Native Bees of the Lower Rio Grande Valley

This book was published in partnership with the
Department of Entomology at Texas A&M University.

Native Bees

of the Lower Rio Grande Valley

A Photographic Guide

Paula Sharp

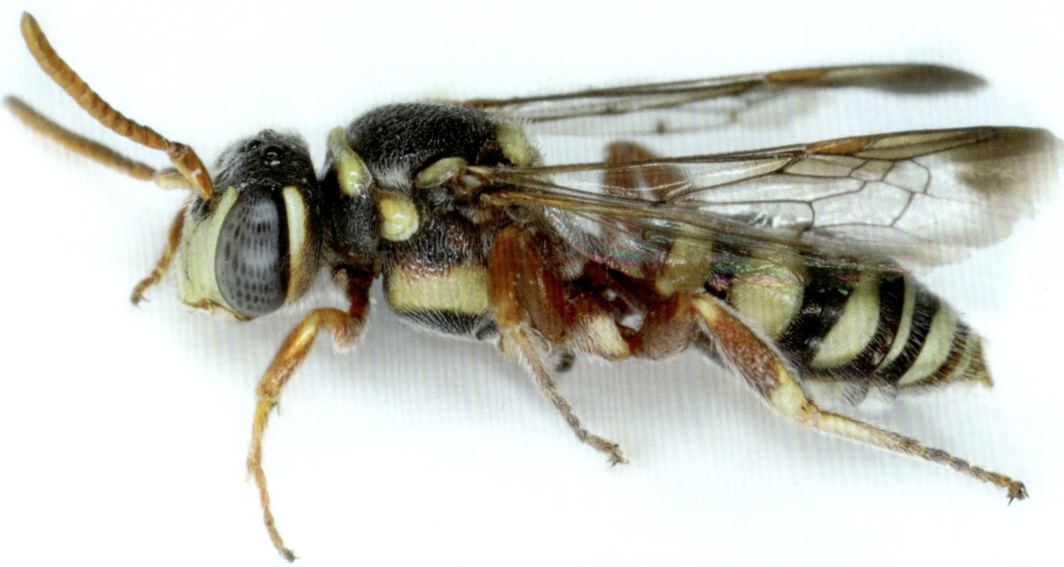

TEXAS A&M UNIVERSITY PRESS
COLLEGE STATION

First edition

∞ This paper meets the requirements of ANSI/NISO Z39.48–1992 (Permanence of Paper).
Binding materials have been chosen for durability.
Manufactured in China by Martin Book Management.

Library of Congress Cataloging-in-Publication Data

Names: Sharp, Paula, author.
Title: Native bees of the Lower Rio Grande Valley: a photographic guide / Paula Sharp.
Other titles: Gideon Lincecum nature and environment series.
Description: First edition. | College Station: Texas A&M University Press, [2024] | Series: Gideon Lincecum nature and environment series | Includes bibliographical references and index.
Identifiers: LCCN 2024013634 (print) | LCCN 2024013635 (ebook) | ISBN 9781648432002 | ISBN 9781648432019 (ebook)
Subjects: LCSH: Bees—Texas—Lower Rio Grande Valley—Identification. | Bees—Ecology—Texas—Lower Rio Grande Valley. | LCGFT: Field guides.
Classification: LCC QL567.1.L8 S63 2024 (print) | LCC QL567.1.L8 (ebook) | DDC 595.79/9097644—dc23/eng/20240328
LC record available at https://lccn.loc.gov/2024013634
LC ebook record available at https://lccn.loc.gov/2024013635

Unless otherwise indicated, illustrations and photographs are by the author.

This book is dedicated to Ross Eatman.

Contents

Acknowledgments

John L. Neff, director of the Central Texas Melittological Institute, helped with every aspect of this book. His contributions were enormous: he identified scores of species and supplied a continuous current of detailed information about the behavior, distinctive physical traits, floral preferences, and geographic ranges of Texas bees. He recommended entomological resources and responded to a relentless stream of queries from the author, which he tended to answer within hours of receiving. His generosity in imparting his knowledge, his gracefully oblique criticisms, and his patience and stamina are all gratefully acknowledged.

Camille Rich, owner of El Mesteño Ranch of Puerto Rico, Texas, accompanied the author on countless bee-searching expeditions. She made the research for this book entertaining, adventurous, geographically expansive, and safe. Her kindness, curiosity, knowledge of the Valley's terrain, flora, and fauna, and her gift of snake chaps are all deeply appreciated.

The author is indebted to the National Butterfly Center for opening its grounds to three years of research. Many thanks in particular to the NBC's plant nursery and gardens manager, Stephanie Lopez, for her expertise and assistance in identifying Rio Grande Valley native plants; to Executive Director Marianna Treviño-Wright for her help and resourcefulness; and to NBC founder Jeffrey Glassberg for his support.

This book would not have been possible without the assistance and cooperation of Texas Parks and Wildlife, whose parks are glorious, beautiful, and large. Special thanks are due to Jose Uribe, superintendent of Falcon State Park, whose gracious welcome and assistance enabled many days of research that produced spectacular results. The author is also indebted to the US Fish and Wildlife Service for providing entrance to the myriad federal refuges and tracts essential to the preservation of natural fauna and flora in the Valley.

Many thanks to William Rich, master gardener of McAllen, Texas, who enabled the author to observe and photograph the elusive feather-tongued bee, a personal favorite. Thanks also to John Brush of Quinta Mazatlán for the invaluable information he provided about local bees and flora; to biologist Ernesto Herrera for his help in pursuing bees of Hidalgo County; to Javier Gonzalez for information on *Bombus pensylvanicus* at South Padre Island Nature and Birding Center; to Raziel Flores for his help with botanical identifications; and to the McAllen Nature Center for introducing the author to the immensely capable young scientists and naturalists who work there. And many thanks also to Hardy Jackson of Starr County's Campos Viejos Ranch for allowing the author to glimpse the ranch's breathtaking array of unique fauna and flora; and to Seth Welliver for making Oleander Acres available to research.

Dr. John S. Ascher of the National University of Singapore, and Hadel Go of the American Museum of Natural History, collaborated with John L. Neff in helping to identify *Megachile (Tylomegachile)* cf. *toluca*. Dr. Karen Wright of Texas A&M University identified *Svastra sabinensis*. I am thankful for their assistance. I also feel deeply indebted to the two peer reviewers of the manuscript, whose thoughtful, detailed, smart, and always helpful comments greatly enhanced the book.

This book's conception and completion were made possible by the exceptional quality of TAMU Press's production crew. I am indebted to the expertise of Managing Editor Katie Duelm and the aesthetic eye of Senior Book Designer Kristie Lee for meeting the considerable visual and design challenges presented by *Native Bees of the Lower Rio Grande Valley*. Finally, I wish to thank Laurel Anderton for her dazzling copyediting skills.

The following photographers generously contributed pictures to this guide: Victor Engel (*Ceratina cobaltina*), Ernesto Herrera (*Halictus ligatus*), Cin-Ty Lee (*Xylocopa strandi*), Beth Willoughby (*Xylocopa griswoldi*), Jeff Winget (*Melissodes tristis*), John Yochum (scrum of *Diadasia*), Laura Langlois Zurro (*Megachile* cutting leaf, and sleeping aggregation of *Melissodes*), and Ross Eatman (*Apis mellifera, Augochlora azteca, Coelioxys edita, Megachile policaris*, and *Svastra sabinensis*).

....

A note about common names for bees appearing here. Common names for bees are tricky, as there is no unified system for authorizing particular English names for North American bees. Nonetheless, many bee genera already have established names in English—among them, honey bees, bumble bees, carpenter bees, and sweat bees. In addition, several superb general bee guides have made valiant attempts to hammer out common names for bee genera and species—including well-traveled databases that have proven invaluable to citizen scientists, such as BugGuide and iNaturalist. Lower Rio Grande Valley bees are unusual, however, and many Valley genera and species names have never been translated from the original Latin by those sources.

In this book, common names, where possible, have been translated literally from the scientific name. Where direct translation fails because the original Latin meaning is too obscure or tangled to provide a useful English handle, the scientific genus names have sometimes been preserved (as in, e.g., *Exomalopsis* and *Mesoxaea*), with the unfounded hope that these might become household words in the style of *Tyrannosaurus rex*. In some cases, English common names have been selected that highlight the physical attributes or preferred flora of particular species in a way that might aid naturalists in identifying them (e.g., barrel-cactus longhorn, pale-gold longhorn).

The name "American bumble bee" was adopted from *Bumble Bees of North America*.[1] John L. Neff suggested the epithets "ignored" for *Perdita ignota*, "woolly" for *Eucera paenalbata*, and "ribbed" for *Stelis costalis*. The following common names were adopted from BugGuide: the species names barrel cactus longhorn (*Svastra duplocincta*), globemallow bee (*Diadasia diminuta*), common little leafcutter (*Megachile brevis*), silver-tailed petal cutter (*Megachile montivaga*), and sunflower chimney bee (*Diadasia enavata*); and the genus names cellophane bee, chimney bee, cuckoo leafcutter, digger and oil-digger bee, fairy bee, furrow bee, mason bee, mining bee, pebble bee, and resin bee.

1. Williams et al. (2014).

Native Bees of the Lower Rio Grande Valley

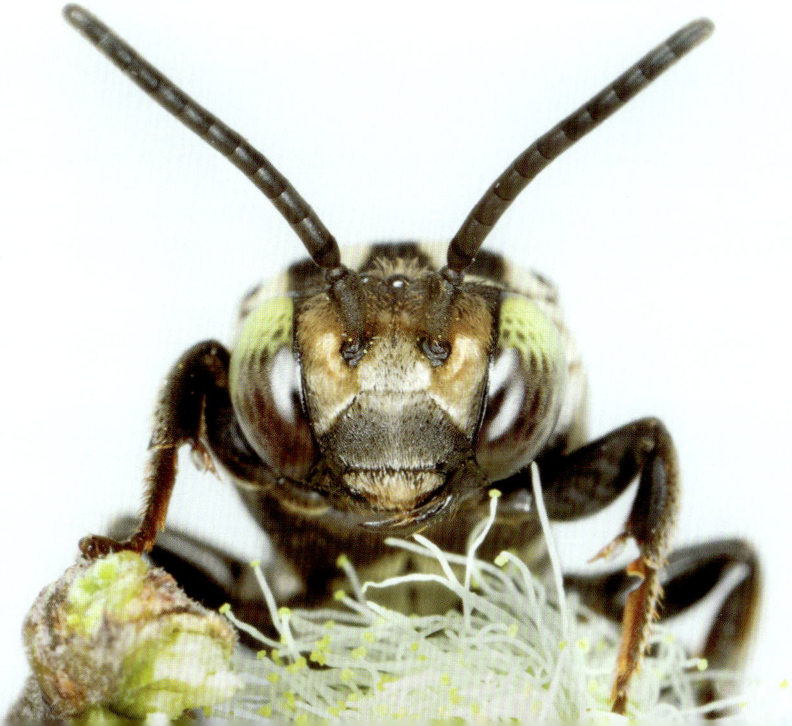

Introduction

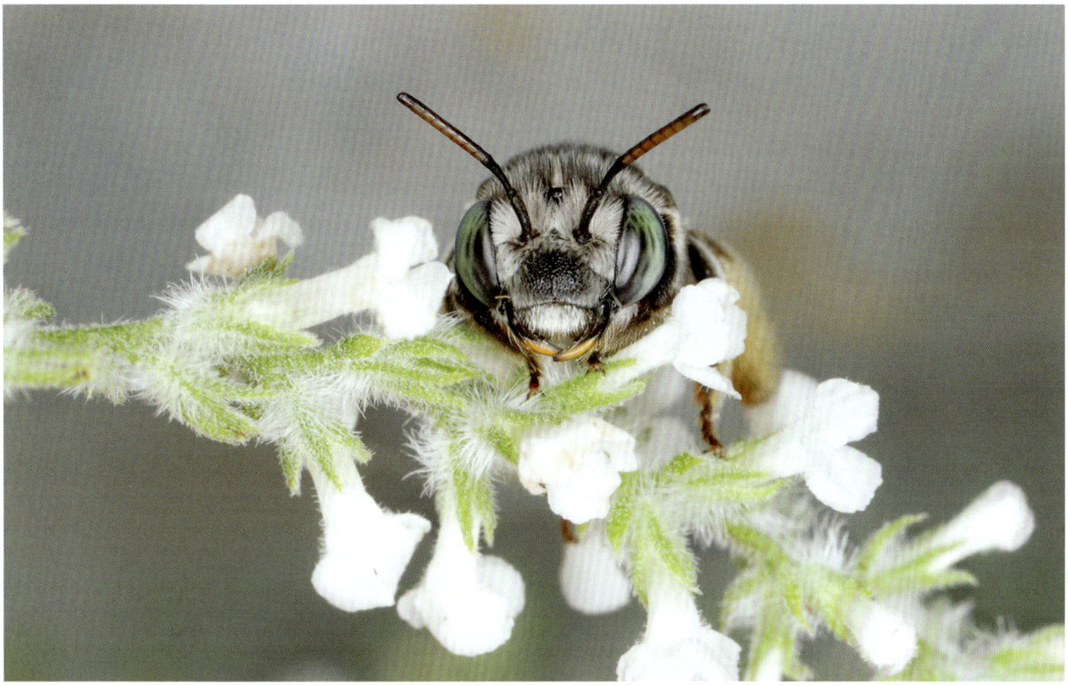

A worthy longhorn bee (*Florilegus condignus*) on whitebrush

The state of Texas is a remarkable hub for wild bees. In the United States, East and West Coast bee populations differ greatly from each other—and Texas lies at the intersection of their two pollinator universes. Some of the species shown here appear as far east as New York. Others are common in California and arid habitats of the Southwest. Texas offers naturalists a rare opportunity to glimpse bee species from both sides of the continent.

Texas, in addition, spans a wide and varied geographic terrain that incorporates, at its southern extreme, a subtropical climate that harbors bee species unknown in other parts of the country. The Texas Lower Rio Grande Valley, located on the Mexican border in the floodplain of the Rio Grande, contains a multitude of distinct habitats acclaimed for their biodiversity. The Valley harbors unique mammals and birds, more than 1,200 varieties of plants, and hundreds of species of butterflies and bees.

Many native bee species of the Lower Rio Grande Valley are rarely seen north of Mexico. Others are found nearly exclusively in Texas or near the Mexican border. Among these are bees belonging to genera seldom featured in popular North American pollinator guides—fascinating creatures such as the crooked-legged bee, *Melissoptila* and *Florilegus* longhorns, the beautiful compact anthophorula, and the glorious protoxaea. Even those bee genera recognizable to many casual naturalists and gardeners—carpenter bees, sweat bees, and leafcutters, for example—tend to appear in a surprising array of species and exotic colors specially adapted to the Lower Rio Grande Valley's long growing season and remarkably varied flora.

Many of the Lower Rio Grande Valley's native bees were first recorded in the late 1800s by prominent entomologists such as E. T. Cresson and T. D. A. Cockerell, who described bee species in detailed written accounts unaccompanied by illustrations. A number of these species have been photographed rarely or not at all. Others remain underresearched, and their habits are largely unknown.

More than 100 bee species and 45 genera are covered in this guide. Species descriptions summarize existing knowledge about little-known native bees of the Lower Rio Grande Valley and supplement that information with detailed photographs and notes on bee behavior. Specific data on where and when individual species appeared between 2018 and 2021 is provided in the guide's appendix. My hope is that students of entomology use this book as a springboard for further research into the novel and diverse bee species of the Lower Rio Grande Valley.

FIELD IDENTIFICATION OF BEES

This guide is addressed to anyone interested in native bees, particularly citizen scientists and naturalists—as these are the people who, in the end, will fight to preserve natural lands where our native fauna and flora thrive. Accordingly, to the extent possible, the guide avoids highly technical scientific terminology. Any esoteric terms are explained in the illustrations of bee anatomy and definitions provided at the end of the introduction. For rare or unusual species particular to

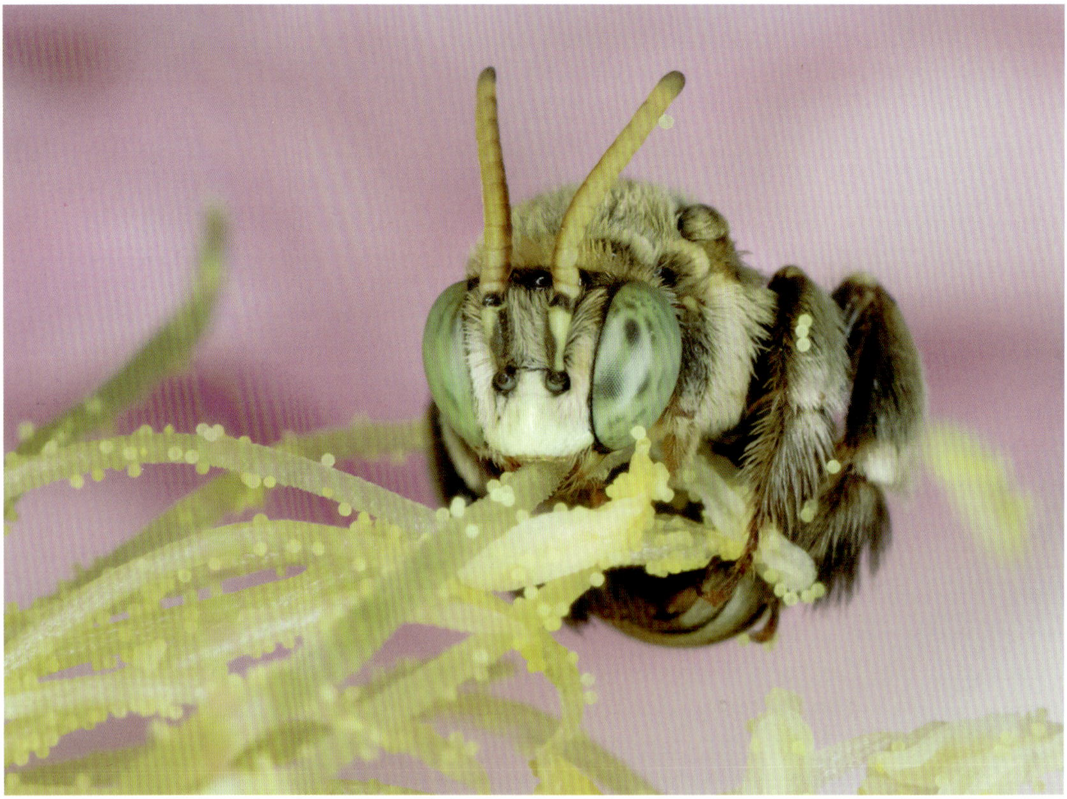

A compact anthophorula (*Anthophorula compactula*) on twisted-rib cactus

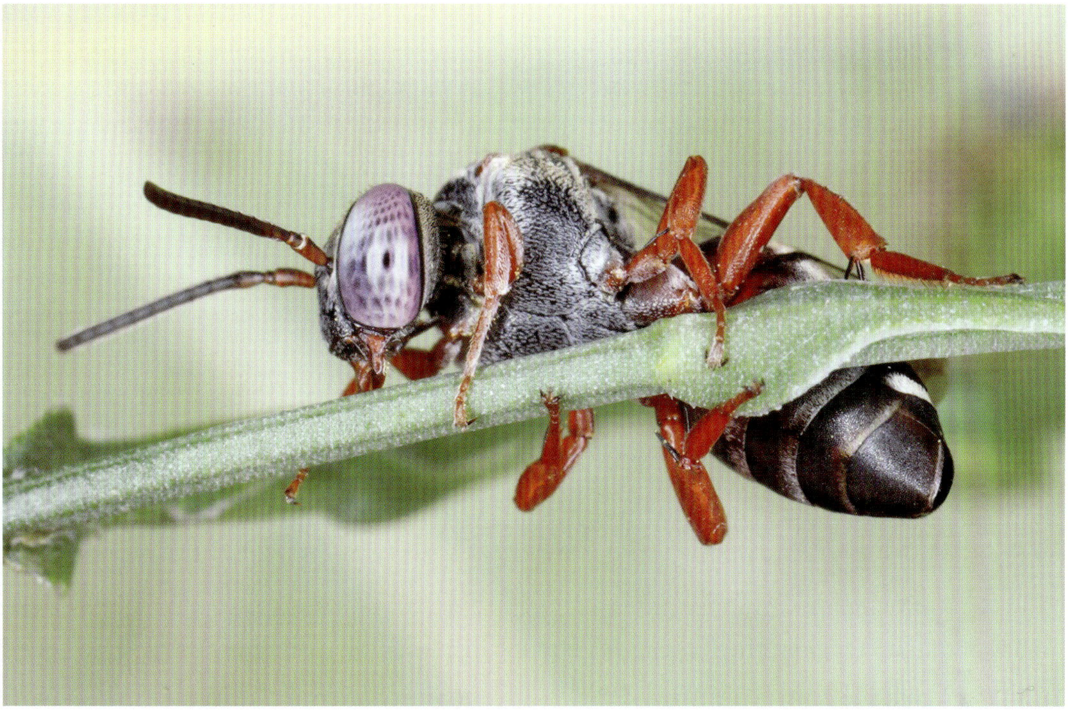

A dwarf epeolus (*Epeolus pusillus*), with red legs and violet eyes

the Lower Rio Grande Valley, guide entries are longer and more detailed. Entries are shorter for species that are better known and more widespread.

To date, few field guides for native bees exist. This is because bees, unlike birds or butterflies or mammals, are small creatures whose species differentiation often turns on minute characteristics undetectable to the naked eye. Traditionally, bee identification by entomologists involves the use of complex taxonomic keys that require extensive knowledge of the nuances and vocabulary of bee anatomy. Such keys typically rely on the examination of dead specimens preserved in collections. For nonexperts seeking to identify live bees in the field, wrestling with taxonomic descriptions and keys can be immensely challenging.

The widespread availability of digital photography has made the recording of bee characteristics simpler and sometimes allows easier field identification. Among other advantages, photographs preserve bee color, a trait that makes many live bees readily recognizable but that often plays no part in taxonomic keys, since dead specimens tend to lose or change color and to relinquish the play of light evoked through iridescence. Thus, Ellen Ordway, an entomological pioneer who described the dazzlingly iridescent bees of the genus *Augochlorella*, lamented that the specimens she examined in museum collections had turned from green to orange in their trays, leaving her at a loss to characterize their color.

For these reasons, wherever possible, the colors of bees, including eye color, are noted for species included in this guide. Eye color in particular can be immensely helpful in identifying Lower Rio Grande Valley bees in the field. The compound eyes of subtropical bees are frequently their most arresting trait—they may be red, denim blue, lime green, or the opaque black of apple seeds. The violet eyes of the Valley's dwarf epeolus cuckoo bee distinguish it from almost all other native bees of the region. Color is also an essential general trait of many subtropical species: the red thorax of Aztec cuckoo leafcutters sets them apart from others of their genus in the region; and the electric blue green of the Aztec sweat bee is the very feature that lures casual naturalists to follow it for an extra glimpse of its splendor.

Close observation of insect behavior also aids in field identifications of Lower Rio Grande Valley bees. The time of year a bee emerges, the flowers it prefers, and its method of carrying pollen are all characteristics that help narrow down species. The idiosyncratic and entertaining antics of bees in action also help reveal species identities. Some bees are noisy buzz pollinators; others are territorial and pursue intruders; and some hover like hummingbirds in flight or dart in jittery figure eights above the ground. Some Lower Rio Grande Valley bees sleep in poppies, or they build turrets outside their nest entrances, or they nest in vast aggregations numbering hundreds of bees.

THE TEXAS LOWER RIO GRANDE VALLEY REGION

The Lower Rio Grande Valley is a distinct cultural and environmental region located near the mouth of the Rio Grande. The Rio Grande floodplain cuts across Cameron, Hidalgo, and Starr Counties in southern Texas, as well as the Mexican state of Tamaulipas. A scattering of medium-sized cities such as Brownsville, McAllen, Rio Grande City, and Roma dots the Valley's Texas border, but much of the surrounding terrain is sparsely populated and relatively rural.

The subtropical climate of southernmost Texas is characterized by short winters that rarely dip below freezing, and intensely hot summers. The Valley's ecological biomes and habitats include, among others, woodlands, scrub forest, and expanses of cactus-laden landscape. Much of the land is semiarid, although some woodlands bordering the Rio Grande are humid by day and full of mist in the mornings.

Research for this guide focused on the three Texas border counties of Cameron, Hidalgo, and Starr and encompassed a variety of environmental settings that included Texas state parks, federal refuges and land tracts, city parks, nonprofit nature preserves, private ranches, agricultural tracts, and gardens—as well as hardscrabble semiurban landscapes, road shoulders, railroad rights-of-way, trailer parks, and vacant lots. Entries for individual bee species in this guide provide information about the habitats in which bees were found, wherever relevant to a bee's appearance, lifestyle, floral associations, or nesting habits.

MAP OF THE TEXAS LOWER RIO GRANDE VALLEY

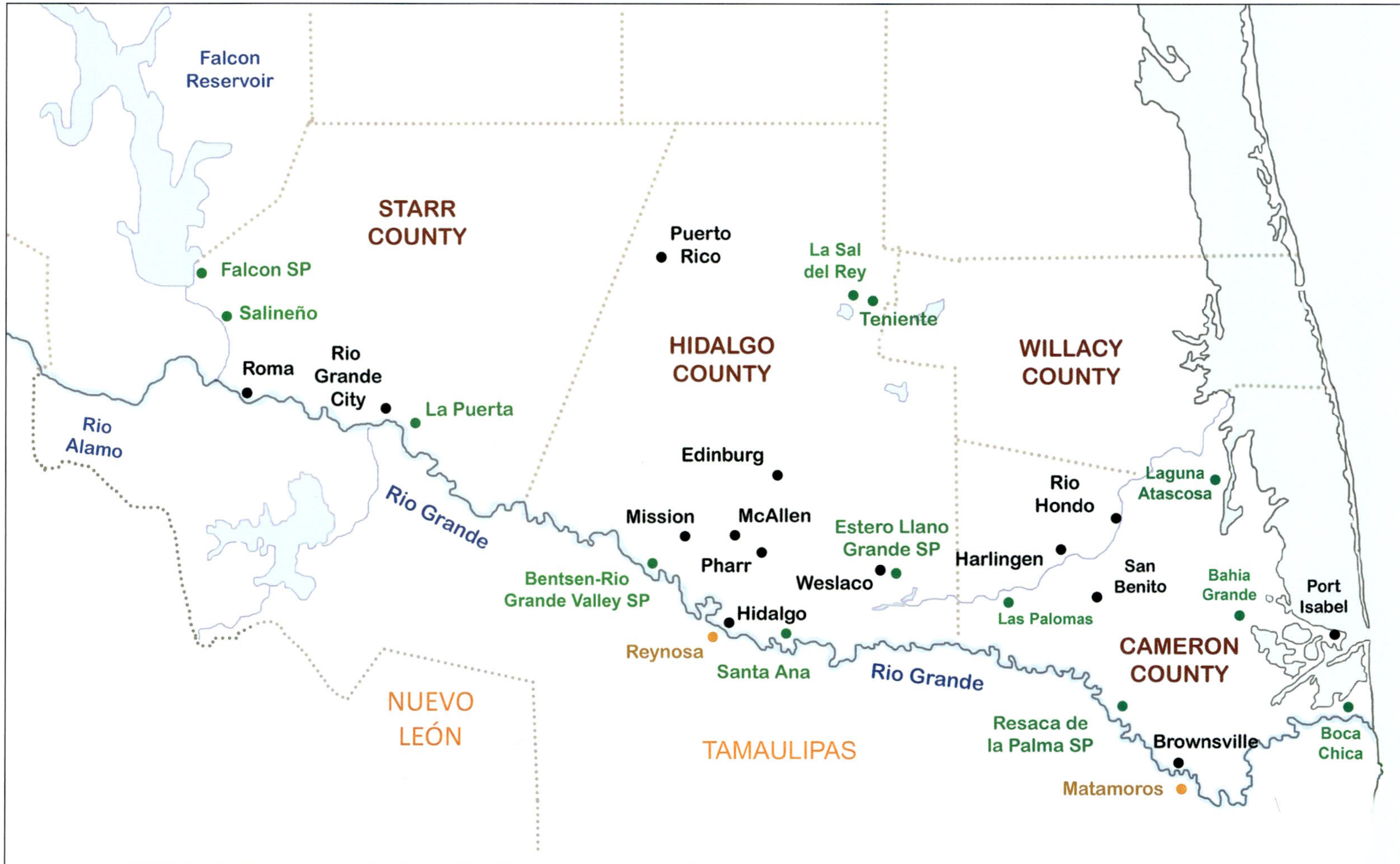

Definitions

NESTING BEHAVIOR

Solitary: The solitary female bee constructs her own individual nest, lays her own eggs, and gathers pollen and stores it in the nest without the aid of other bees of her species. Solitary bees often nest in the ground—less commonly, they carve out nests in rotted logs, or they nest in preexisting holes in plant matter or human-made structures. Their nests often consist of tunnels that connect to individual egg chambers. The majority of solitary bees produce a single generation per year; the parent bee usually dies before her young hatch. Some solitary bees may have two or more generations between spring and summer, particularly in warm climates with long flight seasons. Most of the native bees of the Lower Rio Grande Valley are solitary.

Gregarious: Solitary bees often nest gregariously—that is, close to one another. Gregarious bees may form **aggregations** that consist of a few to thousands of individual nests. Usually, aggregation inhabitants do not work cooperatively to construct nests or gather pollen, and they do not swarm or participate in similar collective action. Sometimes, however, aggregations may encourage bees to take advantage of one another's company to shift from solitary to communal behavior. Examples of gregarious bees that nest in aggregations are crooked-legged bees (*Ancyloscelis apiformis*), Texas mesoxaea (*Mesoxaea texana*), and Rincon chimney bees (*Diadasia rinconis*).

Communal: Sometimes multiple female bees cohabit a single nest but do not share basic duties: each nest occupant still gathers her own food and constructs and provisions her own egg chambers. Such bees may work together in limited ways, however. For example, the nest tunnels leading to individual bees' egg cells may share a common entrance tunnel. This economizes on labor and makes nest defense easier. Such bees are said to live communally. In communal bee groups, interactions usually occur only among adult bees of the same generation. Adult bees die or leave the nest before their eggs hatch. Examples of bees that live communally are the green metallic sweat bee (*Augochloropsis metallica*), the nightshade exomalopsis (*Exomalopsis solani*), the Sabine longhorn bee (*Svastra sabinensis*), and the four-banded nomia (*Nomia tetrazonata*).

Social: "Social" is a term used to describe various kinds of cooperative behavior among bees.

Subsocial: In a subsocial colony, adult females share the same nest with their offspring, tending, feeding, and guarding them until they mature. The female parent usually dies before the offspring become adults. Examples include some large carpenter bees of the genus *Xylocopa* and some small carpenter bees of the genus *Ceratina.*

Semisocial: In a semisocial colony, adult females cohabit a nest in which there is some division of labor into a simple caste system. For instance, some females have fully developed ovaries, mate, and produce eggs, while others have undeveloped ovaries, remain unmated, and perform worker activities such as nest building, collecting pollen, and provisioning egg cells. The females of semisocial colonies usually belong to the same generation. Examples include various species of *Augochlorella* and *Augochlora* sweat bees.

Primitively eusocial: In primitively eusocial colonies, multiple generations of adult bees work cooperatively. Their society may be divided into castes that perform different tasks. In the United States, primitively eusocial colonies usually persist for one year. Examples include some *Augochlora*, *Augochlorella*, *Halictus*, and *Lasioglossum* sweat bees.

Eusocial: Eusocial bees form complex societies in which female castes, such as queens and workers of multiple generations, perform separate tasks and frequently differ from one other in size and physical attributes. Egg-laying queens usually are larger and survive longer than other colony members. Colonies are highly adapted for survival because of the functions performed by versatile workers, which build nests, gather food, and feed offspring. An example is bumble bees (*Bombus*). **Highly eusocial** colonies may persist beyond a year and contain tens of thousands of individuals. An example is honey bees (*Apis mellifera*).

Bivoltine and multivoltine: These terms describe bees that pass through two (bivoltine) or several (multivoltine) generations in a single year.

POLLINATION

Oligolectic or specialist pollinators: Some bees specialize in collecting pollen from plants of a particular species, genus, or family. Such bees are called oligolectic or specialist pollinators. Examples are Wilmatte's longhorn bee (*Eucera wilmattae*), which collects pollen principally from aster-family flowers, and the Rincon chimney bee (*Diadasia rinconis*), which is a cactus specialist.

Polylectic or generalist pollinators: Bees that collect pollen from an array of flowers are referred to as generalist or polylectic pollinators. Examples of broad generalists are honey bees (*Apis mellifera*), the common little leafcutter (*Megachile brevis*), and the mournful longhorn bee (*Melissodes tristis*).

Vibratile or buzz pollinators: Bees known as vibratile or buzz pollinators have the ability to vibrate their flight muscles while grasping a flower: this shakes pollen loose from the anthers in a manner commonly compared to jiggling a salt shaker to dislodge salt. Vibratile pollinators exist in several bee families.

Scopal hairs or scopae: Most female bees have hairs called scopae that are specially designed to carry pollen. Such hairs are usually located on the bee's upper hind legs or sternum, or less commonly, on the propodeum. Some bees lack scopae: masked bees of the genus *Hylaeus*, for example, carry pollen internally. Cuckoo bees, which lay their eggs in other bees' nests, do not gather pollen and thus have no need for scopae; their offspring consume pollen stored by their hosts. Male bees lack scopal hairs.

PARTS OF THE BEE

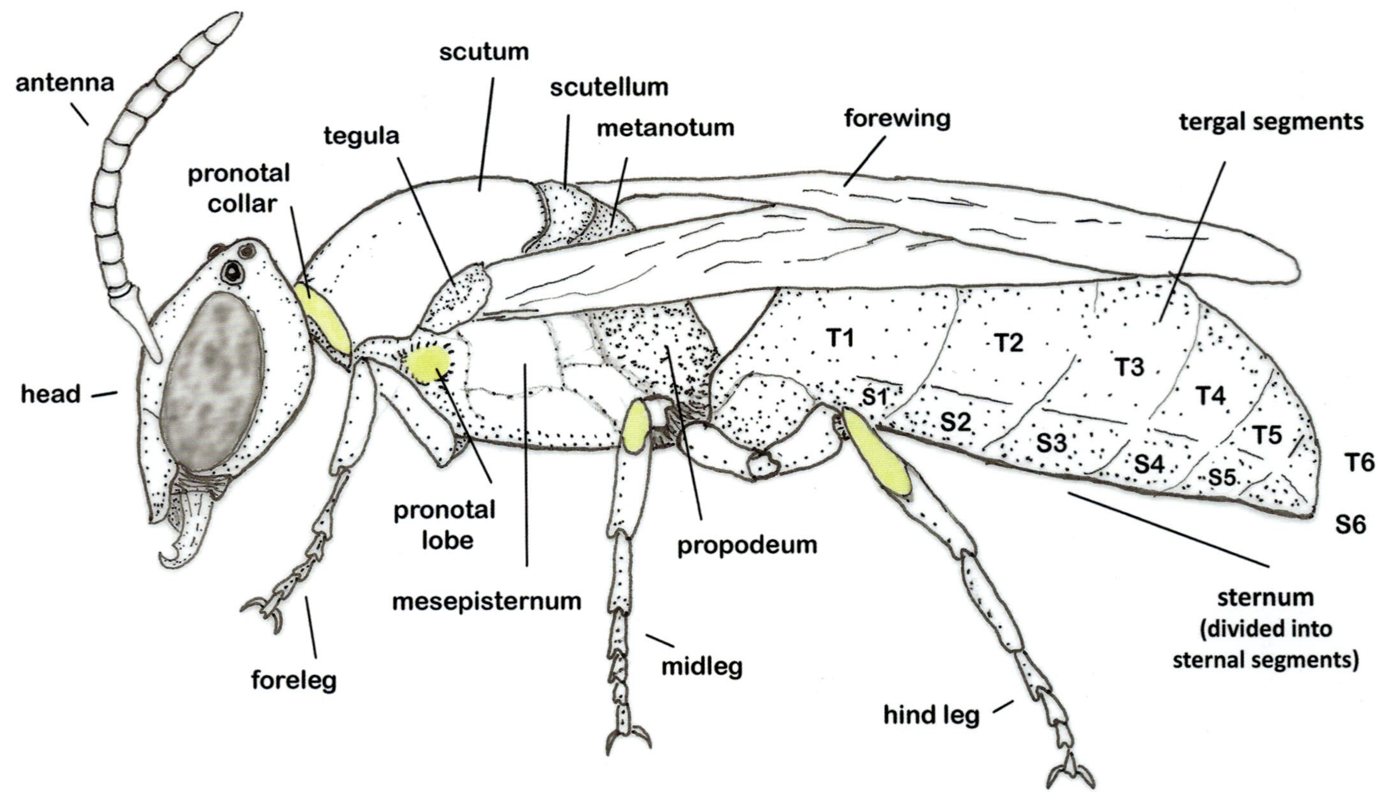

BEE ANATOMY

Body: The bee's body is divided into three parts: the **head, thorax,** and **abdomen**. The bee's outer skin, or exoskeleton, which houses the internal organs within these three sections, is referred to as the **integument**. Bees have two sets of wings attached to the thorax: the **forewings** and **hind wings**. Small, shield-like plates called **tegulae** cover the bases of the wings. Bee wings have distinctive vein patterns that are sometimes used to identify genera and species.

Thorax and Abdomen: The top surface of the thorax is divided into sections. From front to rear, these include the **scutum, scutellum,** and **metanotum**. Flanking the rear edge of the scutellum are toothlike or rounded projections known as **axillae**. The front of the scutum is called the **pronotum**. Some bees have pale markings on the front rim of the pronotum, called the **pronotal collar**, or on small structures located ahead of the tegulae, called **pronotal lobes**. Behind the metanotum is the **propodeum**, a part of the abdomen that is fused to the back of the thorax. The combined thorax and propodeum are called the **mesosoma**.

The abdomen is divided into segments. The top portions of these segments are called **terga**, which are numbered from front to back as T1 to T6 (or T1 to T7 on males). The abdomen's underside is called the **sternum** and is divided into sternal segments numbered S1 to S6 (S1 to S7 on males). The term **metasoma** is used to describe the abdomen minus the propodeum.

Some bees have a **pygidial plate** on the tip of the abdomen (T6 on females and T7 on males). The pygidial plate is a hard, flat, roughly triangular structure. In some genera, such as *Nomia*, the female alone has a pygidial plate. In others, both males and females have pygidial plates. They are most common among ground-nesting bees. Female bees may use their-pygidial plates to smooth waterproofing secretions onto egg chamber walls and to tamp down soil when digging ground nests.

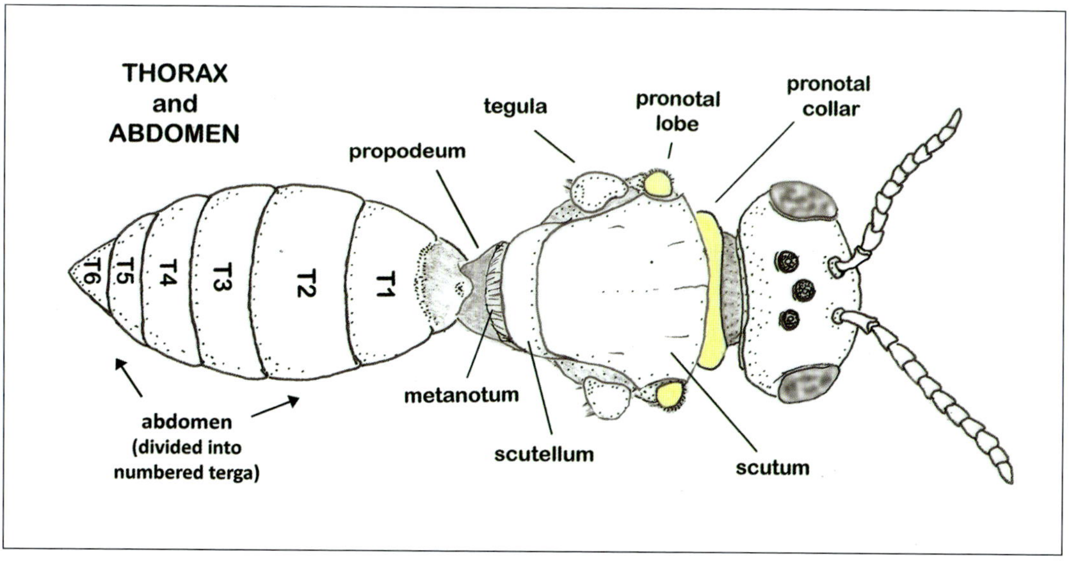

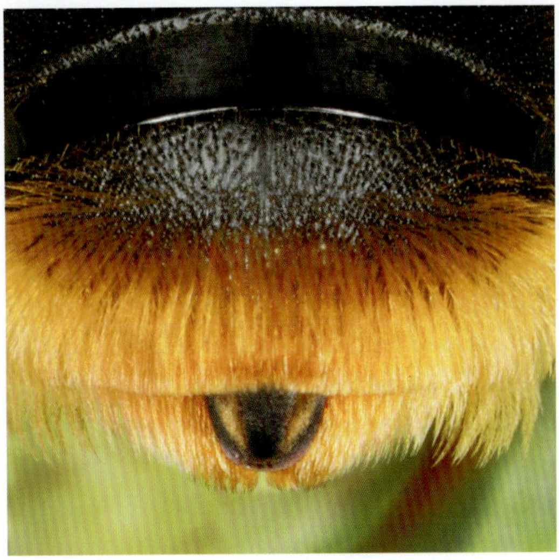

The pygidial plate of a *Protoxaea*.

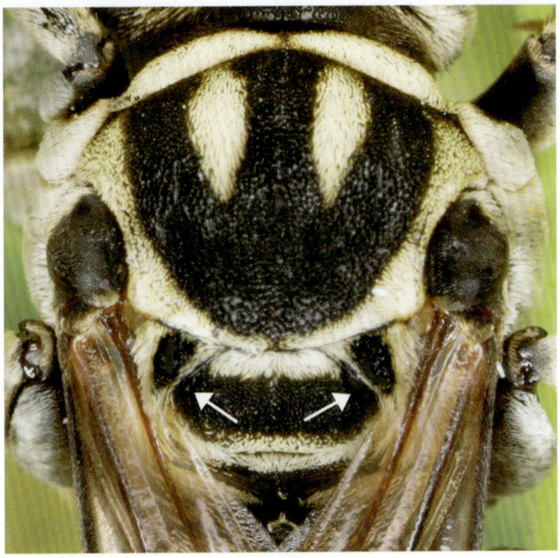

The axillae of a male *Triepeolus*.

Head and Face: All bees have mandibles, antennae, and two sets of eyes: a pair of large ones called **compound eyes**, and three smaller, simple eyes called **ocelli**, which house photo receptors used to detect movement. The side of the bee's face, behind the compound eye, is called the **gena**, or cheek. The top of the head, behind the ocelli, is the **vertex**. The "forehead" area below the ocelli and above the antennae is the **frons**.

The bee's antennae are affixed to its head in nodes called **antennal sockets** and are divided into three sections. The **scape** is located at the antenna's base and is usually relatively long; right above this is the **pedicel**, which is usually short. The remainder of the antenna is called the **flagellum** and is divided into segments called **flagellomeres**. The segments are numbered from base to tip, from F1 to F10 (on females) or from F1 to F11 (on males). Bee mandibles may be simple or have cutting edges or teeth: teeth are numbered from bottom to top.

Legs: A bee's legs attach to its thorax. The legs are divided into segments: the coxa, trochanter, femur, tibia, and tarsus (from top to bottom). The tarsus (plural tarsi) is divided into tarsal segments, or tarsomeres, and terminates in tarsal claws. A small pad called an **arolium** (plural arolia), located between the tarsal claws, is found in some bee genera and may aid in their identification.

Differentiating Female and Male Bees: Bee gender can be determined by counting the number of segments on the abdomen or antennae. Female bees have six abdominal terga, while males have seven. Female bees' antennae have 12 segments: a scape, a pedicel, and 10 flagellar segments. Males' antennae have an extra flagellar segment and a total of 13 segments. Female bees also have stingers, which serve as modified ovipositors. All male bees lack stingers. In many bee genera, males are smaller than females. In some species, facial markings and eye, hair, and body colors of male and female bees also often differ.

FACE AND HEAD OF THE BEE

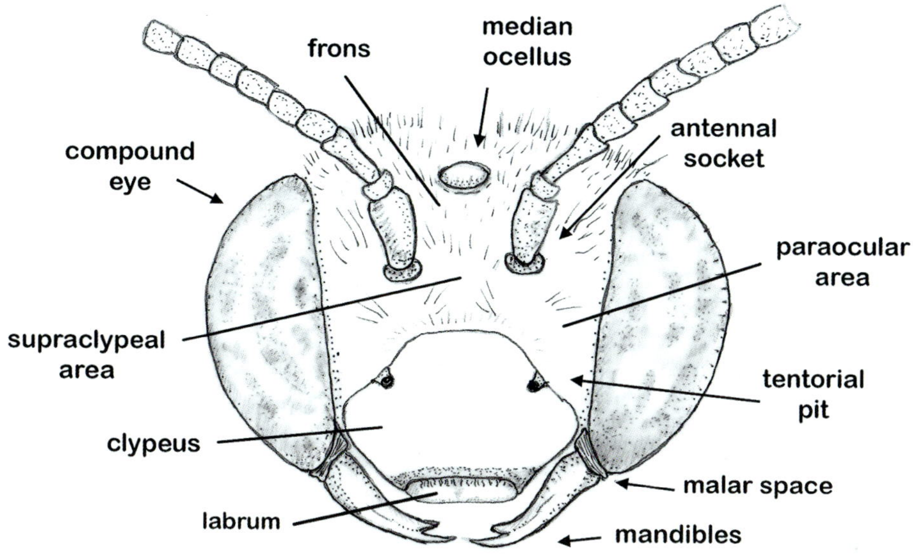

FACE
(frontal view)

HEAD
(dorsal view)

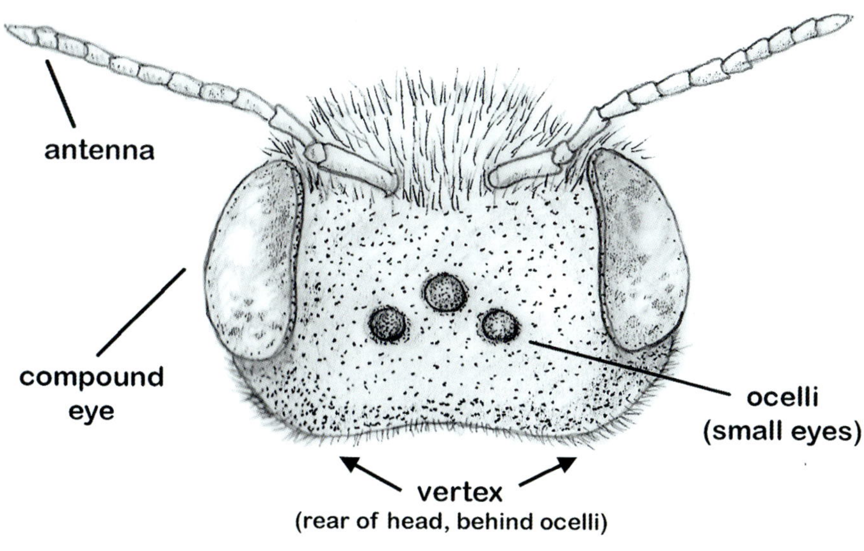

HEAD, MANDIBLES, AND ANTENNA OF THE BEE

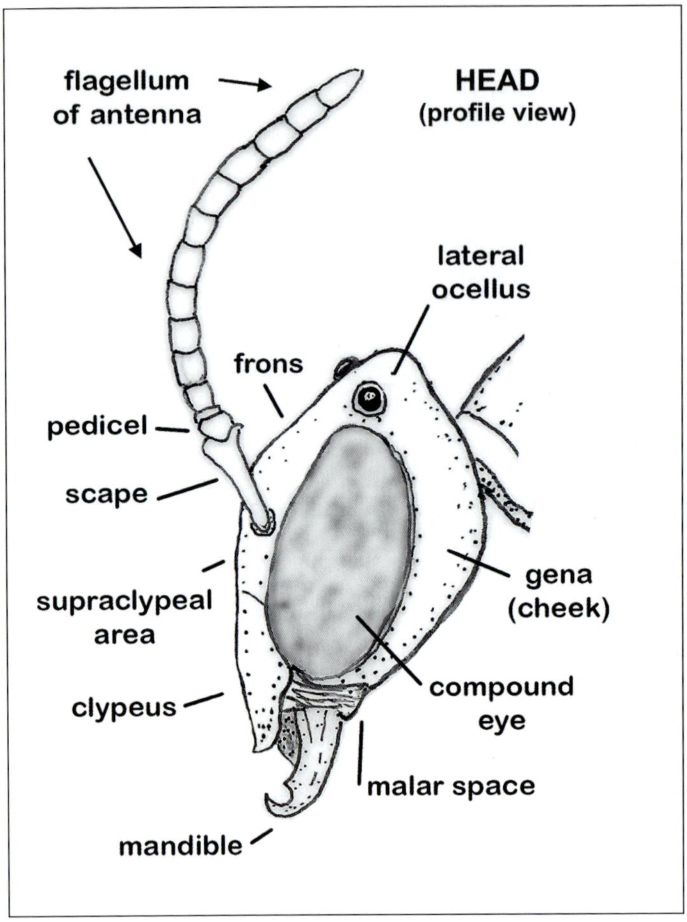

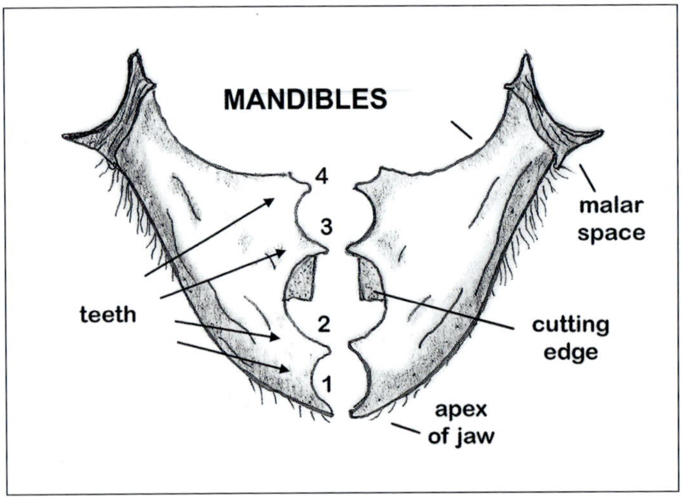

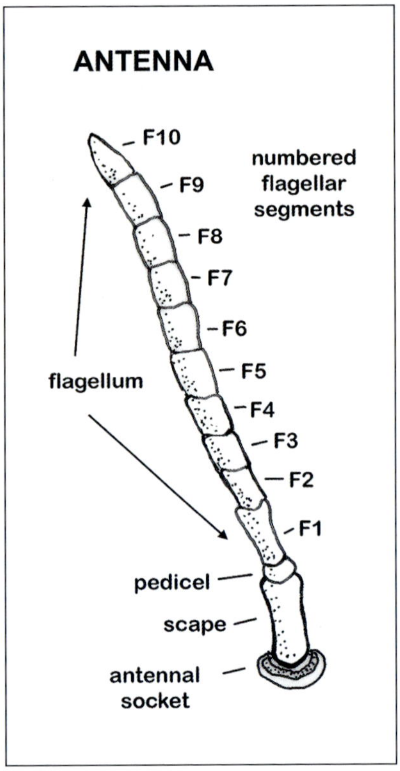

LEG AND WINGS OF THE BEE

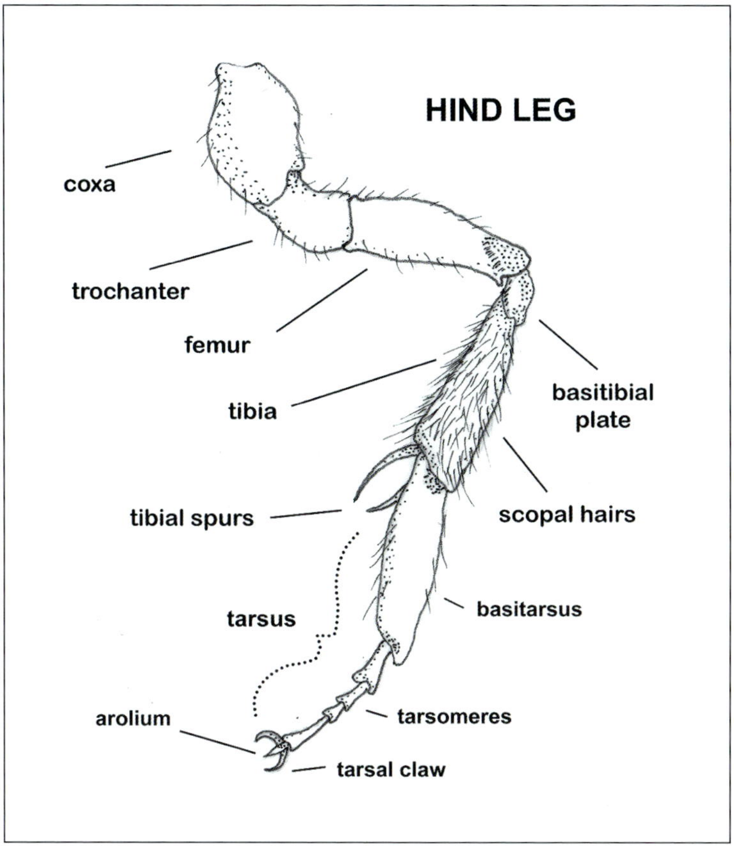

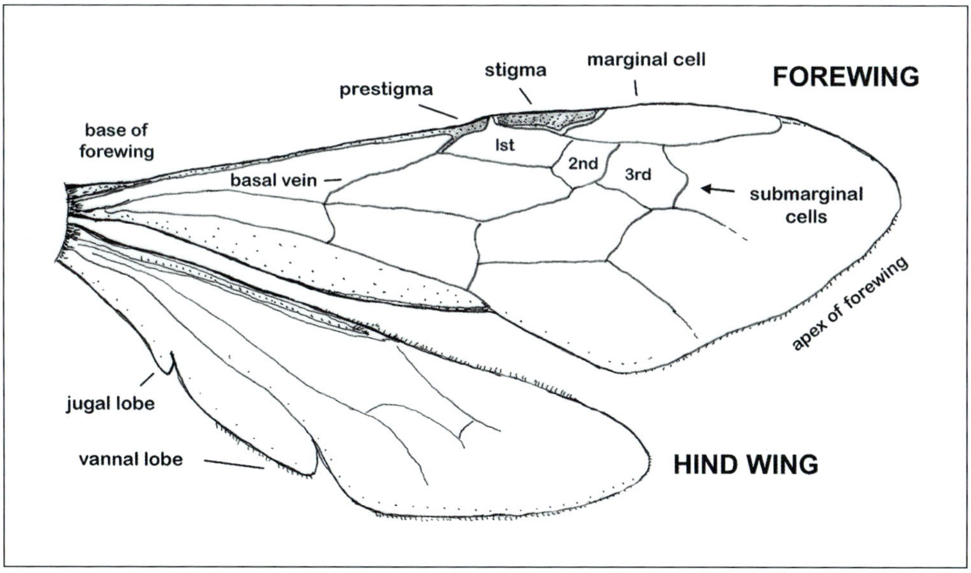

BEE FAMILIES AND GENERA OF THE TEXAS LOWER RIO GRANDE VALLEY

FAMILY ANDRENIDAE

GENUS
- Andrena
- Calliopsis
- Mesoxaea
- Perdita
- Protandrena
- Protoxaea

FAMILY APIDAE

GENUS
- Ancyloscelis
- Anthophora
- Anthophorula
- Apis
- Bombus
- Centris
- Ceratina
- Diadasia
- Epeolus
- Eucera
- Exomalopsis
- Florilegus
- Melissodes
- Melissoptila
- Melitoma
- Nomada
- Svastra
- Triepeolus
- Xylocopa

FAMILY COLLETIDAE

GENUS
- Colletes
- Hylaeus
- Ptiloglossa

FAMILY HALICTIDAE

GENUS
- Agapostemon
- Augochlora
- Augochlorella
- Augochloropsis
- Dieunomia
- Halictus
- Lasioglossum
- Nomia
- Sphecodes

FAMILY MEGACHILIDAE

GENUS
- Ashmeadiella
- Coelioxys
- Dianthidium
- Heriades
- Lithurgopsis
- Megachile
- Osmia
- Stelis

Bee Guide

Family Andrenidae

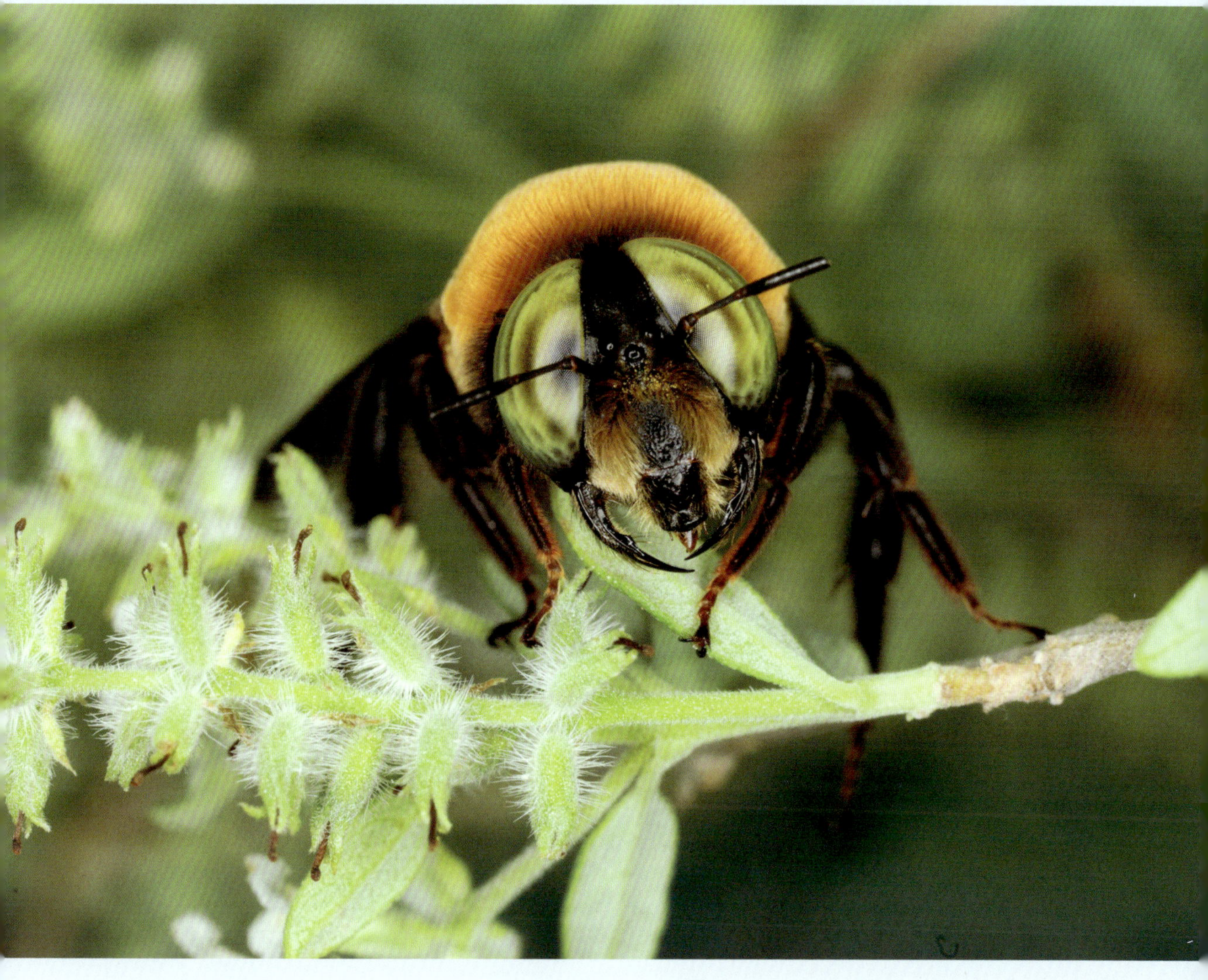

Mesoxaea and Protoxaea
Mining Bees
Protandrena
Fairy Bees
Calliopsis

ANDRENIDAE FAMILY GENERA
OF THE TEXAS LOWER RIO GRANDE VALLEY

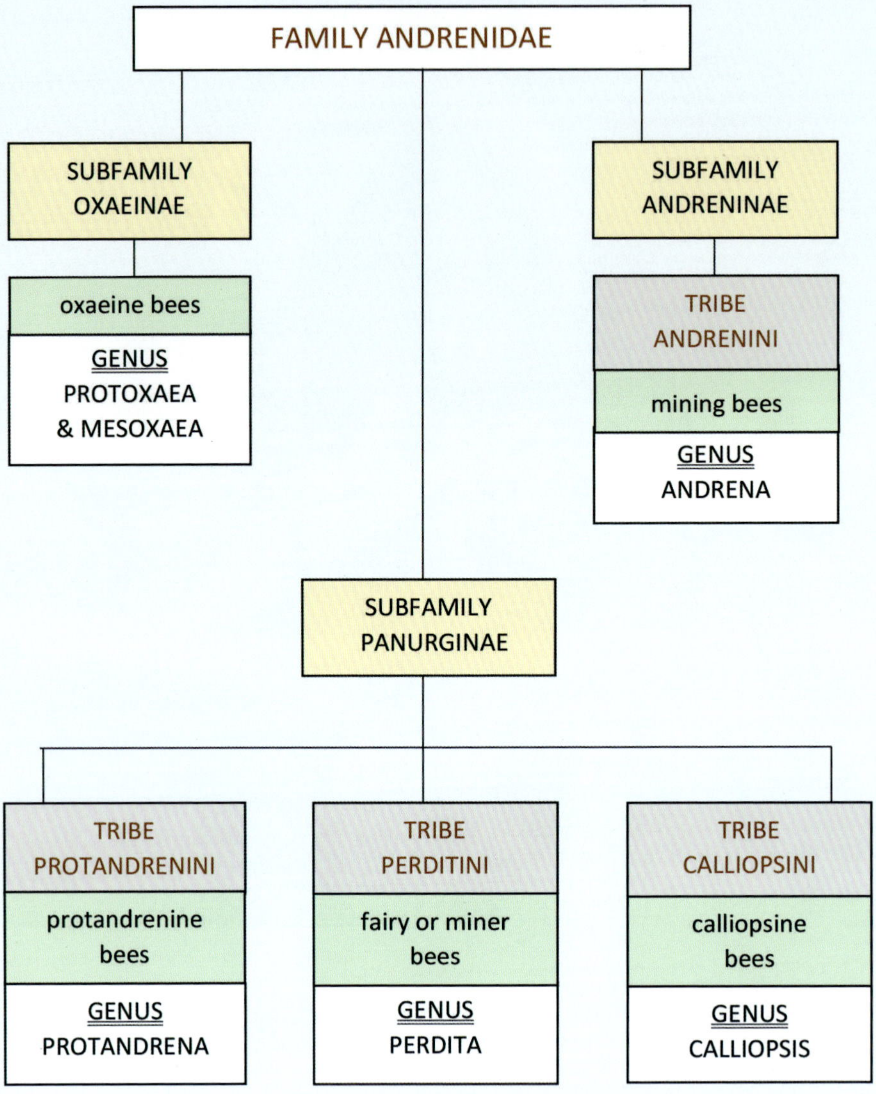

Oxaeine Bees

Subfamily Oxaeinae
Genera *Mesoxaea* and *Protoxaea*

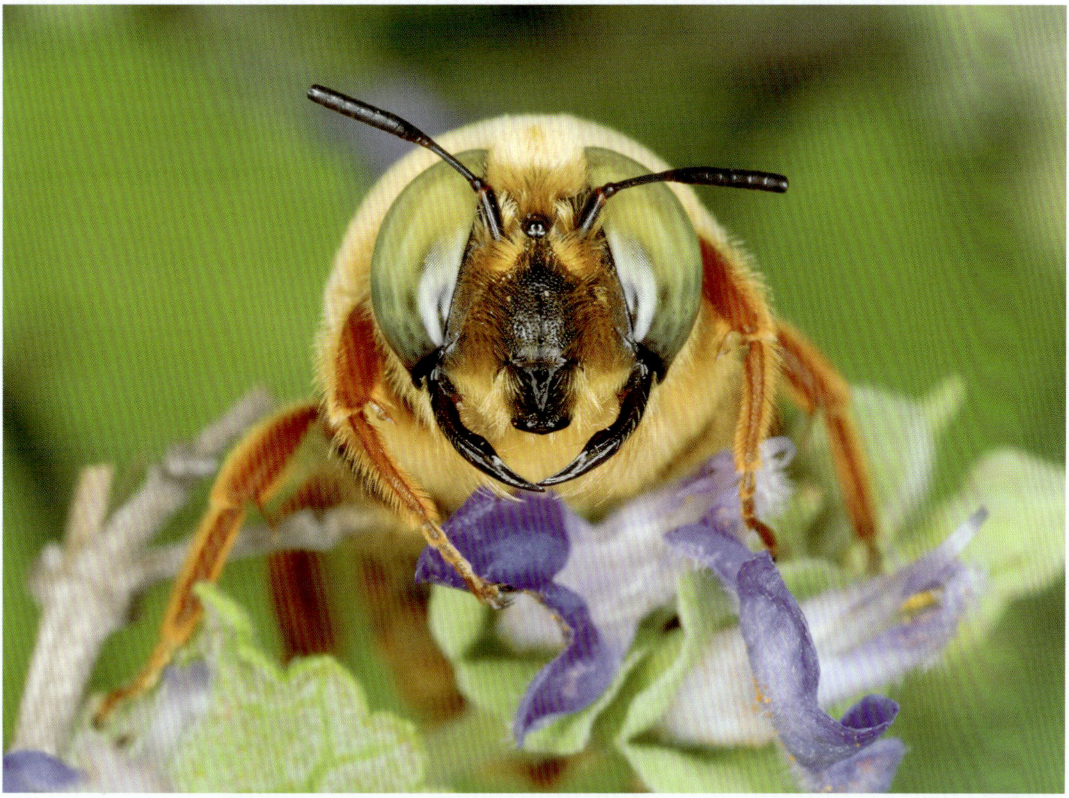

A male *Protoxaea gloriosa*

Bees in the subfamily Oxaeinae are represented in the Lower Rio Grande Valley by two singular species: the Texas mesoxaea and the glorious protoxaea. Oxaeinae are hairy, robustly built, and quite large—usually 13 to 26 mm long (half an inch to an inch). Their faces are unusual: their clypei are protuberant; their ocelli are low set; and the compound eyes of males converge upward, giving them a fly-like aspect. The antennae of regional *Mesoxaea* and *Protoxaea* have short flagellar segments, except for the first segment (F1), which is very long, so that the antennae tend to rest stiffly to the sides of the face, rather than arching gently forward.

These remarkable bees form a small group: they comprise a mere four genera and fewer than 25 species, and they are found only in the New World. Oxaeinae once occupied their own bee family, but they are now considered members of the family Andrenidae. Almost all Andrenidae, including Oxaeinae, share a common characteristic: they have two "sulci" (lines) below each antenna, features often obscured by thick facial hair.

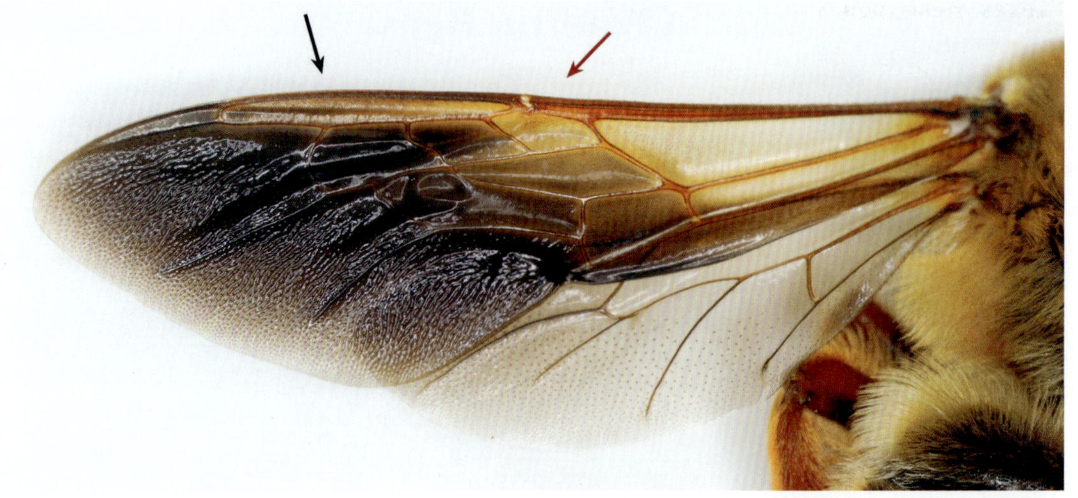

Wing of a male *Protoxaea gloriosa*: note the near absence of a stigma (red arrow), and the elongate, narrow marginal cell (black arrow).

Other traits of Oxaeinae aid in their identification. Their wings are distinctive: on the forewing, the stigma is nearly absent, and the marginal cell is long and narrow and bent away from the wing edge at the tip. On females, the apical end of the hind femur is enlarged and flattened into a plate that joins with a smaller (basitibial) plate on the leg joint below it. Females' scopal hairs are abundant from the very top segment of the hind leg (the trochanter) to the middle of the tibia and form a basketlike structure. The feet of all Oxaeinae have arolia.

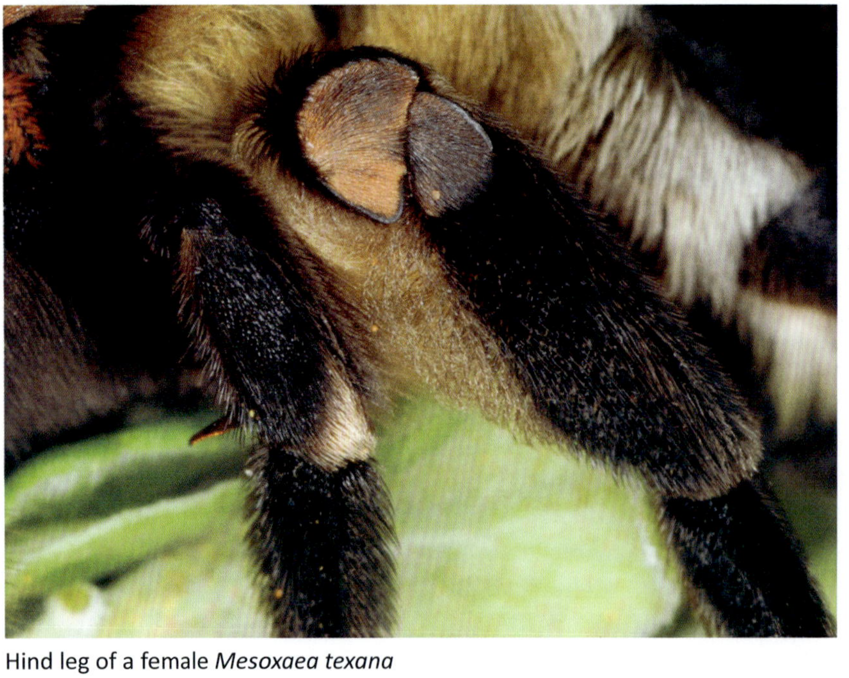

Hind leg of a female *Mesoxaea texana*

Texas mesoxaea ◆ *Mesoxaea (Mesoxaea) texana* (Friese, 1898)

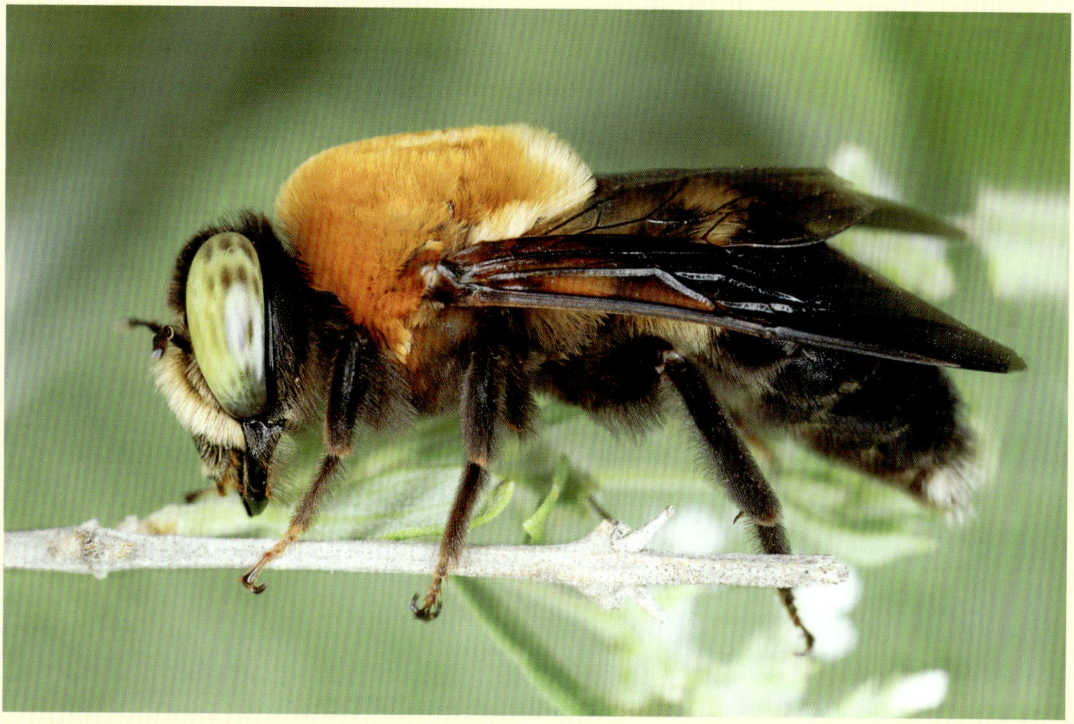

A male *Mesoxaea texana*

Description: Described by T. D. A. Cockerell as "a large and beautiful insect,"[1] the Texas mesoxaea is a fast-flying, hefty bee with prominent green eyes and dark antennae; a boxlike thorax topped with rust-orange hairs; and a dark abdomen with metallic green and purplish tints. The bee's wings are glassy at the base, darkening toward the tips, with reddish-black veins. The tegulae are mostly dark red; on females, they are partly yellowish. Males' faces give them a striking, ferocious appearance: their compound eyes tilt inward at the top, and their reddish-black mandibles, designed to immobilize females during mating, are long and curved. The faces of both males and females are covered with predominantly dark hairs above the antennae, and pale hairs below the antennae.

Dark hairs cover much of the abdomen, and the underside of the thorax. White tufts of hair jut laterally from the tip of the abdomen (T5 on females and T6–T7 on males). The tip of the male's abdomen is notched and translucent orange. The legs of males and females are black or reddish black and covered with mostly black hairs. Females' hind-leg scopal hairs are partly white on the femurs. Females also have short, pale scopal hairs on the sides of the sternum (S1–S2). The lower tarsi of males are reddish and terminate in prominent arolia. Females have inconspicuous arolia. **Size:** females 20–24 mm (4/5–9/10 in.), males 17–20 mm (2/3–4/5 in.).

1. Cockerell (1933), p. 286.

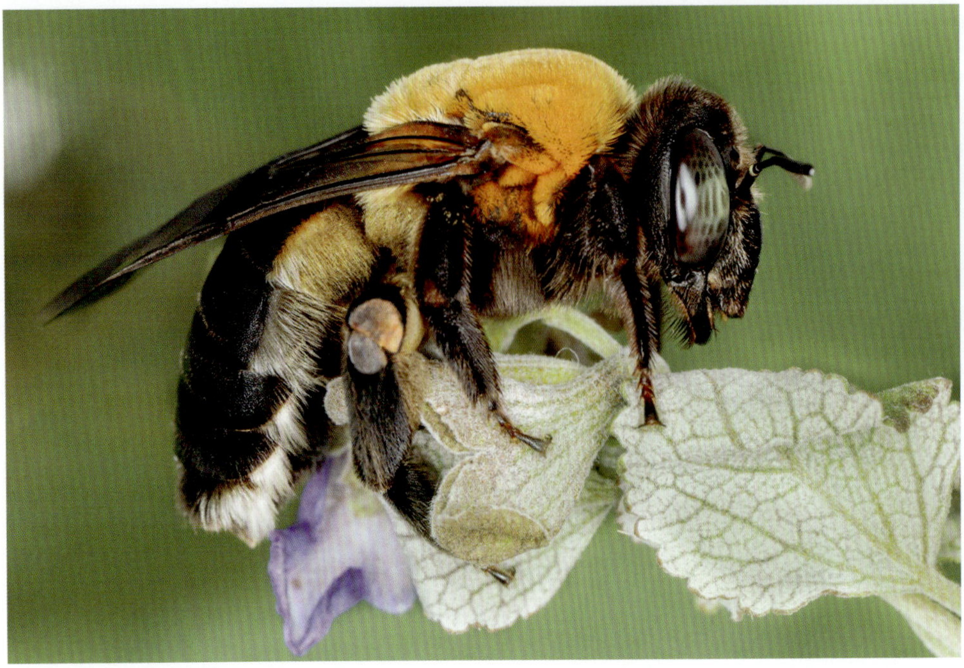

A female *Mesoxaea* texana: note the protuberant clypeus, typical of the subfamily Oxaeinae.

Range: *Mesoxaea texana* is among the rarest bee species of the United States. Most documented sightings of the Texas mesoxaea occurred in various counties of Texas before the 1980s. There are isolated records of this species in western Louisiana and in Oklahoma. *Mesoxaea texana* is also found in the Mexican states of Tamaulipas and Nuevo León, and it occurs as far south as Veracruz and Guanajuato, but even in those locations, official records are few and decades old. The first description of *Mesoxaea texana* appeared in 1898, in an Austrian science journal article written by German bee biologist Heinrich Friese. He based his description on two male specimens, one of which came from Dallas, that formed part of a collection of "exotic" New World bees.

The *Mesoxaea texana* males and females shown here appeared from mid-May through late June 2021 at El Mesteño Ranch in northwest Hidalgo County, and at Starr County's Falcon State Park, which lies near the Tamaulipas border and close to Nuevo León. Both locations contain semiarid Tamaulipan thornscrub habitats with partly sandy soils, and dense growths of shrubby blue salvia (*Salvia ballotiflora*) near abundant water sources.

Associated plants: At both sites in Starr and Hidalgo Counties, male and female *Mesoxaea texana* were found feeding in conjunction with *Protoxaea gloriosa*, shown in the next entry. The *Mesoxaea* females of El Mesteño Ranch appeared by the dozens from early morning through about 9:00 a.m., foraging on shrubby blue salvia growing in red sand. Male *Mesoxaea texana* at El Mesteño visited whitebrush (*Aloysia gratissima*) from 10:00 a.m. to noon, and from 3:00 until 4:30 p.m., avoiding both early morning and the hottest part of the day. At Falcon State Park, both male and female *Mesoxaea texana* visited shrubby blue salvia. Females followed the same morning schedule as those at El Mesteño, but male bees could be found from 10:00 a.m. until 5:00 p.m. even on unusually hot days, vying with one another and foraging for nectar.

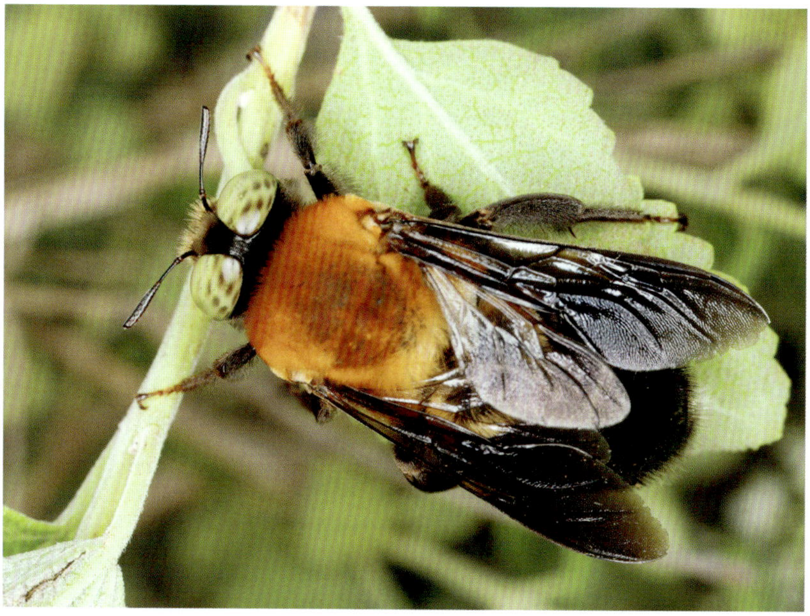

A male *Mesoxaea texana*: its convergent, bulbous eyes lend it a fly-like appearance.

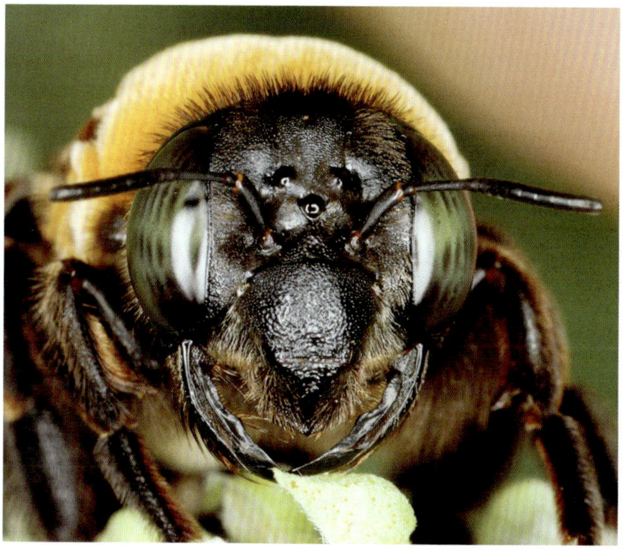

Face of a female *Mesoxaea texana*

Female bees of the subfamily Oxaeinae are thought to collect pollen from a fairly narrow range of plants in the families Fabaceae, Solanaceae, and Zygophyllaceae. In the late spring thornscrub of El Mesteño Ranch and Falcon State Park, plants from all of these families were in bloom, most notably guajillo (*Senegalia berlandieri*), silverleaf nightshade (*Solanum elaeagnifolium*), and guayacan (*Guaiacum angustifolium*). In 1933, Cockerell recorded *Mesoxaea texana* females in the sand dunes of Bexar County nectaring on spotted beebalm (*Monarda punctata*)—which, like salvia, belongs to the plant family Lamiaceae. Texas records from 1928 to 1956 document *Mesoxaea texana* males nectaring on boneset, croton, cotton, knotweed, ironweed, verbena, and whitebrush.

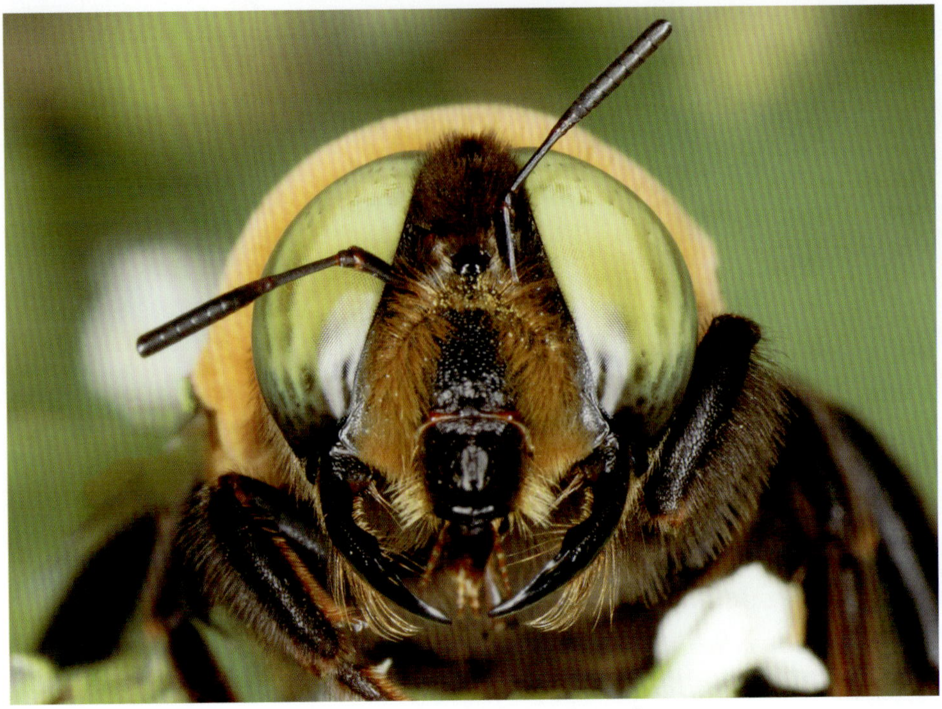

Face of a male *Mesoxaea texana*: note the dark hairs above the antennae.

ID tips: Bees of the genera *Mesoxaea* and *Protoxaea* are similar; at one time, *Mesoxaea texana* was classified as a species of *Protoxaea*. In 1976, entomologists Hurd and Linsley erected the genus *Mesoxaea*, and *Mesoxaea texana* was rechristened under its current name. *Mesoxaea* and *Protoxaea* still might be confused in the field, however, because of their similar size and facial structure. *Mesoxaea* can be distinguished from *Protoxaea* by two traits: (1) the tufts of white hairs on the abdomen tips of *Mesoxaea* males and females; and (2) the notched tip of the male *Mesoxaea*'s abdomen. *Protoxaea* lack these traits. Moreover, the single *Protoxaea* in the United States, *P. gloriosa*, is covered more extensively with golden-brown hairs, and its legs are predominantly red.

Mesoxaea texana is the only species of its genus in Texas. A handful of other *Mesoxaea* are found in Mexico (*M. arizonica*, *M. clypeata*, *M. nigerrima*, *M. rufescens*, *M. tachytiformis*, and *M. vagans*). Two of these, *Mesoxaea arizonica* and *M. rufescens*, sometimes cross the border into Arizona. The male *Mesoxaea texana* can be distinguished from all others of its genus by the dark hairs covering the area from the antennal sockets to the hind margin of the head.

The male *Mesoxaea texana*'s size, large mandibles, and convergent eyes set it apart from almost all other Texas bees. Nonetheless, the Texas mesoxaea's black abdomen, rust-brown thorax hairs, and dark wings might make it confusable in the field with the hefty male southern carpenter bee, *Xylocopa micans*, which has bulbous, tilted green eyes, a black abdomen, and orangish thorax hairs, and which engages in territorial behavior and hovering maneuvers not unlike those of male *Mesoxaea*. The southern carpenter bee, however, is smaller, its forewings differ, and, most notably, it has a conspicuous pale yellow mark covering much of its lower face, visible even from several feet away.

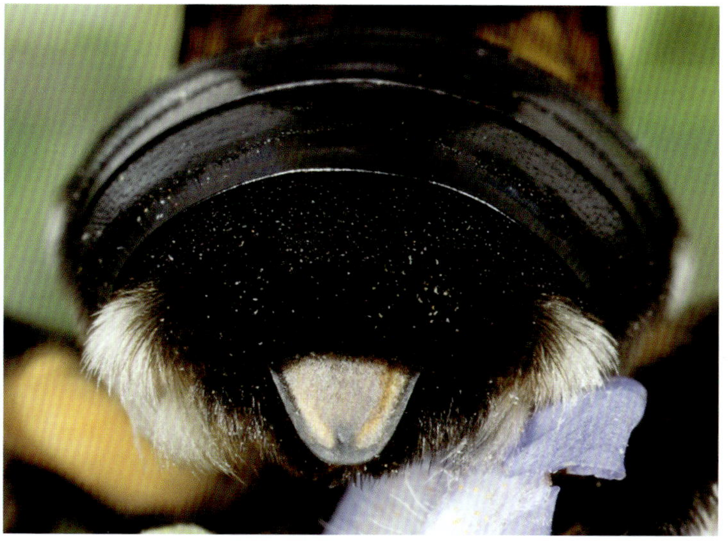

Rear view of a female *Mesoxaea texana* abdomen: note the tufts of white hair on T5. The well-developed pygidial plate is a trait of Oxaeinae generally.

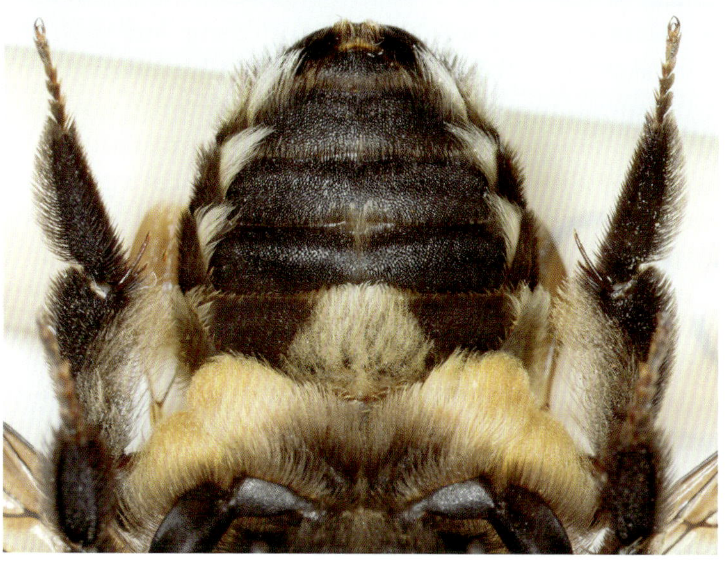

Sternum of a female *Mesoxaea texana*: note the pale scopal hairs on the inner hind legs.

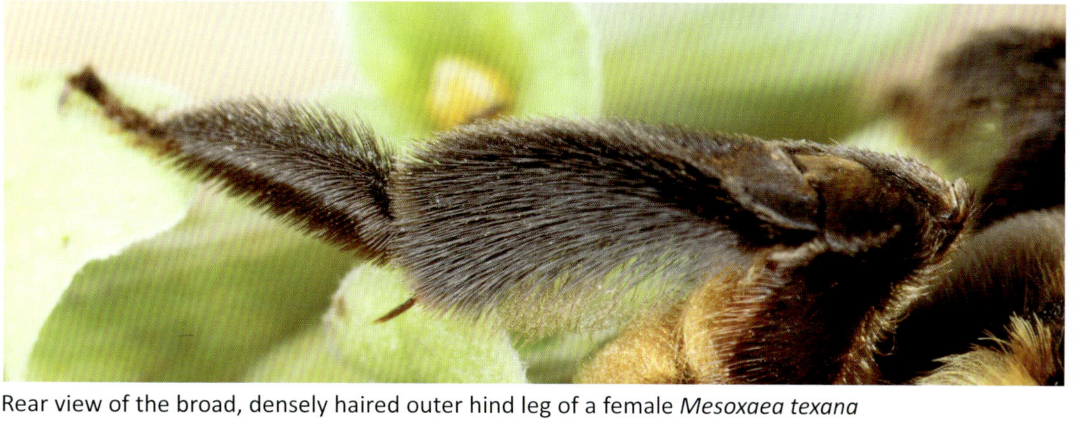

Rear view of the broad, densely haired outer hind leg of a female *Mesoxaea texana*

Behavior: Although generally rare, in those areas of the Valley where it does occur, *Mesoxaea texana* asserts itself vigorously and conspicuously. At El Mesteño Ranch in late spring 2021, male bees zealously guarded their territories, challenging the ranch owner and guests as they walked along pathways bordering vegetation where the bees were foraging. The bees would hover in the air in the middle of a pathway and then zoom back and forth over it to discourage passersby from continuing forward.

Mesoxaea texana is known to nest in large aggregations. In 1933, Cockerell described Pastor Gotthilf Birkmann's encounter in Fedor, Texas, with "a great swarm of wild bees." Birkmann had discovered a 700-by-230-foot *Mesoxaea texana* aggregation that contained an estimated 80,000 nest tunnels and 100,000 bees. The tunnel openings were about half an inch in diameter and descended as deeply as several feet into the soil. Like the bees from El Mesteño, those observed by Birkmann inhabited an area of sandy red soil.

According to Birkmann, the *Mesoxaea* aggregation was exceptionally loud: the bees made "a great ado and hubbub flying in all directions." Birkmann reported that they were "flying fast and like so many little furies, making all the noise possible to them. You could hear them at some distance, the sound was similar to that heard in the telegraph wires at times, but much louder."[2] Cockerell noted that the nesting *Mesoxaea* bees "flew with such energy and speed that they would not try to avoid any one standing in the way, but struck Mr. Birkmann's body and hat." The "wild buzzing and flying" happened repeatedly thereafter on various sunny days as the bees "had a holiday in the air," although they grew quiet during rainy and overcast periods.[3]

Mesoxaea texana females tend to build their nests in bare earth, the entrances haphazardly exposed rather than hidden by vegetation. The nests branch into lateral tunnels, each containing a single egg cell, which the female fills in with dirt after completing each chamber.

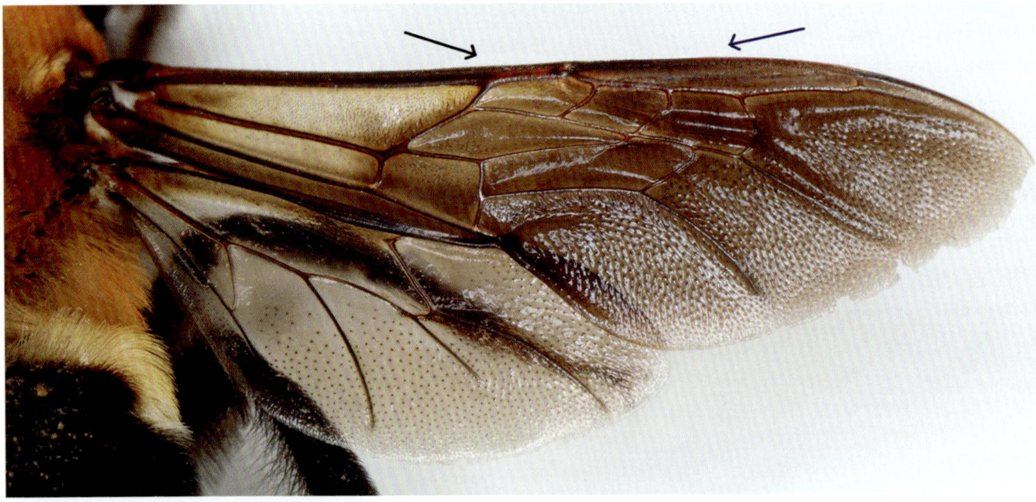

Wings of a *Mesoxaea texana*: the small stigma (black arrow) and long marginal cell (blue arrow) are typical of Oxaeinae: the *M. texana* forewing is clear near the base and darker toward the outer edges.

2. Hurd and Linsley (1976), p. 57.
3. Cockerell (1933), pp. 286–287.

Glorious protoxaea ♦ *Protoxaea gloriosa* Fox, 1893

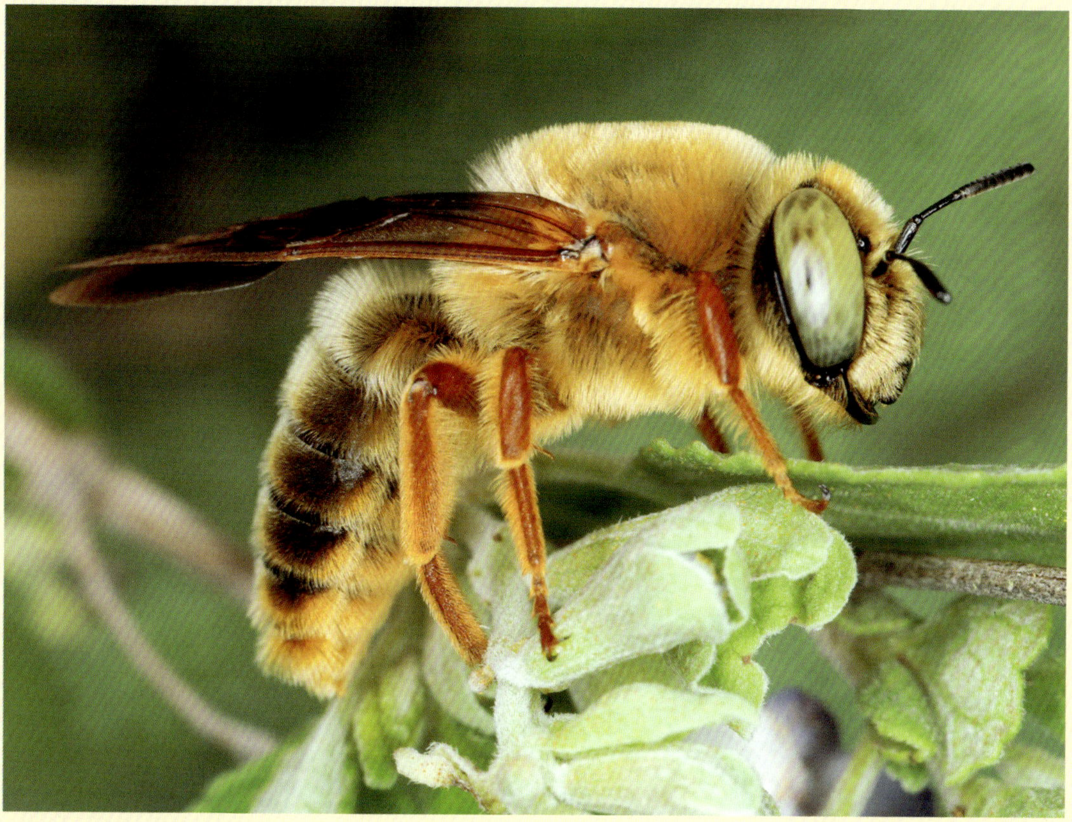

A male *Protoxaea gloriosa*

Description: The glorious protoxaea is a burly, golden-haired bee with a black abdomen, primarily red legs, and large green eyes. Males' wings are glassy near the base and bluish black on the outer edges; those of females are clear or yellowish near the base and bluish black on the outer edges, with strong violet highlights. The bees' tegulae are reddish brown. Like *Mesoxaea* males, *Protoxaea* males have striking and unusual faces: they are fly-like, with convergent compound eyes and low-set ocelli, with the median ocellus close to the clypeus. The female's dark face has a protuberant clypeus and is heavily pitted and fringed with golden hairs. Golden hairs also crown the upper face and vertex of both males and females. Their antennae, clypei, and mandibles are predominantly black, with some reddish areas.

The thorax hairs of both males and females are largely golden brown, although the bees tend to pale with age, so that in older specimens, the thorax hairs may be nearly white. Males' and females' abdomens are also clothed with golden hairs on both the terga and sternum. On males, the hairs are especially long and dense on T1 and T6 and form lateral tufts on T7. On females, white and golden hairs cover the sides of the abdomen and sternum; the upper abdomen is not as densely hairy as the male's; T5 has a thick fringe of long, feathery golden or pale yellow hairs; and T6 has long, reddish-yellow hairs on either side of the pygidial plate. The abdominal hairs

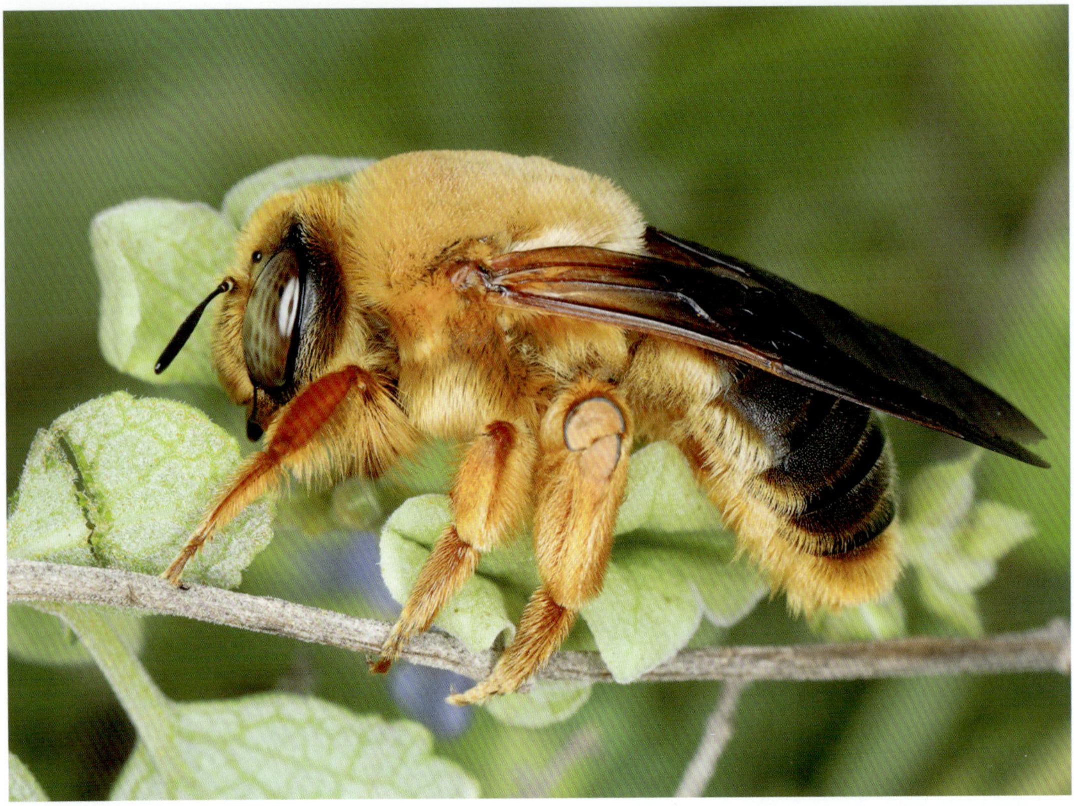

A female *Protoxaea gloriosa*

of both males and females, like those on the thorax, may pale with age. As is typical of Oxaeinae, *Protoxaea gloriosa* females have large, conspicuous knee plates on the apical end of each hind femur that join with smaller plates on the tibiae. The female's hind-leg scopal hairs are pale gold or white, as are additional scopal hairs on the first segments (S1 and S2) of the sternum. **Size:** females 17–22 mm (2/3–7/8 in.), males 13–16 mm (1/2–5/8 in.).

Range: *Protoxaea gloriosa* is principally a Mexican species: it ranges from the west to east coast of Mexico, at least as far south as Querétaro. In the United States, *Protoxaea gloriosa* is relatively uncommon, although it can be abundant locally. It is found mainly along the border in Arizona, New Mexico, and Texas. *Protoxaea gloriosa* was first identified in 1893 by a librarian at the Academy of Natural Sciences of Philadelphia, William J. Fox, after he examined a bee found in Las Cruces, New Mexico. T. D. A. and Wilmatte Porter Cockerell erected the genus *Protoxaea* in 1899 after viewing specimens of *P. gloriosa* from New Mexico and Arizona.

Protoxaea gloriosa males and females appeared during late May and early June 2021 in the Tamaulipan thornscrub habitats of El Mesteño Ranch (northwest Hidalgo Co.) and Falcon State Park (Starr Co.). John Brush, resident urban ecologist of Quinta Mazatlán, also reports seeing *Protoxaea gloriosa* nest aggregations at La Puerta Tract (Starr Co.) in late May 2020. Much of La Puerta Tract is a semiarid barretal habitat.

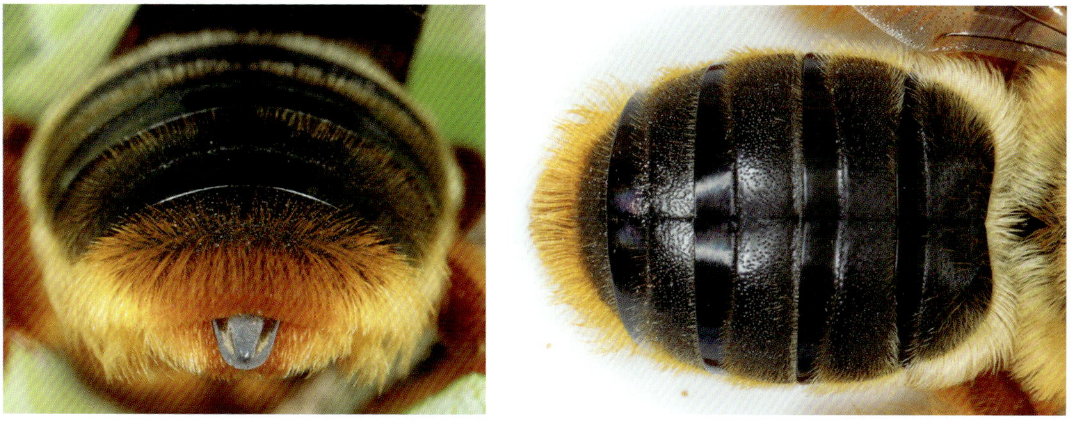

Rear and dorsal views of the abdomen of a female *Protoxaea gloriosa*

Associated plants: *Protoxaea gloriosa* forages for nectar on a fairly wide range of plants, although females tend to collect pollen most frequently from flowers in the families Fabaceae, Solanaceae, and Zygophyllaceae. Most of the information on the feeding habits of this species comes from studies conducted outside Texas, principally in Arizona, New Mexico, and Mexico. Female *Protoxaea gloriosa* have been recorded gathering pollen from Arizona poppy (*Kallstroemia grandiflora*), senna (*Senna bauhinioides*), and nightshades (*Solanum elaeagnifolium* and *S. rostratum*).

The female bees shown here from El Mesteño Ranch and Falcon State Park were feeding on shrubby blue salvia (*Salvia ballotiflora*) growing in sandy soil. Male *Protoxaea gloriosa* in the same location alternated between visiting shrubby blue salvia and whitebrush (*Aloysia gratissima*). As noted in the previous entry, the *Protoxaea* appeared together with *Mesoxaea texana. Protoxaea gloriosa* females followed a schedule like that of the *Mesoxaea*, appearing early in the morning and disappearing around 9:00 a.m. The male *Protoxaea* emerged shortly afterward and foraged through the day until late afternoon.

There are records of *Protoxaea gloriosa* feeding on salvia in Sabinal, Texas, in April 1910. *Protoxaea gloriosa* males and females have been documented in other states and in Mexico visiting plants of diverse families, among them horsetail milkweed (*Asclepias subverticillata*), seepwillow (*Baccharis salicifolia*), holdback (*Pomaria jamesii*), creosote bush (*Larrea tridentata*), French tamarisk (*Tamarix gallica*), sweetclover (*Melilotus officinalis*), and alfalfa (*Medicago sativa*).

ID tips: *Protoxaea gloriosa* is the sole member of its genus in the United States. Its size, exuberant golden hair, and largely red legs make it difficult to confuse with any other Texas bee. Differences between this species and *Mesoxaea texana* are discussed in the preceding entry. Most notably, *Protoxaea gloriosa* lacks tufts of white hair near the end of its abdomen, and the tip of the male's abdomen is blunt with gently curved edges rather than notched.

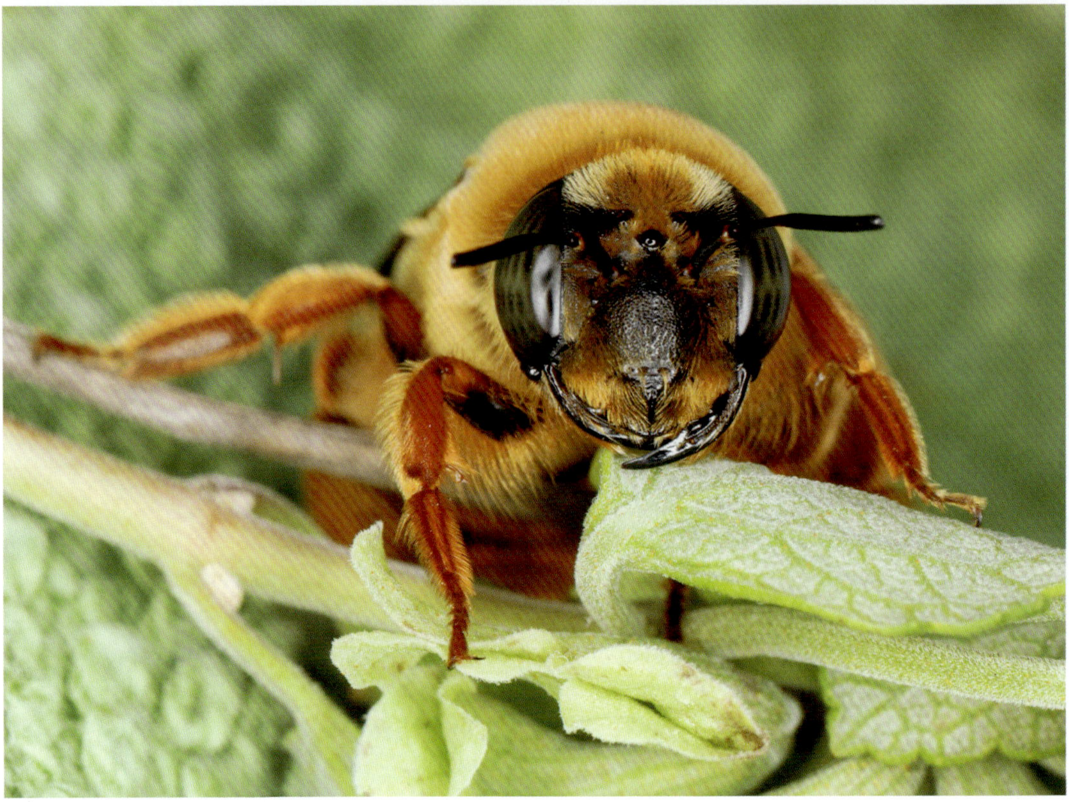

Face of a female *Protoxaea gloriosa*

Behavior: Like *Mesoxaea*, *Protoxaea gloriosa* tend to nest in large aggregations, with their nest entrances exposed rather than hidden under vegetation. Males emerge in spring a few days before the females and are known for their tumultuous behavior. Males stake out territories near food plants, hovering in place a few inches above the ground and frequently darting away from their guarded positions to chase one another through the air. Sometimes, *Protoxaea gloriosa* males mount mating pairs and attempt to dislodge other males by buzzing while clutching them tightly. *Protoxaea gloriosa* males also charge other, larger insects such as butterflies, and occasionally even birds.

During May 2021 at El Mesteño Ranch, male *Protoxaea gloriosa* tended to park in midair beside *Salvia ballotiflora* thickets, sometimes at the eye level of visitors to prevent them from nearing claimed shrubbery. Females are also boisterous: entomologists Cazier and Linsley described female *Protoxaea* butting observers in the head when they trespassed too close to the bees' nests.

Mining Bees

Subfamily Andreninae ◆ Tribe Andrenini
Genus *Andrena*

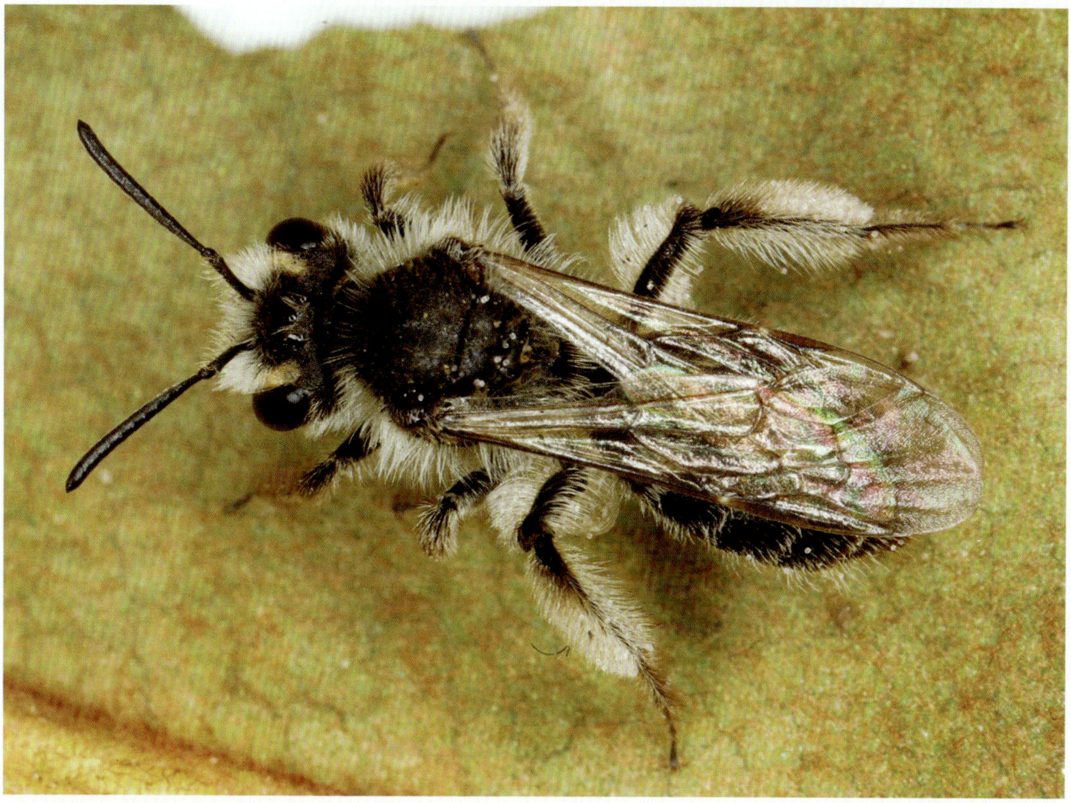

A female *Andrena*

Andrena is one of the largest bee genera, comprising more than 500 species on this continent and nearly 1,500 worldwide. In the Western Hemisphere, *Andrena* appear only in North and Central America. They flourish in temperate areas and seem immune to cold: they range far north, and even in northern Canada, they emerge just after winter ends. In the Lower Rio Grande Valley, many *Andrena* species appear in the cooler months of spring or fall and are rarely seen at other times of year. Fewer than a dozen species have been recorded in the Valley in the last century, although more than 100 species occur in Texas.

Andrena of the Western Hemisphere are solitary. *Andrena* are commonly referred to as "mining bees" because they construct underground nests, although this trait is shared by many solitary bees. *Andrena* females often nest close to one another, sometimes forming aggregations numbering in the tens of thousands. *Andrena* tend to favor sandy soils or bare ground lightly covered with leaf litter and other plant debris.

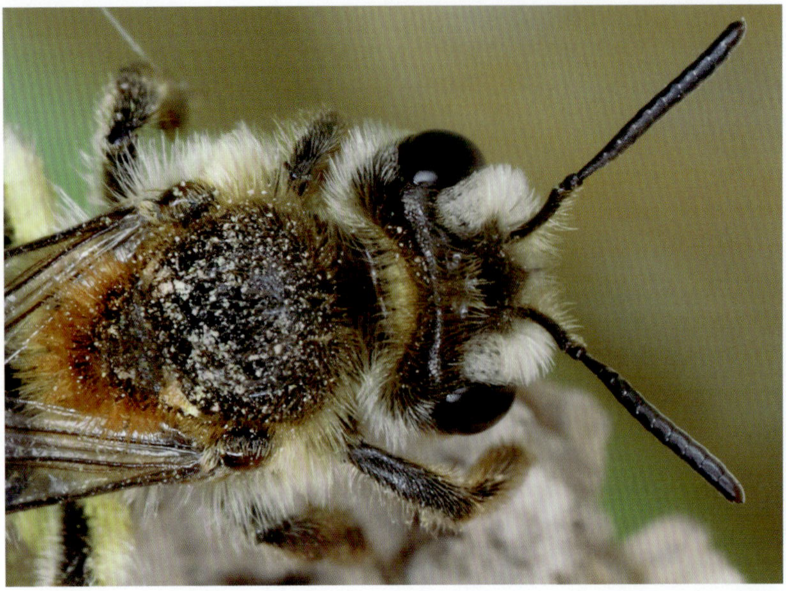

A female *Andrena* with dense hairs covering its facial foveae

Most *Andrena* are the size of honey bees or smaller. They often look "furry"—they have hairy thoraxes, faces, and legs. Their abdomens are usually dark and striped with pale hairs, and they are typically somewhat oval in shape rather than parallel sided. On all female (and some male) *Andrena*, facial depressions called *foveae* extend from just below the vertex to or beyond the antennae. The foveae are covered with hairs sometimes described in popular literature as looking like "sideways eyebrows." Female *Andrena* carry pollen tucked under their hind femurs and trochanters, on their tibiae, and on the propodeum. *Andrena* males tend to be smaller and slenderer than females. Males' faces are frequently covered with shaggy patches of hair and are sometimes adorned with pale or yellow markings.

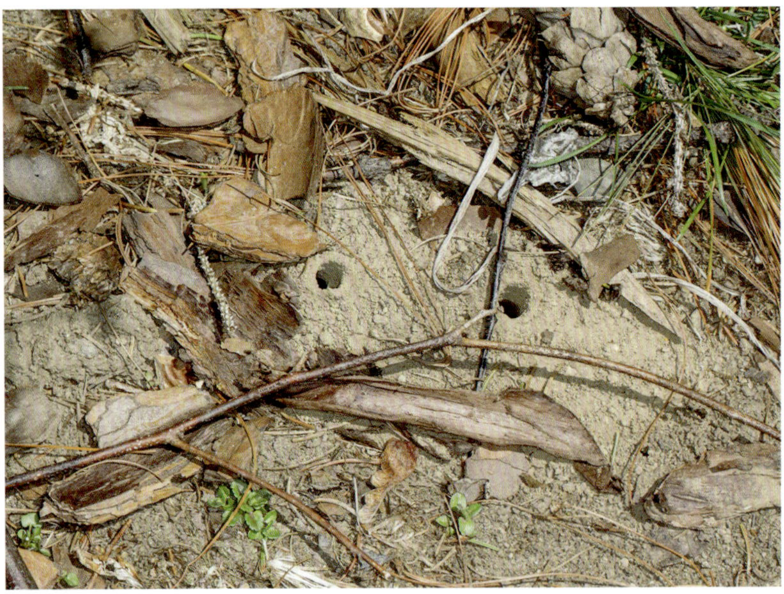

Andrena nests dug by two female bees: the nests are close to one another but lack connecting tunnels below the soil.

Miserable mining bee ◆ *Andrena (Larandrena) miserabilis* Cresson, 1872

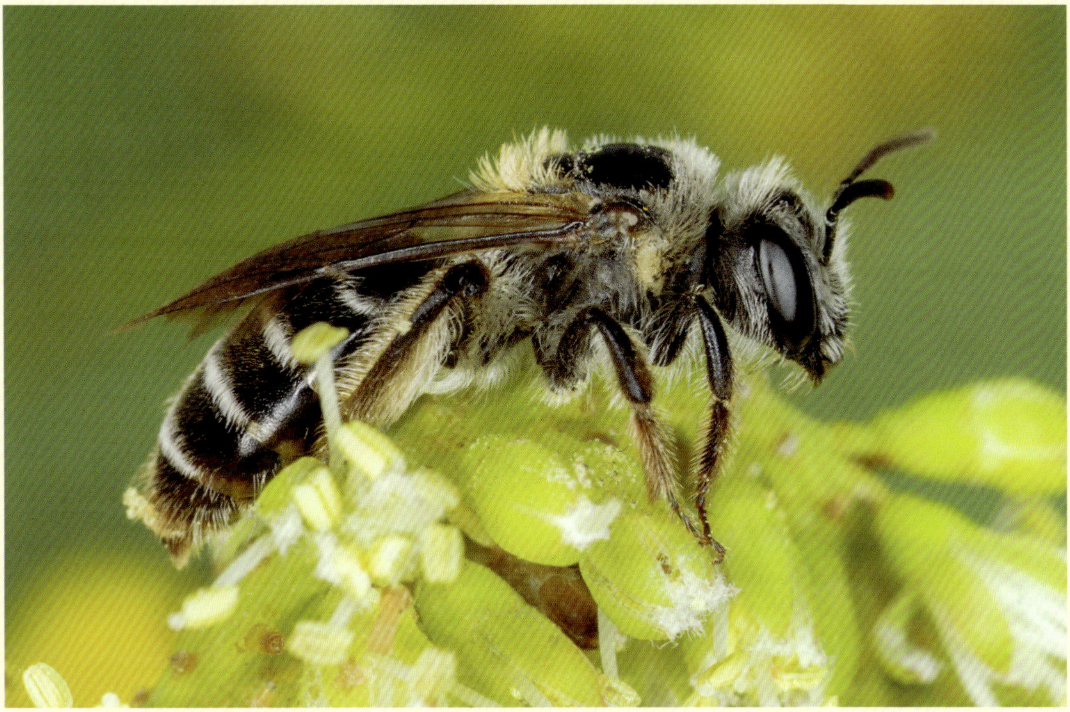

A female miserable mining bee

Description: *Andrena (Larandrena) miserabilis* is a hairy bee with a dark head and a long, dark body. The abdomen is blackish brown, smooth, and encircled with bands of pale hair. The female's thorax is sparsely haired in the center but is rimmed with long, rust-colored hairs along the hind edge of the scutellum. The female's face is squarish, with dark mandibles and antennae. Long strips of pale hair cover the female's facial foveae, extending from each antennal socket to just above the clypeus. On females, the abdominal hair bands on T2–T4 tend to be white and complete; T5 is covered with pale brownish hairs. The female's hind-leg scopal hairs are largely white and especially long on the tibiae. The forewings of both male and female miserable mining bees are glassy brown with brown veins, and their tegulae are brown to dark brown.

E. T. Cresson first described *Andrena miserabilis* in his 1872 publication "Hymenoptera Texana" after examining male and female bees found in Texas. He assigned most of the males, however, to a different species than the females. This result sprang in part from the notable sexual dimorphism of *Andrena miserabilis*. Males are small with narrow bodies; the pale hair bands on their abdomens are weakly defined; their faces are broad and their clypei pale yellow; and their heads and bodies look "fuzzy"—they are covered with long, pale brown to grayish hairs. Males' antennae are generally dark, but often reddish underneath the flagellum, and their mandibles sometimes have amber tips. **Size:** females 8–9 mm (3/10–1/3 in.), males 7 mm (1/4 in.).

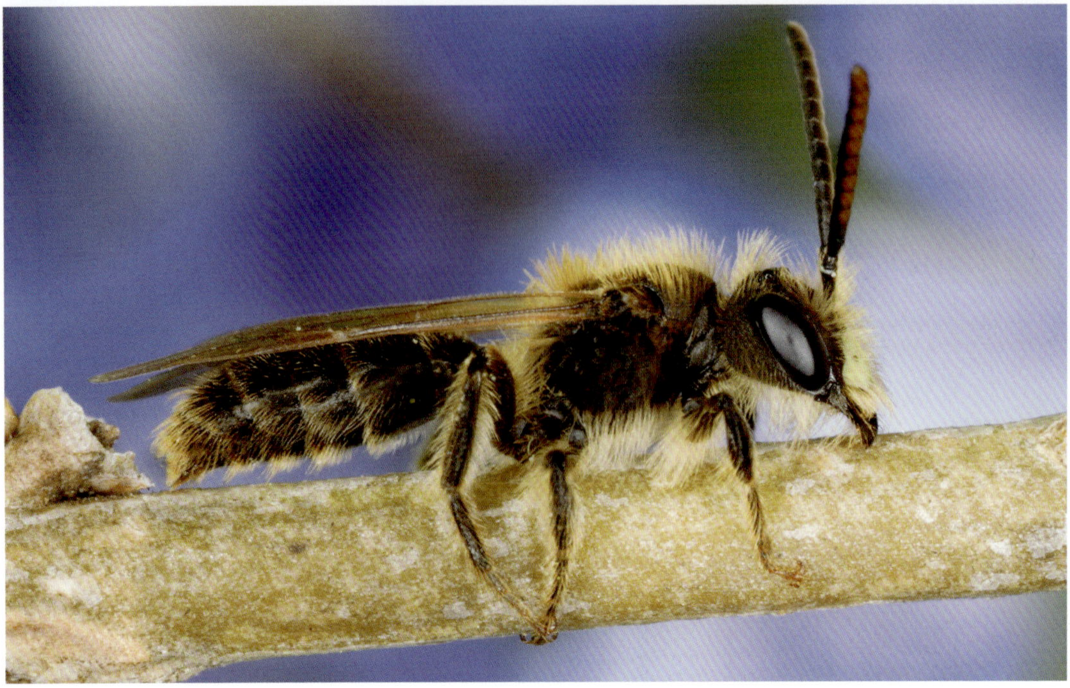

A male *Andrena miserabilis*: note the wide cheek and the long hairs covering the face and thorax.

Range: The miserable mining bee has an impressively wide range—this species is found throughout North America, from the East to West Coast, as far north as Canada, and as far south as Tamaulipas. The bees shown here appeared in April 2019 and 2021 in Mission, Texas (Hidalgo Co.).

Associated plants: Miserable mining bees are considered important pollinators of commercial fruit trees, particularly apples and stone fruits. They are, however, versatile generalist pollinators that visit a remarkable array of flora—they have been documented visiting plants from more than 20 different families. *Andrena miserabilis* emerges in the Lower Rio Grande Valley in early April, just as honey mesquite trees (*Prosopis glandulosa*) begin to bloom. The female bees shown here inhabited a small aggregation of *Andrena miserabilis* nests in sandy, bare soil lining the trails of an area overrun with prickly pear cactus. The *Andrena* females ignored the cactus, possibly relying on blossoming honey mesquite in the immediate area for pollen and nectar.

ID tips: *Andrena (Larandrena) miserabilis* is distinctive: Wallace E. LaBerge erected the subgenus *Larandrena* for *Andrena miserabilis*, and the miserable mining bee remains the sole New World member of its subgenus. Distinguishing traits of *Andrena miserabilis* include the following: (1) the female's clypeus has a broad, shiny area in the middle; (2) the yellow mask on the male's clypeus has two black spots that look a little like false nostrils, and the male's cheeks (behind the compound eyes) are wide; (3) the *Andrena miserabilis* forewing has a large stigma and three submarginal cells, with the second submarginal cell-much shorter than the third; and (4) in the field, the female *Andrena miserabilis* can be distinguished from other regional *Andrena* by the tuft of rust-red hair lining the rear rim of its thorax.

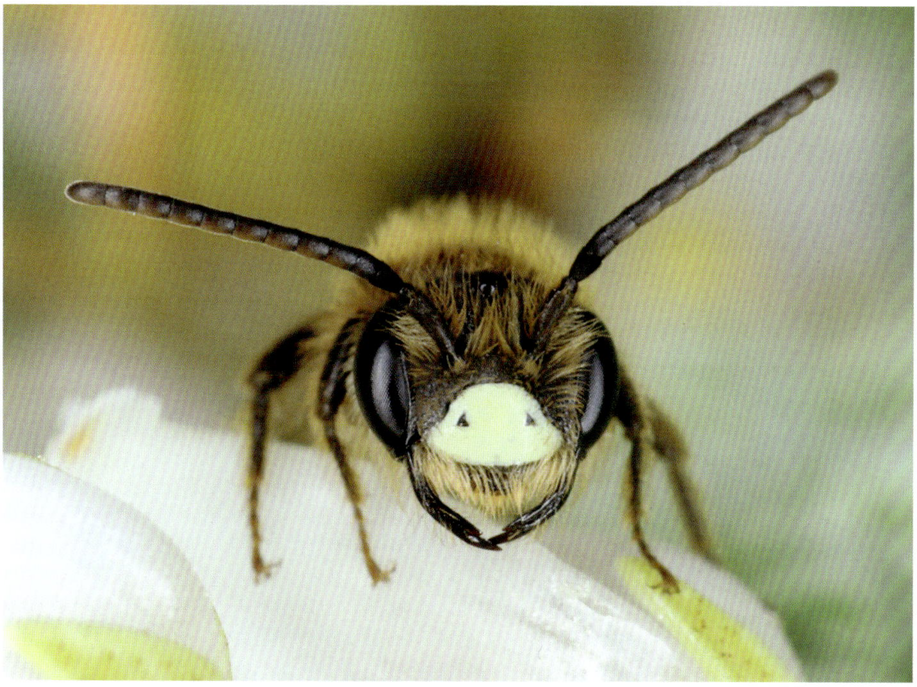

Face of a male miserable mining bee

Behavior: In 1978, entomologists Norden and Scarbrough published a groundbreaking article on *Andrena miserabilis* behavior, based on observations of nesting areas in Highlands County, Florida. Norden and Scarbrough described *Andrena miserabilis* nests as consisting of vertical tunnels running as as deep as 47 cm (18.5 in.) and terminating in lateral tunnels where egg cells were housed. Females built circular mounds of dirt outside their nest entrances. They lined their nests with a waterproofing substance and provisioned their egg chambers with moist, spherical pollen balls. Male *Andrena miserabilis* lurked outside the tunnel openings, waiting for mates—in two instances, males entered nest tunnels when seeking shelter from rain.

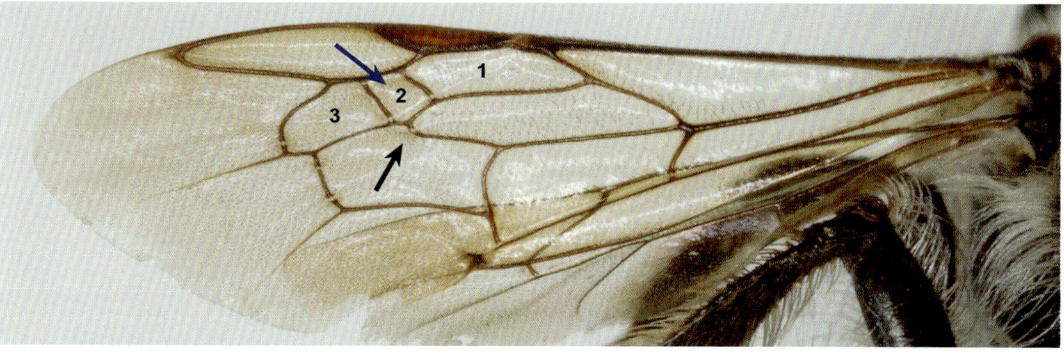

Forewing of an *Andrena miserabilis*: the second submarginal cell (blue arrow) is much shorter than the third, and the first recurrent vein (black arrow) meets the second submarginal cell more than halfway from its inner margin. The brown stigma is relatively large.

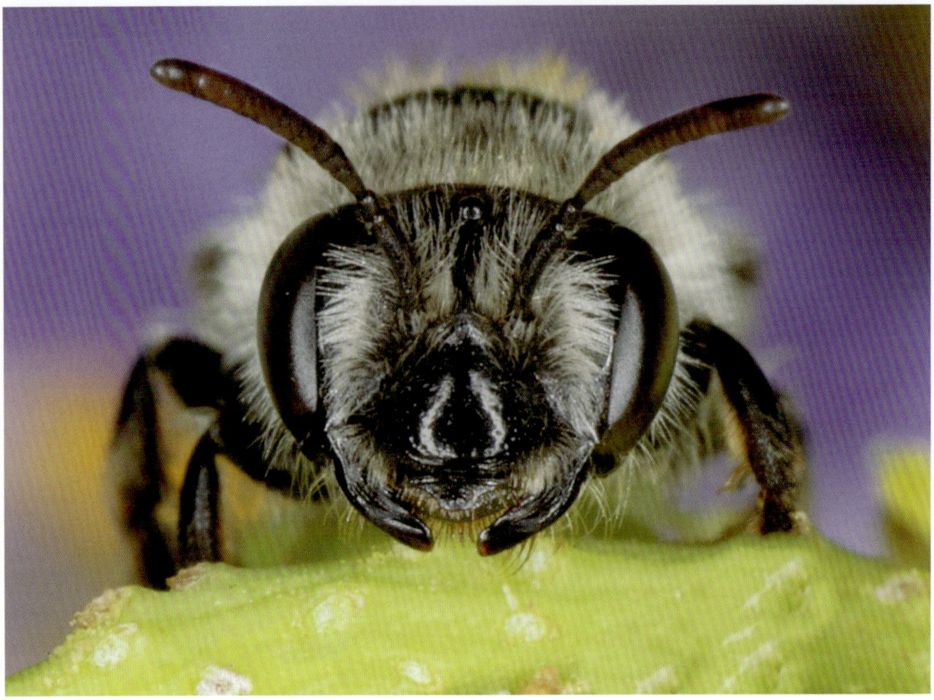

The face of a female miserable mining bee, with a shiny clypeus and pale hairs covering the long facial foveae

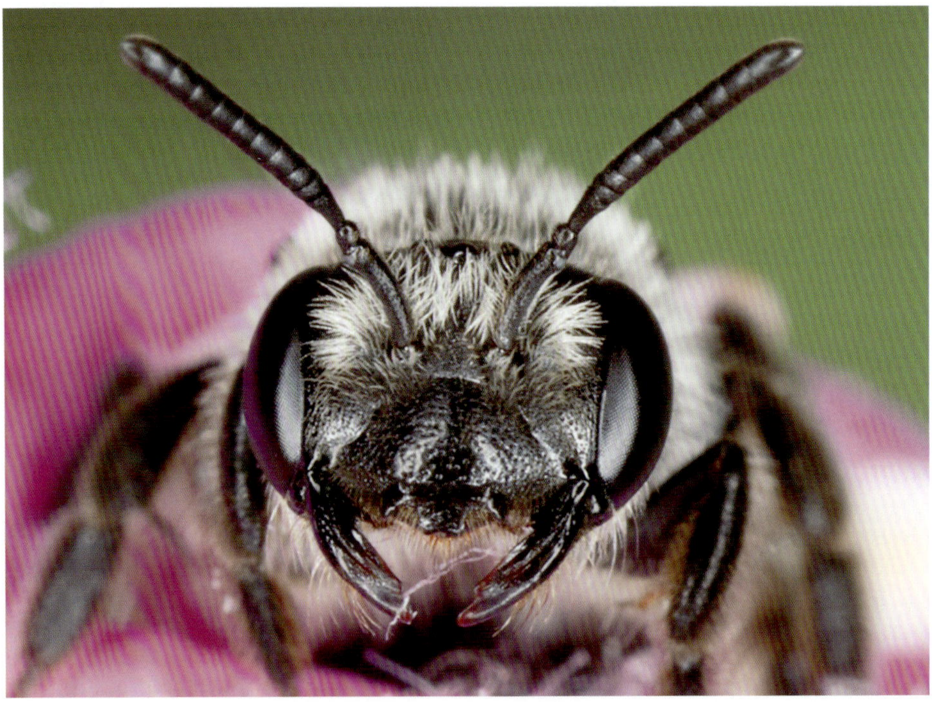

The broad face an *Andrena (Opandrena)*: note the pitted clypeus. The pale hairs that cover the foveae barely extend below the antennal sockets.

Unnamed Andrena ◆ *Andrena (Opandrena)* sp.

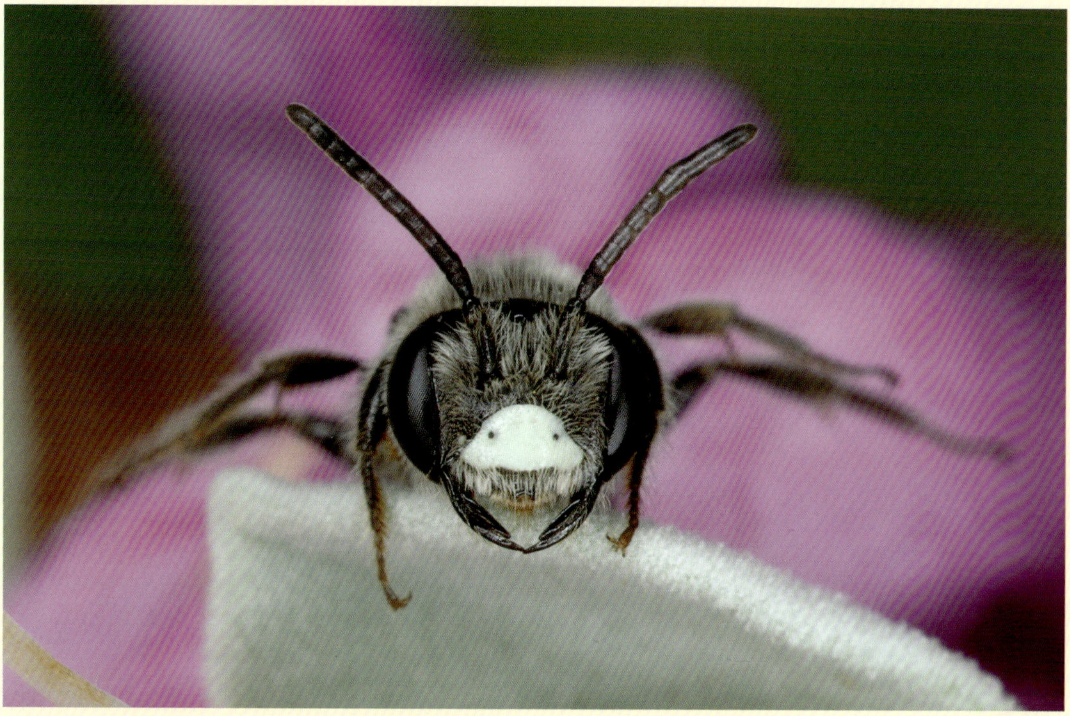

A male *Andrena (Opandrena)* from Starr County

The Lower Rio Grande Valley continues to harbor bee species that are as yet undocumented and unnamed. The mining bee shown here is placed tentatively in the subgenus *Opandrena*. It was identified by John L. Neff, director of the Central Texas Melittological Institute, who for years has studied the association of this species with the native Texas shrub known as cenizo (*Leucophyllum frutescens*).

Description: This is a medium-sized bee with a broad face and a black abdomen encircled with white hair bands (T1–T4 on females and T1–T5 on males). Long white hairs clothe parts of the face, thorax, and legs of both males and females. On females, there are long, dark brown hairs on T5 and T6, and the hind-leg scopal hairs are both brown and white. The female's face is broad, with a pitted clypeus and narrow vertex. White hairs cover the facial foveae, barely extending below the antennal sockets. The male's clypeus has an ivory mask with two dark spots on the tentorial pits, and the male's upper and lower face are partly covered by long pale hairs. The tegulae of females and males are brownish black and their wings glassy light brown with smoky outer edges and brown veins. **Size:** females 8 mm (3/10 in.), males 7.5 mm (< 3/10 in.).

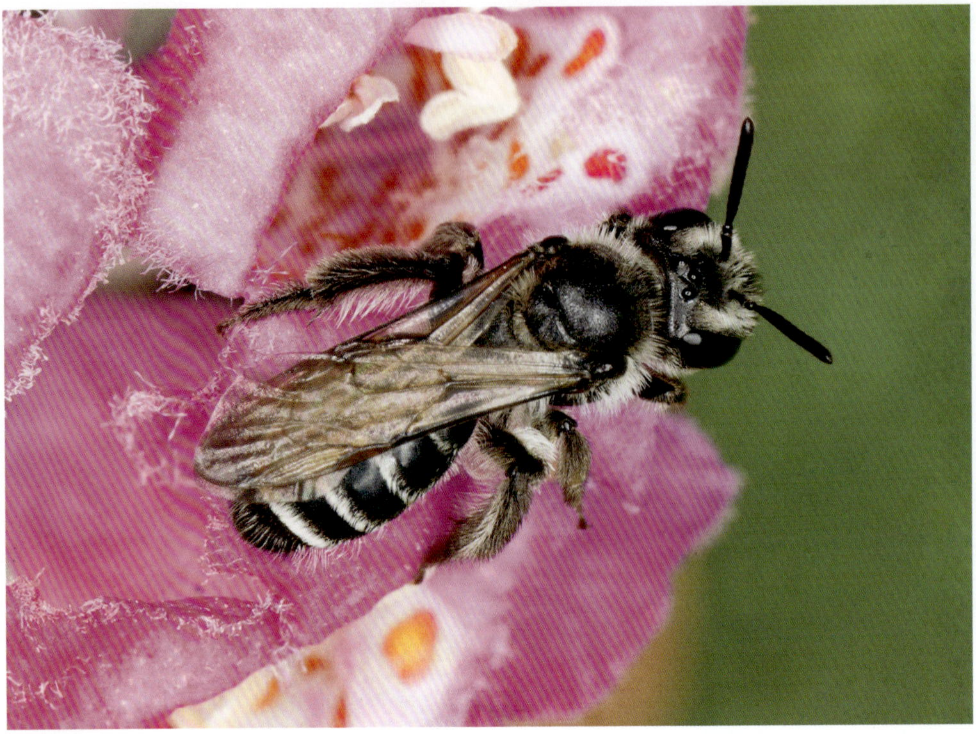

A female *Andrena (Opandrena)* on cenizo

Range: This unnamed *Andrena* species appeared in late May 2021 at Falcon State Park (Starr Co.). Neff has documented the same *Andrena* species previously in Texas and Mexico, feeding on cenizo during the fall, as far north as Kimble County and as far south as Coahuila.

Associated plants: The *Andrena* shown here were among hundreds of female and male bees gathered on three cenizo bushes that were roughly six feet high and covered with abundant blossoms. The cenizo had bloomed immediately following a rainy week preceded by a long dry period. The bees remained on the cenizo for several days, foraging and mating on the leaves and blossoms until the cenizo stopped blooming. Both male and female bees thereafter disappeared entirely from sight. None of the *Andrena* were found on other neighboring flora either during or after the cenizo's bloom period, although flowers from an array of plant families were blooming near the cenizo shrubs.

ID tips and additional information: This bee is best distinguished from other, similar *Andrena* by its association with cenizo. Sometimes called Texas sage or Texas barometer bush, cenizo is a shrub of the figwort family (Scrophulariaceae), with silvery-green leaves and small purple-to-white flowers. Cenizo favors arid and semiarid habitats with calcareous soils and ranges across the southwestern United States, Texas, and much of Mexico. It is common in the Lower Rio Grande Valley and the neighboring Mexican states of Tamaulipas and Nuevo León. In the Valley, periods of cenizo florescence seem largely unconnected to time of year—cenizo blooms in Cameron, Hidalgo, and Starr Counties throughout the spring, summer, and fall.

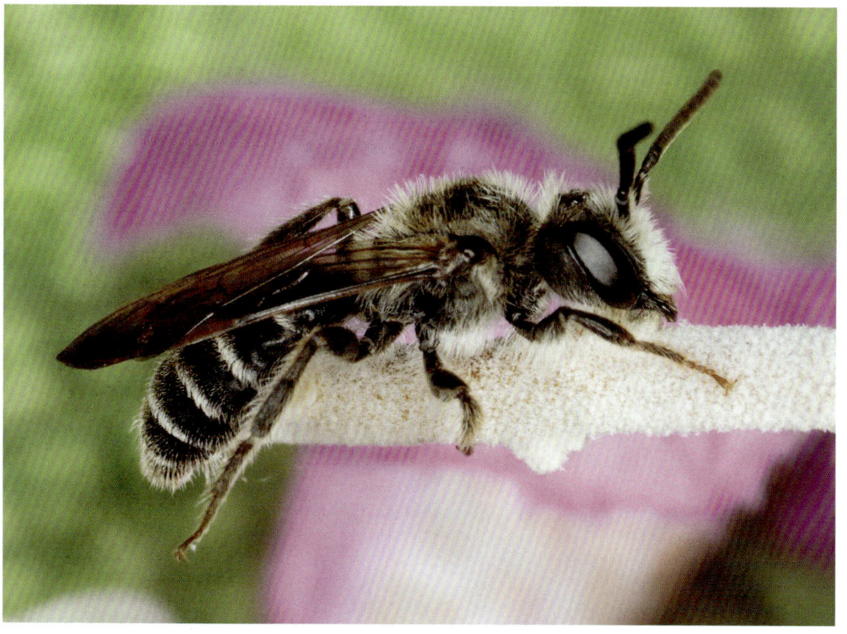

A male *Andrena (Opandrena)*

Cenizo's bloom cycles are largely geared to the appearance of rain. The name "Texas barometer bush" arises from the plant's tendency to bloom in response to increased humidity. Because cenizo has the uncanny tendency to bloom ahead of rain, Valley residents sometimes describe it as a reliable foreteller of rainstorms. The existence of *Andrena* that specialize on cenizo is similarly mysterious, given that the bees' periods of emergence hinge on their ability to accurately predict when cenizo will flower. Presumably, the *Andrena* accomplish this by sensing humidity in the soil.

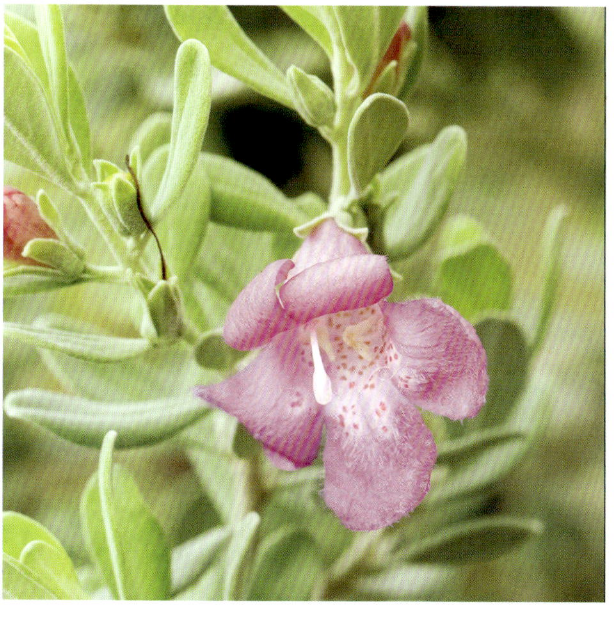

Leucophyllum frutescens, commonly known as cenizo, or Texas barometer bush

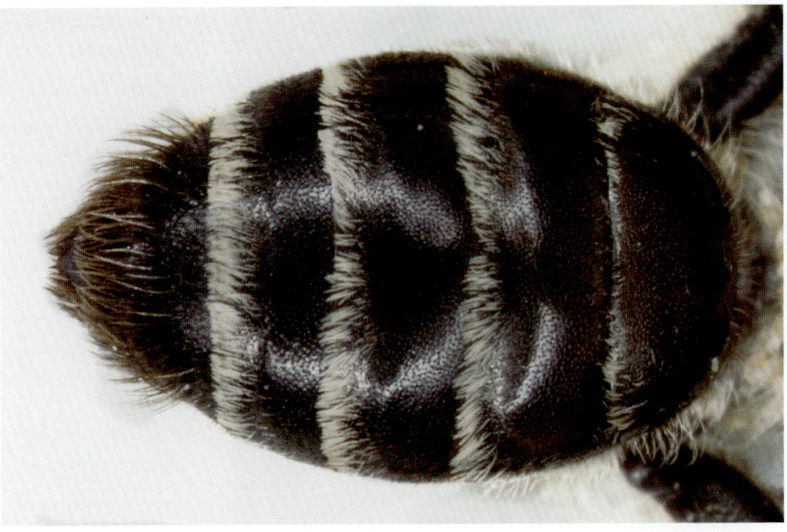

Abdomen of a female *Andrena (Opandrena)*: pale hairs appear on T1–T4; the hair band is weaker on T1 than the bands on T2–T4. T5 and T6 are covered with long brown hairs. The roughly oval shape of the abdomen is a trait shared by most *Andrena*.

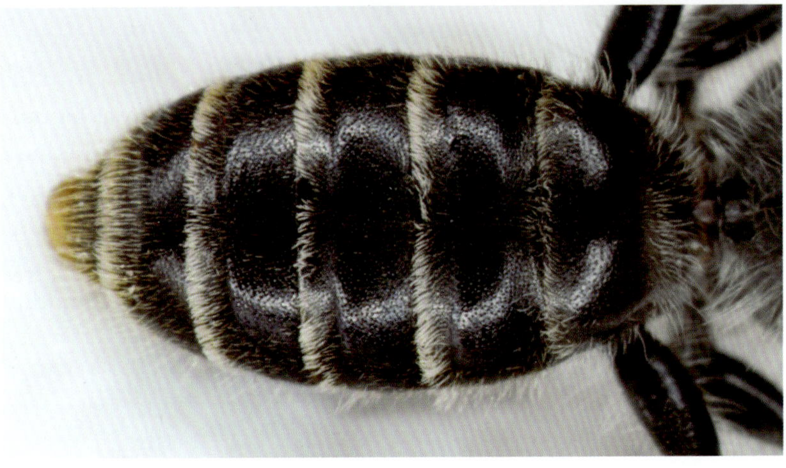

Abdomen of a male *Andrena (Opandrena)*: the abdominal hair bands are narrower and not as well defined as the female's.

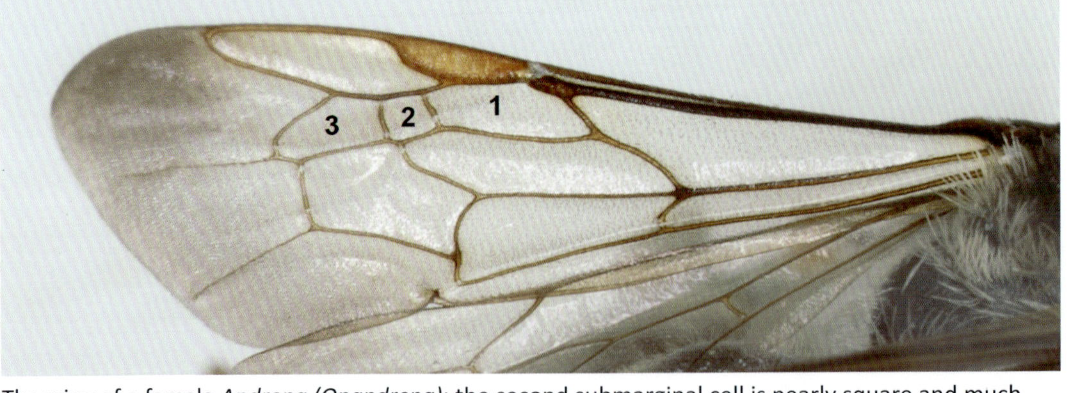

The wing of a female *Andrena (Opandrena)*: the second submarginal cell is nearly square and much shorter than the third, and the outer edges of the wing are dusky. The brown stigma is relatively large.

Protandrenine Bees

Subfamily Panurginae ◆ Tribe Protandrenini
Genus *Protandrena*

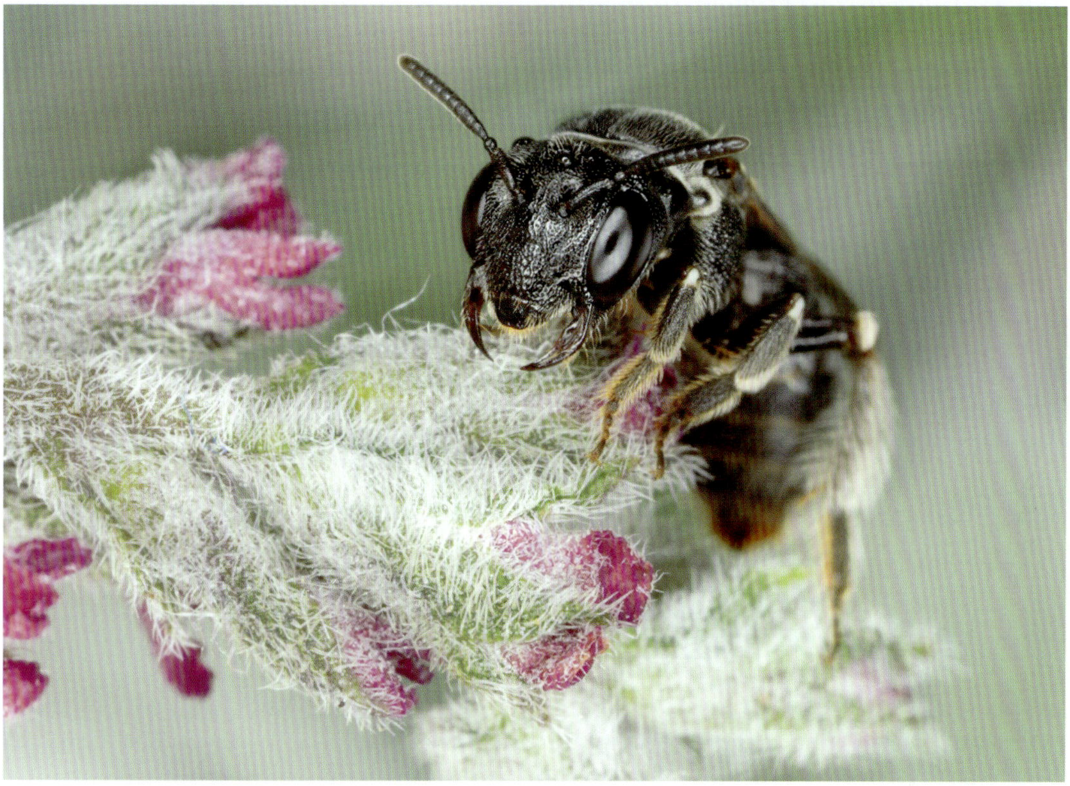

A female Texas protandrena

The subfamily Panurginae contains an array of small bees most diversely represented in temperate and subtropical regions of the New World. In North America, panurgine bees are divided into four tribes: Protandrenini, Perditini, Calliopsini, and Panurgini. Samples of the first three are shown here.[4] Unlike oxaeine bees and *Andrena*, which transport pollen dry, panurgines are "agglutinators"—that is, they add nectar to dry pollen and carry moist or partly moistened pollen loads.

Panurgine bees tend to be dark colored and relatively hairless. Their wings are distinctive and aid in separating them from other bees: the marginal cell of the panurgine forewing is lopped off at the tip and strongly bent away from the edge of the wing. The wings of Panurginae may have two or three submarginal cells.

4. A single member of the tribe Panurgini has been documented in the Lower Rio Grande Valley—*Panurginus polytrichus*. Entomologists Michener, Beamer, Rozen, and Stephen reported finding the species in Cameron County in 1950. *Panurginus polytrichus* is also the sole species of its genus found in Texas (Neff 2003).

The tribe Protandrenini contains the genus *Protandrena*: a group of small, usually dark colored, nonmetallic bees that often have pale markings on their faces, thoraxes, and legs. The prefix *prot-* means "earliest form of," and *Protandrena* look like no-frills, tougher versions of *Andrena*. *Protandrena* have pitted heads and bodies that give them an armored appearance, and their facial foveae are smaller than those of *Andrena* and hairless. Male *Protandrena* lack the fuzzy look typical of many male *Andrena*.

Protandrena appear only in the Western Hemisphere. They are diversely represented in Mexico and the western United States, and they are widespread throughout temperate North America. *Protandrena* females are solitary ground nesters that tend to build in sunny locations with sparse plant cover. Like *Andrena*, *Protandrena* waterproof their egg chambers and often nest in aggregations.

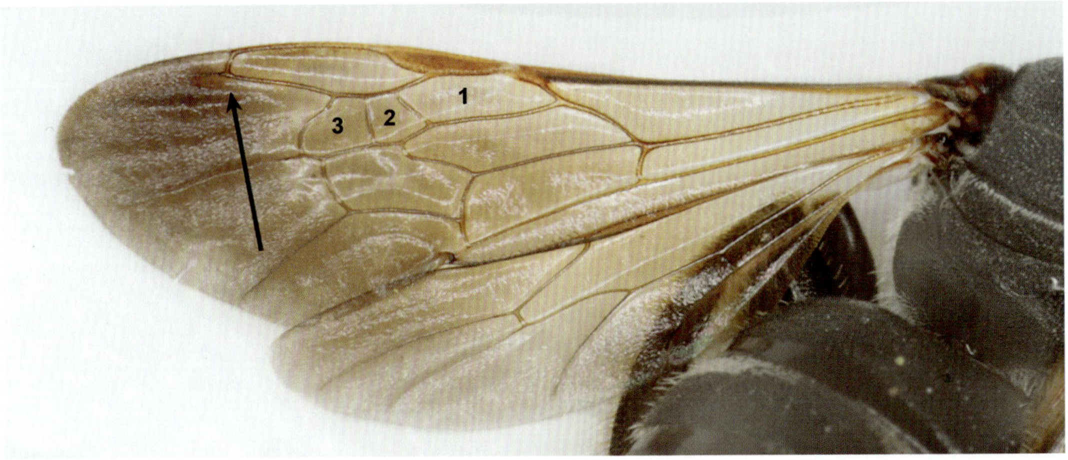

Forewing of a female *Protandrena (Protandrena) bancrofti*: the marginal cell is lopped off at the tip (black arrow), and there are three submarginal cells.

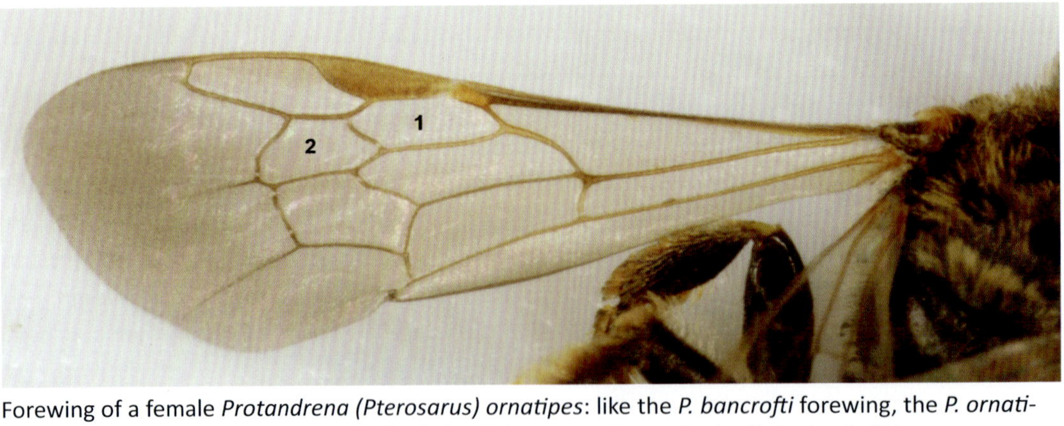

Forewing of a female *Protandrena (Pterosarus) ornatipes*: like the *P. bancrofti* forewing, the *P. ornatipes* forewing has a truncated marginal cell, but it has two submarginal cells instead of three.

Bancroft's protandrena ◆ *Protandrena (Protandrena) bancrofti* Dunning, 1897

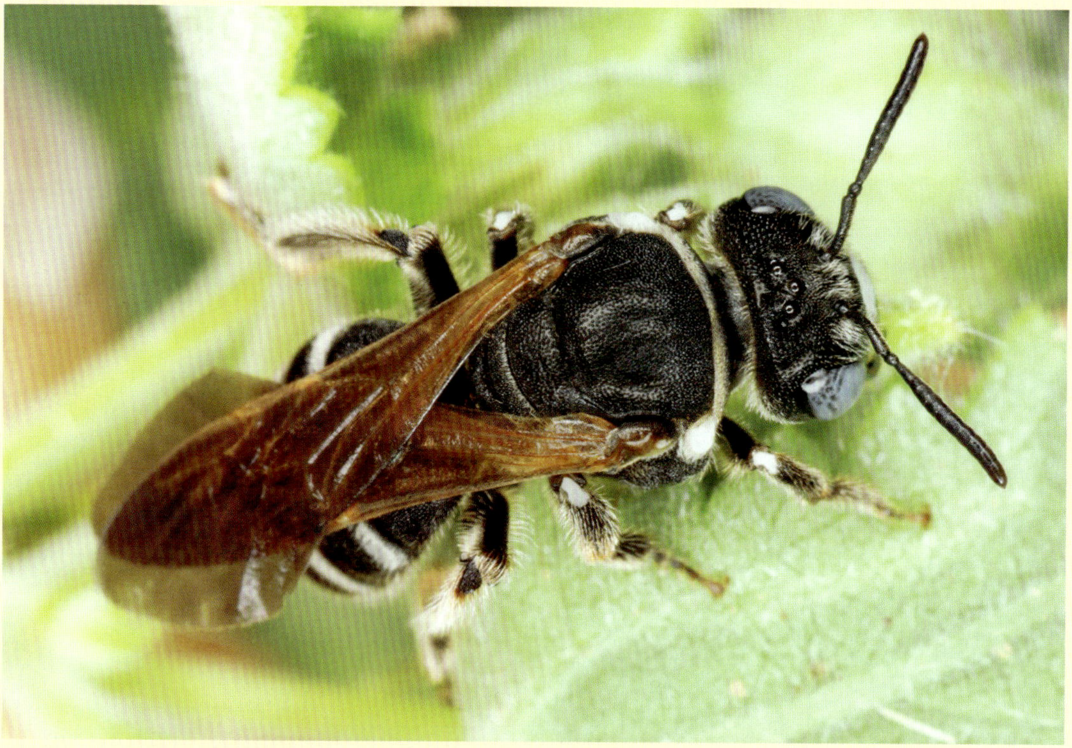

A female *Protandrena bancrofti*

Description: Bancroft's protandrena is a black, medium-sized bee with ivory facial markings and an abdomen banded by pale hairs. The bee's thorax is sparsely haired, revealing a black, pitted surface. The front of the female's thorax is rimmed with short, pale hairs. The outer rims of the bees' abdominal segments are distinctive: they are smooth, unpitted, and somewhat glassy. On males, the abdominal hair bands appear on T2–T5; on females they appear on T2–T4, and sparse brown hairs cover T5–T6. The abdomens of males are narrower than those of females. The legs of both female and male bees are predominantly dark and covered with fine, pale hairs. Their tegulae are dark brown to pale reddish brown, and their forewings have three submarginal cells and are light brown, darkening toward the tips, with reddish-brown veins. The stigma of each forewing is reddish brown, with a small pale spot on its inner angle where it joins the prestigma.

Both male and female *Protandrena bancrofti* have pale markings on the pronotal lobes and tegulae. Females may have an additional small pale spot on the small plate behind each tegula, and yellow spots on the upper tibiae of the front and middle legs. On males, most of the tibiae and some of the tarsi are white or pale yellow; the pale markings are most extensive on the hind legs.

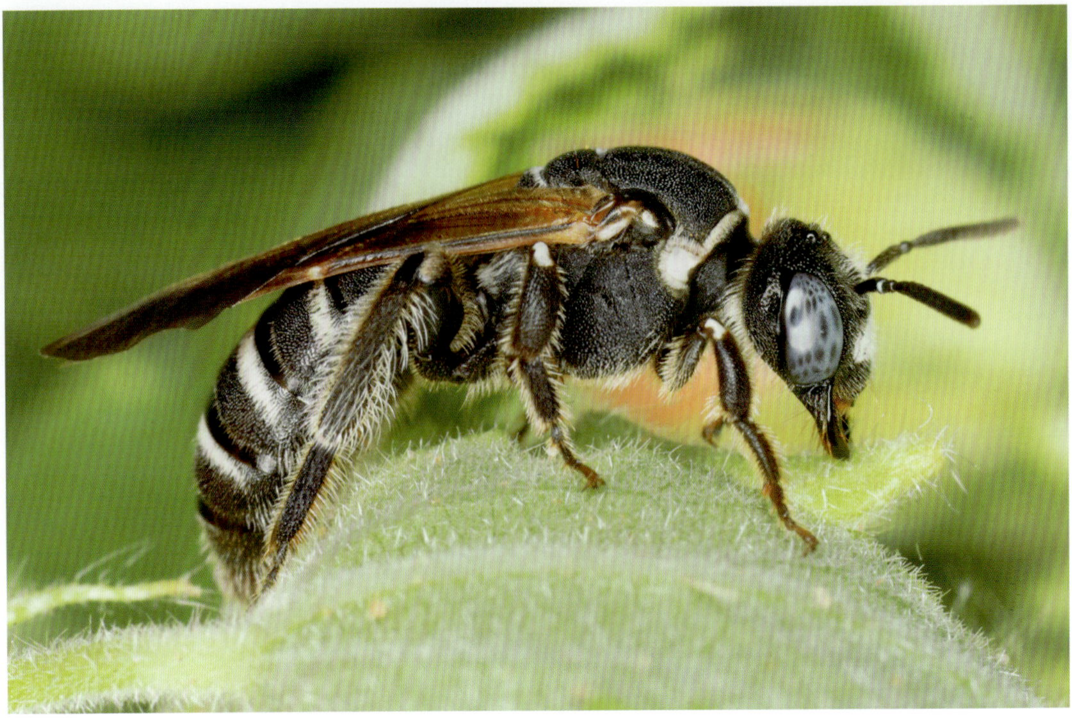

A female *Protandrena bancrofti*

The mandibles, clypeus, and antennae (including the scapes) of *Protandrena bancrofti* are predominantly black on both males and females, and the faces of both are deeply pitted. Females have black labrums and distinctive pale markings on the upper clypeus, described in more detail below. Males often have pale labrums. On males, pale facial markings cover the lower clypeus, and the pale areas on either side of the clypeus expand upward to just short of the base of each antenna and outward, sometimes touching the inner orbits of the compound eyes. **Size:** 9–11 mm (1/3–2/5 in.). *Protandrena bancrofti* are large for members of their genus.

Range: Bancroft's protandrena appeared in early June 2021 at El Mesteño Ranch (northwest Hidalgo Co.). There appear to be no prior records of this species in Hidalgo County, but it is widespread in Texas, although relatively uncommon. *Protandrena bancrofti* occurs throughout the central United States, from the Mexican border to Canada. It ranges south as far as central Mexico.

Associated plants: The female Bancroft's protandrena shown here was found foraging on copper globemallow (*Sphaeralcea angustifolia*). *Protandrena* may be either generalist or specialist pollinators. *Protandrena bancrofti* has been documented feeding on a spectrum of flora, including snowberry (*Symphoricarpos occidentalis*); pea-family plants such as milkvetch and melilot; buffalobur nightshade (*Solanum rostratum*); mustard; milkweed; and aster-family plants such as sunflower and *Baccharis*. *Protandrena bancrofti* is a vibratile ("buzz") pollinator, an unusual trait among Andrenidae.

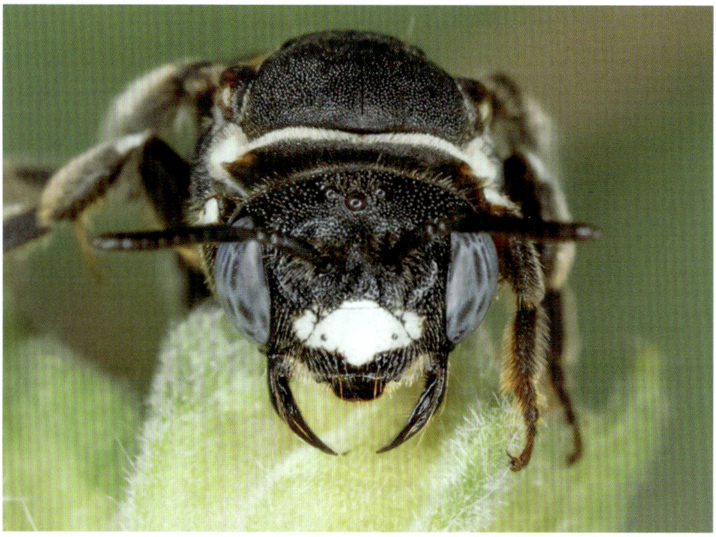

Face of a female *Protandrena bancrofti*

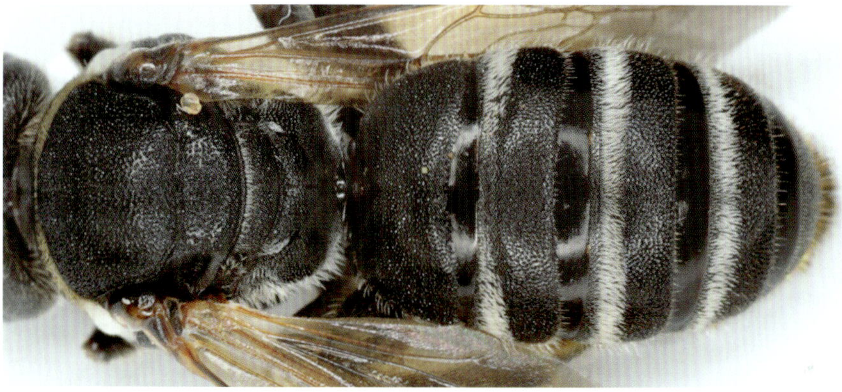

Dorsal view of the thorax and abdomen of a female *Protandrena bancrofti*: note the depressed centers of the terga and their smooth outer rims.

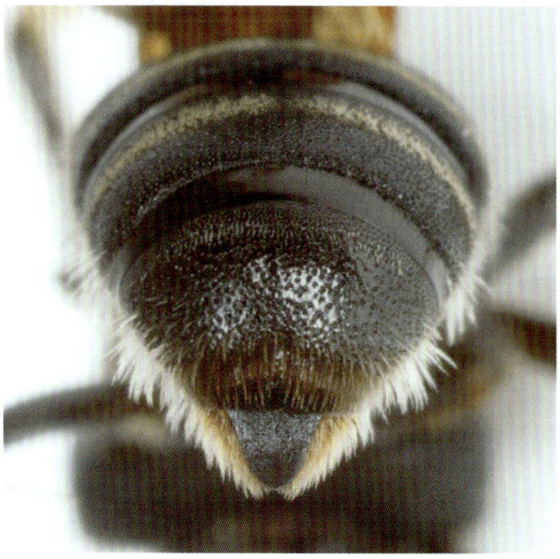

Rear view of a female bee's abdomen

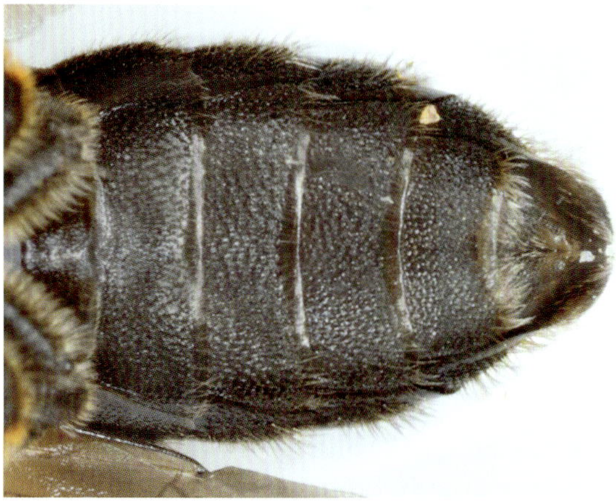

Sternum of a female bee

ID tips: *Protandrena* are often differentiated from one another by the markings on their faces and bodies, the sculpting of their abdomens, or the color of their wings and abdomens. (1) The female *Protandrena bancrofti* is easily distinguished by its idiosyncratic facial markings, which Cockerell described as "trilobed, the lobes acute, like a leaf."[5] (2) The extensive yellow areas on the male's face distinguish it from similar *Protandrena* species. The often-yellow labrum of *Protandrena bancrofti* males is also distinctive. (3) An unusual trait of this species is that the central portions of T2–T5 are depressed; the depressions are especially conspicuous near the middle of the base of each segment. (4) Equally useful in identifying this species are the smooth outer rims of these terga. (5) Finally, both male and female bees have richly colored blue-gray eyes, a conspicuous trait that aids in field identifications.

5. Cockerell (1921b), p. 13.

Texas protandrena ◆ *Protandrena (Pseudopanurgus) texana* (Timberlake, 1976)

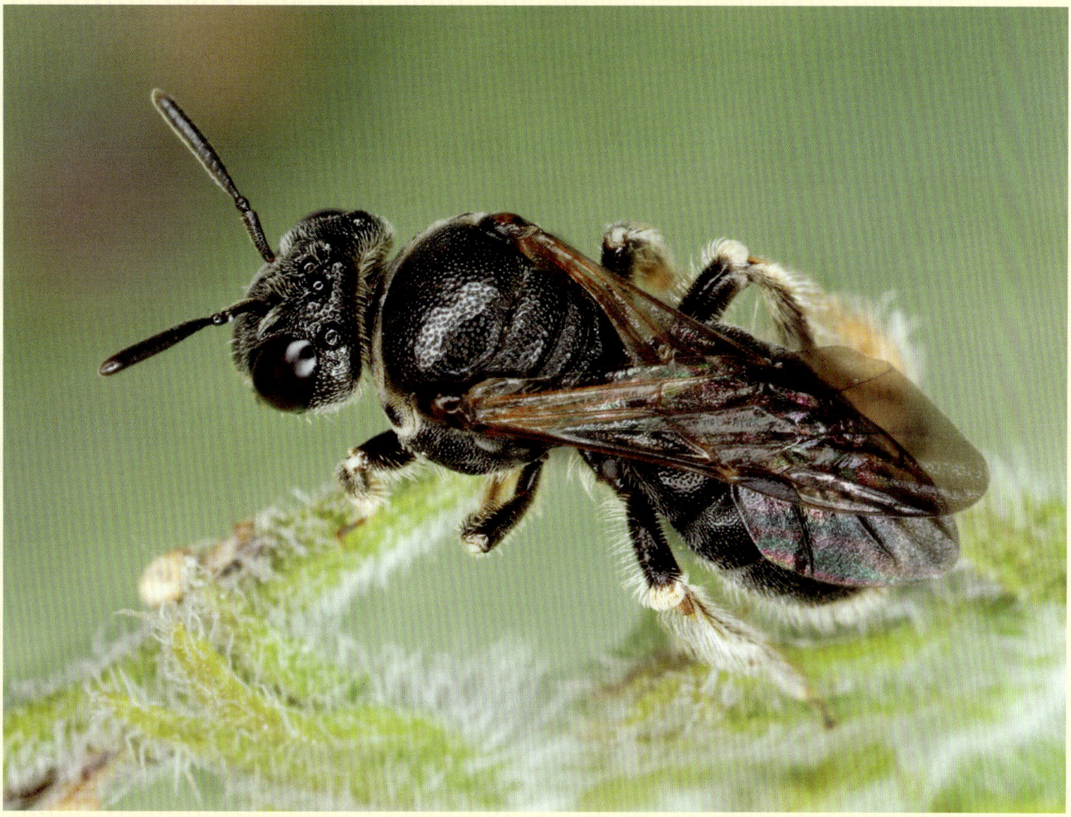

A female *Protandrena texana*

Description: *Protandrena texana* was known until fairly recently as *Pseudopanurgus texanus*. The former genus *Pseudopanurgus* was defined by characteristics still useful in identifying this species: *Pseudopanurgus* were described as small but robustly built, with relatively hairless black bodies; broad faces; thoraxes with a deeply and densely pitted central area; abdomens frequently with weakly defined or absent hair bands; and forewings with two submarginal cells. Females formerly classified as *Pseudopanurgus* typically have black or mostly black faces and large, hairy spines on the front coxae; males have primarily yellow faces and some pale markings on their legs and pronotal lobes. All of these qualities are true of *Protandrena texana* and, taken in combination, aid in distinguishing it from *Protandrena* of other subgenera.

The female *Protandrena texana* has an entirely black face. The male's face is ivory below the antennae, except for two small black areas near the mandibles. Both males and females have distinctive rounded bosses (bumps) behind the top inner edges of each compound eye.

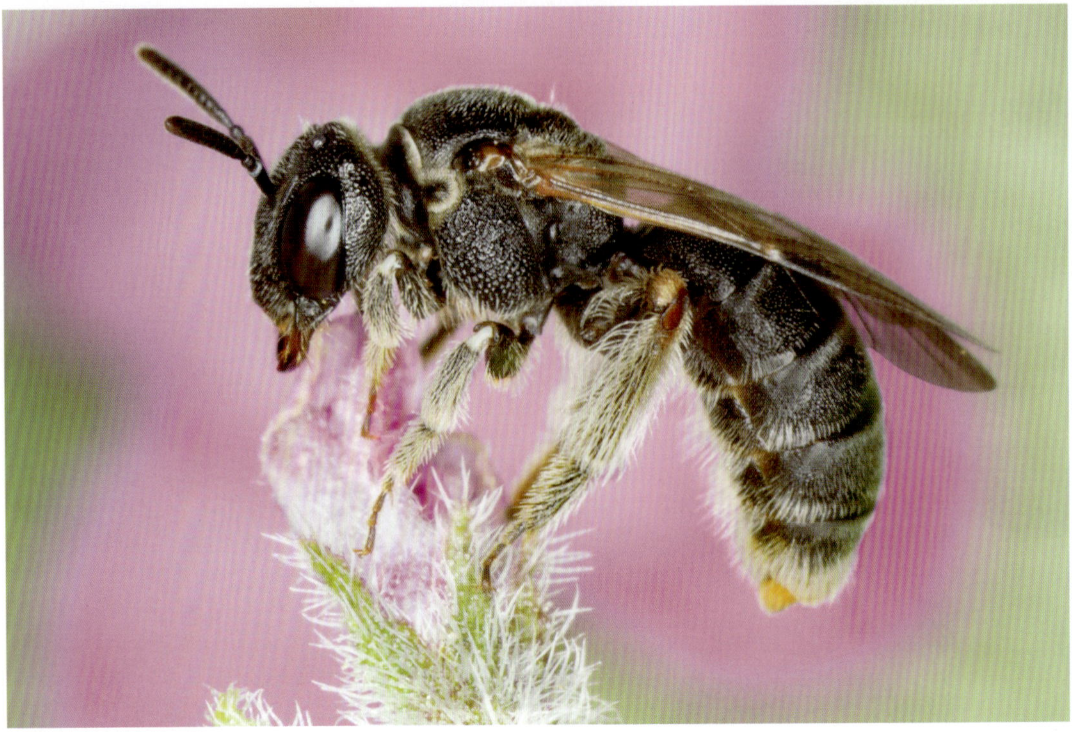

A female *Protandrena texana*

The tegulae of *Protandrena texana* are reddish brown, and their wings are glassy brown, becoming dusky toward the outer edges, with dark reddish-brown veins. The legs of both males and females are primarily black and partly covered with pale hairs. Females carry pollen on long, feathery, dense white scopal hairs located on their hind tibiae and basitarsi. Females' legs have ivory markings on the base of each front and middle tibia, and the apex of the hind femur is covered by a dense tuft of short white hairs. Males have extensive pale yellow areas on all tibiae; their tarsi are pale yellow and reddish.

Both male and female *Protandrena texana* appear hairless to the naked eye. Some pale, thin, very short hairs cover parts of the face and thorax. Narrow collars of short white hairs thinly edge the female's pronotal lobes and the front of the thorax. Sparse white hairs appear on the female and male bees' abdomens. Viewed from the side, the hairs form the suggestion of bands on T2–T5 of the female's abdomen; T6 of the female is covered with fine reddish hairs. The male's sternum is clothed with short, dense white hairs; S6 is notched. On both males and females, the centers of the tergites are depressed; this feature is especially evident on the outer rims of T2–T3. **Size:** females 7 mm (1/4 in.), males 6 mm (1/5 in.).

Range: Entomologist P. H. Timberlake originally described this species in 1973 after examining bees found in Matamoros, Tamaulipas, and the Texas counties of Jim Wells, Karnes, Maverick, and Zavala. *Protandrena texana* is widespread in southern Texas, although uncommon. It has been found as far west as New Mexico. *Protandrena texana* appeared at Falcon State Park (Starr Co.) in late May and early June 2021.

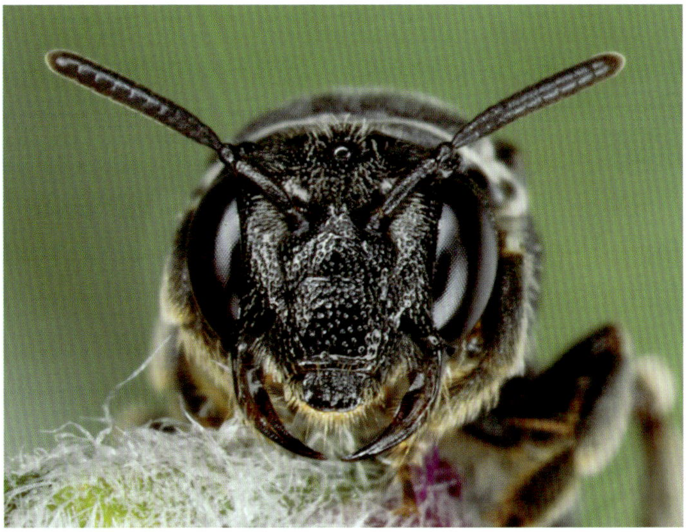
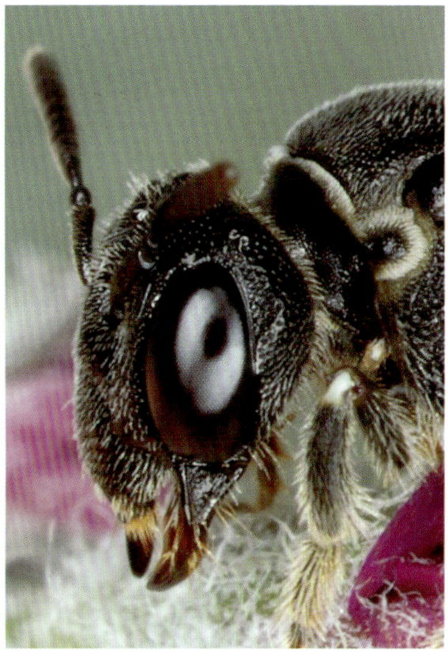

The deeply pitted face of a female *Protandrena texana*: the clypeus and labrum are entirely black, and the clypeus is somewhat protuberant. The jaws are reddish black near the base.

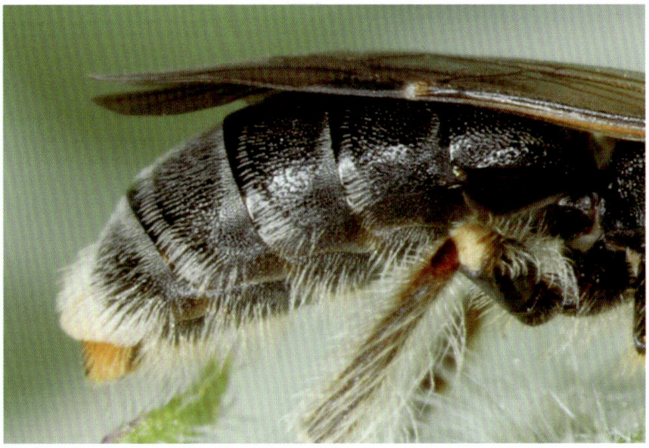
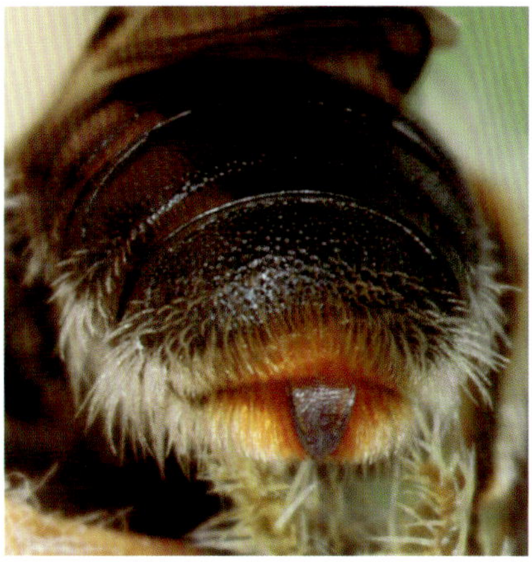

Lateral and rear views of the abdomen of a female *Protandrena texana*

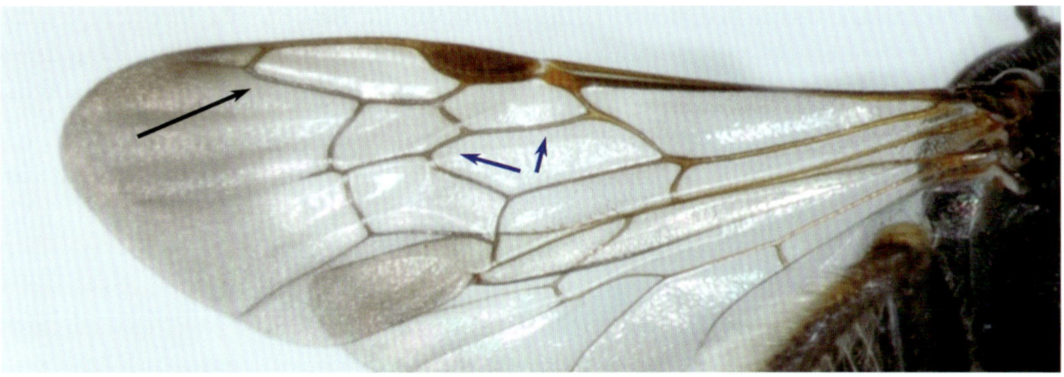

Forewing of a *Protandrena texana*: the marginal cell is lopped off at the tip (black arrow), and there are two submarginal cells (blue arrows).

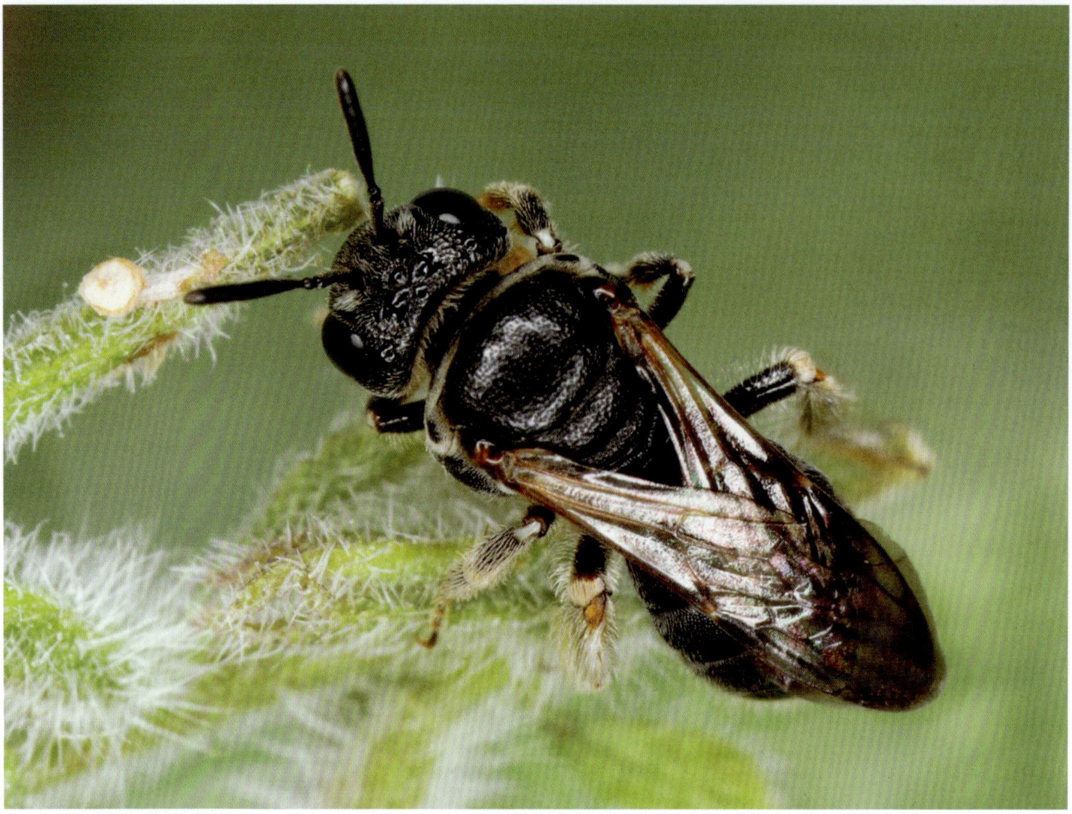

Protandrena texana has a robust, armored appearance. Its body is largely hairless and deeply pitted.

Associated plants: The female bees shown here were found nectaring on the verbena-family plant Rio Grande bee brush (*Aloysia macrostachya*), and gathering pollen from the aster-family plants Texas palafox (*Palafoxia texana*) and hairy wedelia (*Wedelia acapulcensis* var. *hispida*). *Protandrena* formerly classified as *Pseudopanurgus* often specialize in pollinating flowers of the Asteraceae family, but little published information is available on the specific foraging habits of *Protandrena texana*.

ID tips: *Protandrena texana* is difficult to distinguish from similar bees that once occupied the genus *Pseudopanurgus*. Species identification may require the assistance of an expert and frequently turns on such minute traits as the pitting on the bees' thoraxes and abdomens, and the size and quality of the bossed areas behind the compound eyes.

Fancy-footed protandrena ♦ *Protandrena (Pterosarus) ornatipes* (Cresson, 1872)

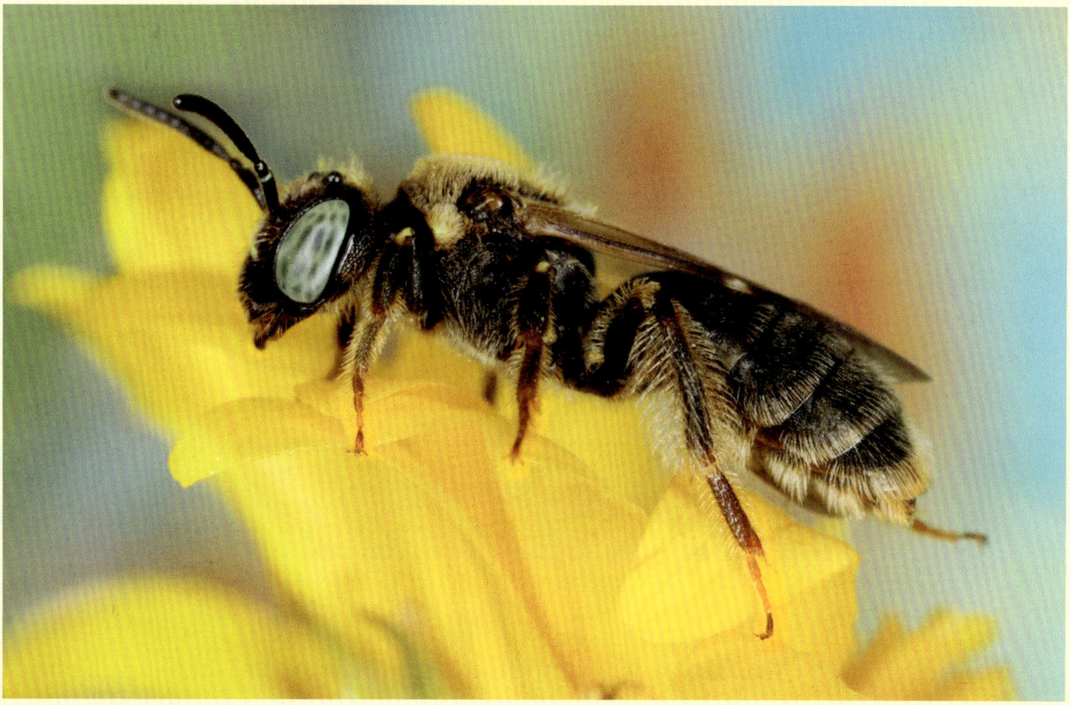

A female *Protandrena ornatipes*

Description: Like the Texas protandrena of the previous entry, *Protandrena ornatipes* was formerly classified as a *Pseudopanurgus.* The fancy-footed protandrena is a small, dark, green-eyed bee with yellow markings and fine, white and pale brown hairs on its head, body, and legs. Both male and female bees have a broad, ridged collar on the front of the thorax, and on both, T1 has a large median notch. The bees' tegulae are mostly brown, and the forewings are glassy, iridescent, and dusky toward the outer edges, with two submarginal cells and a blunt-tipped marginal cell. (A *Protandrena ornatipes* forewing is shown in the preceding entry introducing the genus *Protandrena.*) **Size:** 5–6.5 mm (1/5–1/4 in.).

Although this species is common in Texas, the female bee has never been described. For this reason, a somewhat detailed description is provided here. The female *Protandrena ornatipes* is robustly built, with a dark brown head and body. Its face has a distinctive small, yellow, boomerang-shaped mark: the mark straddles the base of the seam where the clypeus joins the upper face. The face is coarsely pitted and the clypeus somewhat protuberant. The mandibles are mostly dark brown and reddish at the base, and ridged along the outer rims. The apical edge of the clypeus is also ridged, and the labrum is smooth, curved on the apical edge, and fringed with light orange hairs. The antennae are dark brown with reddish front surfaces.

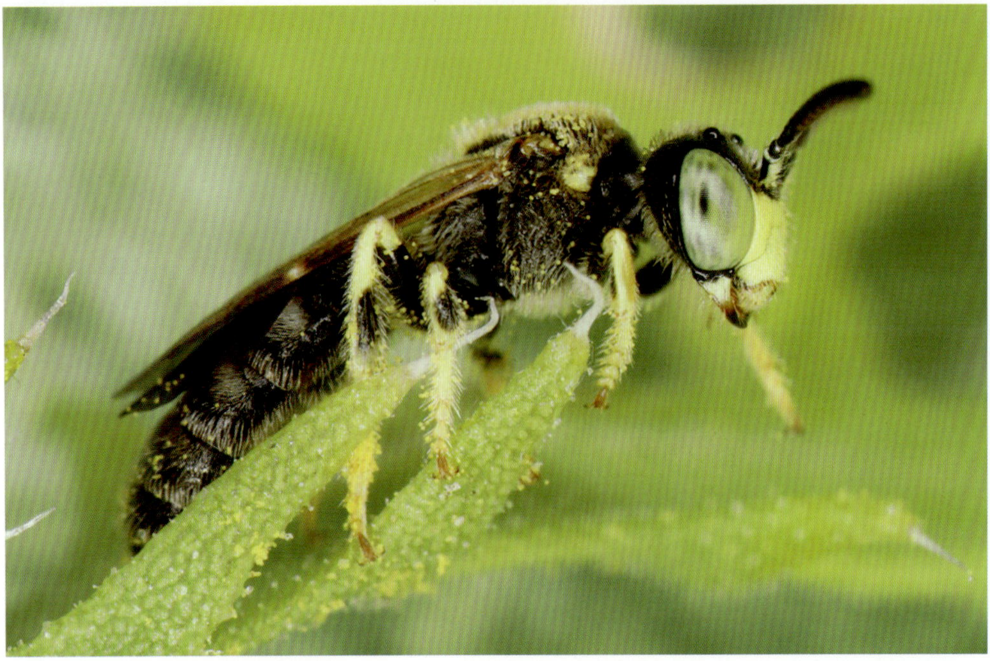

A male *Protandrena ornatipes*

The female's vertex, scutum, and scutellum are finely pitted and covered with short pale hairs. There are yellow markings on the pronotal lobes. The outer rims of T2–T5 are translucent brown, unpitted, and lined with white hairs above and laterally. Long orange hairs cover T6 above and laterally. The pygidial plate is triangular with gently curved, ridged sides. Horizontal grooves run across the top half of the pygidial plate. The sternum is dark brown and less finely punctured than the terga. White hairs rim the sides of S1–S4 and the apical rim of S5.

The female's front and middle tibiae and basitarsi are fringed with white hairs and its tarsi with light brown hairs. The long, pale tibial scopal hairs are feathery and branched, a trait typical of the subgenus *Pterosarus*. The basitarsi and tarsi are reddish brown. There are yellow markings on the base of each of the front and middle tibiae.

Male *Protandrena ornatipes* are blackish brown with finely pitted heads and thoraxes and extensive lemon-yellow markings. On males, the pronotal lobes, the face below the antennae, the front surfaces of the scapes, and the labrum are yellow. The antennae are otherwise black, with dark brown front surfaces on the flagella. The mandibles are yellow at the base, reddish in the middle, and black at the tip. The tibiae and tarsi (except the apical segment) are also largely yellow, and the bees' "knees" (the apical portions of the femurs) are also yellow. There are elongate black markings on the otherwise yellow tibiae. The name of this species may derive from the appearance of the male's lower tarsi, which are wide and conspicuous.

Short white hairs cover the male's head and thorax; the mesothorax hairs are brownish. The male's abdomen is elongate, finely pitted, and covered with brownish hairs dorsally; the apical portions of the terga are depressed, smooth, and polished. The bases of T1–T3 are reddish and rimmed with short pale hairs laterally. The hairs on the remaining segments are yellowish brown laterally. T1–T3 are constricted at the base.

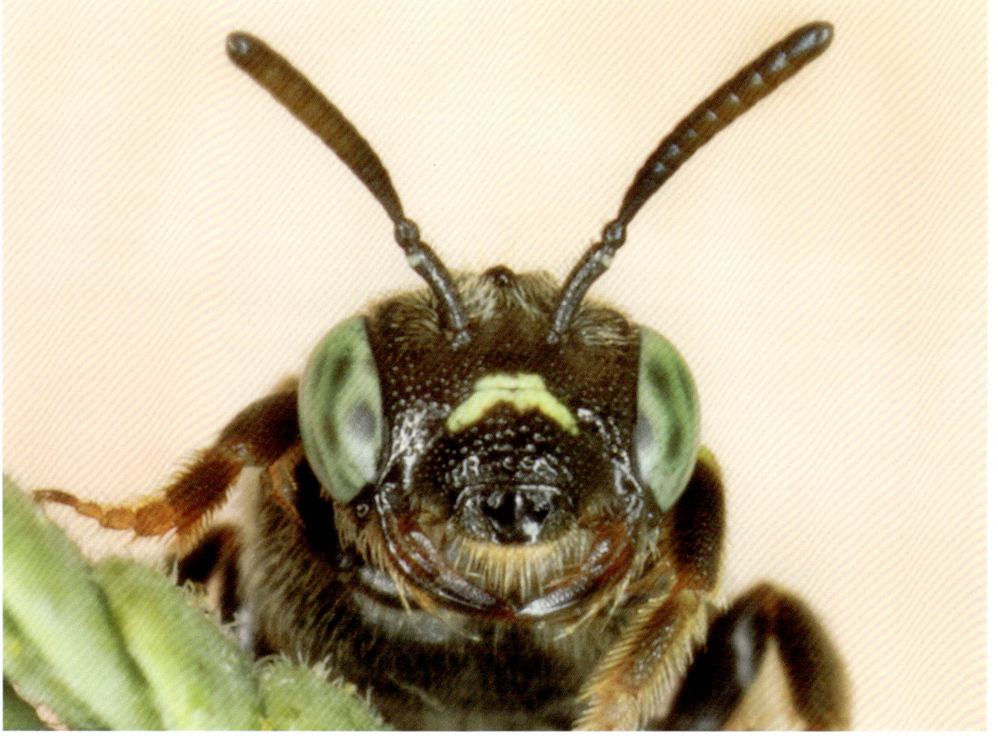

Face of a female *Protandrena ornatipes*

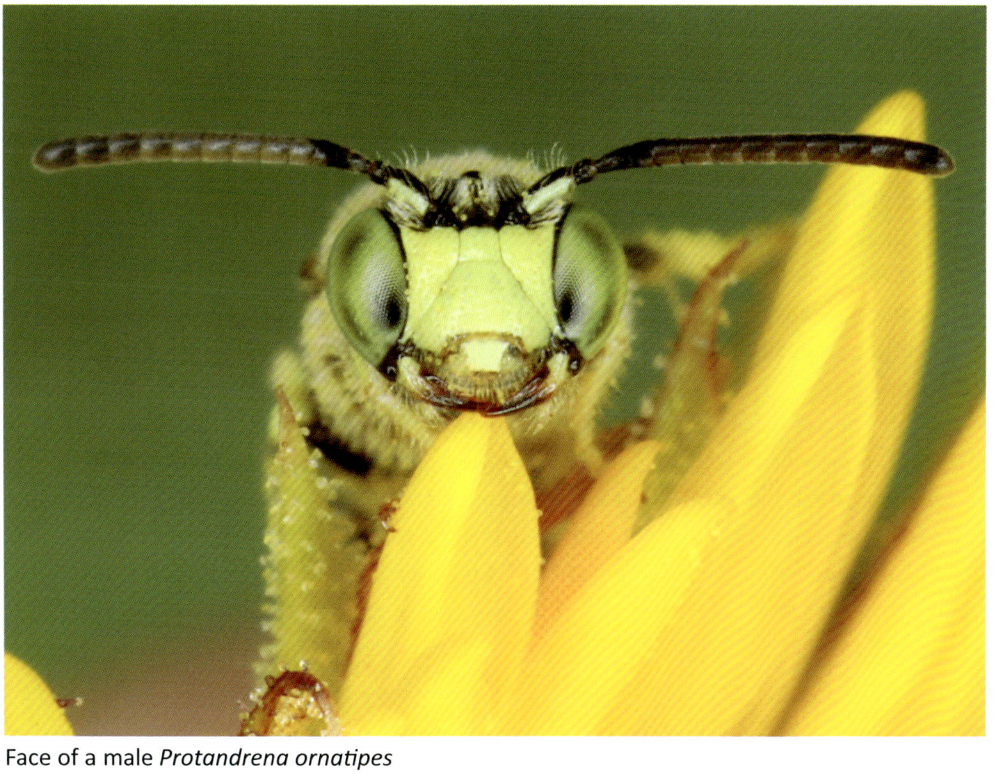

Face of a male *Protandrena ornatipes*

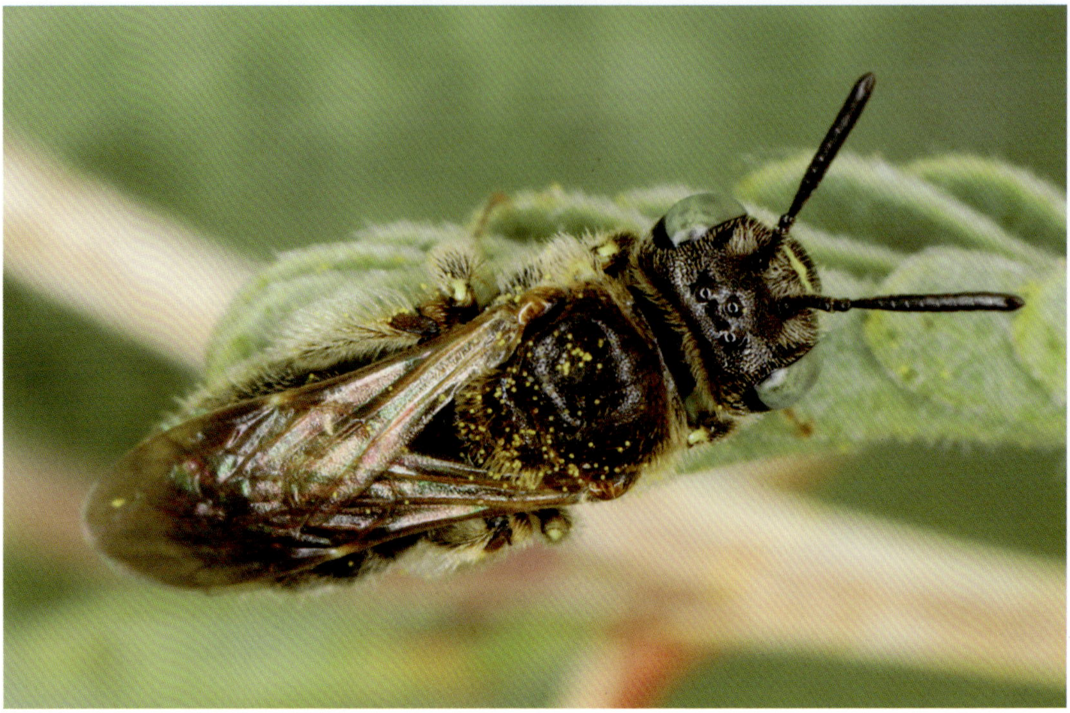

A female *Protandrena ornatipes*

Range: Cresson first described males of this species in his 1892 "Hymenoptera Texana" after examining specimens collected in Texas. The fancy-footed protandrena remains widespread in Texas. This species has been reported as far west as New Mexico and as far north as Colorado and Nebraska.

Associated plants: The short, branched scopal hairs characteristic of female *Protandrena ornatipes* are a trait associated in Andrenidae with aster-family flower specialists. *Protandrena ornatipes* is an Asteraceae specialist that has both a spring and fall season in Texas. The bees shown here appeared in large numbers feeding on camphor daisy (*Rayjacksonia phyllocephala*) in the Bahia Grande wildlife refuge (Cameron Co.) in mid-November 2021.

ID tips: This species is best recognized in the field by the distinctive yellow markings on both the male and female bees' faces and legs. *Protandrena ornatipes* is sometimes misidentified as *P. renimaculatus*, which has similarly small, although not identically shaped, pale facial markings, and whose range is more northern. Close inspection under magnification and expert assistance may be required to verify identifications of this small bee. Differences in the genitalia of *Protandrena (Pterosarus)* often aid in their identification.

Fairy Bees

Subfamily Panurginae ◆ Tribe Perditini
Genus *Perdita*

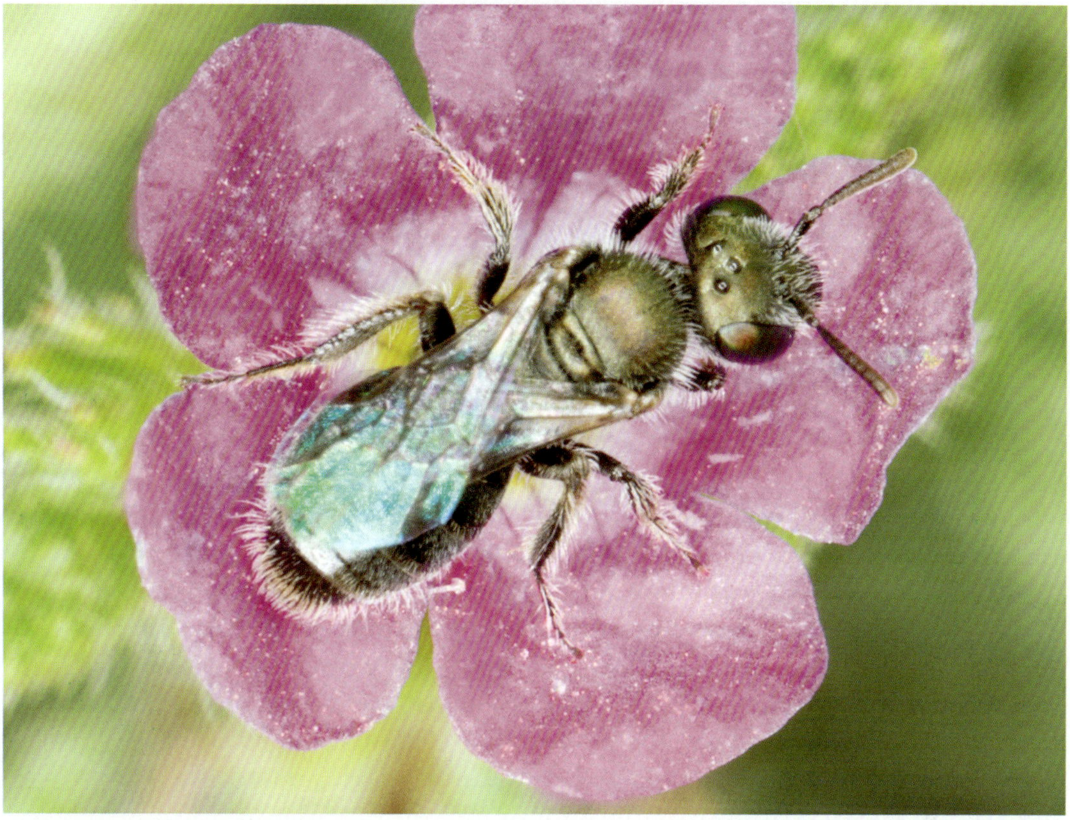

A female *Perdita (Hexaperdita)*

The family Andrenidae contains some of the Valley's largest bees—*Mesoxaea* and *Protoxaea*—and also its smallest, *Perdita*, known by the common name "fairy bee." The genus *Perdita* harbors more than 650 known species, with many yet uncataloged. As the name ("lost thing" in Latin) might imply, *Perdita* are easy to miss; the smallest known bee species is a *Perdita* that measures a mere 2 mm (1/13 in.).

Perdita occur predominantly in arid climates of northern Mexico and the southwestern United States. *Perdita* tend to be sparsely haired and black, or metallic blue or brassy green, often with pale markings on their faces, bodies, and legs; in some *Perdita* species, the pale markings are so expanded that the entire bee appears pale green or yellow. Some *Perdita* have dark reddish-orange abdomens or reddish markings. Most *Perdita* of the Valley are ant sized.

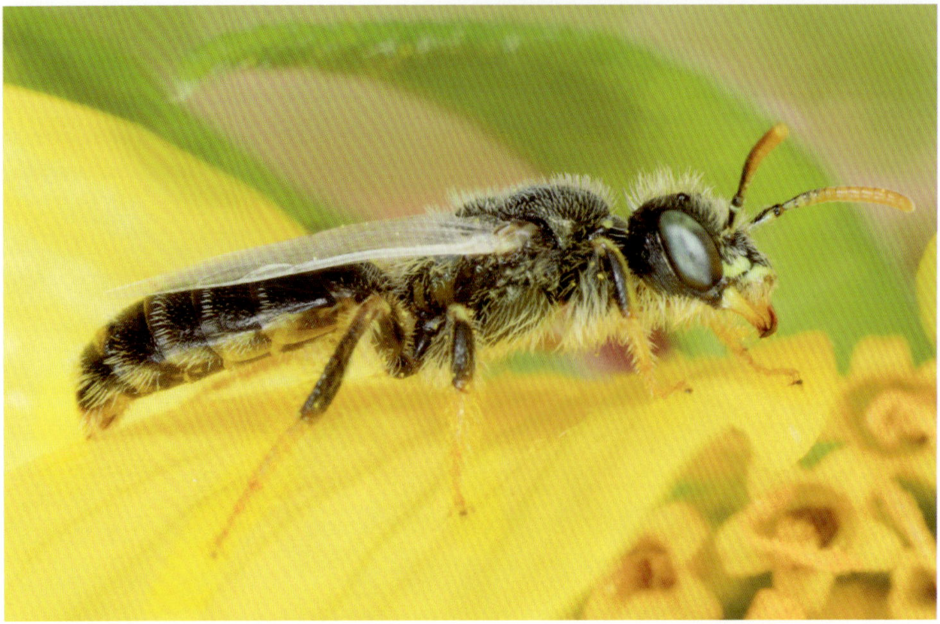

A male *Perdita scopata*: the flat abdomen is a trait of bees of the genus *Perdita*.

Perdita are solitary bees that nest in the ground. They sometimes nest near one another in aggregations; others live communally, with several females sharing a common burrow or tunnel entrance. Although *Perdita* may waterproof provisions stored in egg chambers, they ordinarily do not line their nests with waterproofing materials, a trait unusual for ground-nesting bees. Most *Perdita* are oligolectic.

Perdita abdomens appear flattened when viewed from the side; this character proves helpful in identifying the genus in the field. Close examination of the wings is also useful: the forewings of *Perdita* typically have two submarginal cells, and a marginal cell that is relatively short, lopped off at the end, and bent away at nearly a right angle from the wing edge.

Timberlake devoted much of his career to describing hundreds of *Perdita* species: he dedicated eight long volumes to the genus. He differentiated *Perdita* species by such traits as color; the presence and pattern of pale markings on their heads and bodies; head size and relative breadth; characteristics of their mandibles; the relative hairiness of their faces and bodies; differences in male genitalia; qualities of the females' hind-leg scopal hairs; and wing transparency and venation. Given that *Perdita* of the Valley run quite small, almost all these qualities are far beyond the reach of the naked eye. The time of year in which *Perdita* emerge, and the flora on which they appear, aid in guessing species identities in the field, but strong magnification and expert assistance are usually required to pinpoint species identifications.

A sampling of Lower Rio Grande Valley *Perdita* species representing three subgenera is shown in this section. *Perdita (Cockerellia)* tend to run larger than *Perdita* of other subgenera and have yellow or yellow-banded abdomens. All collect pollen from large-flowered aster-family plants. *Perdita (Hexaperdita)* are relatively small and tend to be oligolectic on small-flowered Asteraceae. *Perdita (Perdita)* contains the majority of species within the genus. Members of this subgenus vary widely in appearance and forage on a diverse range of flowers.

Scopate fairy bee ♦ *Perdita (Cockerellia) scopata* Timberlake, 1953

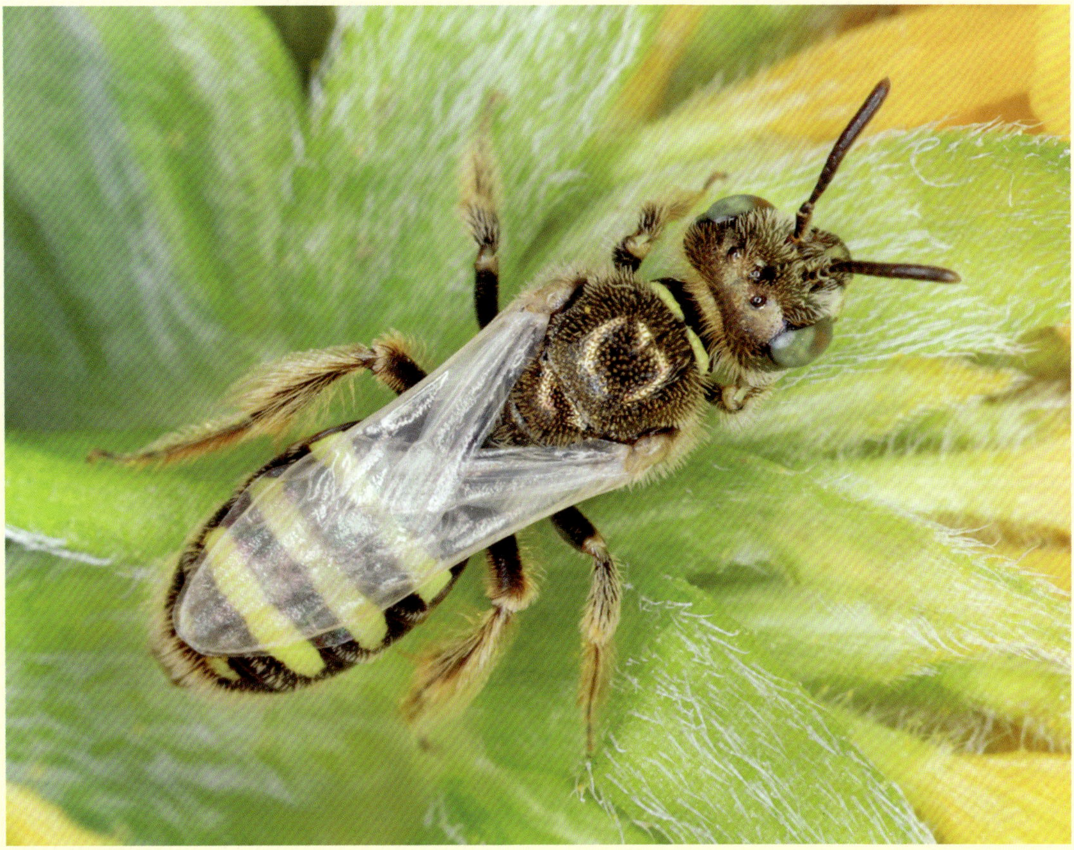

A female *Perdita scopata*

Description: The scopate fairy bee is one of many *Perdita* of the subgenus *Cockerellia* that inhabit the Lower Rio Grande Valley. Under magnification, *Perdita (Cockerellia)* often reveal a singular beauty: many have shimmering bronze heads and thoraxes, dark abdomens striped with electric yellow bands, partly orange mandibles, and vibrant green eyes.

Male and female *Perdita scopata* have bronze heads and thoraxes. The female's abdomen is banded boldly with yellow, while the male's is predominantly black, usually lacks yellow markings, and is striped with faint bands of pale hair; the tip of the male's abdomen (T7) is brown. The female has additional yellow markings on the pronotal collar and on the front of each pronotal lobe. Males sometimes have small yellow lateral spots on the hind margin of the pronotum. Both the male and female have bright rust-orange mandibles that are yellowish at the base and dark at the tip. Their antennae are mostly dark brown; the front surfaces of the flagella are light brown. The bees' wings are a glassy or translucent white with pale yellow-orange veins, and the tegulae are reddish brown. Females have dark legs; the tarsi and basitarsi of males are pale yellow. **Size:** females 7–9 mm (1/4–1/3 in.), males 6–8 mm (1/5–3/10 in.).

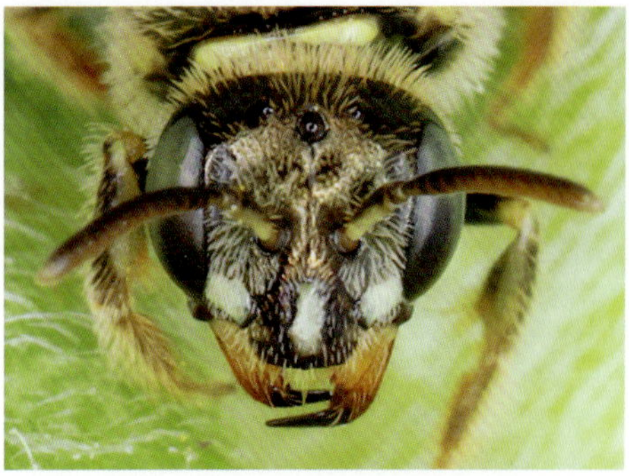 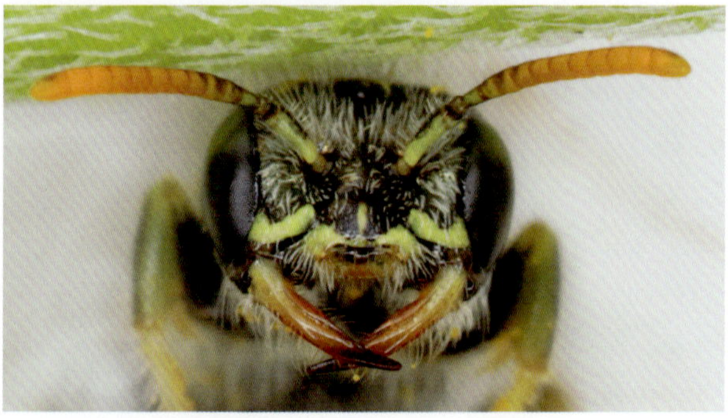

Face and mandibles of a female (*left*) and male (*right*) *Perdita scopata*

Range: Timberlake first described *Perdita scopata* in 1953 after examining bees from Willacy, Cameron, and Hidalgo Counties. He noted in 1954 that *Perdita scopata* was found in coastal regions of Texas from Galveston to Brownsville. This species appeared at multiple locations in Hidalgo County in spring 2021. *Perdita scopata* also appeared in mid-June 2021 at Campos Viejos Ranch in Starr County

Associated plants: The *Perdita scopata* that Timberlake described in the early 1950s were discovered feeding on coreopsis, sunflower, and other yellow composites; as noted, *Perdita (Cockerellia)* are oligolectic on large-flowered Asteraceae. The bees from Hidalgo and Starr Counties shown here were found between late February and mid-June feeding on common sunflower (*Helianthus annuus*), cowpen daisy (*Verbesina encelioides*), and Mexican hat (*Ratibida columnifera*).

ID tips: In the Valley, *Perdita (Cockerellia)* are represented by a continuum of similar species that sometimes cohabit the same area and intermingle on the same flowers. The female scopate fairy bee can be distinguished from others of its subgenus by the following features: (1) the clypeus is usually dark with a median yellow vertical strip; (2) the mandibles are hooked, at nearly a right angle to each base; and (3) the yellow bands on the abdomen (T2–T5) are broad, uninterrupted, and of even width from side to side; they stop just short of the abdomen's lateral edges.

The facial characteristics of male *Perdita scopata* are variable and less prominent than on their female counterparts. The yellow areas on the male's face frequently consist of a small mark in the middle of the clypeus, flanked by pale spots on either side of the clypeus. Sometimes the median mark is absent, or the clypeus lacks yellow markings altogether. Unlike the female, the male has mandibles that are long, tapering, and curved—a trait typical of males of the subgenus *Perdita (Cockerellia)*.

Mission fairy bee ◆ *Perdita (Perdita) missionis* Timberlake, 1958

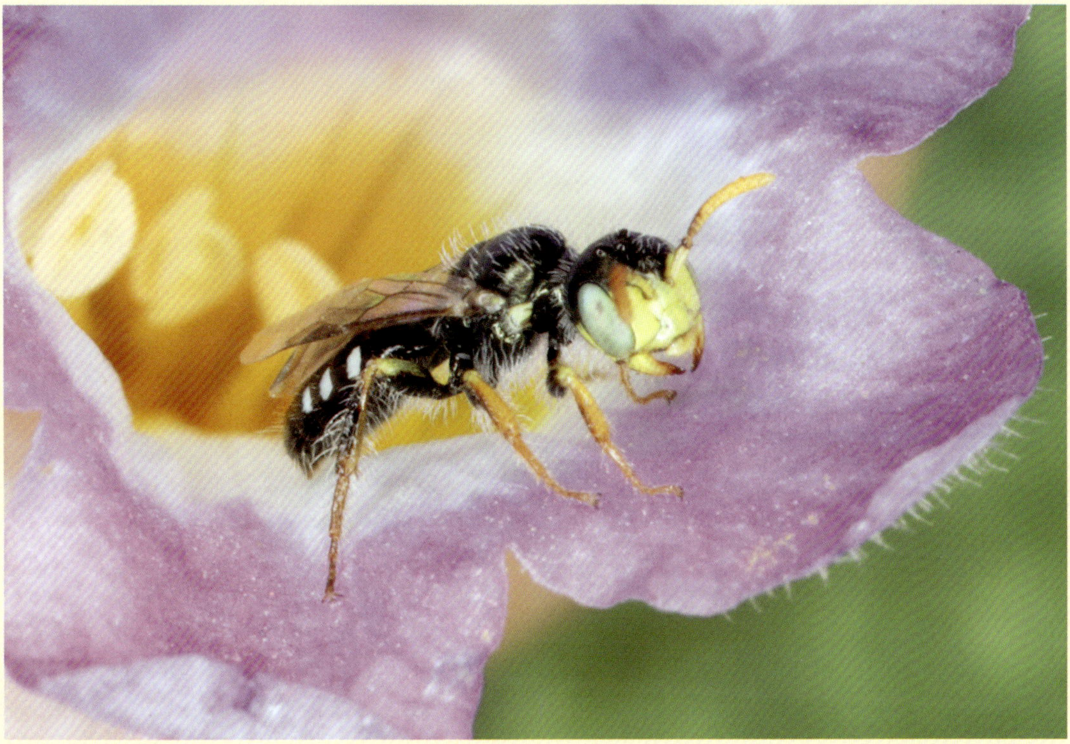

A male *Perdita missionis* on bristly nama

Description: *Perdita missionis*, or the Mission fairy bee, is a tiny insect with a dark head and thorax and an abdomen adorned with ivory spots arranged in pairs. The face of the female bee is black with blue eyes, and its mandibles are mostly reddish, black at the tips, and yellow at the base. Males have bright yellow markings on the lower half of the face, green eyes, and yellow mandibles with red-to-black tips. Females' abdomens have six spots; those of males have six to eight spots. **Size:** females 4–5 mm (1/6–1/5 in.), males 2.75–3.75 mm (1/10–1/7 in.).

Range: Timberlake named *Perdita missionis* after the town of Mission, Texas, because the first specimen of the species he examined had been collected in that Hidalgo County town. In 1958, Timberlake described the range of *Perdita missionis* as extending as far west into Texas as Terrell County, as far north as Mason County, and as far south as San Luis Potosí, Mexico.

Associated plants: *Perdita missionis* is associated with bristly nama (*Nama hispidum*), a low-growing spring-blooming annual with pink-to-purple funnel-shaped flowers, native to Cameron, Hidalgo, and Starr Counties. The bees shown here were found at the McAllen Nature Center (Hidalgo Co.) in March 2019 and April 2021, feeding on both bristly nama and wavy-leaf nama (*Nama undulatum*). The Mission fairy bee also appears on varieties of ground cherry, such as *Chamaesaracha coniodes* and *Quincula lobata*.

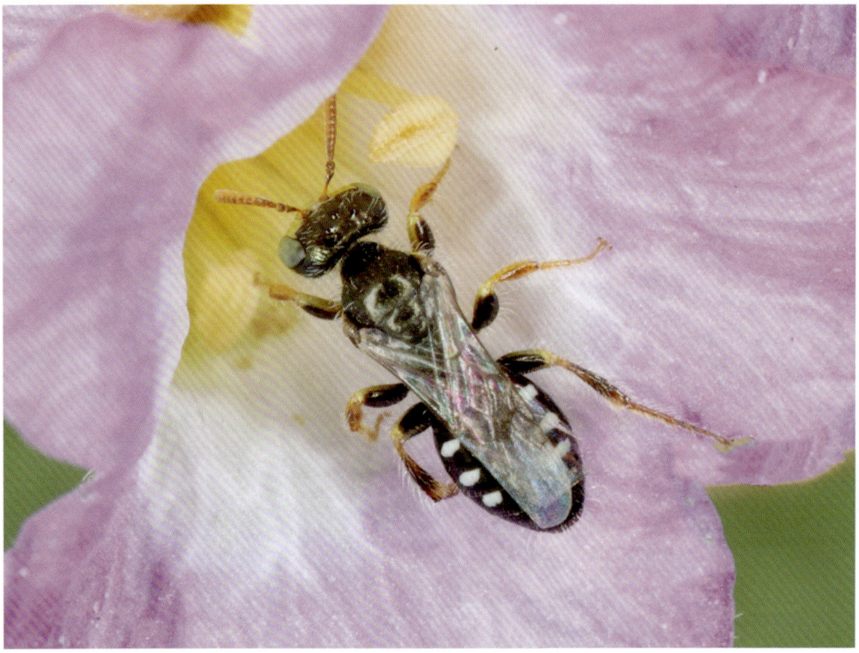

A male *Perdita missionis* entering a bristly nama blossom

ID tips: The flat-bodied shape characteristic of *Perdita* and the presence of six or more spots on the abdomen of *Perdita missionis* help distinguish it from other small bees. In some parts of Texas, populations of this species overlap with those of the very similar and far more widespread six-spotted fairy bee (*Perdita sexmaculata*). Timberlake noted that *Perdita missionis* and *P. sexmaculata* differ only in minor details, none visible to the naked eye. (They include the number of joints in small mouthparts known as maxillary palpi and characteristics of the genitalia). Differentiating between the two species usually requires examination by an expert. In the field, the Mission fairy bee sometimes can be sorted from the six-spotted *Perdita* by plant choice: *P. missionis* is more likely to be found on nama than is *P. sexmaculata*.

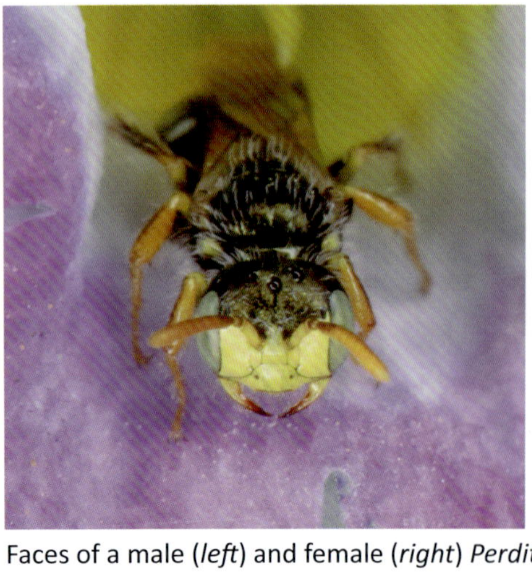

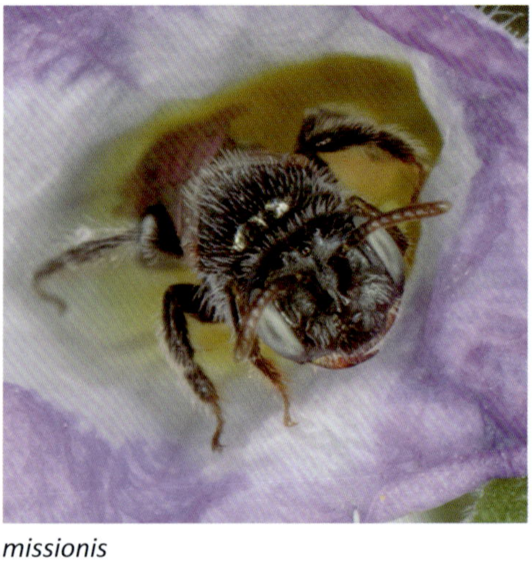

Faces of a male (*left*) and female (*right*) *Perdita missionis*

Ignored fairy bee ◆ *Perdita (Hexaperdita) ignota* Cockerell, 1896

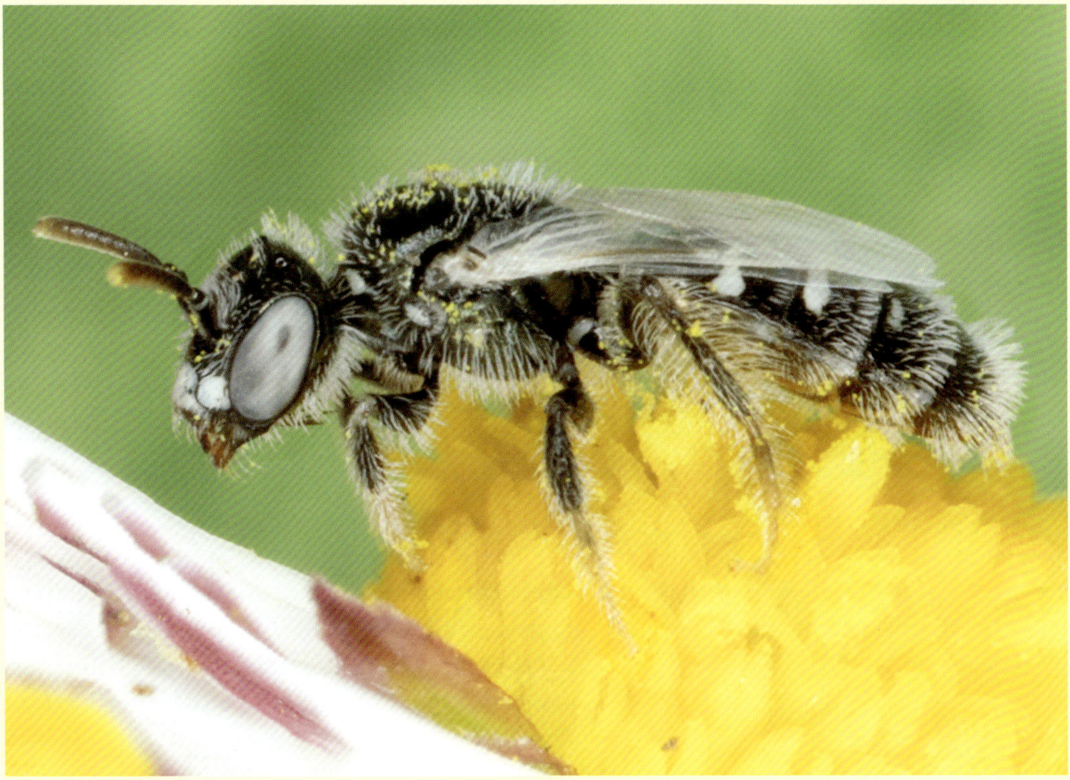

A female *Perdita ignota*

Description: *Ignota* means "unknown" or "ignored," a name that seems to refer to this bee's diminutive size, which makes it easy to overlook. *Perdita ignota* is a milky-winged, dark bee that glints gold in sunlight. The female bee has two pale broken bands on the second and third segments of its abdomen (T2–T3) and may have smaller pale dots on T4. The abdomen of the male bee is almost entirely dark except for the last segment, which is yellowish.

The female ignored fairy bee has milky-blue facial markings: a large bell-shaped mark on the clypeus, flanked by two smaller spots. On males, the pale facial markings are smaller and usually confined to the basal half of the mandibles and labrum and a single dot in the middle of the clypeus flanked by two roughly triangular shapes. The heads and bodies of males and females are covered with pale hairs. The legs of females are dark brown, with brown basitarsi and yellowish-brown lower tarsi. Males' lower legs have extensive pale yellow markings. **Size:** females 4–4.5 mm (1/6–1/5 in.), males 3–3.5 mm (1/9–1/8 in.).

Range: The ignored fairy bee is found throughout the central United States, from Texas to Montana. It ranges as far west as Nevada, and southward into Mexico. This species is common in Texas. It appears in the Lower Rio Grande Valley from April through November.

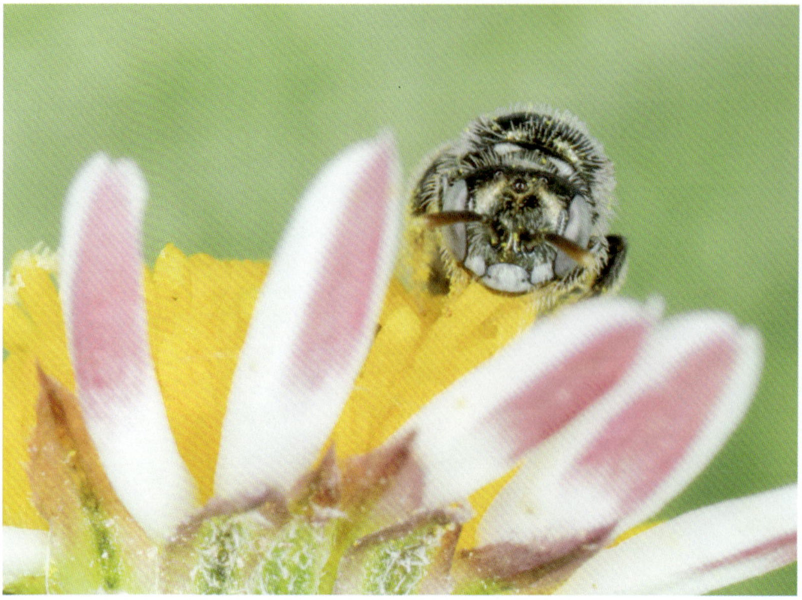

A female *Perdita ignota* on dozedaisy

Associated plants: *Perdita ignota* has been recorded foraging on a range of Asteraceae, including daisies, grindelia, gutierrezia, goldenaster, and goldenrod. In the Valley, it can be found feeding on such composite flowers as camphor daisy (*Rayjacksonia phyllocephala*), cowpen daisy (*Verbesina encelioides*), plains dozedaisy (*Aphanostephus ramosissimus*), and coatbuttons (*Tridax procumbens*).

ID tips: Although the minute size of this species makes it difficult to identify in the field, its distinctive milky wings, visible even to the naked eye, greatly aid in distinguishing *Perdita ignota* from other small bees of the Valley.

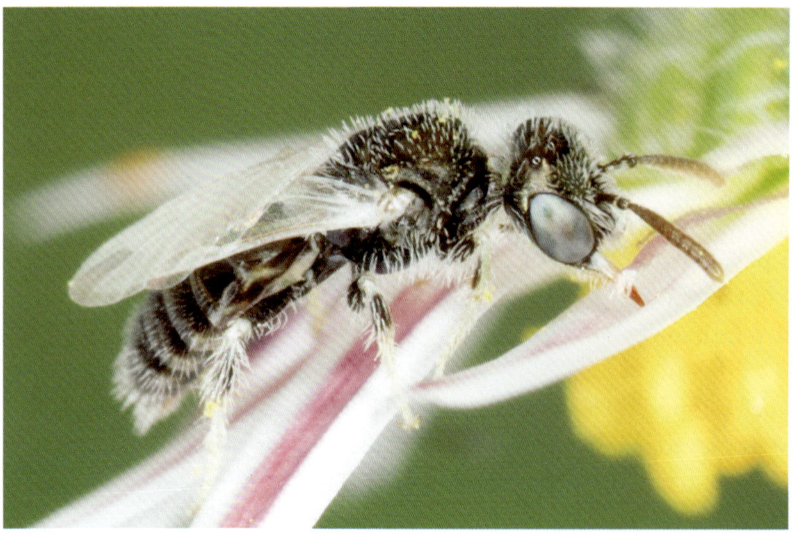

A 3 mm male *Perdita ignota*: Note the milky-white wings.

Calliopsine Bees

Subfamily Panurginae ◆ Tribe Calliopsini
Genus *Calliopsis*

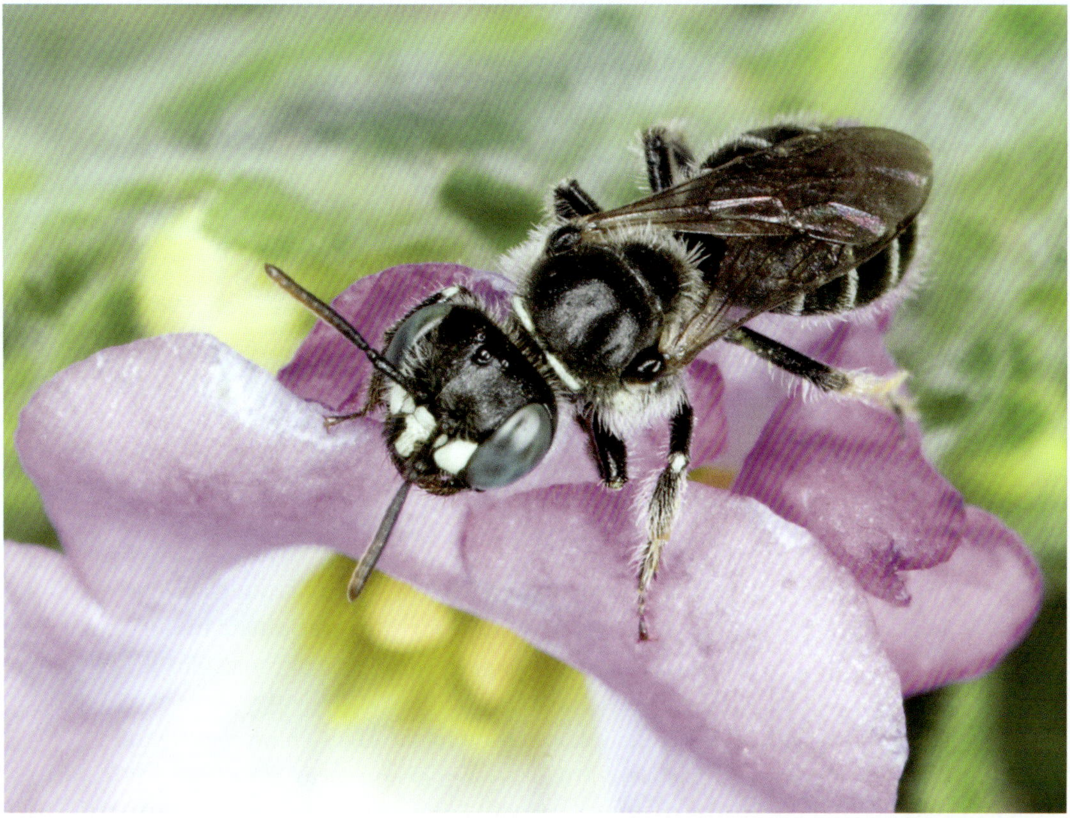

A female *Calliopsis hondurasica*

Genus information: *Calliopsis* is the sole genus of the tribe Calliopsini in North America. Calliopsini are robustly shaped, small to very small bees. They tend to have dark heads and bodies, and abdomens that are banded by pale hairs. Both males and females have pale markings on their faces, and they often have white or yellowish coloration on the pronotum or legs as well. *Calliopsis* forewings have two submarginal cells. Their forewings differ from those of *Perdita*, as illustrated below.

The genus *Calliopsis* is special to the Western Hemisphere. In the United States, *Calliopsis* are the most common and diverse in the Southwest and West. They are usually roughly oligolectic, gathering pollen from a narrow range of plants.

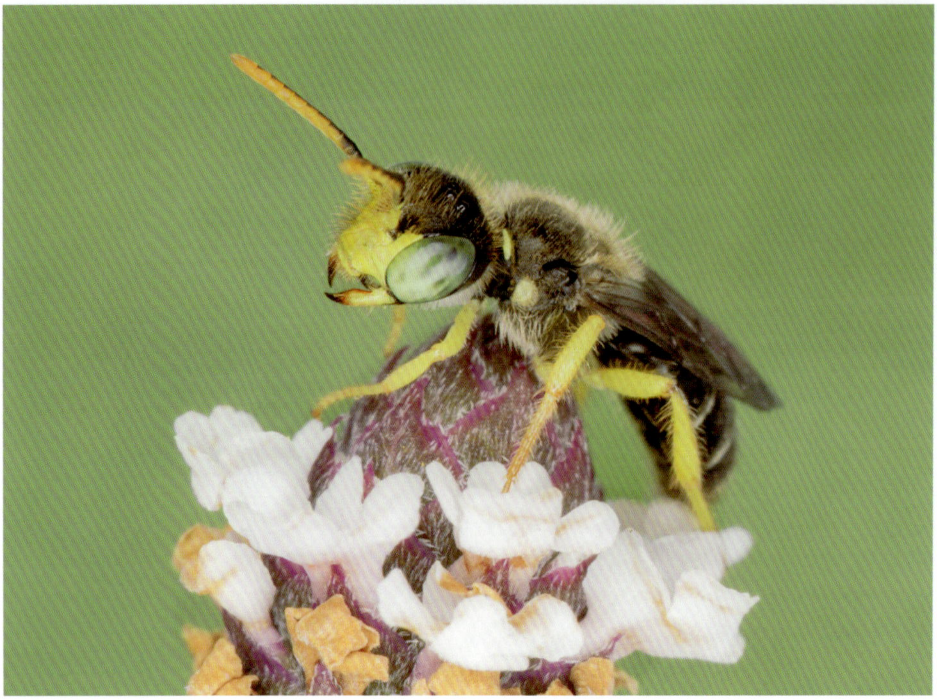

A male *Calliopsis hondurasica*

Like *Andrena* and *Perdita*, bees of the genus *Calliopsis* are solitary ground nesters that sometimes form aggregations. *Calliopsis* nests consist of tunnels that terminate in horizontal cells partly lined with a waxlike substance. Female *Calliopsis* shape pollen provisions into spherical waterproofed masses stored within egg chambers.

Some *Calliopsis* females exude a smell that entomologist Alvin Shinn described in 1967 as reminiscent of lemongrass. He discovered that the tunnel entrances of one species, *Calliopsis andreniformis*, were imbued with the same odor, and he speculated that the bees used the scent to mark and locate their nest entrances. Such entrances may be well hidden: *Calliopsis* females have been observed plugging their nest holes with soil or sand and entering their tunnels by "swimming" through the plug to reach the main burrow.

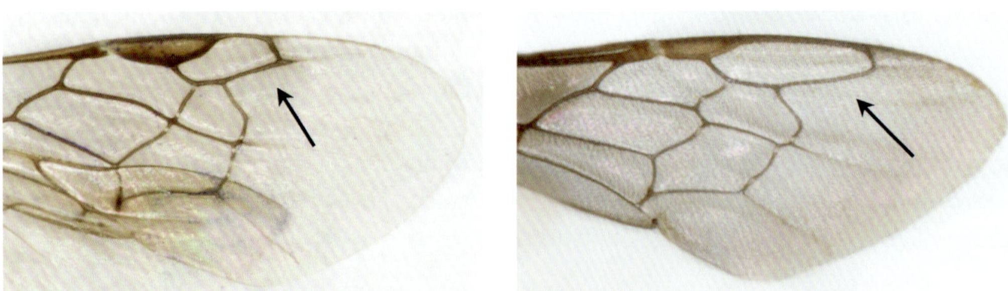

Details of the forewing marginal cells of *Perdita* (*left*) and *Calliopsis* (*right*). Most *Perdita* forewings have a very short, sharply lopped-off marginal cell and an enlarged stigma, while the marginal cell of the *Calliopsis* forewing is longer and somewhat rounder at the tip, and the stigma is reduced.

Honduran calliopsis ◆ *Calliopsis (Calliopsis) hondurasica* Cockerell, 1949

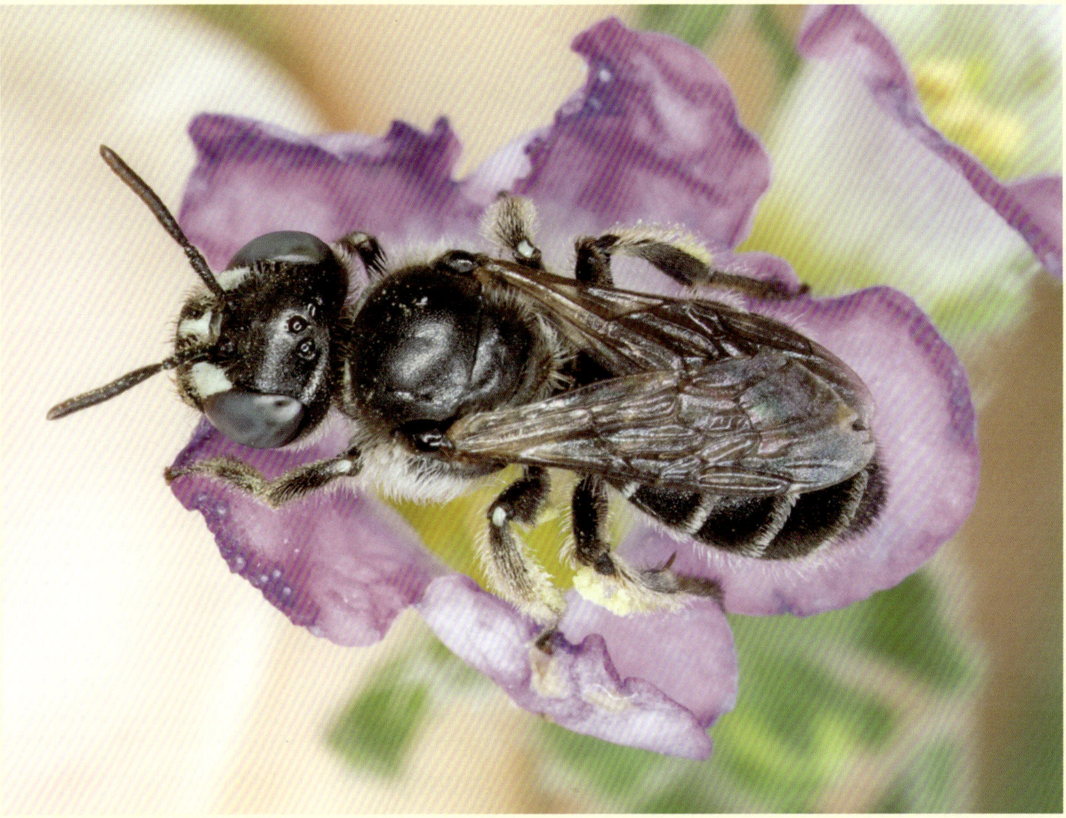

A female *Calliopsis hondurasica*

Description: Honduran calliopsis males and females are small and dark, with abdomens ringed by pale hair bands. The thoraxes have two narrow oblong markings on the pronotum, and the faces sport distinctive masks and jutting clypei. Females' legs are dark and covered with fine white hairs, while males' legs are relatively hairless and mostly yellow. When squeezed slightly, females give off a distinctive lemony smell.

The female *Calliopsis hondurasica* has a predominantly dark head, dark mandibles and antennae, dark blue eyes, and ivory facial markings—one on the center of the clypeus and smaller ones on either side of it. The male has green eyes and a bright yellow mask that covers most of the face below the antennae, including the clypeus and labrum. The flagella of the male's antennae are mostly yellow, with some brown on the hind surface; the antennal scapes are also mostly yellow, usually with some dark markings. The male's mandibles are yellow at the base, partly brownish red toward the middle, and black at the tip. **Size:** females 7 mm (1/4 in.), males 6 mm (1/5 in.).

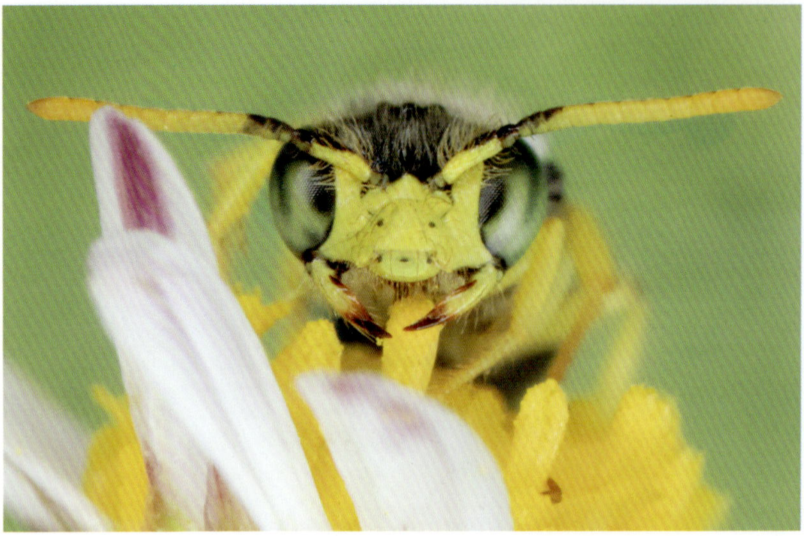

Face of a male *Calliopsis hondurasica*

Range: As its name suggests, the Honduran calliopsis is principally a subtropical bee: it is the only *Calliopsis* species in Central America. It is also the most widely distributed *Calliopsis*—it ranges from the Texas-Mexico border to Panama. In the United States, *Calliopsis hondurasica* extends into Texas and southern Louisiana. It flies from April through November in the Lower Rio Grande Valley.

Associated plants: *Calliopsis hondurasica* forages on a range of flora. In the Valley, this species shows a preference for varieties of nama. The bees featured here were found feeding on bristly nama (*Nama hispidum*), Texas frogfruit (*Phyla nodiflora*), small coastal germander (*Teucrium cubense*), and plains dozedaisy (*Aphanostephus ramosissimus*). The Honduran calliopsis has been documented elsewhere feeding on *Cassia*, *Kallstroemia*, and *Sphaeralcea.*

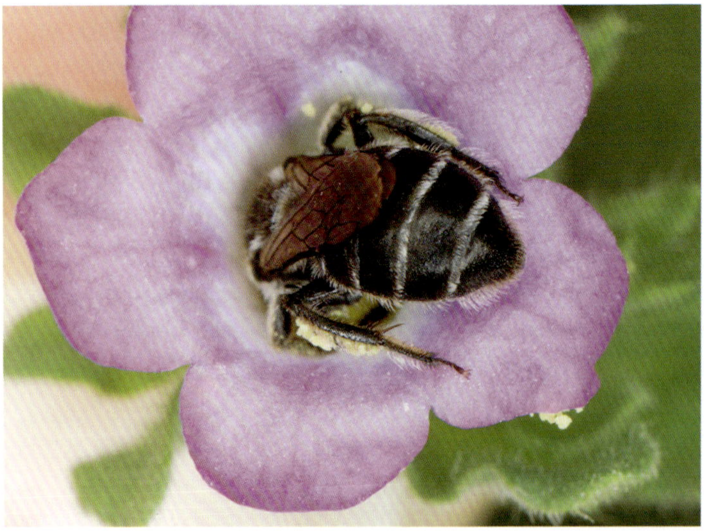

A female *Calliopsis hondurasica* foraging on nama

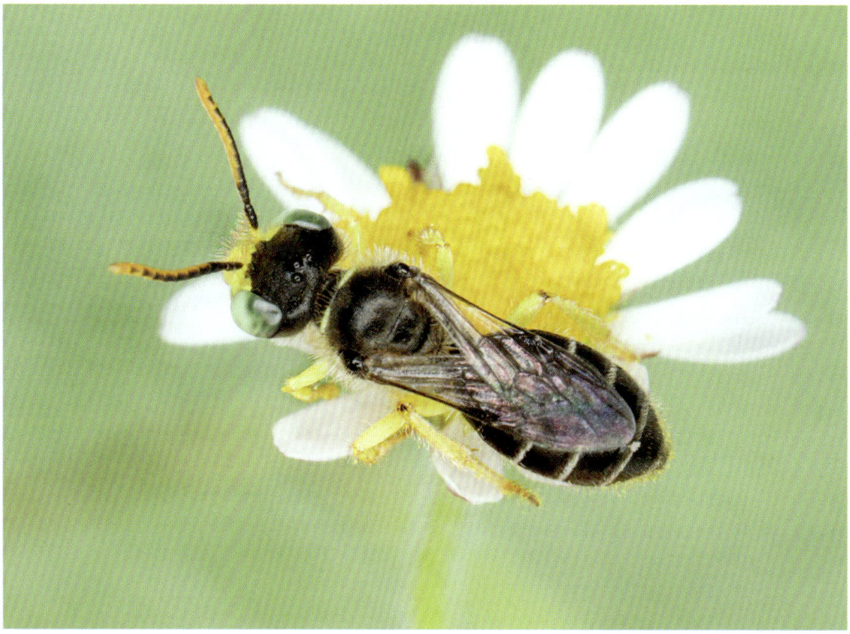

Dorsal view of a male *Calliopsis hondurasica*

ID tips: *Calliopsis* males of the Lower Rio Grande Valley might be confused with local male *Perdita* that also have yellow faces and dark, banded bodies. Notably, however, male *Calliopsis* species of the Valley lack pale abdominal markings or bands that form part of the integument; instead, their abdominal bands are formed by hairs.

Facial markings sometimes aid in differentiating among *Calliopsis* species. Nonetheless, the facial markings of the Honduran calliopsis are nearly identical to those of the widespread eastern calliopsis (*Calliopsis andreniformis*). Males of these two species are somewhat easier to distinguish from each other than are females. The male *Calliopsis andreniformis* has all-yellow antennal scapes, while the scapes of male *C. hondurasica* tend to have some brown on the rear apical surfaces. In addition, the flagella of *Calliopsis hondurasica* males are slighter longer than the bee's head; those of the male *C. andreniformis* are shorter—about as long as the head or less.

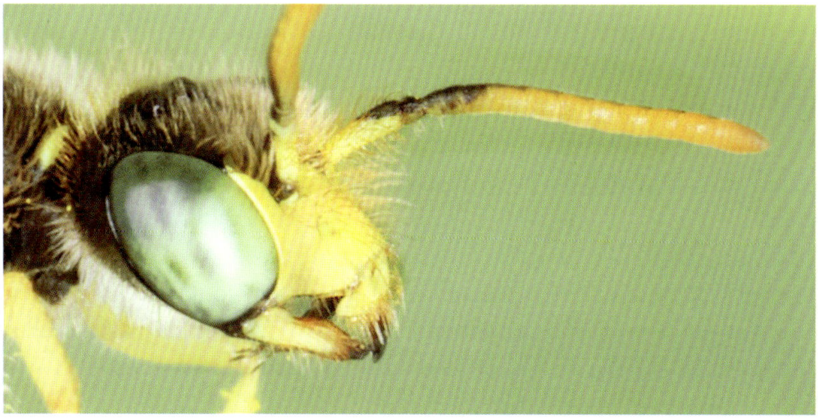

Profile of the head of a male *Calliopsis hondurasica*: note the long flagellum, and the dark brown markings on the top, rear portion of the scape.

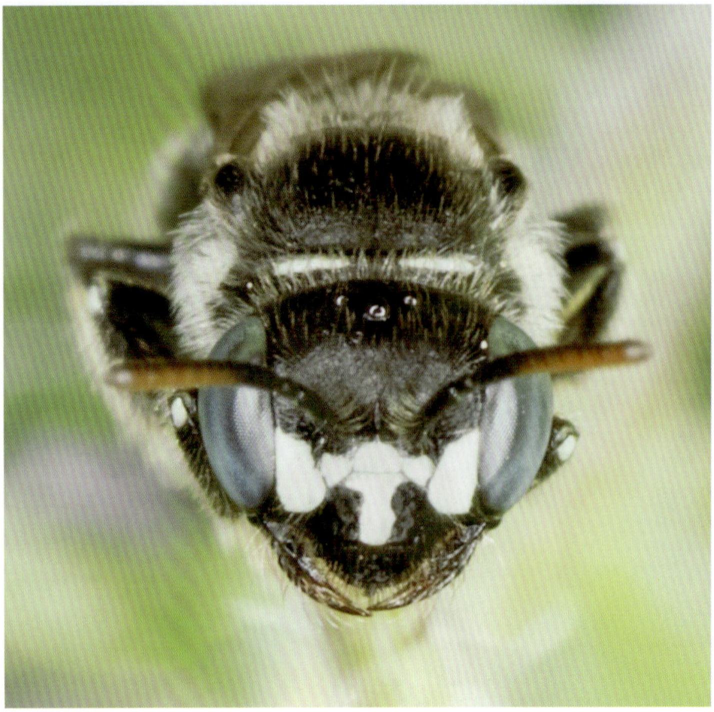

Face of a female *Calliopsis hondurasica*

Distinguishing female *Calliopsis hondurasica* from female *C. andreniformis* requires the assistance of an expert and hinges on characteristics visible only under great magnification. Three field traits sometimes help separate females: (1) the frons (or "forehead") of *Calliopsis andreniformis* tends to have a brassy metallic glint, while *C. hondurasica* lacks this quality; (2) in the Lower Rio Grande Valley, *Calliopsis hondurasica* is more likely to be associated with nama than is *C. andreniformis* (which forages on a broad range of flora but tends to prefer pea-family plants such as sweetclover); and (3) while *Calliopsis hondurasica* is largely subtropical, the distribution of *C. andreniformis* is centered north of the border, extending throughout much of the central and eastern United States, as far north as Canada.

Behavior: In 2000, entomologist William T. Wcislo, then stationed at the Smithsonian Tropical Research Institute of Panama, published a detailed account of the behavior of *Calliopsis hondurasica* on Panama's central Pacific coast. Wcislo reported that male *Calliopsis hondurasica* sleep in shallow burrows, which they dig near areas where female bees are nesting. The males sometimes mate in midflight, using their jaws and legs to bite females and pull at their wings.

Males also stake out territories near nests and chase away other *Calliopsis hondurasica* males, and sometimes other insects as well. In 1967, Shinn reported watching *Calliopsis* males "dog-fighting." Wcislo described male *Calliopsis hondurasica* engaging in a practice he called "spiral flight": pairs of males faced off, jostling each other and flying rapidly in an increasingly tight vertical spiral to a height of about one meter. Eventually, one male would grab the other, making both crash to the ground, where they tumbled and wrestled, sometimes biting each other on the legs until the bee defending the territory prevailed and the intruder caved and fled—or until the intruder successfully evicted the territory holder and took up residence.

Family Apidae

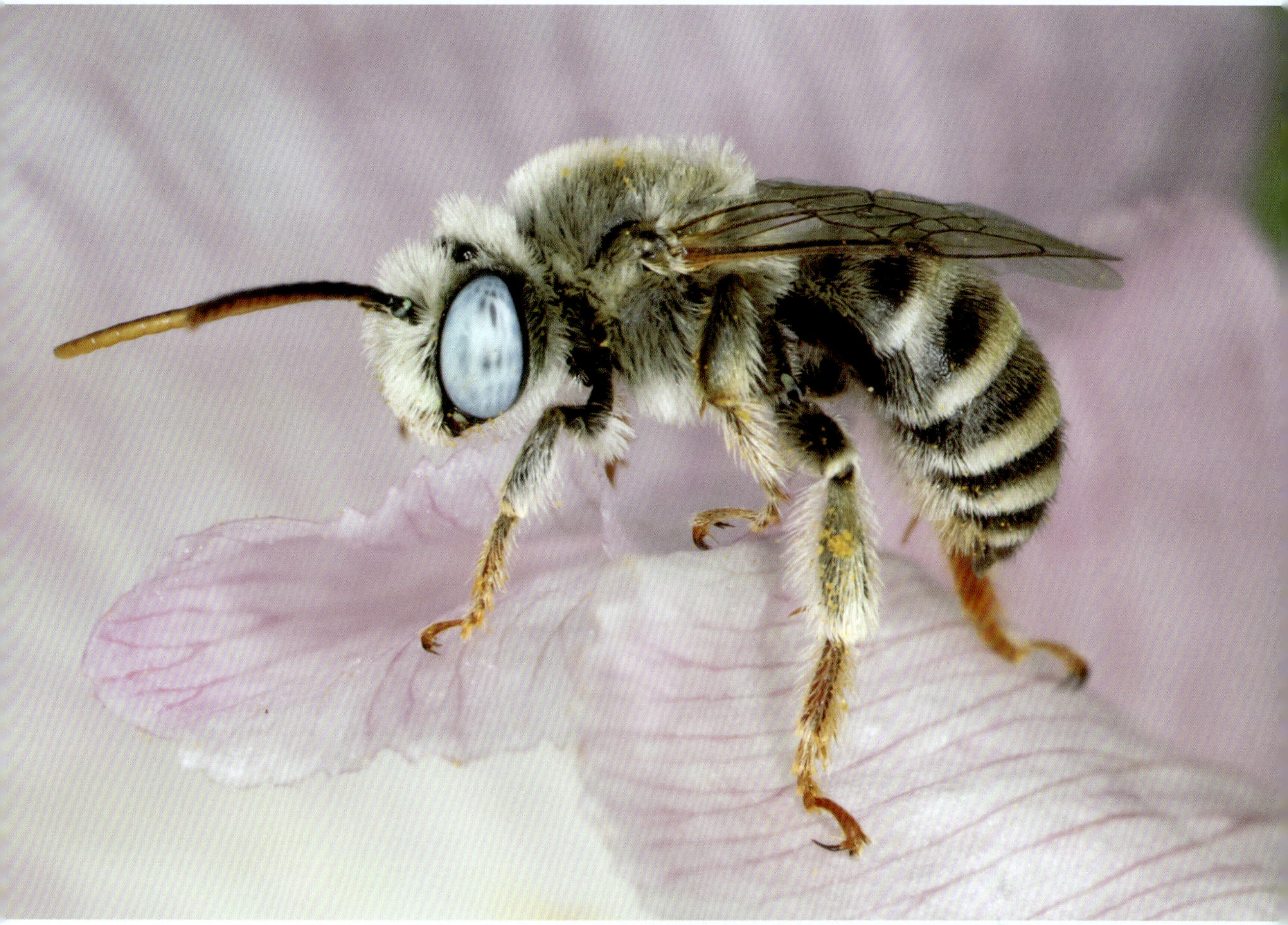

Chimney Bees
Crooked-Legged Bees
Longhorn Bees
Anthophorula and *Exomalopsis*
Digger Bees and Oil-Digger Bees
Large and Small Carpenter Bees
Cuckoo Bees
Bumble Bees
Honey Bees

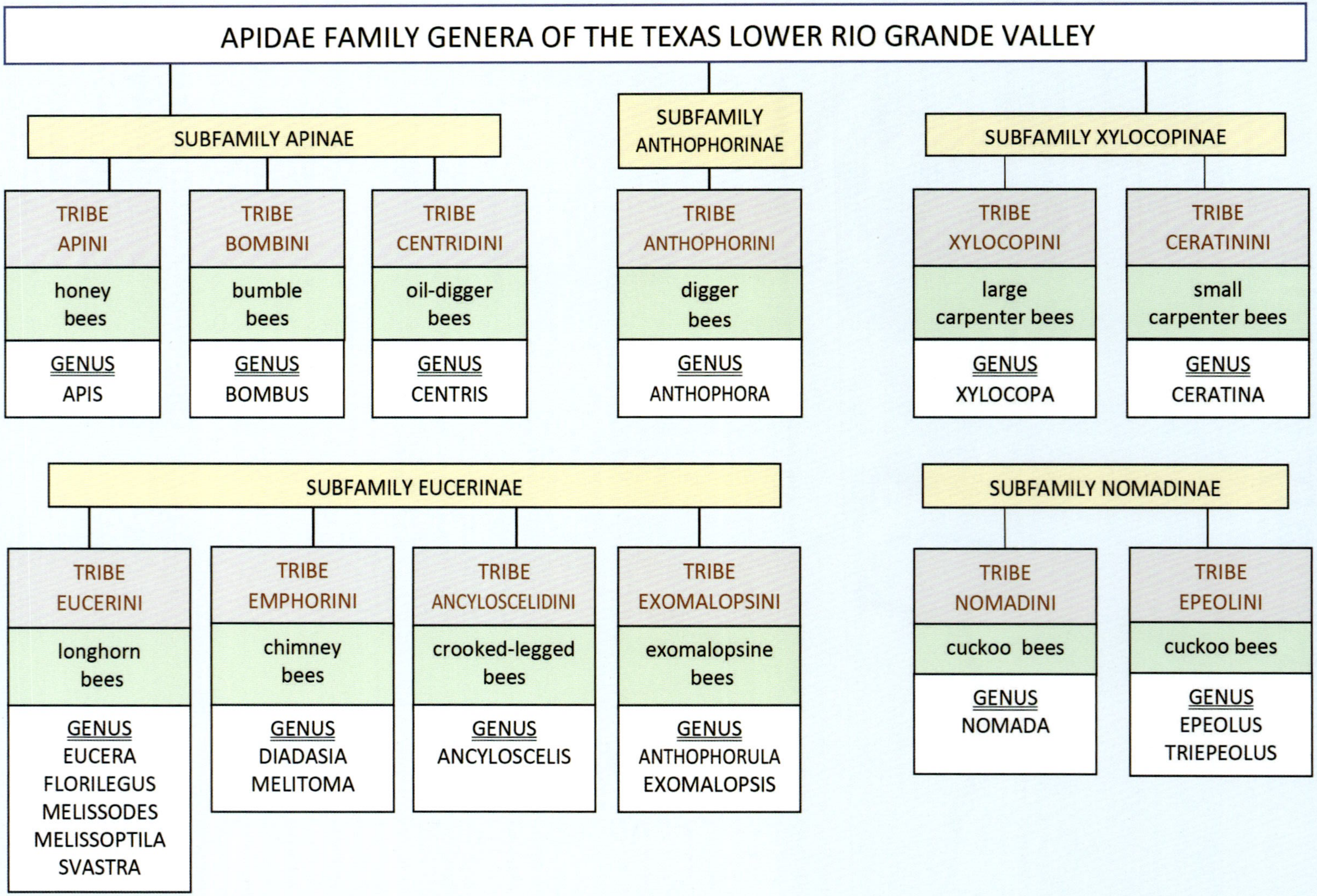
APIDAE FAMILY GENERA OF THE TEXAS LOWER RIO GRANDE VALLEY
SUBFAMILY APINAE
TRIBE APINI
honey bees
GENUS
APIS
TRIBE BOMBINI
bumble bees
GENUS
BOMBUS
TRIBE CENTRIDINI
oil-digger bees
GENUS
CENTRIS
SUBFAMILY ANTHOPHORINAE
TRIBE ANTHOPHORINI
digger bees
GENUS
ANTHOPHORA
SUBFAMILY XYLOCOPINAE
TRIBE XYLOCOPINI
large carpenter bees
GENUS
XYLOCOPA
TRIBE CERATININI
small carpenter bees
GENUS
CERATINA
SUBFAMILY EUCERINAE
TRIBE EUCERINI
longhorn bees
GENUS
EUCERA
FLORILEGUS
MELISSODES
MELISSOPTILA
SVASTRA
TRIBE EMPHORINI
chimney bees
GENUS
DIADASIA
MELITOMA
TRIBE ANCYLOSCELIDINI
crooked-legged bees
GENUS
ANCYLOSCELIS
TRIBE EXOMALOPSINI
exomalopsine bees
GENUS
ANTHOPHORULA
EXOMALOPSIS
SUBFAMILY NOMADINAE
TRIBE NOMADINI
cuckoo bees
GENUS
NOMADA
TRIBE EPEOLINI
cuckoo bees
GENUS
EPEOLUS
TRIEPEOLUS

Chimney Bees

Subfamily Eucerinae ◆ Tribe Emphorini
Genera *Melitoma* and *Diadasia*

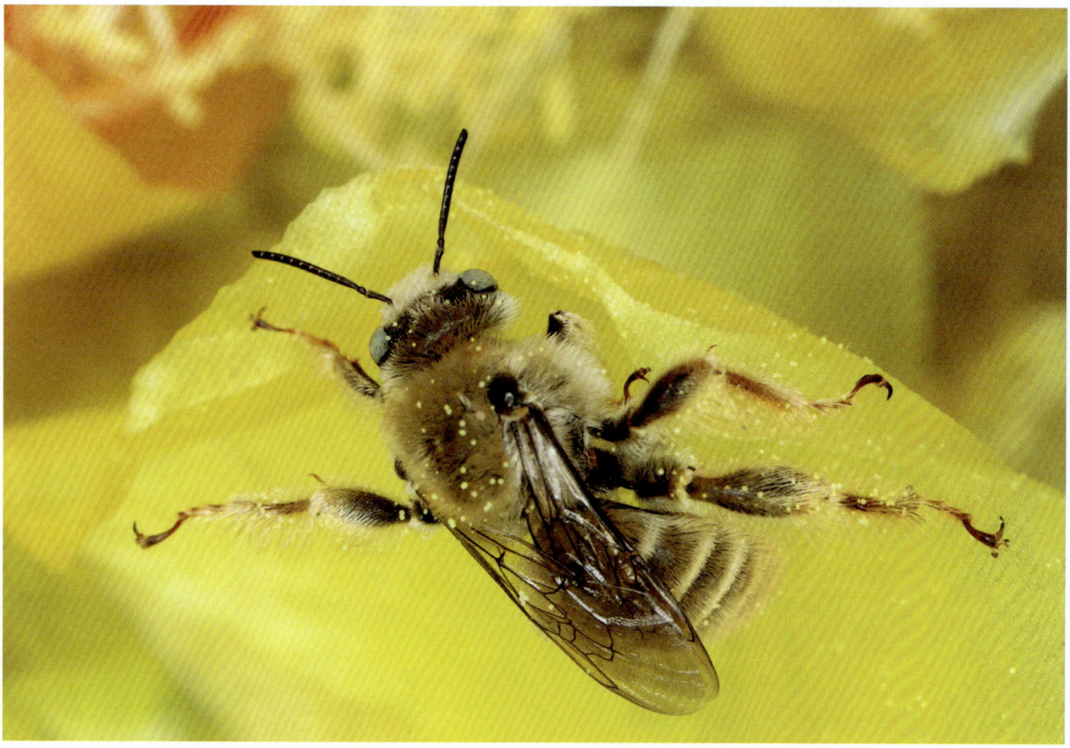

A long-legged male *Diadasia rinconis* negotiating the edge of a prickly pear blossom

Chimney bees belong to the tribe Emphorini, within the subfamily Eucerinae. They are hairy bees, robustly built, with abdomens striped with pale bands or covered entirely with pale hairs. Males often have long hind legs. Males also tend to have short antennae, a trait that sets them apart from males of similar bee tribes.

Chimney bees are found only in the Western Hemisphere. Within the Lower Rio Grande Valley, Emphorini are divided into two genera: *Melitoma* and *Diadasia*. A third chimney bee genus, *Ptilothrix* (hibiscus bees), is found in Texas north of the Valley. Some of the chimney bees that appear in the Valley are relatively rare in the United States—among these are *Melitoma marginella* (the marginate chimney bee) and *Diadasia tropicalis* (the tropical chimney bee). Chimney bees are usually oligolectic: *Melitoma*, for example, favor plants of the morning glory family (Convolvulaceae). Many *Diadasia* are cactus, mallow, or Asteraceae specialists.

Chimney bees are solitary ground nesters: each provisions its own individual tunnel-shaped nest. Nonetheless, female *Melitoma* and *Diadasia* often form aggregations. Chimney bee nests tend to be shallow and tunnel shaped, sometimes with branching tunnels terminating in brood cells. The nests are often constructed in hard-packed soil. The name "chimney bee" derives from the small, tubular, turret-like structures that many Emphorini build at the entrances to their nests. Female chimney bees carry water from nearby areas to their nests, using it to moisten the soil near tunnel entrances. They then reshape the mud into pellets, which they use to construct the "chimneys."

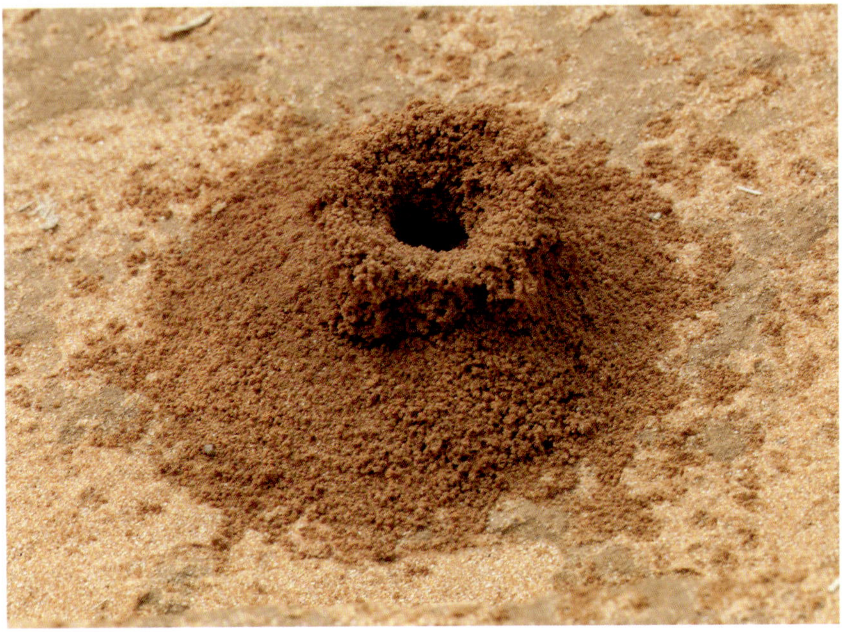

The chimney-like entrance of a chimney bee nest

Marginate chimney bee ◆ *Melitoma marginella* (Cresson, 1872)

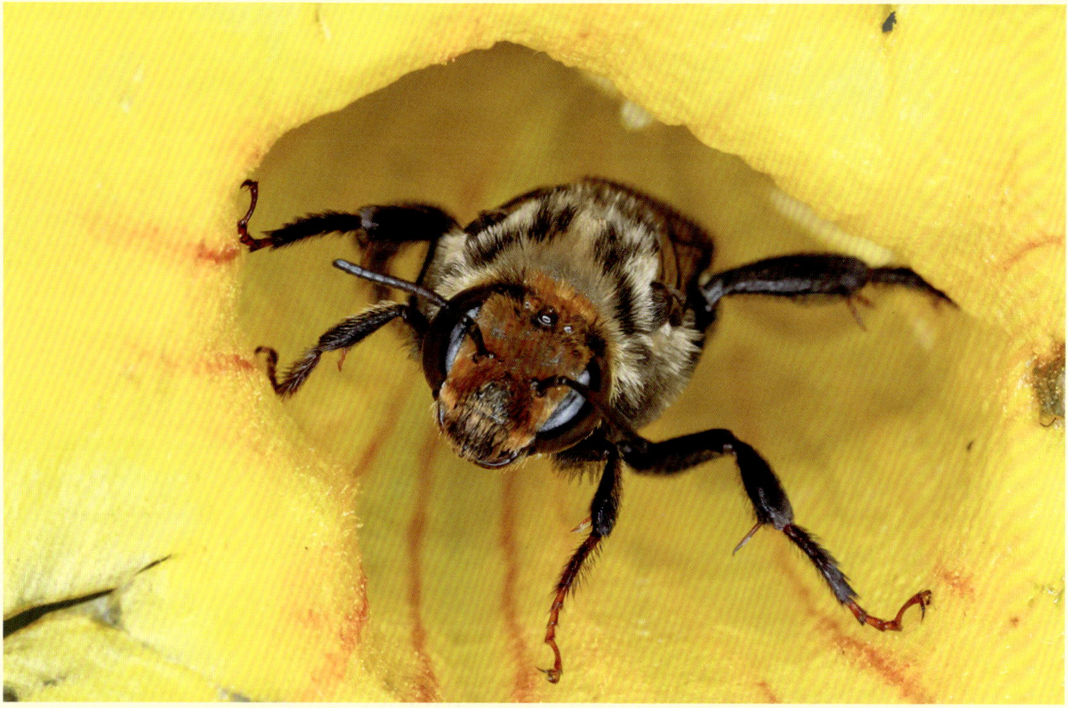

A male marginate chimney bee peering from an esperanza blossom

Description: *Melitoma marginella* is a black bee with bright rust-colored hairs on its face; a body covered with black, gray, and white hairs; and elongated legs. The bee's abdomen is banded with pale hairs that sometimes have a yellow-orange tint. The legs are mostly black, with red lower tarsi and partly red hind basitarsi and femurs. Females' legs are shorter than males' and fringed with long black hairs. Females' hind-leg scopal hairs are long and feathery. Males and females have black eyes and antennae. **Size:** 9–12 mm (1/3–1/2 in.).

Range: The marginate chimney bee is considered a Neotropical species. It is recorded as far south as Ecuador. In the United States, this species is relatively uncommon and occurs principally along the southern Texas border; even in Texas, there are few published photographic records of *Melitoma marginella*. It was first described by Cresson in his 1872 "Hymenoptera Texana": he discovered *Melitoma marginella* in Comal County. This species appeared in substantial numbers at the National Butterfly Center (Hidalgo Co.) during spring and fall of 2018, 2019, and 2021.

Associated plants: *Melitoma marginella* is commonly associated with the red-center morning glory (*Ipomoea amnicola*) and the pink morning glory (*Ipomoea carnea*). Nonetheless, in the Valley, this bee is often found ensconced in the trumpet-shaped yellow flowers of esperanza (*Tecoma stans*), an ornamental shrub of the family Bignoniaceae, native to northern Mexico and South Texas.

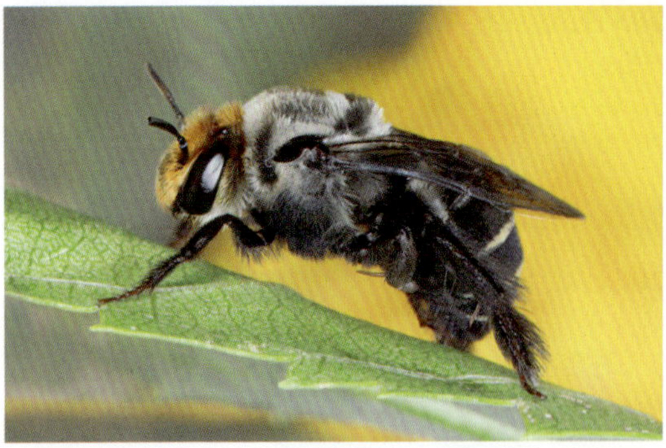

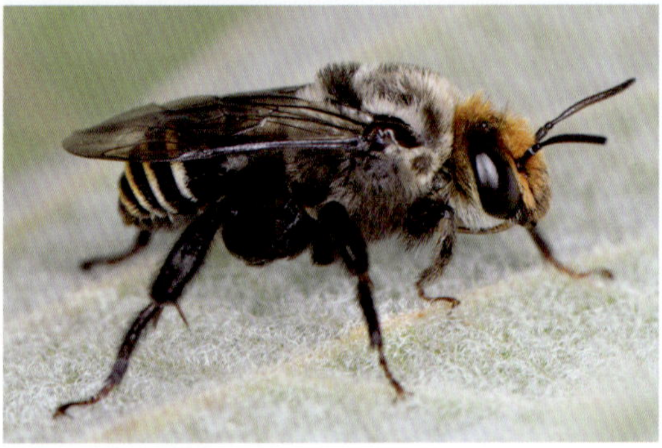

Female (*left*) and male (*right*) marginate chimney bees

ID tips: This is a striking species, difficult to confuse with Texas bees of other genera. *Melitoma marginella* is easily identified by its orange facial hairs and long black legs. In addition, dark and light hairs on the bee's thorax form a distinctive U shape that angles downward over the bee's "shoulders." A similar species, *Melitoma segmentaria*, occurs south of Texas from Mexico through Argentina. *Melitoma segmentaria* can be differentiated in the field in part by these traits: (1) the middle and hind legs of *M. segmentaria* have more extensive red areas than those of *M. marginella*; and (2) the antennal scapes of female *M. segmentaria* are often reddish rather than black.

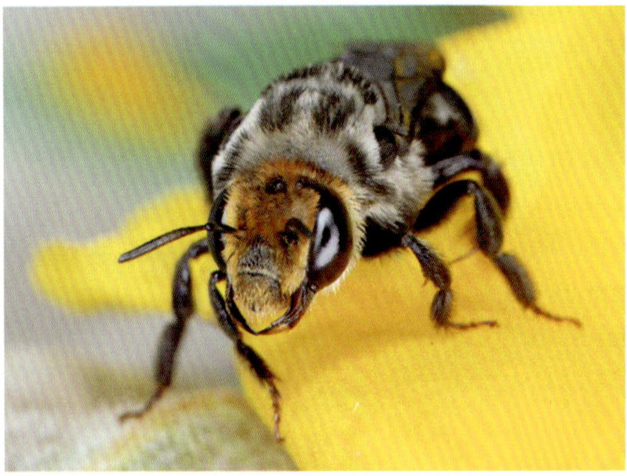

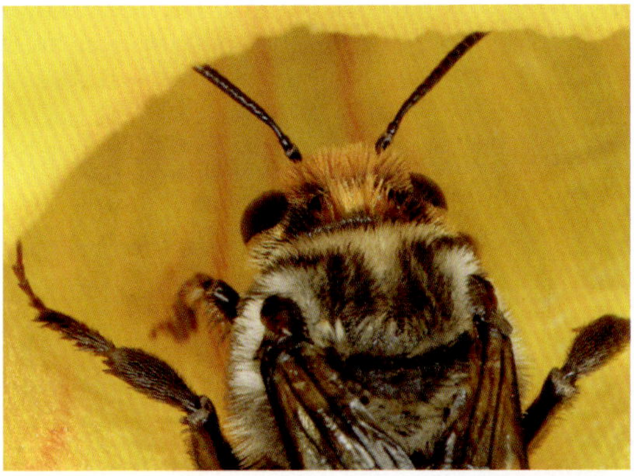

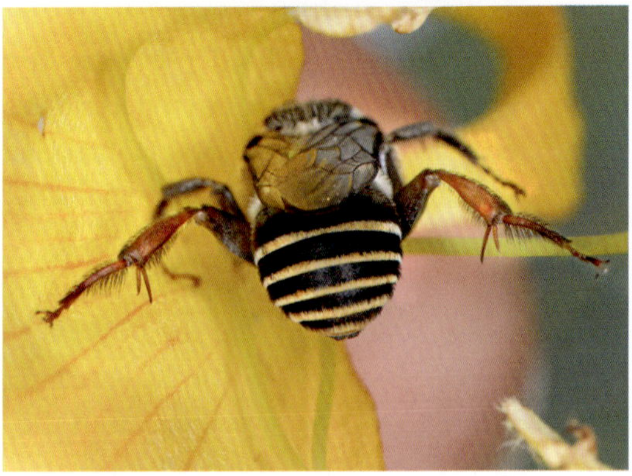

Top left: The rust-orange face of a female *Melitoma marginella*

Top right: A close-up of the U-shaped pattern of dark-and-light hairs on the thorax of a male *Melitoma marginella*

Bottom: Rear view of the elongated and reddish hind legs of a male *Melitoma marginella*

Behavior: Marginate chimney bees seek out dry riverbanks, clayey soils, or adobe structures, frequently near water sources and stands of morning glory. Mexican, Central American, and South American *Melitoma* occasionally take advantage of preexisting structures rather than building their own nests. Studies have documented *Melitoma* females settling into abandoned termite nests; porous brick walls; the eaves of wattle-and-daub houses; the overhangs of mine entrances; and outdoor clay ovens. *Melitoma* have even been observed occupying nests originally built by apiform crooked-legged bees (*Ancyloscelis apiformis*), whose life cycles and floral preferences often coincide with those of *Melitoma.*

Female marginate chimney bees emerge in the Lower Rio Grande Valley in April to begin gathering nectar and pollen from flowers. After mating, female *Melitoma marginella* lay eggs in urn-shaped brood cells with smooth earthen walls. New generations emerge in late summer and occupy the Valley through late November.

Male *Melitoma marginella* are entertaining to watch. They dedicate their daylight hours to zipping from blossom to blossom in search of mates, or plumbing the depths of trumpet-shaped flowers to waylay females gathering pollen. As is true of males of most ground-nesting native bee species, *Melitoma marginella* males do not overnight in nests. Instead, they brave the elements, sleeping outside and seeking shelter in flowers. In late fall of 2019 at the National Butterfly Center, throngs of *Melitoma marginella* males stationed themselves in the deep-throated blossoms of esperanza shrubs during cold days and thunderstorms. The wide mouths of the down-turned blossoms served as umbrellas, keeping the bees dry during the rain. During mid-November, when temperatures in the Lower Rio Grande Valley plummeted into the 30s and 40s, the cold

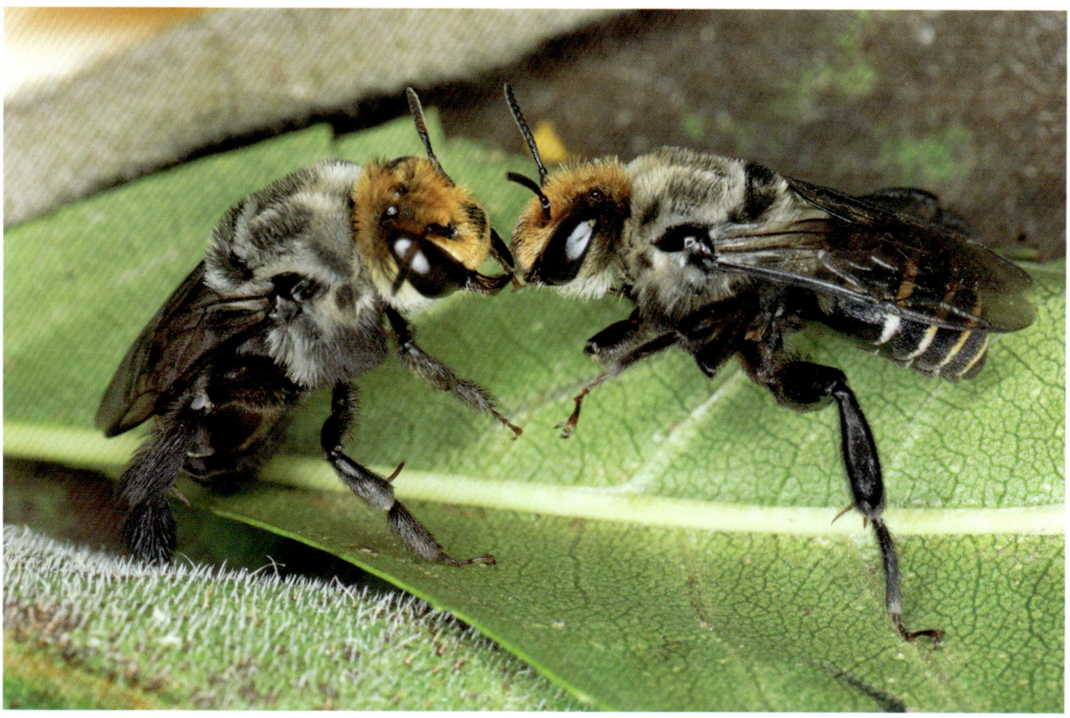

Female (*left*) and male (*right*) *Melitoma marginella* meeting on a leaf

snap caused esperanza bushes to wilt and drop their yellow flowers. Female bees crawled into their underground nests. Male *Melitoma* clung to their holds inside the esperanza blossoms as the flowers floated through the air and parachuted to earth like escape pods.

After the cold snap, *Melitoma marginella* males could be found in large numbers hunkered on the ground inside fallen esperanza flowers, peering from their deep throats with bright rust-colored faces. When the cold rains and weather recommenced, the *Melitoma* males remained housed inside the cast-off blossoms, drinking what remained of the esperanza nectar. As the cold snap gave way to warmer weather, the male bees drifted back to visiting new blossoms until November gave way to winter.

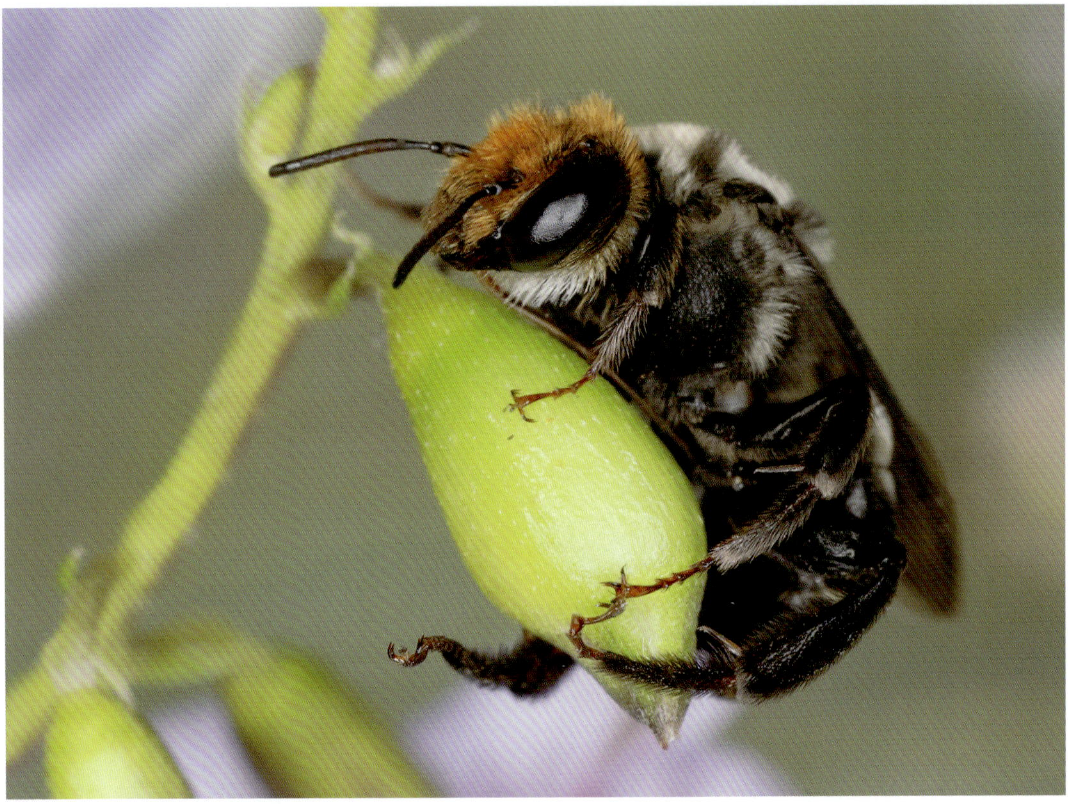

A male *Melitoma marginella*

Bull chimney bee ◆ *Melitoma taurea* Say, 1837

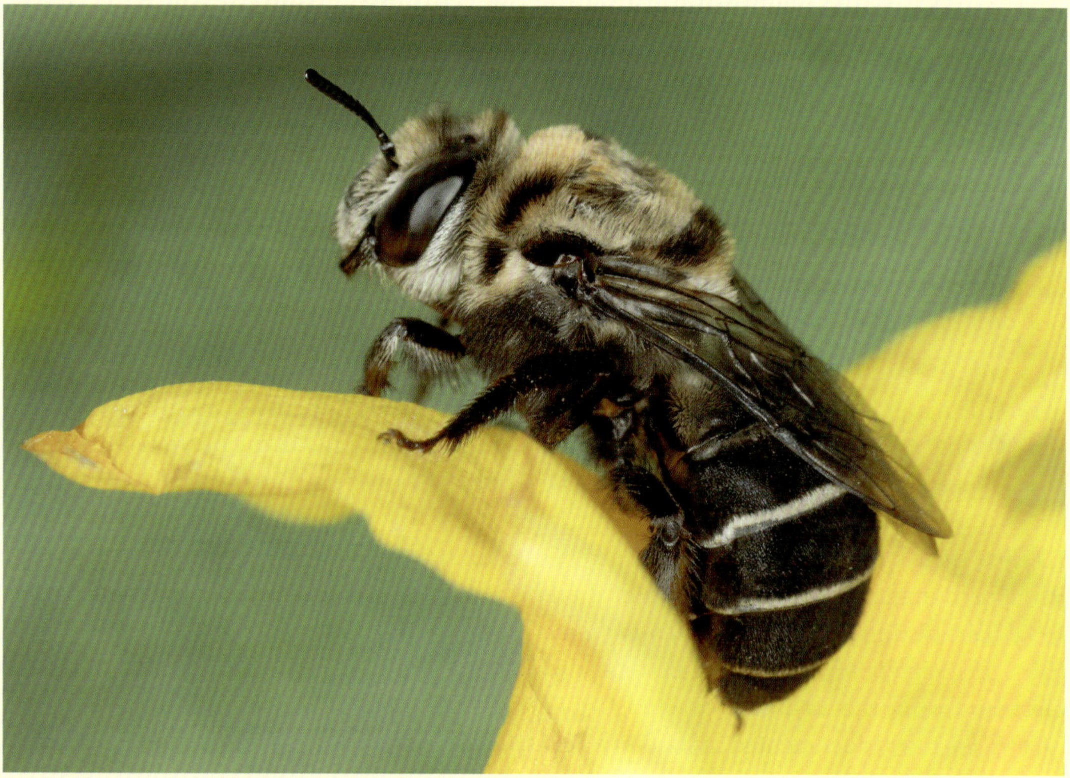

A female *Melitoma taurea*

Description: The name *taurea*, Latin for "bull," most likely refers to the stocky appearance of this chimney bee. *Melitoma taurea* has a hefty black abdomen striped with narrow white bands. Patterns of predominantly dark and white hairs, intermingled with sparse pale orange hairs, cover the black face, head, and thorax. The legs and leg hairs are black. Females have long, feathery, black hind-leg scopae. The hind legs of both males and females are somewhat elongated. **Size:** females and males 9.5–12.5 mm (3/8–1/2 in.).

Range: *Melitoma taurea* is rarely seen in the Valley. This species occurs principally in the eastern United States. Most documented sightings in Texas have occurred near the Louisiana border. Nonetheless, a moderate-sized population of *Melitoma taurea* appeared in Falcon State Park (Starr Co.) during June 2021, evidencing a range extension for this species.

Associated plants: *Melitoma taurea* is associated with flowers of the family Convolvulaceae, in particular wild potato (*Ipomoea pandurata*) and beach morning glory (*Ipomoea pes-caprae*). The long legs and feathery scopal hairs of females enable them to carry the large pollen grains typical

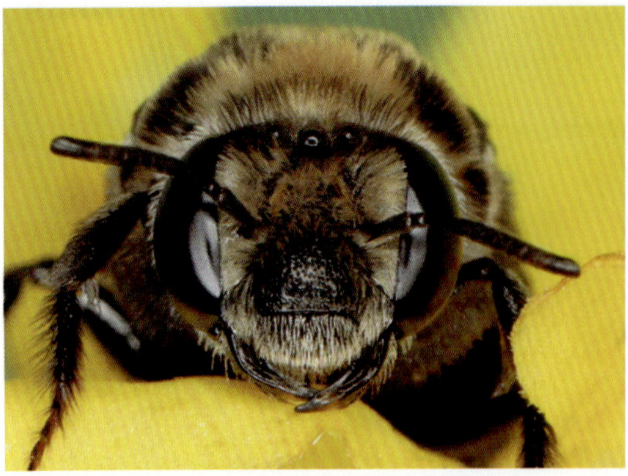 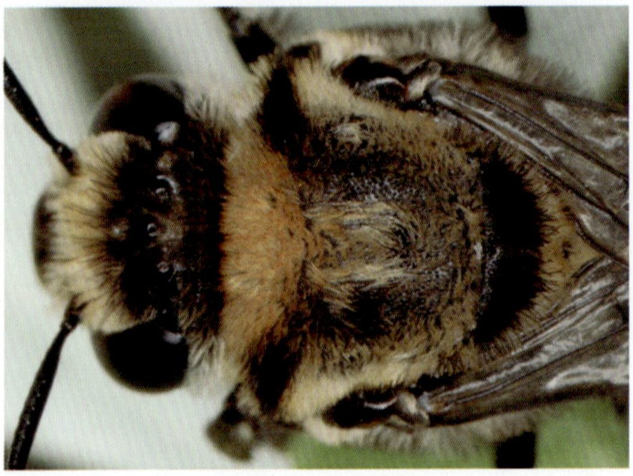

Face of a female *Melitoma taurea* (*left*), and close-up of the thorax (*right*).

of such plants and to plumb their deep-throated blossoms. The Falcon State Park bees shown here were visiting esperanza (*Tecoma stans*), like the *Melitoma marginella* in the previous entry. *Melitoma taurea* has also been documented visiting strawberries, caneberries, melons, and hibiscus.

ID tips: Three species of *Melitoma* chimney bee occur in the United States. *Melitoma marginella* is the predominant *Melitoma* species of the Lower Rio Grande Valley. Its bright rust-orange facial hairs distinguish it from *Melitoma taurea*, which has a dark face covered mostly with pale hairs. *Melitoma taurea* is also a more robustly built bee, and it lacks the well-defined pattern of black-and-white thorax hairs characteristic of *Melitoma marginella*. The third *Melitoma* species, *M. grisella*, rarely documented in southern Texas, is a generally paler bee than either of the other two species: its head and thorax are covered with predominantly white hairs.

Behavior: Bull chimney bees tend to nest in aggregations in compact, clayey soil near water. Some evidence suggests that the bees prefer nesting in shady areas under the protective cover of roots and vegetation. *Melitoma taurea* males emerge in the summer before females and can be found gathering outside their nest holes. In 1925, entomologist Philip Rau recorded a colorful description of male *Melitoma taurea* "quarreling" and "fussing" around a nesting area at the top of a clayey embankment in Missouri. The males fought with one another at nest entrances, rolled down the embankment clasped in struggle, and used their mandibles to drag one another by the legs, "so fiercely," Rau wrote, "that I thought surely they would be snipped off."[1]

Rau observed female *Melitoma taurea* carrying water from puddles to moisten and work the clayey ground for use in constructing their vertical tunnel-shaped nests. The females lined their brood cells with waterproofing materials and papery coverings. When digging nest burrows, they pushed excavated soil behind them, their heads still buried in the earth as they deftly used their back legs to construct complicated turrets.

1. Rau (1929), pp. 171–172.

Rincon chimney bee ◆ *Diadasia (Coquillettapis) rinconis* Cockerell, 1897

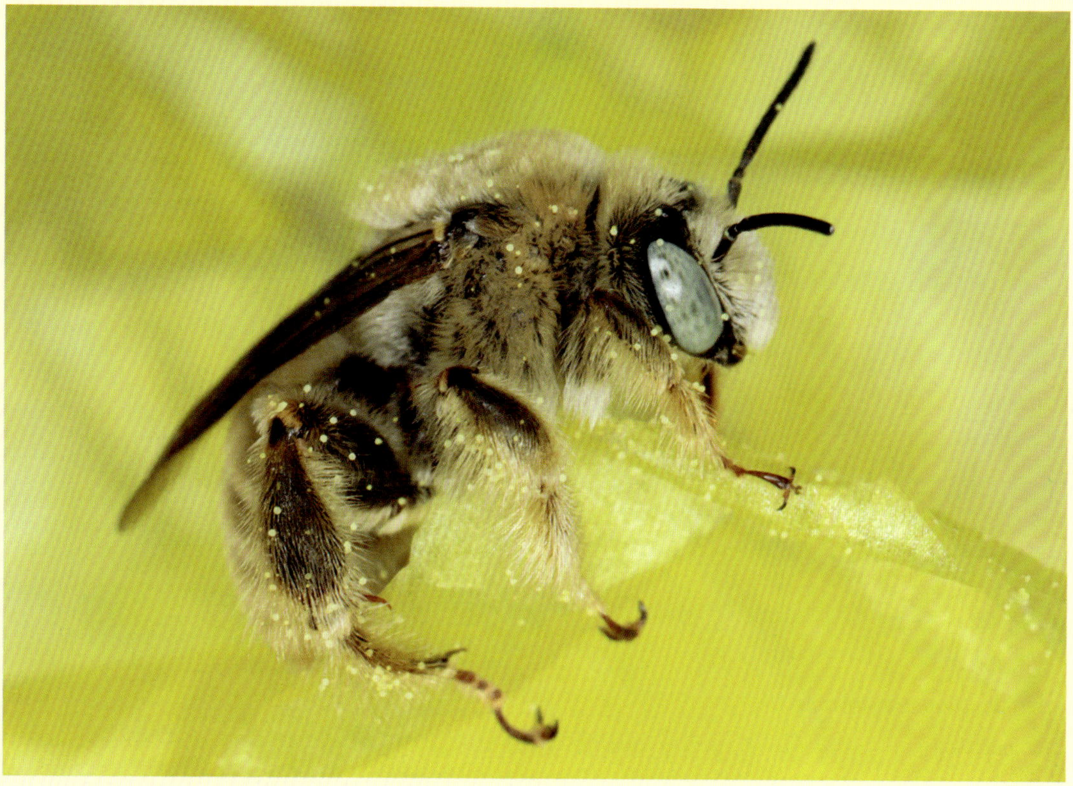

A male Rincon chimney bee

Description: *Diadasia* are small to medium-sized, furry-looking chimney bees. Most appear golden or beige, with abdomens that are either striped with pale bands or covered with light gold hairs. Female *Diadasia* have feathery scopae on their hind legs.

Diadasia (Coquillettapis) rinconis is a robust chimney bee that shows notable sexual dimorphism. Males are covered entirely with tawny hairs and have greenish eyes. Females have dark abdomens girded by broad bands of pale hair; light gold hairs covering their heads and thoraxes; bushy beige scopal hairs on their hind tibiae; and blue-green eyes. The male and female bees' tegulae are dark brown, and their wings dusky with dark veins. **Size:** females and males 11–12 mm (2/5–1/2 in.).

Range: The genus *Diadasia* occurs exclusively in the Western Hemisphere, and *Diadasia* generally favor arid climates. The Rincon chimney bee was named after Rincon, a small desert town in New Mexico, where Cockerell first found the species. *Diadasia rinconis*, however, occurs throughout much of the western United States, from Texas to California, and northward to Montana and Alberta. This species also ranges southward through Mexico to Querétaro. *Diadasia rinconis* is found in all three border counties of the Lower Rio Grande Valley, from March through the end of June.

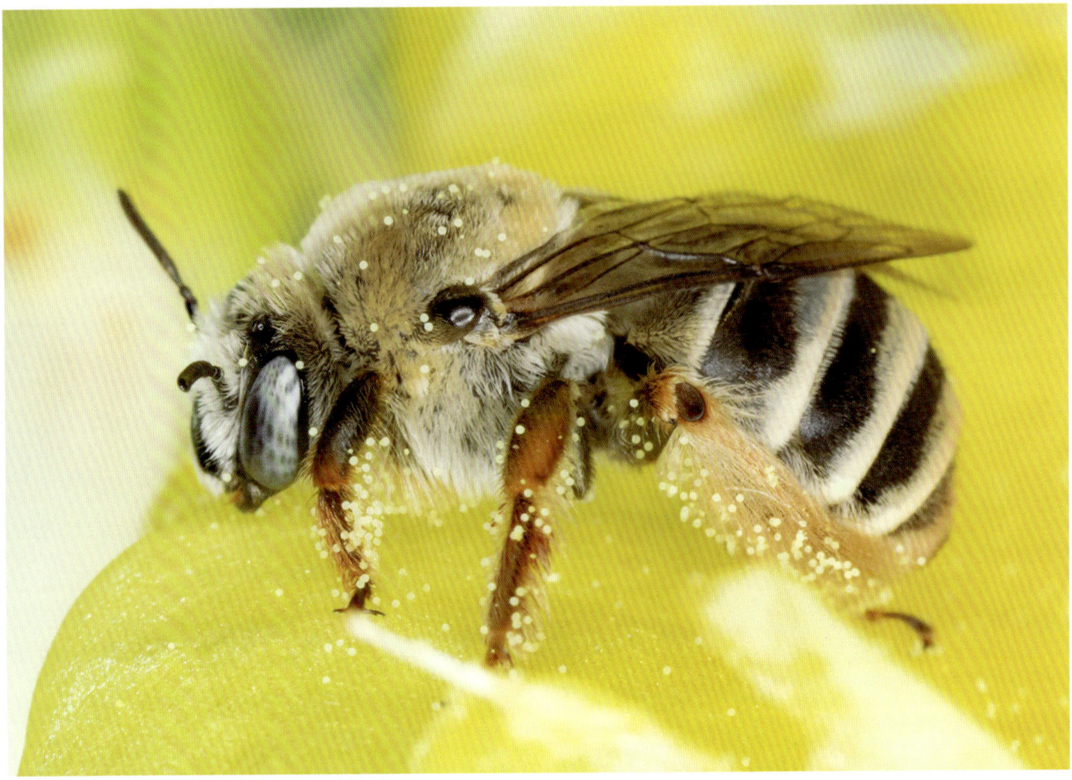

A female *Diadasia rinconis*

Associated plants: *Diadasia rinconis* is considered an oligolege of Cactaceae, and its emergence in the Valley is associated with spring cactus blooms. In the Valley, *Diadasia rinconis* females are most often found gathering pollen from Texas prickly pear cactus (*Opuntia engelmannii* var. *lindheimeri*), tasajillo (*Cylindropuntia leptocaulis*), and prickly poppies (*Argemone*). In late June 2021, this species appeared at Campos Viejos Ranch (Starr Co.) visiting candy barrel cactus (*Ferocactus wislizeni*) and star cactus (*Astrophytum asterias*). Males and females drink nectar from plants as diverse as wild olive (*Cordia boissieri*), Texas paloverde (*Parkinsonia texana*), esperanza (*Tecoma stans*), Texas thistle (*Cirsium texanum*), and mesquite (*Prosopis juliflora*).

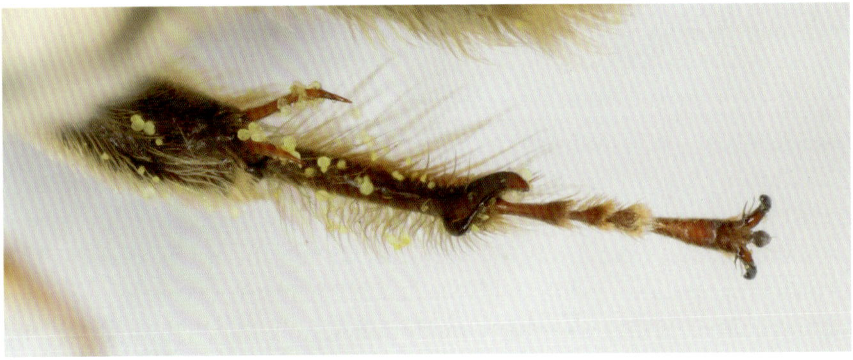

Hind leg of a male *Diadasia rinconis*: the enlarged, comma-shaped projection on the basitarsus extends half the length of the tarsomere below it.

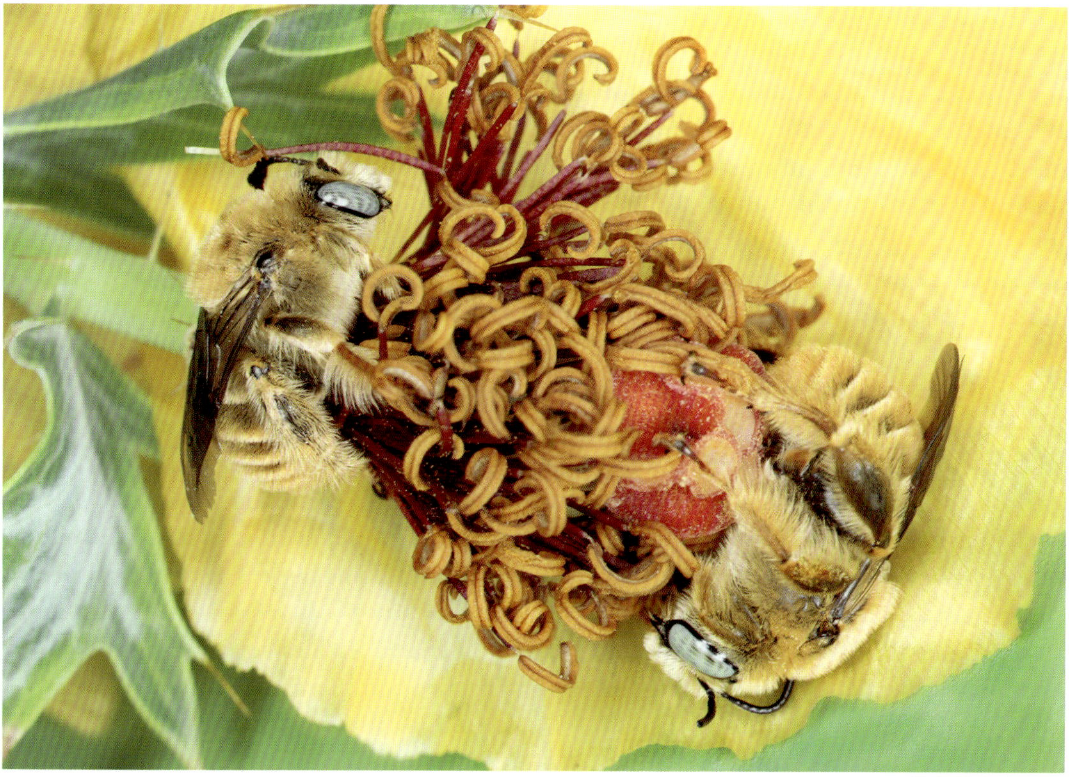

Male Rincon chimney bees resting inside a prickly poppy (*Argemone aenea*)

ID tips: The Rincon chimney bee is the most common *Diadasia* found on prickly pear cactus in the Lower Rio Grande Valley. Its comparatively large size makes it fairly easy to distinguish from other *Diadasia* in this guide. Notably, however, female *Diadasia rinconis* are nearly identical to female *D. australis*, which also pollinate prickly pear and tend to commingle in Texas with *D. rinconis*. Females of these two species can be identified most easily by their association with more easily differentiated males. Male *Diadasia rinconis* can be distinguished from male *D. australis* by distinctive projections on the lower hind legs, shown on the opposite page.

Behavior: During spring in the Lower Rio Grande Valley, male Rincon chimney bees emerge before females and often gather in yellow and red prickly poppies (*Argemone aenea* and *A. sanguinea*). Prickly poppies are common in the Valley—they ornament vacant lots, the edges of railroad tracks, and untended fields. Often three to four feet tall, they have multiple stems bearing chalice-shaped, tulip-sized blossoms that provide protection from wind and rain. Intricate spines arm the leaves and pods, preventing trespass by predators. Prickly poppies provide abundant pollen but not nectar, and thus they are not a food source for males. Instead, males seek out the poppies for shelter from wind or inclement weather.

Once the cactus bloom commences, *Diadasia rinconis* males visit prickly pear blossoms, as well as other flora, for nectar. At night, however, the males return to the *Argemone* poppies; on early mornings during cactus blooms in Hidalgo County, males can be found sleeping inside poppy blossoms.

Female Rincon chimney bees overwinter in the soil and then emerge shortly after males to begin building new ground nests during the spring cactus blooms. Although they gather pollen from *Argemone* poppies, female *Diadasia rinconis* do not seek shelter in them.

Female *Diadasia rinconis* sometimes nest in aggregations that number in the hundreds. Like other Emphorini, Rincon chimney bees erect turret-like structures at their nest entrances. The function of *Diadasia* chimneys is debatable. Female *Diadasia* have been observed disassembling the chimneys after egg laying and using them to plug nest entrances. Various sources hold that the chimneys prevent flooding, keep debris out of the tunnels, and even discourage invaders such as velvet ants.

A 1992 study of *Diadasia rinconis* by Neff and Simpson described the behavior of male Rincon chimney bees as they waited at nest entrances for virgin females to emerge and vied for the opportunity to mate. Males attempted to drive one another from nest holes, to break up mating pairs, and even to mate with other males. As soon as a female bee emerged from her hole, a male guarding the nest entrance would spring on her. Where males were abundant, a group of them would pile together into a scrum and onto the female, who would surface between 30 and 90 seconds later with a male riding on her back, buzzing and rubbing his legs against hers. Known colloquially as "bee balls," such scrums of male Rincon chimney bees are a common sight in the Valley during the prickly pear cactus bloom. Once the mated females begin provisioning their new nests, males patrol flowers rather than linger at active nest sites.

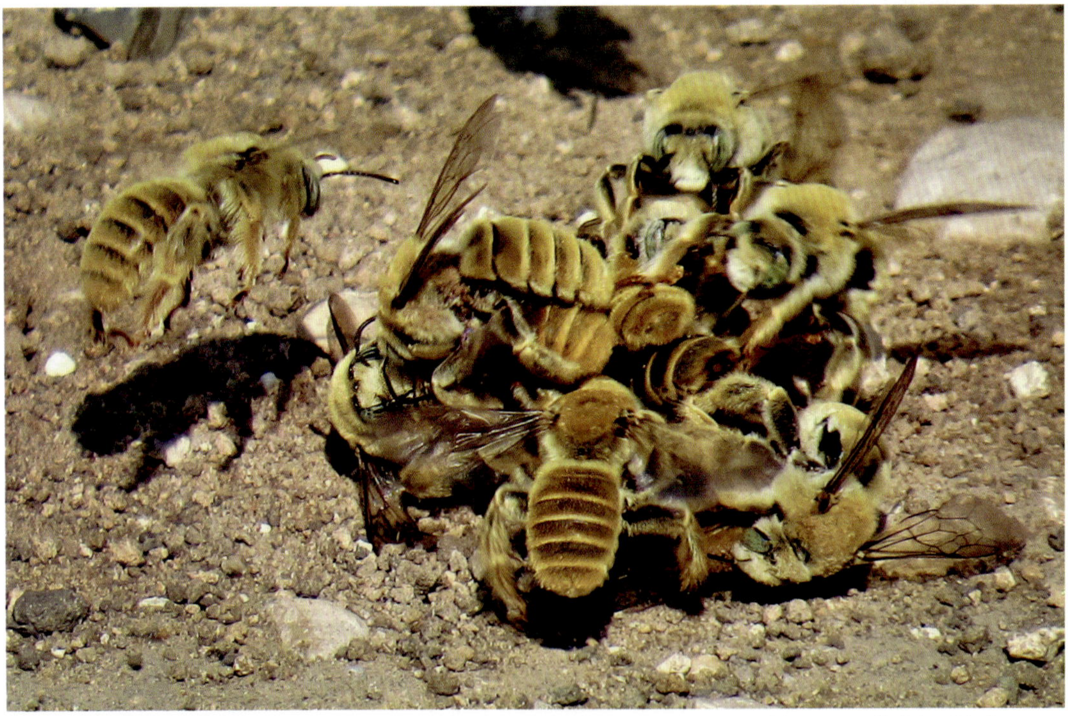

A scrum of male *Diadasia (Coquillettapis)* piling on top of a nest entrance. Photo credit John Yochum.

Globemallow bee ◆ *Diadasia (Coquillettapis) diminut**a*** (Cresson, 1878)

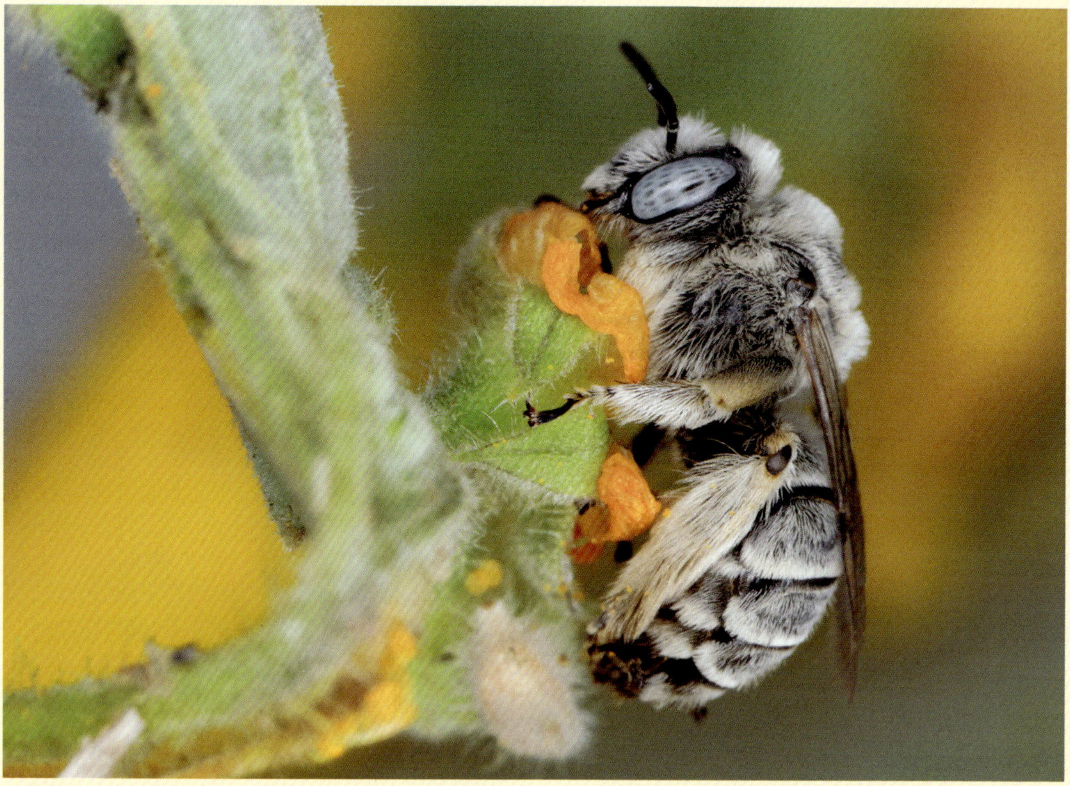

A female globemallow bee on spiked malvastrum

Description: As its name implies (*diminuta* means "tiny" in Latin), the globemallow bee is a small *Diadasia*—the smallest of all the *Diadasia* in this guide. In flight or at rest, *Diadasia (Coquillettapis) diminuta* females look fuzzy and grayish, with vaguely defined pale abdominal hair bands. Males appear pale beige. Females have long pale hairs on their hind legs; the hind-leg scopal hairs are partly black or reddish orange. The lower segments of the males' legs are reddish. Both males and females have dense white hair on their faces, black clypei, black antennae, and bright blue eyes. **Size:** females 7–9 mm (1/4–1/3 in.), males 6–8 mm (1/5–3/10 in.).

Range: The globemallow bee occurs throughout the western United States, as far north as Alberta. It ranges southward from Texas through Mexico and Central America to El Salvador. It is common in Hidalgo County in April and May.

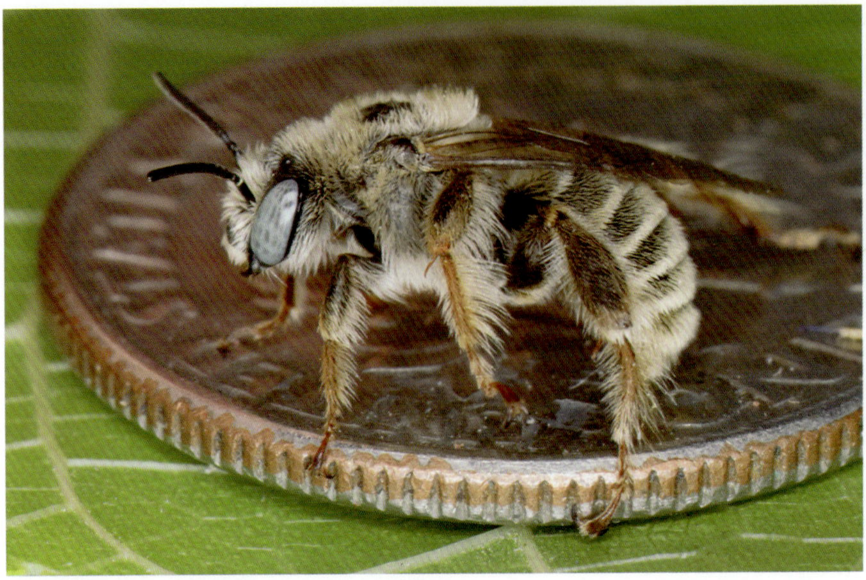

A male globemallow bee resting on a dime

Associated plants: In Hidalgo County, *Diadasia diminuta* chimney bees are frequent visitors of the mallow-family flowers spiked malvastrum (*Malvastrum americanum*) and teabush (*Melochia tomentosa*). Elsewhere, *Diadasia diminuta* is closely associated with globemallow (*Sphaeralcea*). This bee also visits a range of other plants, among them aster-family flowers, *Argemone*, *Kallstroemia*, prickly pear cactus, spurges, and vetch. In late May 2021, *Diadasia diminuta* appeared at the Mesteño Ranch (northwest Hidalgo Co.) feeding on tasajillo (*Cylindropuntia leptocaulis*).

ID tips: *Diadasia diminuta* is best recognized by its small size. In the field, the bright blue eyes of the male globemallow bee make it easily distinguishable from the slightly larger male *Diadasia tropicalis*, which has green or greenish eyes. Both male and female globemallow bees also have a patch of white hair covering the labrum, visible to the naked eye.

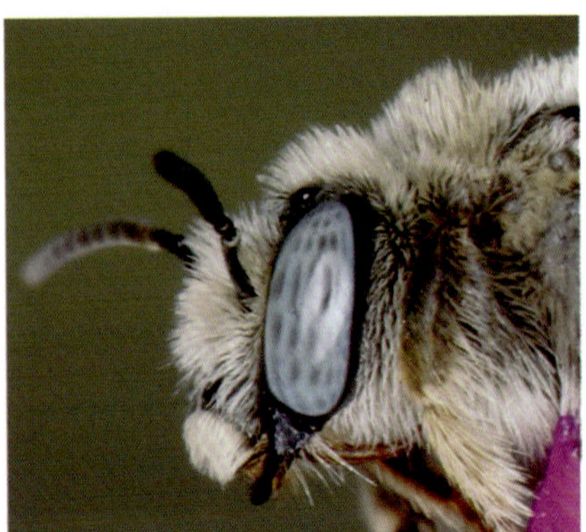

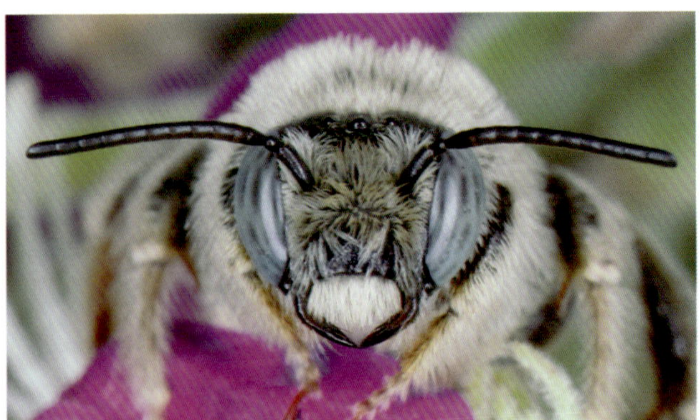

Face of a male globemallow bee

The grayish appearance of female globemallow bees aids in distinguishing them in the field from other regional *Diadasia*. The banded effect of females' abdomens is produced by rows of dense pale hairs alternating with rows of sparsely arranged pale hairs that reveal the dark surface of the integument below. This trait leaves the edges of the bands hazily demarcated and contributes to the bee's overall grayish look.

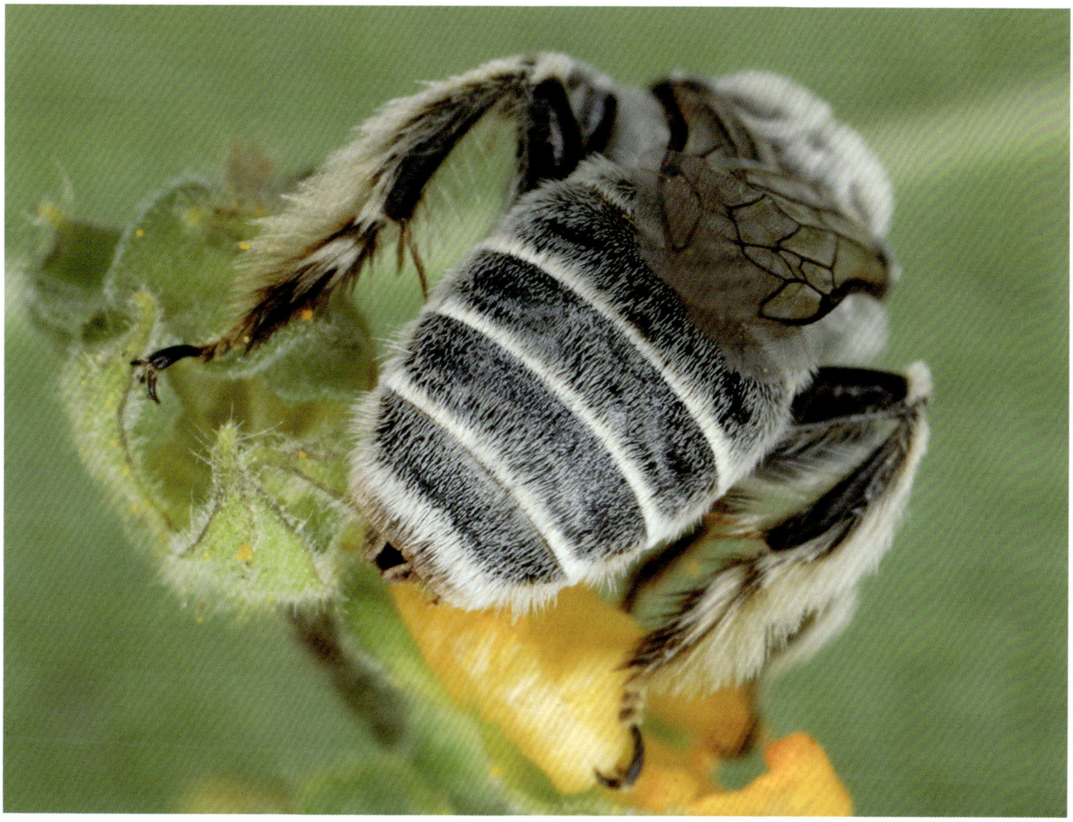

Rear view of the hair bands on the abdomen of a female *Diadasia diminuta*

Behavior: Female globemallow bees tend to nest in bare soil, often with their tunnel entrances fully exposed rather than concealed under vegetation or debris. Like other chimney bees, *Diadasia diminuta* construct turrets at their tunnel entrances. A study of Arizona globemallow bees in 1977 by entomologists Eickwort, Eickwort, and Linsley described *Diadasia diminuta* fashioning turrets by backing partway out of their burrows until their bodies were partly exposed, and then rotating while vibrating their abdomens in order to pack the soil around them into chimney shapes. When finished, the turrets ran horizontal to the ground like fallen chimneys, rather than upright.

In order to provision a nest with pollen for offspring, a female globemallow bee may make more than a dozen trips per day. Nonetheless, globemallow bees often leave their nests after staying in them for only a week or two; they seal each finished nest with dirt and move on to start a fresh nest. *Diadasia diminuta* females have been observed making as many as three nests in one season. Globemallow bees also tend to choose new sites for nest aggregations each year, rather than return to old nesting areas. Eickwort and colleagues hypothesized that this practice may enable the bees to avoid parasites and predators.

Globemallow bees commonly nest in groups of 10 to 100 bees. A casual observer of such an aggregation might notice a single female bee flying from nest hole to nest hole; when this happens, the female is probably searching for her own nest and mistakenly entering the wrong holes. At times, however, individual female globemallow bees occupy empty nests abandoned by other females.

Ochraceous chimney bee ◆ *Diadasia (Dasiapis) ochracea* (Cockerell, 1903)

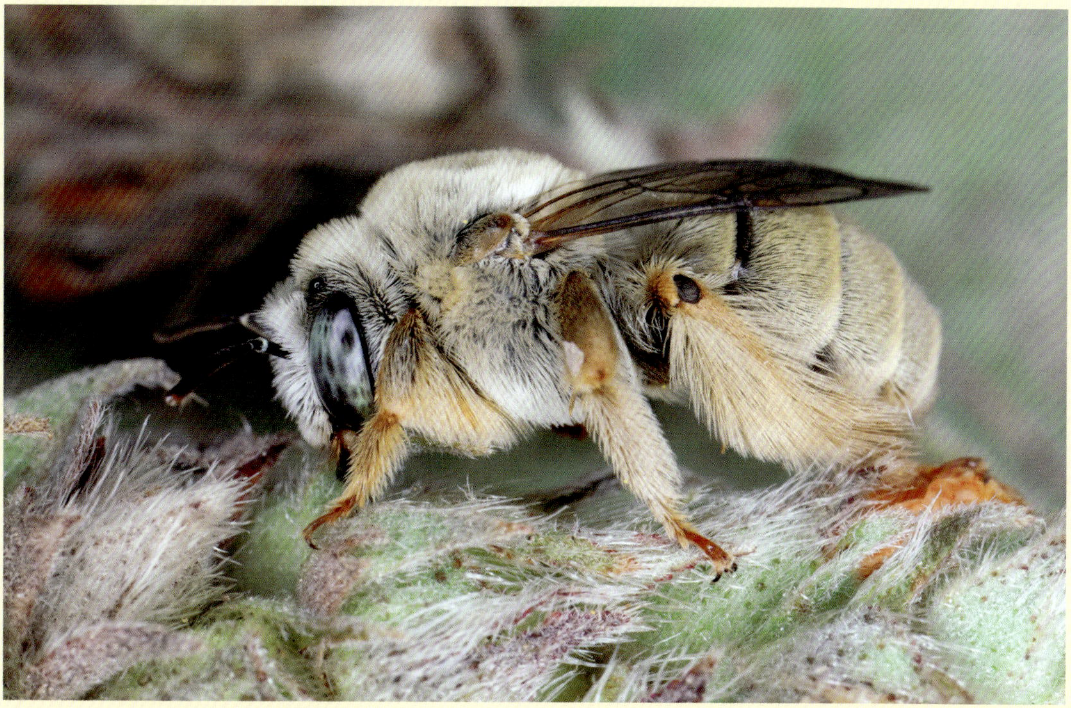

A female ochraceous chimney bee

Description: *Diadasia (Dasiapis) ochracea* is a medium-small bee covered with lush, pale gold hairs. Females have exuberant blond scopal hairs on their hind legs; the inner scopal hairs on their hind tibiae may be black or dark reddish orange. Females also have brushes of long hair on their front-leg tibiae, a distinctive trait of all bees of the subgenus *Dasiapis*. The tegulae and the integument of the upper legs of male and female *Diadasia ochracea* are brown or dark brown. Their wings are light brown with brown veins. Females have dark clypei obscured by pale hairs. The males' clypeus is pale yellow, and the labrum is also sometimes pale. Males and females have green or blue-green eyes. **Size:** females 8 mm (3/10 in.), males 7 mm (1/4 in.).

Range: Ochraceous chimney bees range from Texas through the southwestern United States, and westward to California and Washington. They are common in the Lower Rio Grande Valley in spring and fall, and they occur as far south as the Mexican state of Puebla.

Associated plants: In the Valley, *Diadasia ochracea* is commonly found on native mallows such as spiked malvastrum (*Malvastrum americanum*), Rio Grande abutilon (*Abutilon hypoleucum*), and bladdermallow (*Herissantia crispa*). Males and females nectar on such plants as Texas thistle (*Cirsium texanum*) and wild petunia (*Ruellia nudiflora* var. *runyonii*).

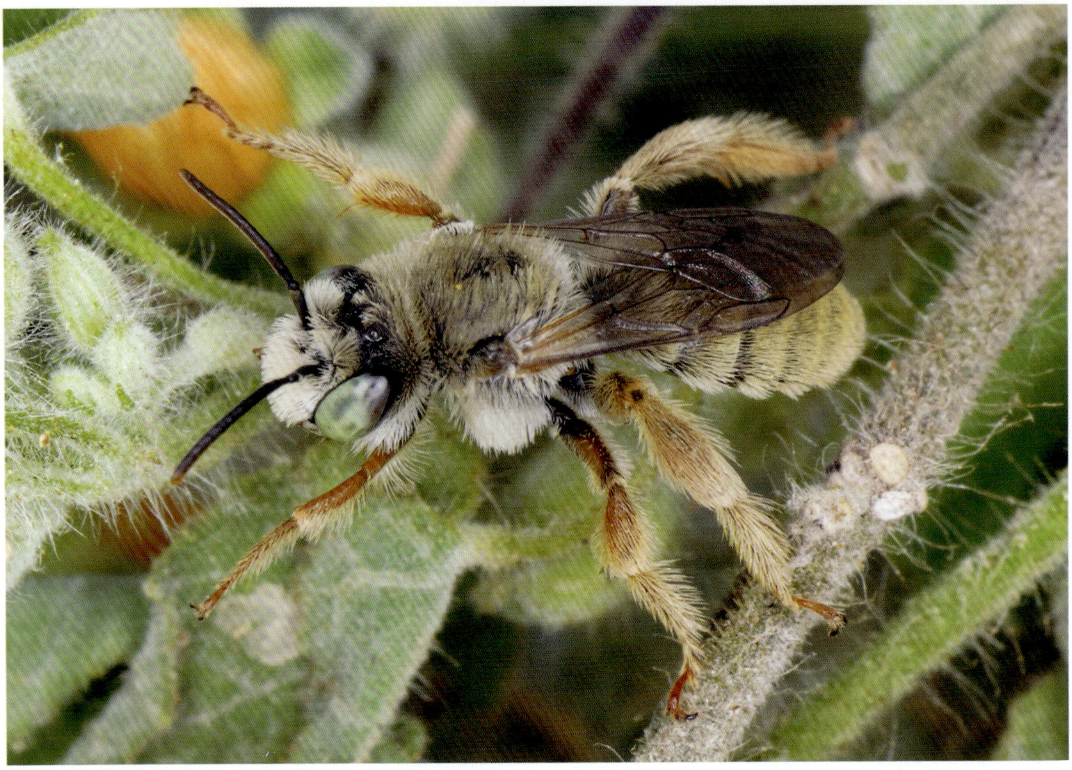

A male *Diadasia ochracea* on spiked malvastrum

ID tips: Michener noted that *Diadasia* can be separated into two groups by appearance. Those that are specialist pollinators of cacti and sunflowers tend to be large and (at least among females) to have banded abdomens. Those that favor mallows and evening primrose are mostly small, with a uniform covering of hairs on the abdomen. The ochraceous chimney bee fits into the latter category. Thus, in the field, *Diadasia ochracea* can be distinguished from other furry, beige *Diadasia* of the Lower Rio Grande Valley—such as cactus-loving *D. rinconis*, shown earlier in this guide—by its relatively small size and its preference for mallows.

Equally notable, North American *Diadasia* of the subgenus *Dasiapis* share a common trait: males have yellow markings on the bases of their mandibles. This feature is visible to the naked eye and helps differentiate *Diadasia ochracea* from similar bees, such as *D. diminuta* and *D. rinconis*, which belong to the subgenus *Coquillettapis* and lack this feature.

On the other hand, *Diadasia ochracea* may be difficult to differentiate from *D. tropicalis* chimney bees, which also belong to the subgenus *Dasiapis*. Differences between these two species are examined in the next entry.

Tropical chimney bee ◆ *Diadasia (Dasiapis) tropicalis* (Cockerell, 1918)

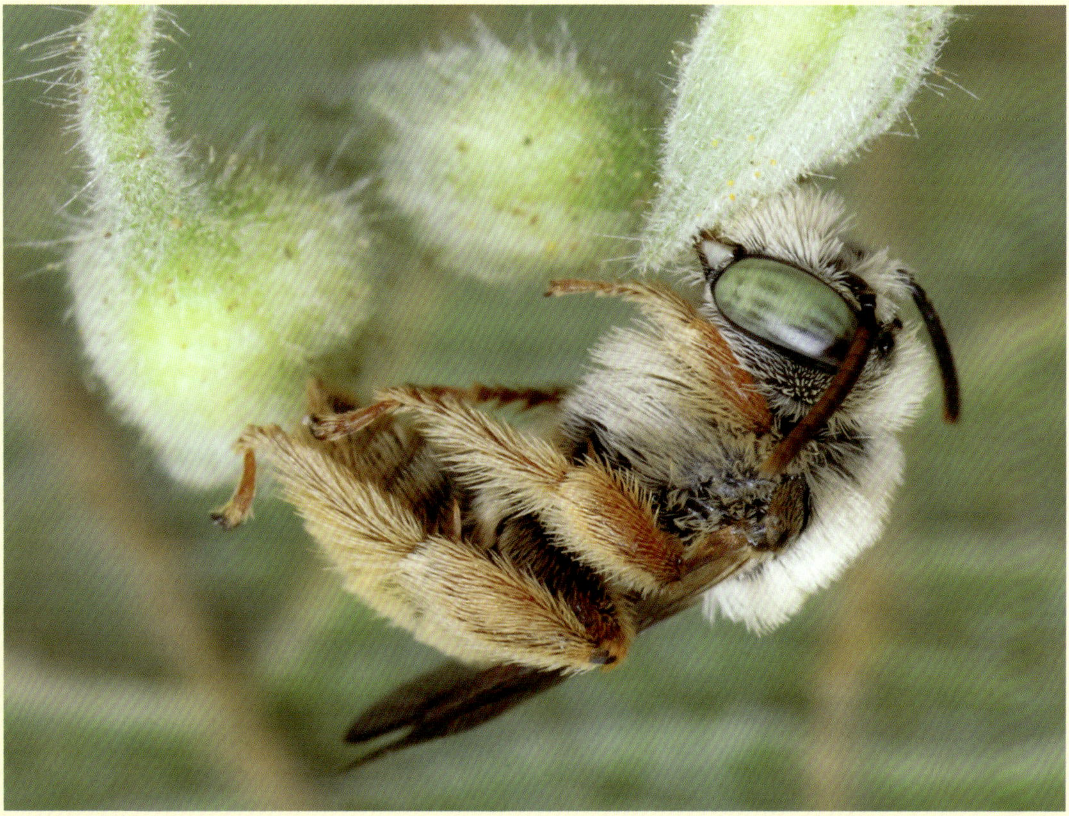

A male tropical chimney bee

Description: *Diadasia (Dasiapis) tropicalis* is a medium-small bee with a generally beige appearance, reddish legs and tegulae, and green or blue-green eyes. The face, thorax, and abdomen of both males and females are black and clothed in dense white and tawny hairs. The female's hind-leg scopae are mostly pale gold, with some rust-red hairs on the inner lower legs. The integument of the upper legs of male and female *Diadasia tropicalis* is reddish. Females have black faces and antennae that are bright red on the undersides. The male bee has a pale clypeus and labrum. Its antennae are black with reddish-brown undersides. **Size:** females 8–10 mm (3/10–2/5 in.), males 7–8 mm (1/4–3/10 in.).

Range: The tropical chimney bee is found in Mexico, Central America, Colombia, and possibly Venezuela. In the United States, *Diadasia tropicalis* has been documented only in the Lower Rio Grande Valley and along the Arizona-Mexico border. A substantial population of *Diadasia tropicalis* appeared during the spring of 2018 and 2019 at the National Butterfly Center (Hidalgo Co.). This species is found in all three border counties of the Valley.

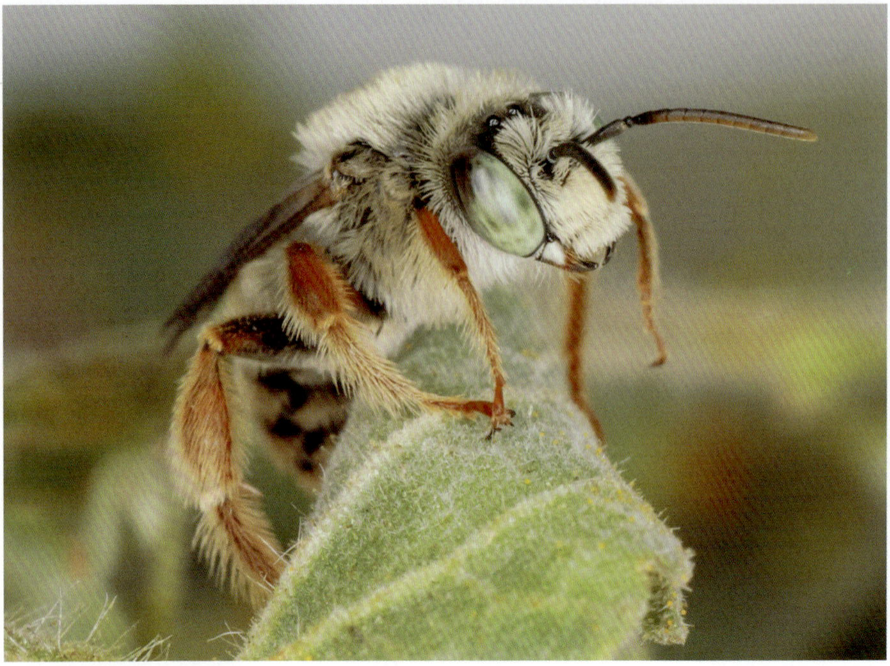

A male *Diadasia tropicalis*

Associated plants: Within the Lower Rio Grande Valley, tropical *Diadasia* are found most commonly on mallows such as Rio Grande abutilon (*Abutilon hypoleucum*), spiked malvastrum (*Malvastrum americanum*), and teabush (*Melochia tomentosa*). In mid-May 2021, male *Diadasia tropicalis* appeared feeding on shrubby Indian mallow (*Abutilon abutiloides*) at Campos Viejos Ranch in Starr County.

ID tips: Tropical chimney bees have an overall reddish appearance. They also bear the hallmark of their subgenus *Dasiapis*—males have pale spots on the bases of the mandibles, a trait visible in the field. As noted in the previous entry, *Diadasia tropicalis* closely resembles the ochraceous chimney bee, which shares this trait. The species are generally distinguished in the field by color—the femurs and tegulae of *Diadasia ochracea* are brown, while those of *D. tropicalis* are reddish.

The ochraceous versus tropical chimney bee: *Diadasia tropicalis* was first described in 1918 by Cockerell in an article quaintly titled "Neotropical Bees, Principally Collected by Professor Bruner in Argentina." Cockerell's description was based on a pinned *Diadasia* specimen from the Mexican state of Veracruz. He proclaimed the bee a new species but questioned whether it might simply be a redder subspecies of the ochraceous chimney bee. Seventy-six years later, entomologist Roy Snelling produced a study of *Diadasia* of the subgenus *Dasiapis*, in which he too opined that *Diadasia tropicalis* "might be nothing more than a southern variant of *Diadasia ochracea*."[2] He concurred with Cockerell that color alone should not be enough to differentiate

2. Snelling (1994), p. 6.

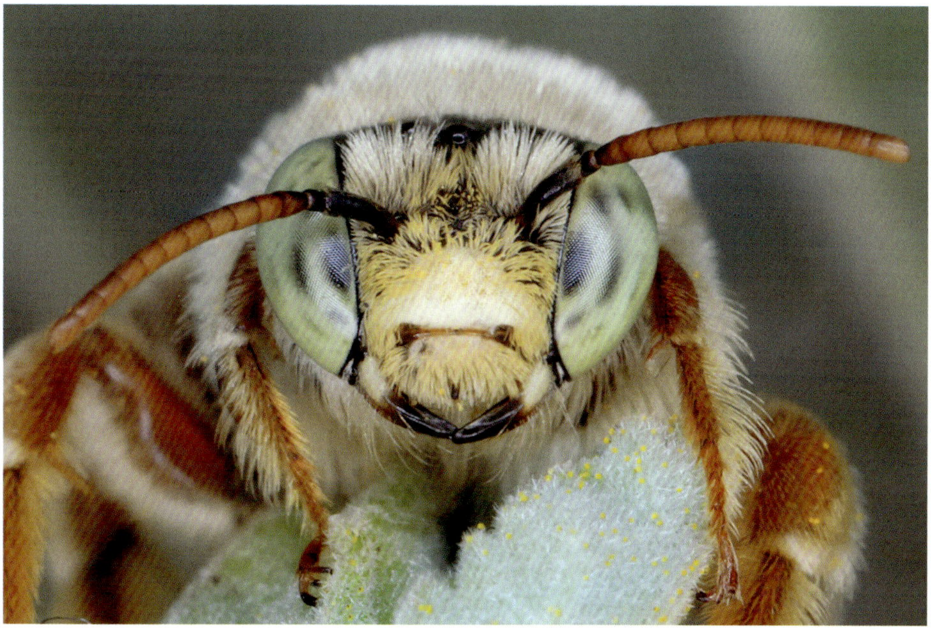

Close-up of the face of a male *Diadasia tropicalis*

species. Nonetheless, Snelling retained the separate species names, noting that *Diadasia ochracea* had a more northerly range and in Mexico tended to be confined to the cooler Central Highlands. He also noted a minute structural difference between the two species that went beyond color: the thorax of *Diadasia tropicalis* has well-defined, dense pits on the scutum, a trait absent or muted in *D. ochracea*.

Further study is probably required before this issue is put to rest. During 2018–19 at the National Butterfly Center in Hidalgo County, bees fitting the descriptions of both species appeared in tandem—as did a continuum of *Diadasia* sharing combinations of their traits (for example, red upper legs with dark brown tegulae). All foraged on mallows.

There is little published data on the floral preferences of *Diadasia tropicalis*. Anecdotal evidence based on observations at the National Butterfly Center in 2019 and 2021 suggests that *Diadasia tropicalis* is oligolectic on mallows. Among the broad array of flora available at the NBC, *Diadasia tropicalis* was found most often, and in large numbers, on teabush (*Melochia tomentosa*). Variations of this bee that lacked red coloration and that fell definitively under the species label *Diadasia ochracea* were not found on this plant. Teabush is a subtropical shrub whose range mirrors the recorded range of *Diadasia tropicalis*: it is found along the border between Hidalgo County and Tamaulipas, and in Mexico, Central America, and northern South America. Formerly placed in the cacao family, Sterculiaceae, teabush is now considered a member of the family Malvaceae.

Diadasia tropicalis specimens that Snelling examined were found principally on mallows such as fanpetals (*Sida*), globemallows (*Sphaeralcea*), and viscid mallow (*Bastardia viscosa*). A few were collected on other flowers, among them orange cosmos, spurge, *Kallstroemia*, morning glories, coral vine (*Antigonon leptopus*), feverfew (*Tanacetum parthenium*), verbena, lantana, and plants of the family Zygophyllaceae.

Comparison of *Diadasia ochracea* and *Diadasia tropicalis*

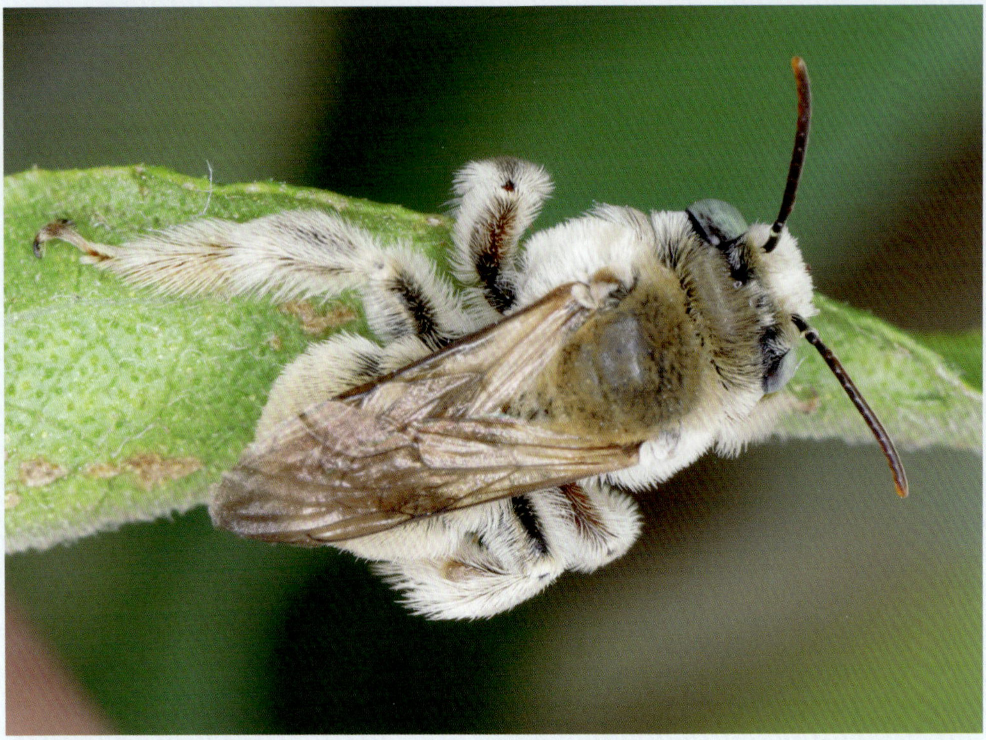

The upper legs of the male *Diadasia ochracea* are dark brown.

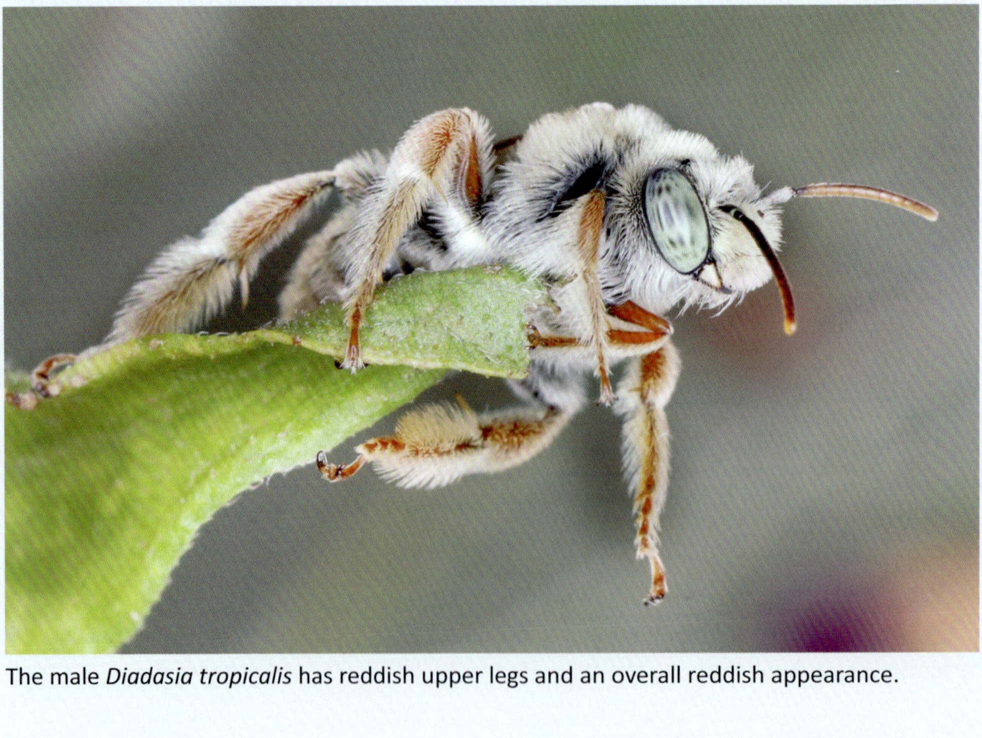

The male *Diadasia tropicalis* has reddish upper legs and an overall reddish appearance.

Sunflower chimney bee ◆ *Diadasia (Diadasia) enavata* (Cresson, 1872)

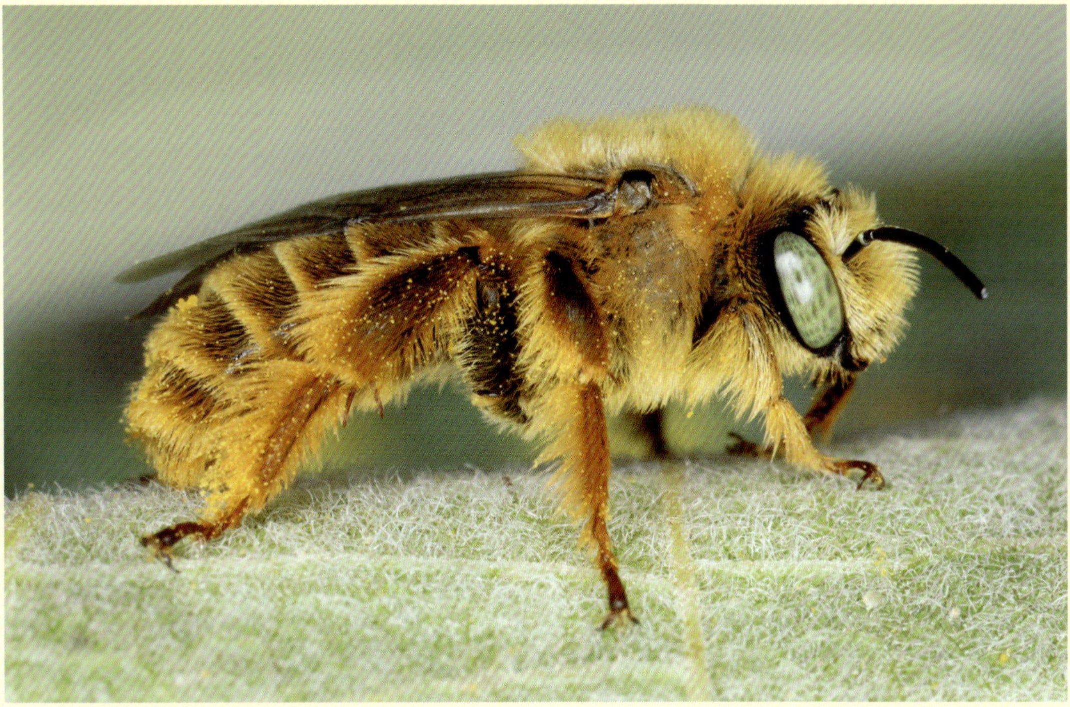

A male sunflower chimney bee

Description: *Diadasia (Diadasia) enavata* is a medium-large bee with a black head, body, and legs, all covered with dense golden-brown hairs. Female bees have yellow-orange scopae on their hind legs. Both males and females have black antennae and green eyes. **Size:** females and males 12–15 mm (1/2–3/5 in.).

Range: The sunflower chimney bee is a widespread North American *Diadasia* species. It ranges throughout the western United States, eastward into Nebraska and Mississippi, northward to the Canadian province of Saskatchewan, and as far south as the Mexican state of Tamaulipas. It appears in the Lower Rio Grande Valley from early June through September.

Associated plants: *Diadasia enavata* is most often associated in the Valley with common sunflower (*Helianthus annuus*) and cowpen daisy (*Verbesina encelioides*). This species is considered oligolectic on flowers of the aster family.

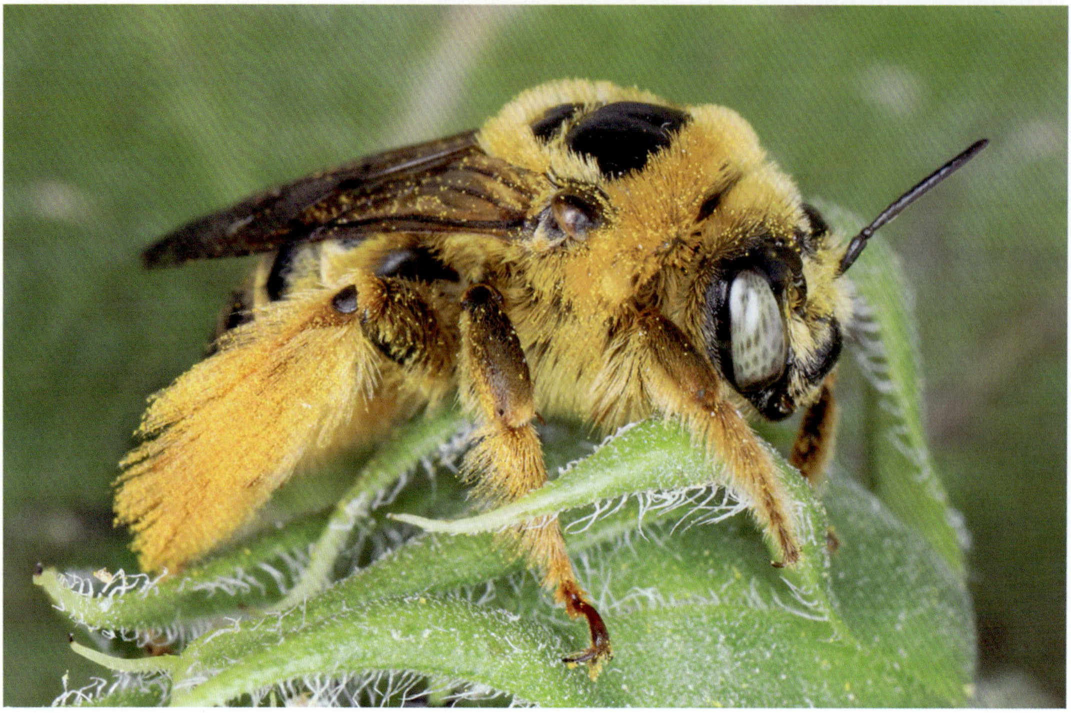

A female sunflower chimney bee

ID tips: The sunflower chimney bee's comparatively large size and coat of golden hairs make it easy to distinguish from other *Diadasia* of the region. The close association of this species with sunflowers also aids in its identification.

Behavior: *Diadasia enavata* is the sole member of its genus in the Valley that specializes in sunflowers and related aster-family flowers. Because of its prowess as a sunflower crop pollinator, *Diadasia enavata* has been more thoroughly researched than most *Diadasia.*

A seminal 1980 study by Hurd, LaBerge, and Linsley reported that *Diadasia enavata* females are well equipped to gather sunflower pollen: they have fine, feathery hind-leg scopal hairs specially adapted to carry the fine-grained pollen of sunflowers. In addition, while many bees concentrate pollen-gathering activity in the morning or late in the day, female *Diadasia enavata* are energetic bees that start collecting sunflower pollen in early morning and continue to do so until sunset, as long as pollen is still available.

Sunflower chimney bees are solitary ground nesters that tend to favor sandy and hard-packed soils. Their nests consist of long vertical tunnels that branch into egg chambers. The bees construct half-inch turrets at nest entrances. Sunflower chimney bees vary in their nest habits. They sometimes dig individual nests far from one another, but they also form large aggregations, often near stands of sunflowers or other Asteraceae.

Crooked-Legged Bees

Subfamily Eucerinae ◆ Tribe Ancyloscelidini
Genus *Ancyloscelis*

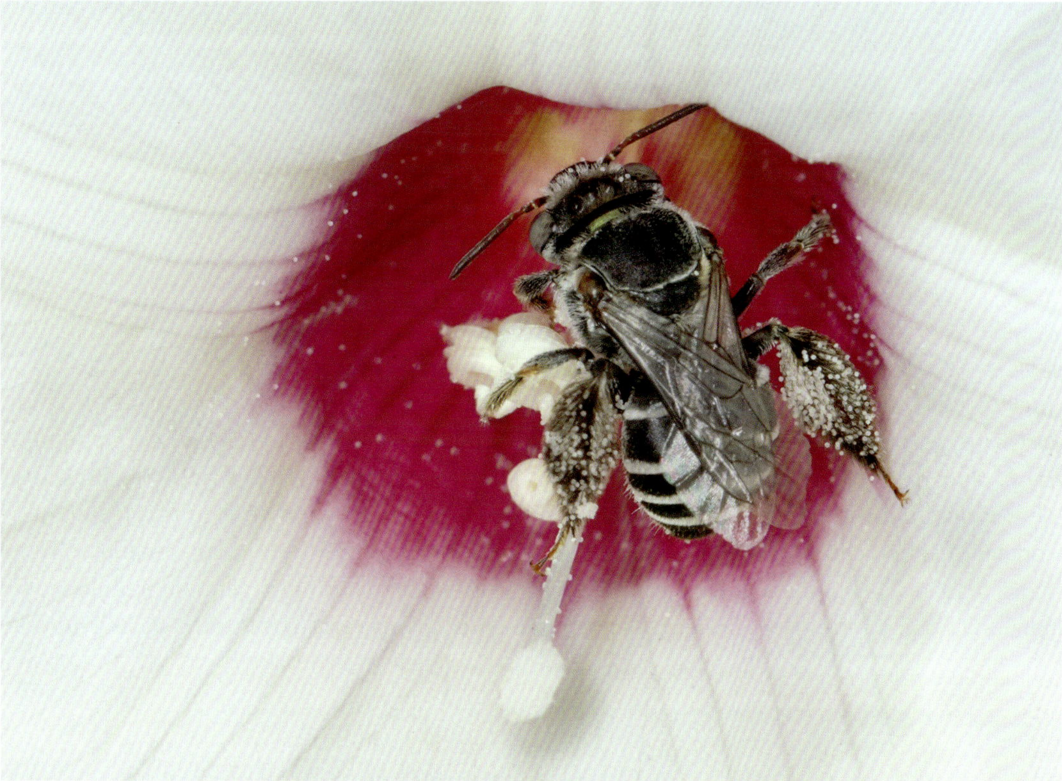

A female *Ancyloscelis apiformis* entering an alamo vine blossom

Crooked-legged bees are hairy, small to medium-sized bees with distinctly banded abdomens and protuberant clypei that give them a Roman-nosed appearance. *Ancyloscelis* males have long, ornately curved hind legs. The genus name *Ancyloscelis,* which means "crooked legged," derives from this trait. Females have dense, coarse scopal hairs on the hind-leg tibiae and basitarsi.

In the recent past, these singular bees were classified as a type of chimney bee. The genus *Ancyloscelis* now occupies its own tribe, Ancyloscelidini, within the subfamily Eucerinae. Unlike *Melitoma* and *Diadasia,* crooked-legged bees do not build turrets at their nest entrances. *Ancyloscelis* are solitary bees that often nest in large aggregations. Michener described *Ancyloscelis* nests as consisting of branched burrows leading to single egg cells or short series of cells, their inner surfaces "beautifully smooth but without visible linings."[3]

---

3. Michener (2007), p. 703.

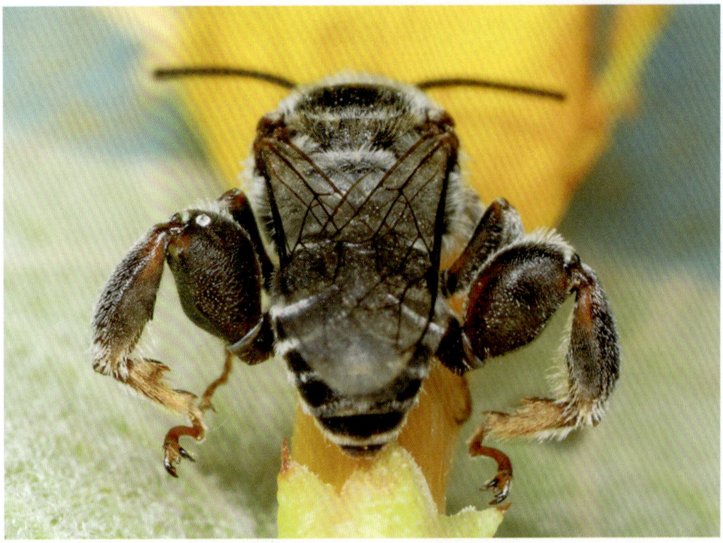

The crooked legs of a male *Ancyloscelis apiformis*

***Ancyloscelis* behavior:** In the Valley, female *Ancyloscelis apiformis* tend to congregate in shady areas and banks abutting canals or other waterways, often close to *Melitoma marginella* populations. Male *Ancyloscelis apiformis* can be found buzzing around nests or nearby flowers. Males use their elongated hind legs for leverage when mating or feeding. They spread their rear legs to keep from falling into wide-mouthed flowers as they sip nectar, or as they peer down into blossoms in search of mates. Males fly in quick figure-eight patterns when patrolling vegetation frequented by females. At night, males sleep hanging by their jaws from plants visited by females during the day.

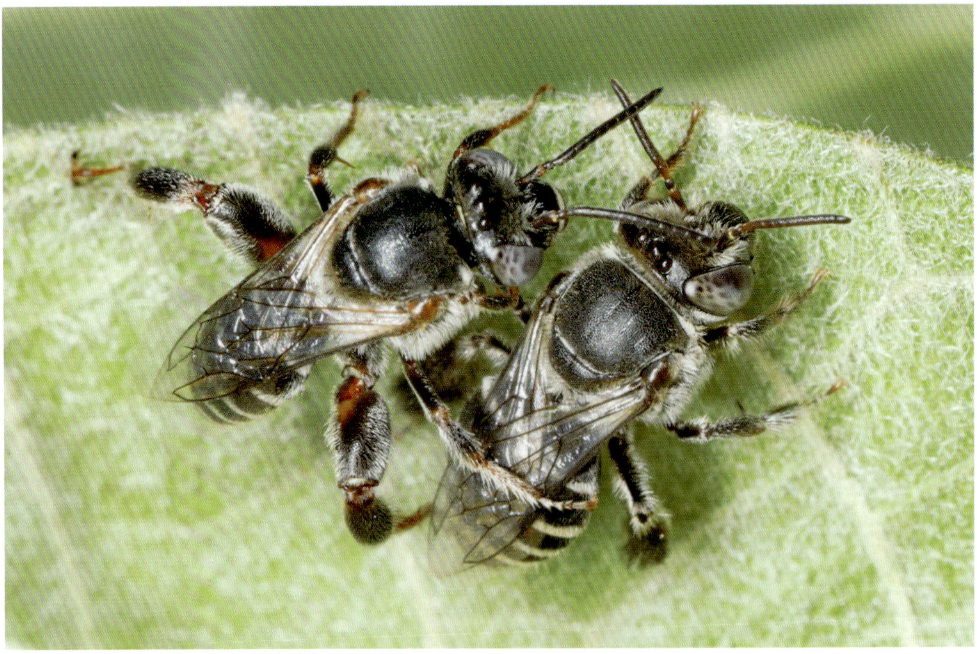

Male (*left*) and female (*right*) *Ancyloscelis apiformis* during mating

Apiform crooked-legged bee ◆ *Ancyloscelis apiformis* (Fabricius, 1793)

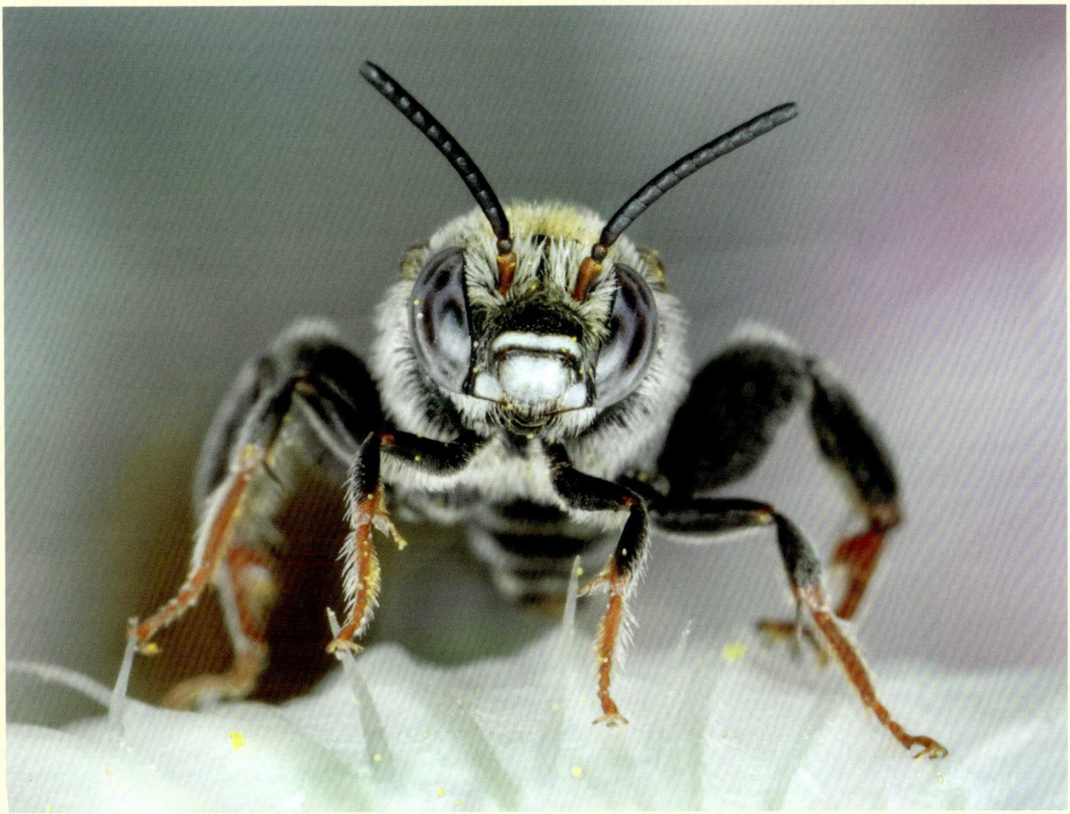

A male *Ancyloscelis apiformis*

Description: *Ancyloscelis apiformis* is a small, dark, stoutly built bee with an abdomen girded with crisply defined bands of white hair. Males have the long, curved hind legs typical of their genus. Females' hind legs are proportionately short and straight and sport stiff, widely spaced, feathery scopal hairs. Females have reddish tarsi; on males, the tarsi, basitarsi, and apical ends of the tibiae are red. In other ways, male and female *Ancyloscelis apiformis* are similar. Their antennae are relatively short and their eyes blue gray. Their faces are oval and predominantly dark, and their mandibles are pale. The antennae are mostly black or reddish black; males have red scapes. **Size:** females 7–8 mm (1/4–3/10 in.), males 6 mm (1/5 in.).

Range: Bees of the genus *Ancyloscelis* occur only in the Western Hemisphere. They range from Texas, Colorado, and Arizona through Mexico and Central America, and as far south as central Argentina. There are approximately 20 species; only three have been found in the United States (*Ancyloscelis apiformis*, *A. melanostoma*, and *A. sejunctus*). Of these, *Ancyloscelis apiformis* is the species most likely to be found in the Valley: it is native to Cameron, Hidalgo, and Starr Counties and has been found as far north in Texas as Medina County.

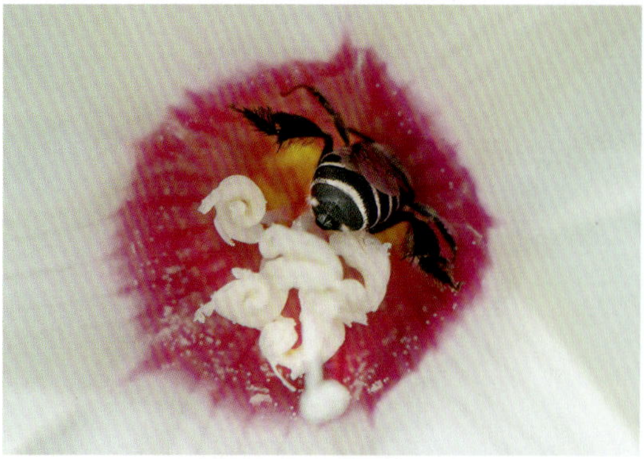 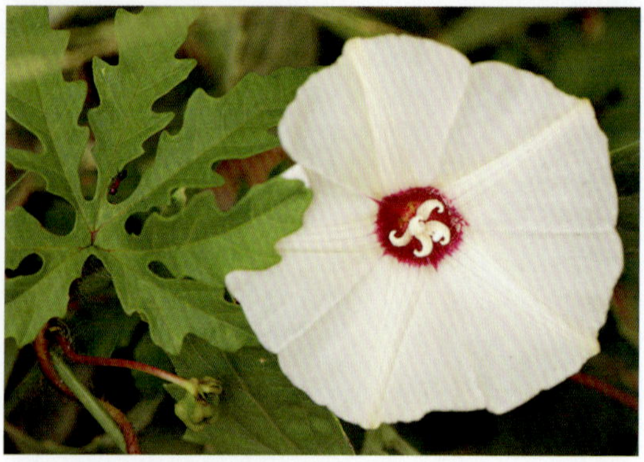

A female *Ancyloscelis apiformis* gathering pollen from alamo vine (*left*). An alamo vine blossom (*right*).

Associated plants: *Ancyloscelis* are oligolectic: all North and Central American and at least some South American species collect pollen from flowers in the morning glory family (Convolvulaceae). Some in South America are oligolectic on flowers of the families Cactaceae and Pontederiaceae. Within the Lower Rio Grande Valley, *Ancyloscelis apiformis* is usually found feeding on alamo vine (*Merremia dissecta*), a plant in the Convolvulaceae family bearing white trumpet-shaped flowers with magenta centers. This bee's close association in the Valley with alamo vine aids greatly in its identification. *Ancyloscelis apiformis* chimney bees are commonly spied with their heads plunged into the deep centers of alamo vine blossoms, with only their striped abdomens and black hind legs visible from above. *Ancyloscelis apiformis* of the Valley also commonly visit esperanza (*Tecoma stans*).

ID tips: Male *Ancyloscelis apiformis* can be identified by their long, hooked legs. Males typically have distinctive facial markings as well: a broad white strip on the apical rim of the black clypeus, a white labrum, and red-tipped white jaws. Females sometimes have pale mandibles, but their faces are predominantly dark. Nonetheless their compact builds, stiff black hind-leg hairs, and boldly striped abdomens make them difficult to confuse with other small, dark bee species of the Valley.

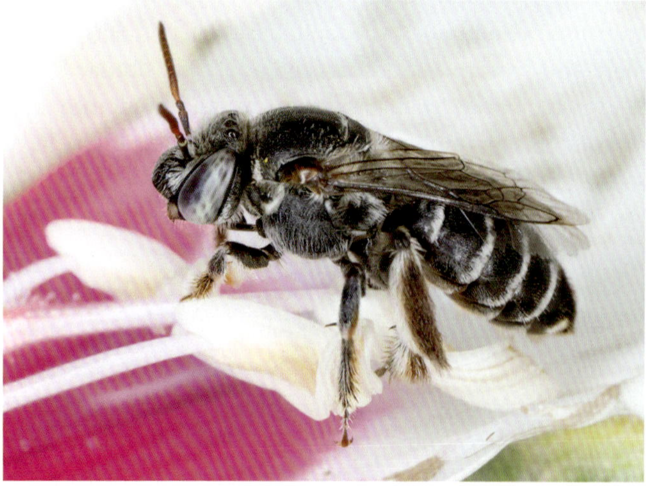

A female apiform crooked-legged bee: note the protuberant clypeus typical of the genus *Ancyloscelis.*

Longhorn Bees

Subfamily Eucerinae ◆ Tribe Eucerini
Genera *Melissoptila, Florilegus, Eucera, Svastra,* and *Melissodes*

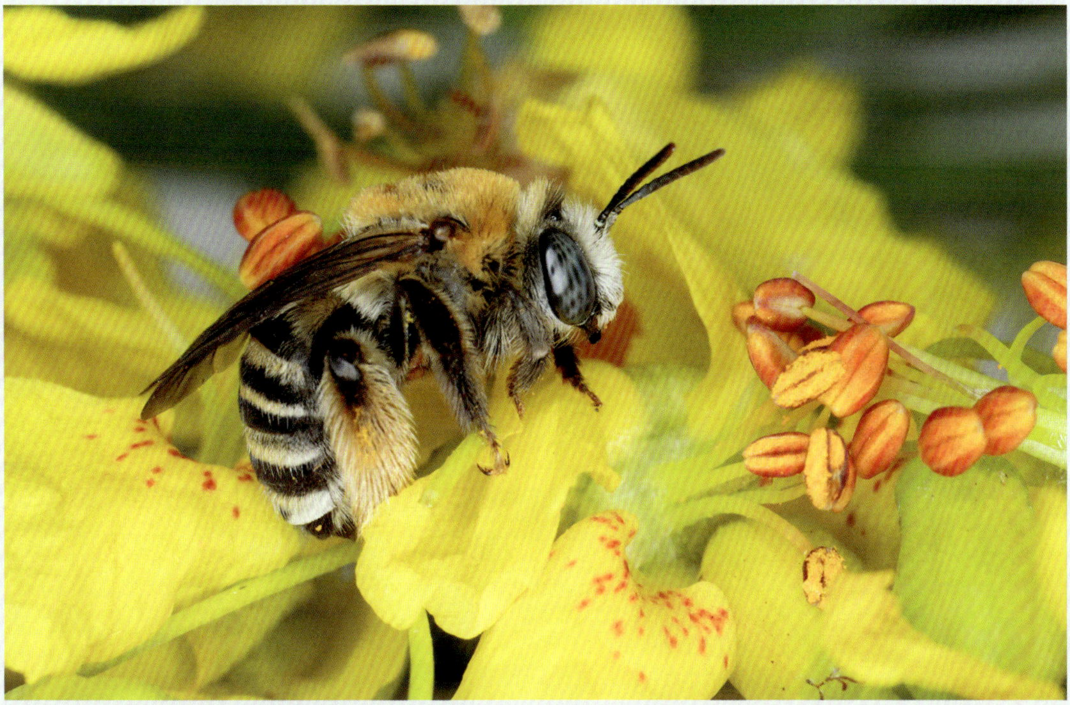

A female *Melissodes tepaneca* on retama blossoms

Longhorn bees belong to the tribe Eucerini, within the subfamily Eucerinae.[4] Eucerini serve as essential pollinators for multimillion-dollar crop industries and include, among others, *Svastra*, instrumental in sunflower production; *Eucera*, which pollinate melons and squash; and *Melissodes*, a genus of irreplaceable pollinators of cotton and sunflower crops, garden flowers, and wildflowers. Less commonly encountered are two additional genera that inhabit the Lower Rio Grande Valley: *Melissoptila* and *Florilegus*, both significant pollinators of native flora.

Bees of the tribe Eucerini are generally robust and have hairy bodies. Males tend to have unusually long antennae—hence the name "longhorn." Female longhorn bees of the Lower Rio Grande Valley usually have bushy hind-leg scopal hairs, dark faces, and shorter antennae than

4. Scientific nomenclature for longhorn bee genera is currently in a state of flux. Recent phylogenetic studies clarifying the evolutionary history of various Eucerini will eventually result in the renaming of some genera and subgenera shown here. A recent analysis by Freitas et al. (2023) reclassifies North American *Svastra* in the genus *Epimelissodes*; absorbs the subgenus *Melissodes* (*Tachymelissodes*) in the larger subgenus *Apomelissodes*; and splits up species of *Eucera* (*Xenoglossodes*), placing some in the newly resurrected genus *Xenoglossa*.

males. Males are often smaller and slenderer than females. Males of all the longhorn species in the Valley tend to have yellow or ivory clypei. A rule of thumb that helps separate longhorn bees from chimney bees (Emphorini) is that a longhorn bee's vertex is straight or slightly concave, while a chimney bee's vertex is rounded (slightly humped).

Longhorn bees may be either pollen generalists or specialists. The mournful longhorn *(Melissodes tristis),* for example, forages on a diverse range of flora, while many longhorn species show preference for aster-family flowers. *Asteraceae* are known as composites: their blossoms are made up of multiple tightly packed florets. A sunflower is a familiar example; although it looks like a single large blossom, a sunflower is made up of many small florets packed into a circular center, surrounded by other petal-like flowers. Bees that specialize in visiting composites are able to feed efficiently, gathering nectar and pollen from several florets without having to fly to a new flower.

Most longhorn bees whose nest behavior has been studied have proven to be solitary ground nesters. There are some exceptions: some *Svastra* and *Melissodes* females, for example, live communally, sharing burrows while provisioning their own individual nests. The nest holes of Lower Rio Grande Valley longhorn bees are often hidden under brush or grass and are more difficult to locate than those of other regional large solitary bees such as *Diadasia* and *Anthophora.*

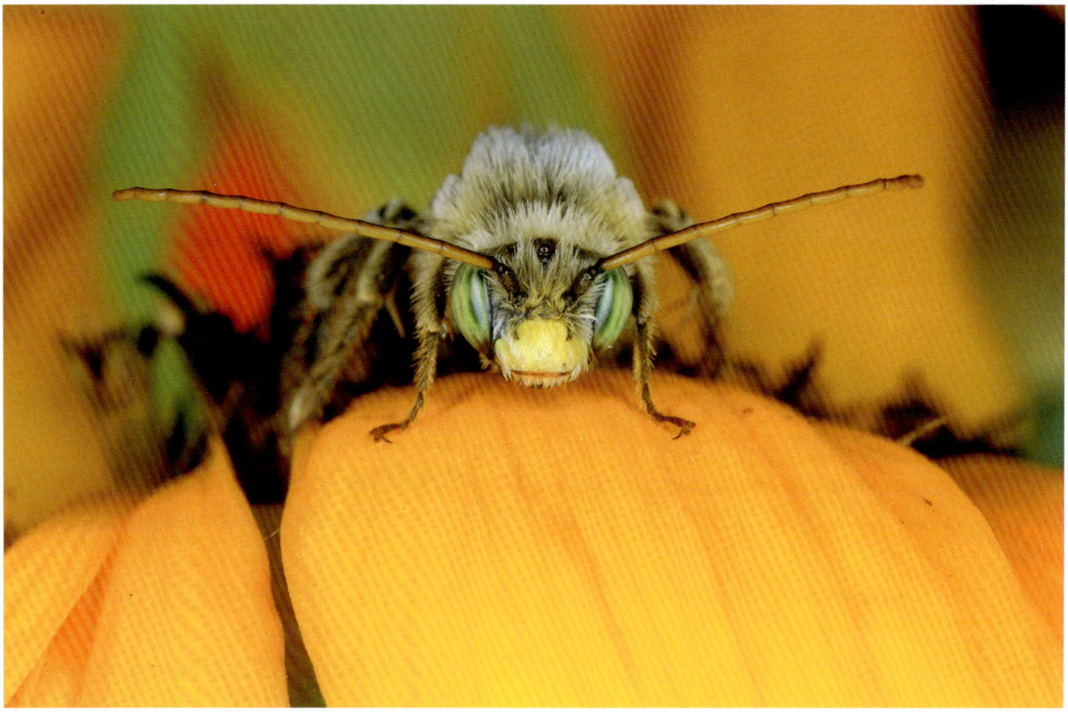

A male *Melissodes* with the long antennae, green eyes, and yellow clypeus common to many male longhorn bees

Portly longhorn bee ◆ *Melissoptila pinguis* (Cresson, 1878)

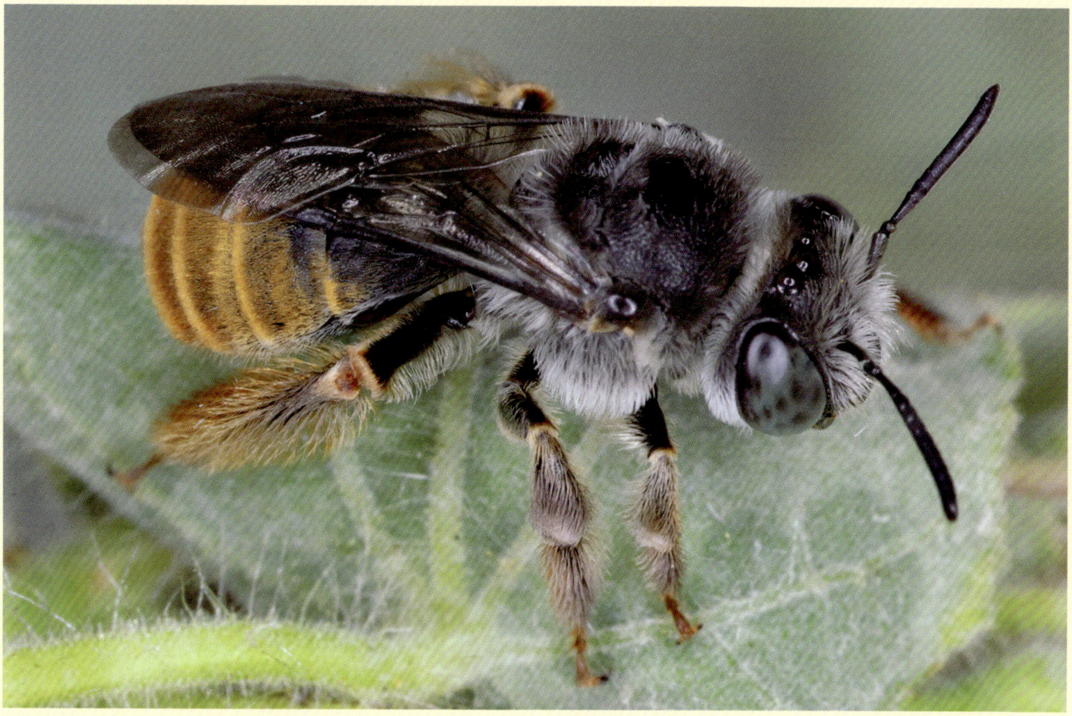

A female *Melissoptila pinguis*

Description: *Melissoptila pinguis* is the only species representing its genus in the United States. It is a handsome black bee with an abdomen covered with dense tawny hairs. White and dark hairs clothe the bee's head and thorax, giving it a slate-gray appearance. The tegulae are black, and the wings glassy with dusky outer margins. The female has long, feathery, golden scopal hairs on its hind-leg tibiae. The clypeus of the male bee is yellow; females have entirely black faces. Both have black antennae and gray-blue eyes. The sixth segment (T6) of the female's abdomen has blunt teeth on either side; on males, these toothlike projections appear on both T6 and T7. **Size:** females and males 9–10 mm (1/3–2/5 in.).

Range: According to Michener, *Melissoptila pinguis* is relatively rare in North America. In the United States, it appears principally in southernmost Texas, although it ranges as far north as Travis County. *Melissoptila pinguis* has also been documented in Mexico, Guatemala, and Honduras. This species appeared in substantial numbers at the National Butterfly Center (Hidalgo Co.) in November 2018, 2019, and 2021.

Associated plants: Anecdotal evidence suggests that *Melissoptila pinguis* is a specialist pollinator of flowers of the mallow family (Malvaceae). At the National Butterfly Center, *Melissoptila pinguis* females were found principally on spiked malvastrum (*Malvastrum americanum*), and the

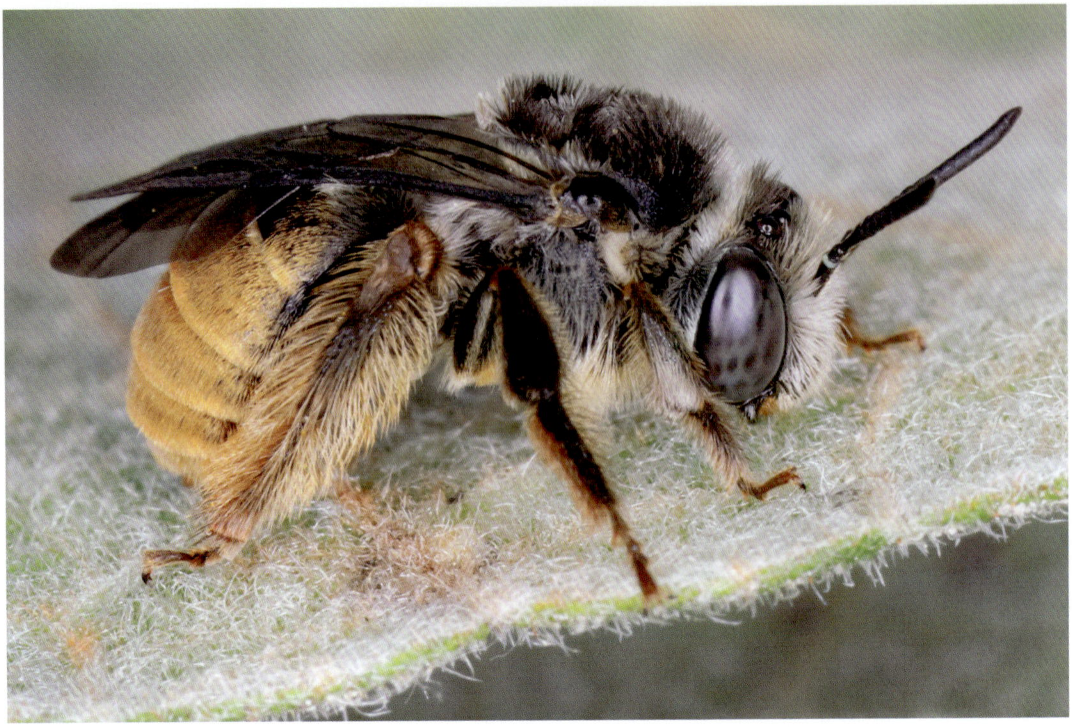

A female *Melissoptila pinguis*

bees' appearance there coincided with the bloom period of that mallow. Neff has recorded Texas specimens of *Melissoptila pinguis* feeding on mallows of the genus *Wissadula*. At the NBC, male *Melissoptila pinguis* occasionally appeared nectaring on cowpen daisy (*Verbesina encelioides*).

Other documented *Melissoptila* species of Central and South America have proven to be specialists. Several show a preference for mallows, particularly those of the genus *Sida*. Others target aster-family flowers; the species *Melissoptila tandilensis* is a significant pollinator of sunflower crops in Argentina.

ID tips: Michener described *Melissoptila* as similar to the more commonly seen longhorn bees of the genus *Melissodes*. Michener noted, however, that *Melissoptila* tend to be smaller and can be differentiated from *Melissodes* in part by the appearance of their abdomens. The abdomens of *Melissoptila* commonly lack the well-defined bands typical of *Melissodes* and are instead covered with a fairly uniform coat of short hairs. *Melissoptila* means "fuzzy bee."

Melissoptila pinguis is a distinctive bee within the Valley, easily recognized in the field by the combination of its modest size, the dense tawny hairs on its abdomen, and its dark thorax. Identification is greatly aided by traits visible with the aid of magnification. These include the toothlike projections on the male bee's abdomen and the bee's wing structure: a diagnostic trait of *Melissoptila* is that the stigmas of the bee's forewings are shorter than the prestigmas.

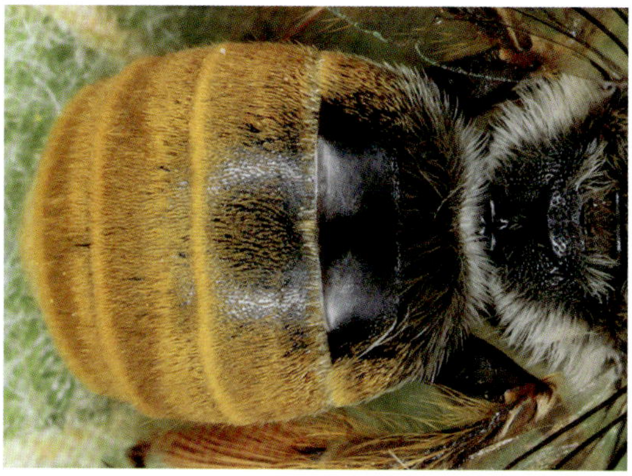

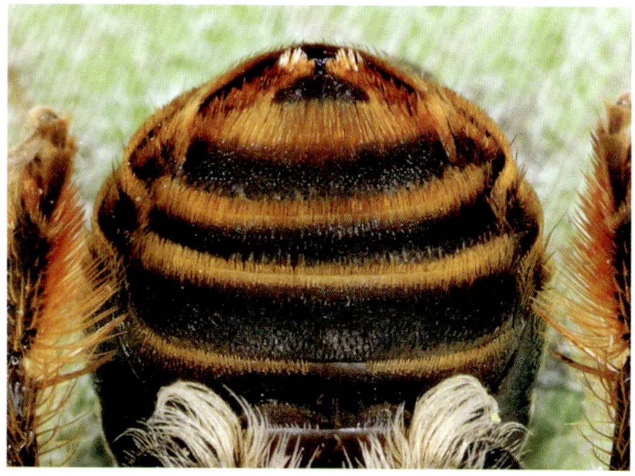

Dorsal view of the abdomen (*left*) and sternum (*right*) of a female *Melissoptila pinguis*

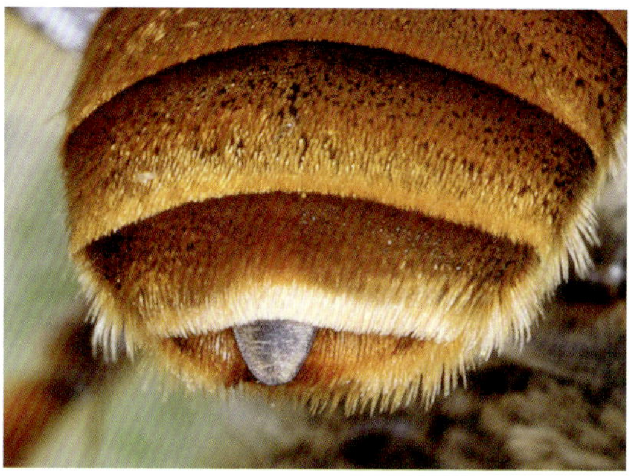

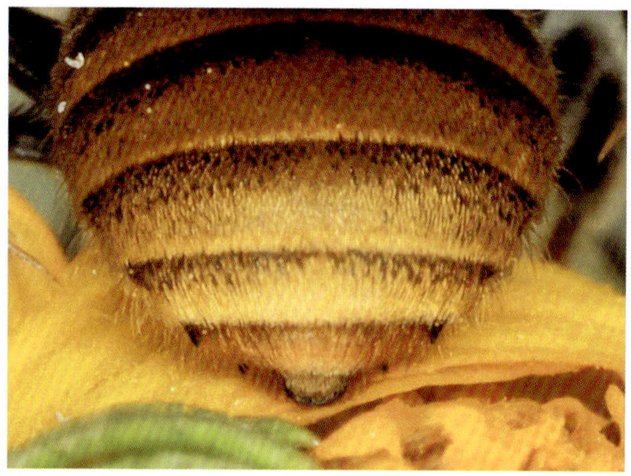

Left: Tip of a female's abdomen and pygidial plate. *Right*: Tip of abdomen of a male *Melissoptila pinguis*, with toothlike projections on T6 and T7.

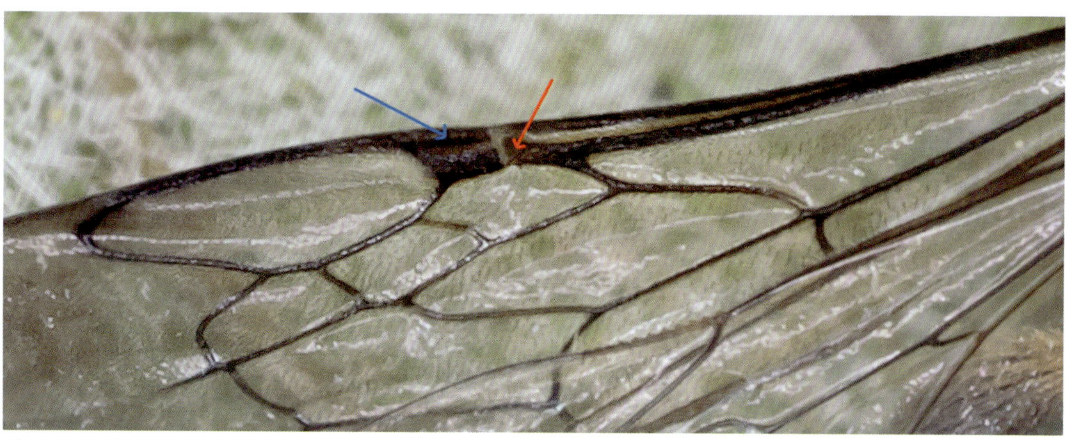

Close-up of a *Melissoptila pinguis* forewing: the prestigma (red arrow) is longer than the stigma (blue arrow).

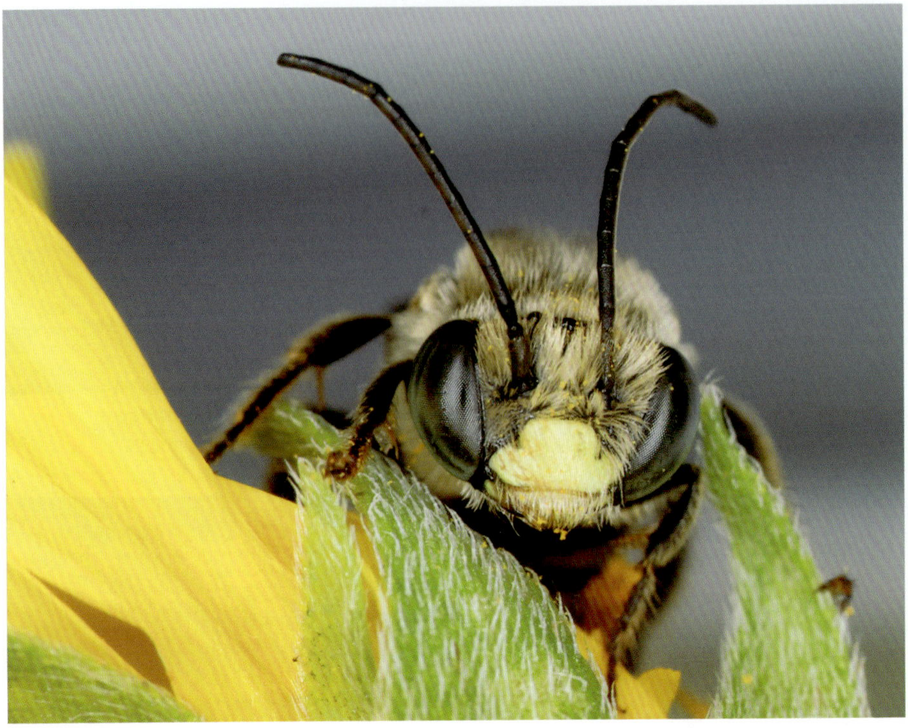

Face of a male *Melissoptila pinguis*: males have yellow clypei and long antennae.

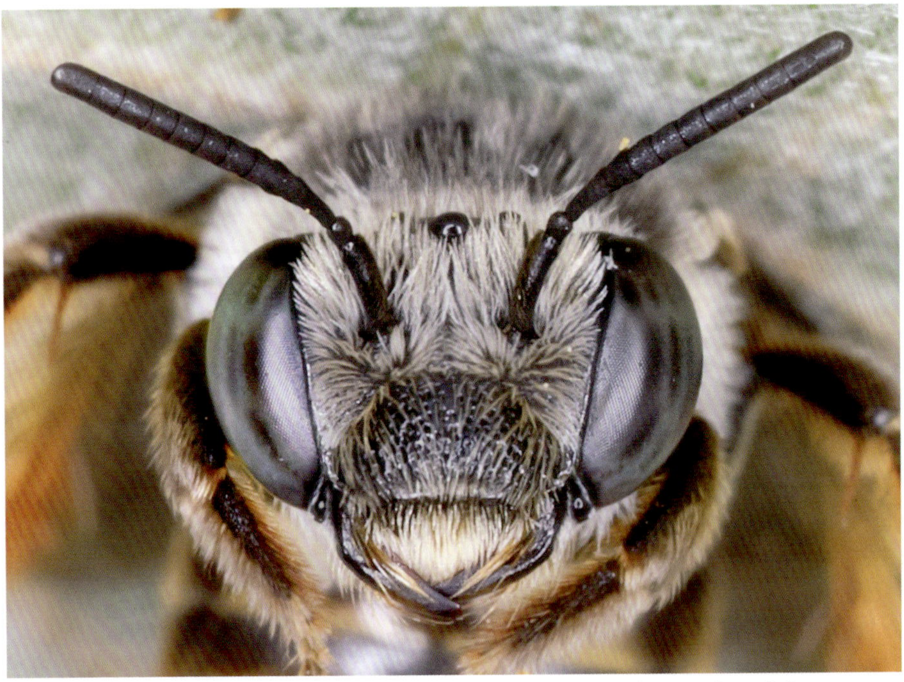

A female *Melissoptila pinguis*: the clypeus is black and coarsely pitted, and the antennae are relatively short.

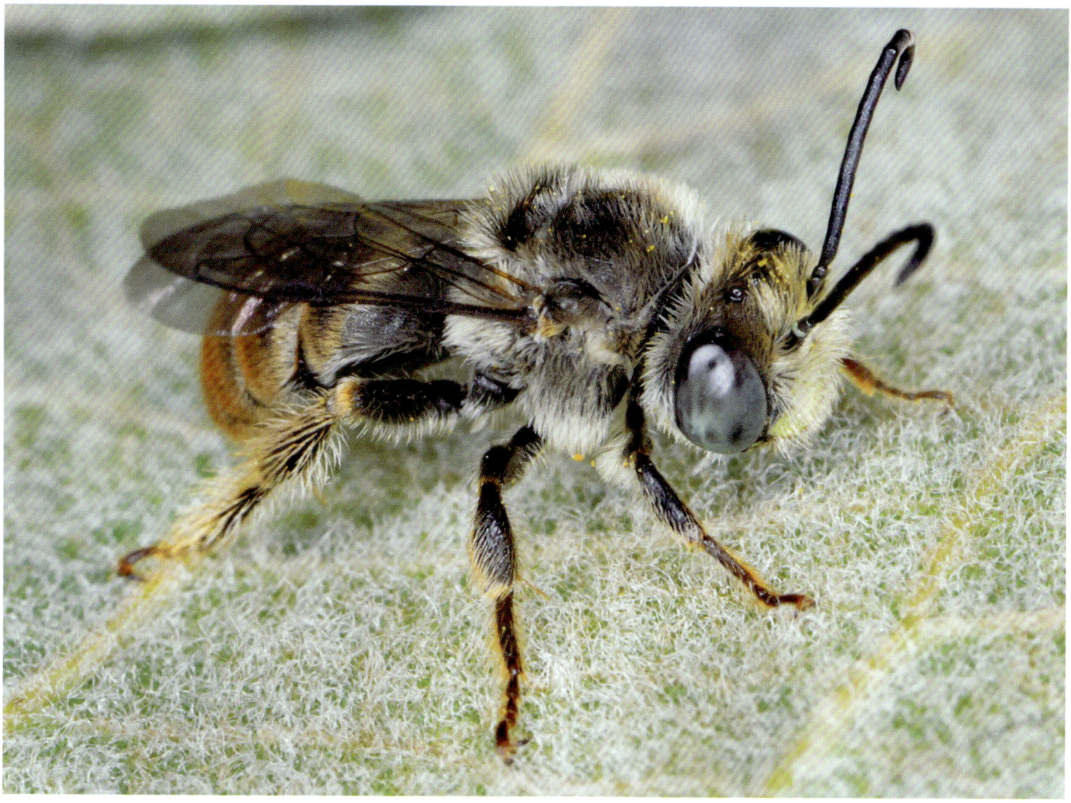

A male *Melissoptila pinguis*

History of the genus *Melissoptila*: In 1998, a prominent Brazilian entomologist criticized the only definitive study of *Melissoptila* longhorn bees available at that time: "The revision," she wrote, "of the genus *Melissoptila* penned by Urban in 1968 was, lamentably, a publication of poor quality, with pages left blank, sentences illegible or cut in half, and endless typographical errors."[5] The entomologist denouncing this earlier publication was none other than Urban herself, returned 30 years later to reconquer the same territory.

A trailblazer for women of science in Brazil, Danuncia Urban discovered and documented approximately 320 new bee species during her 60-year career, founded the distinguished entomology program at the Federal University of Paraná, and achieved recognition as a prominent taxonomist and authority on Neotropical bees. Much of Urban's work has focused on longhorn bees. In 1998, Urban published a comprehensive article on the genus *Melissoptila*, titled "New South American Species of *Melissoptila* Holmberg with Taxonomical Notes." Prior to her 1998 article, 21 species of *Melissoptila* had been described. Urban's 1998 work added 31 more species and remains the most definitive study of the genus to date.

Melissoptila is a Neotropical genus. The 52 *Melissoptila* species identified by Urban are found as far south as Argentinean Patagonia and central Chile, and as far north as the Greater Antilles

5. Urban (1998), p. 1.

and the US Rio Grande border. Since Urban's 1998 study, new *Melissoptila* have been documented in Hispaniola and Cuba.

The species *Melissoptila pinguis* itself has a tortuous history. The first published record of this bee appeared in an 1878 article by Cresson titled "Descriptions of New Species of North American Bees." In this publication, Cresson described a species he christened *Melissodes pinguis*; he based his description on an examination of six female specimens collected in Mexico. Later in the same article, Cresson described the species *Melissodes otomita*, basing his description on a single male specimen, also found in Mexico. Both the males and females had been collected by the Swiss entomologist Adrien Louis Jean François Sumichrast. With time, *Melissodes pinguis* and *M. otomita* would prove to be the same species.

After Cresson's seminal article, *Melissodes pinguis* and *M. otomita* continued to find their way sporadically into the annals of entomological literature. In 1897, Cockerell included *Melissodes pinguis/otomita* in an article he titled "New and Little-Known Bees." He described both male and female specimens, placing them together under the single species name *Melissodes pinguis*. Two years later, he included *Melissodes pinguis* in a catalog of the bees of Mexico.

Sixty years later, in 1957, LaBerge published a study of New World longhorn bees titled "The Genera of Bees of the Tribe Eucerini in North and Central America." In this study, he affirmed that *Melissodes otomita* and *M. pinguis* were a single species, which he placed in the genus *Melissoptila*.

In her 1968 treatise on the genus *Melissoptila*, Danuncia Urban linked Cresson's two descriptions under one name, *Melissoptila pinguis*. Since Cresson had described the female bee first, under entomological rules of naming, *pinguis* technically trumps *otomita*. Nonetheless, this species still appears in English-language literature and databases as *Melissoptila otomita*.

***Melissoptila* behavior:** Scant information has been published regarding the nesting habits of *Melissoptila* bees, and there appears to be no published information whatsoever on *M. pinguis* beyond descriptions of the bee's appearance and apparent floral preferences.

A handful of studies by Latin American entomologists from 1912 through the present have furnished some useful general information about the genus: *Melissoptila* are solitary, and at least some South American *Melissoptila* form large aggregations. A 1992 study by entomologists Moffat and Roig-Alsina reported that the Argentinian *Melissoptila pubescens* nests communally. The authors encountered a nest in which a group of females shared both a common entrance hole and a main tunnel that ran vertically underground before dividing into smaller tunnels leading to individual bees' separate egg chambers.

Males of various genera of solitary ground-nesting bees sleep outside in small aggregations. This behavior is particularly conspicuous among longhorn bees. In 2014, Brazilian biologists Mahlman, Hipólito, and de Oliveira published an account of male *Melissoptila* aff. *bonaerensis* sleeping in an aggregation commingled with the longhorn bee *Melissodes nigroaenea*. Such cross-genus behavior is unusual in the bee world. Whether *Melissoptila pinguis* males engage in this practice has yet to be established.

Worthy longhorn bee ◆ *Florilegus condignus* (Cresson, 1898)

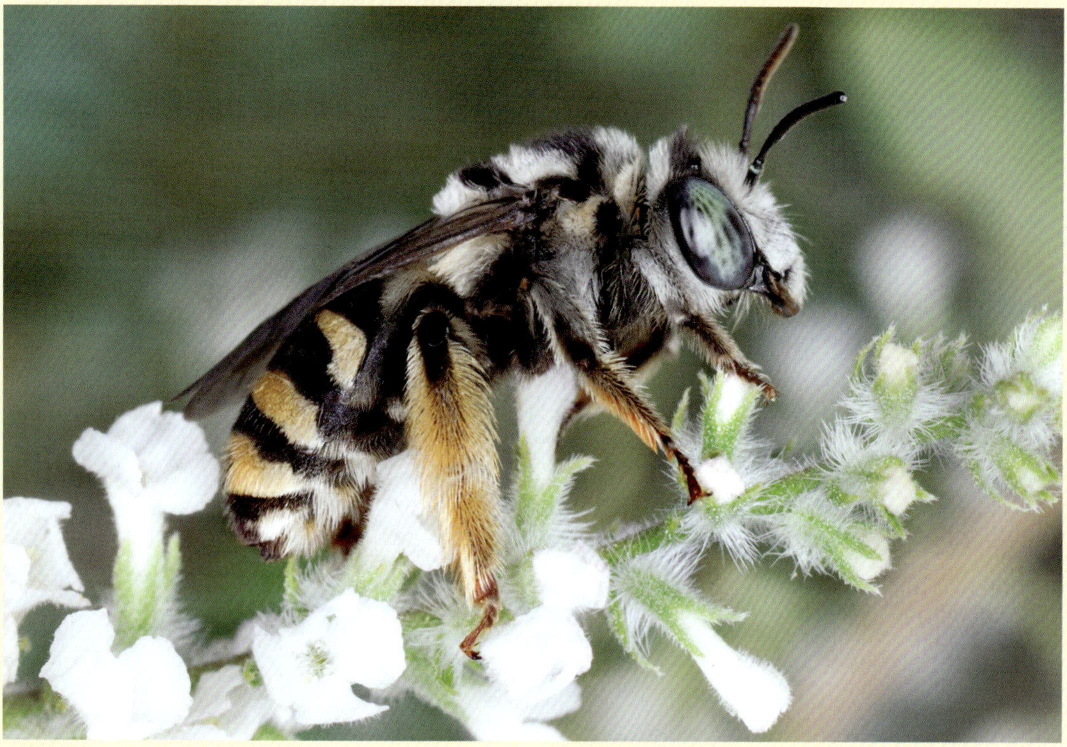

A female *Florilegus condignus*

Description: The genus *Florilegus* contains fewer than a dozen species and is found only in the Western Hemisphere. A single species of *Florilegus* inhabits the United States—the worthy longhorn bee. The female *Florilegus condignus* is a magnificent bee with a plump abdomen girded by golden hair bands, and a thorax adorned with swaths of black and white hair. Females have bulkier builds than males, broadened front femurs, and golden scopal hairs on their hind femurs and basitarsi. Males have a shaggy appearance and differ enough from females to be mistaken for a separate species. Rust-red hairs cover the males' thoraxes, and their abdomens are banded with longish pale hairs; the hair bands continue onto the underside of the bee. The male's clypeus is yellow. Females' faces are black. The antennae of both males and females are mostly black; the flagella have partly gold front surfaces. The eyes are bluish green. **Size:** females 11–12 mm (2/5–1/2 in.), males 9.5–12.5 mm (3/8–1/2 in.).

Range: *Florilegus condignus* has a wide geographic range: it extends throughout Mexico, Central America, and Cuba, and into South America as far as Argentina. In the United States, *Florilegus condignus* is most common in Florida and southernmost Texas but has been documented as far west as Colorado and as far north as New Jersey and Wisconsin. This species is often characterized as rare. Nonetheless, it is abundant in all three border counties of the Lower Rio Grande Valley from midspring through summer.

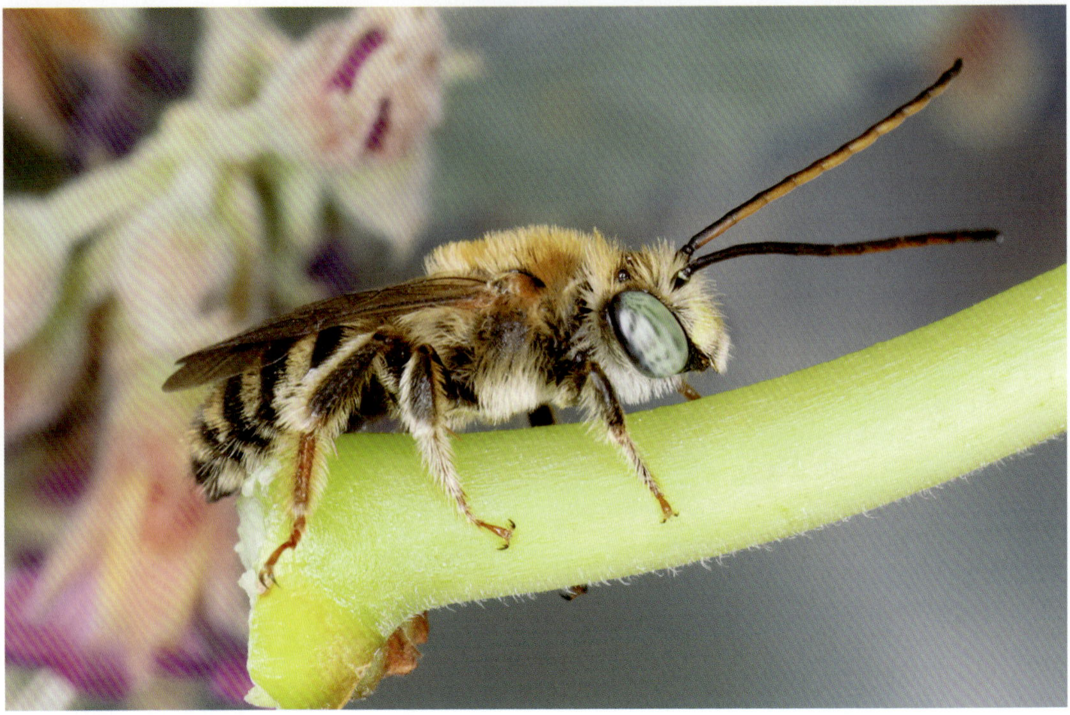

A male *Florilegus condignus*

Associated plants: *Florilegus condignus* is sometimes described in popular literature as a specialist pollinator of pickerelweed, a purple aquatic flower belonging to the family Pontederiaceae. *Florilegus condignus* is probably not a specialist, however. It is often found in association with pea-family plants and was praised by LaBerge and Ribble in 1996 for its prowess as a pollinator of alfalfa. In the Valley, *Florilegus condignus* appears just as the pea-family trees honey mesquite (*Prosopis glandulosa*) and retama (*Parkinsonia aculeata*) begin their spring bloom.

In the Valley, this longhorn bee visits Texas kidneywood (*Eysenhardtia texana*), brushland lantana (*Lantana achyranthifolia*), autumn sage (*Salvia greggii*), shrubby blue salvia (*Salvia ballotiflora*), silverleaf nightshade (*Solanum elaeagnifolium*), golden dewdrops (*Duranta erecta*), Rio Grande bee brush (*Aloysia macrostachya*), teabush (*Melochia tomentosa*), wild olive (*Cordia boissieri*), whitebrush (*Aloysia gratissima*), wild petunia (*Ruellia nudiflora* var. *runyonii*), and diverse flowers of the aster family.

ID tips: The broad front femurs and gold abdominal bands of the female *Florilegus condignus* distinguish it from all other female longhorn bees of the Valley. *Florilegus condignus* males somewhat resemble the common *Melissodes tepaneca* longhorn bee. The following two traits help separate them: (1) the mandibles of male *Florilegus condignus* are entirely dark, while those of *Melissodes tepaneca* have yellow spots at the base; and (2) the tips of the antennae of male *Florilegus condignus* are dark.

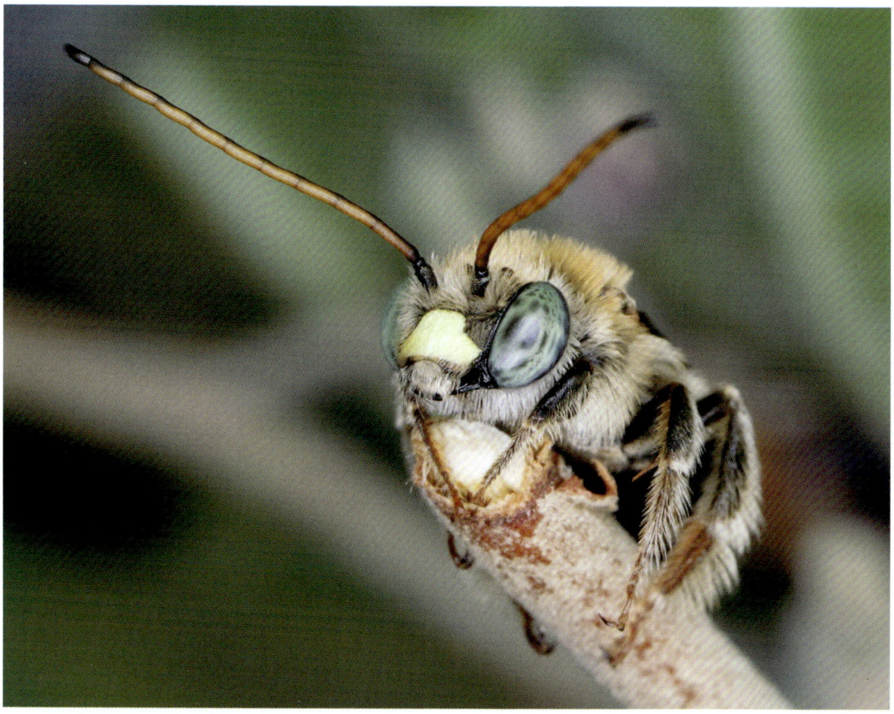

A male *Florilegus condignus*

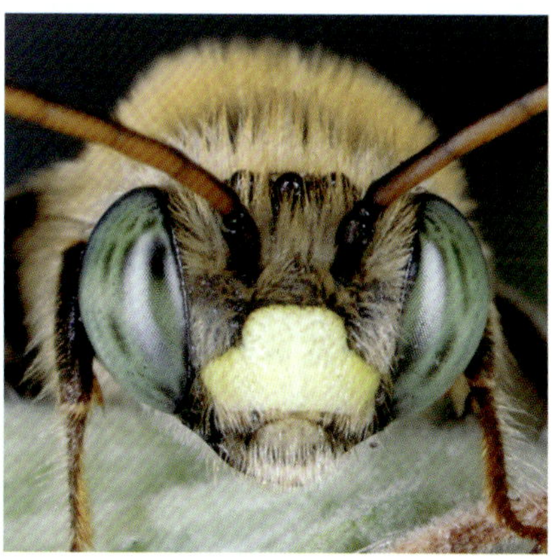

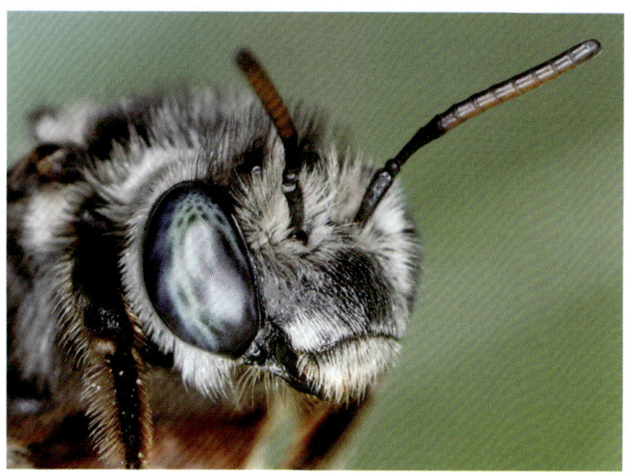

Faces of male (*left*) and female (*right*) *Florilegus condignus*

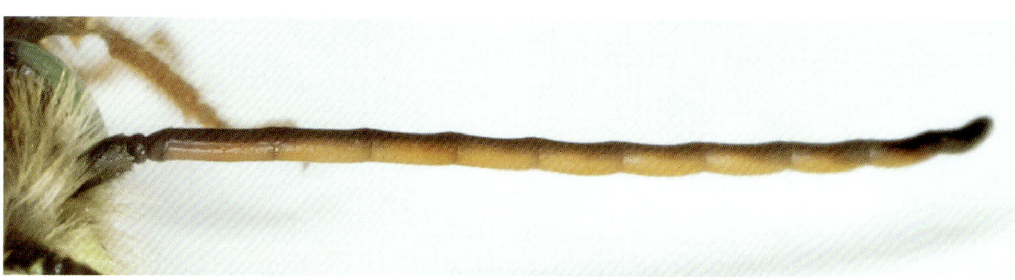

Antenna of a male bee

Additional information: We owe the most comprehensive research on the genus *Florilegus* to Brazilian entomologist Danuncia Urban. In 1970, Urban penned a treatise describing all known *Florilegus* species and announcing two previously unknown species. In South America, members of the genus *Florilegus* are varied in appearance. Some, such as the Ecuadorian purple longhorn bee (*Florilegus purpurascens*), are iridescent. Since Urban's 1970 work, studies of *Florilegus* in English have been few and have focused on the pollen-collecting habits of *F. condignus*.

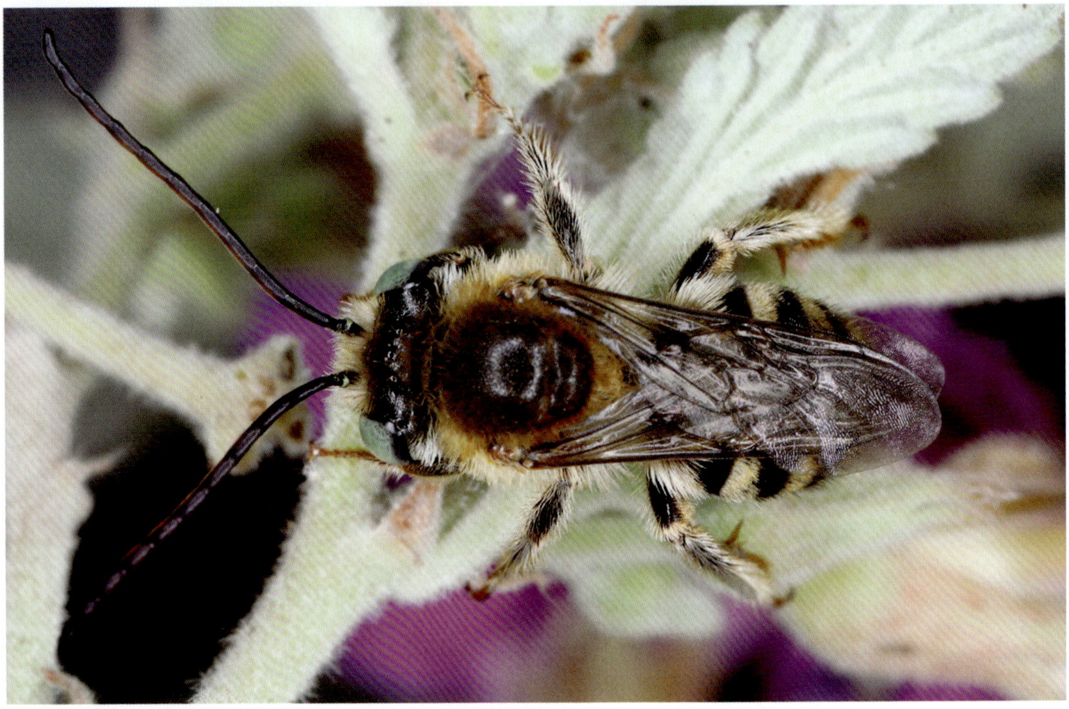

A male *Florilegus condignus*: male bees have a "shaggy" appearance and rust-red thorax hairs.

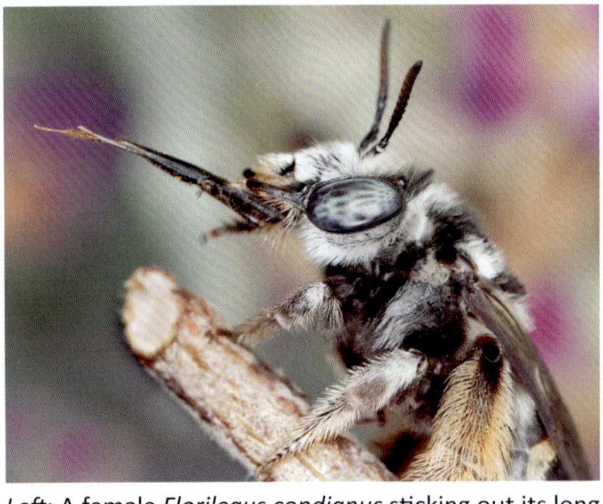

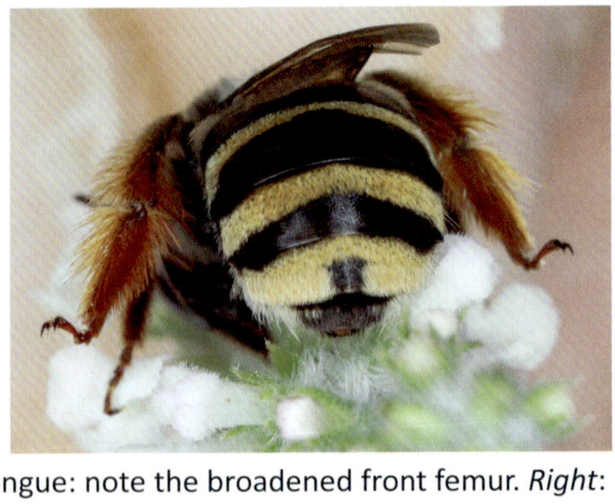

Left: A female *Florilegus condignus* sticking out its long tongue: note the broadened front femur. *Right*: Rear view of a female's abdomen.

Wilmatte's longhorn bee ◆ *Eucera (Xenoglossodes) wilmattae* (Cockerell, 1917)

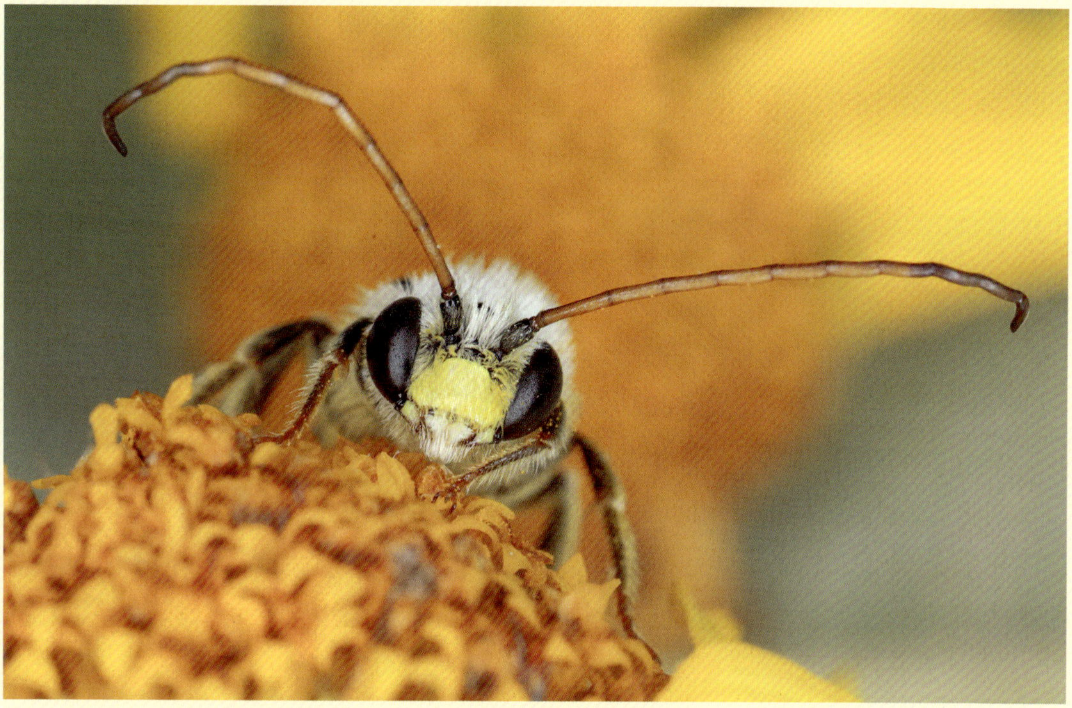

A male Wilmatte's longhorn bee

Description: Wilmatte's longhorn bees are enchanting insects. Males have bright yellow faces and exceptionally long antennae that curl at the tips like the ends of mustaches. Both female and male bees have opaque black eyes. *Eucera wilmattae* females have golden hairs on the head, thorax, and legs, and golden abdominal hair bands. On males, the hairs of the head, thorax, abdomen, and legs are predominantly white or off-white. Females' scopal hairs are bright orange on the hind legs' inner surfaces and tawny on the outer surfaces. Females' faces are mostly dark, with a narrow pale yellow strip on the bottom of the clypeus, and their antennae are much shorter than males'. On male bees, the labrum and most of the mandibles, as well as the clypeus, are yellow. **Size:** females and males 7–10 mm (1/4–2/5 in.).

Range: *Eucera wilmattae* occurs primarily in southeastern Texas and northeastern Mexico. It appears in abundance from late April through November in multiple locations in Cameron, Hidalgo, and Starr Counties. This species was named after biologist Wilmatte Porter Cockerell, who included it in her 1917 work "Collecting Bees in Southern Texas." W. P. Cockerell discovered a single female specimen feeding on a yellow composite flower in Port Isabel, Texas (Cameron Co.). In an appendix to his wife's publication, T. D. A. Cockerell described the bee, christening it *Xenoglossodes wilmattae*.

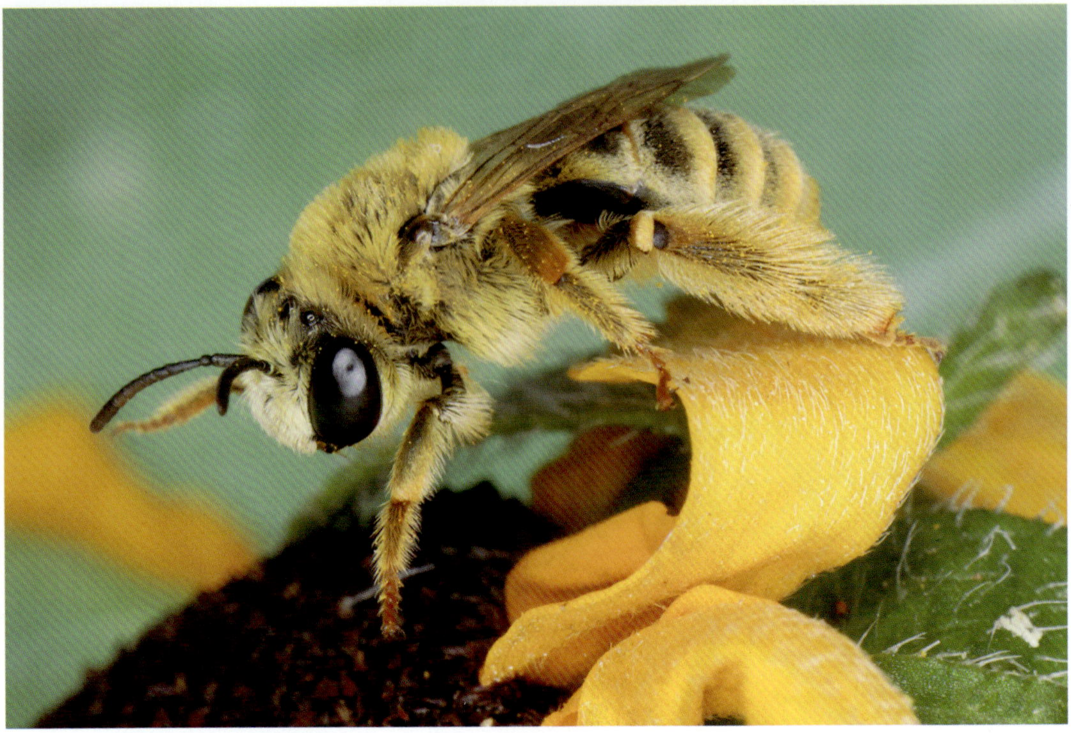

A female Wilmatte's longhorn bee

Associated plants: Wilmatte's longhorn bee is oligolectic on flowers of the family Asteraceae. In the Valley, this species visits, among others, hoary blackfoot (*Melampodium cinereum*), cowpen daisy (*Verbesina encelioides*), blanketflower (*Gaillardia pulchella*), Mexican hat (*Ratibida columnifera*), resinbush (*Viguiera stenoloba*), romerillo (*Bidens alba*), Texas palafox (*Palafoxia texana*), Tiny Tim (*Thymophylla tenuiloba*), and hairy wedelia (*Wedelia acapulcensis* var. *hispida*).

ID tips: *Eucera wilmattae* is sometimes confused with the similar Texas species *Eucera (Xenoglossodes) eriocarpi*. The male Wilmatte's longhorn bee can be distinguished from male *Eucera eriocarpi* in part by the pale hairs clothing the former's head and body. Those on the male *Eucera eriocarpi* are tawnier. In addition, on male Wilmatte's longhorn bees, the tergal hair bands are well defined, while on male *Eucera eriocarpi*, the tergal hair bands are light gold, feathery, and less well defined, and the bands on T3 and T4 commingle with the segments behind them.

With the aid of magnification, traits of the antennae and head also separate these two species. On male Wilmatte's longhorn bees, the last three and a half segments of the antennae are black, a characteristic lacking in *Eucera eriocarpi*. Female Wilmatte's longhorn bees can also be distinguished from female *Eucera eriocarpi* by examination of the face: on female Wilmatte's longhorn bees, the labrum is dark, and the yellow band on the bottom edge of the clypeus is relatively narrow. On *Eucera eriocarpi* females, the labrum is yellow and the band on the bottom of the clypeus is wider.

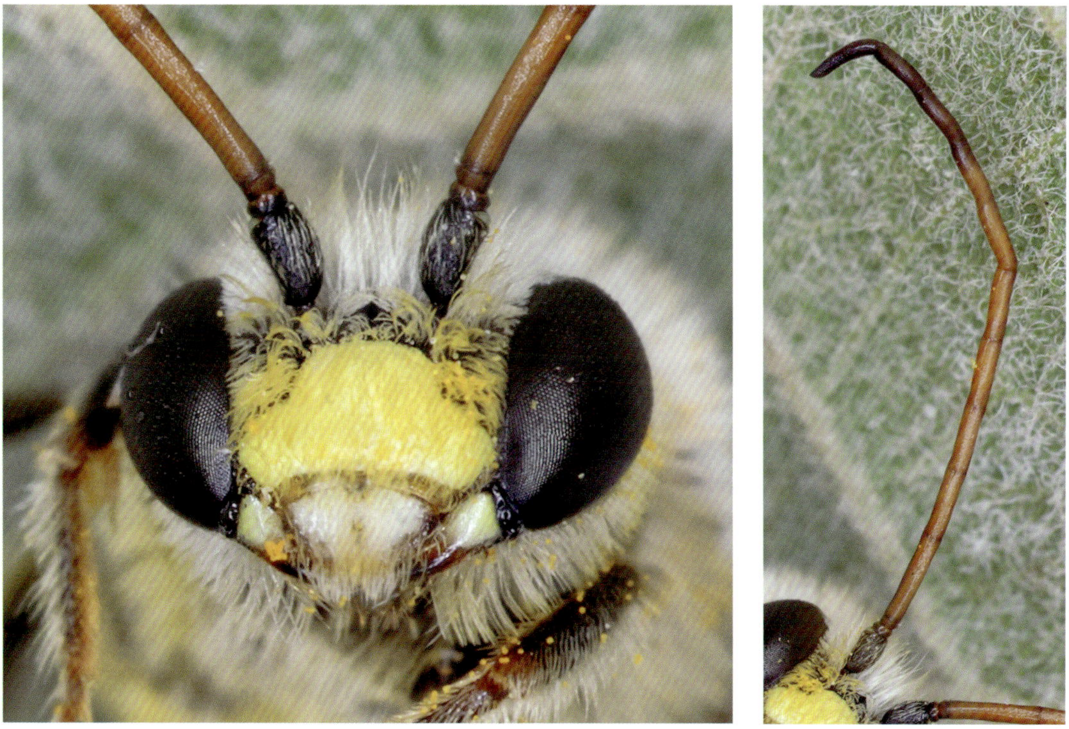

Face and antennae of a male Wilmatte's longhorn bee: note the wide, dark antennal scapes. The last 3½ segments at the tip of the antennae are black.

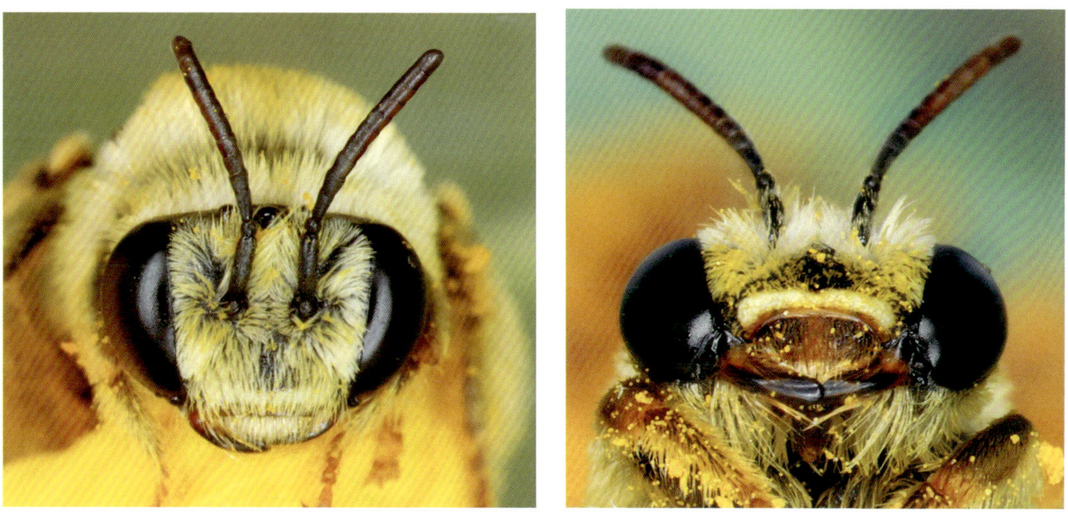

A female *Eucera wilmattae*: note the yellow strip, edged with a narrow brown strip, on the clypeus. Also note the straight vertex: this trait is typical of longhorn bees generally.

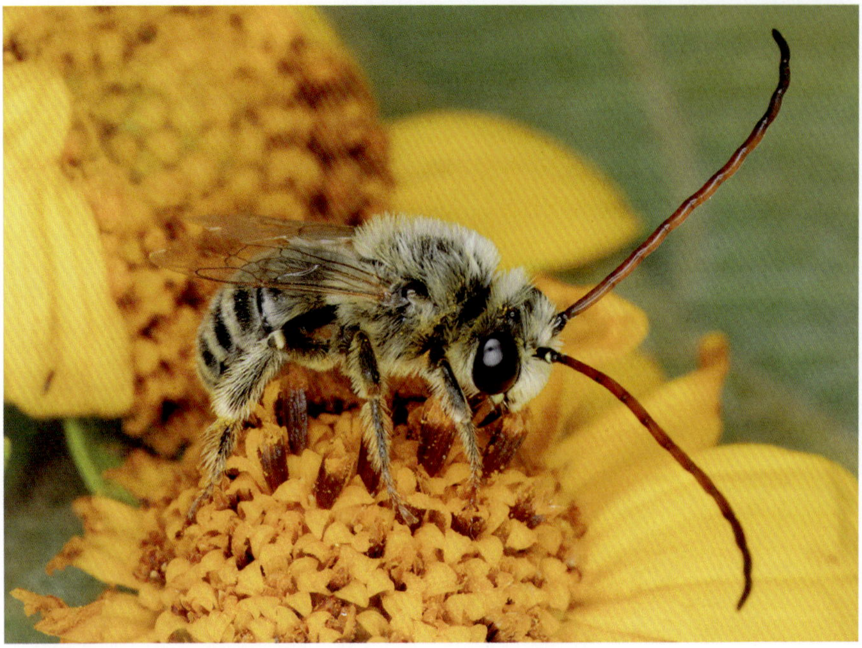

A male Wilmatte's longhorn bee

The genus *Eucera* and the former genus *Tetraloniella*: For many years, Wilmatte's longhorn bee was known as *Tetraloniella wilmattae*. In 2018, changes in longhorn bee taxonomy resulted in the genus *Tetraloniella* being shelved and absorbed into the larger genus *Eucera*.[6] Nonetheless, Wilmatte's longhorn bee continues to appear in widely used internet databases as *Tetraloniella wilmattae*, and its identification is easier if the bee is viewed with an awareness of the defining traits of *Tetraloniella*.

Prior to upheavals in the longhorn bee universe in 2018, the most comprehensive study of *Tetraloniella* was "Revision of the Bees of the Genus *Tetraloniella* in the New World (Hymenoptera: Apidae)," published by LaBerge in 2001. For this 96-page work, LaBerge examined 6,504 specimens of *Tetraloniella*, boiling them down to 35 distinct species that he described in painstaking detail. Among these was *Tetraloniella wilmattae*, which LaBerge removed from the genus *Xenoglossodes*, under which Cockerell had originally classified the bee.

LaBerge generally described *Tetraloniella* as follows: They are small to moderate-sized bees that superficially resemble longhorns of the genus *Melissodes*, with notable differences in their mouthparts and tegulae. The abdomens of *Tetraloniella* are often striped with pale hair bands. Females usually have short antennae and dense, sometimes feathery scopal hairs on their hind legs. Males have long antennae with wide scapes, short first flagellar segments, and long second flagellar segments. The male *Tetraloniella*'s mandibles, clypeus, and labrum are usually pale or yellow. Some males have toothlike projections on either side of T6. Wilmatte's longhorn bee possesses all of these traits.

---

6. See Dorchin, López-Uribe, et al. (2018). The nomenclature status of the three former *Tetraloniella* featured here currently remains uncertain. Freitas et al. (2023) reclassify various *Eucera* (*Xenoglosodes*) as *Xenoglossa*.

Woolly-white longhorn bee ◆ *Eucera (Xenoglossodes) paenalbata* (LaBerge, 2001)

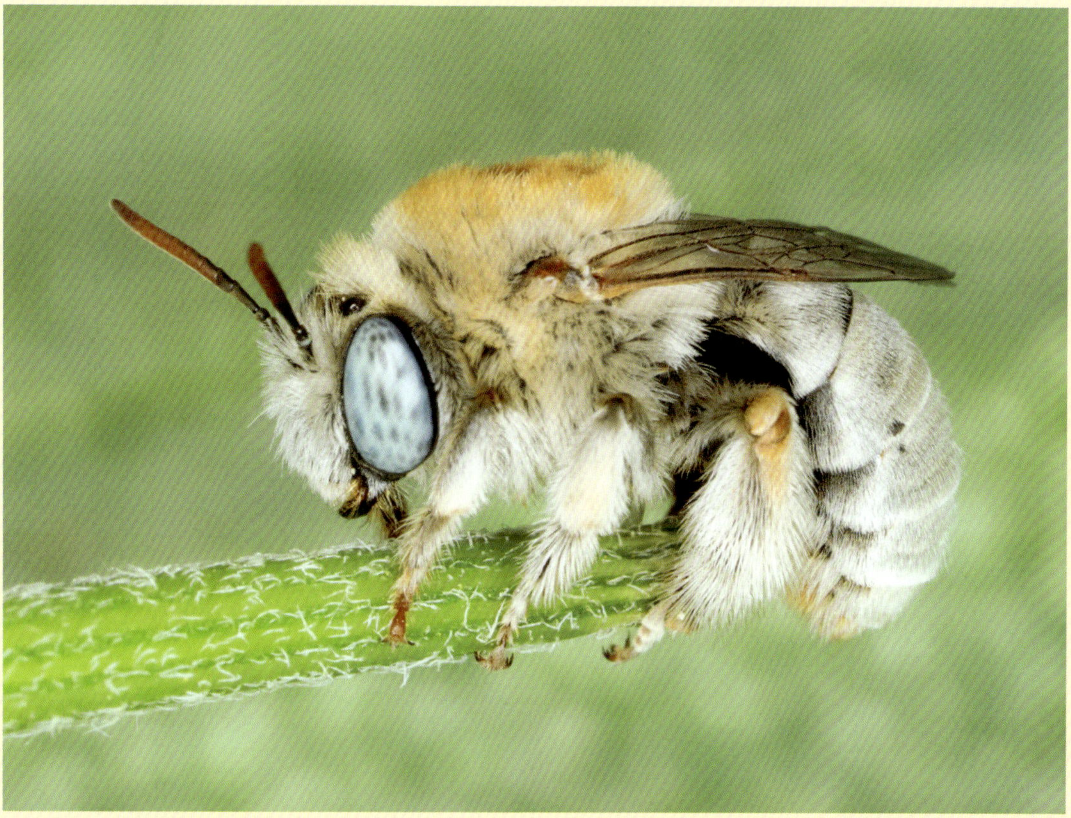

A female *Eucera paenalbata*

Description: First described by LaBerge in 2001, *Eucera paenalbata* is a beautiful bee—petite, with white and pale orange thorax hairs, golden-pink tegulae, and pale blue eyes. The upper abdomens (T3–T5) and legs of both males and females are clothed in long white hairs. Long white and golden-orange hairs intermingle on the female's tibiae and hind basitarsi. The wings of both males and females are glassy with reddish-brown veins. The scientific name of this species (*paen-* means "almost") is a nod to *Eucera albata*, a similar white and reddish-gold bee first described by Cresson in 1872.

Like *Eucera wilmattae* of the previous entry, *Eucera paenalbata* was formerly classified as a *Tetraloniella*, and it exhibits the characteristics of that abandoned genus. The female's hind-leg scopal hairs are dense and plumose. The male *Eucera paenalbata* has a pale clypeus and labrum, and its mandibles are pale at the base. **Size:** females 10–12 mm (2/5–1/2 in.).

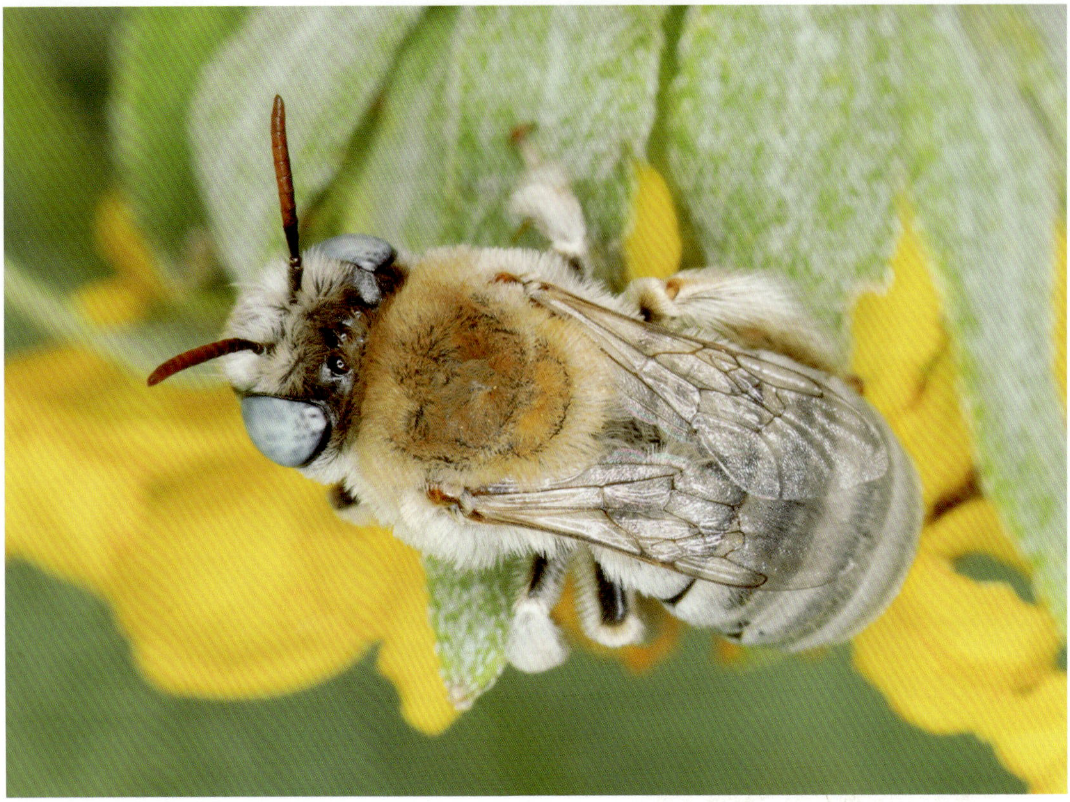

Dorsal view of a female *Eucera paenalbata*

Range: LaBerge first described *Eucera paenalbata* in his 2001 revision of the genus *Tetraloniella*. He called it a "species from the southern prairies" and based his description on bees collected in Colorado, Kansas, Nebraska, and Texas. *Eucera paenalbata*, however, ranges across the Texas border into Tamaulipas, and LaBerge later documented a specimen from as far south in Mexico as Mazatlán. The female *Eucera paenalbata* shown here appeared in late June 2021 at Campos Viejos Ranch (Starr Co.). Prior records exist of this species in Zapata County.

Associated plants: Many of the bees examined by LaBerge were collected on dalea, particularly purple dalea (*Dalea lasiathera*), a pea-family plant. The female bee found at Campos Viejos Ranch was feeding on the aster-family flower Mexican hat (*Ratibida columnifera*).

ID tips: This species closely resembles *Eucera albata*. The two species can be differentiated in part by the color of their antennae. Those of the female *Eucera paenalbata* are orange on the front surfaces above F1. The antennae of the male *Eucera paenalbata* are yellow to red on the front surfaces of F2–F11. On *Eucera albata* males and females, the corresponding flagellar segments are brown on their front surfaces.

Fasciatella longhorn bee ◆ *Eucera (Xenoglossodes) fasciatella* (LaBerge, 1970)

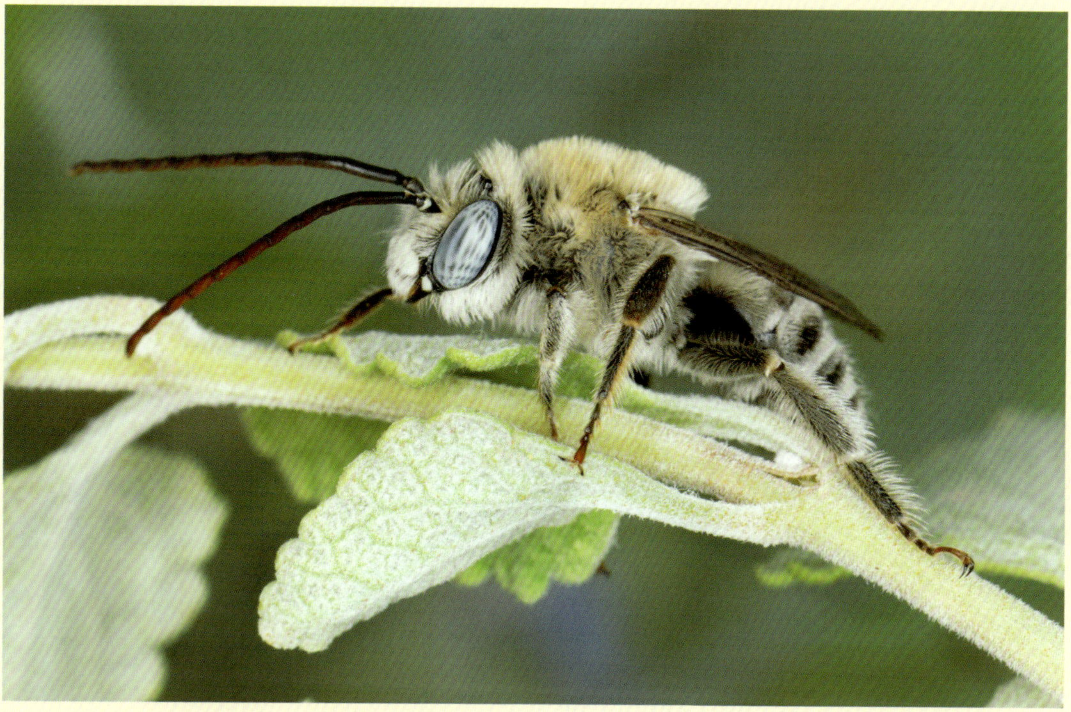

A male *Eucera fasciatella*

Description: *Eucera fasciatella* is a medium-sized, blue-eyed, black bee with white and pale rust-colored hairs covering its face, thorax, and legs. Bands of pale hairs gird its abdomen. The bee's tegulae are pale reddish, the wings transparent reddish brown, and the wing veins dark brown. Males have exceptionally long antennae and distinctive ivory facial markings. **Size:** females and males 11–13 mm (2/5–1/2 in.).

Range: The fasciatella longhorn bee is rarely documented in the United States. It was first described by LaBerge in 1970, based on specimens collected in Nuevo León. The male shown here is one of a handful that appeared in late May and early June 2021 at Falcon State Park (Starr Co.), which is situated near the borders of both Tamaulipas and Nuevo León. No female bees were observed associating with the males. *Eucera fasciatella* has been recorded previously in Val Verde County, but its appearance at Falcon State Park marks a range extension for this species.

Associated plants: The Falcon State Park bees were found foraging on shrubby blue salvia (*Salvia ballotiflora*). LaBerge was unable to ascertain the floral associations of the bees he collected; however, in a 1989 article on the former longhorn bee genus *Pectinapis*, he noted that Neff had found *Eucera fasciatella* nearly a decade before, in Val Verde County, feeding on lemon beebalm (*Monarda citriodora*)—which, like salvia, is a member of the mint family.

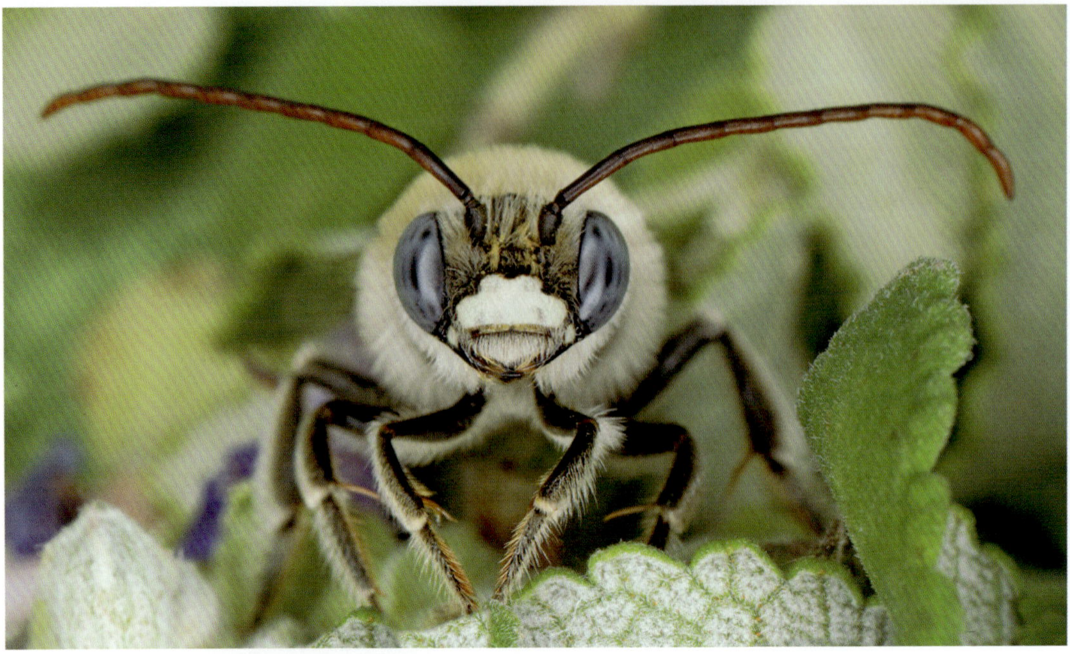

Frontal view of a male *Eucera fasciatella*

ID tips: The shape and color of the ivory mark on the male bee's face distinguish it from all other male longhorn bees of the Lower Rio Grande Valley. Because this bee has been described infrequently, a somewhat more detailed account of the male's appearance is provided here.

Head: The male bee has a wide face with a large ivory area shaped like a crown topped by two rounded projections, which covers the clypeus and extends more than halfway to the antennae. The lower edge of the clypeus is rimmed with brown. The bases of the mandibles each bear a conspicuous ivory spot. The antennal scapes are wide and black, and the pedicels short and black. The flagellum is dark brown generally, but a lighter reddish brown on the front surface. F1 is about the same size as the pedicel. F2 is 1.5 times the width of F3. The remaining flagella are about equal in size to F3. White and rust-colored hairs cover the face and cheeks; the hairs on the rear vertex are pale orange.

Thorax: Intermingled white and pale orange hairs cover the scutum and scutellum and continue laterally to just below the tegulae. Long white hairs clothe the lower half and underside of the thorax.

Abdomen: Long, dense white hairs intermingled with light orange hairs line the base of T1. Four continuous bands of short white hairs rim the apical edges of T2–T5; on the sides, the white hairs of each band are long and dense. The hair bands continue onto the sternum and become sparse medially on S2–S5. T5 is covered with long brown hairs at the base and shorter white hairs on the apical rims and sides. T6 is completely covered with moderately long brown hairs. T2 and T3 additionally have ill-defined bands of short, sparse white hairs rimming the base of each segment.

Legs: The hind legs are clothed in long, plush white hairs. Short yellow hairs rim the inner surfaces of the hind and middle basitarsi. The lower tarsi are reddish.

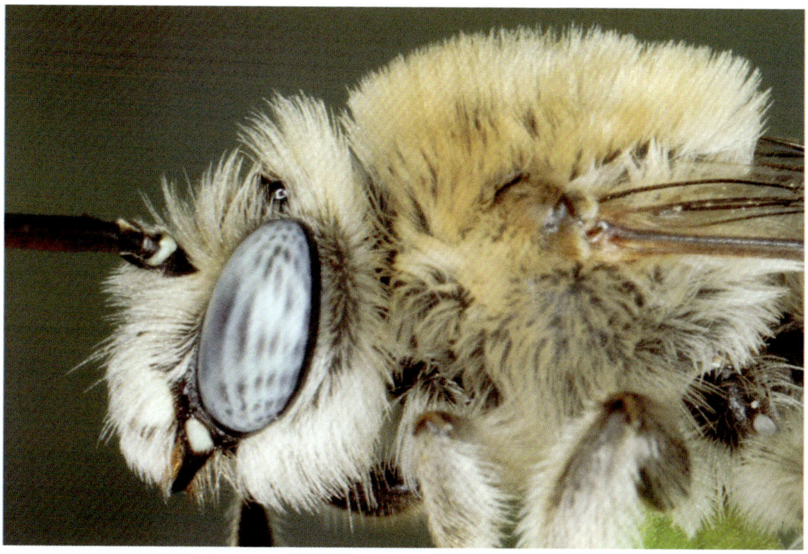

Lateral view of the head and thorax of a male *Eucera fasciatella*

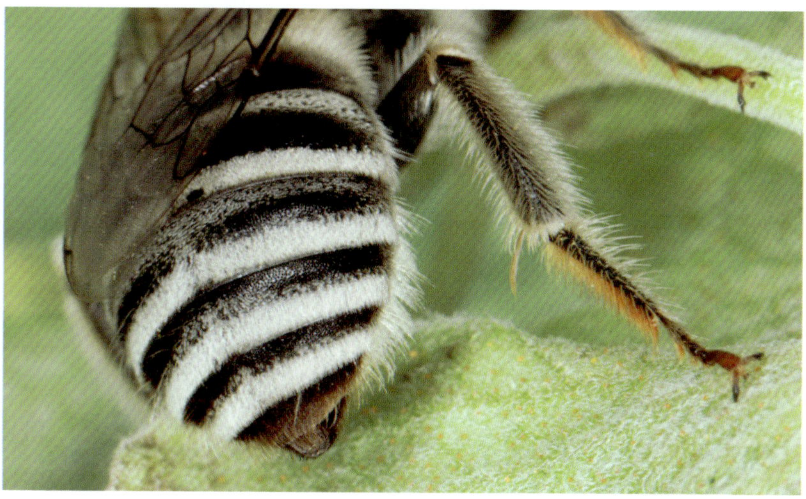

Rear view of a male bee's abdomen

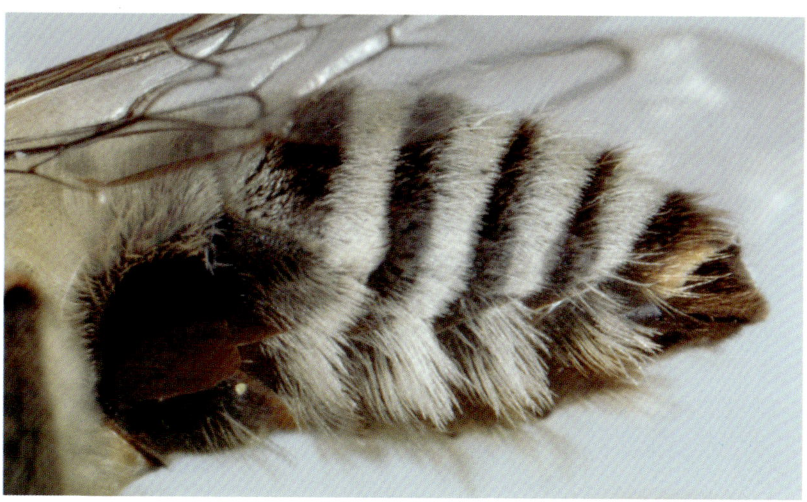

Lateral view of tergal and sternal hair bands

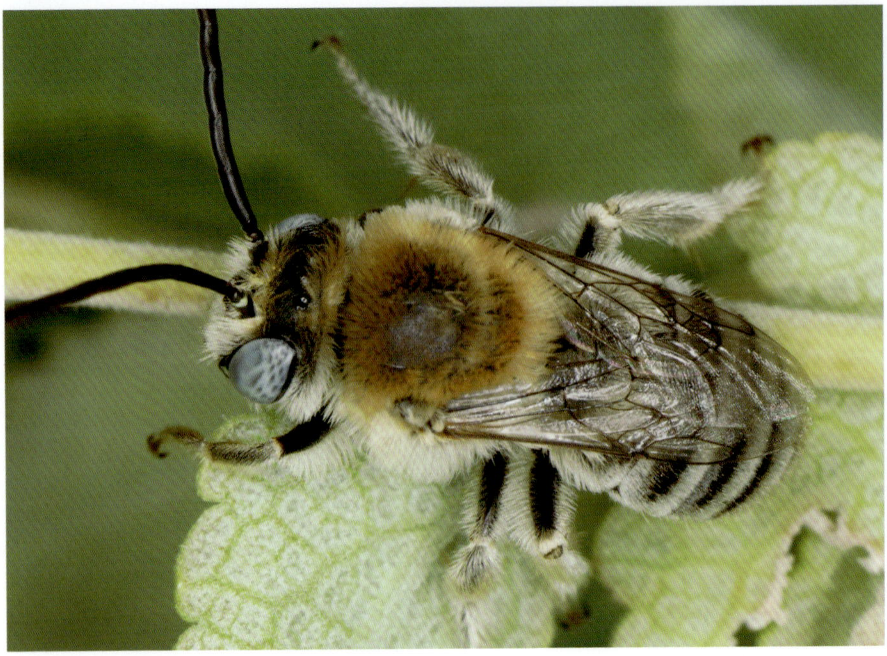

Dorsal view of a male *Eucera fasciatella*

Additional information: In 1970, upon discovering this species, LaBerge placed the fasciatella longhorn bee in a new genus, which he named *Pectinapis.* He located two other Mexican species in the same genus—*Pectinapis auricauda* and *P. fasciata*—which he thought remarkable because of the facial traits of the females. First, they had comblike hairs at the top of the clypeus (*pectin* derives from the Latin for "comb"). Second, the facial areas above and below the comblike hairs were flat or slightly concave and covered with hairs seemingly modified to hold pollen. *Pectinapis fasciatella* females lacked the comblike structure but shared the second trait.

LaBerge ventured that the novel facial apparatus of *Pectinapis* females might have some special use when the bees were digging nests in sand. Additionally, because he found pollen grains on the facial hairs of his specimens, LaBerge speculated that the females used the hairs to rake pollen from flowers. He noted that anthophorine digger bees had similar facial hairs that they used to collect pollen. (The most common Valley species with such clypeal hairs is *Anthophora capistrata*, discussed later in this guide, which collects pollen principally from mints such as salvia.)

LaBerge originally classified *Pectinapis* as a type of Anthophoridae, although he noted that his *Pectinapis* species were similar to *Xenoglossodes* longhorn bees. Michener thereafter renamed *Pectinapis fasciatella* in *The Bees of the World*: he cast it into the same longhorn bee genus as Wilmatte's longhorn bee, rechristening it *Tetraloniella (Pectinapis) fasciatella.* "In various groups of bees," Michener wrote, "specialist collectors of salvia pollen have a flattened clypeus with erect hooked or wavy bristles, as in *Pectinapis.* The females themselves are not elsewhere regarded as generic characters, and I see no reason why they should be generic characters in the Eucerini."[7]

7. Michener (2007), p. 738.

Sabine longhorn bee ◆ *Svastra (Epimelissodes) sabinensis* Cockerell, 1924

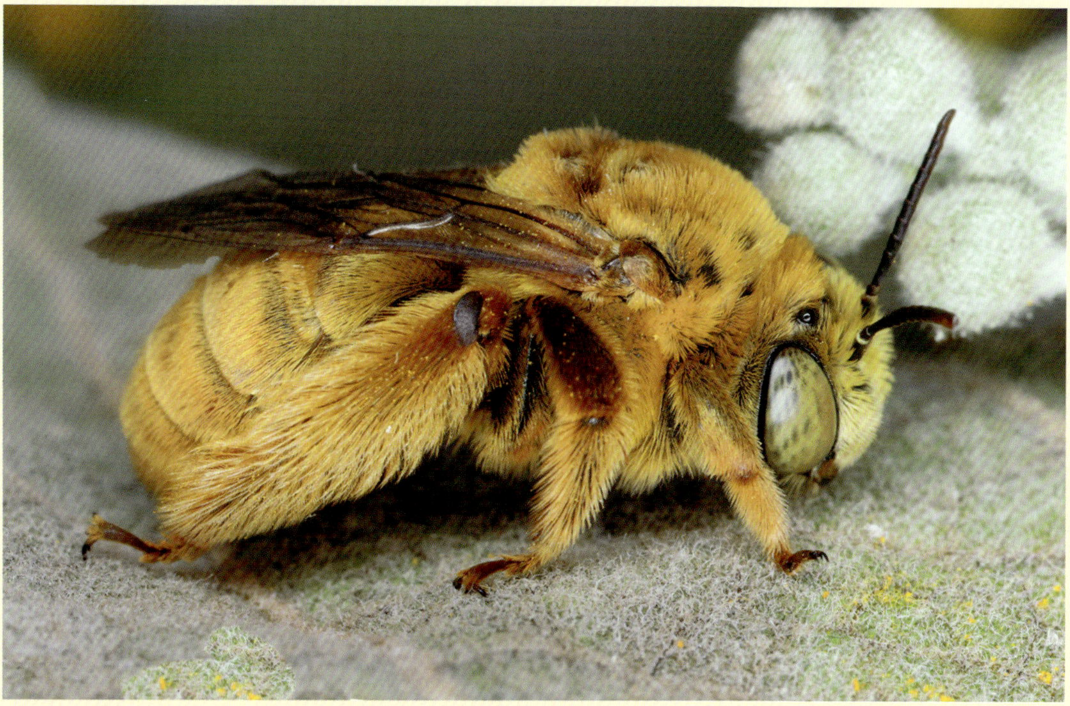

A female Sabine longhorn bee. Photo credit Ross Eatman.

Description: *Svastra (Epimelissodes) sabinensis* is a magnificent longhorn bee, large and leonine, and covered with golden hair. The faces of both male and female Sabine longhorn bees are covered with dense white hairs. Tawny hairs clothe their legs. Females have long golden scopae on their hind-leg tibiae. The bees' tegulae are golden brown, and their wings transparent brown with brown veins. The hind-leg knee plates of females are prominent, black, and ridged on the rims. The male's abdomen has short, toothlike spines on either side of T6. Both males and females have olive-green eyes. Their antennae are relatively short for longhorn bees, a general trait of the genus *Svastra.* Males have yellow clypei. **Size:** females and males 14 mm (1/2–3/5 in.).

Range: *Svastra sabinensis* ranges from Texas to southern California, as far north as Colorado, and as far south as Durango and Baja California. It is relatively uncommon in the Valley. The female bee shown here appeared at the National Butterfly Center (Hidalgo Co.) in mid-April 2019. The male appeared in mid-June 2021 at Campos Viejos Ranch (Starr Co.).

Associated plants: Bees of the genus *Svastra* tend to be oligolectic, most often on flowers of the aster or evening primrose families (Asteraceae and Onagraceae). *Svastra sabinensis* gathers pollen principally from Asteraceae. The bees shown here were foraging on resinbush (*Viguiera stenoloba*) and cowpen daisy (*Verbesina encelioides*).

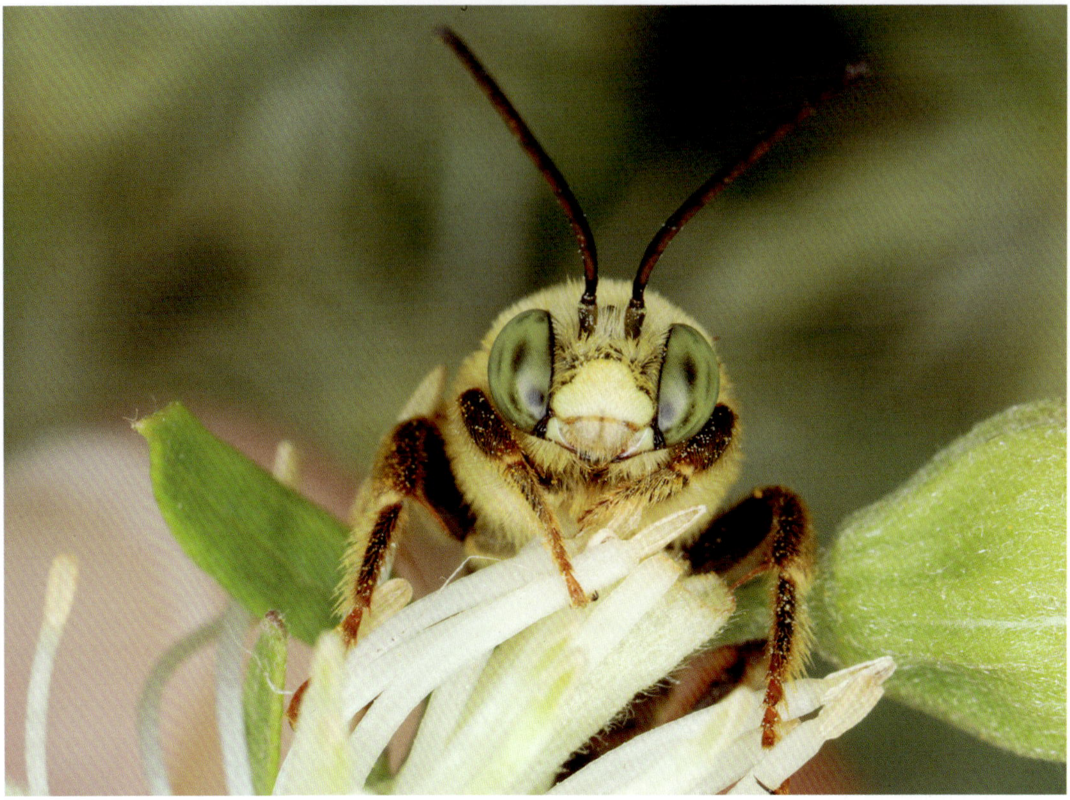

A male Sabine longhorn bee: the clypeus and mandibles of males are pale yellow.

ID tips: *Svastra sabinensis* is fairly easily differentiated from other Texas *Svastra*. The most commonly seen Texas *Svastra* species include *S. obliqua*, *S. compta*, *S. machaerantherae*, *S. petulca*, and *S. texana*. All have dark abdomens with contrasting bands of usually pale hairs; all lack the even covering of short golden hairs characteristic of *Svastra sabinensis*.

General information on the genus *Svastra*: *Svastra* are large to moderate-sized longhorn bees found mostly in the New World. They range from southern Canada through Mexico and into temperate South America. The genus *Svastra* was first documented by the Argentinean naturalist Eduardo Ladislao Holmberg, a prolific science writer, author of science fiction, and director of the Buenos Aires Zoological Gardens. In the late 1800s and early 1900s, Holmberg, at times forsaking his long-standing interest in spiders, undertook a series of bee-collecting expeditions into the mountains of Argentina and the wilds of neighboring nations. He encountered diverse longhorn bee species and named the genus *Svastra* (Sanskrit for "sister"), remarking on its similarity to its sister genus *Melissoptila*, which he had also named.

Svastra are differentiated from other longhorn bees in part by size. In 1960, T. B. Mitchell described *Svastra* as robust bees, noting that some were nearly as large as bumble bee queens (measuring up to 20 mm). Mitchell observed, in addition, that *Svastra* females could be best distinguished from other longhorn bee genera by a single trait, visible to the naked eye: a tuft of long hairs in the middle of the metanotum (at the back of the thorax).

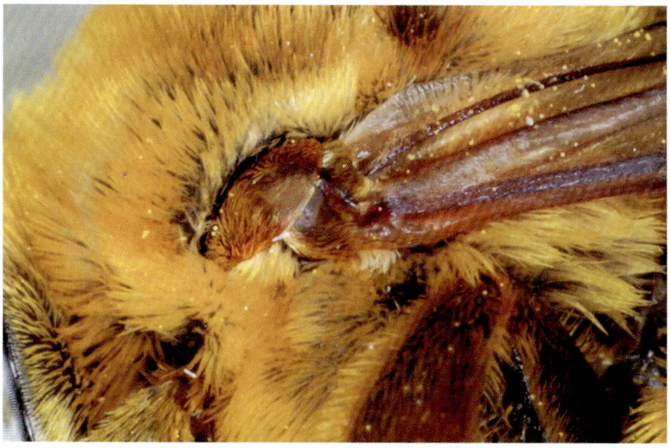
The oval tegula of a female *Svastra sabinensis*

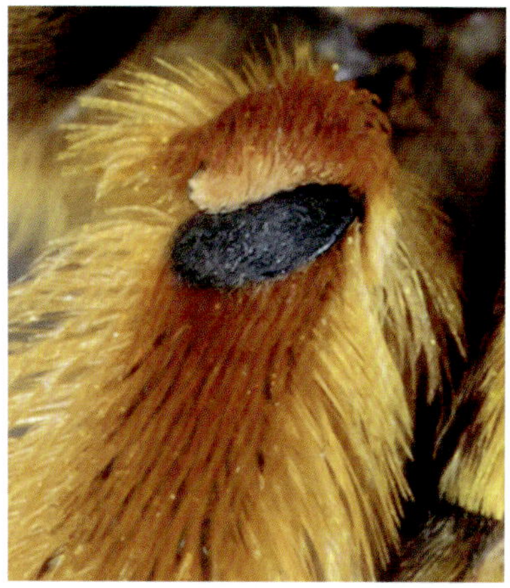
The female's hind-leg basitibial plate

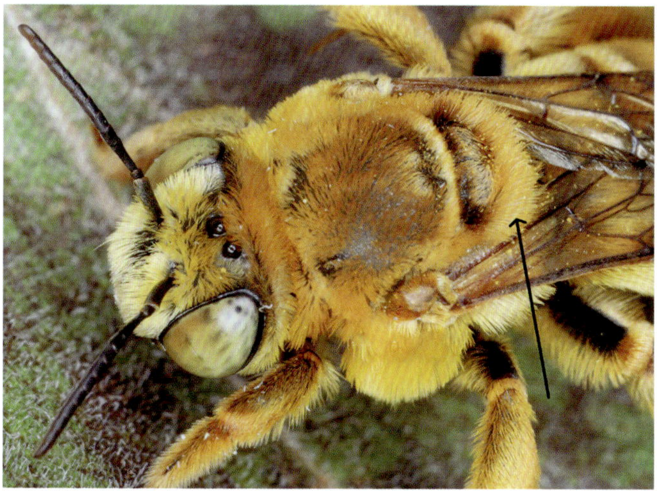
The band of long hairs on the back of a female's scutellum

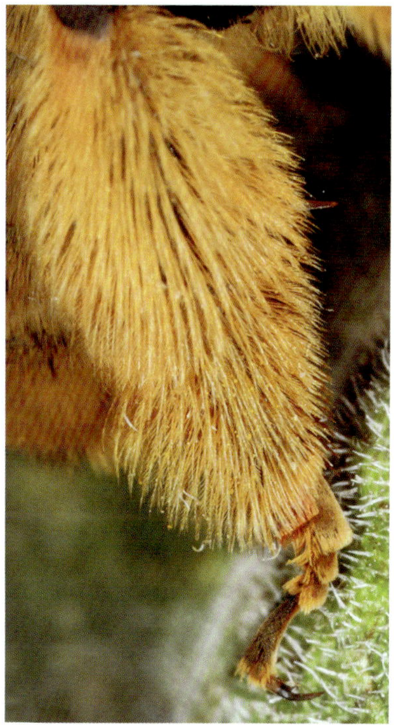
The plumose hind-leg scopal hairs

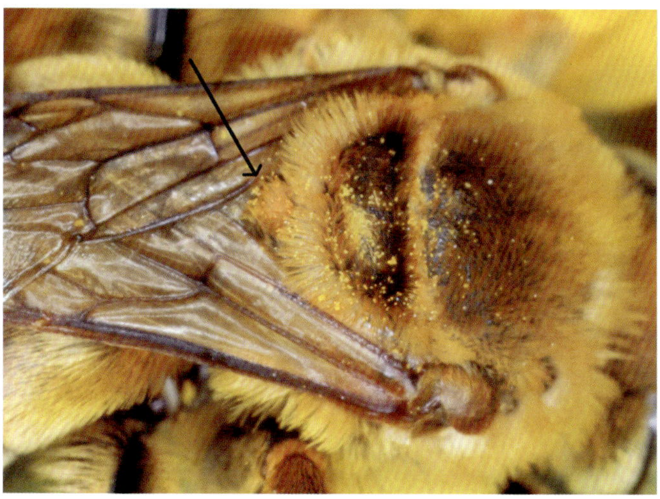
Tuft of hair on the metanotum

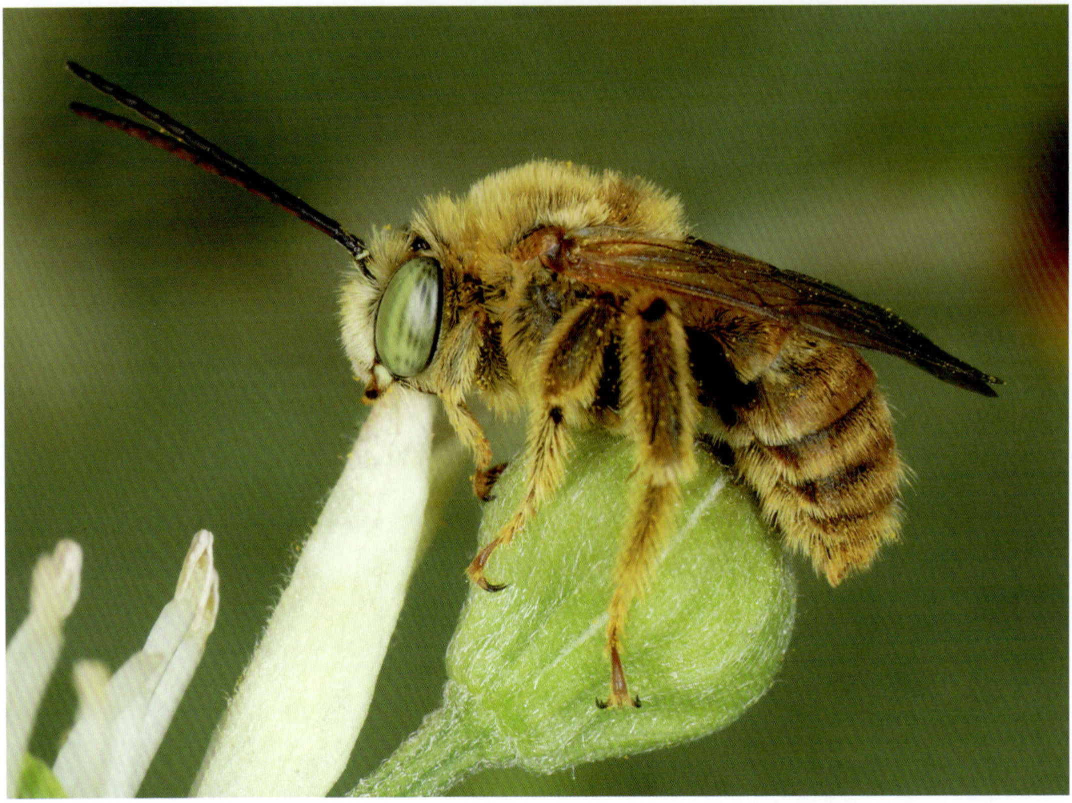

A male Sabine longhorn bee

All members of the genus *Svastra* have "spatuloplumose" (long, flat) hairs between the scutum and scutellum, on the rim of T1, and on the base of T2. This characteristic is beyond the range of the naked eye and most macro lenses, but it is a defining attribute that separates *Svastra* from other current longhorn bee genera. The tegulae of *Svastra* are oval, a trait that also helps separate them from *Melissodes* longhorn bees.

***Svastra* behavior:** *Svastra* nest in the ground. They often construct turrets at their nest entrances and tend to plug the entrances while foraging. Many *Svastra* are solitary bees that build nests close to one another in large groups. Aster specialists of *Svastra (Epimelissodes)*, however, tend to be communal: several females share a single nest and jointly excavate tunnels and nest entrances, but each individual bee builds and provisions her own egg cells.

In August 1982, while on a research expedition 20 miles southeast of Tucson, entomologist Jerome Rozen lit on a nest of *Svastra sabinensis*. He discovered that the nest was communal—it was occupied simultaneously by two female bees. It had a branched and meandering burrow that descended more than 8 cm (3 in.) to a level that housed egg cells. The cells were long and vertical and coated with a waterproof lining, and parts of the nest were plugged with sorted pebbles. Rozen observed a cleptoparasitic cuckoo bee, *Triepeolus penicilliferus*, attempting to enter the nest. Rozen brought the two female *Svastra* and the *Triepeolus* back with him to the American Museum of Natural History in New York.

Frisky longhorn bee ◆ *Svastra (Epimelissodes) petulca* (Cresson, 1878)

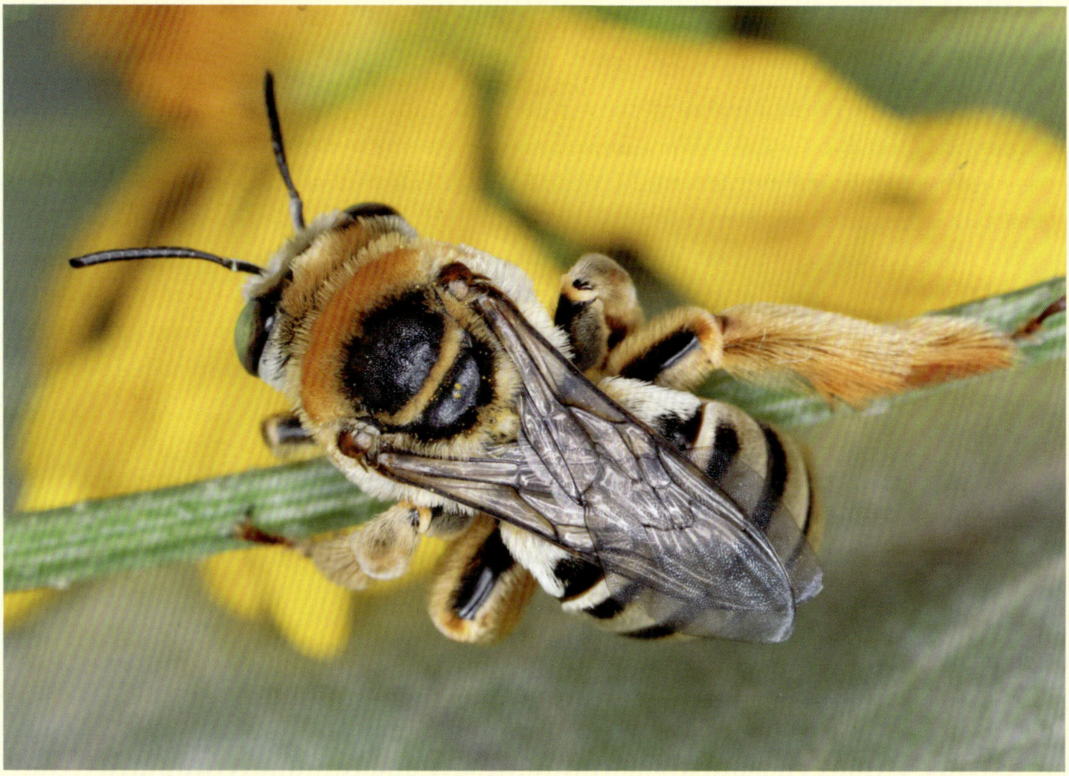

A female frisky longhorn bee

Description: *Svastra (Epimelissodes) petulca* is a hefty bee with an abdomen striped by alternating wide bands of light-colored hair and narrow bands of short black hair. The central areas of the female's scutum and scutellum are covered with dense dark hairs, while the thorax hairs around them are long and rust orange to yellowish white. On males, the dark patches of thorax hairs are less conspicuous, and the thorax hairs around them tend to be all off-white, giving the male a paler appearance overall. Both males and females have bright green eyes, faces thickly covered with white hairs, and pale yellow hairs on the vertex. The female's face is black, and the labrum and the bases of its mandibles yellowish. On males, the clypeus, labrum, and bases of the mandibles are bright yellow.

The tegulae of *Svastra petulca* are reddish gold on females and yellowish on males. The wings are transparent light brown, with brown veins. Female frisky longhorns have pale yellow and orange scopal hairs on their hind tibiae. The hairs on the females' other legs, and on all the legs of male bees, are predominantly white or yellowish white. **Size:** females 12.5–16 mm (1/2–5/8 in.), males 11–13 mm (2/5–1/2 in.).

Range: This is a common *Svastra* in all three border counties of the Lower Rio Grande Valley, where it appears from mid-April through November. *Svastra petulca* ranges south from the border through Tamaulipas, and northward through Texas to Kansas. Its population also extends eastward along the Gulf Coast of the United States, and northward from Florida to New Jersey.

Associated plants: Frisky longhorns show a preference for flowers of the family Asteraceae. Within the Lower Rio Grande Valley, *Svastra petulca* gathers pollen from composites such as cowpen daisy (*Verbesina encelioides*), Mexican hat (*Ratibida columnifera*), resinbush (*Viguiera stenoloba*), common sunflower (*Helianthus annuus*), tropical threefold (*Trixis inula*), and hairy wedelia (*Wedelia acapulcensis* var. *hispida*). In June 2021, male *Svastra petulca* appeared in substantial numbers at Falcon State Park and Campos Viejos Ranch in Starr County, feeding on guajillo (*Senegalia berlandieri*), old man's beard (*Clematis drummondii*), and Texas kidneywood (*Eysenhardtia texana*). In November 2021, a male bee appeared foraging on spiked malvastrum (*Malvastrum americanum*) at Resaca de la Palma State Park (Cameron Co.).

ID tips: Male *Svastra* species are differentiated in part by the length of their flagellar antennal segments. On male *Svastra petulca*, F2 is nearly three times longer than the short side of F1. Males of the subgenus *Svastra (Epimelissodes)* tend to have only moderately long antennae, which rarely extend backward past the first segment of the abdomen.

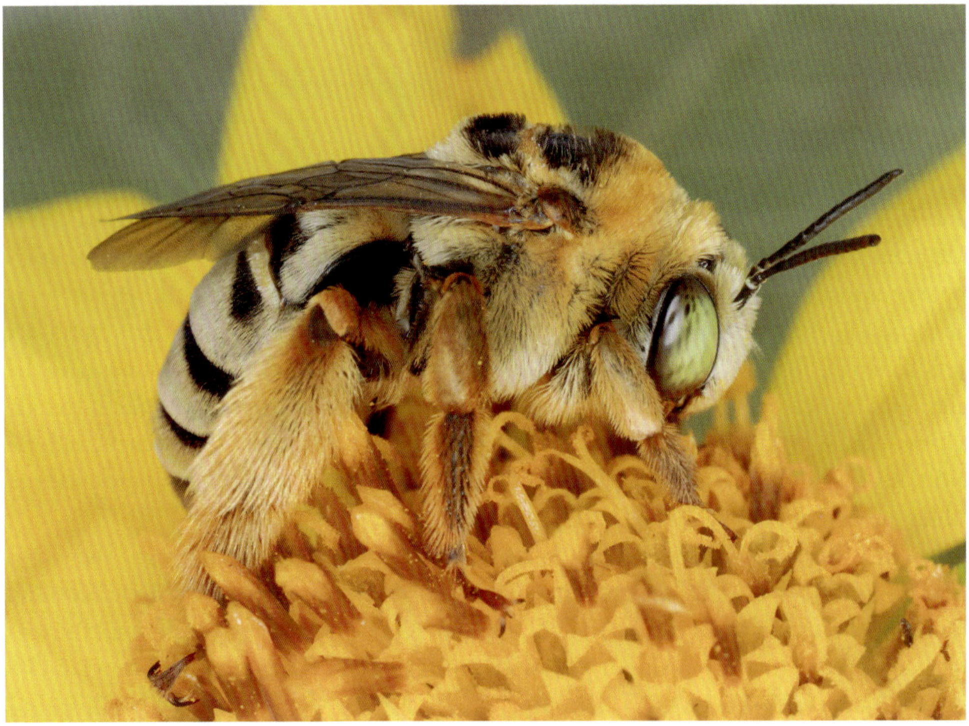

A female *Svastra petulca*

The wide abdominal bands of *Svastra petulca* are the best way to differentiate it from other regional *Svastra*. The species *Svastra petulca* sometimes appears under the common name "broad-banded longhorn bee," although the Latin *petulca* means "frisky." On both males and females, the pale and uninterrupted hair bands on the second and third segments (T2–T3) of the abdomen are at least two-thirds the width of the segment and extend all the way to its hind rim. (The basal third of each segment is covered with black hairs.) This description is sometimes true of T4 as well. On other, similar Texan *Svastra*, these bands are narrower and often interrupted.

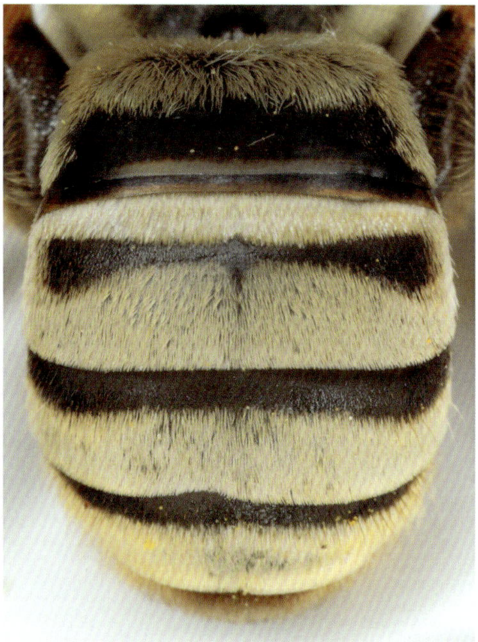

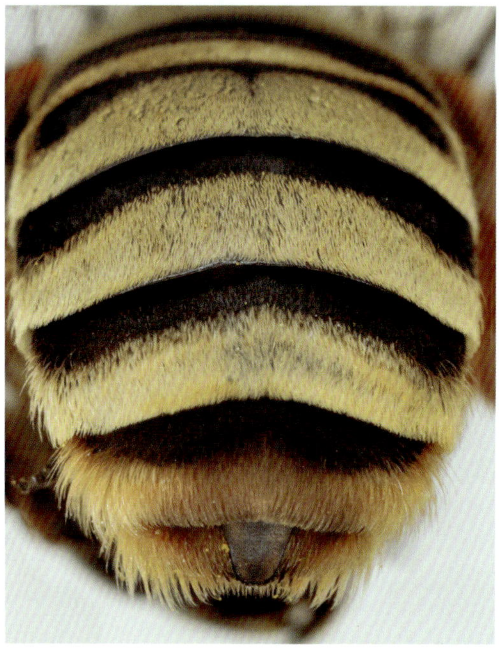

Dorsal and rear views of the wide abdominal bands of a female *Svastra petulc*

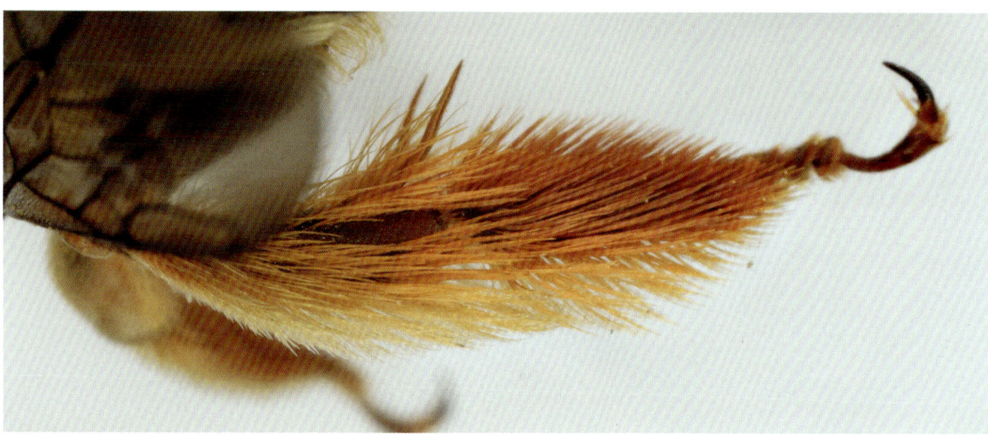

The light yellow and orange scopal hairs on the hind tibia of a female *Svastra petulca*

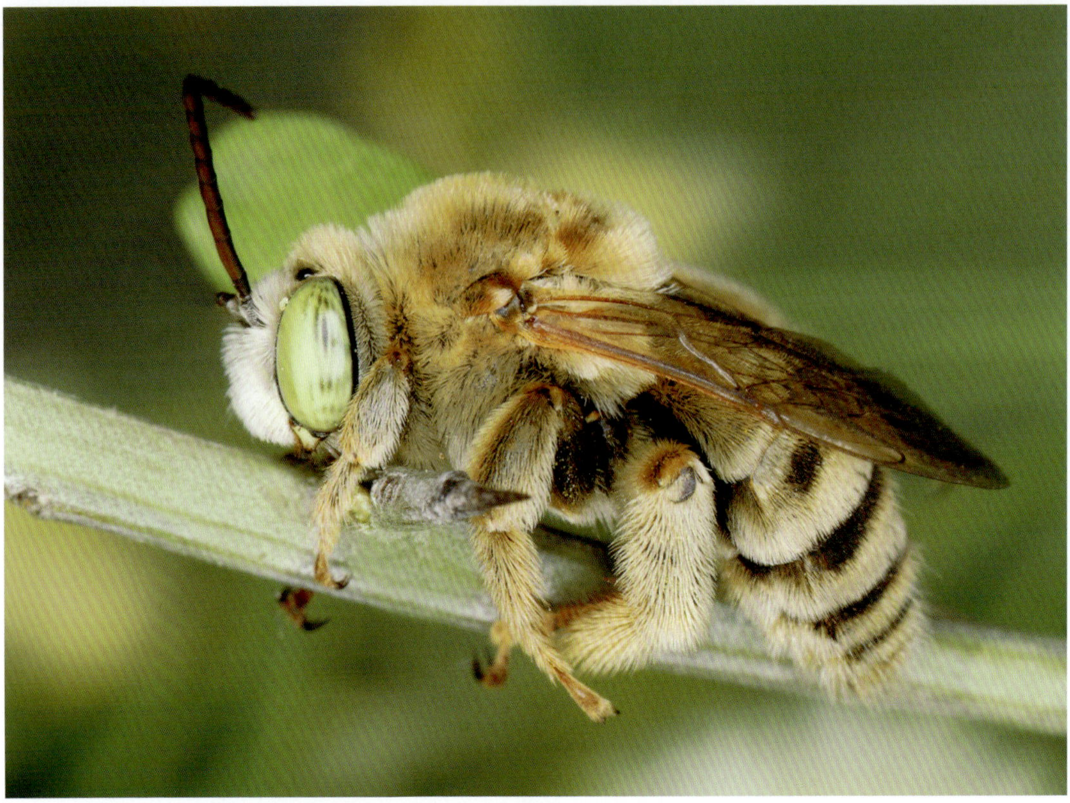

A male *Svastra petulca*: note the light yellow leg hairs and the wide pale bands on its abdomen.

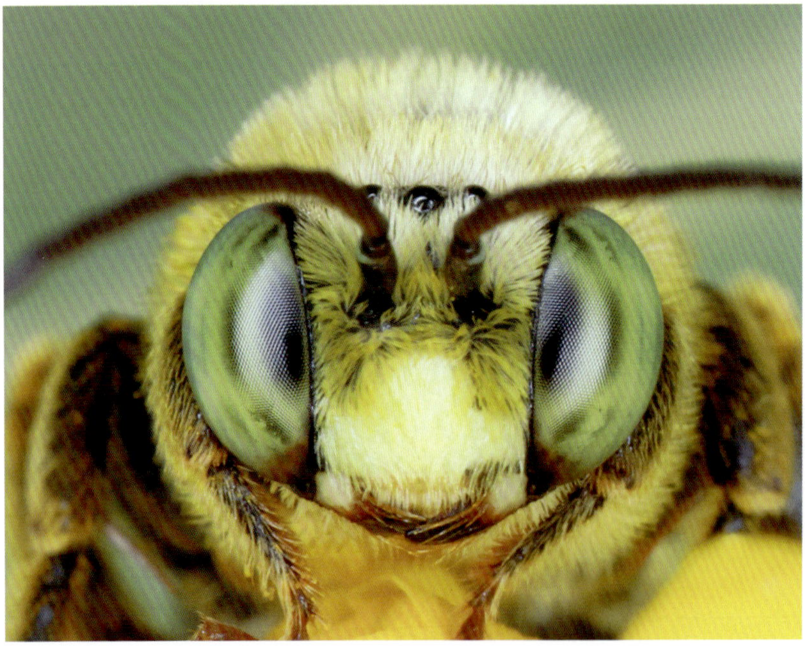

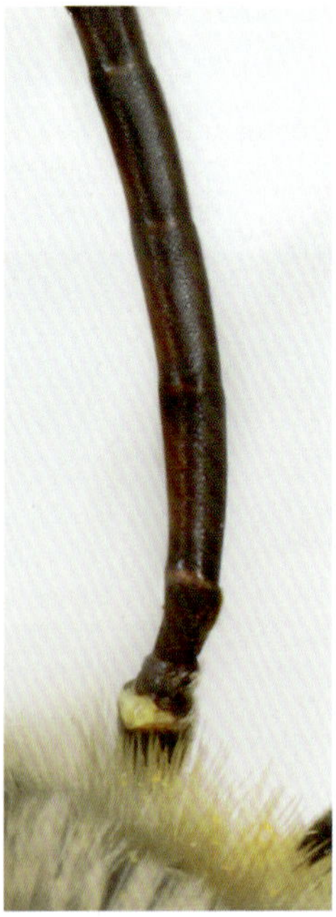

Left: Face of a male *Svastra petulca*. *Right*: Close-up of male's antenna.

Black-legged longhorn bee ◆ *Svastra (Epimelissodes) atripes* (Cresson, 1872)

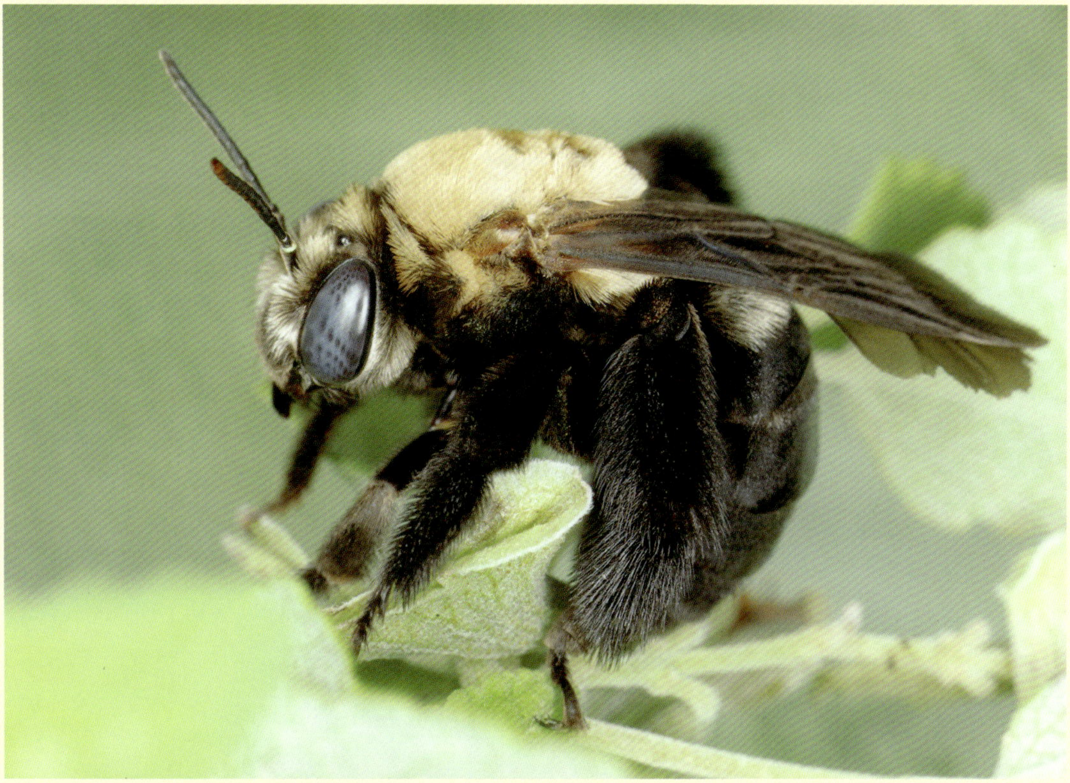

A female dark morph *Svastra atripes*

Description: The female *Svastra (Epimelissodes) atripes* is a large and beautiful black-and-gold bee with smoky wings and deep-blue eyes. The female's head and body are black; a mix of black and pale hairs clothe its dark face and vertex; its thorax is covered largely with pale gold hairs; and its abdomen is clothed with dark hairs, sometimes with pale bands. Exuberant, dense black hairs cover the female's legs; its hind-leg scopal hairs are long and black. Its mandibles are black, with small yellowish or reddish spots near the tips.

On the male *Svastra atripes* shown here, black hairs clothe most of the abdomen except for a pale band along the base of T2, and two widely interrupted narrower bands on T2 and T3. Short, toothlike spines protrude from either side of the male's T6. The male's thorax hairs are pale gold, and there are some light gold hairs on all of the tibiae. The male's tegulae, like the female's, are reddish brown, but its wings are less darkly colored. Males have light-green eyes, and the clypeus, labrum, and bases of the mandibles are yellow. Pale hairs cover the male's face. **Size:** females 15–20 mm (3/5–4/5 in.); males 13–18 mm (1/2–7/10 in.).

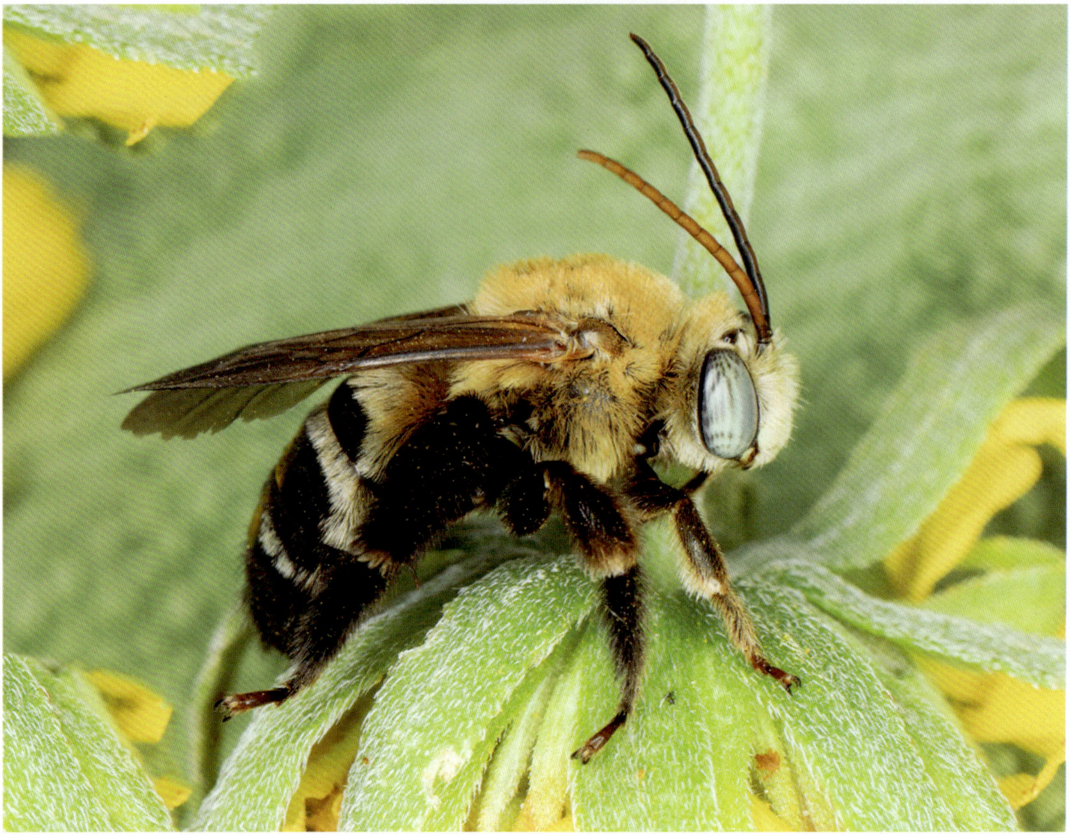

A male *Svastra atripes*

Range: *Svastra atripes* appears in all three border counties of the Lower Rio Grande Valley, but this species is relatively uncommon in the region. The female dark morph variation featured here is quite unusual. The dark morph appeared in mid-June 2021 at El Mesteño Ranch (northwest Hidalgo Co.). The male bee shown here appeared at Campos Viejos Ranch (Starr Co.) in late June 2021. In the United States generally, the various subspecies of *Svastra atripes* range from Utah to the East Coast, as far north as Illinois.

Associated plants: The black-legged longhorn bee is a vibratile ("buzz") pollinator, unlike the many *Svastra* that are Asteraceae specialists. *Svastra atripes* is known to visit a range of plants as diverse as cotton, partridge pea, passionflower, and goldenrod. The female dark morph featured here was found collecting pollen from shrubby blue salvia (*Salvia ballotiflora*). The male bee appeared at Campos Viejos Ranch on devil's claw (*Proboscidea louisianica*).

ID tips: This species is easily distinguished from other *Svastra* of the Valley by the predominantly black hairs covering its hind legs and abdomen.

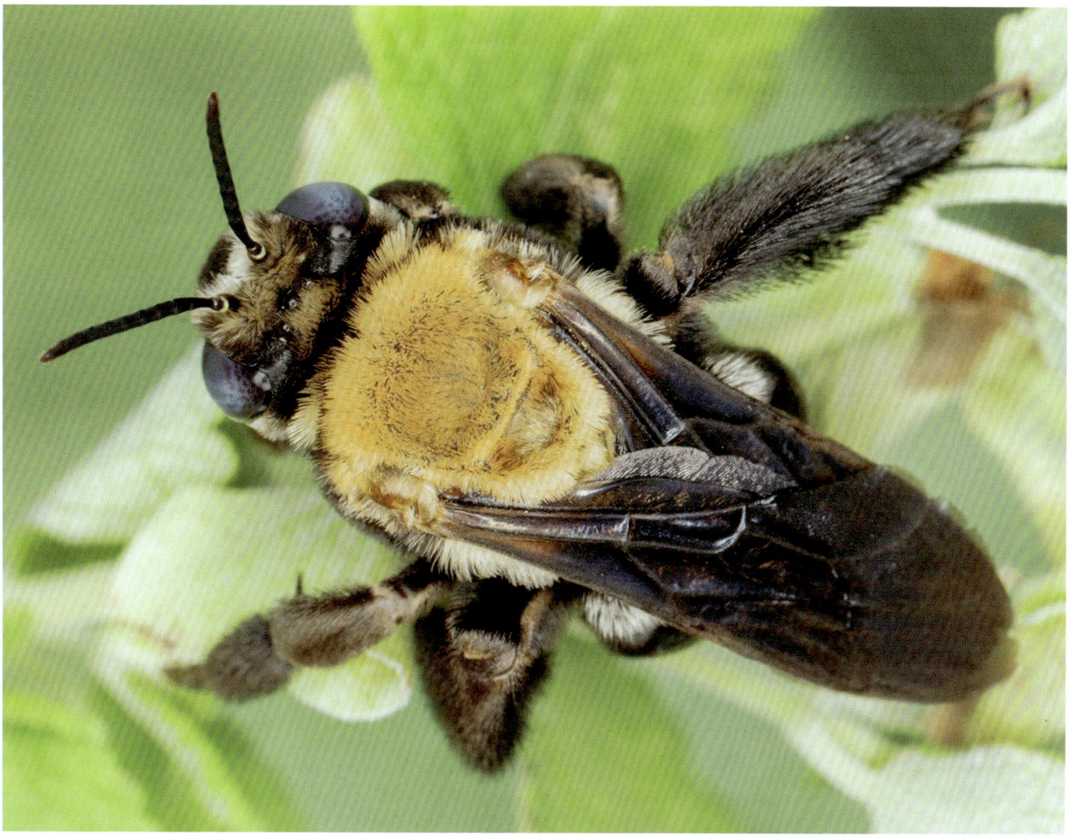

Dorsal view of a female dark morph *Svastra atripes*

Subspecies and morphs: There are several subspecies of *Svastra atripes* in the United States, generally differentiated by hair-color variations and geography. The typical *Svastra atripes atripes* is found primarily in the Southwest and Midwest, from New Mexico and Colorado to Louisiana and Illinois. Females and males of this subspecies have all-black hairs on their hind legs and interrupted white bands on T2 and T3. *Svastra atripes atrimitra* females and males have all-black hairs on the legs and reduced pale-haired banding or no banding on the abdomen. This subspecies is found from Alabama and Georgia to New Jersey. *Svastra atripes georgica* occurs in Georgia and Florida. Females of this subspecies have pale gold hairs on their hind tibiae, and their scopae are less feathery than those of the other subspecies. *Svastra atripes georgica* males have broad pale bands on T2 and T3, and some pale gold hairs on the outer surfaces of their middle and hind tibiae.

None of these subspecies descriptions precisely fit the female dark morph shown here, whose salient characteristic is the dense, dark hair covering the lower half of its thorax—rather than the pale hair typical of all three subspecies listed above. The dark morph has all-dark hairs on all segments of its abdomen except T1, and an absence of any light hairs on its legs. Oddly enough, the description that best approximates the male bee shown here is *Svastra atripes georgica*, although it is well outside the range of that subspecies.

Behavior: In 1995, entomologist James Cane studied aggregations of nesting *Svastra atripes* in Alabama. The females he observed neglected many practices associated with *Svastra*—they did not build turrets at entrances or plug their nests when foraging. Most notably, although the *Svastra atripes* females were gregarious enough to nest close to one another, they did not participate in communal nesting behavior, such as the sharing of nest entrances, typical of *Svastra* females of multiple species, among them *S. sabinensis*, *S. petulca*, and *S. duplocincta*.

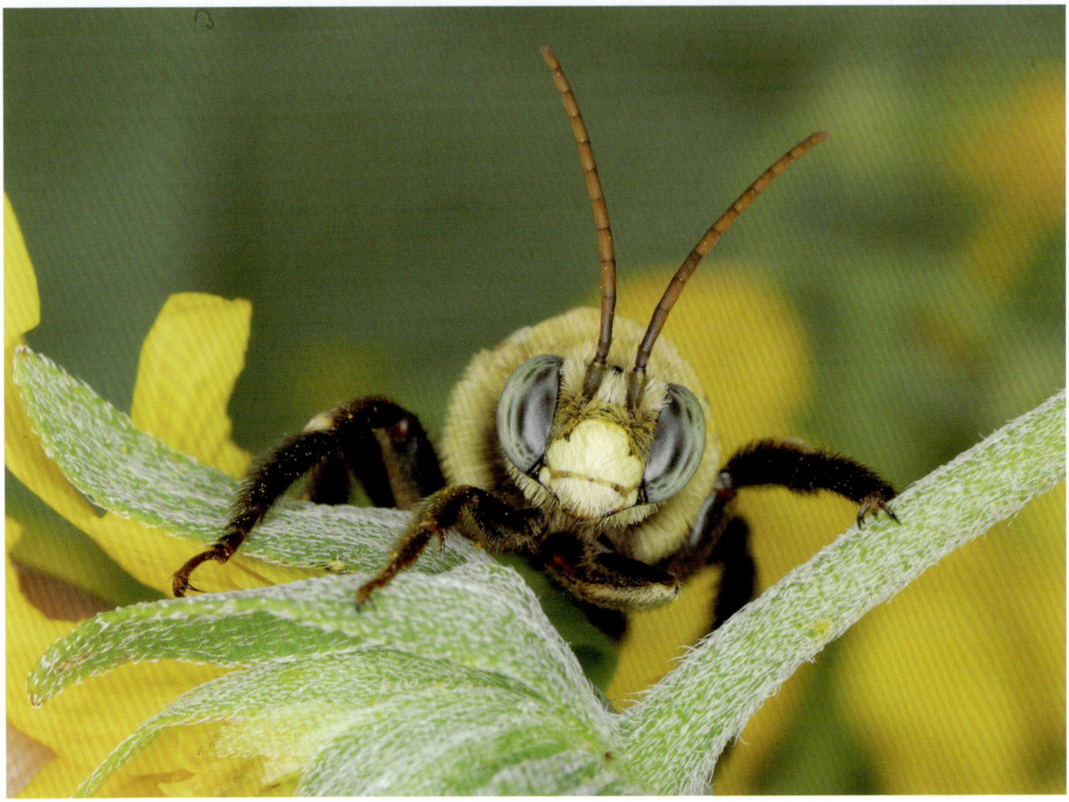

Face of a male *Svastra atripes*

Barrel cactus longhorn bee ♦ *Svastra (Idiomelissodes) duplocincta* (Cockerell, 1905)

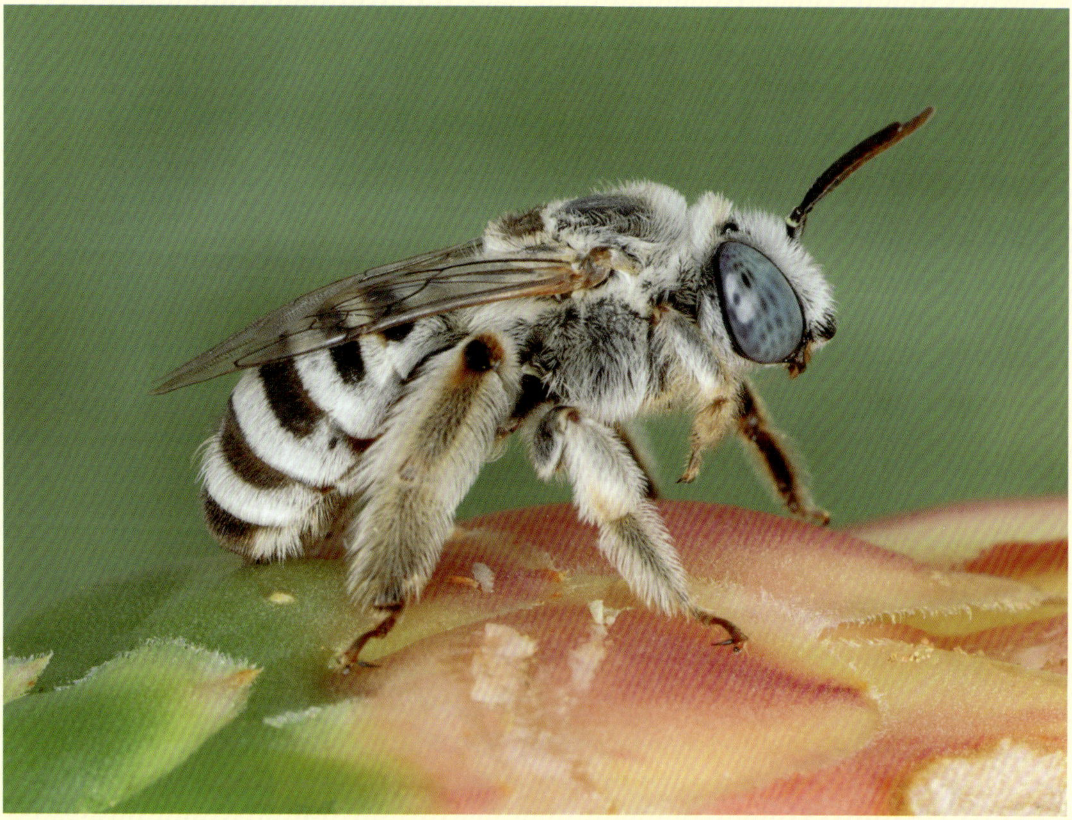

A female barrel cactus longhorn bee

Description: *Svastra duplocincta* is the sole member of its current subgenus *Idiomelissodes.* It is considerably smaller than other *Svastra* in this guide, which belong to the subgenus *Epimelissodes.* The barrel cactus longhorn bee is distinctive, with bold bands of alternating black-and-white hairs on its abdomen. (Females have five white bands and males six.) The head, thorax, and legs are covered predominantly with white hairs, intermingled with some brown hairs. Flattened white hairs line the back of the scutellum and form a tuft on the metanotum. The wings are light brown with dark brown veins, and the tegulae are reddish gold. The female's hind-leg scopal hairs are mostly white and pale brown; the scopae are dark brown on the inner surfaces of the lower legs. Females have black faces, gray-blue eyes, and short antennae. Males have green eyes, bright yellow clypei, yellow markings on the bases of their mandibles, and relatively long antennae for *Svastra.* **Size:** females 10 mm (2/5 in.), males 8–9 mm (3/10–1/3 in.).

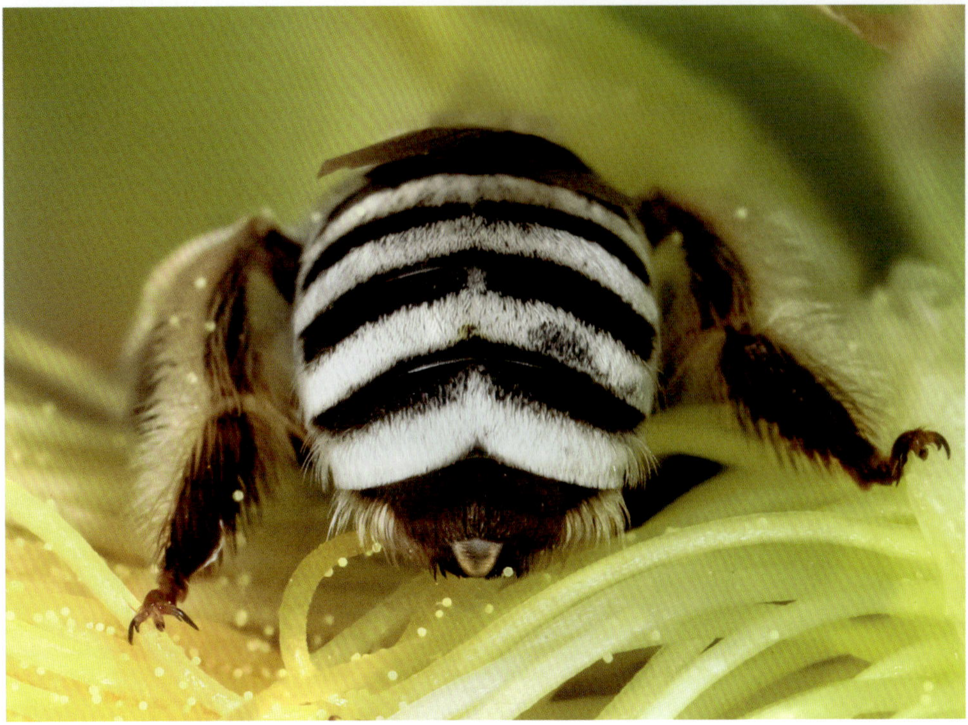

Rear view of a female *Svastra duplocincta* pollinating barrel cactus

Range: In the United States, *Svastra duplocincta* ranges principally along border areas of Texas, New Mexico, and Arizona. It is native to the Mexican states of Tamaulipas, Nuevo León, and Sonora and has been documented as far south as Durango. This species is uncommon in the Lower Rio Grande Valley; it is more likely to be found near the southwestern Texas border. Nonetheless, in mid-June 2021, the barrel cactus longhorn bee appeared in substantial numbers at Campos Viejos Ranch (Starr Co.).

Associated flora: *Svastra duplocincta* is an oligolectic pollinator of cactus: it characteristically emerges in early summer and thus misses most of the prickly pear cactus bloom in the Valley. All the bees discovered at Campos Viejos were feeding on candy barrel cactus (*Ferocactus wislizeni*). There are some records of this species visiting other flora, including creosote bush (*Larrea tridentata*), stinging serpent (*Cevallia sinuata*), flowers in the morning glory family, spurge, and cotton.

ID tips: *Svastra duplocincta* is easily distinguished from all other longhorn bees of the Valley by its relatively small size and its bold white-and-black abdominal banding. On both males and females, the rear abdominal band has a distinctive shape—it peaks in the middle like the roof of a pagoda. Male *Svastra duplocincta* tend to congregate conspicuously in areas where cactus abounds; in early morning, males can often be found sleeping in aggregations that are easily spotted because of the bright banding on the bees' abdomens. According to a 1998 study by entomologist John Alcock, the practice of overnighting in aggregations gives the small *Svastra* some degree of protection against predatory assassin bugs.

Little prickly pear longhorn bee ◆ *Melissodes (Tachymelissodes) opuntiellus* Cockerell, 1911

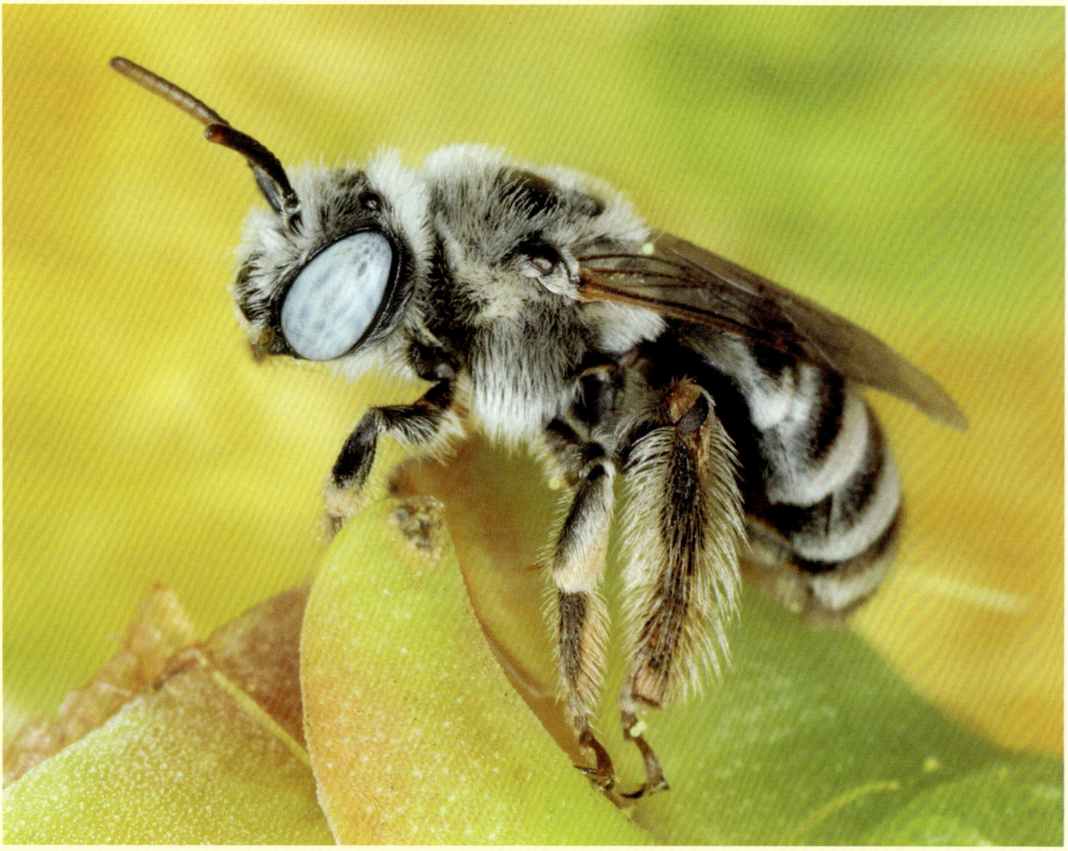

A female little prickly pear longhorn bee on a cactus blossom

Description: *Melissodes (Tachymelissodes) opuntiellus* is a small, attractive longhorn bee with broad bands of white hair striping its abdomen; dense white hairs on its thorax; and bright blue eyes. The tegulae are black, and the wings clear with brown veins. Females' hind-leg scopal hairs are white and loosely separated, an adaptation that allows them to carry prickly pear pollen. The legs of both male and female *Melissodes opuntiellus* are covered predominantly with white hairs. On males, the basitarsi and lower tarsomeres are reddish and partly covered with reddish hairs. On females, the hind-leg tarsi are reddish.

The little prickly pear longhorn bee currently belongs to the small subgenus *Tachymelissodes*. Bees of this subgenus have distinct white banding on their abdomens. The bands on their second through fourth terga (T2–T4) are usually uninterrupted in the middle; on females, each band is of uniform width from one side to the other. On female *Melissodes opuntiellus*, the abdominal hair bands are white on the first, second, and fourth segments (T1, T2, T4); white and grayish on T3; and chocolate brown on T5–T6. On males, the bands on T1–T5 are white, and T6 and T7 have lateral toothlike projections.

The faces of both male and female *Melissodes opuntiellus* are largely hidden behind a covering of copious white hair. Females have entirely black faces, and males have yellow clypei, dark labrums, and mandibles that are yellow at the base. On all *Melissodes (Tachymelissodes)* males, the first flagellar segment of the antennae (F1) is more than half as long as the second (F2). Males tend to have shorter antennae than do male *Melissodes* of other subgenera. **Size:** females and males 8–9 mm (3/10–1/3 in.).

Range: *Melissodes (Tachymelissodes)* are found principally in arid and semiarid regions of the western United States or Mexico. The little prickly pear longhorn bee is somewhat rare in the United States. It is found primarily in southeastern Texas and Mexico. It appears during the spring (March through June) in Hidalgo and Starr Counties. *Melissodes opuntiellus* has been documented as well in Medina and La Salle Counties. There are isolated records of this species in Arizona, Utah, and California.

Associated plants: *Melissodes opuntiellus* was first described by Cockerell in 1911 in a one-page article titled "Some New Bees from Flowers of Cactaceae." The species name *opuntiellus* is a reference to *Opuntia*, or prickly pear cactus. Prickly pear flowers are difficult for pollinators to negotiate: the cactus stamens are tightly packed and their pollen is sticky. Nonetheless, female *Melissodes opuntiellus* push into the cactus flowers with ease and disappear entirely within the stamens, only to emerge minutes later covered with pollen.

During 2019, in multiple locations in Mission, Texas, female *Melissodes opuntiellus* appeared in abundance on the flowers of Texas prickly pear cactus (*Opuntia engelmannii* var. *lindheimeri*) from mid-April until the end of the spring cactus bloom in late May. Males emerged before females, in late March, just before regional prickly pear began flowering. Before and during the bloom, males could be found sleeping on red prickly poppy (*Argemone sanguinea*). The male bees used the poppies for shelter from wind and rain.

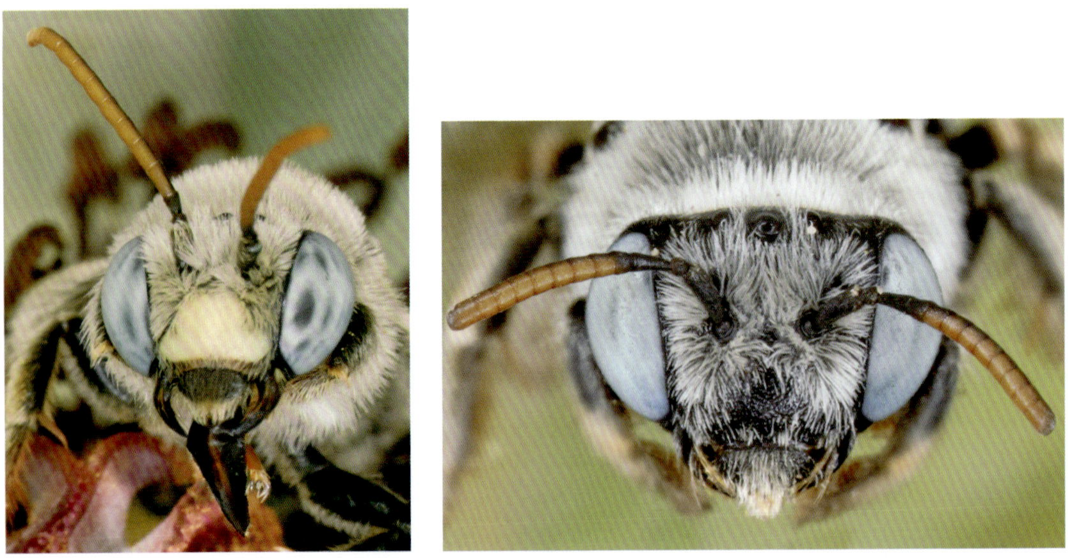

Faces of male (*left*) and female (*right*) *Melissodes opuntiellus*

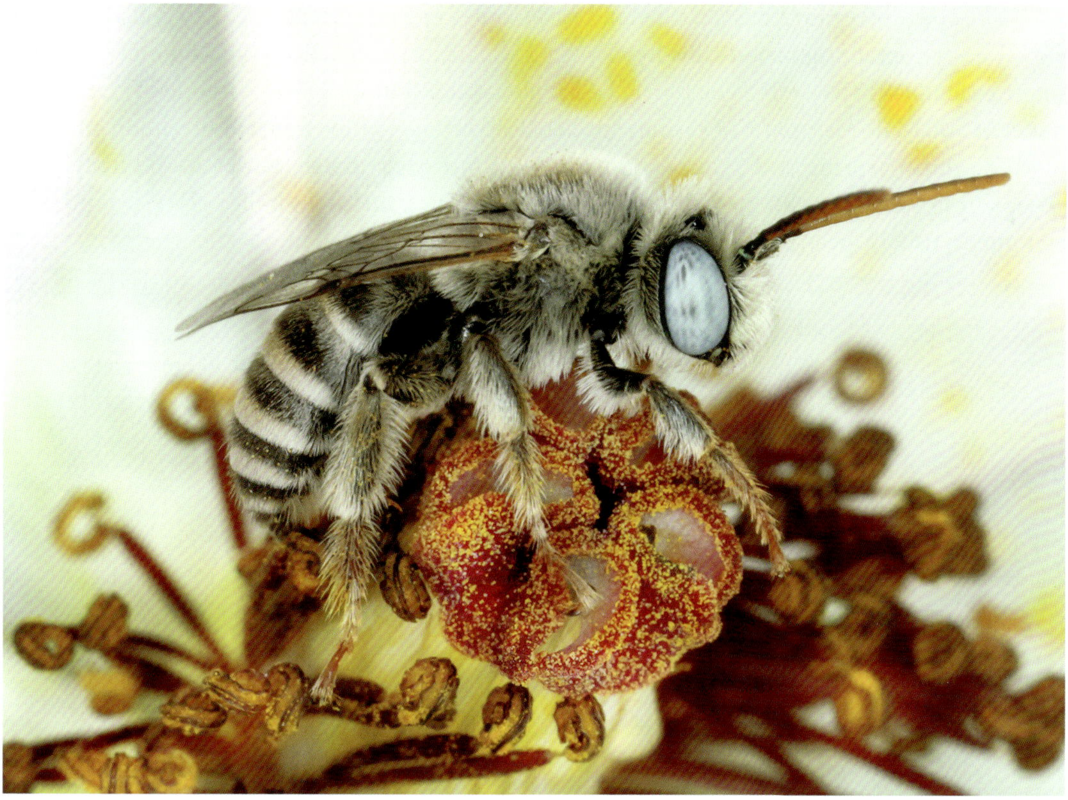

A male little prickly pear longhorn bee on red prickly poppy

Despite their name and adeptness in negotiating cactus blossoms, female *Melissodes opuntiellus* collect pollen and drink nectar from plants other than prickly pear—LaBerge recorded little prickly pear longhorn bees on various composite flowers, including sunflowers. During the spring of 2021, this species appeared in Starr County at La Puerta Tract, visiting Texas paloverde (*Parkinsonia texana*), and at Campos Viejos Ranch, foraging on Texas paloverde, shrubby Indian mallow (*Abutilon abutiloides*), and sticky florestina (*Florestina tripteris*).

ID tips: *Melissodes opuntiellus* is easily distinguished from other regional longhorn bees by the combination of its relatively small size, pale abdominal bands, pale thorax hairs, bright blue eyes, and association with cactus.

The genus *Melissodes*: *Melissodes* generally are robust, medium-sized longhorn bees with hairy heads and bodies. Females carry pollen on bushy hind-leg scopal hairs. Males have narrower builds than females and usually sport much longer antennae. The tegulae of *Melissodes* narrow apically, a trait that distinguishes them from all other longhorn bee genera (but one that, unfortunately, is usually obscured by hairs and requires magnification to perceive). Within the Lower Rio Grande Valley, both female and male *Melissodes* tend to have brightly colored blue or green eyes; eye color within a given species often varies according to gender. At least three distinct *Melissodes* subgenera are found within the Valley: *Tachymelissodes*, *Melissodes*, and *Eumelissodes*.

All *Melissodes* are ground nesters. Females carve holes in the earth where they construct individual nests containing brood cells lined with a waxlike material, each holding one egg and a single pollen ball. The nests, which may be isolated or built in groups, are frequently concealed under brush. Male *Melissodes* of various species sleep in aggregations. In late afternoon and early morning during spring and summer in the Valley, *Melissodes* males can be found hanging in groups from stems or grass stalks, suspended by their jaws, their legs pulled close around their abdomens, usually near flowers frequented by females during the day.

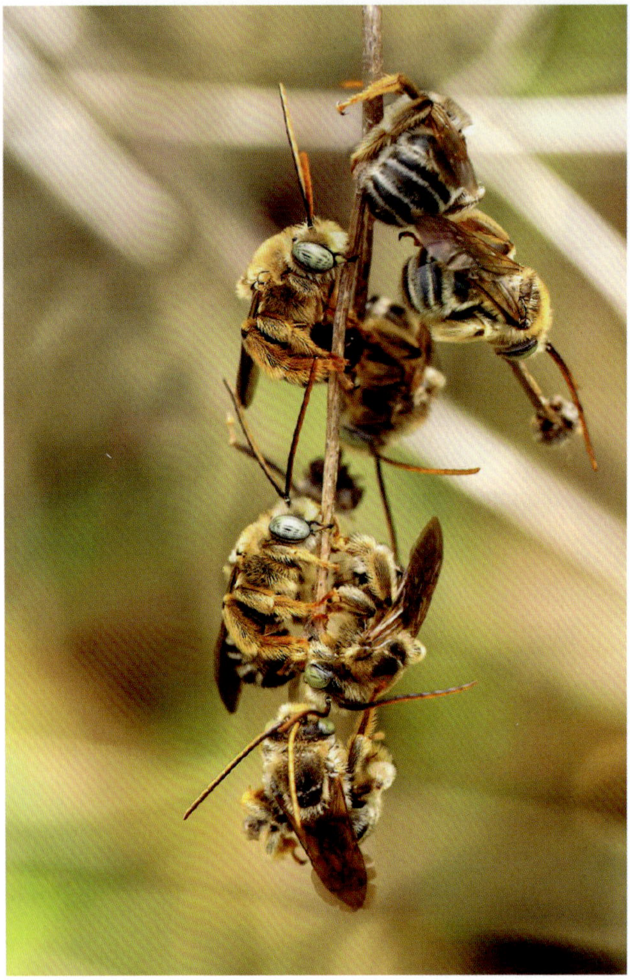

An aggregation of *Melissodes* males sleeping on a twig. Photo credit Laura Langlois Zurro.

Tepanec longhorn bee ◆ *Melissodes (Melissodes) tepaneca* Cresson, 1898

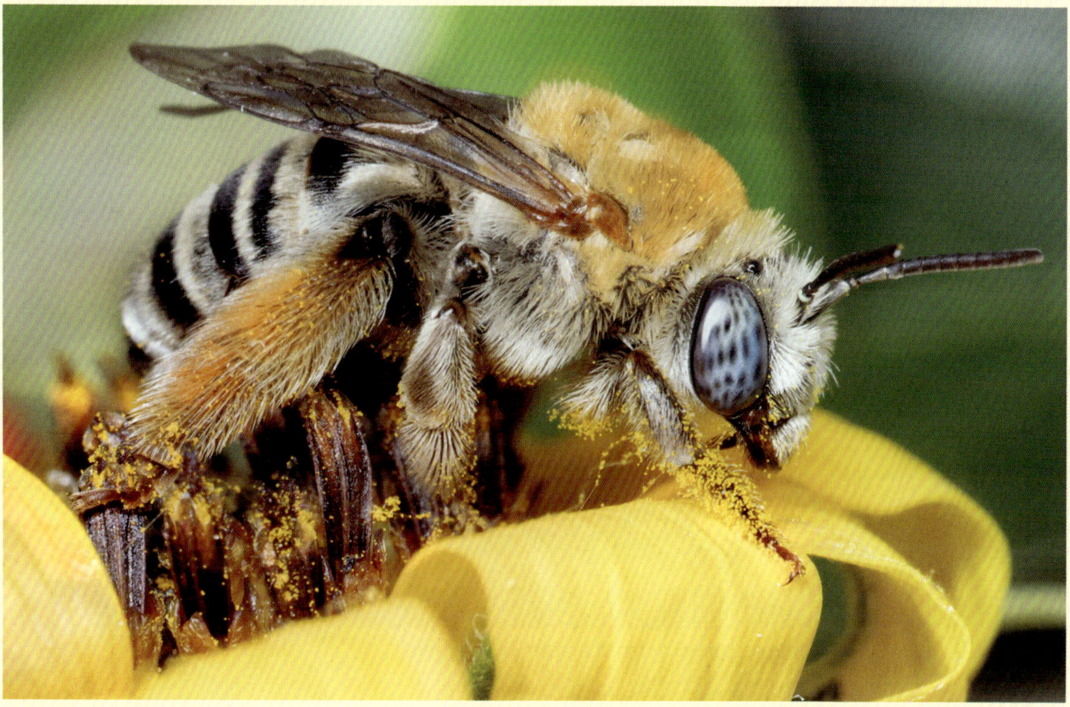

A female Tepanec longhorn bee

Description: *Melissodes (Melissodes) tepaneca* is a burly bee with rust-colored thorax hairs and a black abdomen banded by pale hairs. Females have dark antennae, dark clypei, and blue-gray eyes. The male Tepanec longhorn bee has green eyes and a yellow clypeus, and its antennae are brown on their front surfaces and otherwise black. Males of the subgenus *Melissodes (Melissodes)* typically have pale labrums and very long antennae (that reach at least to the front of their abdomens). On all male *Melissodes (Melissodes)*, the first flagellar segment of the antenna (F1) is one-third to one-tenth the length of the second (F2).

Female *Melissodes tepaneca* have plush blond and orange hind-leg scopal hairs. On both females and males, the lower legs (the tarsi and some or all of the tibiae) are red. The wings are glassy yellow with brown veins; the tegulae are also glassy yellow or (on some males) reddish. **Size:** females 12–13 mm (1/2 in.), males 9.5–11 mm (3/8–2/5 in.).

Range: *Melissodes (Melissodes)* range widely, from Canada to Brazil and the Caribbean. *Melissodes tepaneca* is found throughout Texas, Mexico, and Central America as far south as Panama. This species also ranges throughout the eastern United States, as far north as Maryland and as far south as Florida. It is uncommon west of Texas and Kansas. *Melissodes tepaneca* is the most common longhorn bee species of the Valley and is found throughout Cameron, Hidalgo, and Starr Counties.

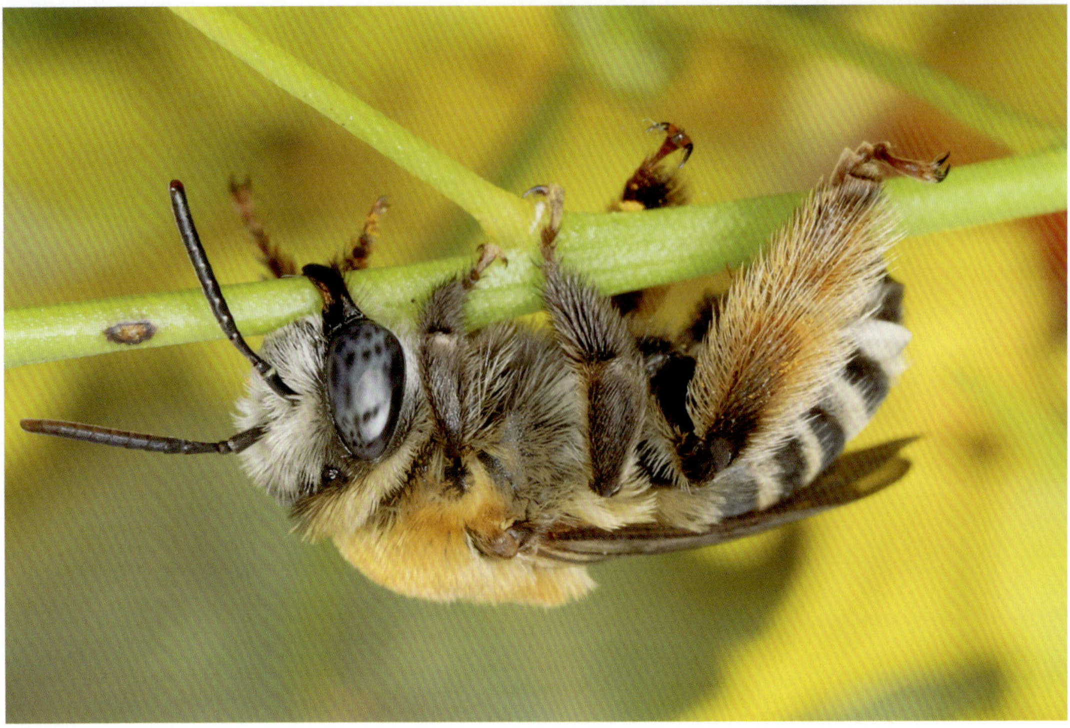

A female Tepanec longhorn bee hanging from a stem by its jaws

Associated plants: Tepanec longhorn bees emerge in April in the Lower Rio Grande Valley and persist through October. They are generalist pollinators that forage on flora from diverse plant families, including, for example, esperanza (*Tecoma stans*), retama (*Parkinsonia aculeata*), Texas paloverde (*Parkinsonia texana*), crucita (*Chromolaena odorata*), resinbush (*Viguiera stenoloba*), common sunflower (*Helianthus annuus*), and, occasionally, Texas prickly pear (*Opuntia engelmannii* var. *lindheimeri*). This species is also an important pollinator of cotton.

ID tips: *Females*: Species identification of *Melissodes* females often hinges on characteristics of their abdominal bands and variations in hair color. The following traits, in combination with the glassy yellow color of the wings, distinguish female Tepanec longhorn bees from similar regional species: (1) there are dark hairs just behind the ocelli—all other hairs on the vertex are pale; (2) the thorax is covered with rust-colored hairs and lacks dark hairs on the top or sides; (3) dark and light hairs clothe the front and middle legs; (4) on T2 and T3, the hair bands are set back from the hind rim; (5) the band on T2 is of uniform width from one side of the tergum to the other, and there are small pits in the bare, depressed portion of the segment's rim; and (6) there is also a dark patch of hair in the center of T4; the hair on the rim, base, and sides of T4 is pale.

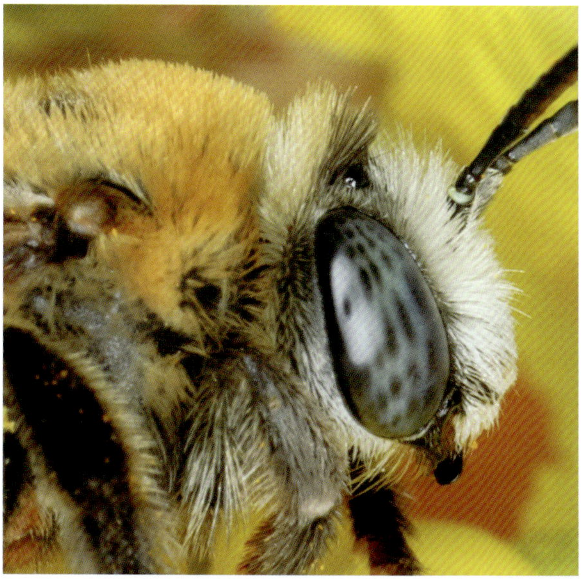

Close-up of the head of a female Tepanec longhorn bee, showing dark hairs behind the ocelli

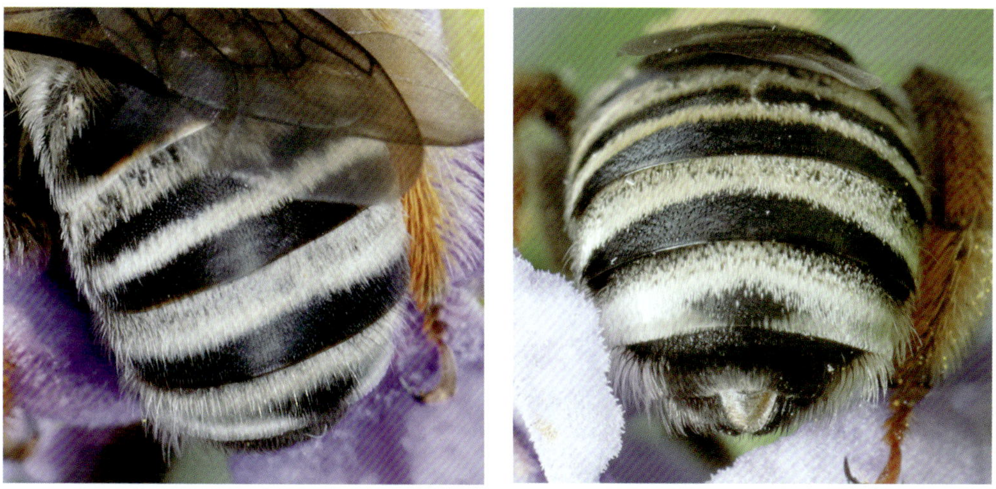

The hair bands on the abdomen of a female Tepanec longhorn bee

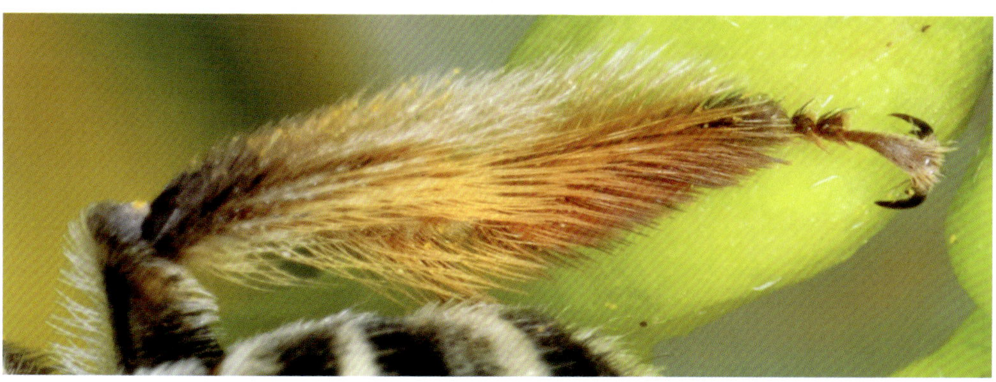

The hind-leg scopal hairs on the tibia of a female *Melissodes tepaneca*

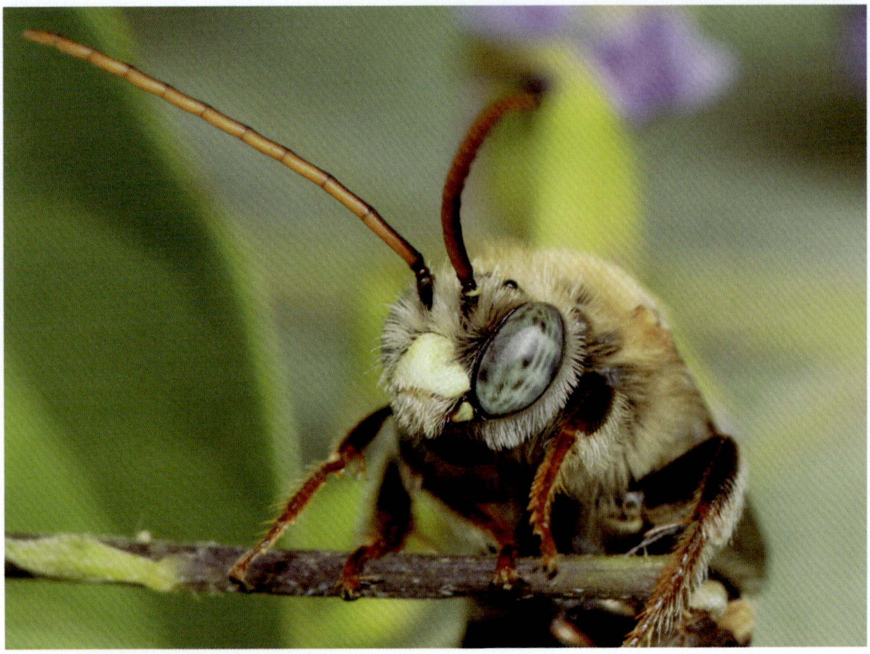

A male Tepanec longhorn bee

Males: The following attributes, taken together, separate the male *Melissodes tepaneca* from similar male longhorn bees of the Valley: (1) the hair on the vertex, thorax, and legs is entirely light; (2) the first flagellar segment of the antennae (F1) is longer than the pedicel; and (3) the yellow spots on the bases of the mandibles, and the lack of dark coloration on the antennal tips, distinguish *Melissodes tepaneca* males from *Florilegus condignus* males.

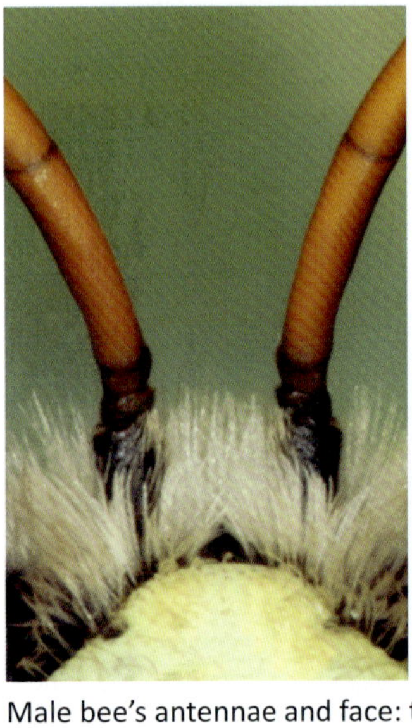

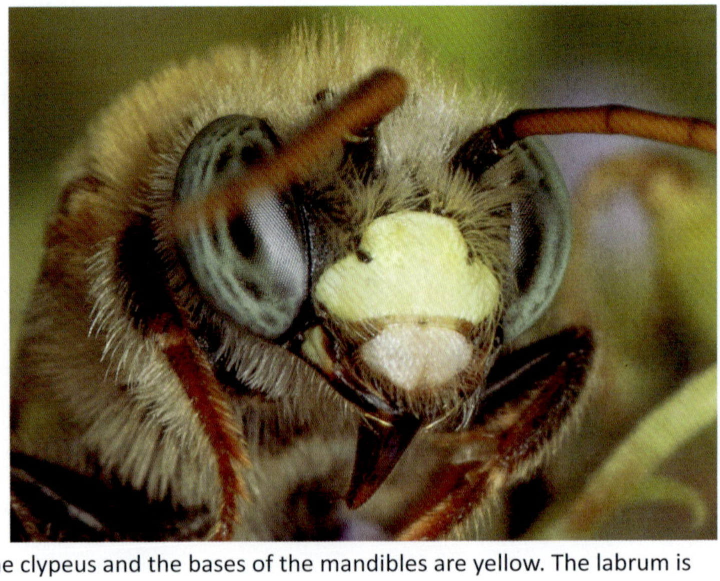

Male bee's antennae and face: the clypeus and the bases of the mandibles are yellow. The labrum is white. F1 of each antenna is much shorter than F2, but longer than the pedicel.

Mournful longhorn bee ◆ *Melissodes* (*Eumelissodes*) tristis Cockerell, 1894

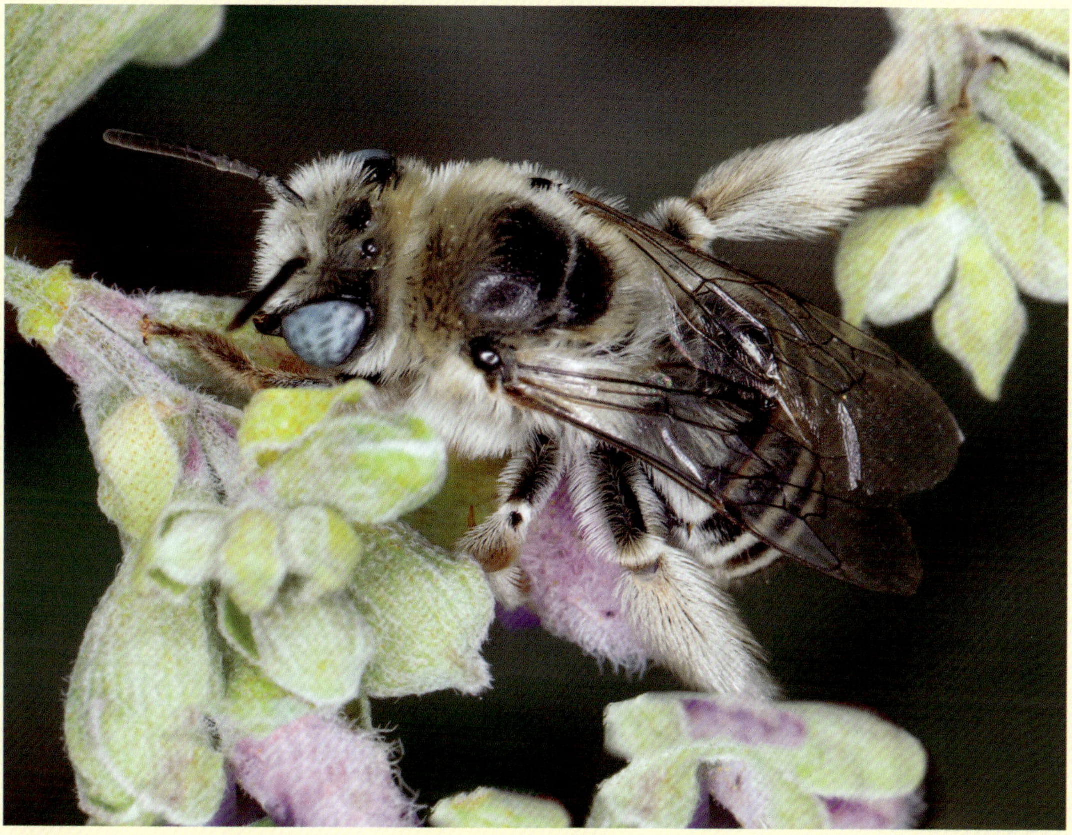

A female *Melissodes tristis*

Description: *Melissodes (Eumelissodes) tristis* is a robust, medium-sized bee, striking in the field because of the lush white hair covering much of its body, and its intensely colored eyes, which are denim blue. Males have unusually long antennae. Female mournful longhorn bees have wide faces; black, pitted clypei; and short, dark antennae. Males' antennae are mostly dark; above the first flagellum, the front surfaces of the antennae are golden.

On both male and female *Melissodes tristis*, white hairs cover most of the head and thorax and band the abdomen. Females have partly brown hairs on the thorax and sternum, and both males and females may have some pale gold hairs on both the vertex and thorax. Some dark hairs line the bases of the third through sixth segments of the abdomen (T3–T6).

The leg hairs of both male and female *Melissodes tristis* are predominantly white. The scopal hairs on the female's hind-leg tibiae are long and bushy and mostly pale. Reddish-brown or brown hairs clothe the female's basitibial plates and basitarsi. On males' basitarsi, the hairs are yellow. The tegulae of both male and female *Melissodes tristis* are brown, and their wings glassy to milky, with brown or black veins. **Size:** females 10–14 mm (2/5–3/5 in.), males 9–12 mm (1/3–1/2 in.).

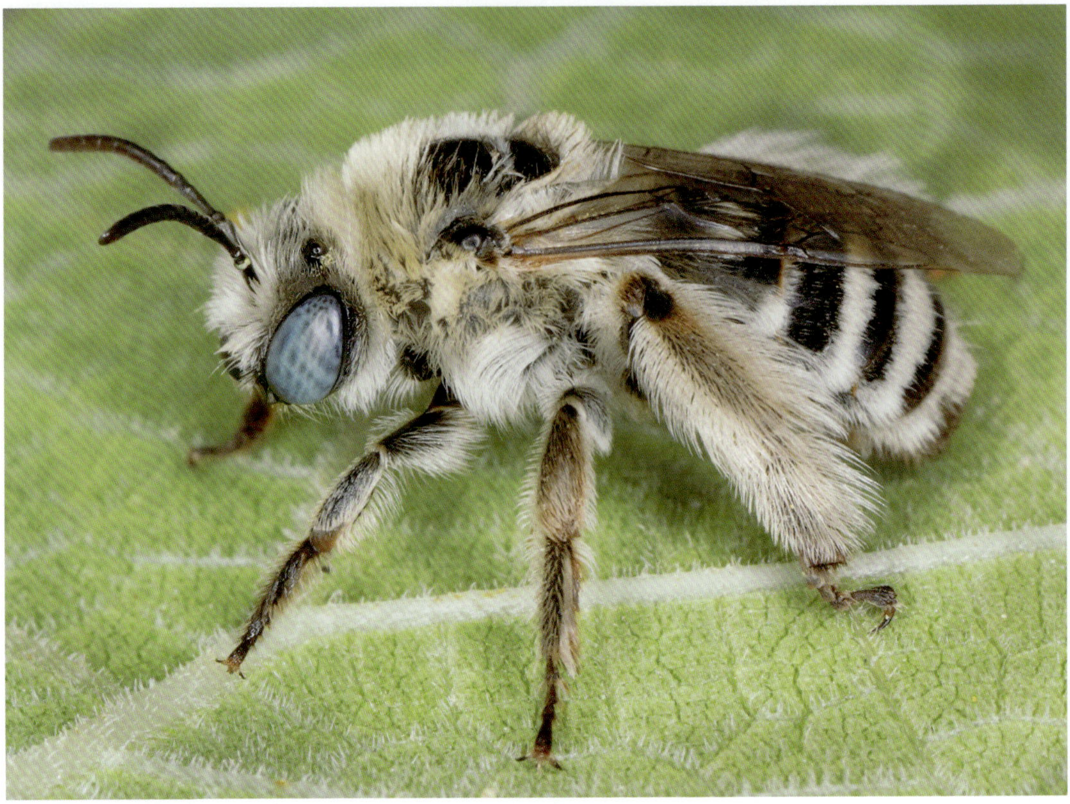

A female *Melissodes tristis*

Range: *Eumelissodes* is the largest subgenus of the genus *Melissodes*, occurring from British Columbia to Panama and Cuba. *Melissodes (Eumelissodes) tristis* is found throughout much of Texas, and its population extends to south-central Mexico. This species also ranges through New Mexico and Arizona and extends from western Oklahoma, Kansas, and Nebraska through Colorado, Nevada, Utah, and southern Wyoming into California and Oregon. *Melissodes tristis* appeared at Quinta Mazatlán and the National Butterfly Center (Hidalgo Co.) in midspring of 2019 and 2021.

Associated plants: *Melissodes (Eumelissodes)* frequently specialize in pollinating aster-family flowers. The mournful longhorn bee, however, is a prodigious generalist that has been documented visiting the flowers of more than 20 plant families. In a comprehensive 1961 treatment of the genus *Melissodes*, LaBerge wrote that the mournful longhorn bee is the most polylectic species of its subgenus. It has a remarkably long flight season in the Valley, from March through November, and is able to adapt to an array of habitats with very different flora. LaBerge believed that *Melissodes tristis* produces as many as three generations per year in Texas. The female bees shown here were found gathering pollen from honey mesquite (*Prosopis glandulosa*) and shrubby blue salvia (*Salvia ballotiflora*).

ID tips: *Melissodes (Eumelissodes)* tend to share the following traits: (1) they are medium large, generally 8 to 16 mm (3/10 to 3/5 in.); (2) females have black clypei and copious feathery scopal hairs on their hind legs; (3) F1 of the male's antennae is much shorter than F2 (often one-third its length or less, but at the very most one-half the length); and (4) on males, the clypeus is usually yellow or white. *Melissodes tristis* possesses all of these traits except the last.

Males: Mournful longhorn bees sometimes appear under the alternative common name "dark-faced longhorn bee" because males of this species have entirely black clypei (and labrums) rather than the pale clypeus typical of their subgenus. This feature is often obscured under a thatch of white facial hairs. Usually, however, the male bee's very long antennae and its saturated blue eyes suffice to identify it in the field.

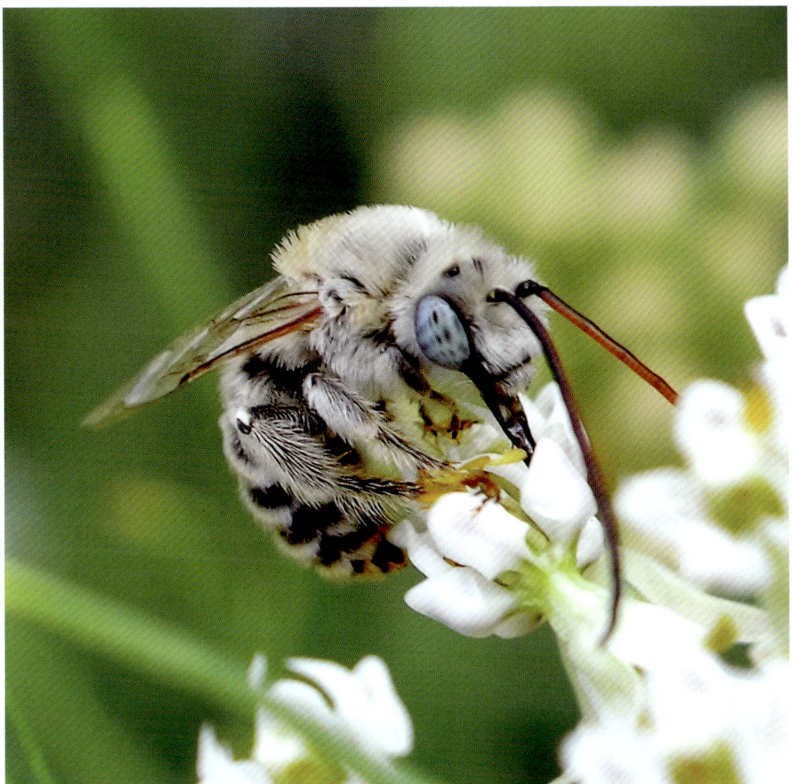

A male *Melissodes tristis* drinking nectar. Photo credit J. A. Winget.

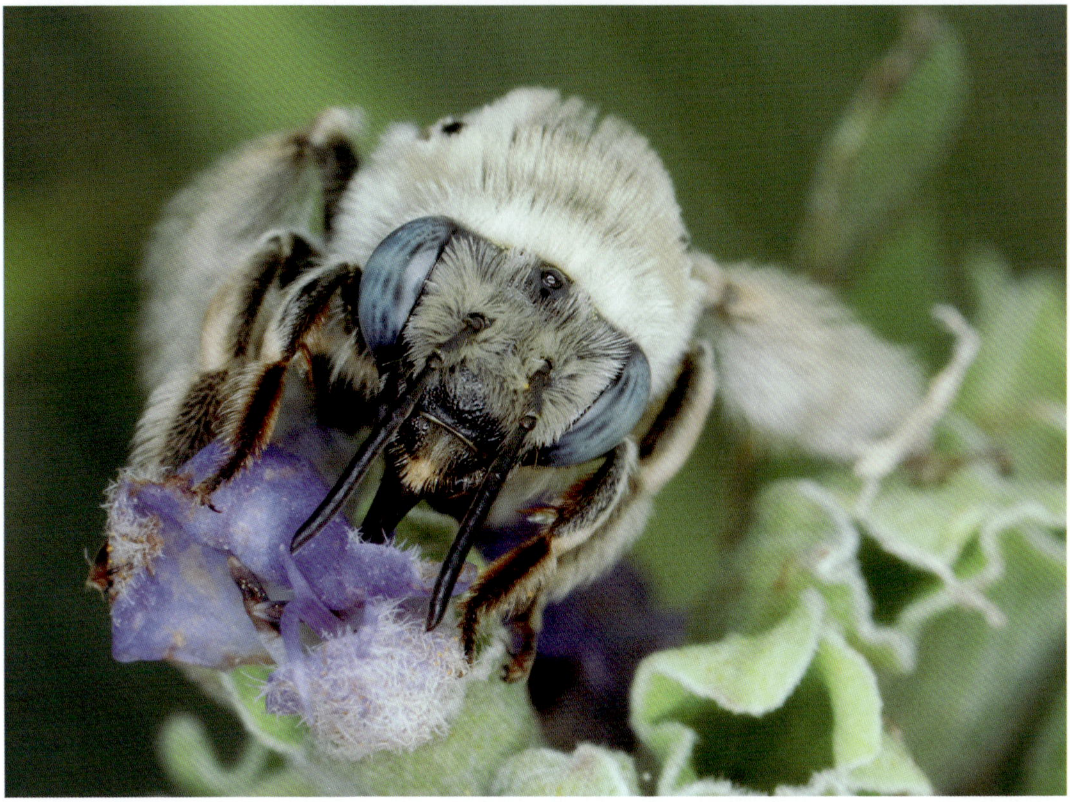

A wide-faced female *Melissodes tristis* drinking nectar from shrubby blue salvia

Females: Minute traits helpful in distinguishing female *Melissodes tristis* include the following: (1) the female's clypeus is unusually wide, with a pronounced ridge on the bottom edge; (2) the rim of the first abdominal segment (T1) is colorless and glassy; and (3) the hairs on the inner surfaces of the hind basitarsi are dark.

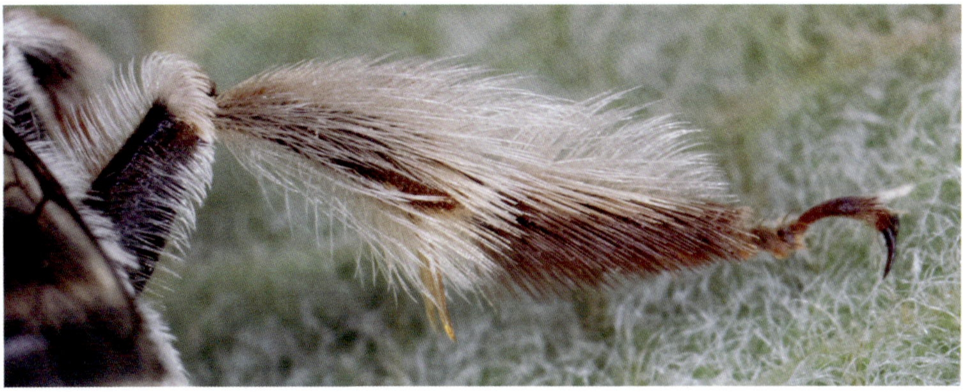

Hind leg of a female *Melissodes tristis*: note the dark hair on the inner basitarsus.

Exomalopsine Bees

Subfamily Eucerinae ◆ Tribe Exomalopsini
Genera *Anthophorula* and *Exomalopsis*

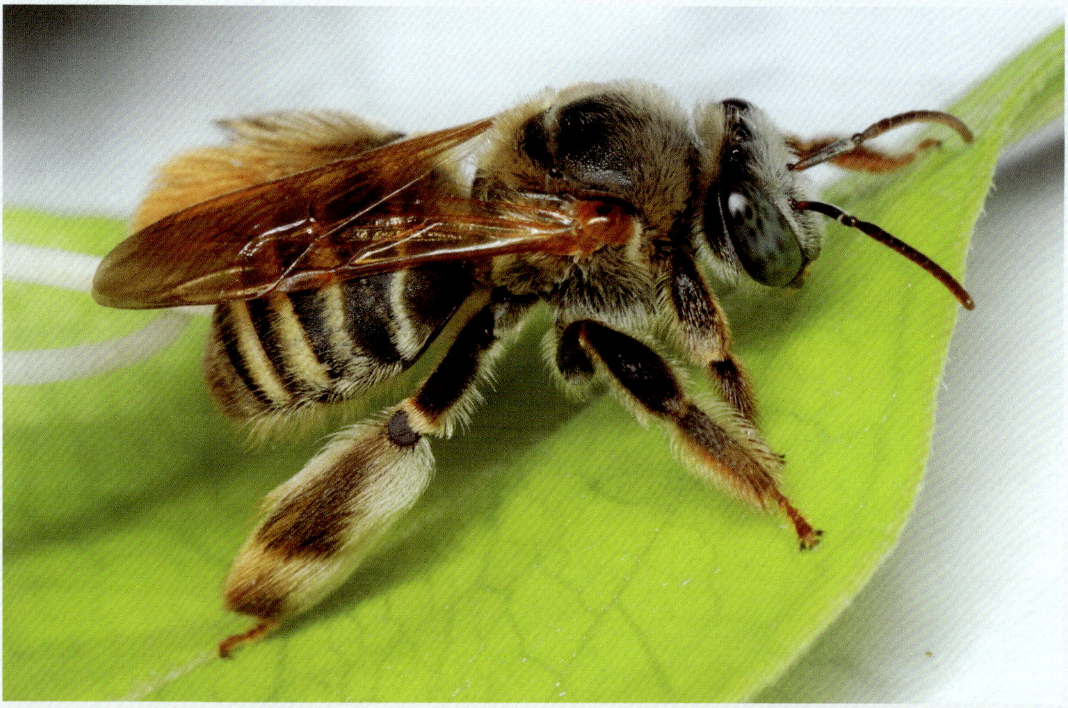

A female honey-footed exomalopsis

Anthophorula and *Exomalopsis* belong to the tribe Exomalopsini of the family Eucerinae. They are found only in the New World, predominantly in subtropical and tropical areas. In the United States, both *Anthophorula* and *Exomalopsis* are relatively uncommon, although *Anthophorula* range from California to New York and *Exomalopsis* occur in much of the West and Southwest as well as Florida, Georgia, and Texas.

Anthophorula and *Exomalopsis* are small but striking bees: their abdomens are dark and ringed with intricate patterns of pale or golden hair bands; their legs are covered with rust-orange or multicolored hairs; and their eyes are brilliant green or blue gray. Females have luxuriant scopal hairs on their hind legs evocative of the groomed fur of champion show dogs. Males have relatively long antennae.

Anthophorula and *Exomalopsis* are now considered two distinct genera but at one time were grouped together in a single genus. They are generally similar in appearance. Bees of both genera share a notable trait: they have a long row of erect, well-separated hairs lining the inside edge of each compound eye orbit.

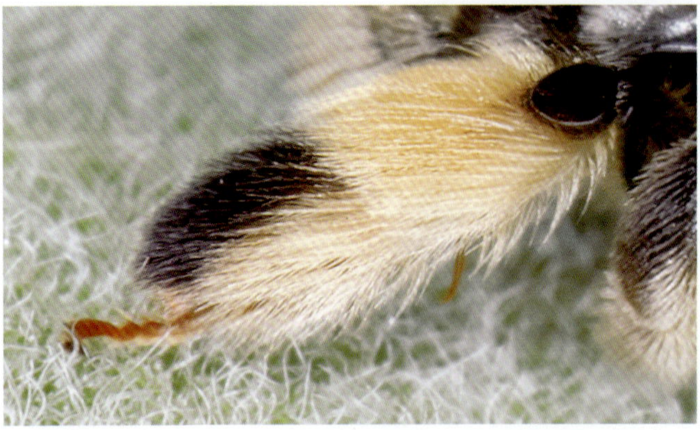

Close-up of the copious hind-leg scopal hairs and large knee plate of a female *Exomalopsis*: *Anthophorula* females have similarly lush scopal hairs, but their knee plates are comparatively smaller.

***Anthophorula* versus *Exomalopsis*:** *Exomalopsis* tend to be larger than *Anthophorula*. Michener noted that the following minute differences help separate the two genera: (1) the hind knee plates are proportionately larger on *Exomalopsis* females than on *Anthophorula* females; (2) with some exceptions, the male *Anthophorula* has a yellow clypeus and labrum, while *Exomalopsis* males have dark clypei (females of both genera have dark faces); and (3) the stigma of the forewing is usually proportionately larger in *Exomalopsis*; on *Anthophorula*, the stigma tends to be less than half the length of the marginal cell.

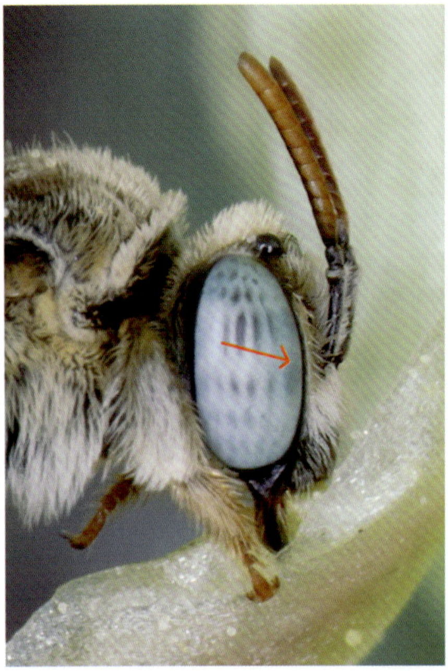

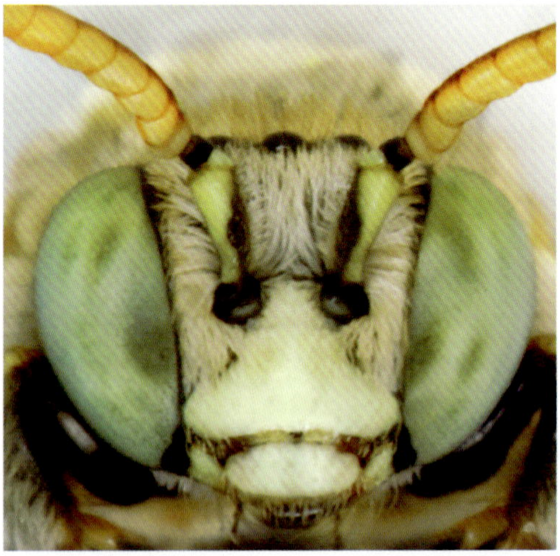

Left: An *Anthophorula* face with a row of erect, well-separated hairs lining the inner compound eye. This trait is shared by *Exomalopsis*. *Right*: The pale yellow clypeus and labrum of a male *Anthophorula*.

Individual *Anthophorula* and *Exomalopsis* species: Ten species of *Exomalopsis* inhabit the United States. At least seven have been documented in Texas: *Exomalopsis analis*, *E. birkmanni*, *E. mellipes*, *E. snowi*, *E. solani*, *E. solidaginis*, and *E. tibialis*. All of these except *E. solidaginis* have been found in the Lower Rio Grande Valley.

There are approximately 40 species of *Anthophorula* in the United States. Most of these inhabit California or the Southwest, although one species, *Anthophorula pygmaea*, occurs throughout most of the country. Fourteen *Anthophorula* species have been found in Texas; of these, *A. compactula* is by far the most common in the Valley.

Identifying *Exomalopsis* and *Anthophorula* species in the field can be challenging because the bees are small and differences among them subtle. Factors considered in diagnosing species include the time of year when a bee emerges; its floral preferences; the smoothness, pitting, and hair patterns of its abdominal segments; and the color of the hairs on the legs and lining the scutellum. Explanations of the distinguishing traits of seven Lower Rio Grande Valley *Exomalopsis* and *Anthophorula* species are provided in the next pages.

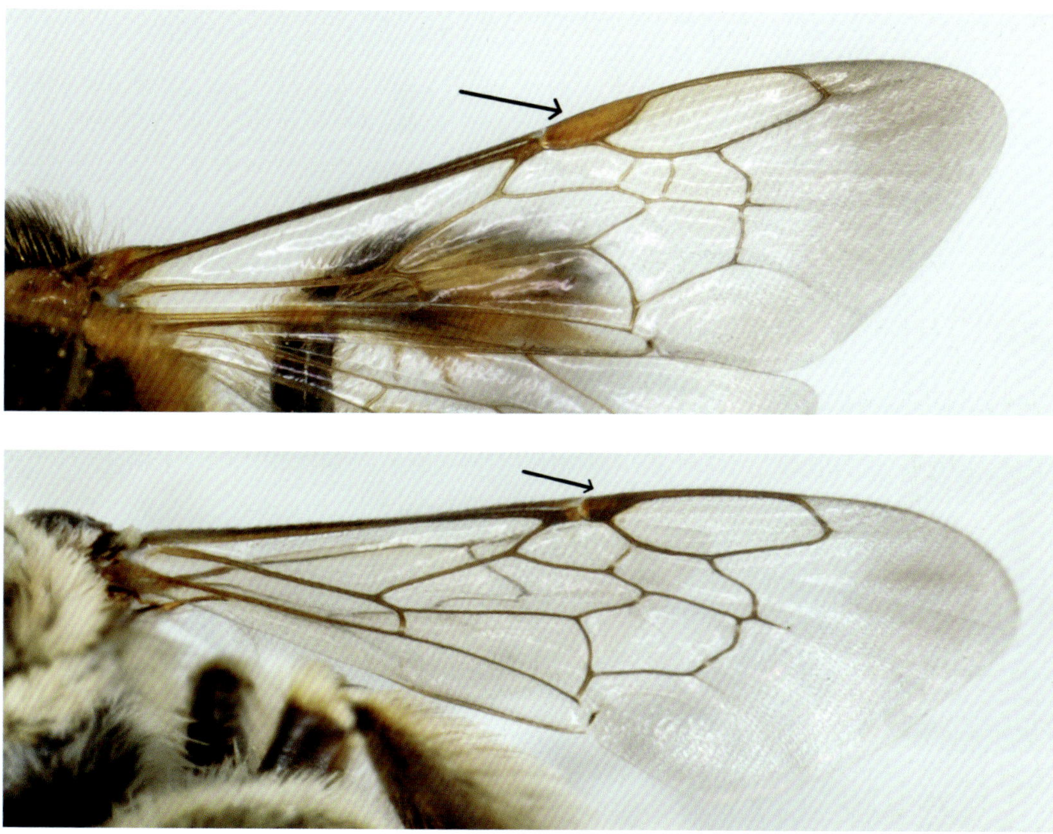

Forewings of an *Exomalopsis* (*above*) and an *Anthophorula* (*below*): the stigma of the *Exomalopsis* wing is proportionately larger. Both genera may have forewings with two or three submarginal cells.

Behavior: The nesting habits of most *Exomalopsis* and *Anthophorula* species remain unknown. Those that have been studied nest in the ground and are mostly communal, sharing tunnels and nest entrances while provisioning their own egg chambers. Female *Exomalopsis* and *Anthophorula* dig out oval egg chambers and coat them with a thin waterproof lining. Females store pollen for offspring on a "foot," a small mound of dirt slightly elevated above the nest floor, to protect food provisions against moisture.

Anthophorula may be oligolectic or polylectic. A few *Anthophorula* species specialize in pollinating such plants as gumweed and leafy spurge or exhibit a preference for aster-family flowers or buckwheat. In the Valley, *Anthophorula* are closely associated with cacti, while *Exomalopsis* are most likely to be found on small-flowered plants and nightshades. All *Exomalopsis* are generalist pollinators.

Compact anthophorula ◆ *Anthophorula (Anthophorula) compactula* Cockerell, 1897

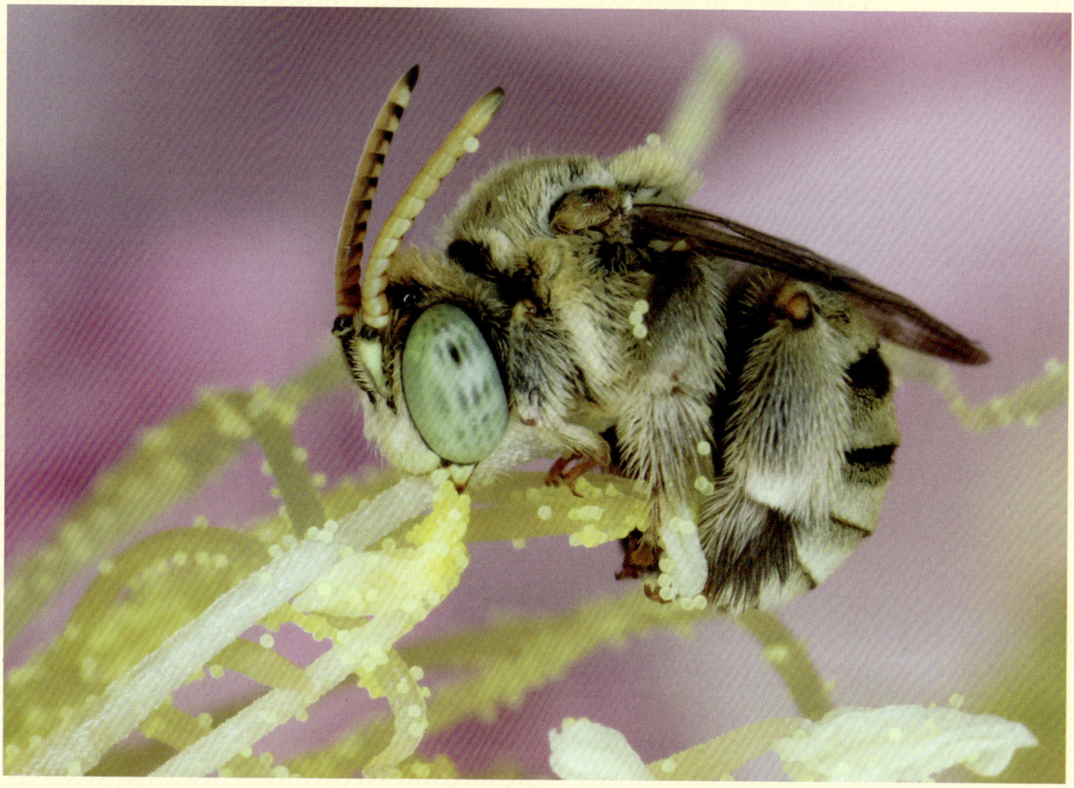

A male *Anthophorula compactula* on twisted-rib cactus

Description: The compact anthophorula is a robustly built, furry-looking bee with wide, pale hair bands on its abdomen. As its name implies, *Anthophorula compactula* is small—it fits easily on a dime. Males have yellow facial markings, bright green eyes, and long antennae. Females have dark faces covered with pale hairs, light blue eyes, and relatively short antennae. **Size:** females 7 mm (1/4 in.), males 4.5 mm (1/6–1/5 in.).

Range: *Anthophorula compactula* appears in the Valley in early April, inhabits central and northern Mexico, and is found throughout Texas and the southwestern United States. It ranges as far west as southeastern California. *Anthophorula compactula* is common from early April through mid-June in Hidalgo and Starr Counties.

Associated plants: Although broadly polylectic, this species emerges in the Valley during the spring cactus bloom and is most likely to be found there on Texas prickly pear (*Opuntia engelmannii* var. *lindheimeri*) and twisted-rib cactus (*Thelocactus setispinus*). After the cactus bloom, males and females visit diverse flora, including whitebrush (*Aloysia gratissima*), Texas paloverde (*Parkinsonia texana*), bracted fanpetals (*Sida ciliaris*), and Tiny Tim (*Thymophylla tenuiloba*).

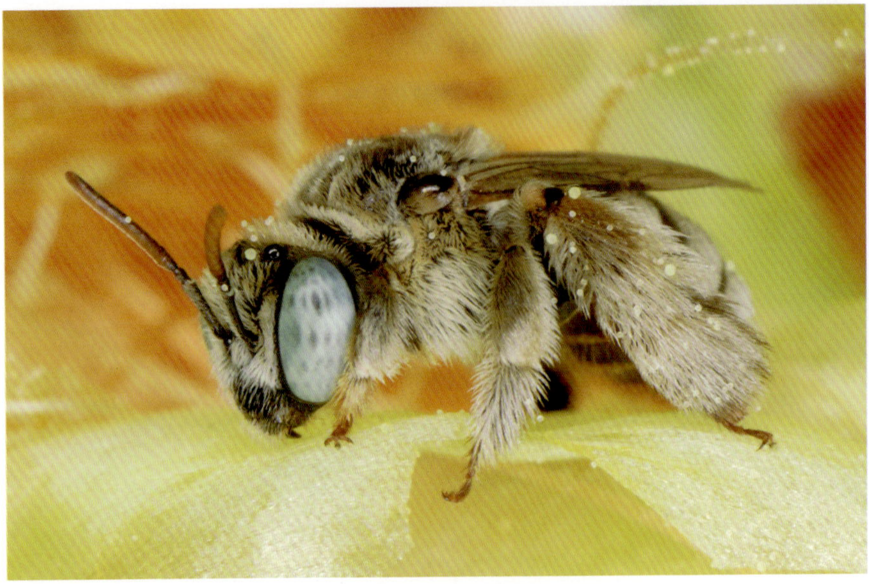

A female *Anthophorula compactula* on twisted-rib cactus

ID tips: The compact anthophorula is easily distinguished in the field from other cactus pollinators by the combination of its diminutive size, broadly striped abdomen, hairy legs, and vibrant green or light blue eyes. The male bee has a distinctive yellow-and-black pattern on its antennal scapes and yellow-and-black striping on the underside of each flagellum. *Anthophorula compactula* forewings have two submarginal cells, a trait that distinguishes them from all other Texas *Anthophorula*.

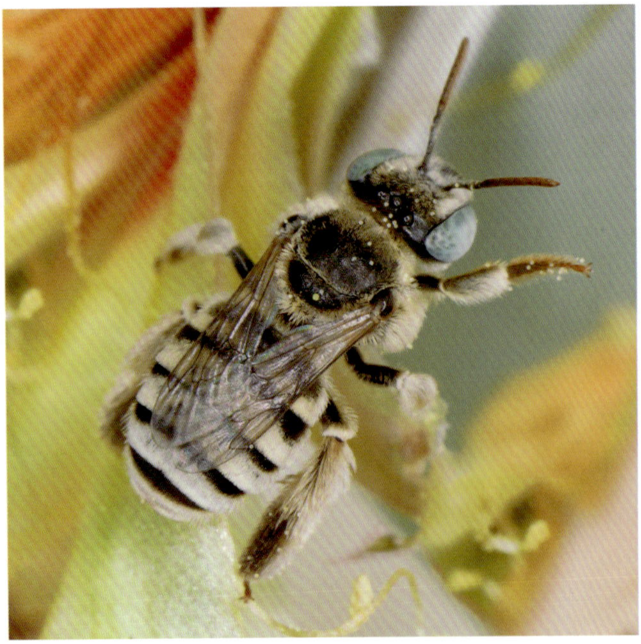

Dorsal view of a female *Anthophorula compactula*

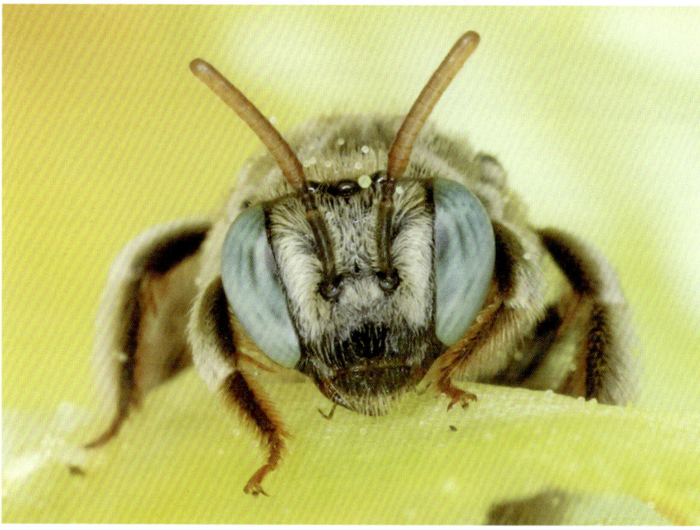

Face of a female *Anthophorula compactula*

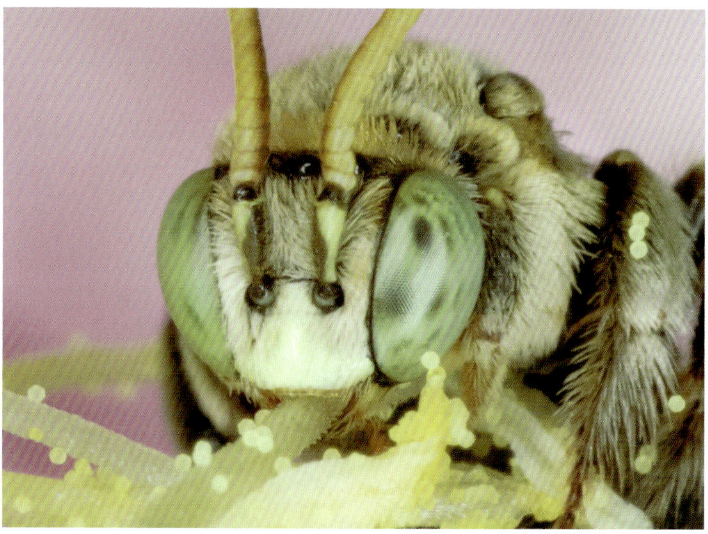

Face of a male *Anthophorula compactula*

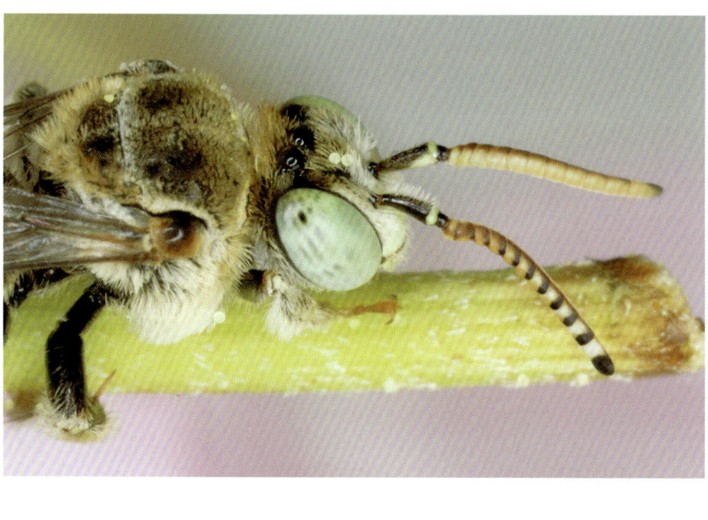

Striped antennae of a male *Anthophorula compactula*

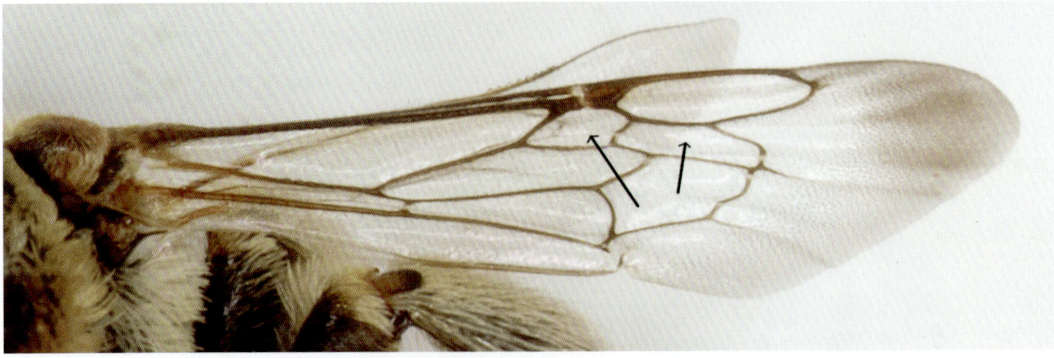

The two submarginal cells of an *Anthophorula compactula* forewing

Behavior: True to their genus, *Anthophorula compactula* are ground nesters that waterproof their nests and store pollen for offspring on an elevated "foot" within each chamber. During the spring 2019 cactus bloom in Hidalgo County, female compact anthophorula commonly appeared in small nest aggregations grouped around low-growing twisted-rib cactus. Many of the females built within a foot of the plants from which they gathered pollen. Shortly before and during the cactus bloom, male *Anthophorula compactula* could be found resting nearby inside red prickly poppies (*Argemone sanguinea*), often sharing flowers with male *Diadasia rinconis.* Male *Anthophorula compactula* occupied themselves by patrolling the ground in search of females' nests, or by visiting cactus blossoms frequented by the female bees.

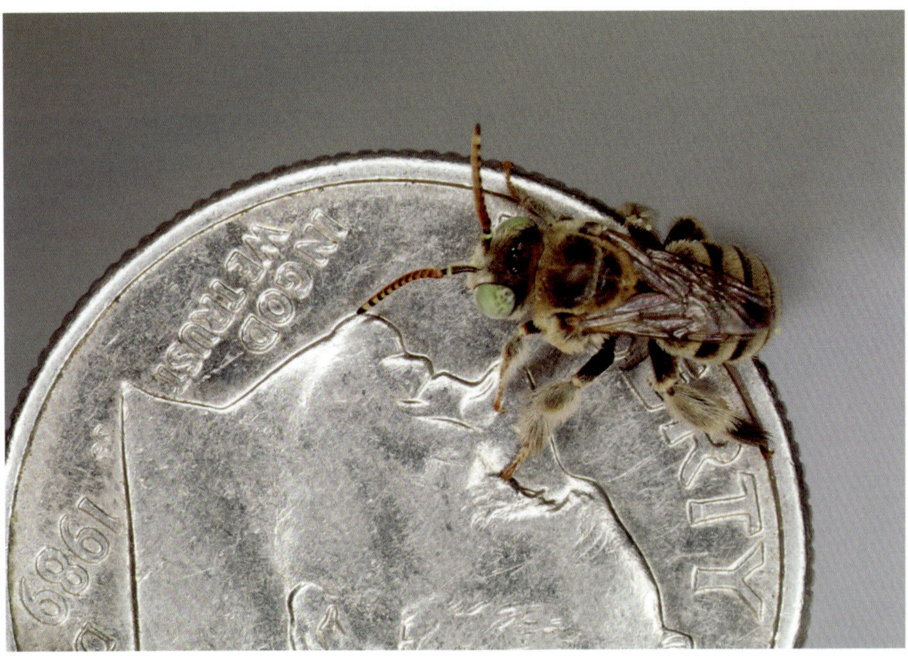

A male *Anthophorula compactula* resting on a dime: *Anthophorula* are generally smaller than *Exomalopsis*.

Snow's exomalopsis ◆ *Exomalopsis (Phanomalopsis) snowi* Cockerell, 1906

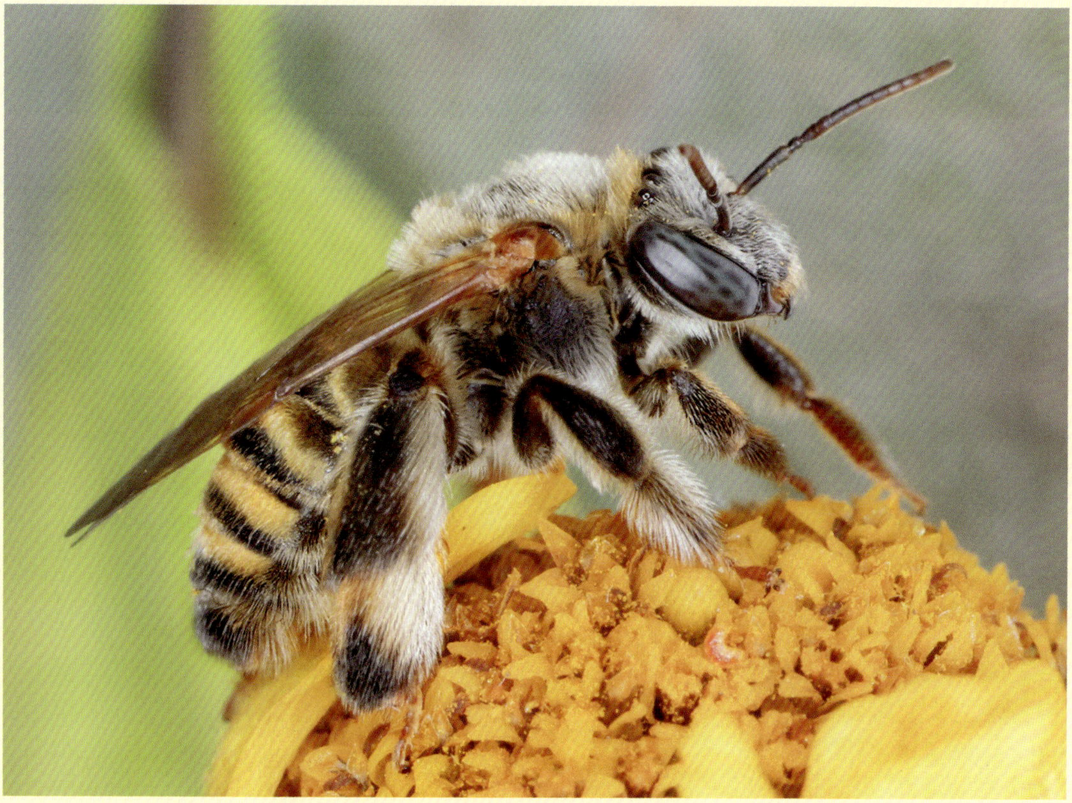

A female *Exomalopsis snowi*

Description: Snow's exomalopsis is a little bee with gold bands of hair on its abdomen. Both male and female *Exomalopsis snowi* have reddish-orange tegulae. Females have bushy scopal hairs that are pale and black on the outer surfaces of the hind legs and orange on the inner surfaces. Males are more narrowly built than females, with legs and faces that are covered with white hairs. The eyes of male and female bees are blue gray. **Size:** females 8 mm (3/10 in.), males 7 mm (1/4 in.).

Range: In the United States, *Exomalopsis snowi* is found primarily along the southern Texas border. The range of this species extends southward through Mexico and Central America to Venezuela. Snow's exomalopsis is relatively uncommon in much of the Valley. It appeared at the National Butterfly Center (Hidalgo Co.) in November 2018 and 2021.

Associated plants: *Exomalopsis snowi* feeds on plants from diverse families. At the National Butterfly Center, females and males were found foraging on whitebrush (*Aloysia gratissima*), Texas ebony (*Ebenopsis ebano*), crucita (*Chromolaena odorata*), seaside goldenrod (*Solidago sempervirens*), and romerillo (*Bidens alba*).

ID tips: Snow's exomalopsis closely resembles the honey-footed exomalopsis (*Exomalopsis mellipes*): both are singular in having reddish tegulae, a trait visible to the naked eye. The pale bands on *Exomalopsis snowi*, however, are distinctly golden, while those on *E. mellipes* are off-white. The ornate coloring of the hind-leg scopal hairs of these two *Exomalopsis* species additionally aids in identifying females. As shown here, the hairs on the inner back legs of the female *Exomalopsis snowi* are predominantly orange. An additional definitive trait of *Exomalopsis snowi* females is visible only with magnification: the surface of the scutum and of the top of the first abdominal segment (T1) are shiny, with few punctures. The scutum and T1 of the female *Exomalopsis mellipes*, by contrast, are densely pitted.

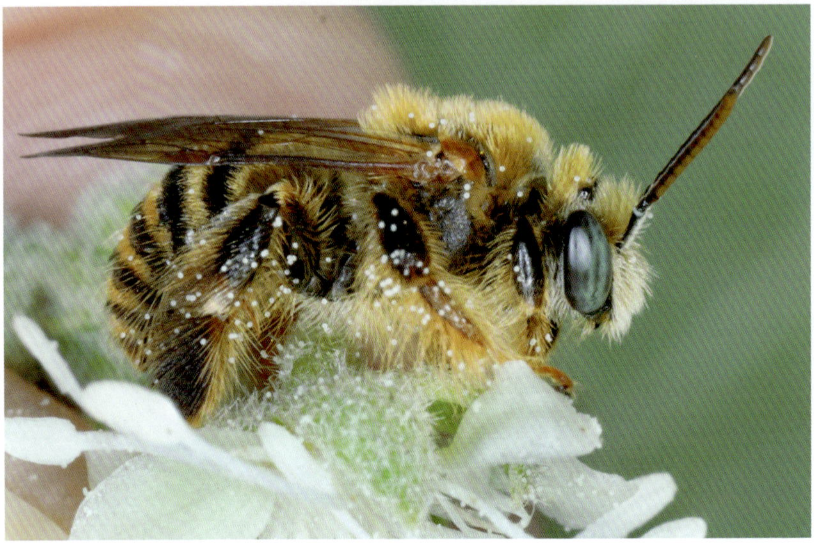

A male *Exomalopsis snowi*

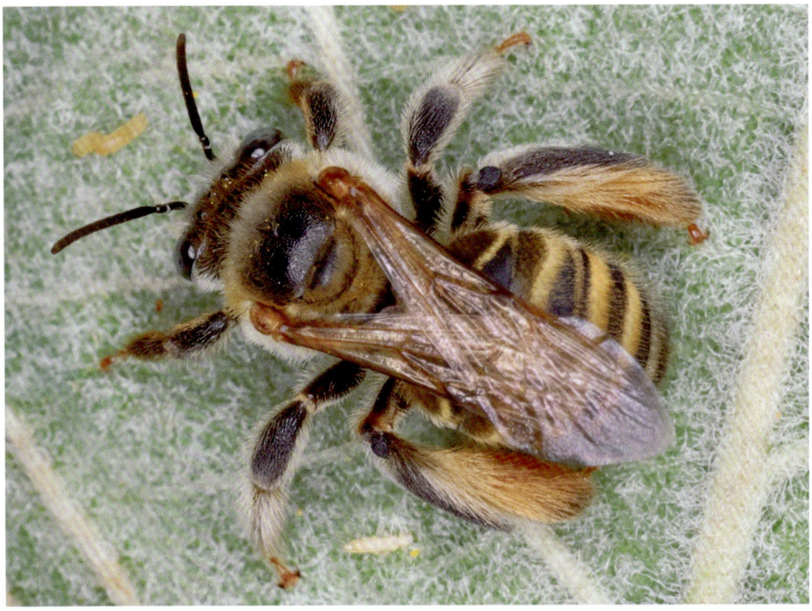

A female *Exomalopsis snowi*, viewed from above. The hind-leg scopal hairs are mostly orange on the inner tibiae. Note the reddish tegulae.

Honey-footed exomalopsis ◆ *Exomalopsis (Exomalopsis) mellipes* Cresson, 1878

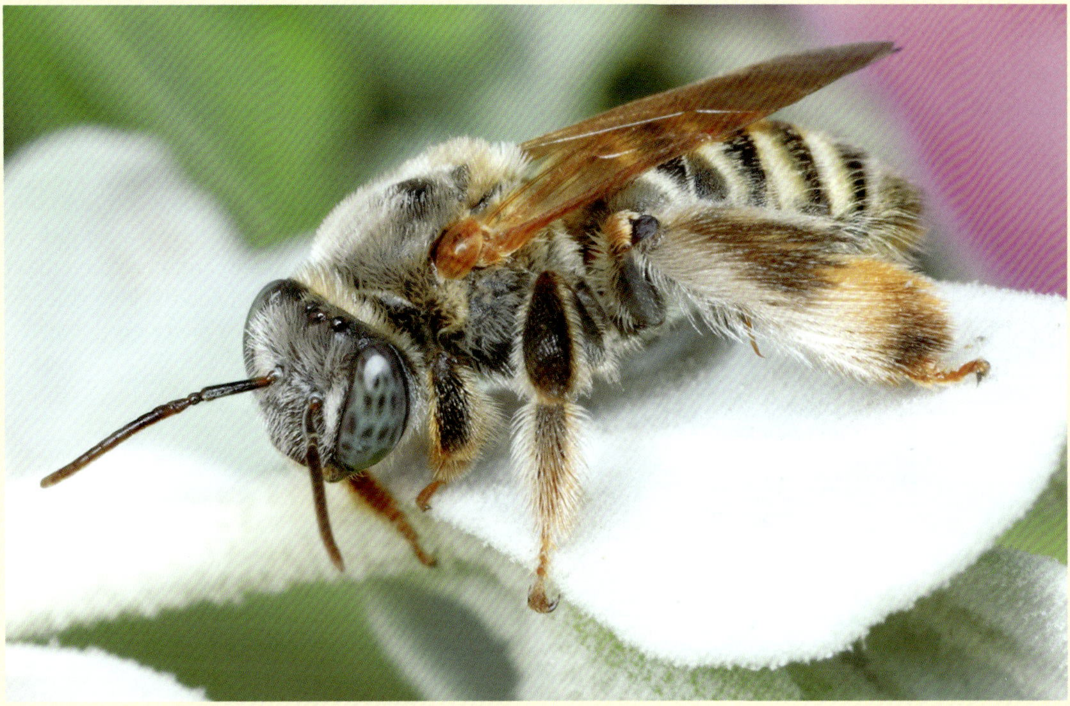

A female *Exomalopsis mellipes*

Description: The honey-footed exomalopsis is a medium-small bee with ivory bands of hair on its abdomen. Females have bushy hind-leg scopal hairs that are predominantly black and ivory on the tibiae (with some orange hairs on the outer surfaces). Males' legs are covered with copious golden hairs. The tegulae of both females and males are reddish orange. Males have red legs and partly red abdomens. The scapes and front surfaces of males' antennae are also red. Females have gray-blue eyes, and males have green eyes. **Size:** females 10 mm (2/5 in.), males 8–9 mm (3/10–1/3 in.).

Range: In the United States, *Exomalopsis mellipes* appears primarily along the Texas border. It occurs in Cameron, Hidalgo, and Starr Counties, and its range extends southward through Mexico and Central America to Panama.

Associated plants: *Exomalopsis mellipes* is found from May through November in the Valley. It forages on plants as diverse as buffalobur nightshade (*Solanum rostratum*), crucita (*Chromolaena odorata*), Texas ebony (*Ebenopsis ebano*), old man's beard (*Clematis drummondii*), three-lobed false mallow (*Malvastrum coromandelianum*), seaside goldenrod (*Solidago sempervirens*), Texas crownbeard (*Verbesina microptera*), whitebrush (*Aloysia gratissima*), Rio Grande bee brush (*Aloysia macrostachya*), cenizo (*Leucophyllum frutescens*), and shrubby blue salvia (*Salvia ballotiflora*).

ID tips: The male *Exomalopsis mellipes* is distinctive because of its overall reddish appearance and the luxurious golden hairs on its head and legs. As noted in the previous entry, the female *Exomalopsis mellipes* closely resembles the female *E. snowi*. The female honey-footed exomalopsis, however, tends to run larger than Snow's exomalopsis, and, as noted, the former's abdominal hair bands are ivory rather than golden. The scopal hairs on the females' hind tibiae also differ, as shown here: on *Exomalopsis mellipes*, they are black and pale, rather than orange.

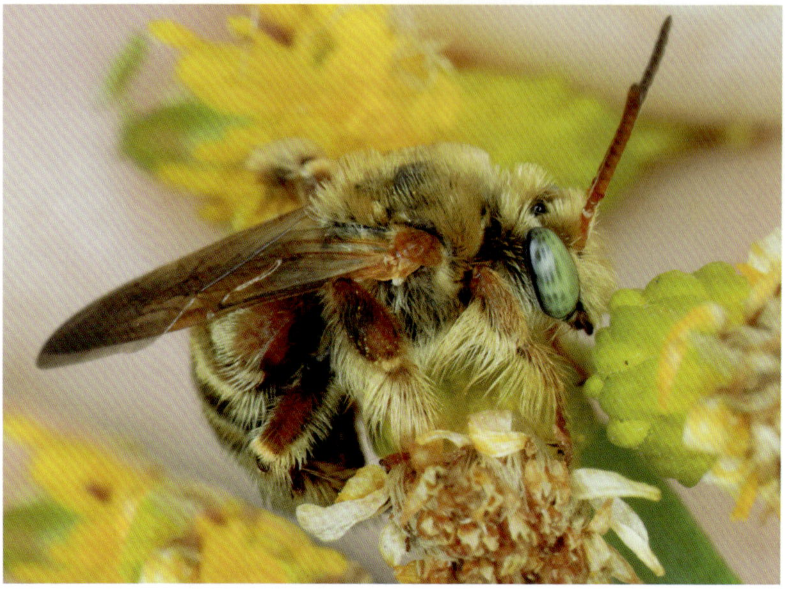

A male *Exomalopsis mellipes*: note the reddish legs, antennae, and tegulae, and the long golden hairs on its legs.

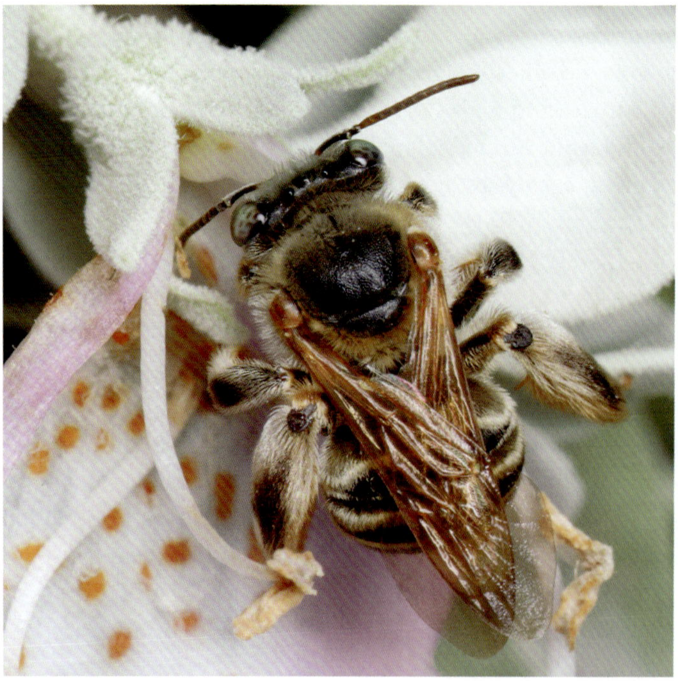

A female *Exomalopsis mellipes*, viewed from above: note that that the scopal hairs on the inner hind tibiae are predominantly black and pale. The top surface of the thorax is relatively hairless, a trait common to many *Exomalopsis*.

Ringed exomalopsis ◆ *Exomalopsis (Exomalopsis) analis* Spinola, 1853

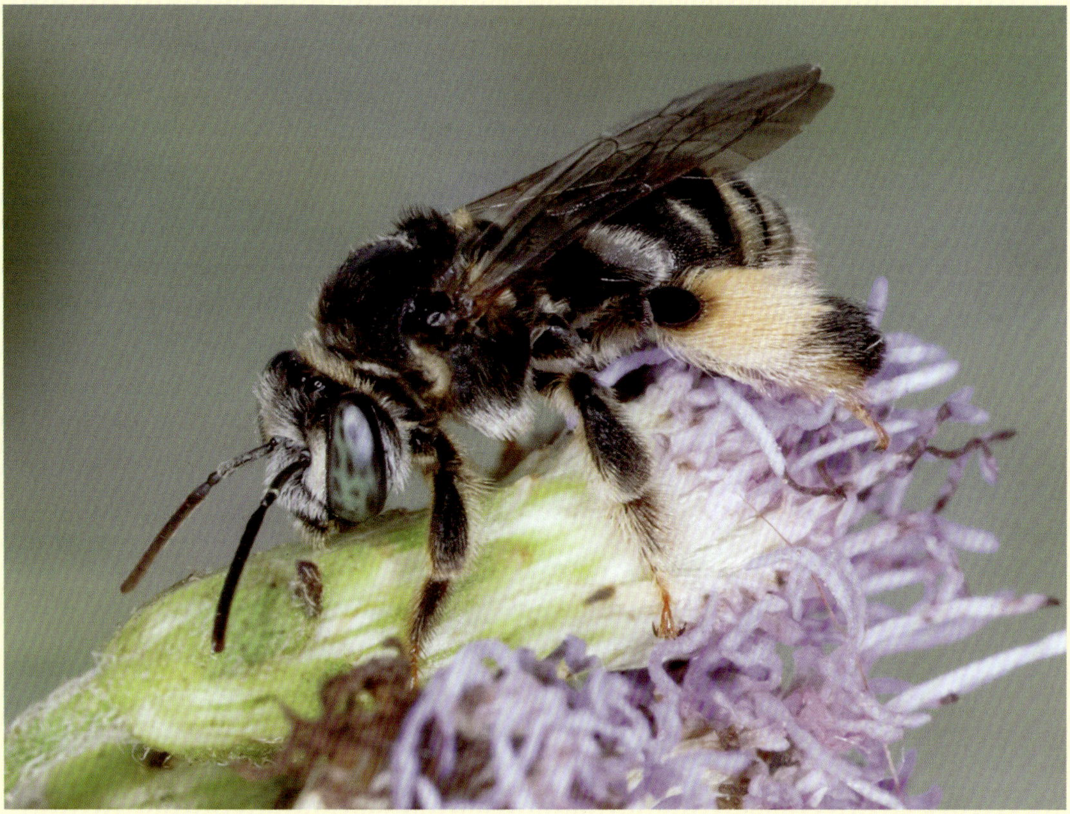

A female *Exomalopsis analis*

Description: The ringed exomalopsis is a small bee with a dark head and thorax, and a dark abdomen striped with pale hairs. The bee's tegulae are black, and its wings brown with dark brown veins. The female's hind-leg scopae are black and pale on the outer surfaces, and orange on the inner surfaces. Males have pale hairs on all legs. Both males and females have dark green eyes. **Size:** females and males 8–9 mm (3/10–1/3 in.).

Range: In the United States, *Exomalopsis analis* occurs principally in the Lower Rio Grande Valley. This species ranges though Mexico, Central America, the Caribbean, and South America as far south as Argentina.

Associated plants: *Exomalopsis analis* appears from May through November in Hidalgo County, feeding on diverse plants, among them crucita (*Chromolaena odorata*), scorpion's tail (*Heliotropium angiospermum*), whitebrush (*Aloysia gratissima*), and Rio Grande bee brush (*Aloysia macrostachya*). Females also collect pollen from silverleaf nightshade (*Solanum elaeagnifolium*).

ID tips: The hairs on the vertex and scutellum of *Exomalopsis analis* are dark, and the bee looks much blacker generally to the naked eye than the other *Exomalopsis* species in this guide. As shown here, the hair colors of the female's scopae and basitarsi are distinctive. The female's abdomen also has a singular pattern of alternating dark hairless patches and pale hair bands (shown at the end of the *Exomalopsis* section).

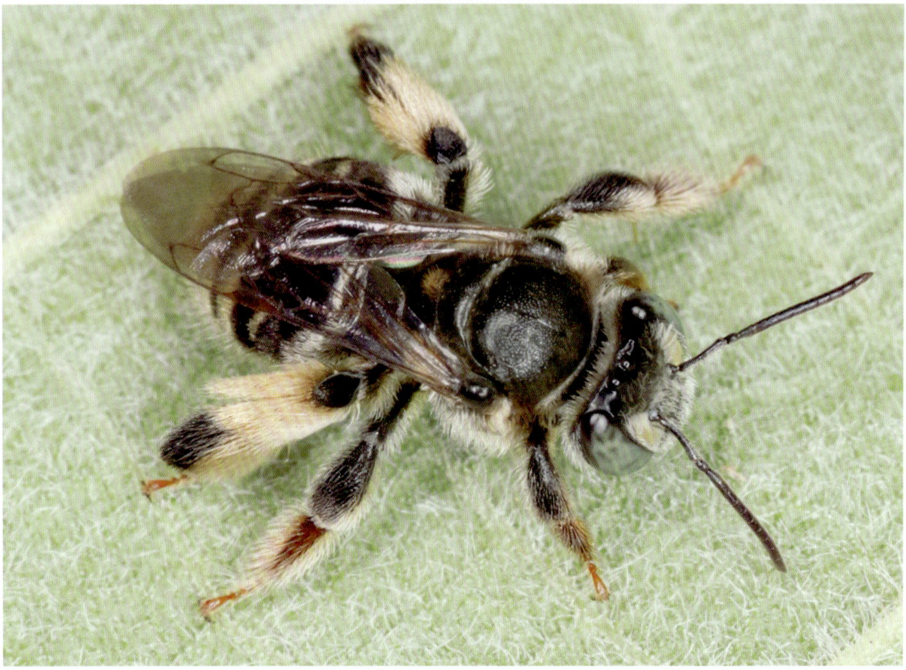

A female *Exomalopsis analis*

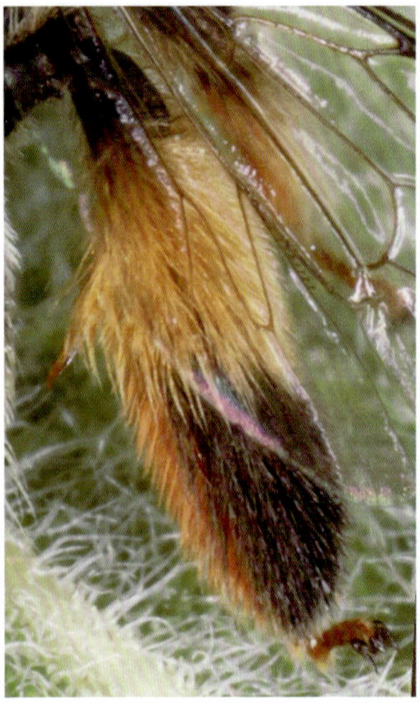

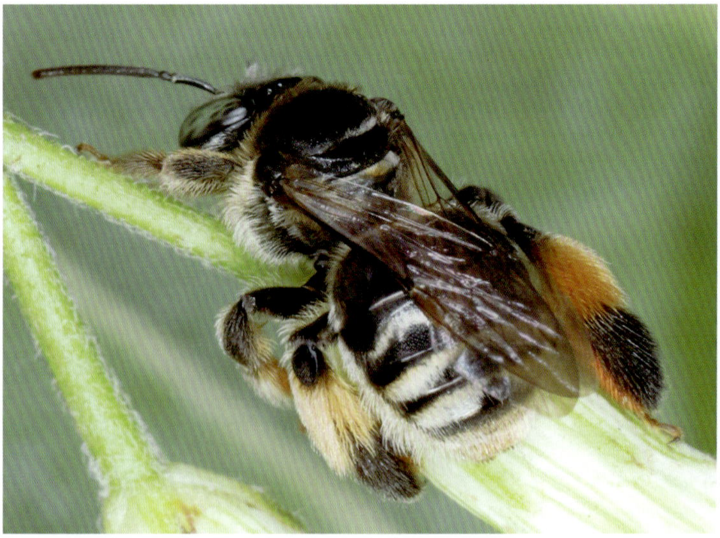

The inner hind-leg scopal hairs of the female *Exomalopsis analis* are black and rust colored. A strip of deep orange hair lines the inner edge of the hind basitarsus.

Similar exomalopsis ◆ *Exomalopsis (Exomalopsis) similis* Cresson, 1865

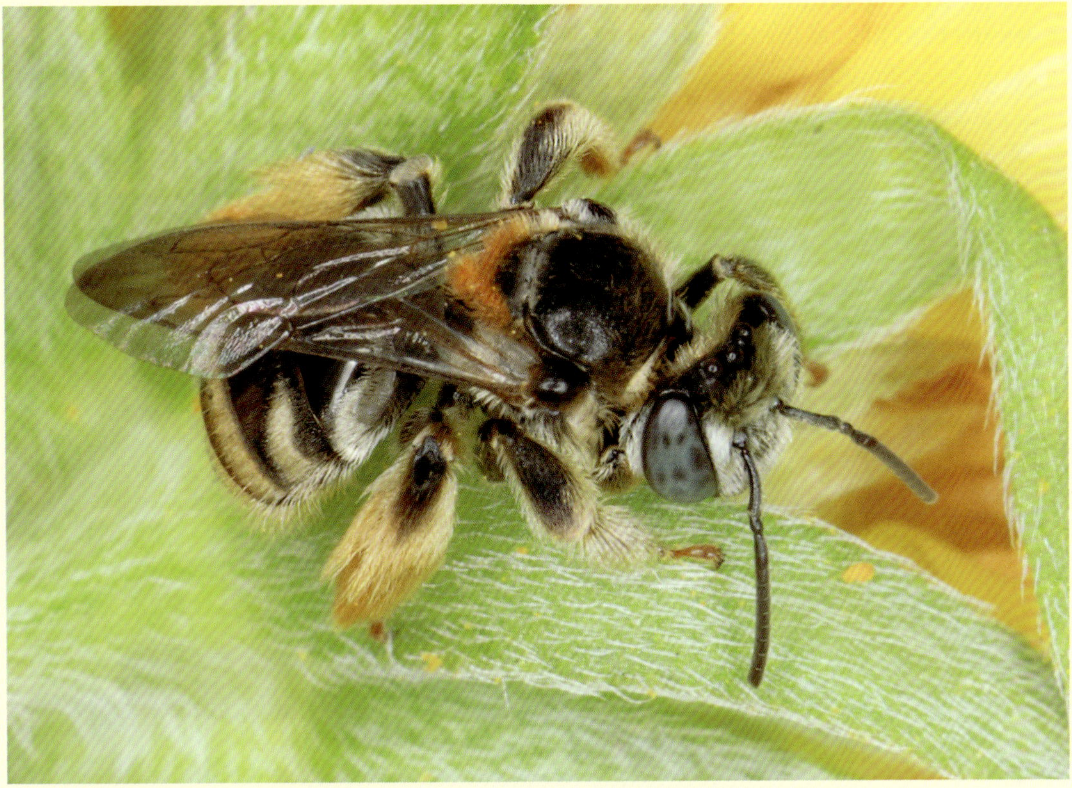

A female *Exomalopsis similis*

Description: The similar exomalopsis is a small to very small, dark bee with a rust-orange tuft of hair rimming the back edge of its scutellum, and pale orange hairs banding its dark abdomen. The tegulae are black, and the wings brown with dark brown veins. The outer scopal hairs on females' back legs are predominantly black and pale; the inner hairs are pale to deep orange. On males, leg hairs are mostly pale. Females have blue-gray eyes and males have green eyes. **Size:** females 6–7 mm (1/4 in.), males 4–5 mm (1/6–1/5 in.).

Range: In the United States, *Exomalopsis similis* occurs principally in the Lower Rio Grande Valley. This species ranges southward through Mexico and Central America to Costa Rica. It is widespread in the Caribbean and has been found in Florida.

Associated plants: *Exomalopsis similis* appeared at the National Butterfly Center in October and November 2019 and 2021. Males and females were found feeding on plants of the aster family, among them seaside goldenrod (*Solidago sempervirens*), resinbush (*Viguiera stenoloba*), and romerillo (*Bidens alba*).

ID tips: *Exomalopsis similis* tends to run smaller than the other five Rio Grande *Exomalopsis* species in this guide. The bright tuft of orange hair on the rear thorax of the female *Exomalopsis similis* is visible to the naked eye and is the best way to recognize this species in the field. On males, the hairs behind the scutellum may be less brightly colored.

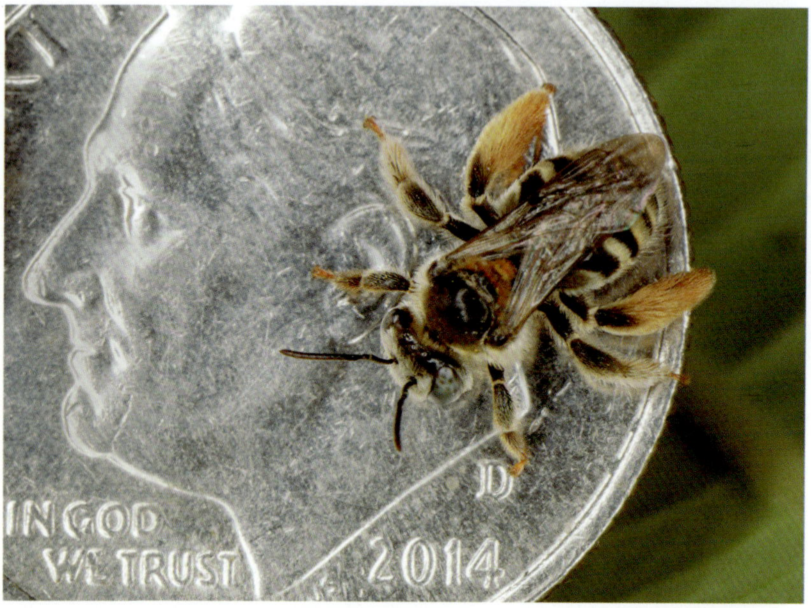

Exomalopsis similis is small for its genus. The female has predominantly orange scopal hairs on its inner hind legs and bright orange hairs rimming its rear thorax.

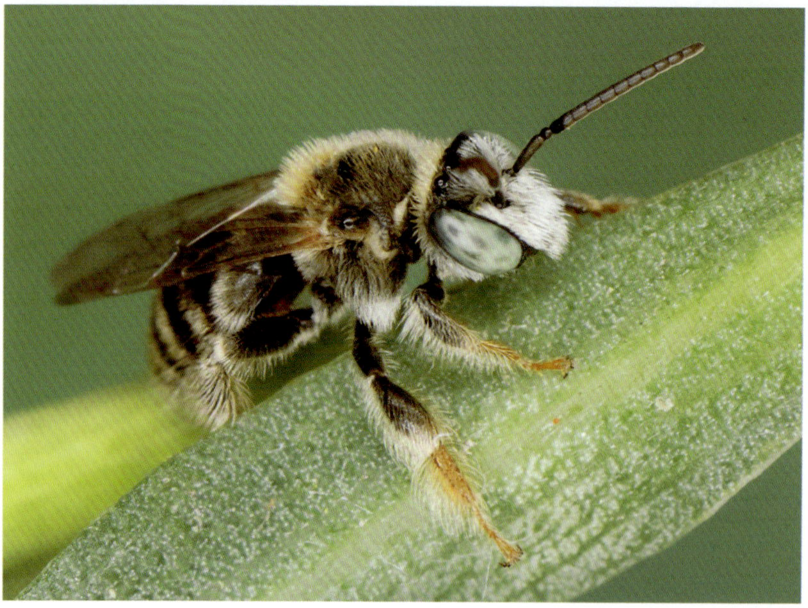

A male *Exomalopsis similis*: males have light hairs on all legs, and pale orange hairs on the thorax.

Birkmann's exomalopsis ◆ *Exomalopsis (Stilbomalopsis) birkmanni* Cockerell, 1922

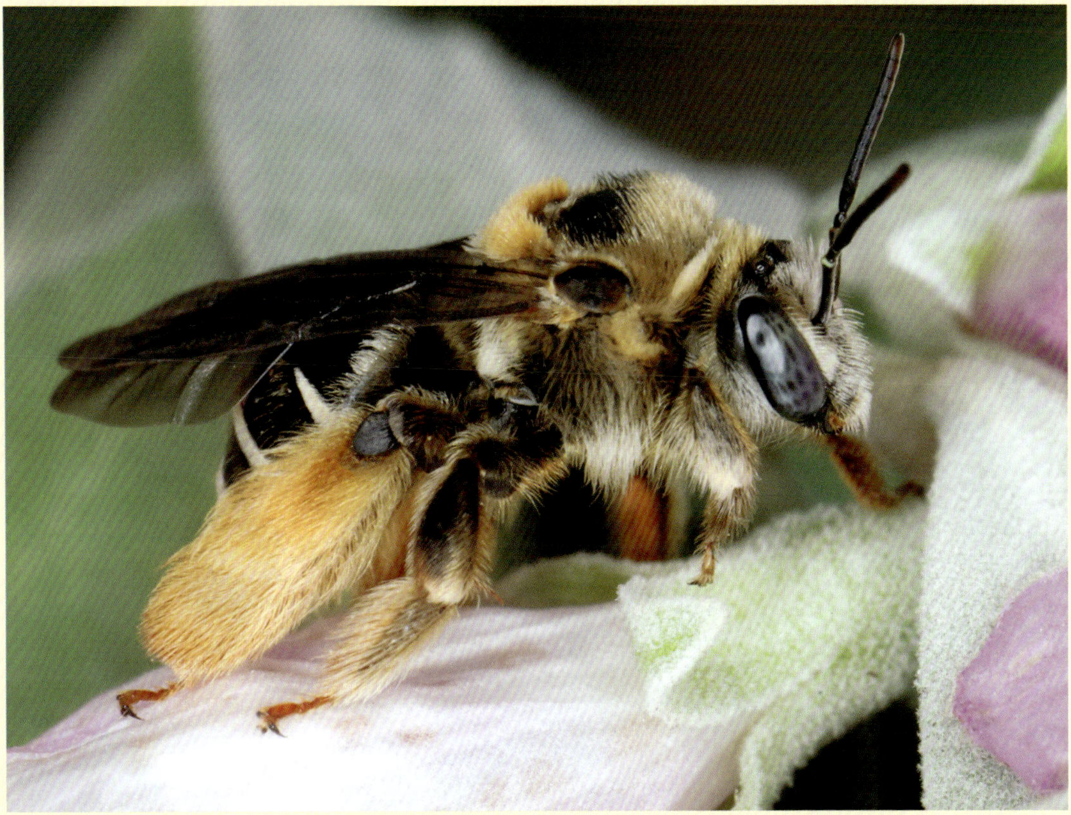

A female *Exomalopsis birkmanni*

Description: Birkmann's exomalopsis is a robust, medium-sized bee with a dark abdomen banded by white hairs. The female's hind-leg scopae are brilliant orange, and dense rust-orange hairs line the back rim of its scutellum. Orange hairs also form a fringe behind the female's ocelli. Pale and dark hairs partly cover the female's scutum. The hairs on the male's thorax, vertex, and legs are a muted pale orange. Females have gray-blue eyes, and males have greenish eyes. **Size:** females 11 mm (2/5 in.), males 9 mm (1/3 in.).

Range: In the United States, *Exomalopsis (Stilbomalopsis) birkmanni* is found principally in the Lower Rio Grande Valley. It has been documented at multiple locations in Hidalgo and Starr Counties during spring and fall. Its range extends into northern Mexico.

Associated plants: In the Valley, female *Exomalopsis birkmanni* commonly gather pollen from cenizo (*Leucophyllum frutescens*), silverleaf and buffalobur nightshade (*Solanum elaeagnifolium* and *S. rostratum*), and shrubby blue salvia (*Salvia ballotiflora*). Birkmann's exomalopsis appeared at Campos Viejos Ranch (Starr Co.) in June 2021 visiting the flowering shrubs brasil (*Condalia* *hookeri*) and coma (*Sideroxylon celastrinum*). This species also forages locally on Asteraceae such as hoary blackfoot (*Melampodium cinereum*), hierba del marrano (*Symphyotrichum subulatum*), coat buttons (*Tridax procumbens*), and Texas crownbeard (*Verbesina microptera*).

ID tips: *Exomalopsis birkmanni* females are fairly easy to identify because of their robust build and bright orange scopal hairs. They are nearly identical to female *Exomalopsis solani*, shown in the next entry, but in that species, the scopae and the hairs rimming the scutellum are much paler. Males of the two species are best differentiated by their proximity to the more distinctly colored females.

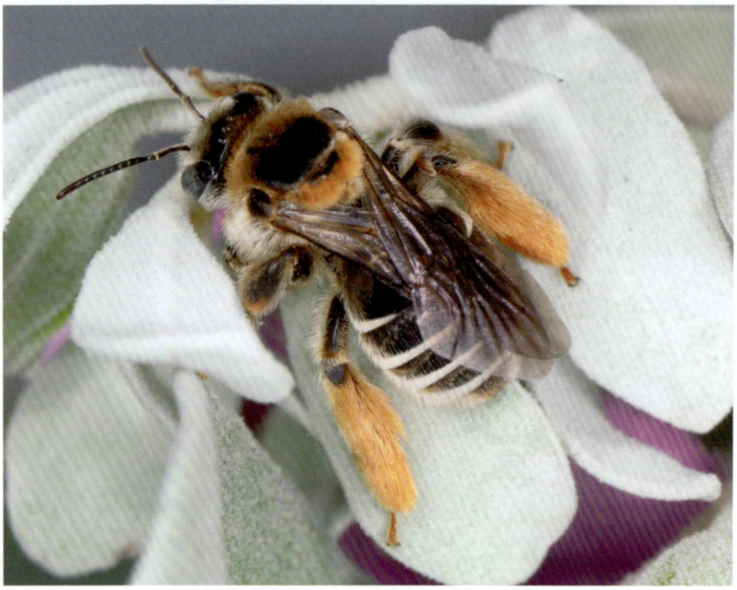

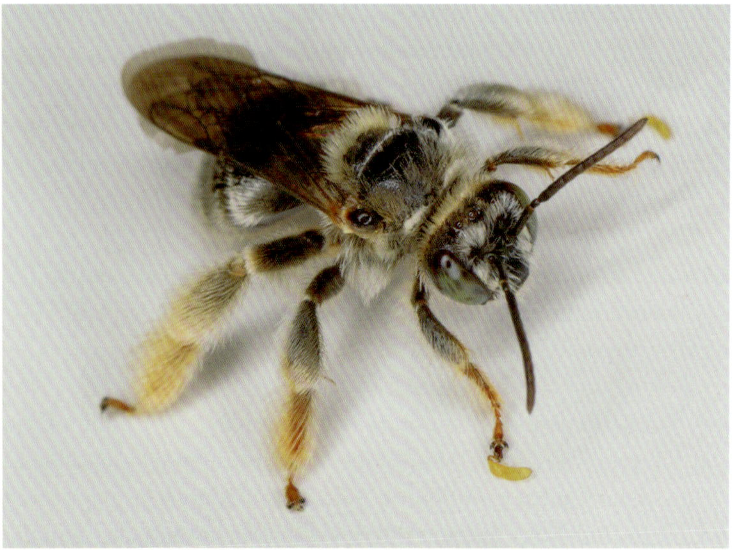

Female (*above*) and male (*below*) *Exomalopsis birkmanni*

Nightshade exomalopsis ◆ *Exomalopsis (Stilbomalopsis) solani* Cockerell, 1896

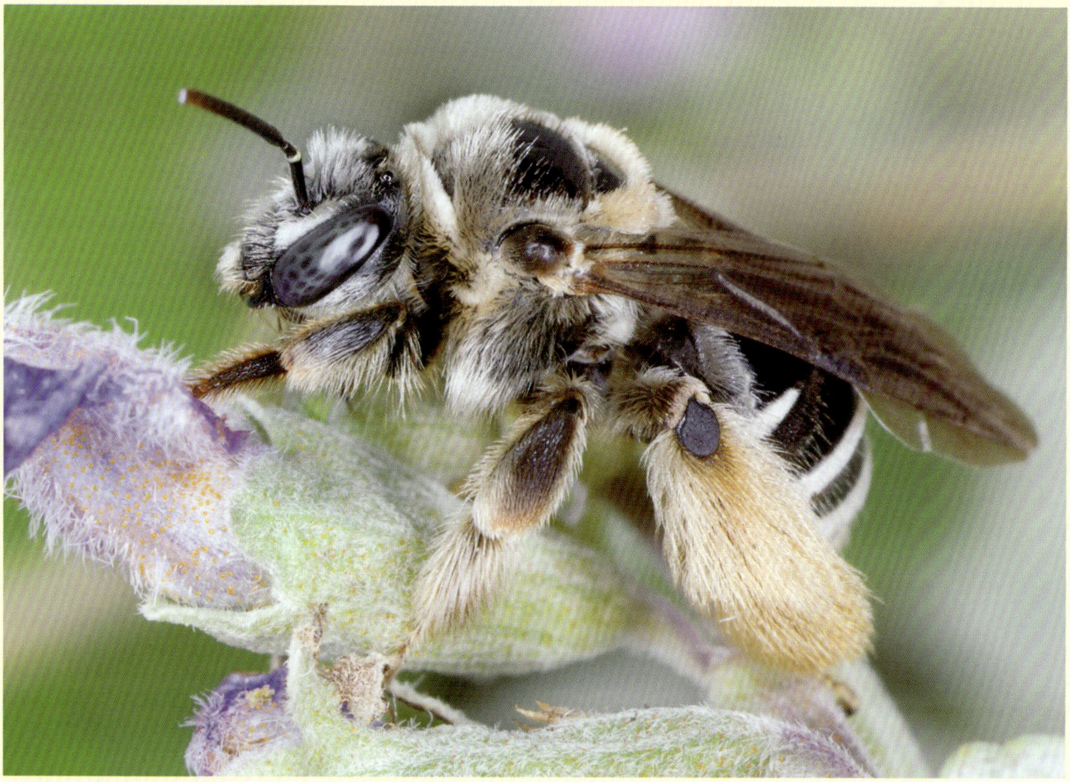

A female *Exomalopsis solani*

Description: The nightshade exomalopsis is a robust, medium-sized bee that resembles *Exomalopsis birkmanni*, described in the previous entry. *Exomalopsis solani*, however, lacks the bright orange hair color: its abdomen is banded by white hairs, and white hairs cover its face and much of its thorax. Females have pale orange hind-leg scopal hairs, and pale orange hairs on the rear scutellum. Hairs behind the female's vertex are mostly white. White and pale orange hairs clothe the male's legs. **Size:** females 10 mm (2/5 in.), males 9 mm (1/3 in.).

Range: *Exomalopsis solani* ranges through much of the southwestern United States from Texas to California, and north to Wyoming. It occurs in Mexico as far south as the state of Puebla.

Associated plants: *Exomalopsis solani* derives its name from its association with plants of the nightshade genus (*Solanum*). This species appeared in October 2019 and November 2021 at the National Butterfly Center (Hidalgo Co.), visiting silverleaf nightshade (*Solanum elaeagnifolium*), shrubby blue salvia (*Salvia ballotiflora*), and crucita (*Chromolaena odorata*).

ID tips: Cockerell first described *Exomalopsis solani* in 1896. In 1922, he described *Exomalopsis birkmanni* as a separate species. As noted, the female *Exomalopsis solani* can be separated from *E. birkmanni* by hair color. Documented specimens of Birkmann's exomalopsis females also run larger than those of *Exomalopsis solani*. Nonetheless, some debate exists over whether *Exomalopsis solani* and *E. birkmanni* are simply color variations of the same species.

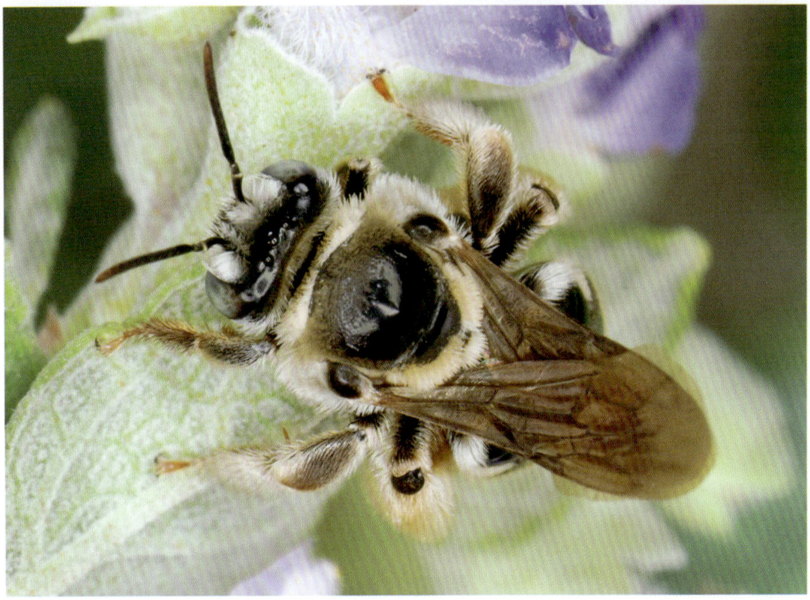

A female *Exomalopsis solani*: note the pale hairs rimming the hind edge of the scutellum.

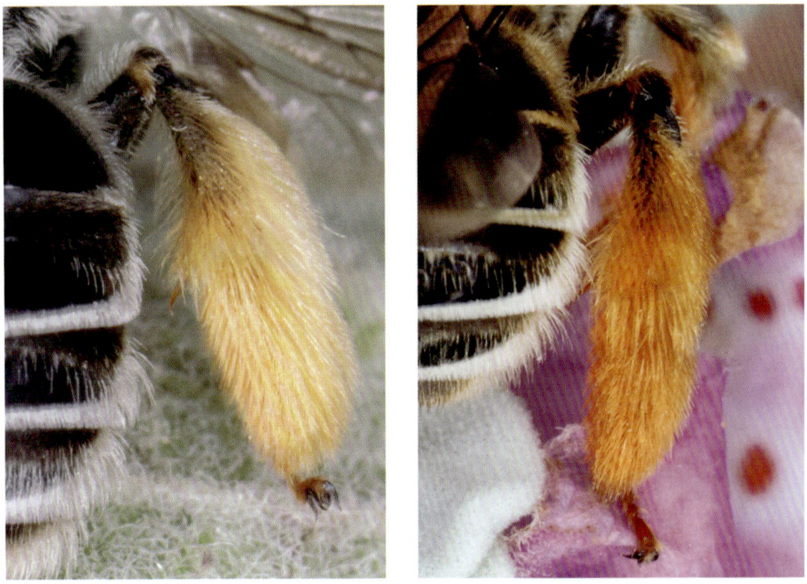

Comparison of scopal hair color of female *Exomalopsis solani* (*left*) and *E. birkmanni* (*right*)

Nesting behavior: According to Michener, among *Exomalopsis*, multiple occupancy of a single nest by several females is fairly common. In 1966, Michener described *Exomalopsis solani* females working together communally to ensure the health and safety of future offspring. Females too immature to develop eggs gathered pollen and brought it to the nest. Michener speculated that the pollen was cached in egg chambers so that mature females could then deposit eggs on top of it. Such cooperative provisioning of nests is relatively rare among solitary bee species: most construct nests that they modify and provision individually.

Additional information: Several *Exomalopsis* species of the Valley occasionally visit nightshades, but *E. solani* is the most likely member of its genus to be found on *Solanum* in the Valley. Plants of the genus *Solanum* often have a complementary-colored "bull's-eye" appearance attractive to pollinators, with bright yellow anthers set against purple or blue-violet petals. *Solanum* pollen also tends to have a high protein concentration particularly nutritious to bees. *Solanum* flowers are difficult to pollinate, however, because their pollen is densely packed within the anthers.

Such pollen can be accessed through vibratile ("buzz") pollination. Large bee species are usually the best candidates for buzz pollination (that is, for vibrating their wing muscles forcefully enough to shake loose pollen). The relatively small *Exomalopsis solani* is nonetheless a proficient buzz pollinator. Vibratile pollination tends to be noisy—one way to locate this species in Hidalgo County is to enter a patch of silverleaf nightshade, the most common wild *Solanum* of the area, and just listen.

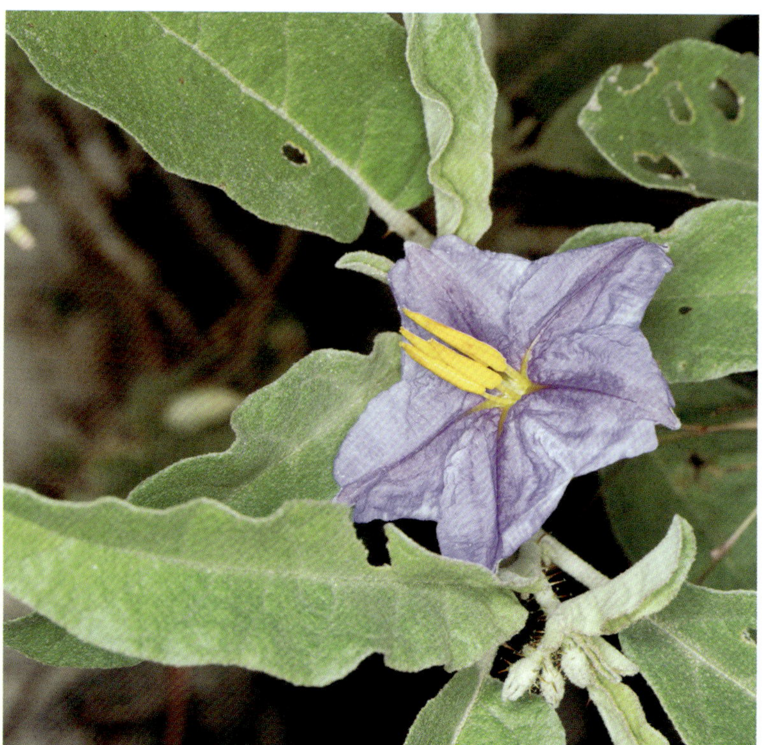

Silverleaf nightshade, a food plant of *Exomalopsis solani*

Hair Patterns on the Abdomens of Six Female *Exomalopsis*

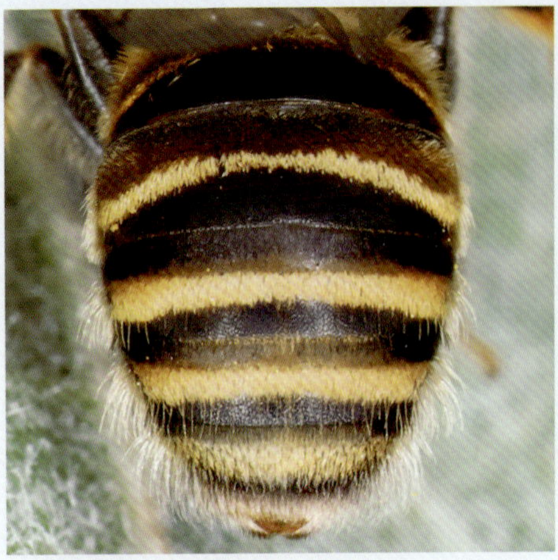

Exomalopsis snowi

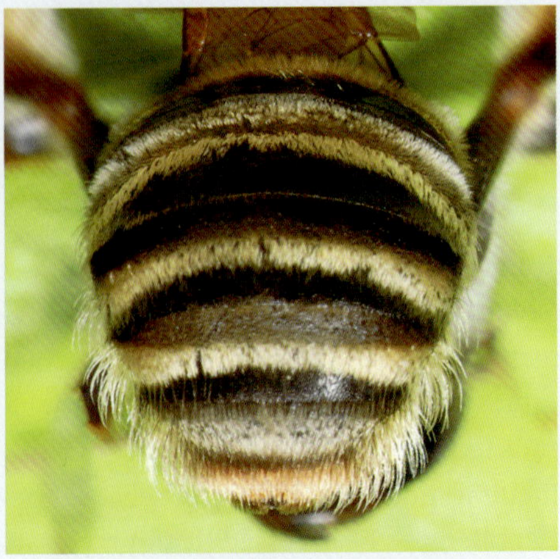

Exomalopsis mellipes

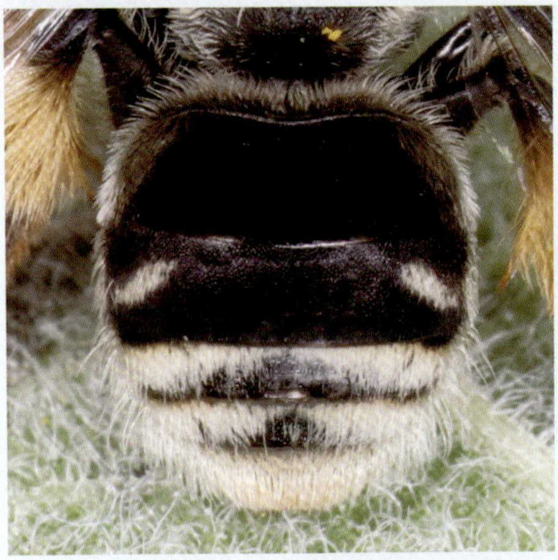

Exomalopsis analis

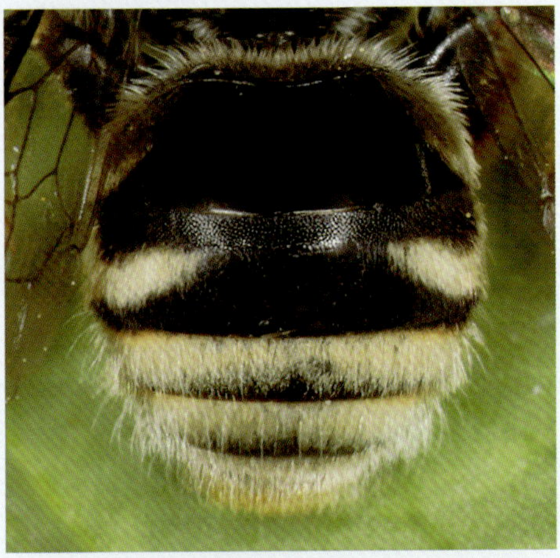

Exomalopsis similis

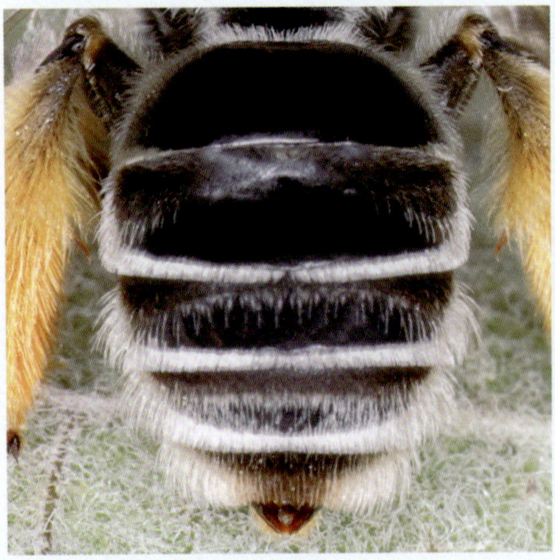

Exomalopsis solani

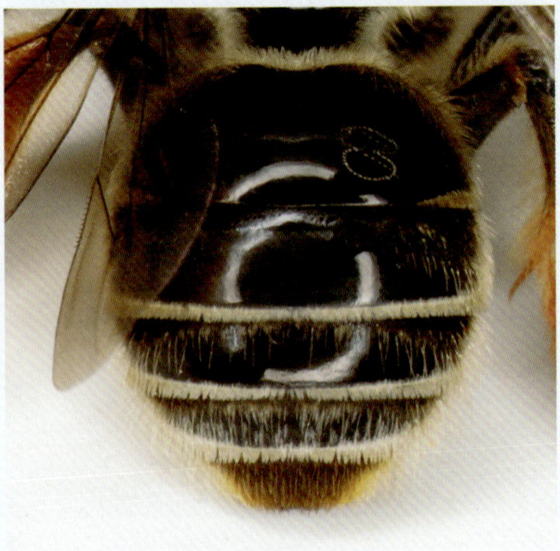

Exomalopsis birkmanni

Digger Bees

Subfamily Anthophorinae ◆ Tribe Anthophorini
Genus *Anthophora*

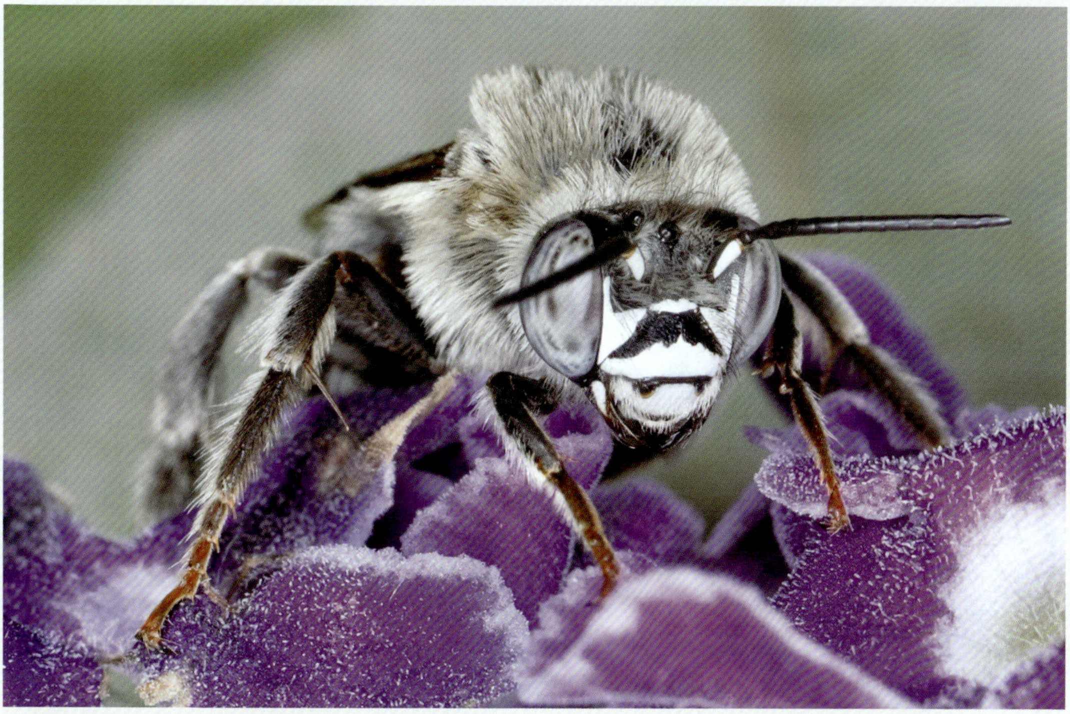

A male *Anthophora capistrata*

Digger bees (*Anthophora*) are robustly built, entertaining, unusually noisy insects with a distinctive way of flying. Buzzing raucously, they zip speedily around plants and then stop abruptly, hovering in front of blossoms and angling their long tongues down the flowers' throats without letting their feet touch the petals. Most North American *Anthophora* are generalist pollinators capable of buzz pollination. They play an important role in maintaining wildflower diversity, in part because their long tongues allow them to pollinate deep-throated and tubular blossoms inaccessible to many bees.

In the Lower Rio Grande Valley, *Anthophora* species tend to have hairy thoraxes, striped abdomens, and brightly colored eyes. Male and female *Anthophora* can differ notably in appearance. Females often have dark faces, while males have elaborate facial markings, usually ivory or yellow. Some male *Anthophora* have long hind legs equipped with wide teeth or ornamented with brushes of hair.

General behavior: As their name suggests, digger bees typically nest in the earth, either in level ground or vertical banks. *Anthophora* females construct solitary ground nests by digging with their front legs and using their mandibles to loosen dirt. In some species, females use their front

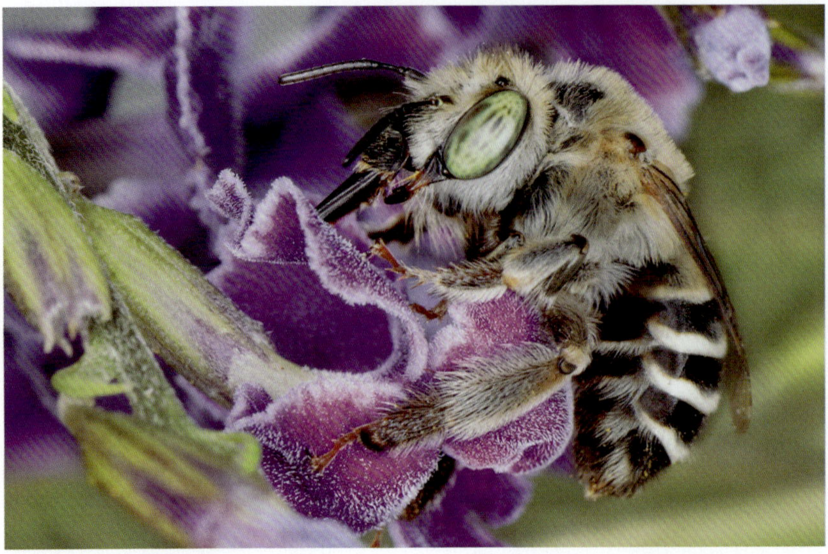

A female California digger bee drinking nectar from *Duranta erecta* blossoms

legs to haul mud through the air; they then work the mud to moisten the dirt of dry nesting areas, softening them up for better digging. Some *Anthophora* females shape excavated soil outside their holes into turrets. Females may vigorously defend their nests against predators: *Anthophora* females have been observed bodily dragging parasitic bees from nest holes.

Various species of *Anthophora* have been documented engaging in boisterous behavior. In 1929, entomologist Phil Rau wrote of *Anthophora abrupta*, a species found in Central Texas: "They are neither timid nor aggressive, but . . . how conspicuous they are as they noisily swing their ponderous bodies to and fro on the wing, arrive home and scramble into their burrows or come tumbling out headlong and dash off into the sunny fields, with all the exuberance of boys just out of school. They have none of the shy, stealthy ways of maneuvering, whereby some of the smaller and daintier varieties of bees and wasps hold their own in a competitive world."[8]

Despite their solitary status, female digger bees tend to be gregarious and sometimes form aggregations that number in the hundreds or even thousands. Bees in such aggregations may appear to behave cooperatively. Entomologist Robbin Thorp recorded in 1969 that if he strolled through a nest colony of Edwards digger bees (*Anthophora edwardsii*), "swinging an insect net about a meter above the ground, the females would aggregate in the plane of motion. Thus, it was possible to capture 25 to 50 females in five sweeps."[9] Thorp also noted that objects thrown over or onto the nesting site were likely to be followed aggressively by large numbers of females. He found this behavior remarkable because solitary bees rarely act collectively. Recent evidence suggests that some female *Anthophora* emit gland secretions containing alarm or aggregation pheromones that attract or alert other females.

Anthophora males often gather outside nest tunnels dug by females. Like the male *Diadasia* discussed earlier, male *Anthophora* sometimes pile together in scrums outside nest entrances, awaiting the emergence of a female. When she appears, they jump on her en masse.

8. Rau (1929), p. 156.

9. Thorp (1969), p. 325.

California digger bee ◆ *Anthophora (Anthophoroides) californica* Cresson, 1869

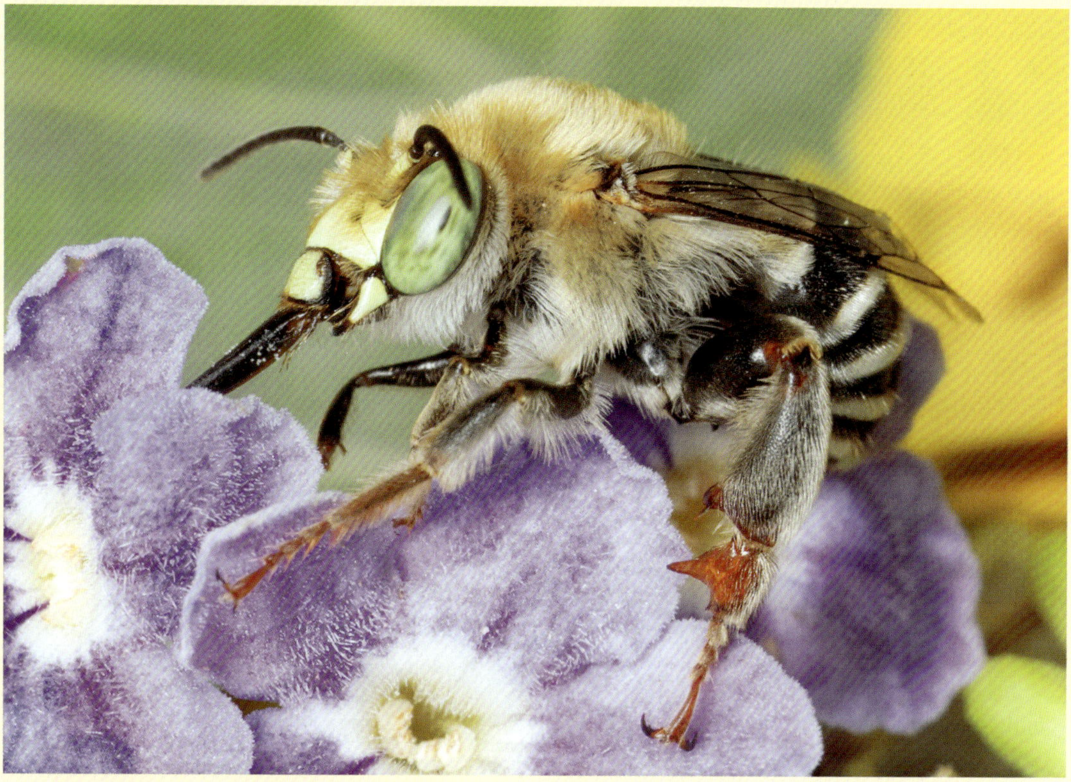

A male California digger bee

Description: *Anthophora californica* is a medium-sized bee with crisply defined white bands on its abdomen, a densely hairy thorax and head, and green eyes. Female California digger bees have dark faces, and their heads and thoraxes are covered with copious grayish-brown hairs. Males have pale yellow faces and mandibles, and golden hairs on their heads and thoraxes. The legs of male California digger bees are elongated, red on the lower segments, and armed with large teeth on the basitarsi. The hind-leg tibiae of females are covered with long, feathery hairs. **Size:** females and males 11–12 mm (2/5–1/2 in.).

Range: This species is widespread through much of the south-central and western United States: it ranges westward from Texas to California and Oregon. *Anthophora californica* occurs throughout Mexico into Central America and has been documented as far south as Costa Rica. This is a common species in Hidalgo and Starr Counties.

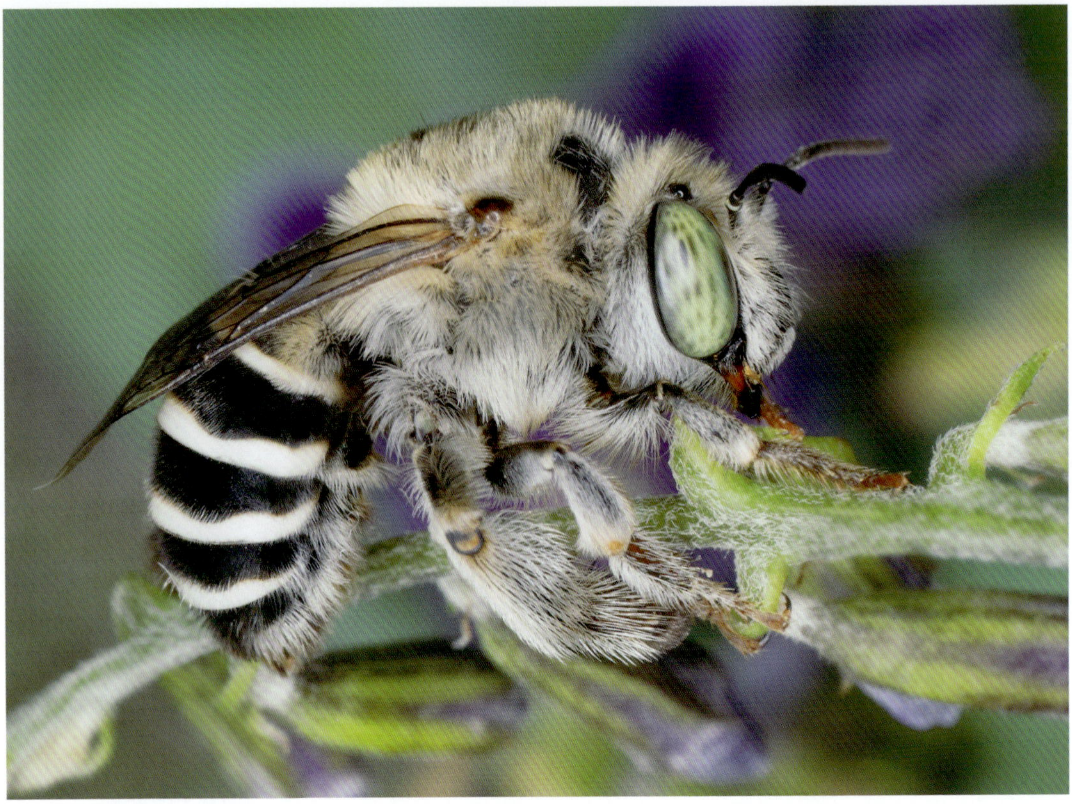

A female California digger bee: note the long, feathery hairs on the hind legs.

Associated plants: The California digger bee forages on flowers in a broad range of plant families. In the Lower Rio Grande Valley, it is frequently associated with Berlandier's fiddlewood (*Citharexylum berlandieri*) and shrubby blue salvia (*Salvia ballotiflora*). Golden dewdrops (*Duranta erecta*), a common ornamental shrub of the Valley, remains steadily in bloom there from April through November and is a magnet for both male and female *Anthophora californica*.

Male *Anthophora californica* are among the first bees to emerge in spring in Hidalgo County. They are visible as early as late February and have the distinction of being one of a mere handful of Valley bee species that visit guayacan (*Guaiacum angustifolium*), an unusual purple-flowered shrub of the family Zygophyllaceae, native to northern Mexico and southern Texas. An evergreen with waxy leaves that have earned it the local common name "soapbush," guayacan produces abundant nectar during its early-spring bloom, when it is visited predominantly by native flies and, where established, feral honey bees.

Female California digger bees emerge about two weeks after males. In Hidalgo County, female *Anthophora californica* can be found in mid-March and April nesting in aggregations on the banks of irrigation canals bordering the Rio Grande. Once females emerge, they are commonly found collecting pollen from shrubby blue salvia.

Additional identification information: California digger bees belong to a different subgenus (*Anthophoroides*) than the masked salvia digger bee (*Anthophora capistrata*), described in the next entry. *Anthophora (Anthophoroides)* digger bees run fairly small for their genus, usually measuring around 14 mm (1/2 in.) or less. On all male *Anthophora (Anthophoroides)*, the hind leg basitarsus sports a tooth; on the California digger bee, this tooth is large enough to be seen with the naked eye when the bee is at rest on a flower.

To the casual observer, female California digger bees closely resemble female *Anthophora capistrata*, although the latter run somewhat larger. Both female and male California digger bees can be distinguished, however, by a trait that is obvious on close inspection: the white bands on the California digger bee's abdomen are part of the integument; they are not formed by bands of pale hairs.

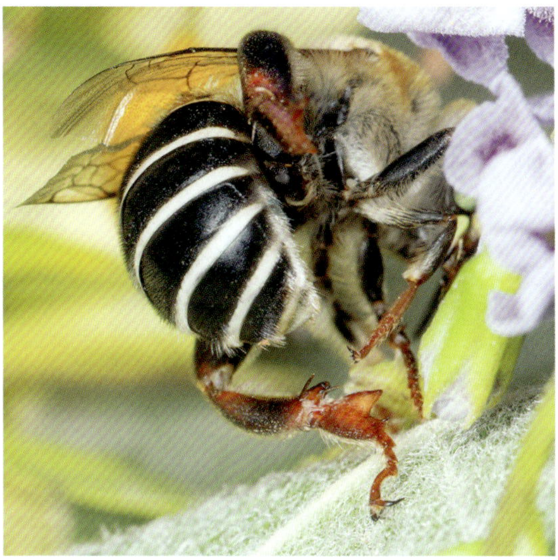

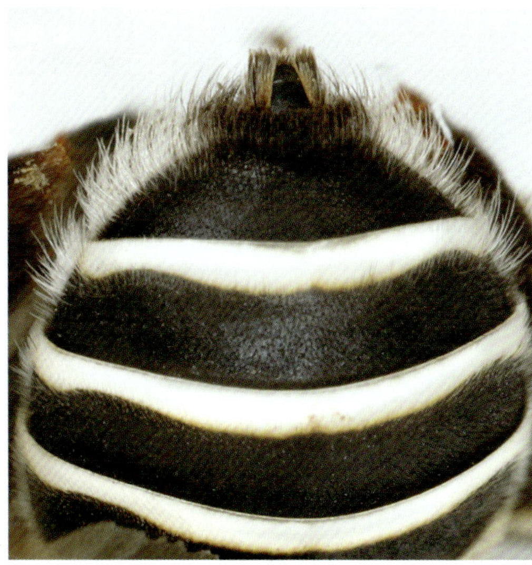

Left: Rear view of a male *Anthophora californica. Right*: Close-up of the abdomen of a female *Anthophora californica*: the white bands are part of the integument.

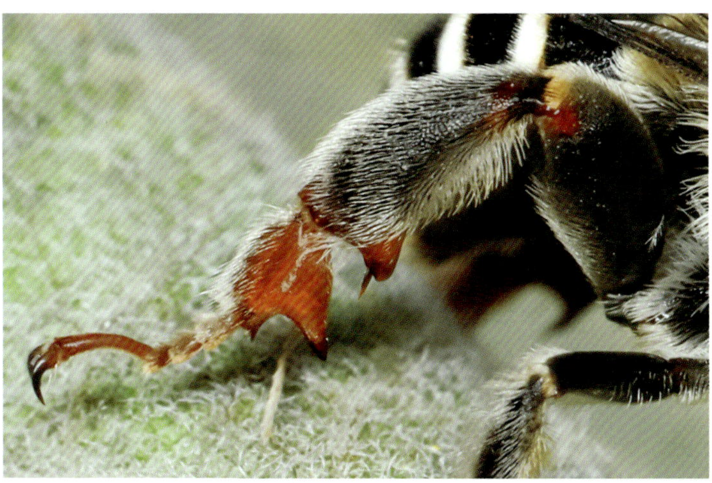

Close-up of the hind leg of a male California digger bee

Behavior: Male *Anthophora californica* are fast-flying and flashy bees, with bright yellow faces and red legs. They are territorial, often stationing themselves near a plant and vigorously defending it by buzzing lustily while patrolling it. Males also commonly gather near nesting sites. They use their ornate legs to grasp females during mating.

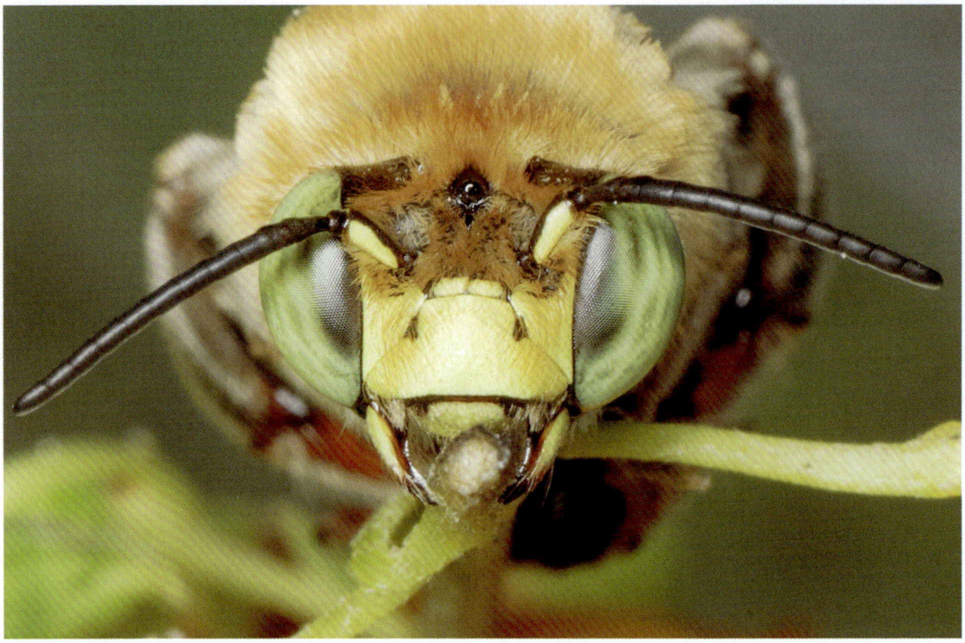

Face of a male California digger bee

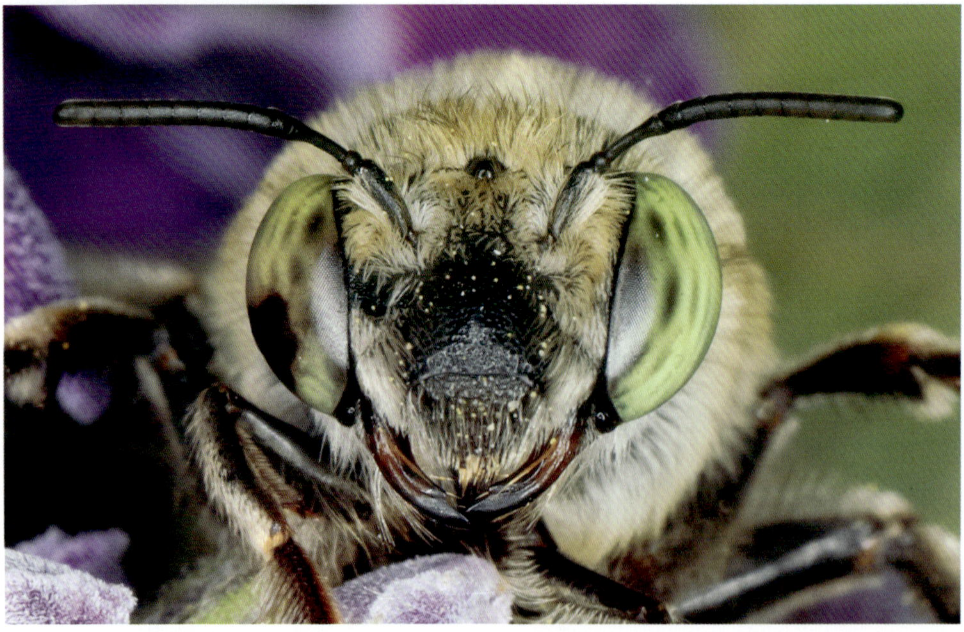

Face of a female California digger bee

Masked salvia digger bee ◆ *Anthophora (Mystacanthophora) capistrata* Cresson, 1878

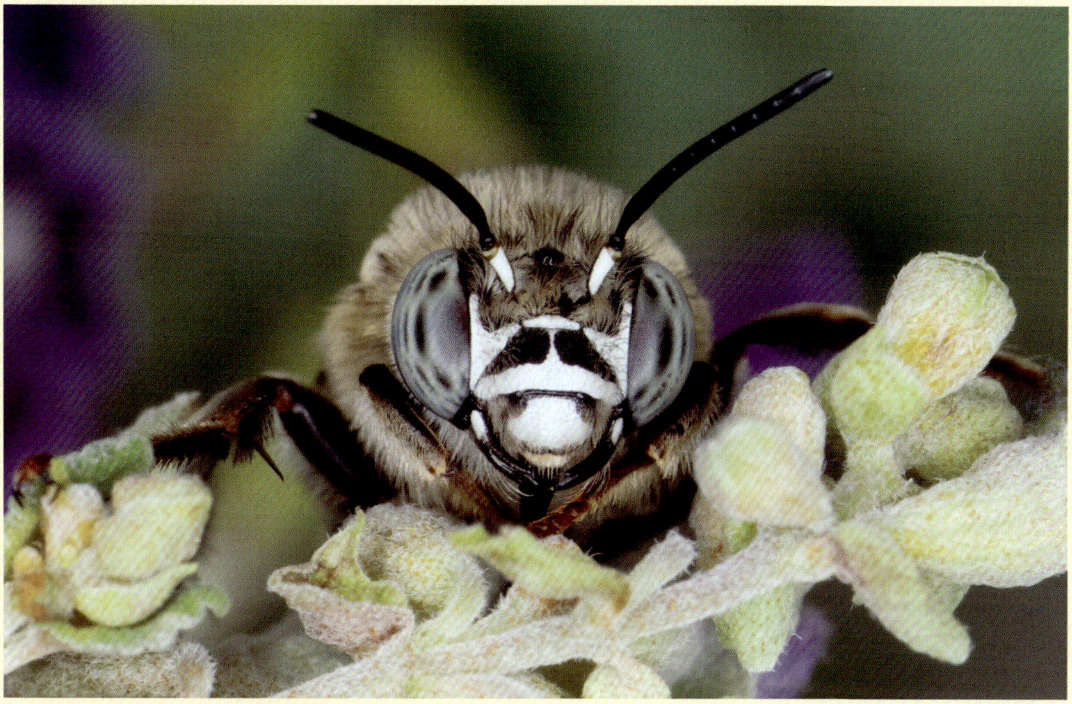

A male *Anthophora capistrata*

Description: *Anthophora (Mystacanthophora) capistrata* is a robust, furry-looking bee. Its head and thorax are covered with pale-and-black hairs that give the bee a grayish cast from a distance. Its abdomen is black and boldly striped with pale hair bands. Male *Anthophora capistrata* have striking and intricate black-and-white facial masks. Female bees have dark faces. The eyes of both males and females are steel blue. **Size:** females and males 12–13 mm (1/2 in.).

Range: The distribution of *Anthophora capistrata* is centered in Mexico, where the species occurs from the border to the southern state of Oaxaca. In Texas, *Anthophora capistrata* occurs principally along the southeast border but ranges at least as far north as Travis County. The bees shown here appeared from September to November 2019 and 2021 at the National Butterfly Center and Quinta Mazatlán (Hidalgo Co.).

Associated plants: In Hidalgo County, female *Anthophora capistrata* are most often found collecting pollen from shrubby blue salvia (*Salvia ballotiflora*), which, as noted earlier, is a member of the mint family, Lamiaceae. In other regions, *Anthophora capistrata* has been recorded visiting Lamiaceae such as azure blue sage (*Salvia azurea*) and canyon sage (*Salvia lycioides*). In the Valley, male bees sometimes nectar on golden dewdrops (*Duranta erecta*).

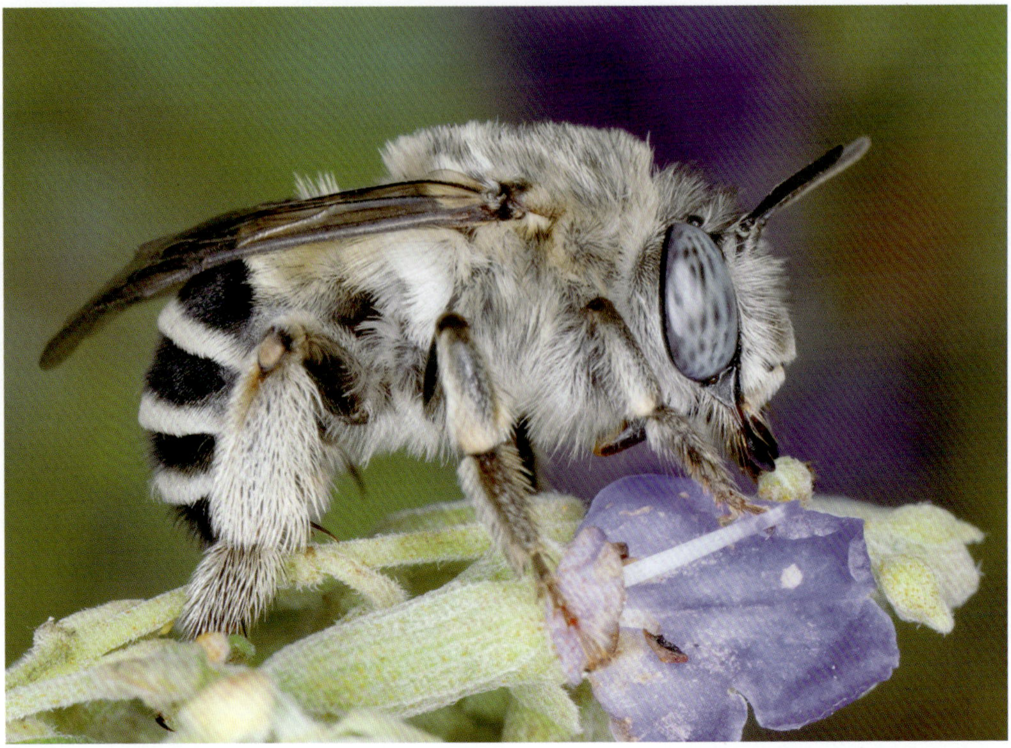

A female *Anthophora capistrata*: females have robust builds. Dense pale hairs cover their heads and bodies, intermingled with darker hairs on the thorax and vertex.

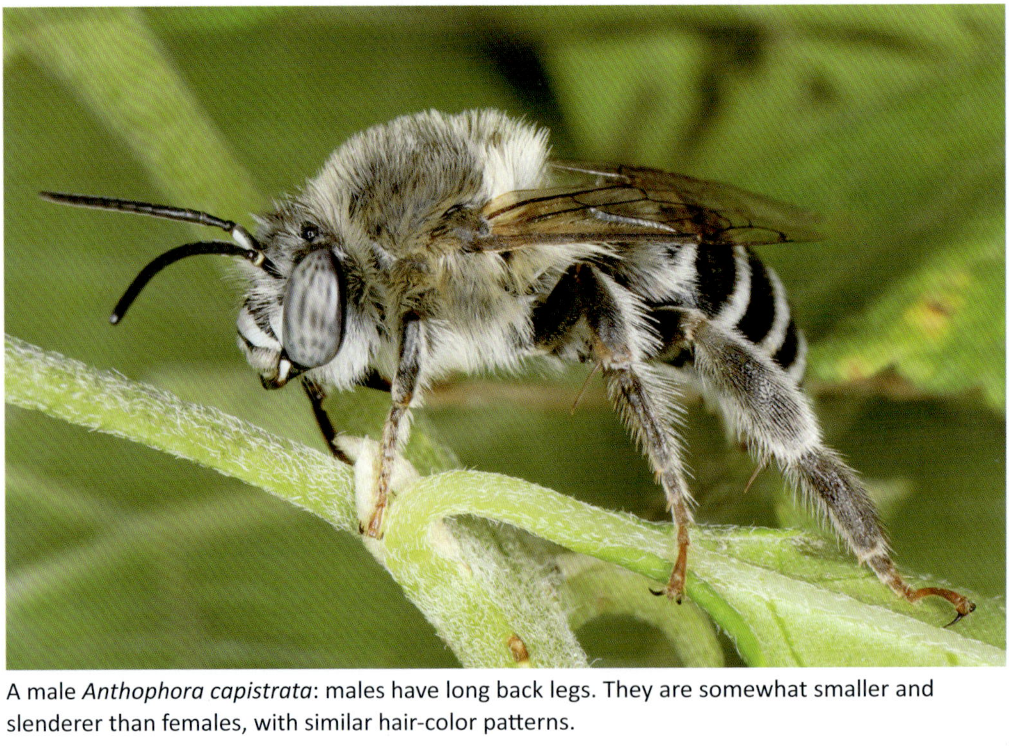

A male *Anthophora capistrata*: males have long back legs. They are somewhat smaller and slenderer than females, with similar hair-color patterns.

ID tips: *Anthophora capistrata* was first recorded by Cresson in 1878. He based his species description on an examination of two male bees found by Swedish entomologist Gustave W. Belfrage during Texas bee-collecting expeditions in the mid-1880s.

Capistrata means "wearing a halter," a reference to the distinctive black-and-white pattern on the male bee's face. Cresson painstakingly described this pattern, a trait that readily identifies the species. He also noted three other attributes of the male *Anthophora capistrata*: it has black areas on the bases of its mandibles; its antennal scapes are white; and its head is densely covered with hair.

The face of the female *Anthophora capistrata* is black and lacks a mask—nonetheless, it is remarkable for another reason: female digger bees of the subgenus *Anthophora (Mystacanthophora)* have a fringe of hairs lining the base of the clypeus, which have hooked or wavy ends. The subgenus prefix, derived from the Greek *mystaco*, or "mustache," relates to this trait. Female *Anthophora capistrata* use such hairs to collect pollen from mints.

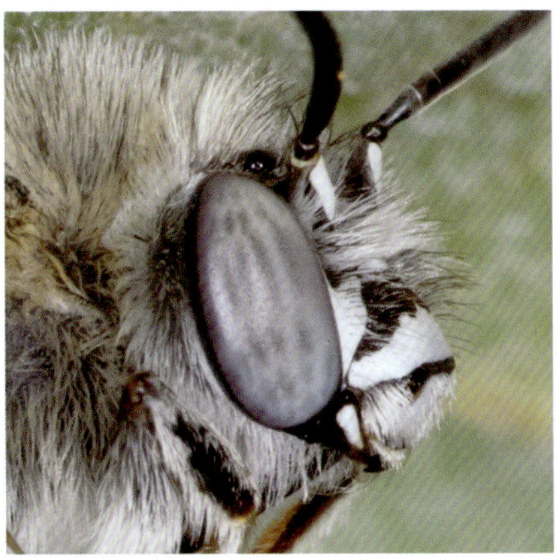

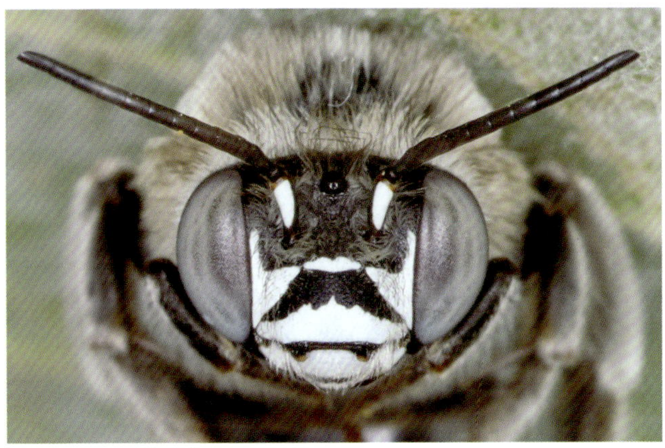

Lateral and anterior views of the head of a male *Anthophora capistrata*

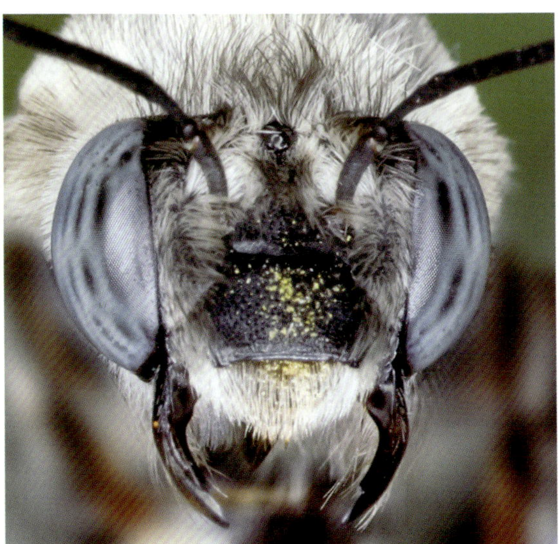

Face of a female *Anthophora capistrata*: females have entirely dark faces, large black mandibles, and dense white hairs around their antennae and fringing their lower faces. Viewed under magnification, the hairs on the base of the female's clypeus appear hooked on the ends.

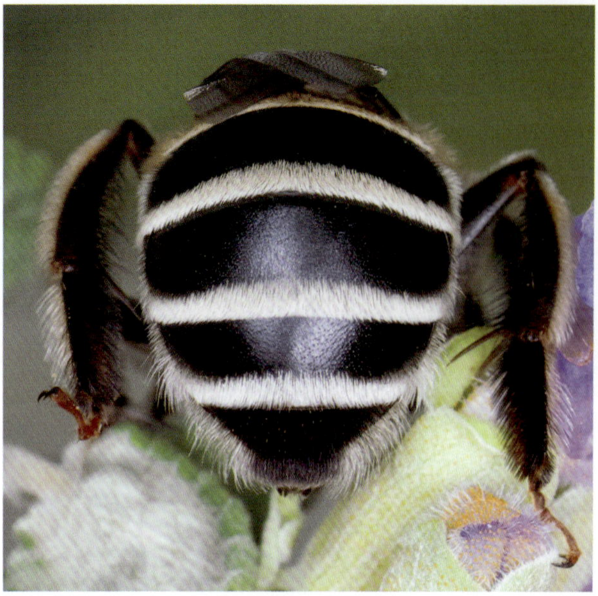

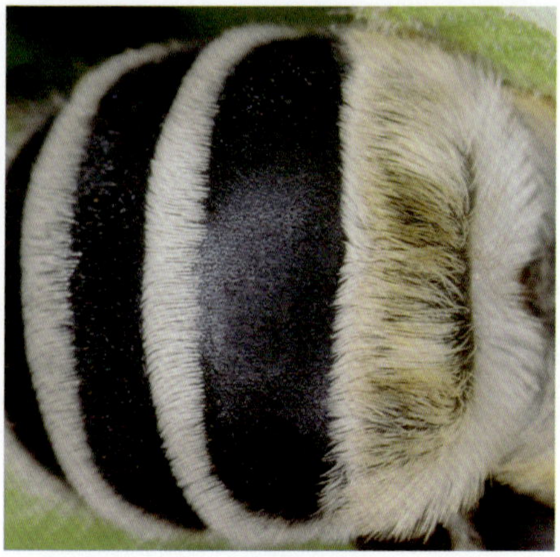

The abdomen of a female *Anthophora capistrata*

Both female and male *Anthophora capistrata* have dark abdomens banded by pale hairs. T1 of both males and females is covered with dense pale hairs. The last segment (T6 on females and T7 on males) is covered with dark hairs.

Additional information: *Anthophora capistrata* is fairly uncommon in the Valley. The author found females exclusively where large shrubby blue salvia bushes (*Salvia ballotiflora*) abounded. This woody shrub is native to the Lower Rio Grande Valley and southwestern Texas. It can grow up to 8 feet tall, is densely branched, and bears numerous blossoms. The flowers are unusually aromatic and bloom dependably from April through late fall. *Salvia ballotiflora* appears to be critical to the presence of several unusual species native to the Valley—among them *Mesoxaea texana*, *Protoxaea gloriosa*, and *Eucera fasciatella*—as well a food source for myriad generalist bees.

Within the Valley, *Anthophora capistrata* tends to appear in areas with water sources nearby. The population inhabiting Quinta Mazatlán during 2021 appeared on *Salvia ballotiflora* near a pond. The *Anthophora capistrata* of the National Butterfly Center accumulated on salvia near an irrigation canal bank. In both locations, *Anthophora capistrata* appeared in conjunction with *A. californica*, described in the previous entry. Entomologist Robert W. Brooks has noted that such distinct *Anthophora* species, and their nest sites, often intermingle.

Oil-Digger Bees

Subfamily Apinae ◆ Tribe Centridini
Genus *Centris*

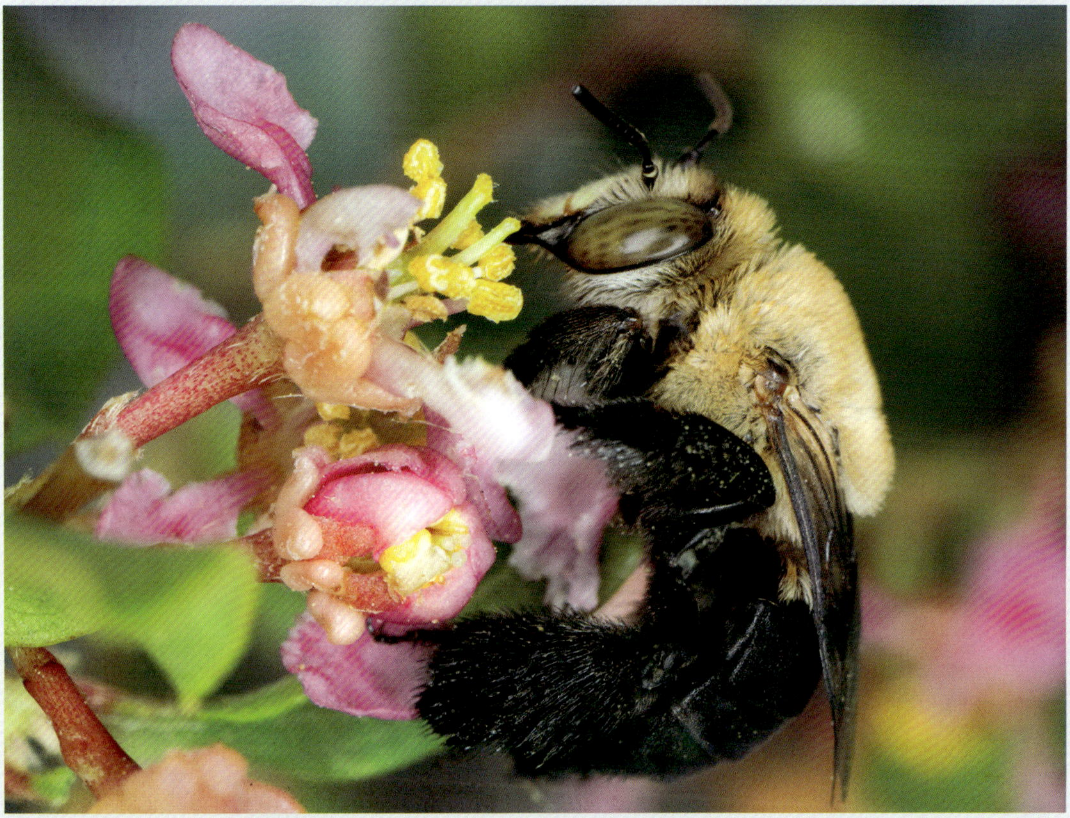

A female *Centris* on *Malpighia*

Fast and adept fliers, *Centris* bees are the quarter horses of the insect world, stopping and changing direction in midflight as they maneuver around plants. In the Lower Rio Grande Valley, *Centris* are difficult to miss. They are hefty, black-and-beige bees that buzz loudly as they zoom through gardens and fields.

In the Valley, *Centris* are usually an indicator of *Malpighia*, a genus of evergreen plants that produce abundant floral oils. Female *Centris* use such oils to line the walls of their nests, plug nest chambers, and provision larvae—a practice that has earned them the common name "oil-digger bees." During spring bloom periods, *Centris* females teem around local *Malpighia* such as wild crapemyrtle (*Malpighia glabra*), making the bushes seem alive with humming. Female *Centris* grasp *Malpighia* blooms with their large jaws while stroking the petals with their legs in order to extract oil. The flowers of oil-bearing plants like *Malpighia* have specialized glands called elaiophores that store oils. The legs of female *Centris* are specially designed to force oil from elaiophores.

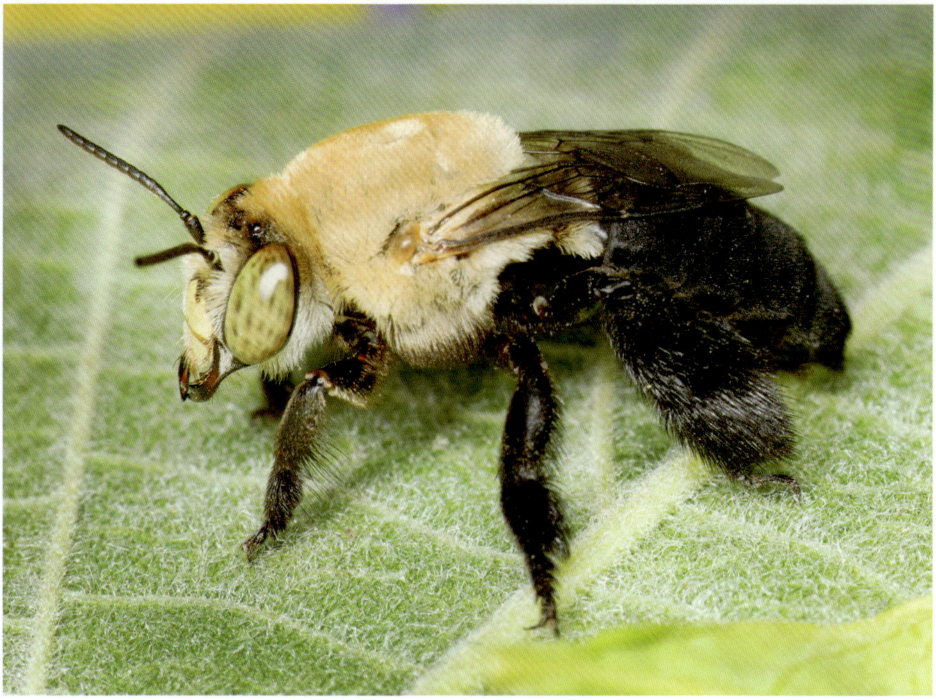

A female *Centris atripes*

A seminal 1981 study of oil-collecting bees by Neff and Simpson described the front and middle legs of most female *Centris* as outfitted with small bristlelike hairs called setae, which form combs (made up of narrow lines of setae) and pads (made up of branched and hooked setae). Some of the setae are bladelike and used to cut open flower receptacles that contain oils.

The *Centris* female's hind legs are covered with bushy pollen-collecting hairs that are sometimes described as looking like "pantaloons." *Centris* females transfer the oils from their front and middle oil-collecting legs to their hind-leg scopal hairs in midflight for transport back to their nests.

Most *Centris* in the Valley nest in the earth, sometimes near the banks of streams or other bodies of water. They are solitary bees that frequently form aggregations at nest sites. Some *Centris* species, such as *C. nitida*, nest in wood or cavities found in trees. *Centris* nests tend to be built in irregular clumps of cells, rather than in series.

Male *Centris* patrol flowers looking for mates, or, in gregarious nest sites, hover around nest entrances waiting for emerging virgin females. Although male *Centris* do not collect floral oils or pollen, their hind legs tend to be nearly as shaggy as those of females. The middle and front legs of male *Centris* also sport comblike arrays of hairs, although these are nonfunctional and less developed than those of females.

Centris males can be difficult to differentiate from females with the naked eye. Males run somewhat smaller but have stout builds, hairy bodies and legs, and coloration patterns that generally mirror those of females. Both female and male *Centris* may have pale or light yellow facial markings, and both frequently have green or greenish-brown eyes. In one Texas species, the red-legged *Centris rhodopus*, females have red eyes, while those of males are green.

Black-legged oil-digger bee ♦ *Centris (Paracentris) atripes* Mocsáry, 1899

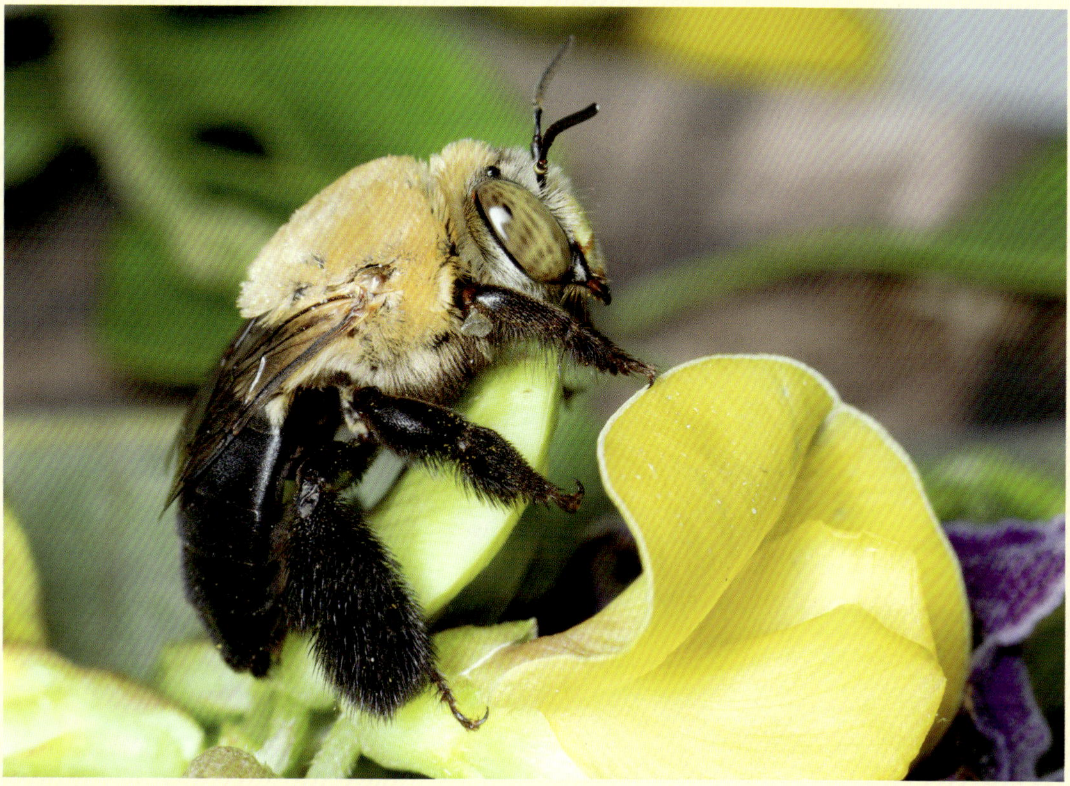

A female black-legged oil-digger bee

Description: *Centris (Paracentris) atripes* is a hefty bee with pale gold hairs on the thorax; black legs; black hairs covering all but the front segment of the abdomen; a pale yellow facial mask; prominent mandibles; and brownish-green eyes. Females and males have shaggy-looking hind legs covered with long, entirely black hairs. Males tend to run slightly smaller than females. **Size:** females and males 15–17 mm (3/5–2/3 in.).

Range: The black-legged oil-digger bee is found throughout Texas and the Southwest, as far north as Oklahoma and Nevada. The distribution of *Centris atripes* reaches through Mexico and Central America to Costa Rica. This species occurs in all three border counties of the Valley.

Associated plants: *Centris atripes* appears in the Valley from May through November. Females commonly gather oils from native Malpighiaceae, most often wild crapemyrtle (*Malpighia glabra*) and big berry manzanita (*Arctostaphylos glauca*), or from nonnative ornamentals such as Barbados cherry (*Malpighia emarginata*). Elsewhere, *Centris atripes* females collect oils from *Krameria* and a variety of other Malpighiaceae. In the Valley, both males and females also visit esperanza (*Tecoma stans*), golden dewdrops (*Duranta erecta*), Texas snout bean (*Rhynchosia senna* var. *texana*), and *Salvia*.

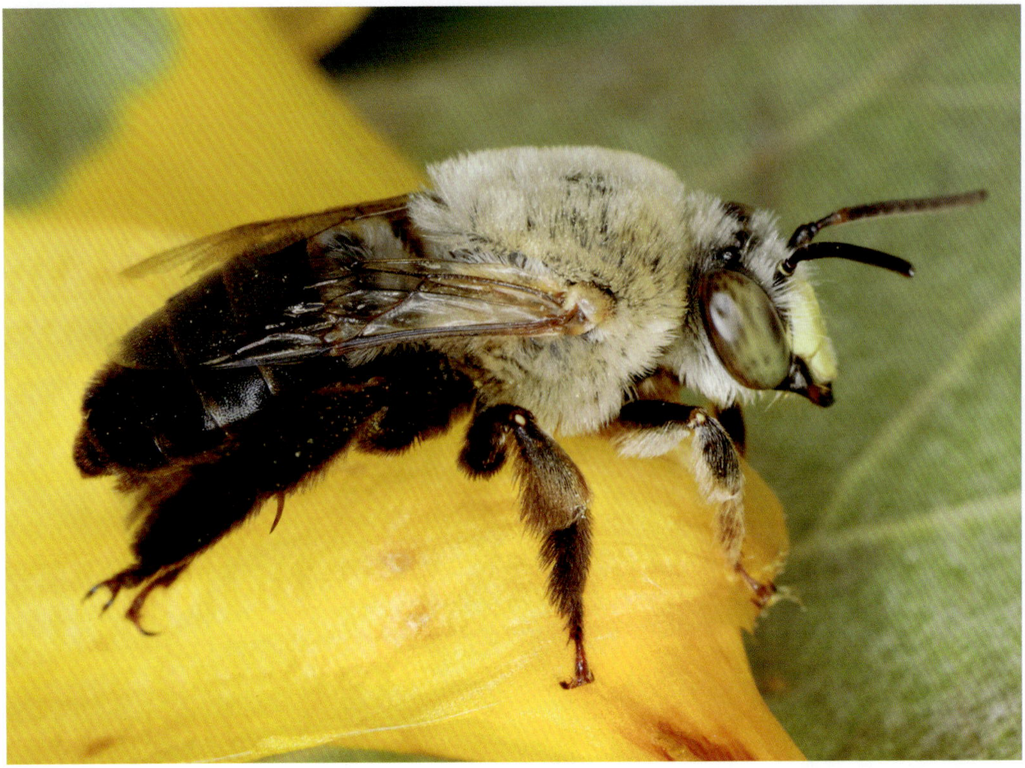

A male *Centris atripes*

ID tips: Distinguishing traits of *Centris atripes* include the following: (1) the hairs on the front, middle, and hind legs are entirely dark; (2) the abdomens, and the hairs covering them, are entirely dark on all segments but T1; (3) T1 is covered with mostly light hairs on males and light mixed with dark on females; (4) males' thorax hairs are all light (off-white, tan, or light yellow brown), while those of females are predominantly light, but dark on the mesepisternum (sides of the thorax); and (5) hairs on the male's face and vertex are light or all light; on the female, dark vertex hair may be intermixed with light hair, or there may be dark patches of hair behind the ocelli.

The facial markings of *Centris atripes* females help distinguish them from digger bees of the genus *Anthophora*, which fly at similar times of year and are also hefty, noisy, swift fliers with dark abdomens and furry-looking thoraxes. Among Texas *Anthophora*, males alone have pale facial markings, while females have dark faces. Female *Centris atripes*, however, have partly pale faces like their male counterparts.

The facial markings of both male and female *Centris atripes* are also the best means of distinguishing them from similar *Centris* species of the Valley. As shown here, the female's clypeus is predominantly yellow, sometimes with a dark lateral and upper border; the paraocular area (the area rimming the inner compound eyes) is entirely dark. On males, the clypeus is nearly entirely yellow, and additional yellow triangular patches extend laterally from the clypeus to the compound eye. The mandibles of both male and female *Centris atripes* are large and primarily dark; those of males may have some dark red or pale areas.

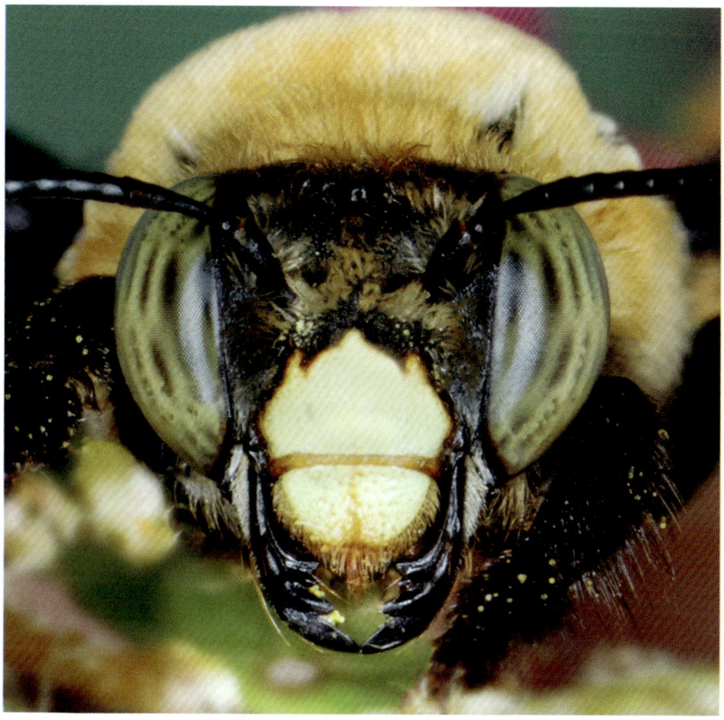

Face of a female black-legged oil-digger bee

Comparison of Faces of Male *Centris nitida* and *Centris atripes*

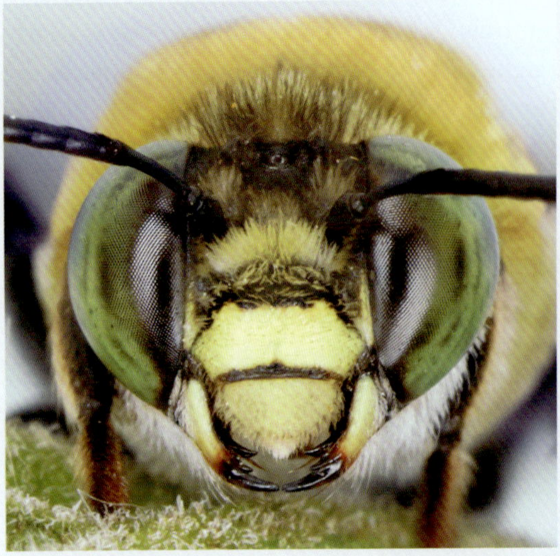
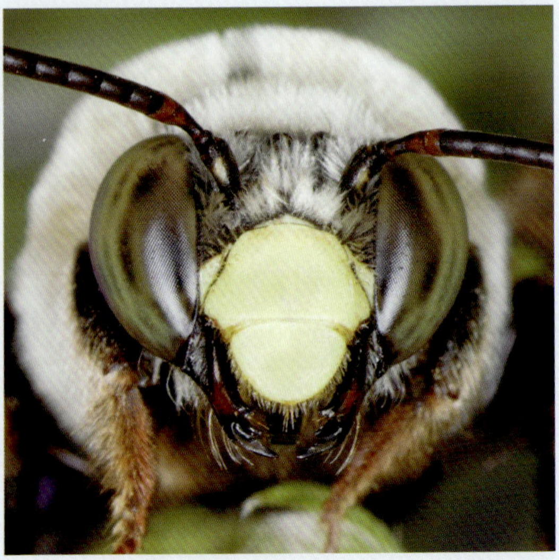

Face of a male *Centris nitida* (*left*) and a male *C. atripes* (*right*): more black intrudes into the clypeus of the male *C. nitida*, its mandibles are predominantly yellow, and its antennal scapes are entirely dark.

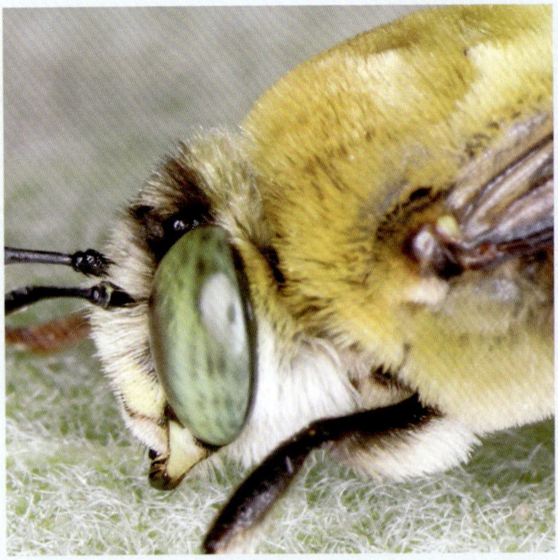
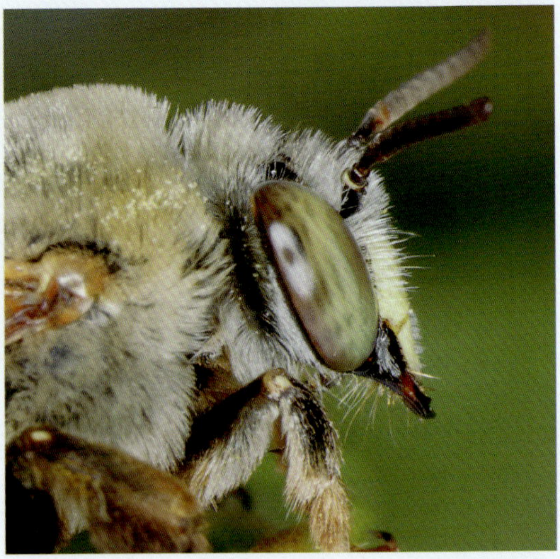

Profiles of the head of a male *Centris nitida* (*left*) and a male *C. atripes* (*right*): dark hairs on the vertex of *C. nitida* commingle with lighter hairs; on *C. atripes*, the vertex hairs are all pale.

Shining oil-digger bee ◆ *Centris (Heterocentris / Hemisiella) nitida* Smith, 1874

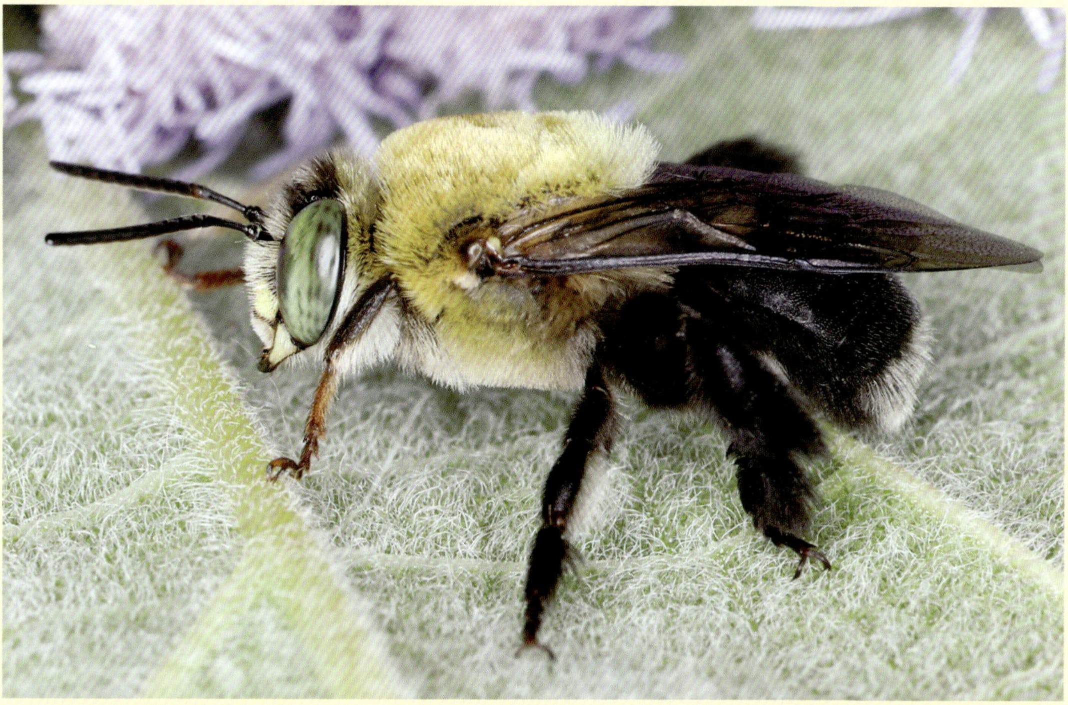

A male shining oil-digger bee

Description: *Centris (Heterocentris / Hemisiella) nitida* is less common in the Valley than the black-legged oil-digger bee, described in the previous entry. The two species appear alike at first glance: shining oil-digger bees are somewhat smaller but have a robust build, a furry-looking appearance, and similar coloring. The abdomens and hind legs of both male and female *Centris nitida* are covered with mostly black hairs. On females, yellow hairs cover the first abdominal segment (T1), and on males the hairs on T1 and T5 are primarily light. Males also have fringes of white hairs on the hind edges of their femurs. As discussed in more detail below, males' rear legs are widely built, with expanded femurs and tibiae. The bees' eyes are green, and both males and females have yellow-and-black facial masks. **Size:** females and males 13–14 mm (1/2–3/5 in.).

Range: This species appeared at the McAllen Nature Center and the National Butterfly Center (Hidalgo Co.) in November 2018. *Centris nitida* is a Neotropical species that generally ranges through Mexico and Central America, and as far south as Bolivia. Records of *Centris nitida* in Tamaulipas extend back to the 1950s. Shining oil-digger bees occasionally stray across the border into the Lower Rio Grande Valley. In 1997, *Centris nitida* began establishing itself as an adventive species in Florida. Although initially greeted with alarm there, it has proven slow spreading and vulnerable to cold snaps and is now viewed as having a relatively harmless impact on native Florida ecosystems.

Associated plants: In the Lower Rio Grande Valley, female *Centris nitida* gather floral oils from wild crapemyrtle (*Malpighia glabra*) and Barbados cherry (*Malpighia emarginata*). Female *Centris nitida* are polylectic and capable of vibratile pollination. They have been recorded elsewhere collecting pollen from nightshade-family plants, and from pea-family plants such as valamuerto (*Senna pendula*). Both male and female *Centris nitida* have been documented outside the Valley nectaring on a wide gamut of flora. The bees shown here were observed visiting crucita (*Chromolaena odorata*).

In more southerly Neotropical areas where this species is widespread, female *Centris nitida* collect floral oils from a range of oil-producing plants in the family Malpighiaceae, such as *Galphimia*, *Stigmaphyllon*, and *Byrsonima*; and from flowers in the family Orchidaceae that mimic oil plants. Entomologists Pemberton and Liu have documented the role of *Centris nitida* as a significant pollinator in Florida of the native long-key locustberry (*Byrsonima lucida*) and the rare cowhorn orchid (*Cyrtopodium punctatum*).

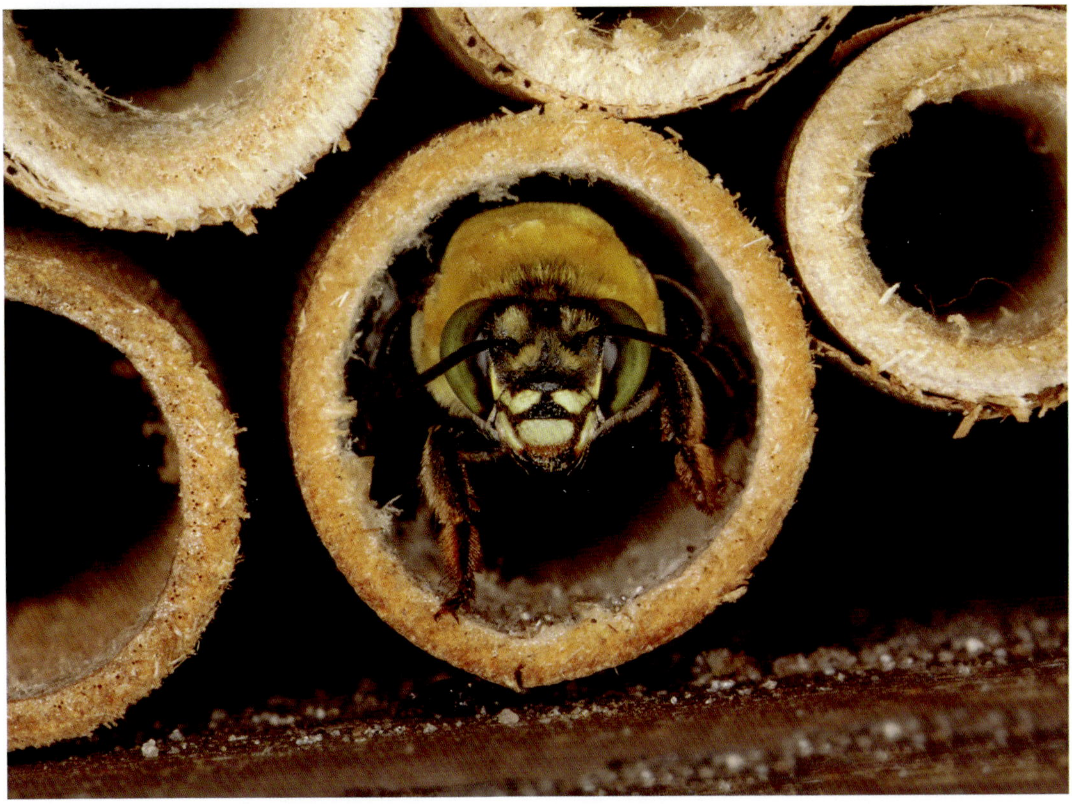

A female *Centris nitida* emerging from her nest in a bamboo cane

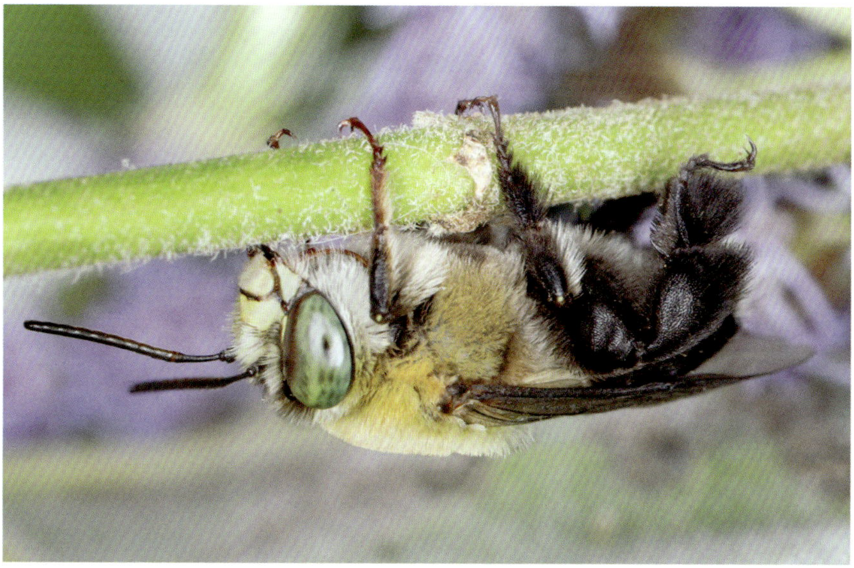

A male shining oil-digger bee

ID tips: The facial masks of both male and female *Centris nitida* are distinctive and are the best means of quickly distinguishing this species locally from the more common *C. atripes*. The male shining oil-digger bee's clypeus is predominantly yellow, with a dark pattern that stretches across the top of the clypeus. The female's face is marked with three distinctive yellow patches separated by a roughly triangular black patch. The mandibles of both males and females are yellow with dark tips; females have three-toothed jaws. Both males and females have antennal scapes that are entirely black.

Behavior: Most Texas *Centris* are ground nesters, but *C. nitida* females seek out cavities in trees and logs as nest sites; they settle readily into bamboo canes used to construct bee houses. In urban areas, *Centris nitida* females sometimes nest in the walls of abandoned buildings. As is generally true of tree hole–nesting *Centris*, shining oil-digger bees tend to select shady areas for nest sites and to avoid open, arid terrain.

Studies of *Centris nitida* in Costa Rica indicate that its preferred habitat there is shady oak forests, although it can adapt to a variety of other habitats. The *Centris nitida* shown here were found in partly wooded areas of Hidalgo County, rather than in the open scrublands and fields more characteristic of the area.

Additional information: Of special note is the unusual structure of the hind legs of male *Centris nitida*, which have enlarged femurs and tibiae. These leg modifications harbor special glands that produce chemicals or "scents" that males use to mark their territories. Some, but not all, male *Centris* species possess such glands, either in their mandibles or legs.

A 1984 study of *Centris nitida* in Costa Rica, headed by entomologist H. J. Williams, reported that male shining oil-digger bees vigorously patrol areas close to food plants. Males' territories are small—usually no more than three or four feet in diameter—but the bees chase away intruders and rival males to a distance of 18 feet.

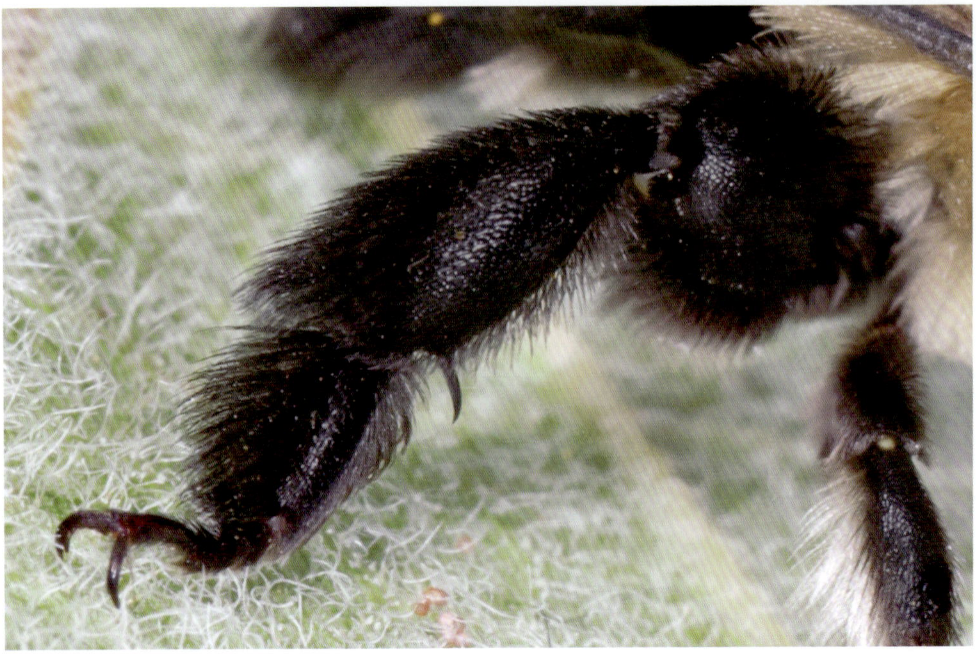

The hind leg of a male *Centris nitida*: the greatly expanded femur and tibia house scent glands. Feathery hairs line the lower legs.

Centris nitida males like to hover downwind from groups of selected flowers, rubbing their legs together in flight. Williams and colleagues suggested that such activity may help disperse chemicals secreted by the scent glands in the bees' expanded femurs and tibiae. After hovering downwind, the bees fly upwind, landing briefly on flowers, twigs, and leaves to mark their surfaces with scents. Such behavior serves to delineate the bees' territories and possibly to attract female bees as well.

Large Carpenter Bees

Subfamily Xylocopinae ◆ Tribe Xylocopini
Genus *Xylocopa*

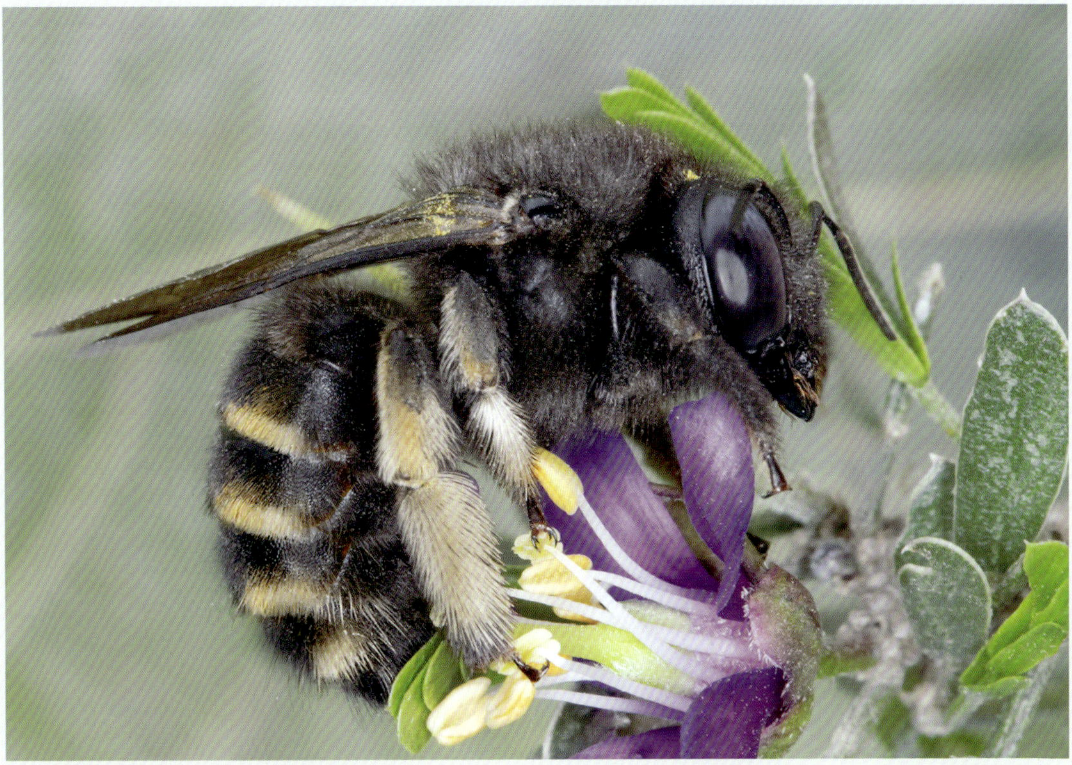

A female parkinsonia carpenter bee

Large carpenter bees of the genus *Xylocopa* are burly, loud, and fast flying. In North America, female large carpenter bees are predominantly black with dark, iridescent wings. Males frequently differ greatly from them in appearance: they may be partly or completely covered with gold, beige, or grayish hairs and sport bulging green eyes or pale facial masks. Male large carpenter bees are harmless to people but often engage in comically aggressive displays of behavior to mark their territories, dive-bombing other bees and buzzing alarmingly at human intruders to make their presence known.

The Lower Rio Grande Valley has a broader array of large carpenter bee species than is typical of most regions of North America. In the 1980s, while surveying Hymenoptera of the Lower Rio Grande Valley, biologist Charles C. Porter recorded three large carpenter bee species of three different subgenera: the striped *Xylocopa (Notoxylocopa) tabaniformis parkinsoniae* (parkinsonia carpenter bee); the more widely known *Xylocopa (Schonnherria) micans* (southern carpenter bee); and the unusually large *Xylocopa (Neoxylocopa) mexicanorum* (Mexican carpenter bee), since reclassified as *Xylocopa griswoldi* (Griswold's carpenter bee). Fifty years later, all three of these species continue to exert a strong presence in the Valley. A fourth species occupying a fourth subgenus, *Xylocopa (Stenoxylocopa) strandi* (Strand's carpenter bee), also frequents the National Butterfly Center in Hidalgo County.

Large carpenter bee behavior: *Xylocopa* are usually solitary, although among some species, nests are multigenerational, with mothers and daughters cohabiting the same nest. Griswold's, parkinsonia, and southern carpenter bees use their strong jaws to excavate nests in wood, while Strand's carpenter bees carve out nests in bamboo. All four species scrape fibers from the inner walls of plant materials to form a kind of particleboard used to partition nests into separate egg chambers. Carpenter bees do not eat wood or plants; they feed exclusively on pollen and nectar.

Despite their key role as crop pollinators, large carpenter bees are often characterized as pests that dig holes in wooden structures such as rafters, doorframes, and fences. Notably, however, none of the Valley's four carpenter bee species typically nest in human-made structures. This behavior is a characteristic of the eastern carpenter bee, *Xylocopa (Xylocopoides) virginica virginica.* Although there have been rare sightings of individuals of this species in and near Cameron County, *Xylocopa virginica virginica* is not considered native to the Lower Rio Grande Valley borderlands.

Large carpenter bees and pollination: Large carpenter bees are of inestimable importance as generalist pollinators of native wildflowers in the Lower Rio Grande Valley. The bees' size and attendant ability to carry prodigious pollen loads make them highly efficient pollinators.

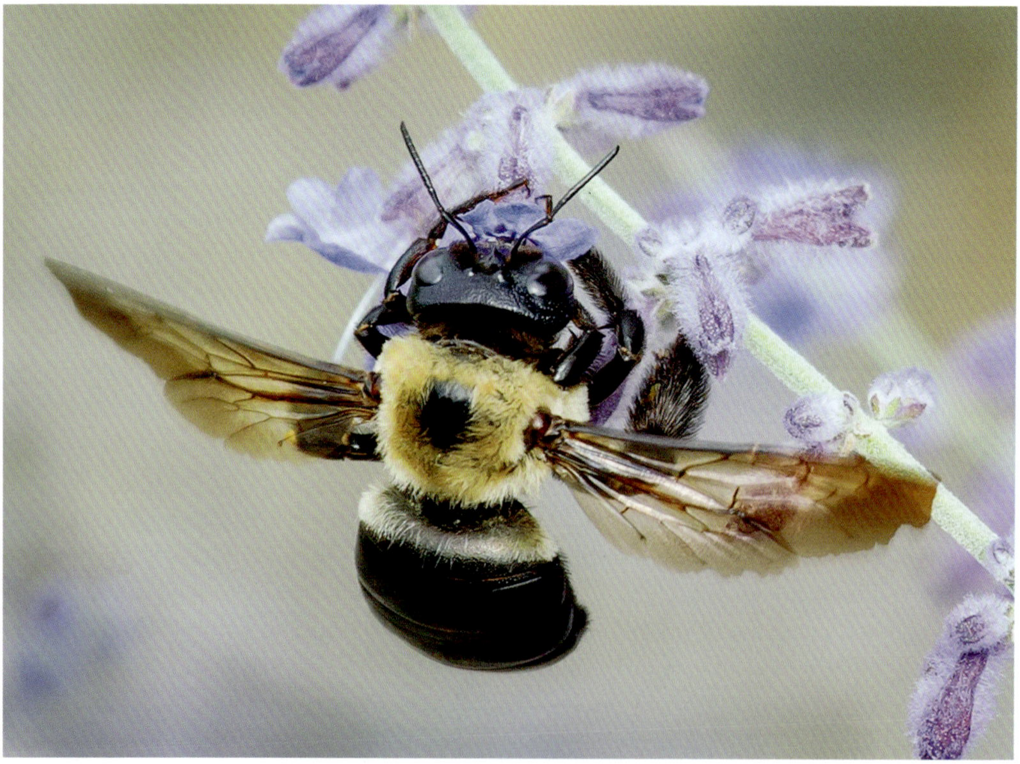

A female eastern carpenter bee (*Xylocopa virginica virginica*)

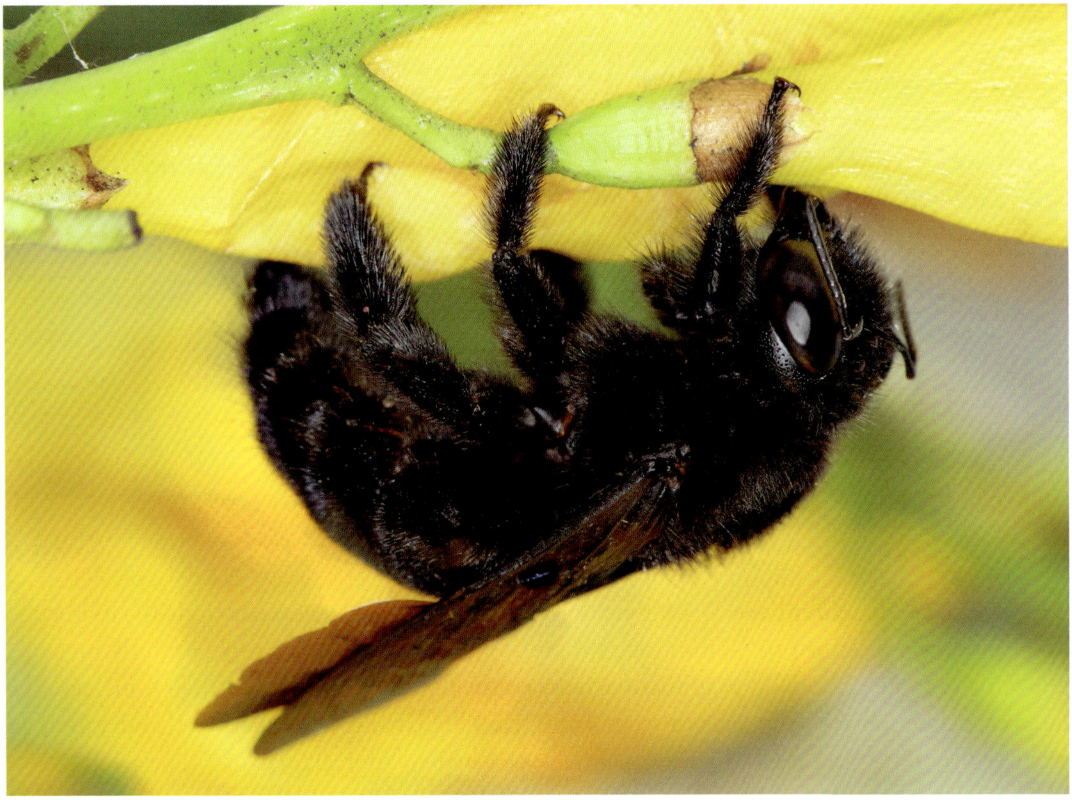

A female carpenter bee robbing nectar from an esperanza blossom

Large carpenter bees are also capable of buzz pollination, and they serve as essential pollinators of commercial crops in the family Solanaceae—tomatoes, eggplants, potatoes, and peppers—whose tightly packed pollen requires violent shaking by vibratile pollinators to be dislodged. Blueberries also require buzz pollination, and their commercial production depends in part on the presence of southern and eastern carpenter bees.

Because large carpenter bees are so hefty, they have difficulty entering a variety of flowers to obtain nectar. In lieu of crawling into blossoms with narrow, tubular necks, large carpenter bees use their sharp mouthparts (massive, pointed galeas) like lances, piercing and slitting the corollas of flowers. The bees then "rob" nectar by lapping it through the holes, thereby avoiding the contact with flower anthers and stigma necessary to pollination.

Critics complain that such behavior additionally impacts flower pollination by other insects, either because other potential pollinators drink nectar through the holes left by carpenter bees or shun the depleted blossoms entirely. Such conduct was observed throughout 2018–20 at the National Butterfly Center: green sweat bees routinely drank nectar from esperanza through slits made in the flowers by carpenter bees. Nonetheless, countervailing studies suggest that at least sometimes, nectar robbing may increase pollination because small bees that slip easily into holes made by carpenter bees are enticed to make multiple trips into the flowers, leading to increased pollen transfer.

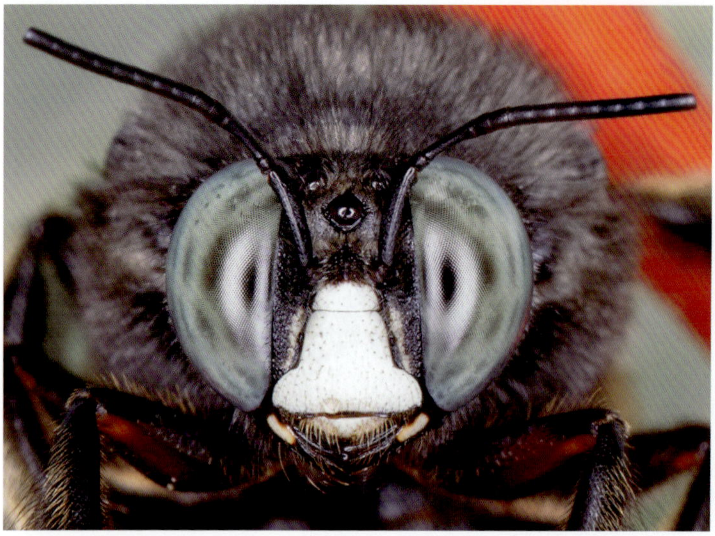

Close-up of the facial markings of a male parkinsonia carpenter bee

Traits of large carpenter bees: Distinguishing characteristics of *Xylocopa* include the following: they are large bodied (13 to 30 mm long); their bodies and heads are shiny or faintly metallic; and females carry dry pollen on stiff scopal hairs on the tibiae and basitarsi of their hind legs. The large carpenter bee forewing lacks a stigma and has a long, narrow marginal cell and a roughly triangular middle submarginal cell sometimes described as "foot shaped."

Large carpenter bees are sometimes confused with bumble bees, which are also hefty and noisy. The two genera can be distinguished by several traits. The heads of *Xylocopa* are disproportionately large for their bodies, their jaws are prominent and forceful, and their faces lack a malar space (the gap between the compound eye and mandible base). Some male carpenter bees of the Valley have ivory facial markings. By contrast, the heads of bumble bees are relatively small, their jaws are less prominent, and their faces lack pale integumental markings.

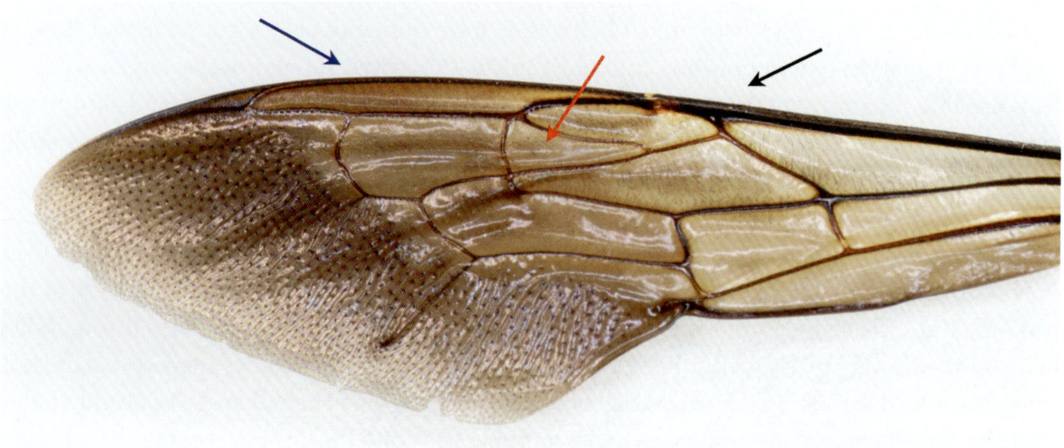

A *Xylocopa* forewing, with a long, narrow marginal cell (blue arrow) and a "foot-shaped" second submarginal cell (red arrow). Note the lack of a stigma (black arrow).

Parkinsonia carpenter bee ◆ *Xylocopa (Notoxylocopa) tabaniformis parkinsoniae* Cockerell, 1917

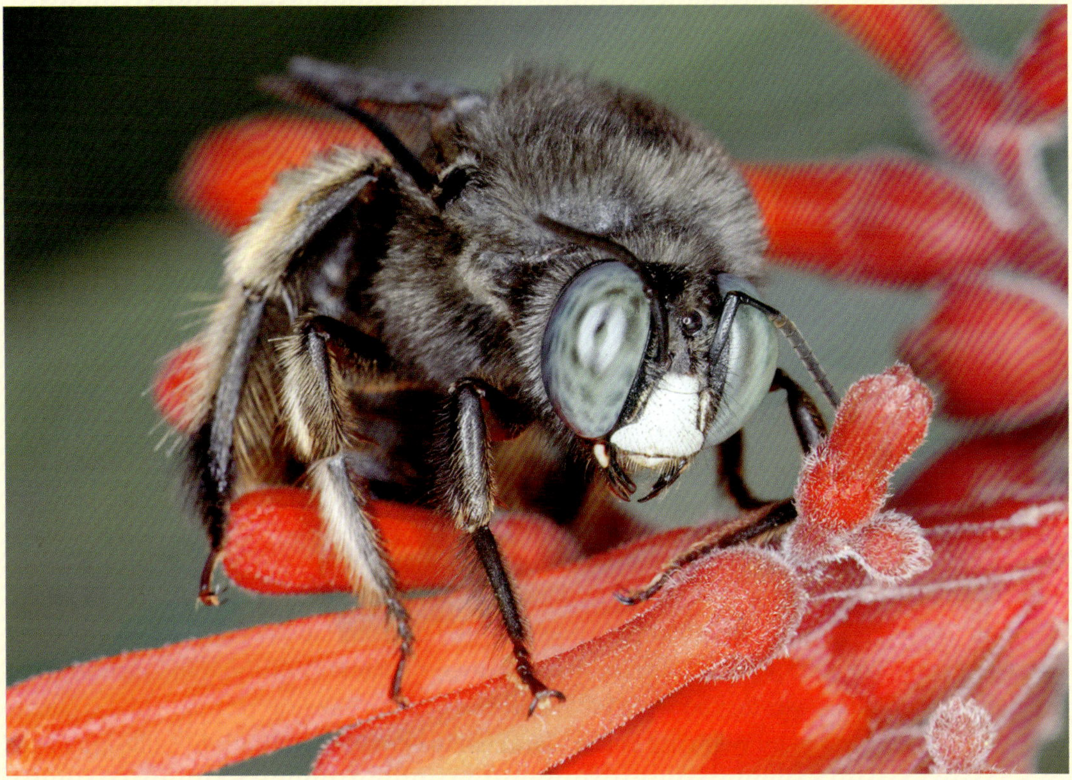

A male parkinsonia carpenter bee

Description: *Xylocopa (Notoxylocopa) tabaniformis parkinsoniae* is a hefty, hairy bee with a dark abdomen girded by bands of pale hairs. Female parkinsonia carpenter bees have dark heads and thoraxes covered with dark hairs, and gray-blue eyes. Males have bulging green eyes, partly ivory faces and mandibles, and grayish hair on their heads and thoraxes. The abdominal bands on both males and females are interrupted in the middle; on males, the sixth abdominal segment (T6) has long patches of white hair on either side. The middle and hind legs of both males and females are partly covered with thick, pale yellow hairs; the front legs are clothed in dark hair. The mandibles of female bees have three teeth on each tip, while males have two. The parkinsonia carpenter bee is currently classified as a subspecies of *Xylocopa tabaniformis*, a species generally referred to as the horsefly-like carpenter bee. **Size:** females and males 17–20 mm (2/3–4/5 in.).

Range: *Xylocopa tabaniformis parkinsoniae* is found in both Hidalgo and Cameron Counties. It ranges into Mexico as far south as the state of Puebla. Within Texas, the parkinsonia carpenter bee occurs at least as far north as Dallas County. Reports of the bee's more western range are muddied because they are commingled with those of other subspecies of *Xylocopa tabaniformis* that extend much farther west than *X. tabaniformis parkinsoniae*.

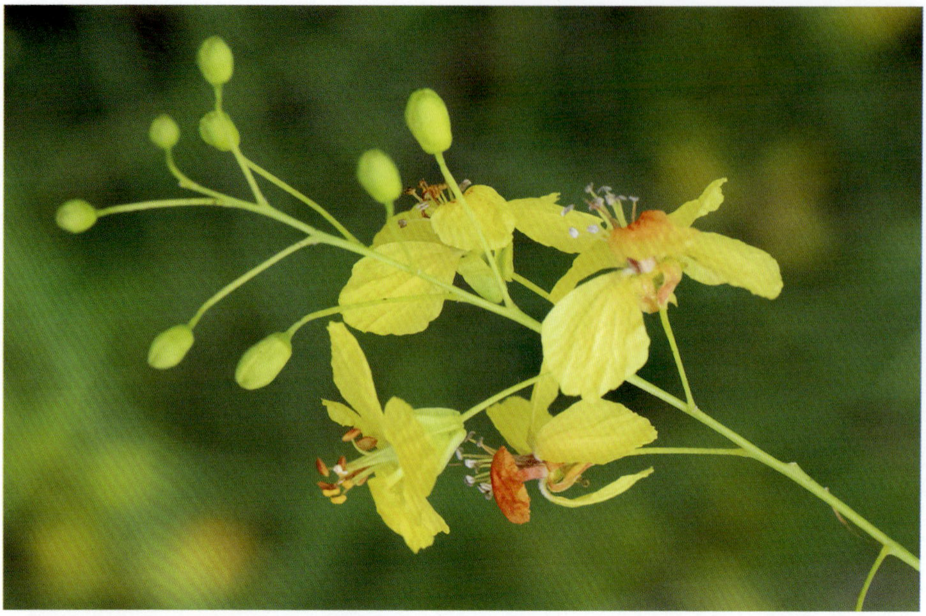

Retama flowers (*Parkinsonia aculeata*)

Associated plants: *Xylocopa tabaniformis parkinsoniae* sometimes appears mislabeled under the common name "Parkinson's carpenter bee," presumably a mistranslation of the bee's Latin name. The name *parkinsoniae* does not derive from the human surname Parkinson, but instead from the plant *Parkinsonia aculeata*, known in the Lower Rio Grande Valley as retama. This carpenter bee was first documented by Wilmatte Porter Cockerell in her 1917 publication "Collecting Bees in Southern Texas." She found the bee on *Parkinsonia aculeata* in San Benito (Cameron Co.).[10]

In a 1981 survey of Hymenoptera of the Lower Rio Grande Valley, Charles C. Porter observed that carpenter bees tend to frequent the topmost flowers of trees. He noted that parkinsonia carpenter bees in particular liked visiting the uppermost flowers of *Parkinsonia aculeata*. Porter also recorded this subspecies feeding on croton, mesquite (*Prosopis juliflora*), huisache (*Vachellia farnesiana*), and fringed twinevine (*Funastrum cynanchoides*). Parkinsonia carpenter bees are now known to visit flowers from a wide range of families. In the Lower Rio Grande Valley, this subspecies is closely associated with esperanza (*Tecoma stans*), scarletbush (*Hamelia patens*), flame acanthus (*Anisacanthus quadrifidus*), and purple passionflower (*Passiflora incarnata*).

ID tips: The species *Xylocopa tabaniformis* contains more than 10 subspecies of varying appearances, all found in North, Central, and South America. Of these, only the parkinsonia carpenter bee is native to the Valley. Two other subspecies occur in the United States—*Xylocopa tabaniformis orpifex* and *X. tabaniformis androleuca*. Both of these are predominantly black and lack the distinctive pale bands of *Xylocopa tabaniformis parkinsoniae*.

10. *Xylocopa tabaniformis parkinsoniae* was first described by T. D. A. Cockerell in an addendum to Wilmatte Porter Cockerell's 1917 article.

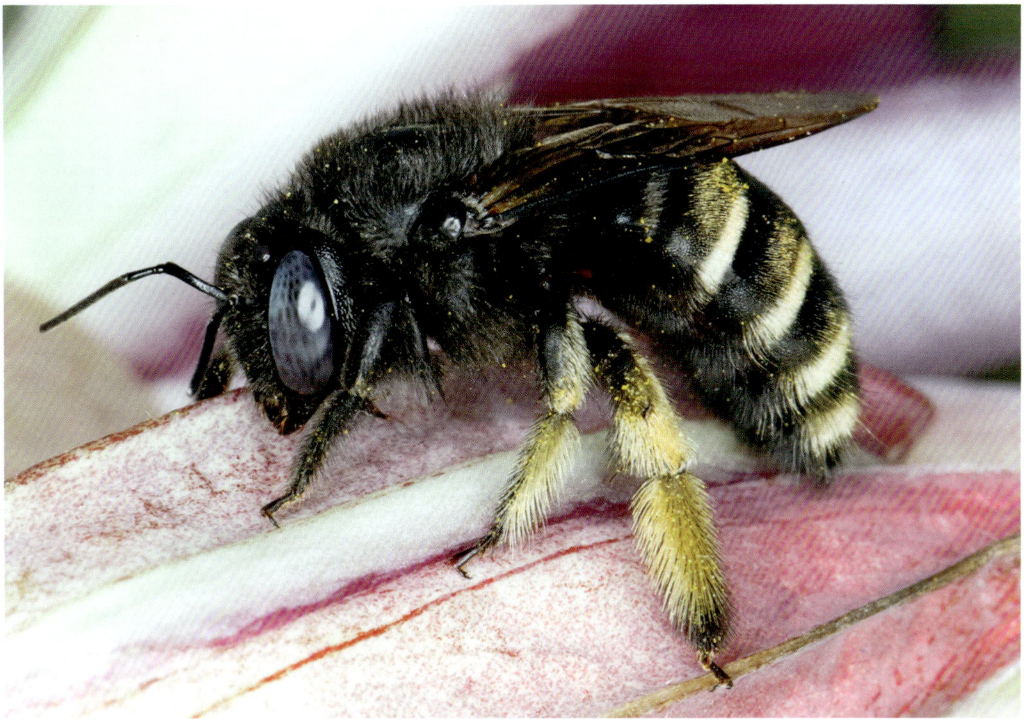

Female (*above*) and male (*below*) parkinsonia carpenter bees: females' heads and thoraxes are covered entirely with dark hairs; commingled light and dark hairs give the heads and thoraxes of males a grayish appearance. Both sexes have interrupted pale hair bands ringing their abdomens, and pale hairs partly covering their middle and hind legs.

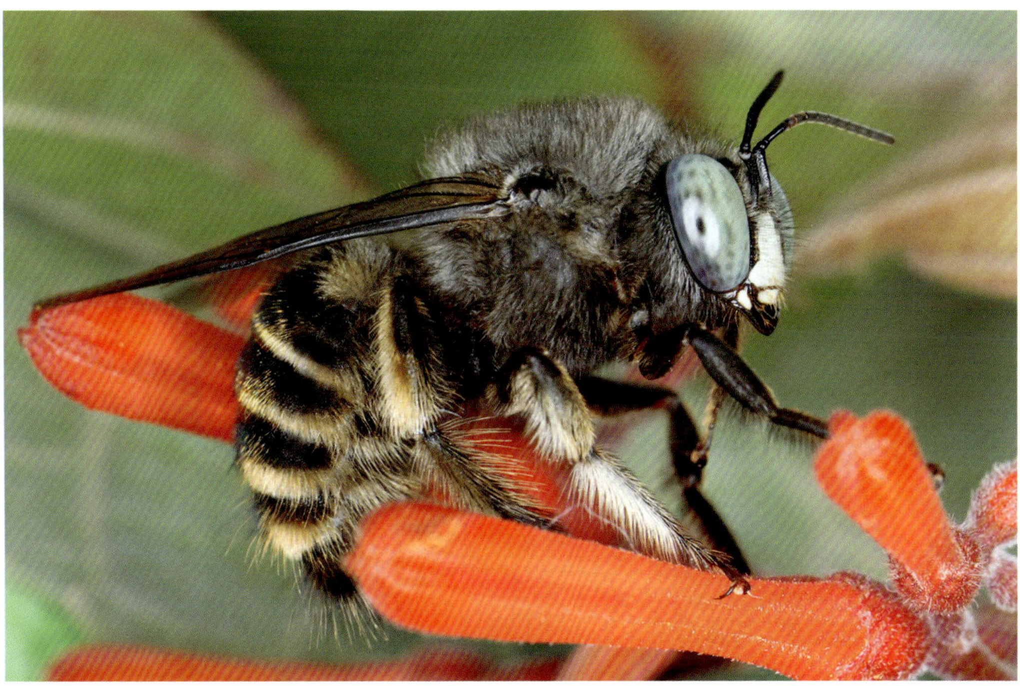

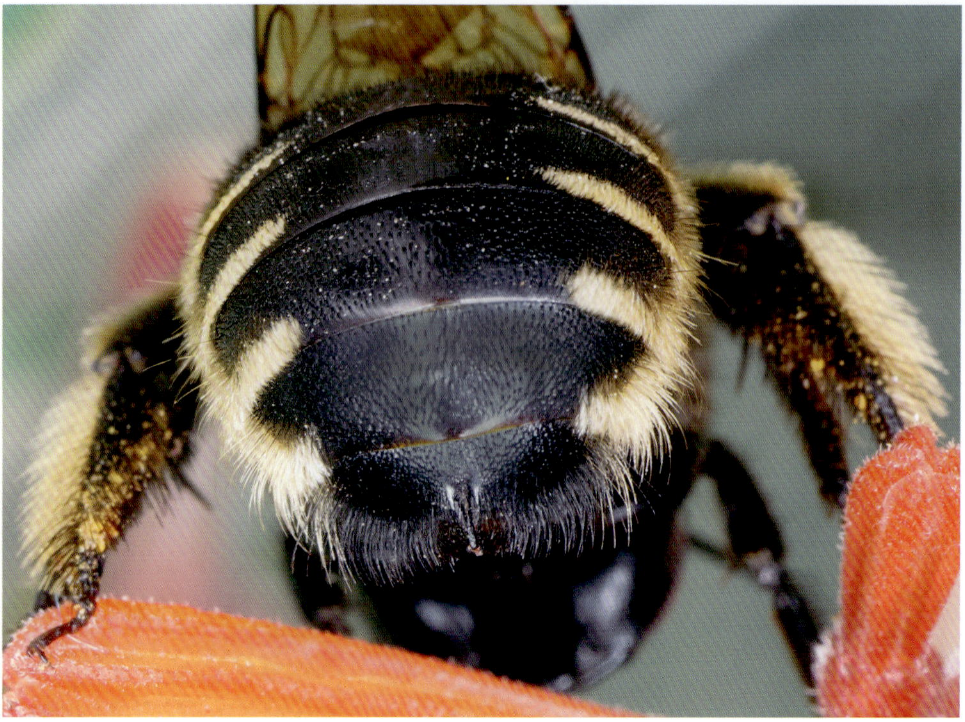

The pale interrupted hair bands on the abdomen of a female (*above*) and male (*below*) parkinsonia carpenter bee: note the long lateral hairs on T6 of the male bee.

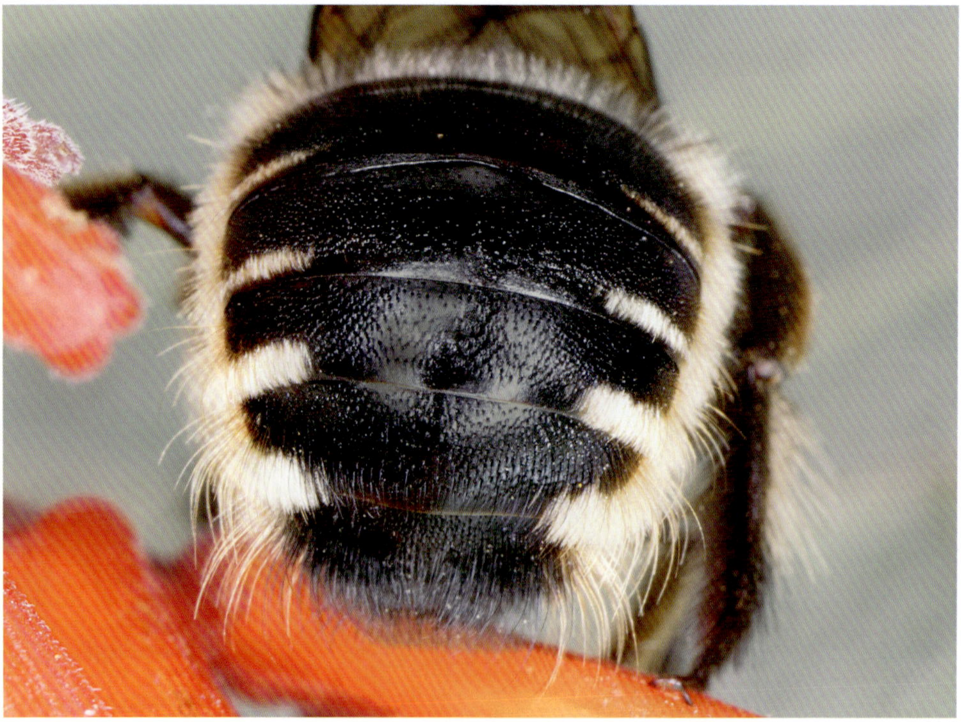

Griswold's carpenter bee ◆ *Xylocopa (Neoxylocopa) griswoldi*
Mérida, Hinojosa-Díaz & Ayala, 2022

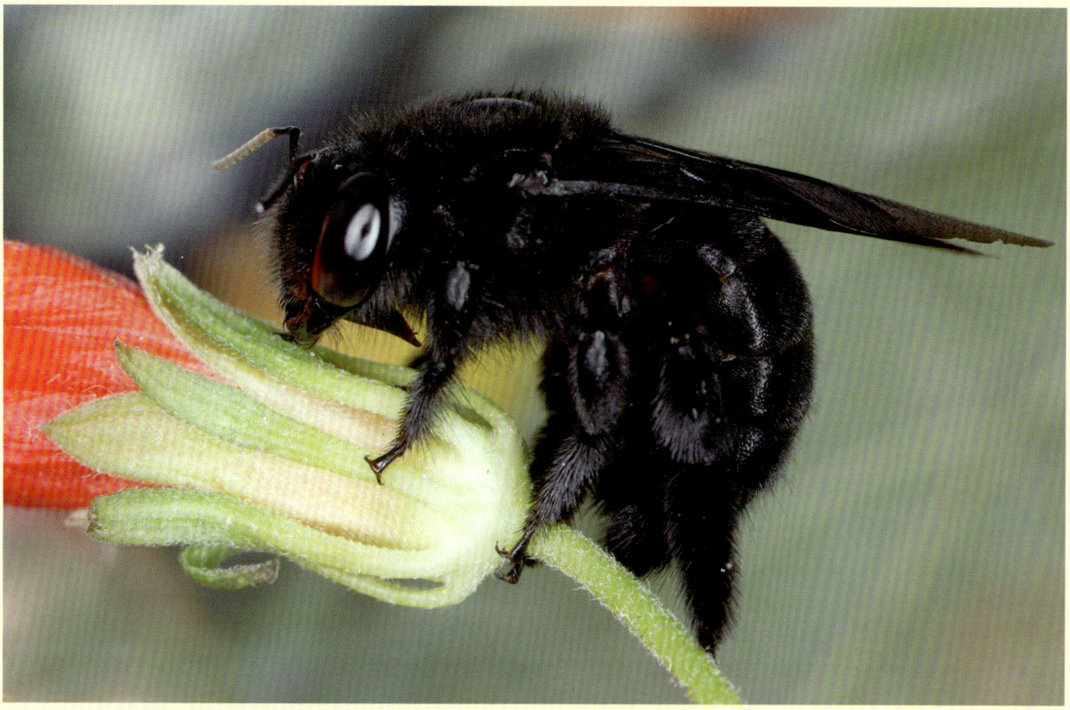

A female Griswold's carpenter bee

Description: *Xylocopa (Neoxylocopa) griswoldi* is the largest bee of the Lower Rio Grande Valley—a visitor to the National Butterfly Center once described Griswold's carpenter bees as "flying black mice." Females are robustly built. Their heads, bodies, and legs are thickly covered with black hair. The integument of females' heads and abdomens is also black. Females' tegulae are black, and their wings are dark brown with a generally purple-to-magenta iridescence that becomes bluer near the bases. Male Griswold's carpenter bees differ greatly in appearance from females. Males' heads, bodies, and legs are covered with lush tawny hairs. The integument of the male's head and body is predominantly yellowish orange, its tegulae are reddish gold, and its wings light brown or orangish brown with brown or reddish veins. Females have black faces, black antennae, and dark eyes; males have predominantly yellowish-orange faces with some brown markings, yellow-and-brown antennae, and green eyes. **Size:** females 24.5 mm (nearly 1 in.), males 25.5 mm (1 in.).

Until recently, this bee was known in the Valley as the Mexican carpenter bee (*Xylocopa mexicanorum*). In 2022, Mexican entomologist Jorge Mérida-Rivas and his colleagues produced a revision of subgenus *Xylocopa (Neoxylocopa)* that unveiled the description of a new species—*Xylocopa griswoldi*. The species *Xylocopa mexicanorum* is now considered endemic to central and southern Mexico, and bees previously identified in the Valley as *X. mexicanorum* are now presumed to be *X. griswoldi*. These two species differ in subtle traits of the clypei and hind tibiae, apparent under magnification. Female Mexican carpenter bees have wings with bluish-green iridescence rather than the purplish iridescence typical of female *Xylocopa griswoldi*.

Range: In the United States *Xylocopa griswoldi* is found principally in southeastern Texas. It ranges through northeastern Mexico and along the slope of Mexico's Gulf Coast. It has been documented in the Mexican states of Nuevo León, Querétero, San Luis Potosí, Tamaulipas, and Veracruz. Griswold's carpenter bee is common in Cameron, Hidalgo, and Starr Counties.

Associated plants: Griswold's carpenter bees appear in the Valley as early as March and linger through November. In the Valley, they are commonly found on retama (*Parkinsonia aculeata*), Texas paloverde (*Parkinsonia texana*), esperanza (*Tecoma stans*), flame acanthus (*Anisacanthus quadrifidus*), Turk's cap (*Malvaviscus arboreus* var. *drummondii*), purple passionflower (*Passiflora incarnata*), and silverleaf nightshade (*Solanum elaeagnifolium*). *Xylocopa griswoldi* is a key pollinator of anaqua (*Ehretia anacua*), a tree in the borage family (Boraginaceae) native to Mexico and South Texas. Griswold's carpenter bees are prolific nectar robbers, perhaps because their extraordinary size prevents them from entering a multitude of flowers.

ID tips: Carpenter bees measuring 24 mm or more that appear in the Lower Rio Grande Valley are nearly invariably *Xylocopa griswoldi*. It is notable nonetheless that the female *Xylocopa griswoldi* can be difficult to distinguish with the naked eye from Strand's carpenter bee (*Xylocopa strandi*), despite the former's superior size. The differences between female *Xylocopa griswoldi* and *X. strandi* are set forth in the next entry.

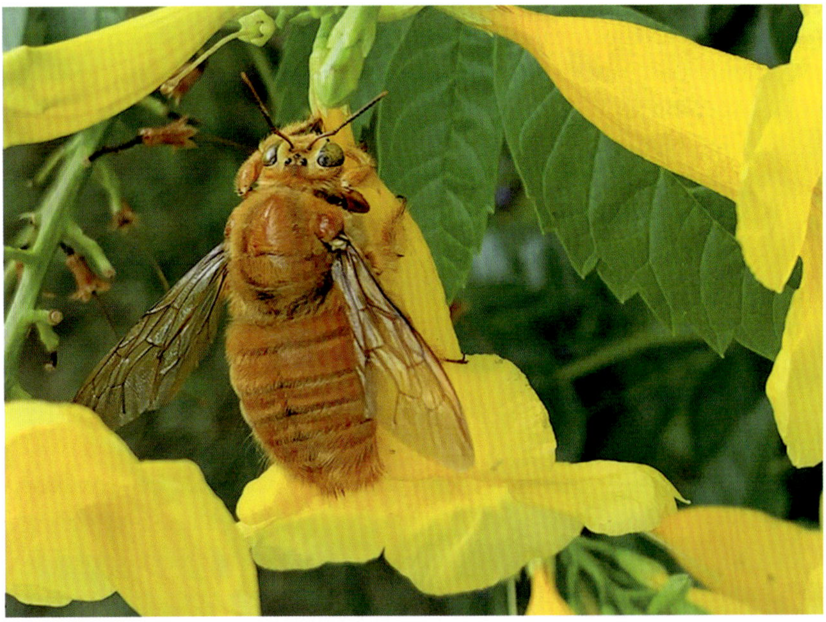

A male *Xylocopa griswoldi* robbing nectar from the base of an esperanza blossom. Photo credit Beth Willoughby.

Strand's carpenter bee ◆ *Xylocopa (Stenoxylocopa) strandi* Dusmet y Alonso, 1924

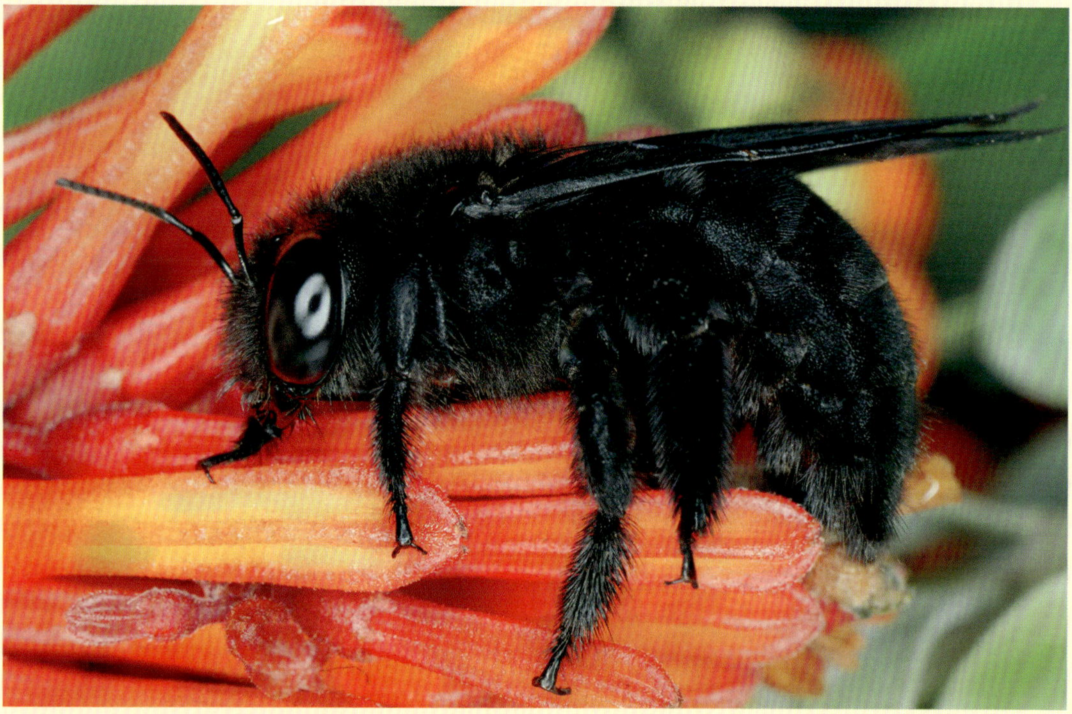

A female Strand's carpenter bee

Description: Female *Xylocopa (Stenoxylocopa) strandi* are robust black bees covered with jet-black hair. Their faces, eyes, and tegulae are black, and their wings are dark, nearly opaque, and tinted with blue and violet iridescence. Male bees have dark heads and bodies covered with brown hairs. Beige hairs commingled with darker hairs cover the thorax of the male bee, making it look grayish in flight. Yellowish-brown hairs form hazily defined bands on the male's abdomen, commingling with some darker hairs toward the front. The male's tegulae are brown, and its wings are transparent brown with gold and purple highlights. Males have green eyes and pale yellow facial markings. **Size:** females and males 19–22 mm (3/4–7/8 in.).

Xylocopa strandi was first documented in 1924 by the Spanish entomologist José María Dusmet y Alonso, who described a single 19 mm male specimen harbored in the collections of Madrid's Museo Nacional de Ciencias Naturales. The specimen had arrived from Costa Rica. Dusmet y Alonso named the species after the Norwegian entomologist Embrik Strand and included it in "Los *Xylocopa* en las colecciones de Madrid," a treatise dedicated to describing 94 distinct large carpenter bee species from Asia, Africa, and the Americas.

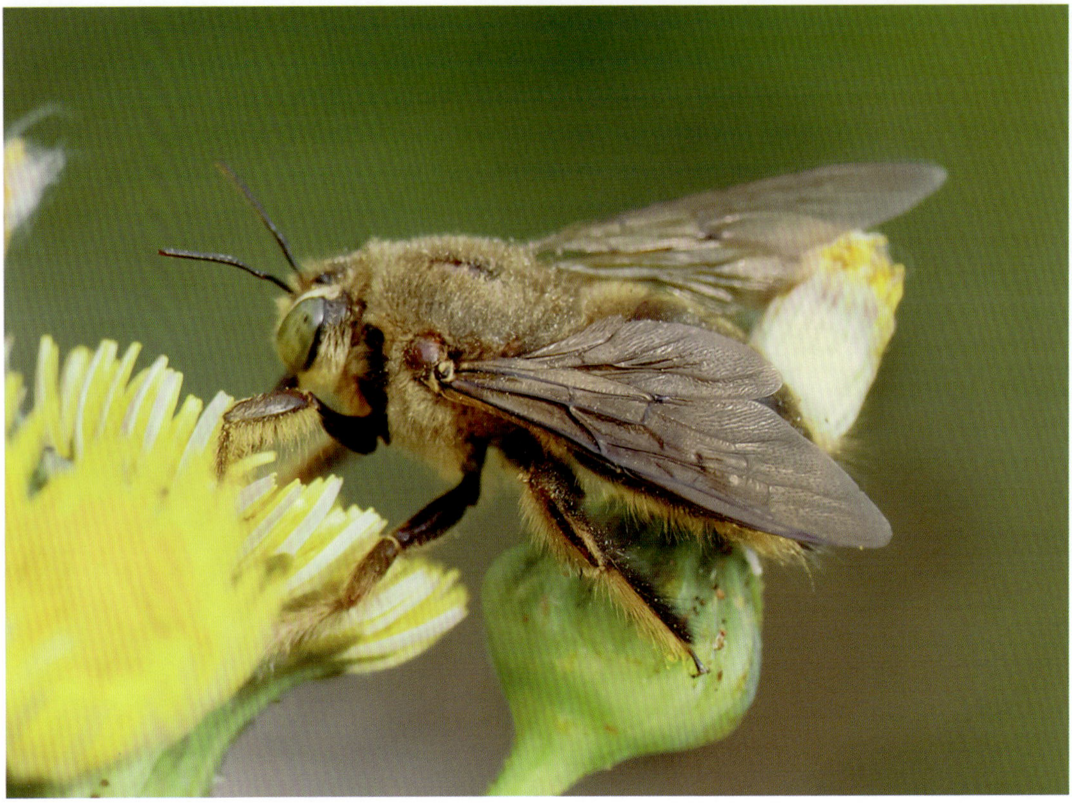

A male Strand's carpenter bee. Photo credit Cin-Ty Lee.

Range: In the United States, Strand's carpenter bees are found nearly exclusively in Texas. Their population extends southward through Mexico to the state of Chiapas. Other than Dusmet y Alonso's specimen, there are few records of this species as far south as Costa Rica. Strand's carpenter bees appeared at the National Butterfly Center in September 2018.

Associated plants: *Xylocopa strandi* nest in bamboo and similar plants. The female bee shown here was foraging on scarletbush (*Hamelia patens*) and red-flowered Turk's cap (*Malvaviscus arboreus* var. *drummondii*). A 2012 study of carpenter bees in Nuevo León documented *Xylocopa strandi* visiting Mexican holdback (*Caesalpinia mexicana*), a yellow-flowered shrub native to the Lower Rio Grande Valley.

ID tips: *Males*: The facial markings of male Strand's carpenter bees are distinctive. On almost all male large carpenter bees of the subgenus *Stenoxylocopa*, two pale yellow bands run from the vertex to the mandibles. On male *Xylocopa strandi*, the bands hug the inner edges of the compound eyes, divided by a dark area. Two roughly triangular pale yellow shapes flank the sides of the lower clypeus.

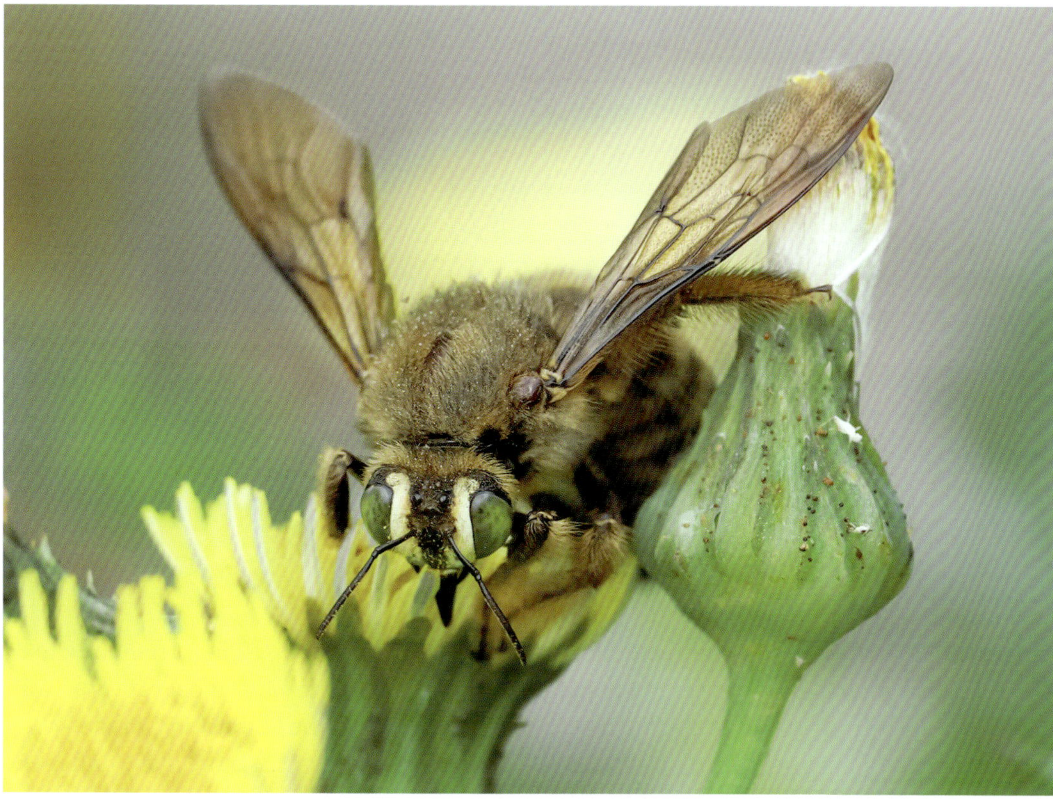

A male Strand's carpenter bee. Photo credit Cin-Ty Lee.

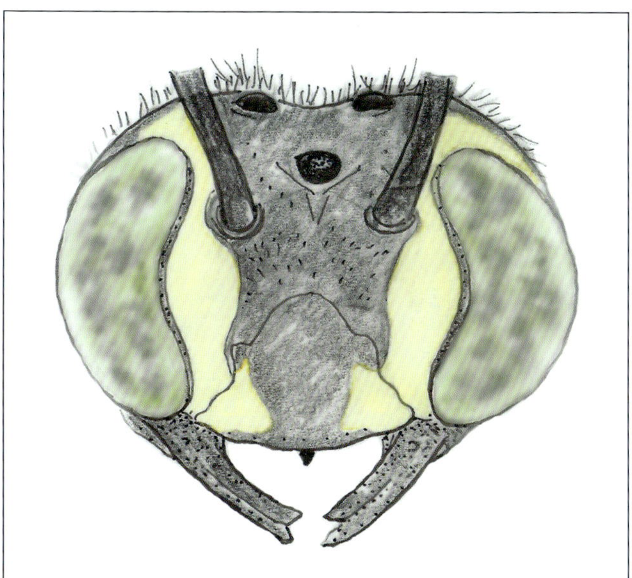

Yellow markings on the face of a male Strand's carpenter bee

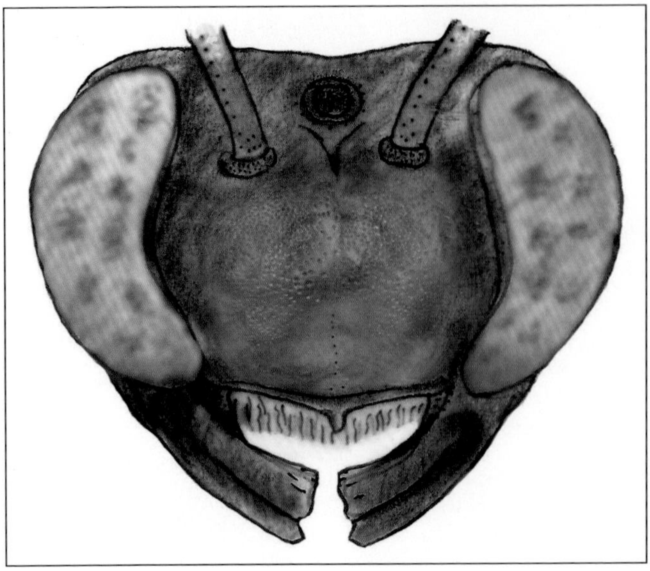

Head and mandibles of a female Strand's carpenter bee: the outer teeth of the mandibles are narrower than the inner teeth.

Females: The female Strand's carpenter bee is difficult to distinguish in the field from the female Griswold's carpenter bee. The latter is larger, but both are hefty, dark-winged bees covered with black hairs. Notably, Griswold's carpenter bees are very common in the Lower Rio Grande Valley, while Strand's carpenter bees are only occasionally documented. Thus, the more likely find in the region is *Xylocopa griswoldi*.

Differentiating between the two species may require close examination of the mandibles. A minute feature related to the nesting habits of *Xylocopa strandi* sets this species apart from other regional carpenter bees: the mandibles of females of the subgenus *Stenoxylocopa* are specially adapted for tunneling in bamboo. The outer teeth on the mandibles of female Strand's carpenter bees are narrower than the inner teeth. This feature allows them to razor deeply into the culms of bamboo to carve out nest chambers.

Southern carpenter bee ◆ *Xylocopa (Schonnherria) micans* Lepeletier, 1841

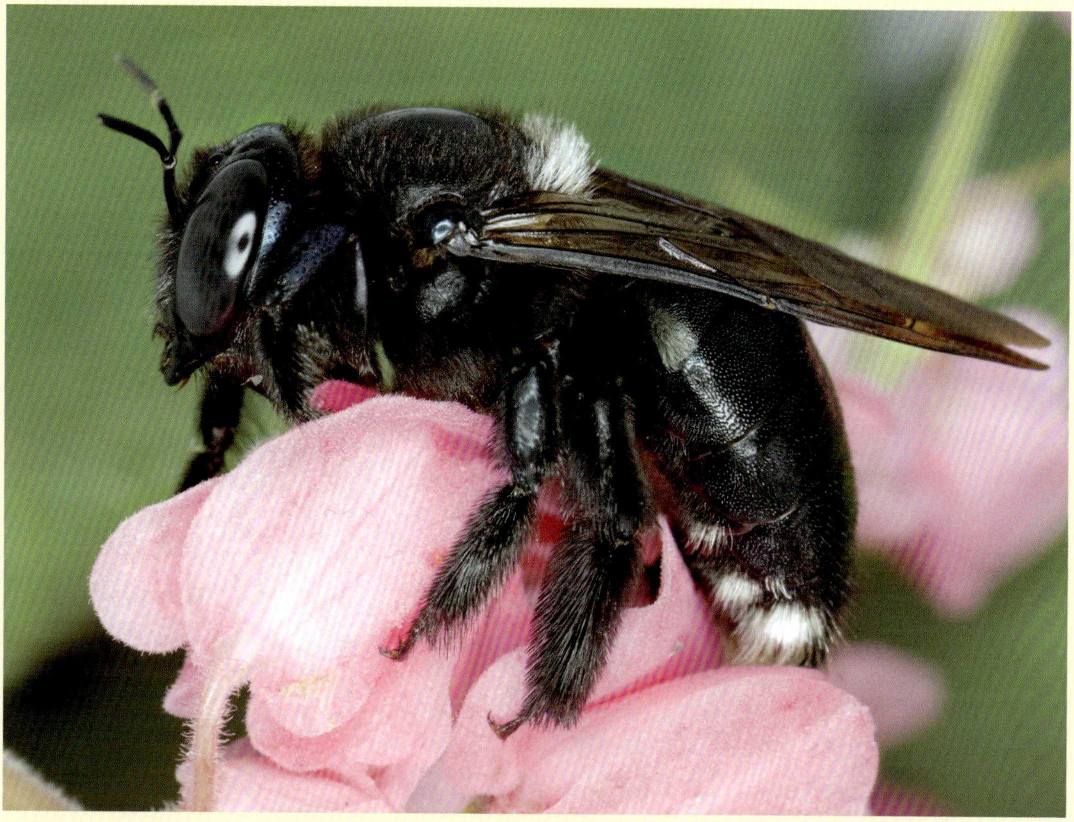

A female southern carpenter bee

Description: Female *Xylocopa (Schonnherria) micans* are purplish-black bees with abdomens that shimmer with blue-violet and green highlights. Females' heads, thoraxes, and legs are clothed in black hair. Their antennae, eyes, and tegulae are black, and their wings are dark with an intense violet and blue iridescence. Tufts of white hair rim the back of the female's thorax and the lower edges of its last three abdominal segments (T4–T6). Male southern carpenter bees look very different: they have thick manes of tawny thorax hair. Their faces sport pale yellow masks, and their eyes are green and bulging. **Size:** females and males 15–19 mm (3/5–3/4 in.).

Range: In the United States, southern carpenter bees are found along the Gulf from southern Texas to southern Florida. *Xylocopa micans* additionally occurs as far west as California, as far east as Virginia, and as far north as Illinois. The species extends southward from Texas into Mexico and Guatemala. *Xylocopa micans* is common in all three border counties of the Valley from spring through fall.

Associated plants: Southern carpenter bees are important generalist pollinators of native and garden flowers. In the Lower Rio Grande Valley, *Xylocopa micans* emerges in May and is frequently associated with flame acanthus (*Anisacanthus quadrifidus*), Mexican shrimp plant (*Justicia brandegeeana*), and purple passionflower (*Passiflora incarnata*).

ID tips: Female southern carpenter bees are easily identified in the Valley by their small size relative to other regional carpenter bee species, and by the distinctive white hairs on their thoraxes and abdomens. Male southern carpenter bees are sometimes confused with male eastern carpenter bees (*Xylocopa virginica virginica*). The latter can be distinguished by their lighter-colored thorax hairs, their somewhat less protuberant eyes, and the lack of yellow stripes on their antennal scapes.

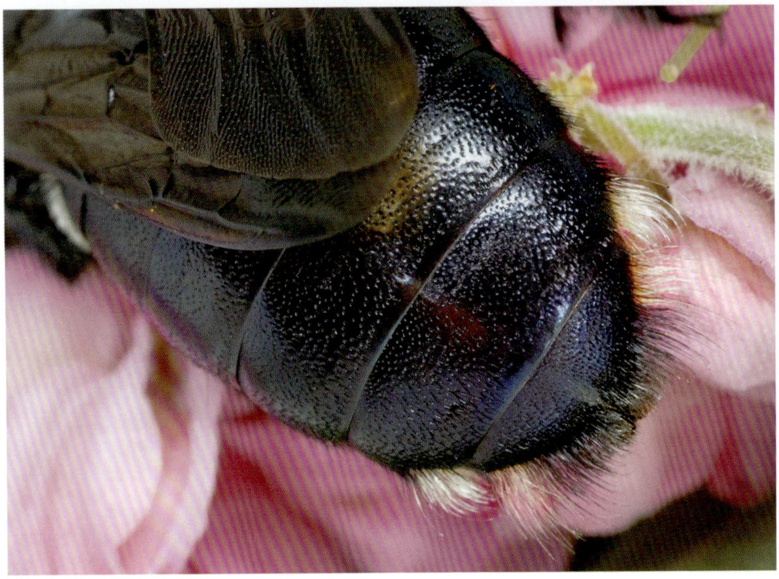

The female southern carpenter bee has distinctive tufts of pale hairs on either side of its fifth and sixth abdominal segments.

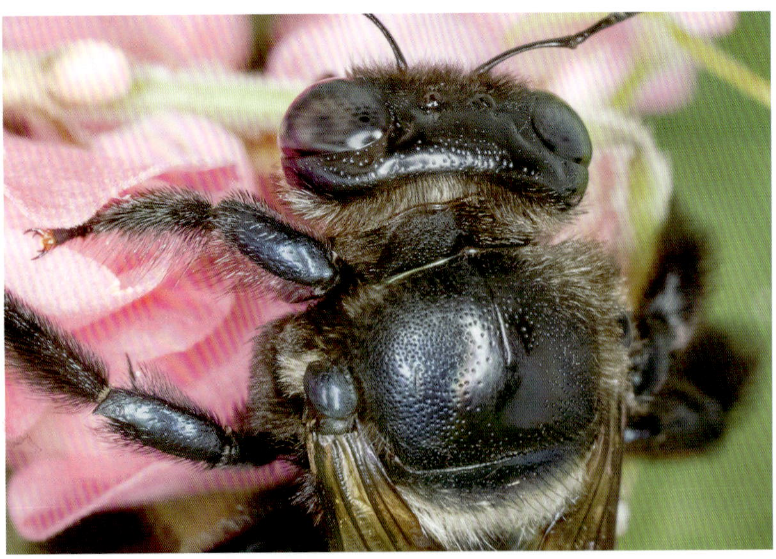

The female bee's thorax is pitted and bisected by a groove on the front half, and rimmed by pale hairs on the back edge.

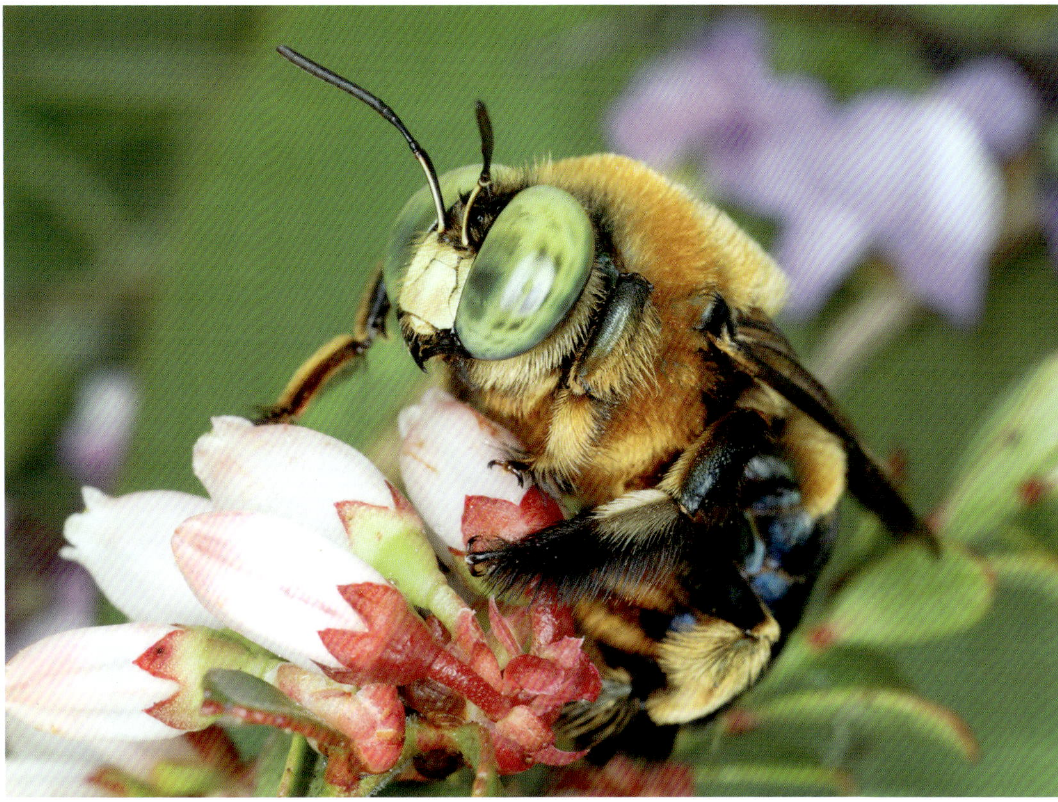

A male southern carpenter bee: males are easily identified by their bulbous green eyes, pale facial masks, and manes of golden thorax hair.

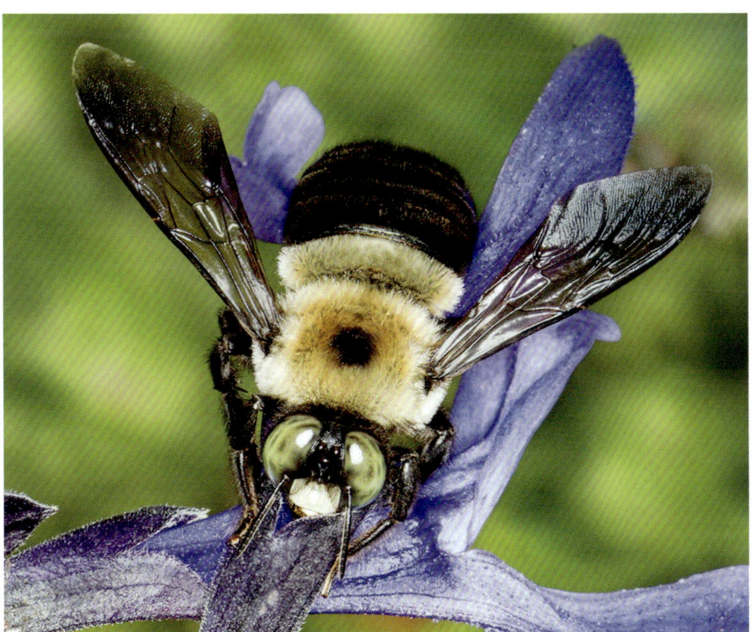

Shown here, the male eastern carpenter bee (*Xylocopa virginica virginica*), found farther north in Texas, somewhat resembles the male southern carpenter bee (*X. micans*). Males of both species have green eyes and pale masks covering their faces just below the antennae. The thorax hairs of the southern carpenter bee, however, are a much richer golden brown and more copious; those of the eastern carpenter bee are trim and beige.

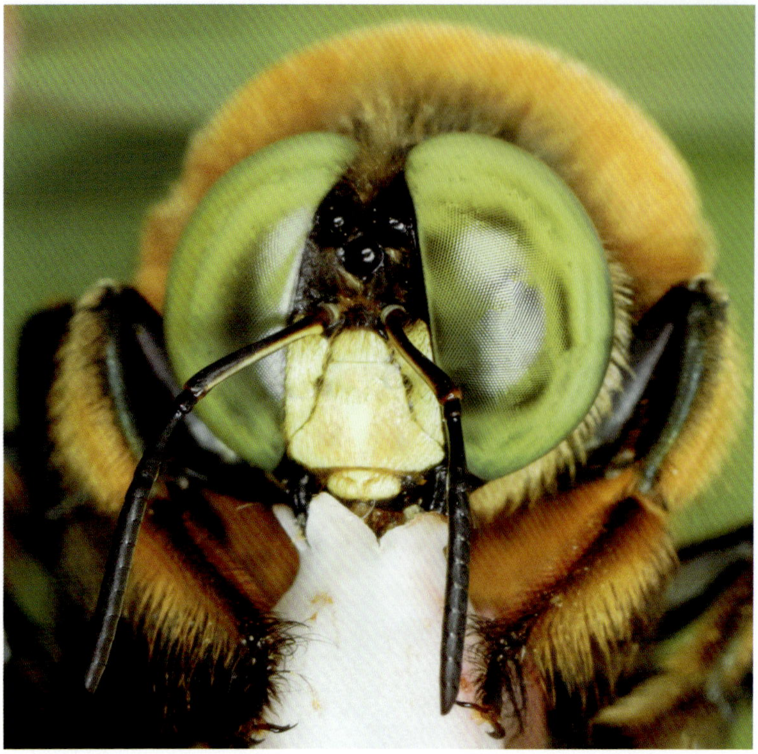

Head of a male southern carpenter bee: the yellow stripes on its antennal scapes help distinguish this species from the male eastern carpenter bee, which lacks this trait.

Behavior: Southern carpenter bees are often targeted misguidedly by pest control companies, despite the fact that this species typically does not damage human-made structures. Southern carpenter bees use dead branches, found principally in woodlands and prairies, for nesting and depend on these materials for survival. When discovered, southern carpenter bees should be left undisturbed because of their key role in pollinating native plants and commercial blueberry and tomato-family crops.

Small Carpenter Bees

Subfamily Xylocopinae ◆ Tribe Ceratinini
Genus *Ceratina*

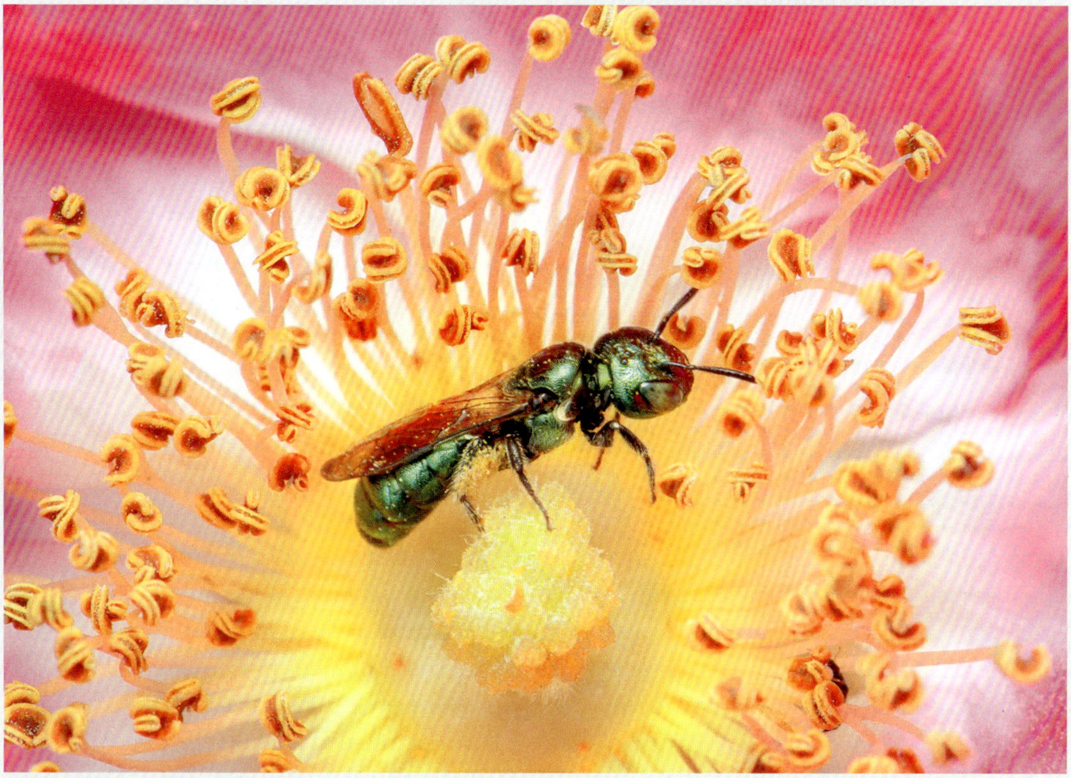

A female small carpenter bee

Small carpenter bees belong to the same subfamily, Xylocopinae, as large carpenter bees. The two groups are so different to the naked eye that few casual observers would think them related: *Xylocopa* are robust and hairy bees, frequently black, while *Ceratina* are small and slender, usually sparsely haired, and dull metallic green. Rau wrote in 1928 that the two genera showed "all the difference of David and Goliath, the one large and powerful in build, the other small and delicate."[11]

Nonetheless, in *The Bees of the World*, Michener noted that the two genera share both physical and behavioral characteristics: among other traits, both large and small carpenter bees have relatively flat faces; usually nest in dead herbaceous or woody plant materials; and store food for adult bees, rather than for young only. Such behavior is unusual in the bee world.

There are approximately 200 species of *Ceratina* worldwide, inhabiting an immense range of habitats. Within the United States, small carpenter bees are represented by roughly 24 docu-

11. Rau (1928), p. 380.

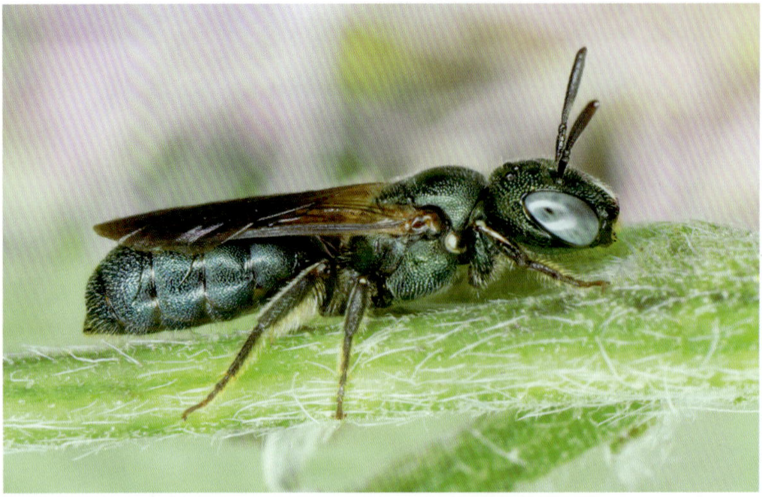

A female *Ceratina*: to the naked eye, small carpenter bees appear ant-like.

mented species—at least a third of these, spanning three subgenera, occur in Texas. *Ceratina* are not common, however, in the Lower Rio Grande Valley. Although small carpenter bees inhabit all continents and can adapt to even the hottest tropical climates, they tend to avoid dry habitats like those found throughout much of the Valley's borderlands.

Genus information: *Ceratina* are easily mistaken for ants: they range from 8 mm (3/10 in.) to less than half that length. The heads and bodies of small carpenter bees are often pitted, faintly iridescent, and greenish. They have relatively large heads and short antennae that are broad at the tips. Their faces and legs often have pale markings. The abdomens of *Ceratina* bell outward and taper abruptly at the ends and are sometimes described as club shaped. *Ceratina* lack a body part common to most bees—the pygidial plate (a flat area, often ridged on the outer rim, frequently found on the sixth abdominal segment of females and the seventh abdominal segment of males). All *Ceratina* are polylectic, and their size allows them to enter small-mouthed flowers unreachable by many larger bees. They are important pollinators of gardens, commercial crops, and wildflowers.

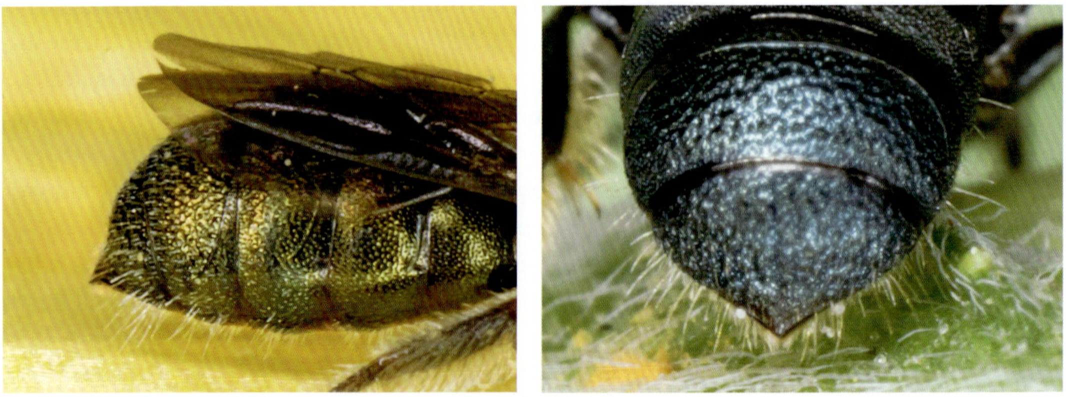

The club-shaped abdomen of a female small carpenter bee: note the lack of a pygidial plate.

Cobalt small carpenter bee ♦ *Ceratina (Calloceratina) cobaltina* Cresson, 1878

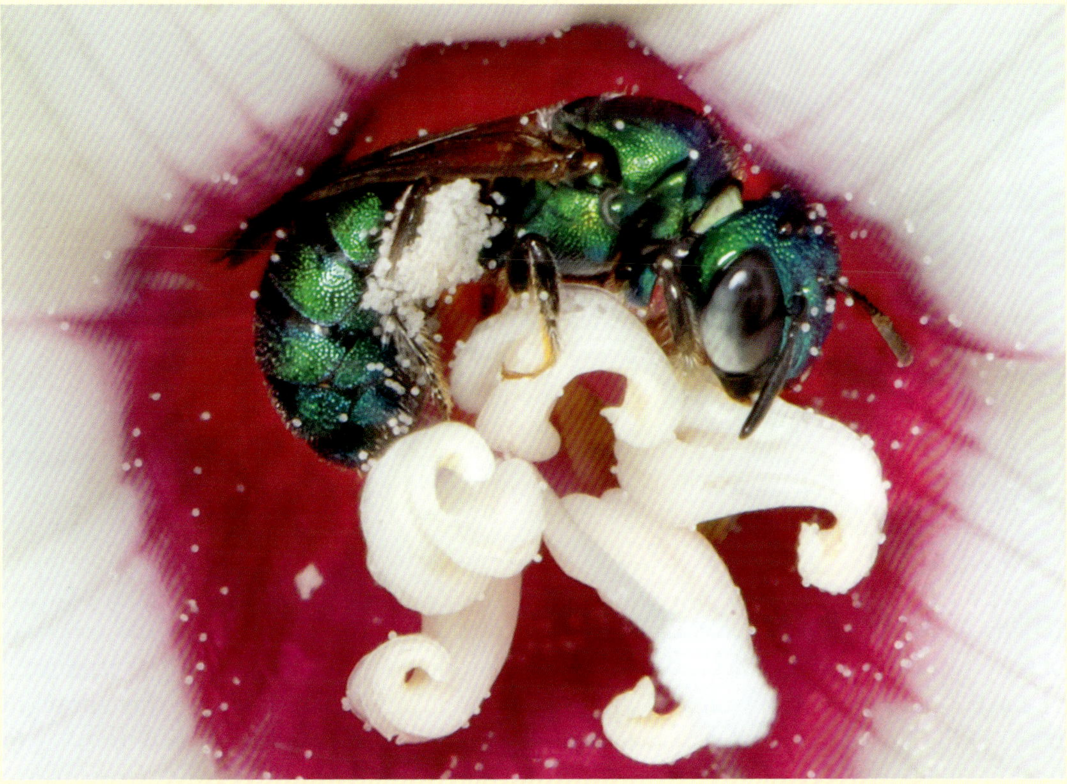

A female cobalt small carpenter bee collecting white pollen from an alamo vine blossom

Description: The Latin prefix *callo-* means "beautiful." *Ceratina (Calloceratina) cobaltina* is a stunning small bee—it is brilliant blue green, with pale markings adorning its face, pronotal lobes, and legs. The mandibles and antennae of *Ceratina cobaltina* are black, and its eyes are grayish green. The bee's head and body are heavily pitted. Its legs and tegulae are dark brown, and the wings are dark colored. **Size:** females and males 7–8 mm (1/4–3/10 in.).

Range: Small carpenter bees of the subgenus *Calloceratina* are primarily Neotropical and range as far south as Brazil. *Ceratina cobaltina* was first identified in 1878 by Cresson, who based his species description on a single male specimen found in Mexico.The distribution of this species is centered in Mexico, although it ranges into southern Texas. It is well established as far north in Texas as Travis County, where it is thought to be an adventive species, possibly transported there on firewood. *Ceratina cobaltina* appears occasionally in Hidalgo County but is generally uncommon in the Valley.

Associated plants: In Hidalgo County, *Ceratina cobaltina* tends to appear in association with alamo vine (*Merremia dissecta*). Locally, it also visits Texas prickly pear (*Opuntia engelmannii* var. *lindheimeri*) and Texas thistle (*Cirsium texanum*).

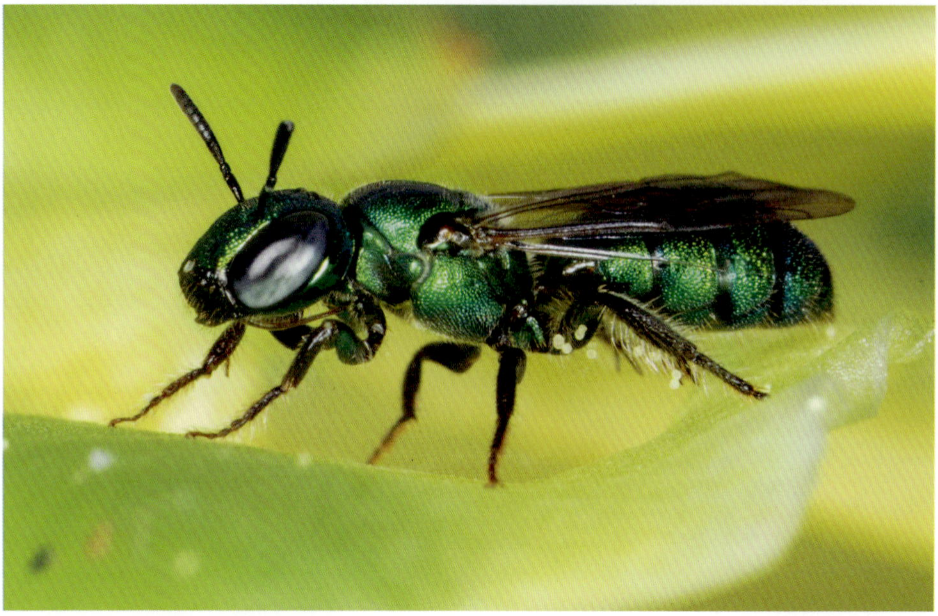

A female *Ceratina cobaltina*

ID tips: The brilliant coloration of *Ceratina cobaltina* is distinctive. The female bee has a coarsely pitted face with black mandibles and black labrum. Its clypeus is metallic green with a black band on the bottom rim marked with a small dot in the middle. The male bee has a three-pronged mark on the lower face, flanked by two small, pale triangles. Males' labrums, and the bases of their mandibles, are ivory.

Additional information: Comparatively little research has been conducted on the behavior of *Ceratina (Calloceratina)* bees. The typical *Ceratina* nest is a linear tunnel excavated in a pithy stem.

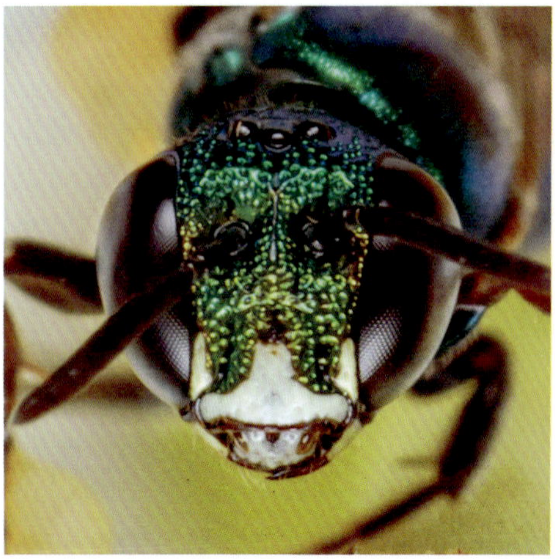

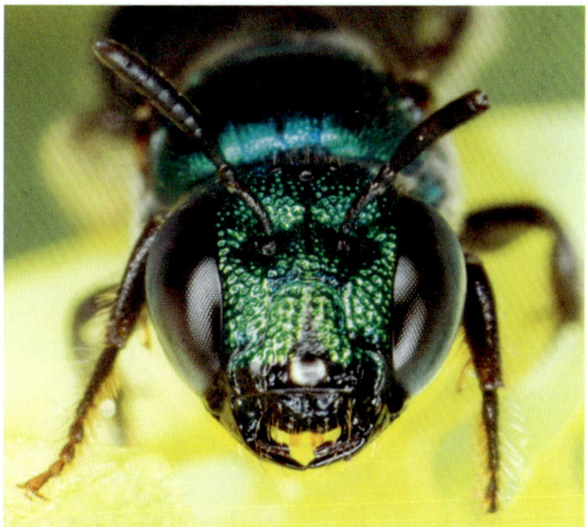

Faces of female (*left*) and male (*right*) *Ceratina cobaltina.* Photo credit Victor Engel.

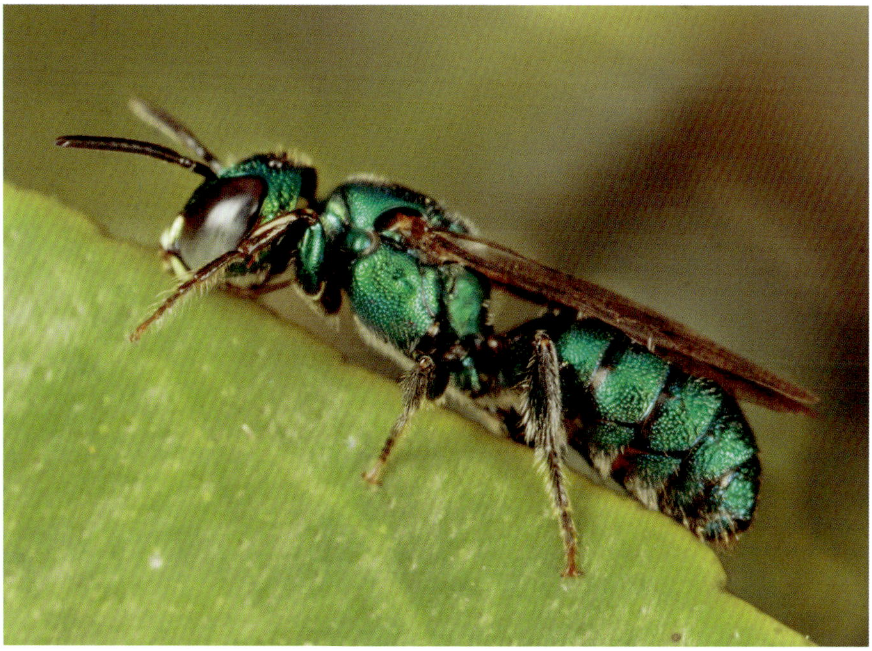

A brightly iridescent male *Ceratina cobaltina.* Photo credit Victor Engel.

In 2004, entomologists Gonzalez, Moreno, and Richards described in detail, with precise illustrations, the nesting habits of *Ceratina (Calloceratina) mexicana currani* (Mexican small carpenter bee). This *Ceratina* nests in the pith of dead sticks, carving out tunnels with rounded, constricted openings. The female bee fills the tunnels with series of small egg chambers divided by partitions made of compacted sawdust, each bearing a small hole in the center.

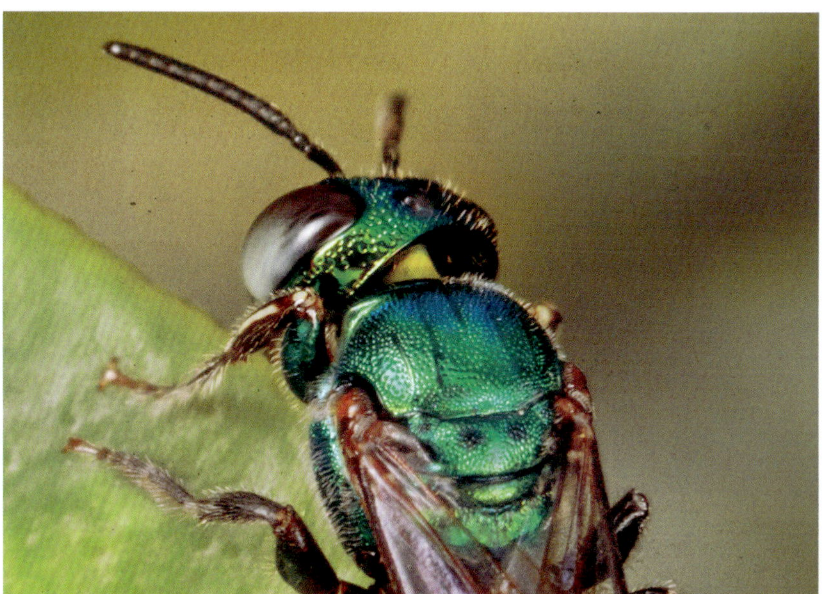

A male *Ceratina cobaltina*: note the carina (ridge) along the back of the head, a trait typical of small carpenter bees. Photo credit Victor Engel.

Behavior of Small Carpenter Bees of the Subgenus *Zadontomerus*

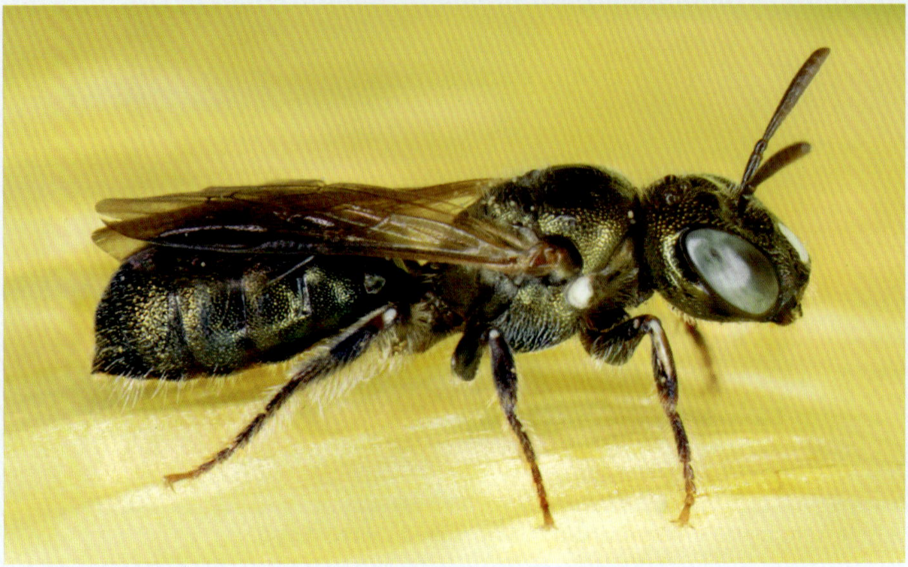

A female Texas small carpenter bee

The Texas small carpenter bee featured in the next entry belongs to the subgenus *Zadontomerus.* This subgenus is widespread, occurring from Canada to Venezuela. In Texas, *Zadontomerus* is represented by an array of species consisting of very similar, faintly metallic bees with pale markings on the face, pronotal lobes, and legs. These include, among others, *Ceratina calcarata*, *C. diodonta*, *C. nanula*, *C. shinnersi*, *C. strenua*, and *C. texana*.

Ceratina are generally described as solitary, but *Ceratina (Zadontomerus)* show a flexibility in behavior that has made them a focus of entomological fascination. Bees are usually characterized as "social" or "solitary" based on the degree to which they act cooperatively. On one end of the spectrum are "eusocial" bees, typified by the highly organized societies of honey bees that work collectively to construct nests, ensure egg production, and rear young. On the other end are solitary bees that build and provision their own nests.

Female *Ceratina (Zadontomerus)* have been described as "subsocial." Although they do not cooperate in building nests or egg laying, they interact in significant ways. In a 2010 study, biologists Rehan and Richards documented long-lived and nest-loyal mother *Ceratina (Zadontomerus) calcarata* caring for their offspring from egg to adulthood, and thus inhabiting a transitional boundary between solitary and social life.

Ceratina (Zadontomerus) females may open brood cells to inspect and groom developing bee larvae, rebuild faulty brood cells, or chew up failed egg chambers and incorporate them into nest walls. As its larvae mature, the mother small carpenter bee may remain in the nest, cohabiting with daughter bees, helping guard the nest against predators and parasites. In the spring, however, the two generations of females disperse. The new females recommence the process of collecting pollen and carving out nest tunnels individually, each setting up its own nest.

Texas small carpenter bee ◆ *Ceratina (Zadontomerus) texana* Daly, 1973

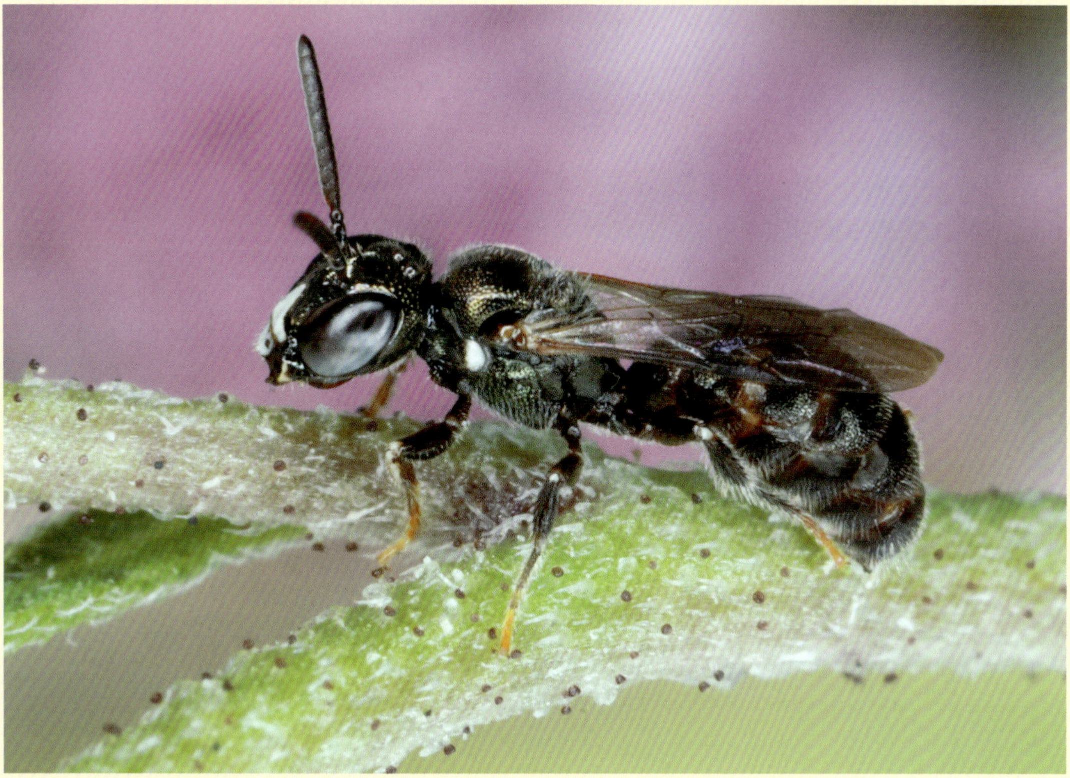

A male *Ceratina texana*

Description: *Ceratina (Zadontomerus) texana* is a diminutive, dark olive-green bee with reddish lower legs and a heavily pitted face and thorax. The head and body of the bee have green and coppery highlights. Both males and females sport pale maculations on the face, legs, and pronotal lobes. The male Texas small carpenter bee has a pale mark on the clypeus described as "hat-shaped" by entomologist Howell V. Daly, who named the species in 1973. The female *Ceratina texana* has a rounded, bell-shaped mark on the clypeus; its mandibles and antennae are partly reddish. Males have predominantly dark antennae, and dark mandibles with pale markings. Males are generally smaller and slenderer than females. **Size:** females and males 5–6 mm (1/5–1/4 in.).

Range: The Texas small carpenter bee is endemic to South Texas. It is common in all three border counties of the Valley.

Associated plants: When defining this species, Daly examined specimens found on Trans-Pecos giant hyssop (*Agastache breviflora*), lemon beebalm (*Monarda citriodora*), blackberries, and raspberries. In the Valley, *Ceratina texana* favors small-flowered plants such as Texas palafox

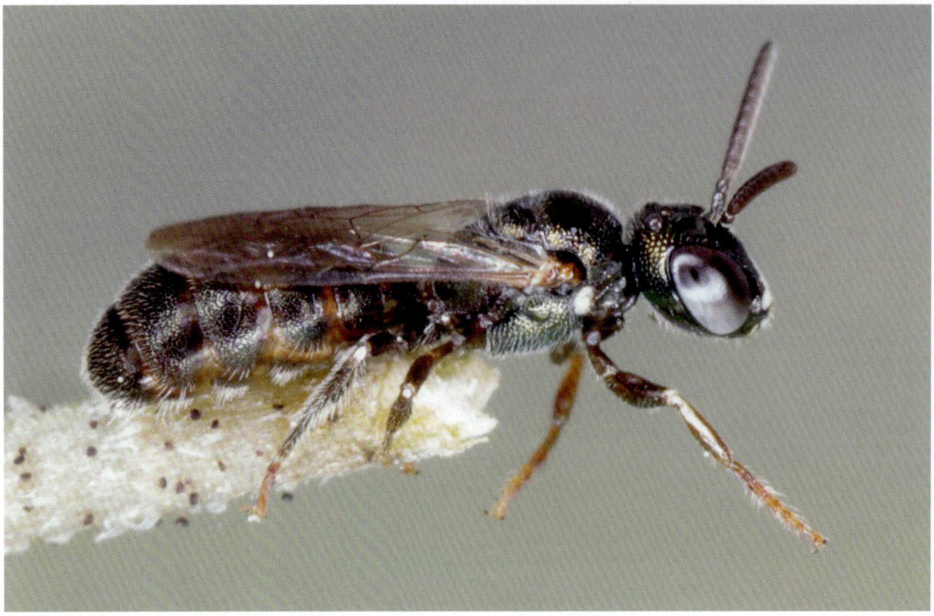

A male Texas small carpenter bee

(*Palafoxia texana*), hoary blackfoot (*Melampodium cinereum*), and wavyleaf nama (*Nama undulatum*). In the spring, the Texas small carpenter bee is among a limited number of Valley bee species that visit golden prickly poppy (*Argemone aenea*) and pitaya (*Echinocereus enneacanthus*).

ID tips: On both male and female Texas small carpenter bees, pale dots mark the base of each tibia. On female bees, an ivory stripe extends more than halfway down each front tibia. The "hat-shaped" mark on the face of the male bee is its most distinctive feature. The male's labrum is also pale, with two dark marks that look like false nostrils.

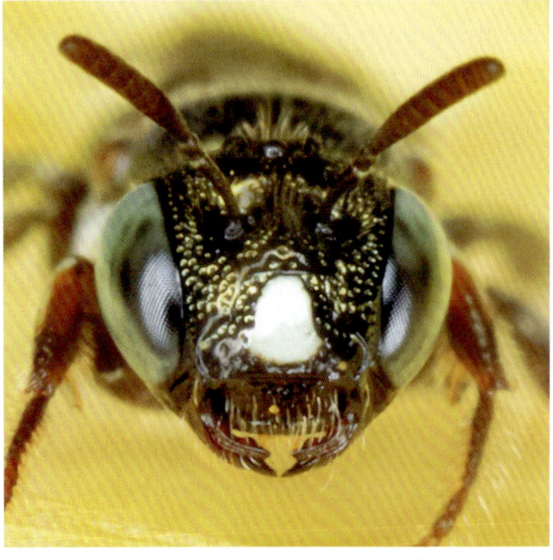

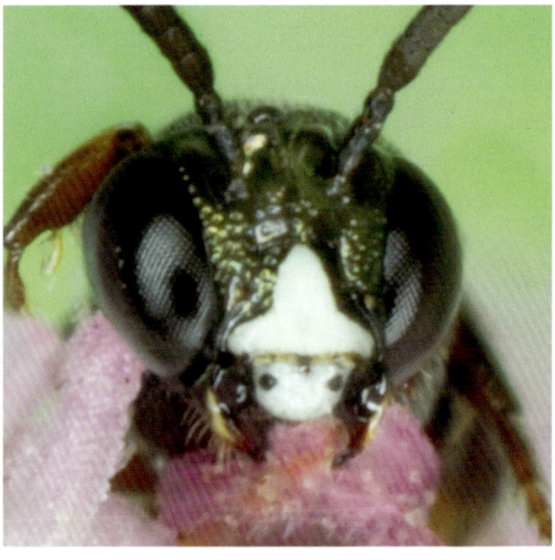

Faces of female (*left*) and male (*right*) *Ceratina texana*

Cuckoo Bees

Subfamily Nomadinae ◆ Tribes Nomadini and Epeolini
Genera *Nomada, Epeolus, Triepeolus*

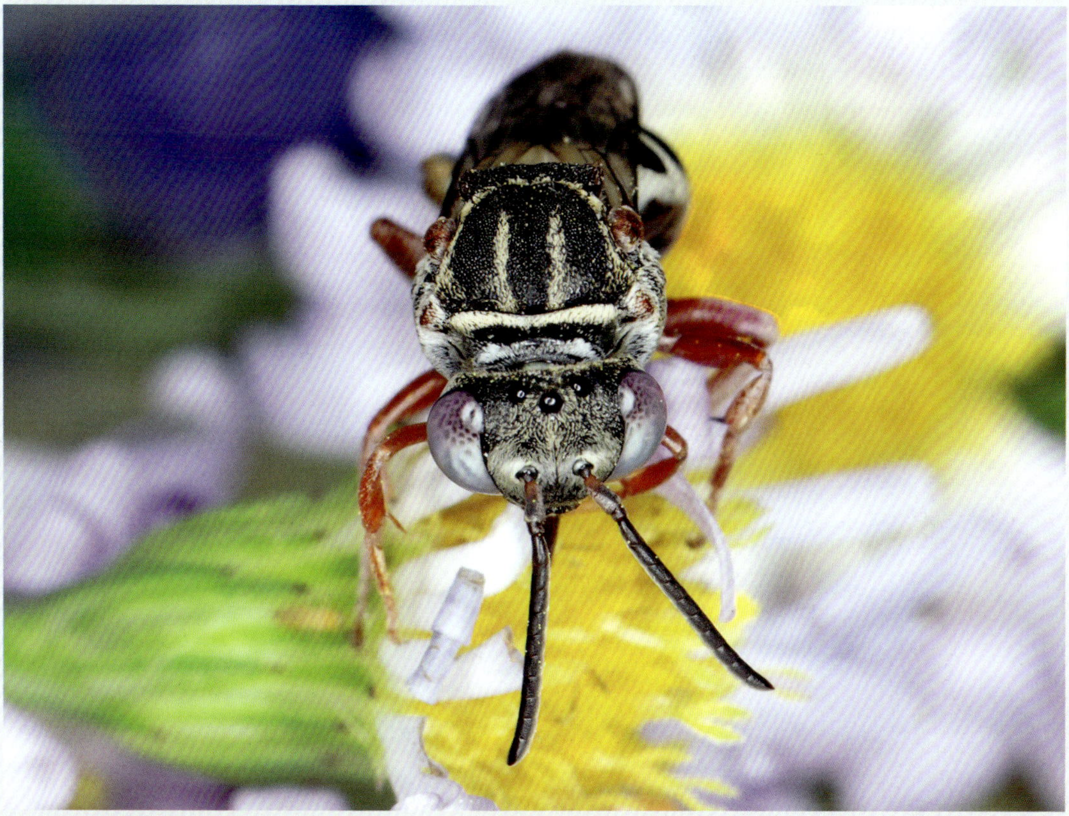

A dwarf epeolus cuckoo bee

The small universe in which wild bees live contains, among other surprises, parasitic cuckoo bees that have evolved over time inextricably linked to the species on which they prey. In the world of entomology, such bees are called cleptoparasites or brood parasites.

Clepto means "theft," and cuckoos are the burglars of the bee world. Cuckoo bees enter other bees' underground nests in order to lay eggs in their hosts' brood cells. Sometimes, the cuckoos destroy their hosts' eggs. More commonly, the cuckoos depart the nest, and when their eggs hatch, the cuckoo larvae annihilate the host bee's eggs or larvae and then feed on the pollen and nectar provisions provided by the host parent for its own progeny.

Because cuckoo bees obtain pollen through such activities, they do not gather it from flowers. Thus, female cuckoo bees lack pollen-collecting scopal hairs. They often look hairless to the naked eye, appearing sleek bodied like wasps. Cuckoo bees sometimes have spade-shaped abdomens or other traits that allow them to dig into the walls of other bees' nests, and cuckoos generally act differently from their hosts—many cuckoo bees spend much of their time skulking around on the ground, looking for their hosts' nests. Cuckoo bees do, however, drink nectar from flowers, and they often gather on plants preferred by their hosts.

Some cuckoos target the nests of bees within their own taxonomic group. Cuckoo leafcutter bees, for example, belong to the family Megachilidae and target the nests of leafcutter bees, which are members of the same family and tribe. *Sphecodes*, which belong to the sweat bee family Halictidae, usually deposit their eggs in the nests of sweat bees of their shared subfamily Halictinae. Both of these varieties of cuckoo bee are described in the sections on the families Megachilidae and Halictidae.

The vast majority of North American cuckoo bees belong to the family Apidae and to the subfamily Nomadinae. Cuckoo bees in this subfamily often cross categories to target bees of other families. Thus, for example, *Nomada* commonly lay their eggs in the brood cells of *Andrena*, which belong to the family Andrenidae. *Epeolus* cuckoos attack the nests of *Colletes*, which belong to the family Colletidae. *Triepeolus* cuckoos target a wide range of host bees from several families.[12]

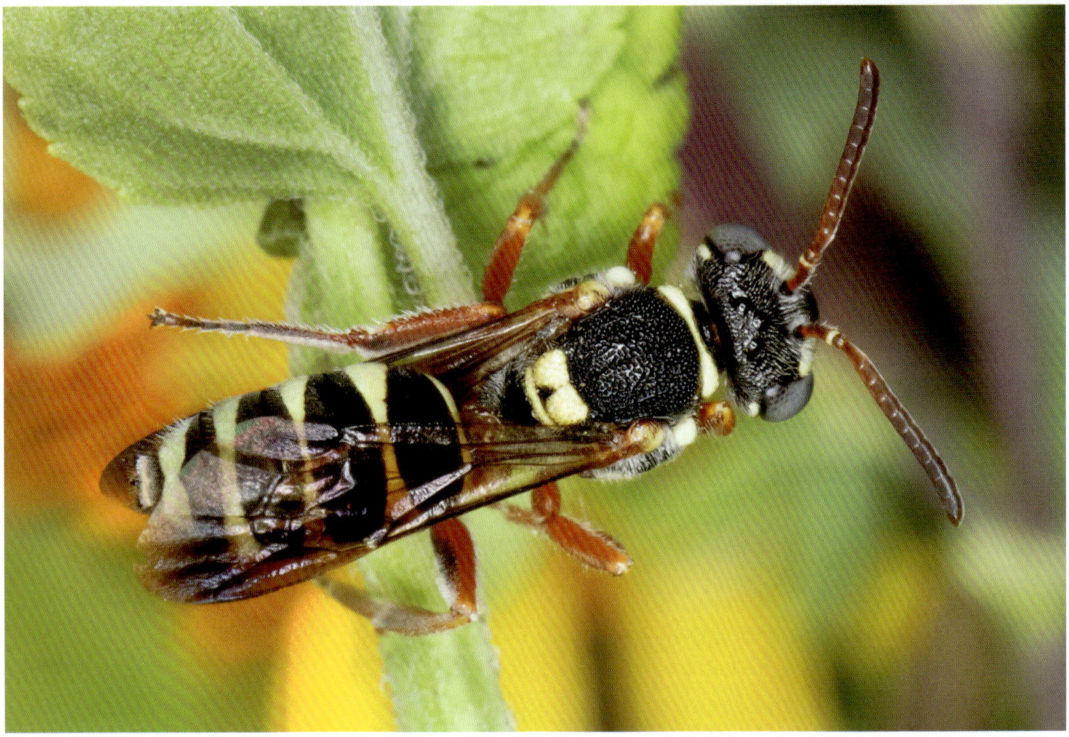

A female Texas nomad bee

12. Nomadinae of the Lower Rio Grande Valley are underdocumented. Less common Nomadinae genera that appear in the Valley, but are not shown here, include *Ericrocis*, *Holcopasites*, *Leiopodus*, *Neolarra*, and *Paranomada*.

Texas nomad bee ◆ *Nomada texana* Cresson, 1872

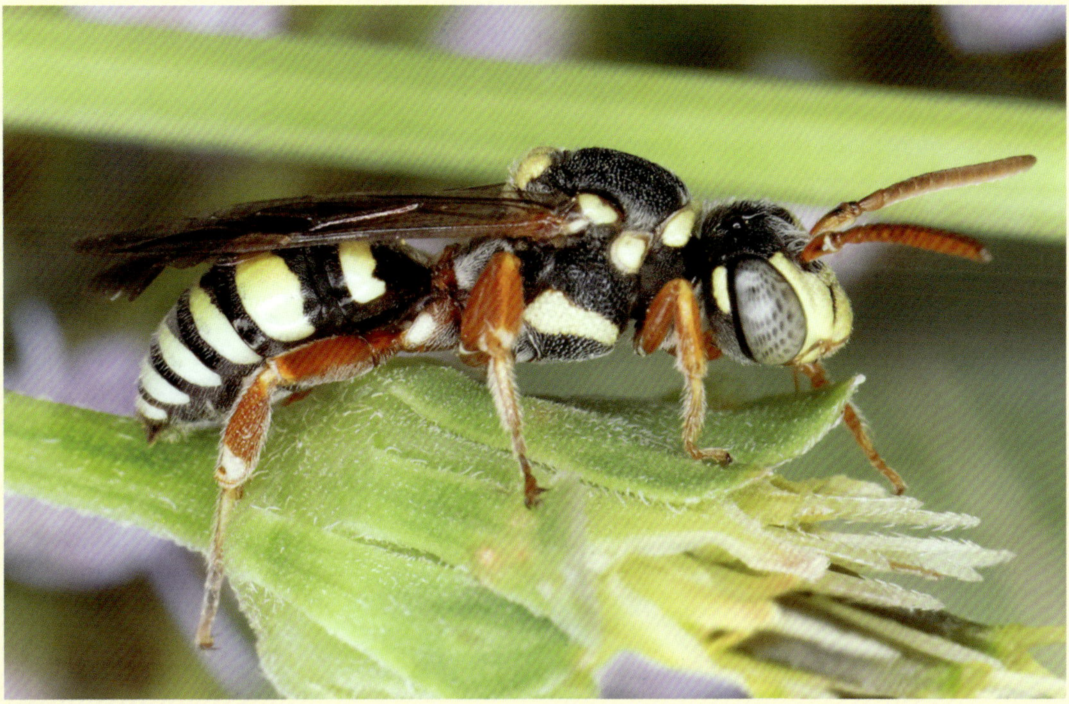

A male Texas nomad bee

Genus description: Nomad bees are slender insects with the flashy appearance of custom-detailed race cars. The heads and bodies of nomad bees are usually black or red and often adorned with yellow or ivory markings. Their legs and antennae may be red or yellow, and their eyes red or green or brown. As is true of cuckoo bees generally, the legs of female nomad bees lack pollen-collecting hairs.

Color and pattern variations among nomad bees are used to identify species. *Nomada texana* has a black body with yellow markings, and reddish legs and antennae. Its abdomen is banded by pale yellow stripes on the first five segments (T1–T5); on male bees, a broken line traverses T6. Both males and females have yellow markings on the front, back, and sides of the thorax. The faces of males are largely yellow, and those of females black with yellow markings. More minute traits, detailed below, help differentiate *Nomada texana* from the very similar Texas species *N. vegana.* **Size:** females 8–11 mm (3/10–2/5 in.), males 7–10 mm (1/4–2/5 in.).

Range: The Texas nomad bee is found throughout Texas and Mexico and as far south as Guatemala. In the United States, *Nomada texana* occurs principally in the southern and western part of the country, but it has been found in such northern locales as Indiana, Massachusetts, and British Columbia. This species is common in Hidalgo and Cameron Counties.

Associated plants: In the Valley, Texas nomad bees can be found during the fall nectaring on aster-family flowers such as blanketflower (*Gaillardia pulchella*), hierba del marrano (*Symphyotrichum subulatum*), Mexican hat (*Ratibida columnifera*), tropical threefold (*Trixis inula*), seaside goldenrod (*Solidago sempervirens*), and romerillo (*Bidens alba*). This nomad bee species has been documented elsewhere visiting an array of plants from several different families.

ID tips for identifying the genus *Nomada*: Nomad bees have several distinctive minute traits that aid in identifying their genus and distinguishing them from wasps. Among them, (1) the pygidial plate (abdomen tip) of males is pronounced and often notched; (2) females have specialized hair patches on the tips of their abdomens; (3) the jugal lobe of the hind wing is inconspicuous; and (4) the thoraxes are heavily pitted.

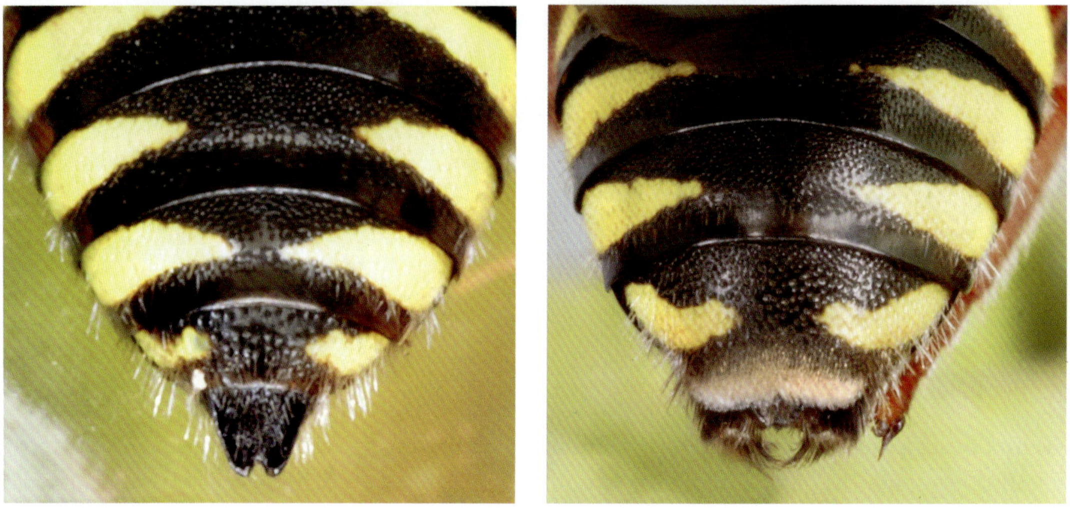

Abdomens of male (*left*) and female (*right*) nomad bees: note the notched pygidial plate of the male bee, and the hair patch on the tip of the female bee's abdomen.

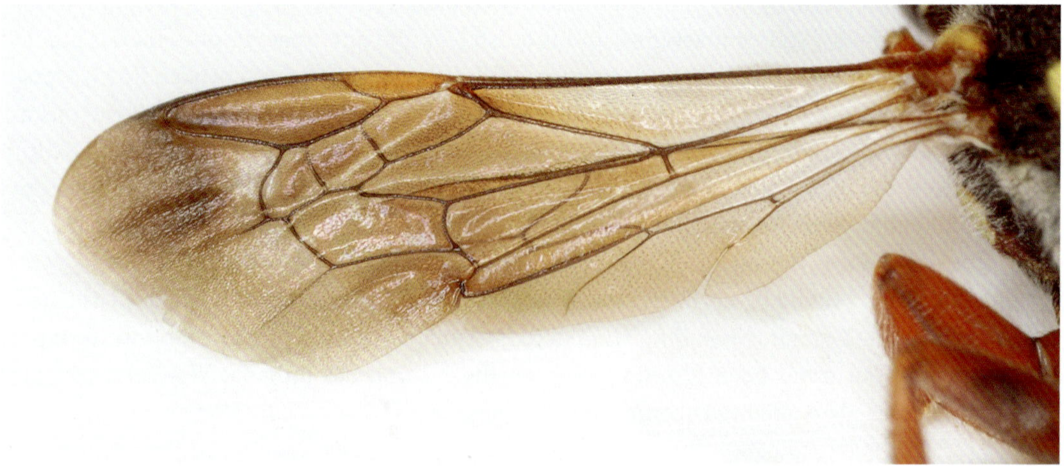

The forewing and partial hind wing of a nomad bee.

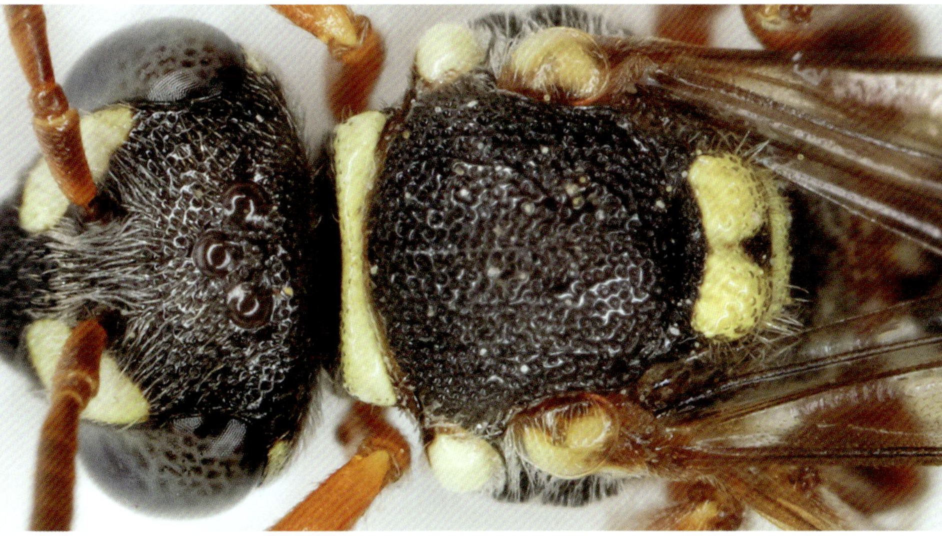

The pitted thorax and vertex of a female Texas nomad bee: the yellow stripe on the pronotal collar, and the yellow marks on the pronotal lobes, tegulae, and scutellum are all characteristic traits of *Nomada texana.*

ID tips for *Nomada texana*: This is the most frequently encountered *Nomada* species in the Valley. The Texas nomad bee is readily distinguished from most other nomad species by the distinctive yellow markings on its face, antennae, thorax, upper abdomen, and sternum, shown in the accompanying photographs. *Nomada texana* can be easily confused, however, with *N. vegana*, which has nearly identical markings and also occurs in Texas. *Nomada vegana* can usually be distinguished by two yellow spots on the propodeum, a feature generally lacking in Texas nomad bees.

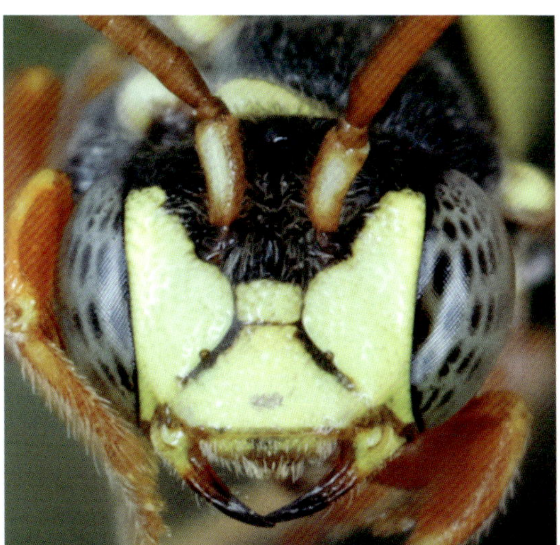

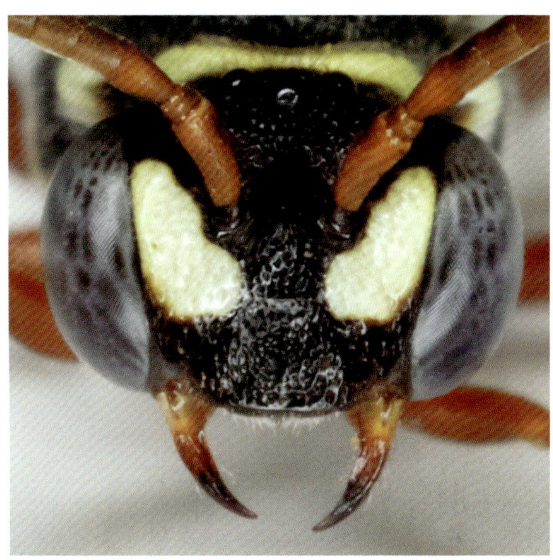

The faces of male (*left*) and female (*right*) *Nomada texana*

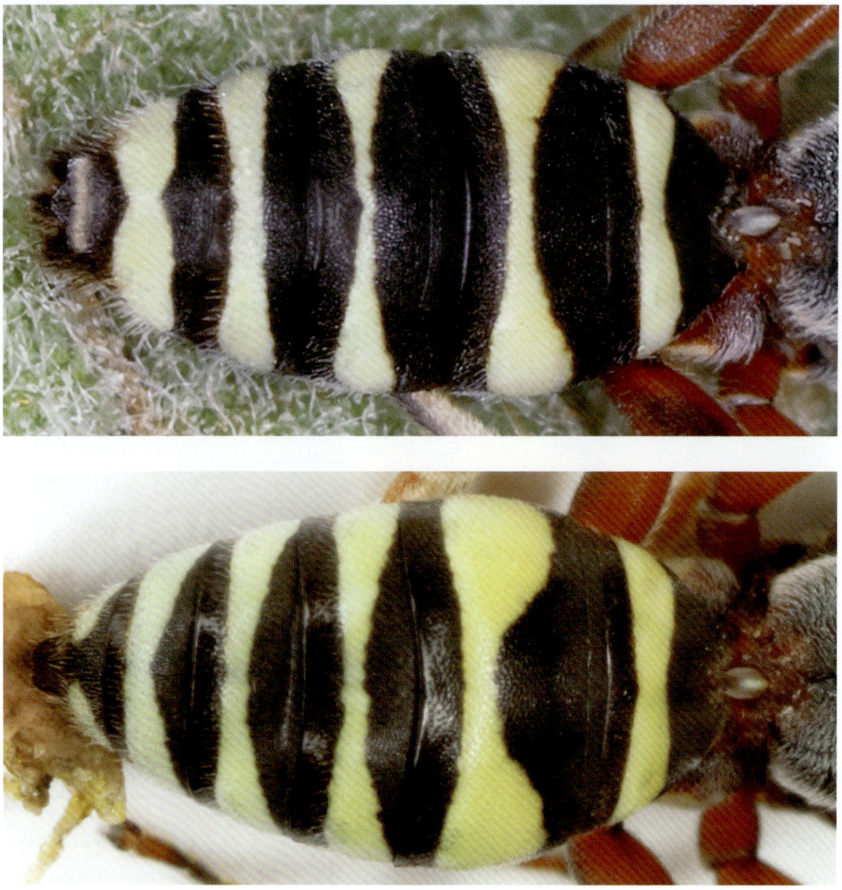

Abdomens of female (*top*) and male (*bottom*) Texas nomad bees

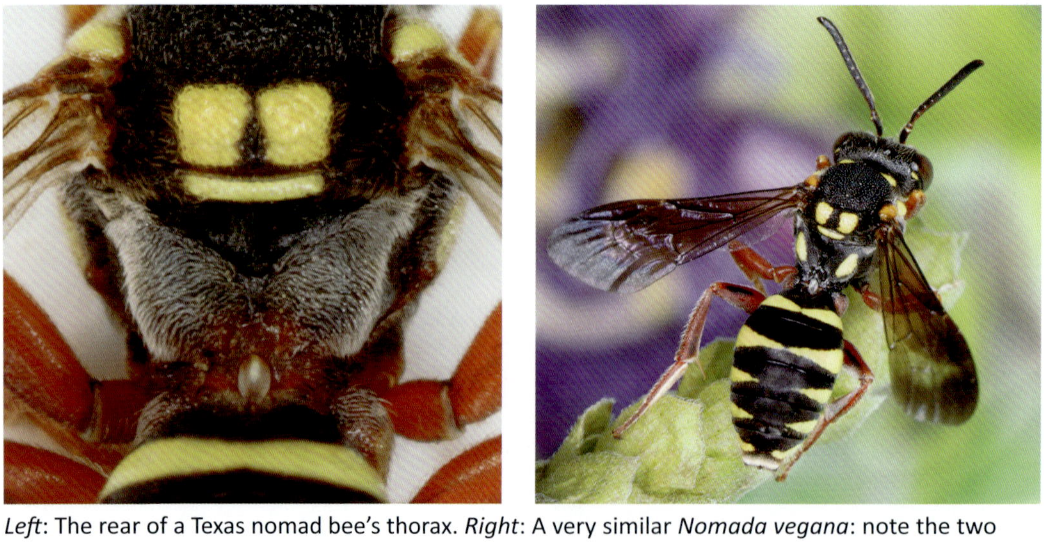

Left: The rear of a Texas nomad bee's thorax. *Right*: A very similar *Nomada vegana*: note the two roughly triangular yellow shapes on the propodeum. Texas nomad bees usually lack this trait.

Nomad bee behavior: As noted above, nomad bees often lay eggs in the nests of *Andrena* mining bees. Newly hatched nomad larvae have large sicklelike mouth parts that they use to kill off the host's larvae before eating the provisions left for them by the host.

The Texas nomad bee belongs to the largely Neotropical *vegana* species group of *Nomada*, whose members target bee genera other than *Andrena*—among them, *Agapostemon*, *Eucera*, *Exomalopsis*, *Halictus*, *Lasioglossum*, *Melitta*, *Nomia*, and *Panurgus*. Michener theorized that *Nomada* originated in Neotropical areas where such diverse hosts abounded, and only later spread northward, encountered *Andrena*, and adopted bees of this genus as preferred hosts.

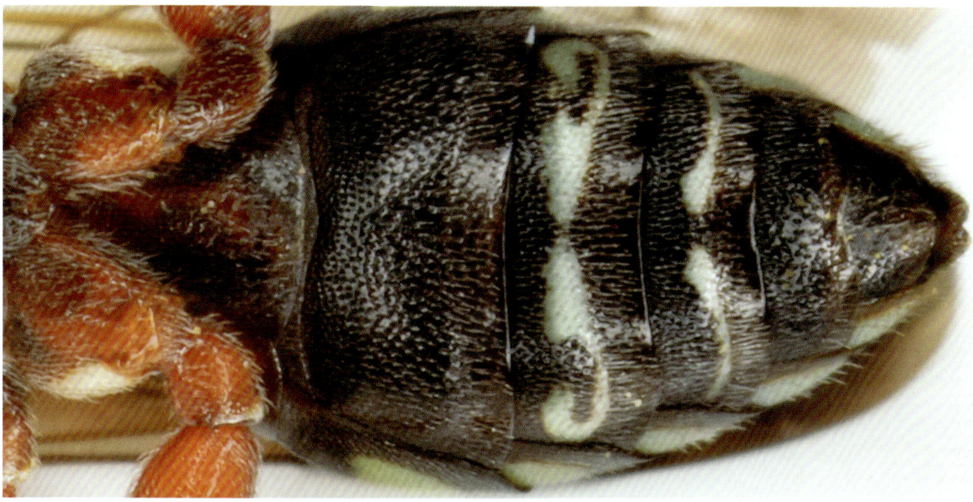

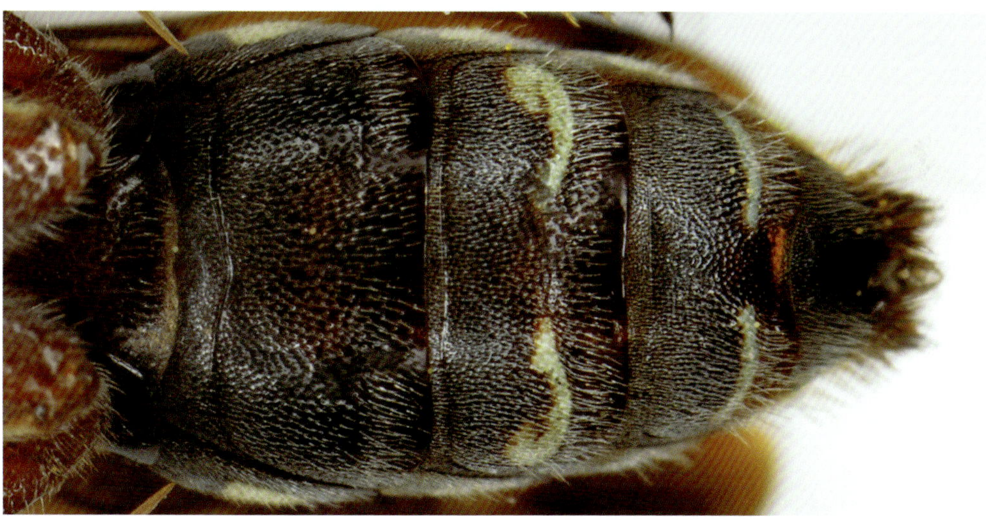

The sterna of male (*top*) and female (*below*) *Nomada texana*: on both, distinctive curved yellow lines traverse S3 and S4.

Little is known about the behavioral habits of *Nomada texana*. It is thought to parasitize the nests of *Agapostemon* sweat bees. In Hidalgo County, Texas, nomad bees are most visible from September through the end of fall. During this period, *Agapostemon melliventris* appear in large numbers on a range of flowering plants throughout the county. This species is often found in tandem with *Nomada texana* on composite flowers.

Viereck's nomad bee ◆ *Nomada vierecki* Cockerell, 1903

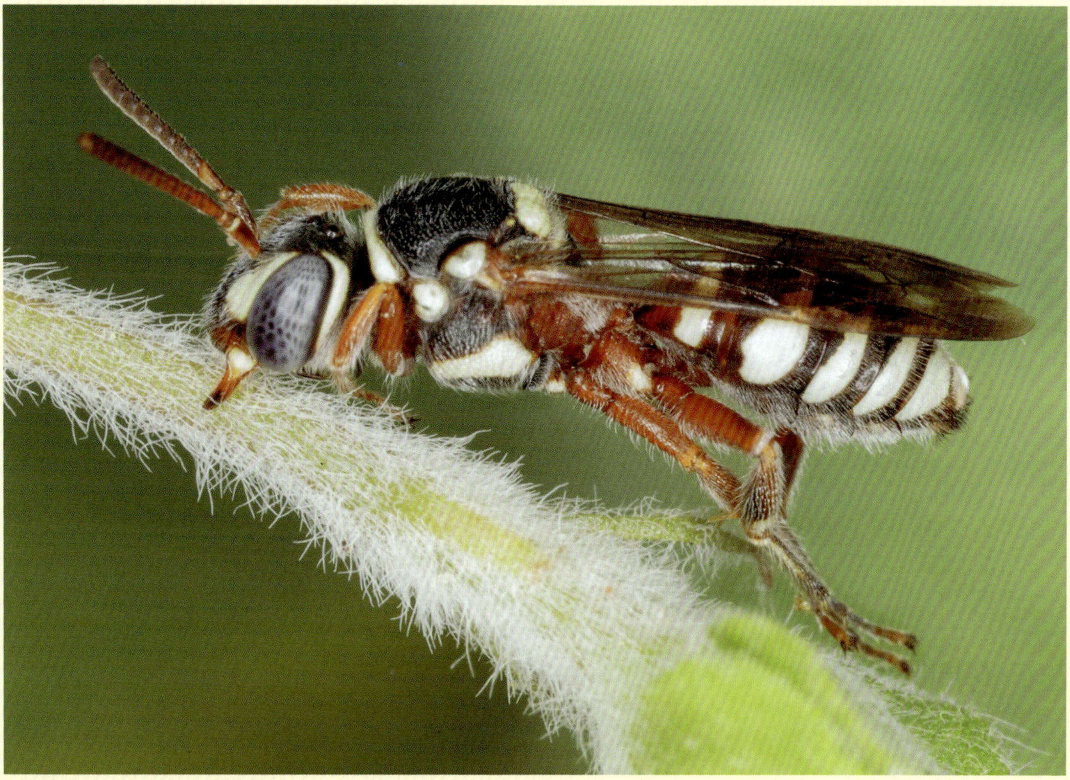

A female *Nomada vierecki*

Description: *Nomada vierecki* is a small red-and-black bee with cream-colored markings on its head and body; red legs with pale markings; predominantly red antennae; and wings with smoky tips and outer edges. The female's abdomen is banded by cream-colored stripes on the first five segments (T1–T5); on male bees, T6 is also banded. On both males and females, the following are often red or partly red: the propodeum, T1, S1–S2, and the undersides of the head and thorax. Females' faces are black and reddish, with large ivory areas that extend from the clypeus to the top of each compound eye. On males, the pale markings are more extensive, covering the entire face below the antennae. The antennae are reddish brown, with some golden brown on the scapes. **Size:** females 7–8 mm (1/4–3/10 in.), males 7 mm (1/4 in.).

Range: Viereck's nomad bee ranges southward from Nebraska and Colorado into New Mexico, Texas, Chihuahua, and Tamaulipas. This species is relatively uncommon in the Lower Rio Grande Valley. Beamer and Michener collected specimens of Viereck's nomad bee in March 1951 in southernmost Cameron County. Sporadic sightings of *Nomada vierecki* were also documented in the 1960s and 1970s in Culberson, Wilbarger, and Val Verde Counties. Male and female *Nomada vierecki* appeared at Falcon State Park (Starr Co.) in May 2021.

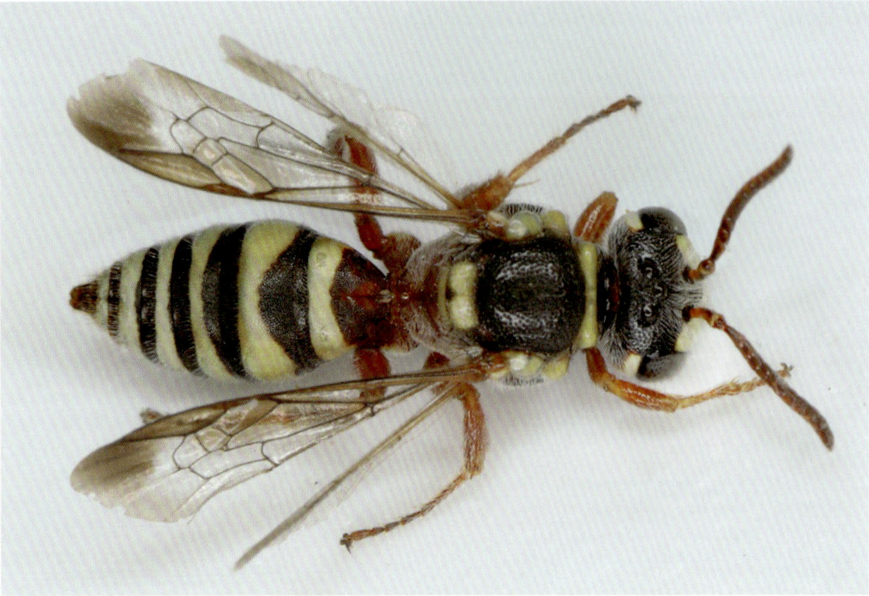

A male Viereck's nomad bee

Associated flora: In 1897, Cockerell and Townsend observed Viereck's nomad bees in New Mexico feeding on poplar, spiny chloracantha (*Chloracantha spinosa*), and touristplant (*Dimorphocarpa wislizeni*). Male and female bees observed at Falcon State Park in 2021 were visiting Arkansas dozedaisy (*Aphanostephus skirrhobasis*), hoary blackfoot (*Melampodium cinereum*), and cenizo (*Leucophyllum frutescens*).

ID tips: The following traits aid in identifying this species: (1) *Nomada vierecki* is smaller than most other *Nomada* of the Valley; (2) the sternum (of both males and females) is banded with curved pale stripes on S3 and S4; and (3) the usually extensive reddish coloration of the head and body help distinguish *N. vierecki* from other Valley *Nomada*; individual specimens of *N. vierecki*, however, show varying amounts of red on the thorax and abdomen.

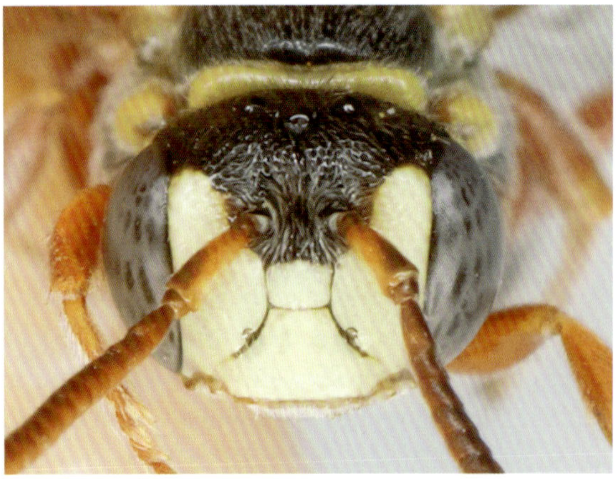

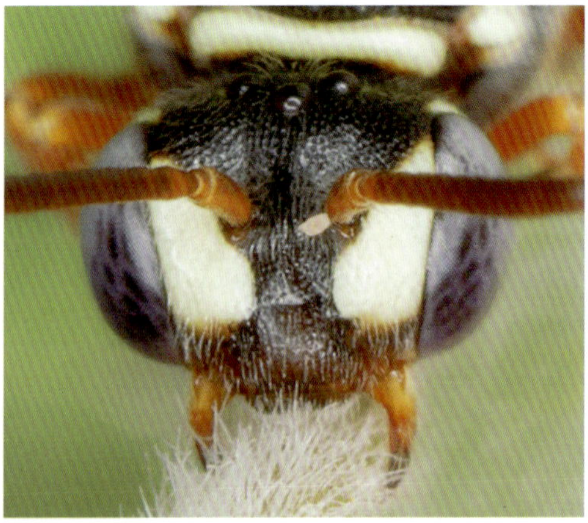

Faces of male (*left*) and female (*right*) Viereck's nomad bees

Behavior: Like the Texas nomad bee, Viereck's nomad bee belongs to the *vegana* species group of the genus *Nomada*, whose members frequently prey on bee genera other than *Andrena*. The smaller Viereck's nomad bee is in the subgroup *Micronomada*, whose members are thought to parasitize the nests of *Exomalopsis* and Halictidae. The male *Nomada vierecki* shown here was found in an area overrun with hundreds of *Exomalopsis birkmanni*.

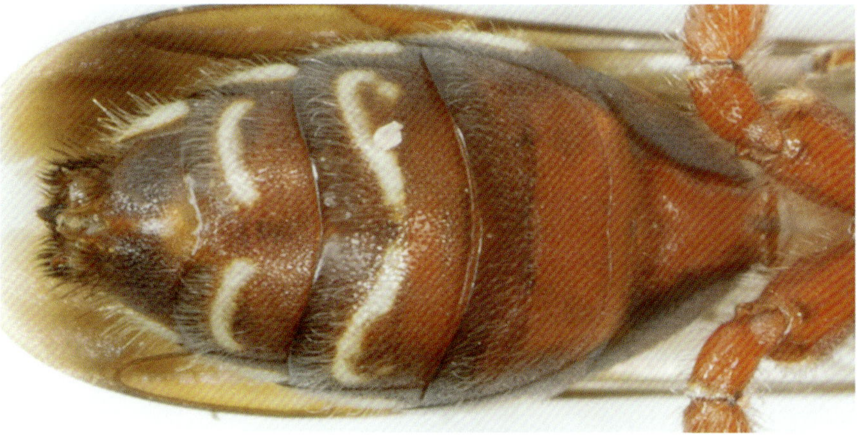

The sternum of a female Viereck's nomad bee

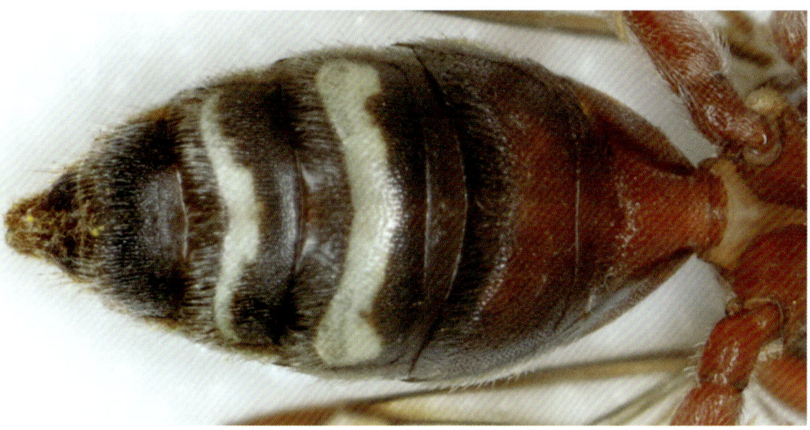

The sternum of a male Viereck's nomad bee

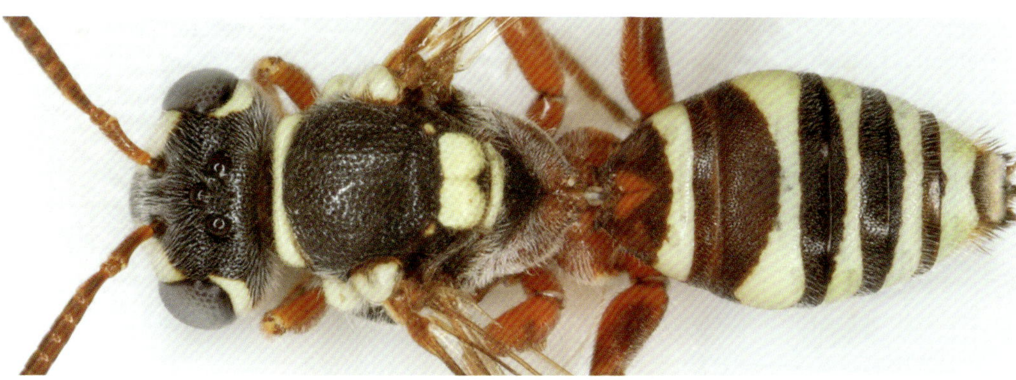

Dorsal view of a female Nomada vierecki

Comparison of *Epeolus* and *Triepeolus*

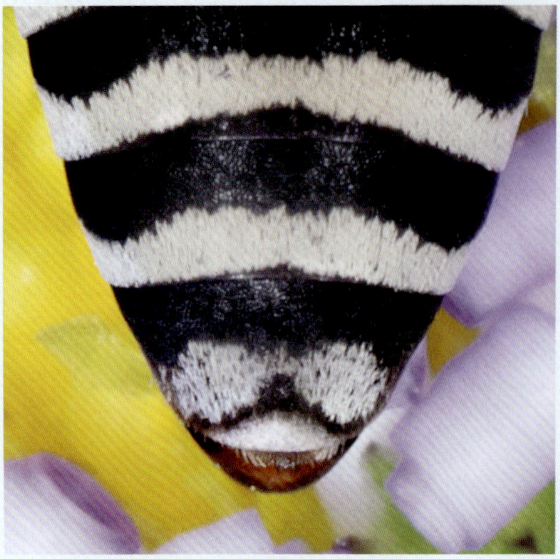

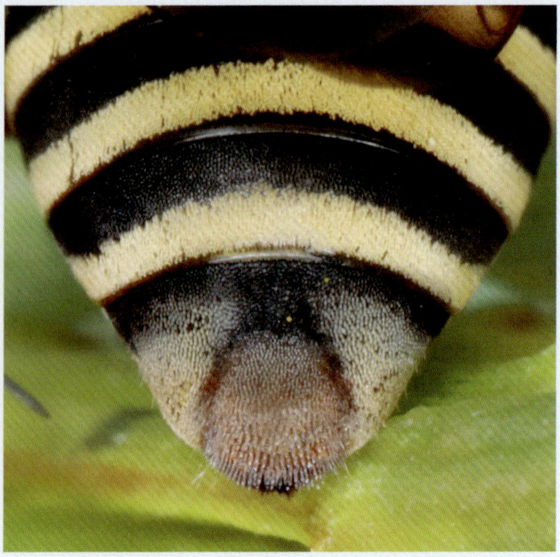

Comparison of the pseudopygidial areas of female *Epeolus pusillus* (*left*) and *Triepeolus rufoclypeus* (*right*).

Epeolus and *Triepeolus* are similar in appearance, although *Triepeolus* tend to run larger. *Triepeolus* cuckoos are more common than *Epeolus* and are represented by a greater diversity of species in the Valley. Male *Epeolus* may be impossible to distinguish from male *Triepeolus* without magnification or the assistance of an expert. Females of the two genera are easier to differentiate.

Epeolus and *Triepeolus* females both have an expanded median area on the fifth tergum called the pseudopygidium, which is covered with specialized setae or bristlelike hairs. Differences in the pseudopygidial area (shown here) can be used to differentiate females of the two genera. On the typical female *Epeolus*, this area is covered with a short, wide band of silvery hairs. On the typical *Triepeolus* female, the pseudopygidial area is narrower and longer, and the bristly hairs covering it usually have a golden cast.

The apparatus on the tips of *Triepeolus* and *Epeolus* females' abdomens also differs in ways that reflect characteristics of the nests each cuckoo genus parasitizes. Female *Epeolus* are specially equipped to attack the nests of *Colletes*: the female *Epeolus* has toothlike projections on the end of its abdomen that allow it to saw through the plastic coverings *Colletes* use to line their egg chambers. *Triepeolus* females, by contrast, have long, narrow, forcepslike projections that they use to dig into the soil walls of ground-nesting bees.

The larvae of *Epeolus* and *Triepeolus* share behavioral characteristics. The early instars (developmental stages) of both *Epeolus* and *Triepeolus* larvae are hospicidal (host killing): among other traits, they have long, sickle-shaped mandibles, and soft bodies lined with leglike tubercules that allow them to maneuver within nest chambers in search of food, newly hatched host larvae, or eggs. After destroying the host eggs or larvae, the cuckoo larvae shed their hospicidal characteristics and develop into adult bees within the nest.

Dwarf epeolus ◆ *Epeolus pusillus* Cresson, 1864

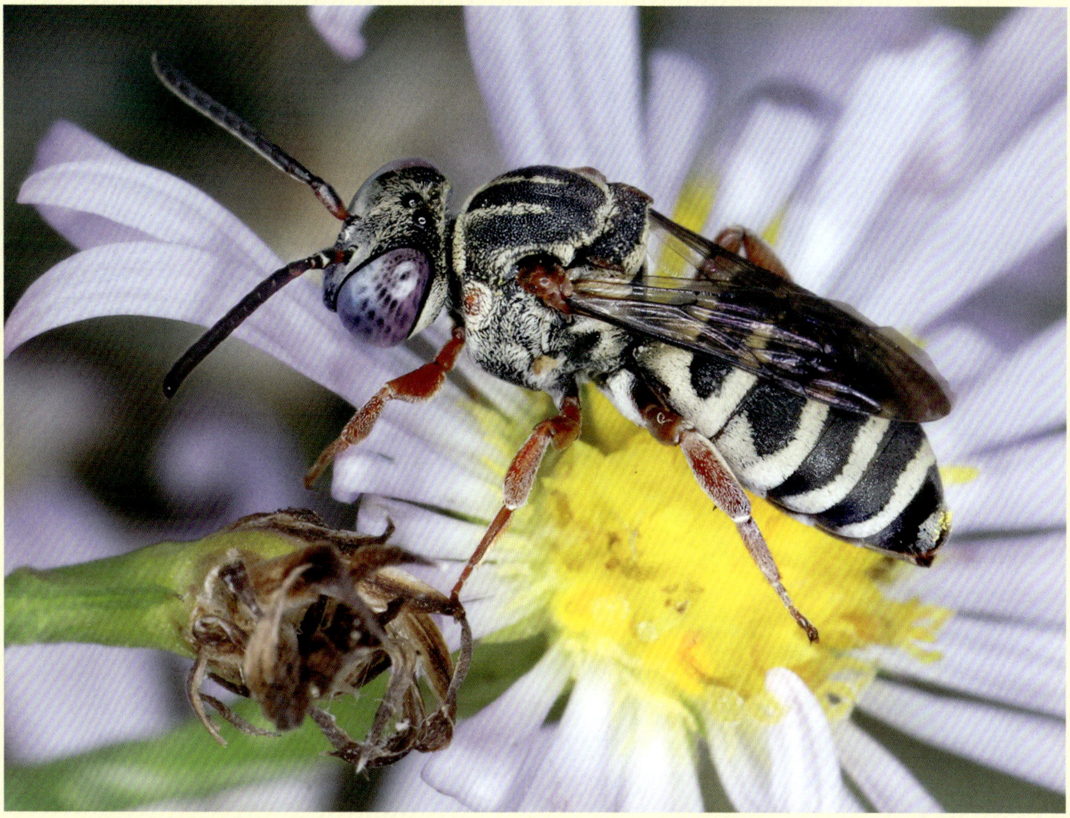

A female dwarf epeolus

The tribe Epeolini contains two similar genera of cuckoo bees found in the United States—*Epeolus* and *Triepeolus*. Bees of both genera have dark bodies, black or red legs, and abdomens and thoraxes adorned with patterns formed by pale, appressed (short and dense) hairs, and less often by yellow or red integumental markings. Like nomad bees, Epeolini lack the furry appearance typical of many bees and are commonly mistaken for wasps.

Species description: The dwarf epeolus is a black bee with distinctive patterns of appressed ivory hairs on its abdomen and thorax. In the Lower Rio Grande Valley, individuals of this species tend to have extensive reddish coloration and eyes that are a striking violet. Red areas of the bee may include its legs, tegulae, axillae, pronotal lobes, the first two segments of the sternum, the bases of the mandibles, and the pedicels and scapes of the antennae. (In more temperate regions of the United States, these parts may appear yellow brown instead of red.) Males closely resemble females but have dense silvery hairs around the antennae, on their hind-leg coxae, and on the sides and underside of the thorax. *Pusillus* means "small," and these bees are small for their genus. **Size:** females and males 8–9 mm (3/10–1/3 in.).

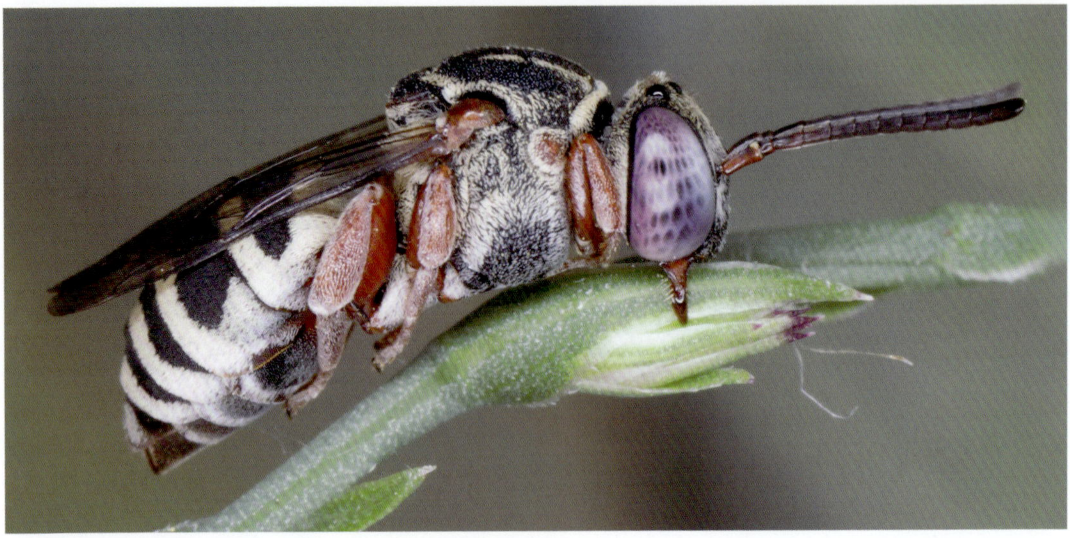

Lateral view of the abdomen of a female dwarf epeolus

Range: The dwarf epeolus is found predominantly in the eastern and south-central United States and the northern half of Mexico. *Epeolus pusillus* has been documented as far north as Ontario, and as far south as Belize. This species appeared at the McAllen Nature Center and the National Butterfly Center (Hidalgo Co.) in November 2018.

Associated plants: Little published research exists regarding the floral preferences of *Epeolus pusillus*. The bees shown here were found visiting the aster-family flowers hierba del marrano (*Symphyotrichum subulatum*) and resinbush (*Viguiera stenoloba*).

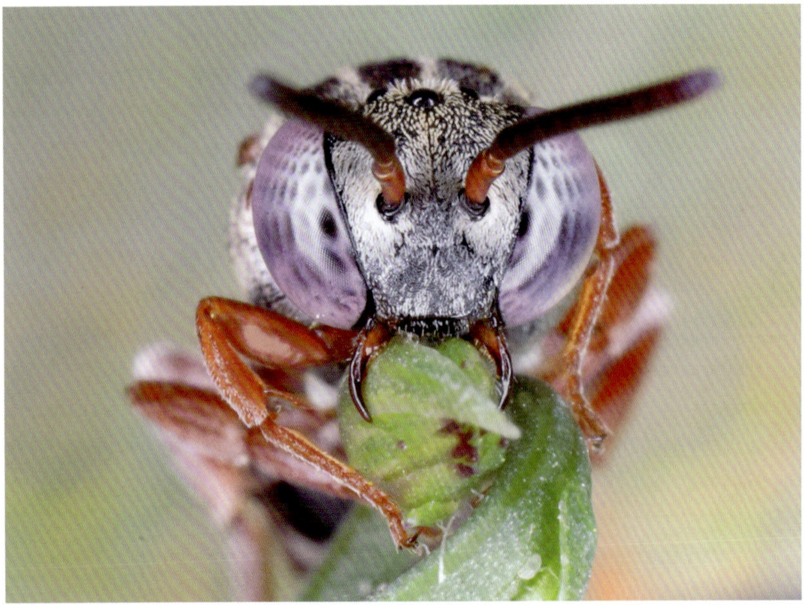

Face of a female dwarf epeolus

ID tips: In the Lower Rio Grande Valley, the small size, red legs, and exquisitely colored violet eyes of *Epeolus pusillus* make it readily recognizable in the field. Generally, *Epeolus* species are differentiated by such features as the black-and-pale patterns on the abdomen and thorax, and by minute traits relating to the sculpting of the thorax, particularly the scutellum and axillae.

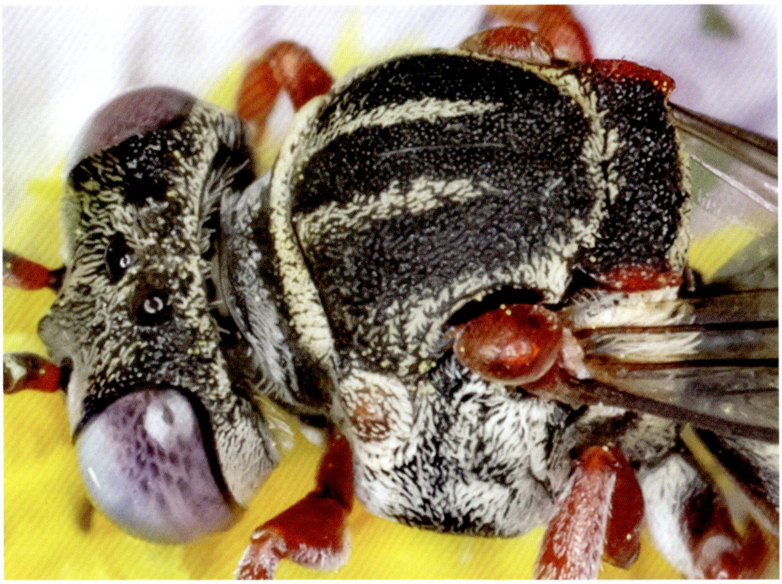

Thorax of a female dwarf epeolus: the axillae are red and slightly curved and extend no farther than the scutellum's rear margin. Two parallel stripes of pale hair join a collar of dense white hairs on the pronotum.

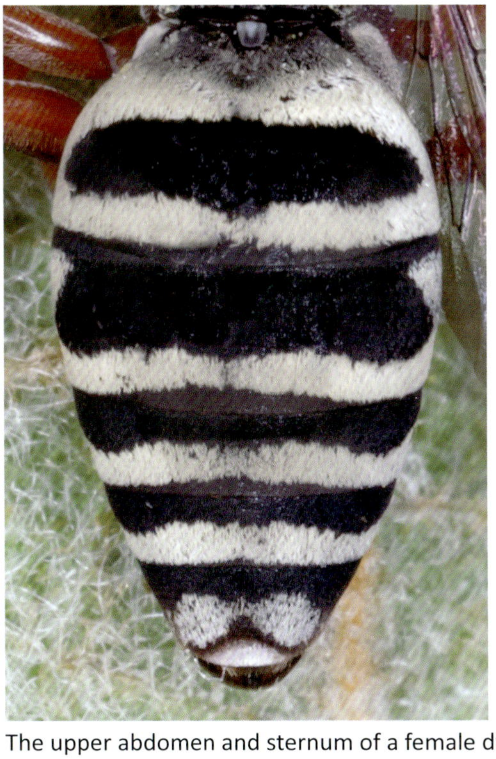

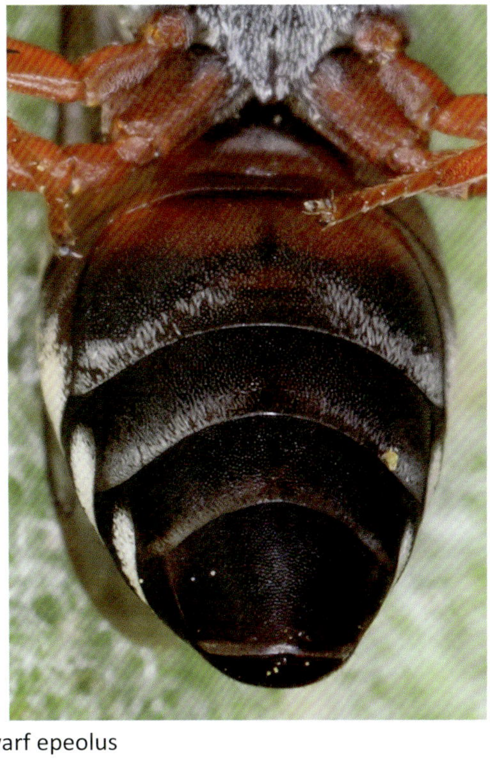

The upper abdomen and sternum of a female dwarf epeolus

Behavior: The dwarf epeolus is known to target the nests of the correspondingly small compact cellophane bee (*Colletes compactus compactus*). In 1968, Rozen and Favreau described dwarf epeolus females darting into the nest tunnels of compact cellophane bees while the hosts were still provisioning their nests and their egg chambers remained open. The cuckoo bees would attach their eggs to the lining of nest cell walls, sometimes near the rear of a chamber so that the emerging cuckoo larvae would have to pass through the cellophane bee's semiliquid pollen-mixture provisions to emerge outside. The cuckoo bees would orient their own eggs so that the front ends pointed toward the nest opening, thus helping direct emerging cuckoo larvae toward open air. After laying eggs, the cuckoos departed the nests to allow the hosts to seal them.

Various studies have conjectured that dwarf epeolus cuckoos parasitize the nests of *Colletes* species other than compact cellophane bees, among them *C. americanus*, *C. ciliatoides*, and *C. deserticola*. None of these species, however—including *Colletes compactus*—has been documented in the Lower Rio Grande Valley. Two known Valley cellophane species, *Colletes birkmanni* and the smaller *C. mandibularis*, are possible but unproven alternative hosts for dwarf epeolus in the Valley. Both *Colletes* fly during periods when *Epeolus pusillus* appears in the Valley, and both frequent aster-family flowers.

Red-faced triepeolus ◆ *Triepeolus rufoclypeus* (Fox, 1891)

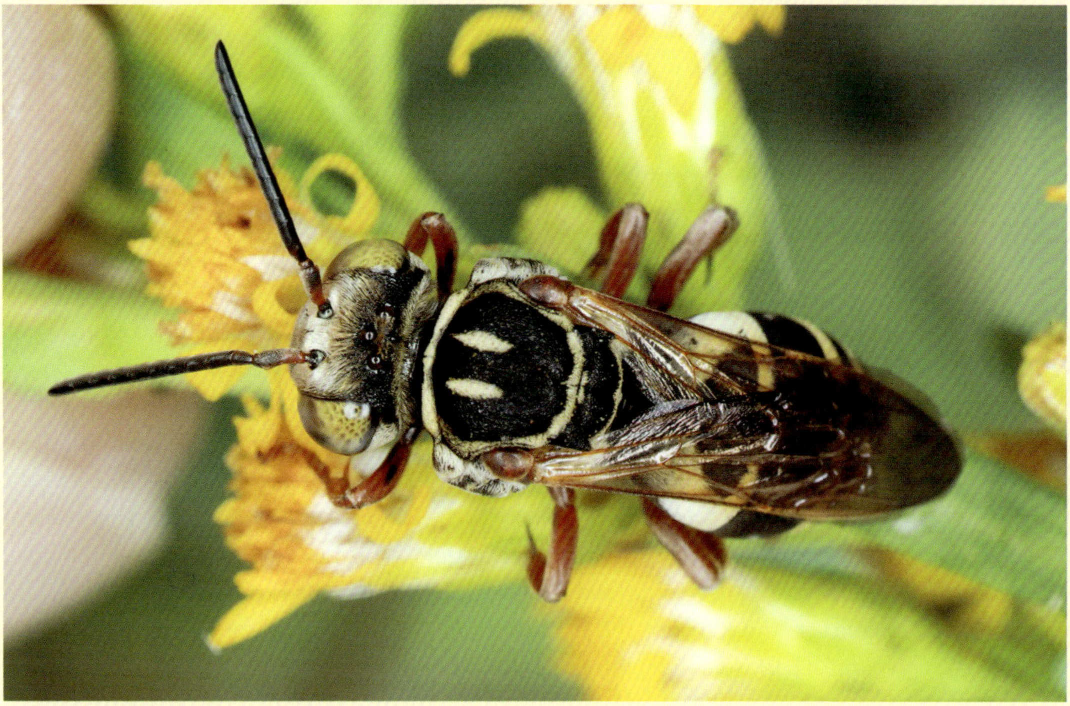

A female red-faced triepeolus

Description: *Triepeolus rufoclypeus* is a black cuckoo bee with red legs and greenish eyes. Dense, short, pale hairs form stripes on its abdomen and curved lines on its thorax. The bee's tegulae and pronotal lobes are red, and its wings are transparent brown with brown veins. Parts of the face, the bases of the mandibles, and the antennal scapes and pedicels are also usually red. The spurs on the legs may be either black or red. **Size:** females and males 8.5–13.5 mm (1/3–1/2 in.).

Range: *Triepeolus rufoclypeus* is a common species in all three border counties of the Valley. Its range extends from Tamaulipas to as far north in Texas as Matagorda County. It has also been documented in Jamaica and Grenada. *Triepeolus rufoclypeus* closely resembles the subtropical *T. cameroni* and may prove to be its northern variation. The two species are now differentiated in large part by their range: *Triepeolus cameroni* is more southern and occurs primarily in southern Mexico, Belize, Honduras, and Costa Rica, overlapping with *T. rufoclypeus* in Tamaulipas.

Associated plants: *Triepeolus rufoclypeus* appears in the Valley from April through November, commonly feeding on composites such as blanketflower (*Gaillardia pulchella*), common sunflower (*Helianthus annuus*), cowpen daisy (*Verbesina encelioides*), resinbush (*Viguiera stenoloba*), and hairy wedelia (*Wedelia acapulcensis* var. *hispida*).

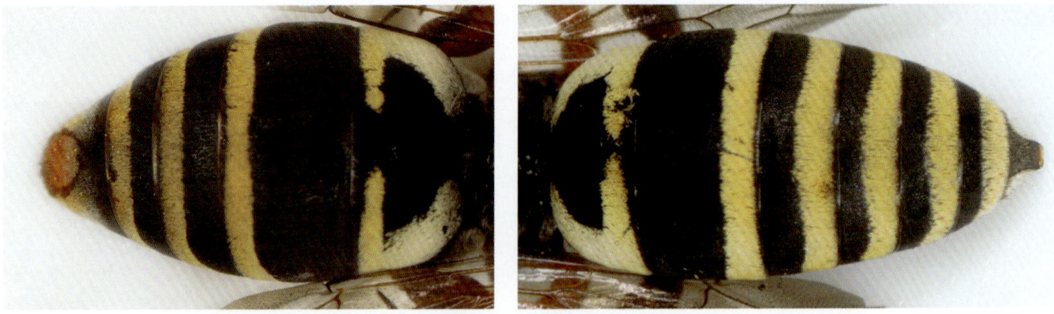

Abdomens of female (*left*) and male (*right*) *Triepeolus rufoclypeus*

ID tips: In 2008, entomologist Molly G. Rightmyer assembled an impressive study of the genus *Triepeolus* that described more than 100 species, among them *T. rufoclypeus*. Rightmyer differentiated species in part by (1) the patterns of pale hairs on the abdomens and thoraxes; (2) the colors of the legs, tegulae, jaws, antennae, and thoraxes; (3) (in females) the shape of the abdominal tip, as seen from above and in profile; and (4) characteristics of the clypeus.

The following are distinctive traits of *Triepeolus rufoclypeus*. (1) *Face*: The scapes, pedicels, and F1 of the antennae are usually partly red. This trait is visible to the naked eye and aids in tentative field identifications. *Rufoclypeus* means "red clypeus," but the clypeus is predominantly black in this species. The labrum and bases of the mandibles may be red, black, or a combination. (2) *Thorax*: The two parallel bands on the scutum do not connect directly with the pale band rimming the pronotum. The axillae are black and pointed and do not extend beyond the scutellum's back rim. (3) *Abdomen*: The bands of short, appressed hairs on both females' and males' abdomens are yellowish, broad, and of even width.

Behavior: The host species of *Triepeolus rufoclypeus* remains unknown. *Triepeolus* cuckoos frequently target longhorn bees such as *Melissodes* and *Svastra*. Less commonly, *Triepeolus* parasitize other ground-nesting bee genera—among them *Anthophora*, *Centris*, *Dieunomia*, *Eucera*, *Melitoma*, *Nomia*, *Protoxaea*, and *Ptiloglossa*.

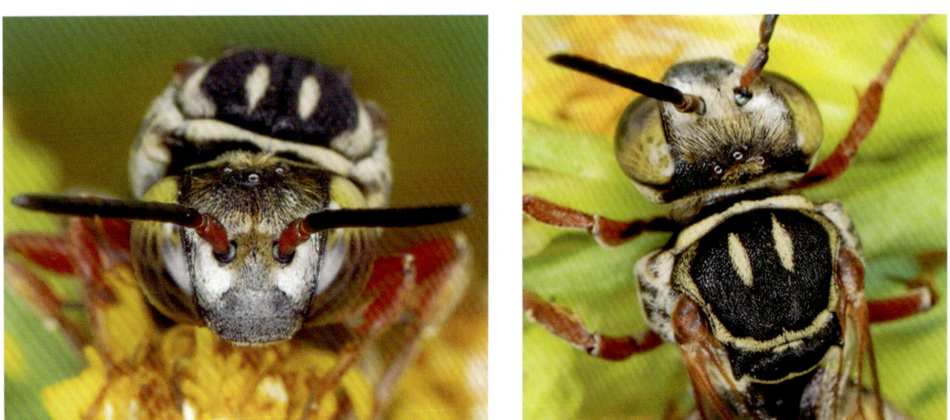

Face (*left*) and thorax (*right*) of a female *Triepeolus rufoclypeus*

Spring triepeolus ◆ *Triepeolus vernus* Rightmyer, 2008

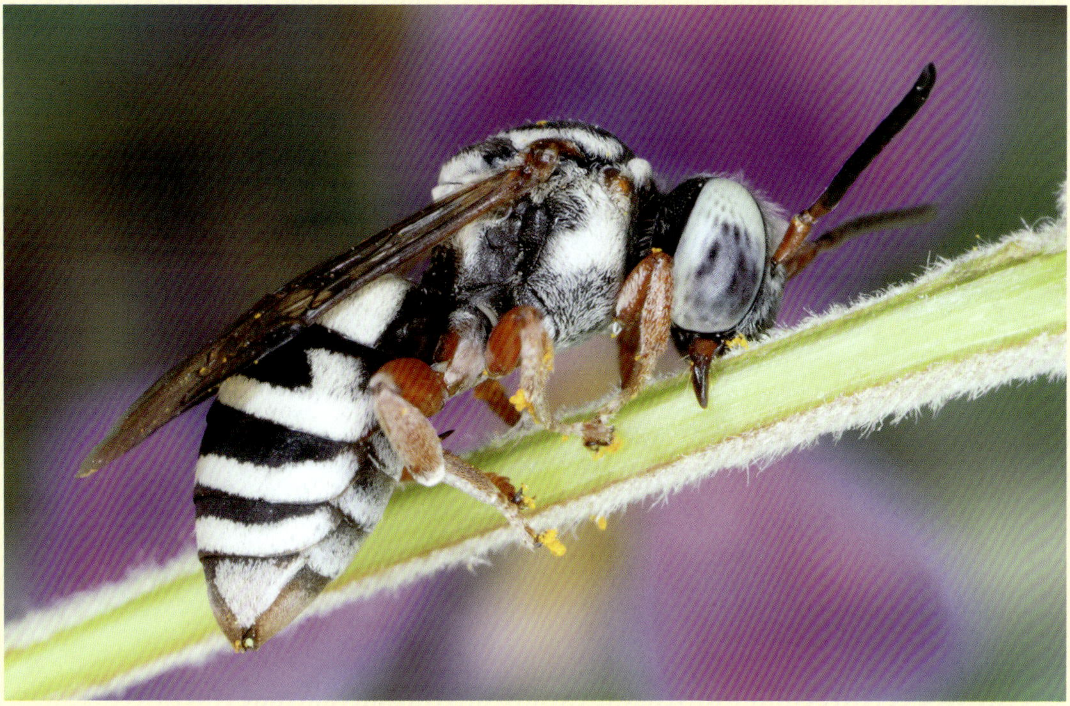

A female *Triepeolus vernus*

Description: *Triepeolus vernus* is a black bee with red legs and pale green eyes. Its body is adorned with dense pale hairs that form broad bands on the abdomen and a pattern of curved and straight lines on the thorax. The tegulae, pronotal lobes, and the last segment of the sternum are red. *Triepeolus vernus* has a predominantly black head, with pale hairs above and around the base of each antenna. The following parts of the head are orange red: the bases of the mandibles, the labrum, the scapes, the pedicels and first flagellar segment (F1) of the antennae, and (sometimes) the bottom half of the clypeus. **Size:** females and males 7–10 mm (1/4–2/5 in.).

Range: *Triepeolus vernus* was first described by Rightmyer. In her 2008 revision of the genus, she noted that *Triepeolus vernus* is found only in Texas, during March and April. This species has been recorded previously in Dimmick, Hidalgo, Maverick, Uvalde, Val Verde, and Zapata Counties. The bee shown here appeared on April 13, 2019, at the National Butterfly Center (Hidalgo Co.).

Associated plants: *Vernus* means "spring." *Triepeolus vernus* cuckoos typically appear in March and April in the Valley, during the spring cactus bloom. The female bee shown here was found foraging on Texas prickly pear (*Opuntia engelmannii* var. *lindheimeri*) and twisted-rib (*Thelocactus setispinus*). Preexisting records document this species feeding as well on *Parkinsonia*, dozedaisy (*Aphanostephus* sp.), and sneezeweed (*Helenium microcephalum*).

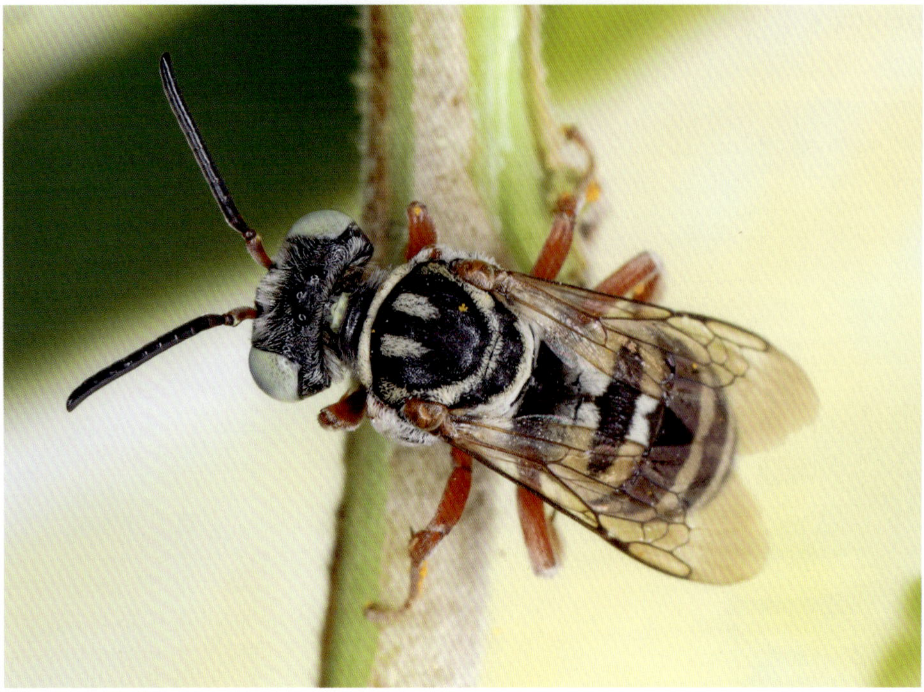

A female *Triepeolus vernus*

ID tips: *Triepeolus vernus* can be distinguished from the more common *T. rufoclypeus* by the following traits: (1) *Triepeolus vernus* is generally smaller and slenderer, and its legs are a more intense orange red; (2) on female *Triepeolus vernus*, pale, diffuse, bristly hairs cover the mesepisternum (lower sides of the thorax); (3) T2–T4 tend to be narrow or interrupted in the middle, while on *Triepeolus rufoclypeus*, T2–T4 are of uniform width and uninterrupted; (4) the axillae on the scutellum of *Triepeolus vernus* are wider than those of *T. rufoclypeus*; and (5) while *Triepeolus vernus* emerges in spring, *T. rufoclypeus* appears in the Valley at all times of year except midwinter.

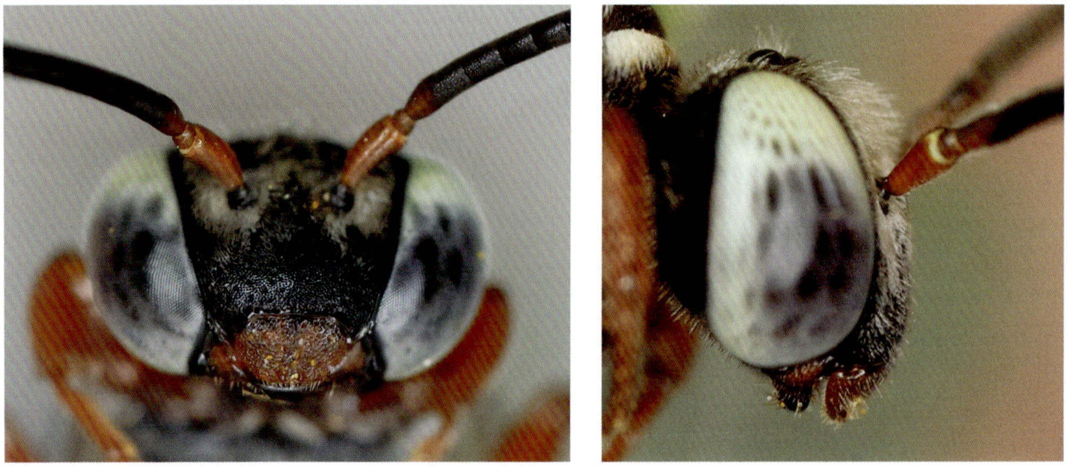

Face of a female *Triepeolus vernus*: the clypeus is relatively short and somewhat pitted.

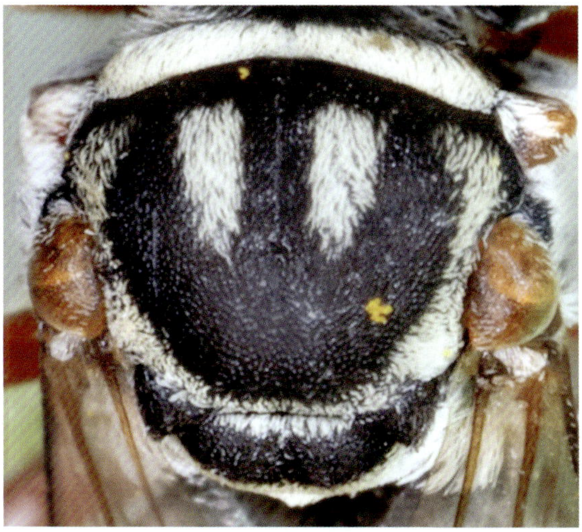

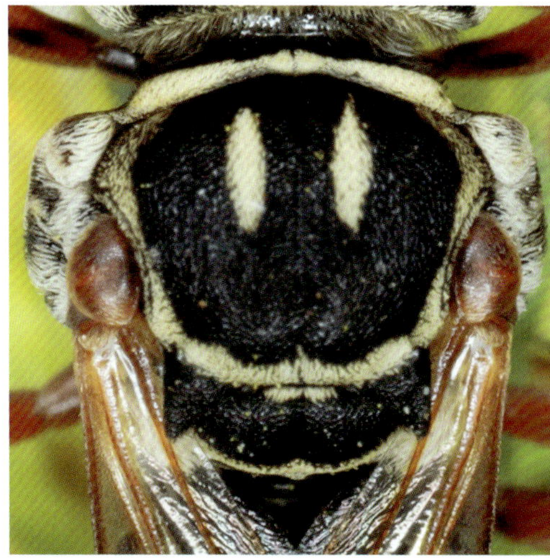

Comparison of thoraxes of a female *Triepeolus vernus* (*left*) and a female *T. rufoclypeus* (*right*)

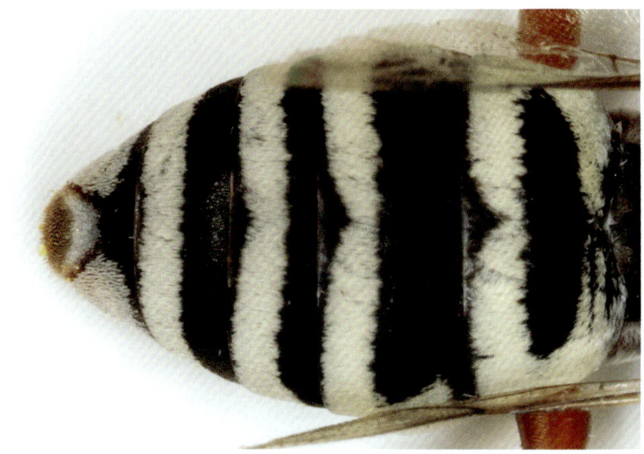

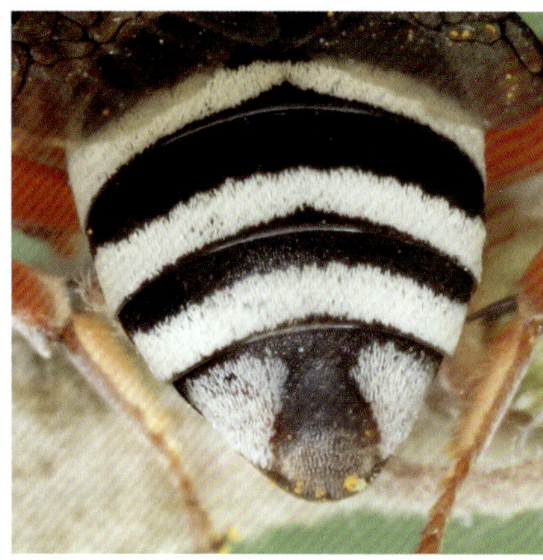

Dorsal and rear views of the abdomen of a female *Triepeolus vernus*

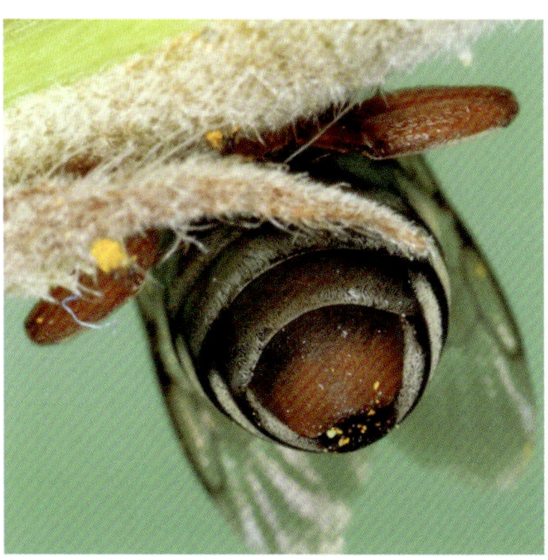

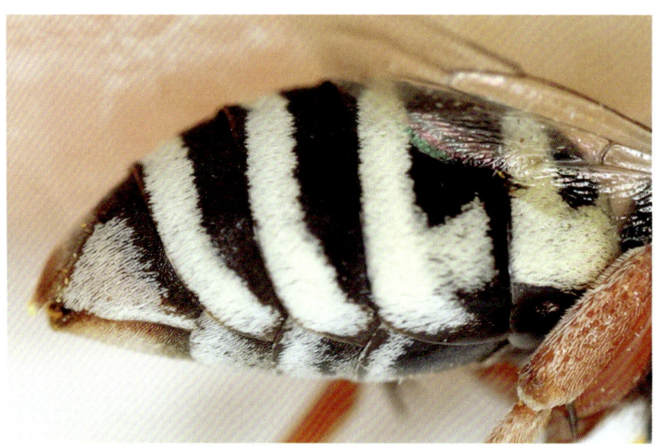

S6 (*left*) and lateral view of the abdomen (*right*) of a female *Triepeolus vernus*

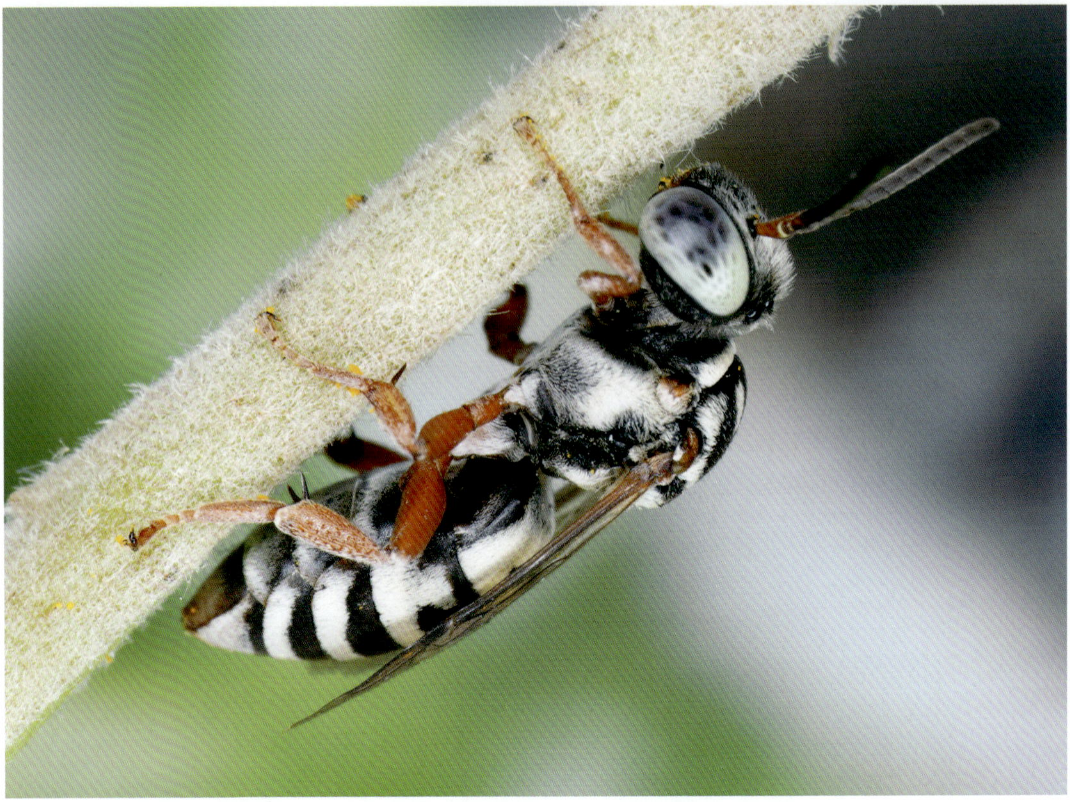

A female *Triepeolus vernus*: the sides of the thorax are covered with pale hairs.

Additional information: Little information is available about this lovely species. The hosts of *Triepeolus vernus* are unknown. The bees shown here were found patrolling the ground in an area where the little prickly pear longhorn bee (*Melissodes opuntiellus*) and the compact anthophorula (*Anthophorula compactula*) were nesting. The flight seasons of these species coincide with that of *Triepeolus vernus*, and all three bees visit Texas prickly pear. Further research might succeed in linking *Triepeolus vernus* with one of these hosts.

Zacatec triepeolus ◆ *Triepeolus zacatecus* (Cresson, 1878)

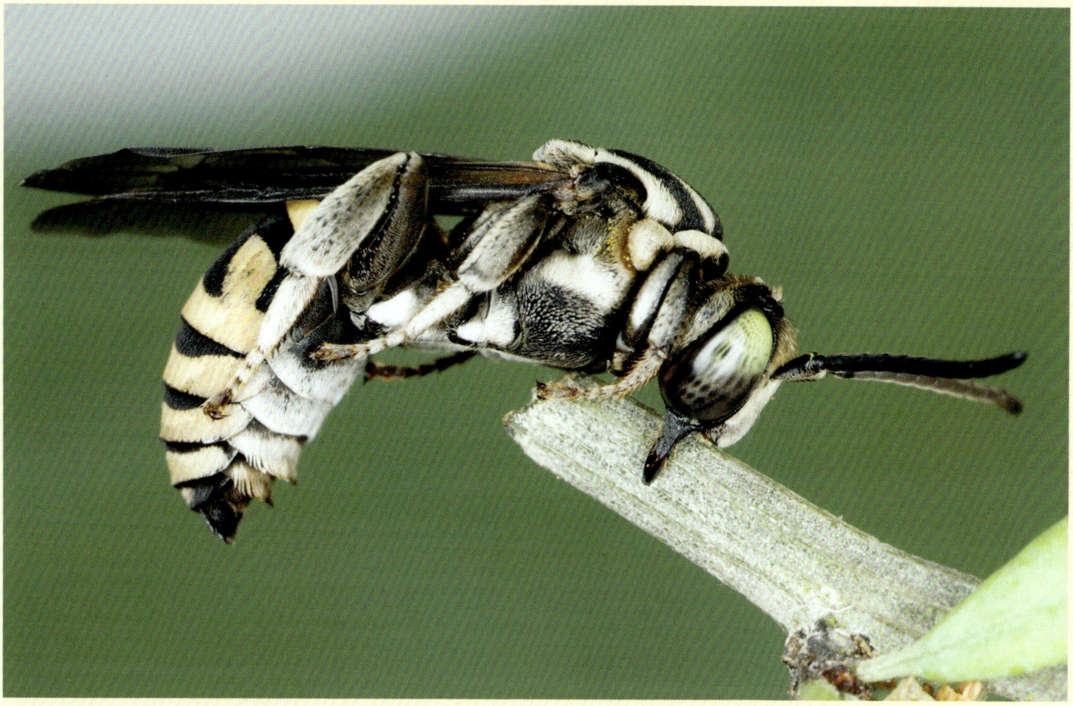

A male *Triepeolus zacatecus*

Description: *Triepeolus zacatecus* is a hefty black bee whose body and legs are covered with patterns formed by appressed white hairs. The bands of appressed hairs on its abdomen are distinctly yellowish. Its tegulae are black and its wings smoky and dark veined. The bee's face is almost entirely dark, with some reddish coloration on the mandibles. The antennae are also dark, sometimes with reddish areas on the outer face of the first segment (F1). This is an exceptionally large *Triepeolus.* **Size:** females and males 15–17 mm (3/5–2/3 in.).

Range: Cresson first described this subtropical species in 1878 after examining a specimen collected by Sumichrast in Mexico, and the distribution of the species is centered in Mexico. (The name *zacatecus* presumably refers to the Zacatec people who once inhabited the Mexican state of Zacatecas). *Triepeolus zacatecus* is uncommon in the Lower Rio Grande Valley. The bee shown here appeared in May 2021 at La Puerta Tract (Starr Co.), just a few miles from the border.

Associated plants: Documentation of the floral associations of *Triepeolus zacatecus* is scant. Rightmyer notes records of *Triepeolus zacatecus* feeding on croton. The male bee featured here was nectaring on guajillo (*Senegalia berlandieri*) in a semiarid barretal habitat.

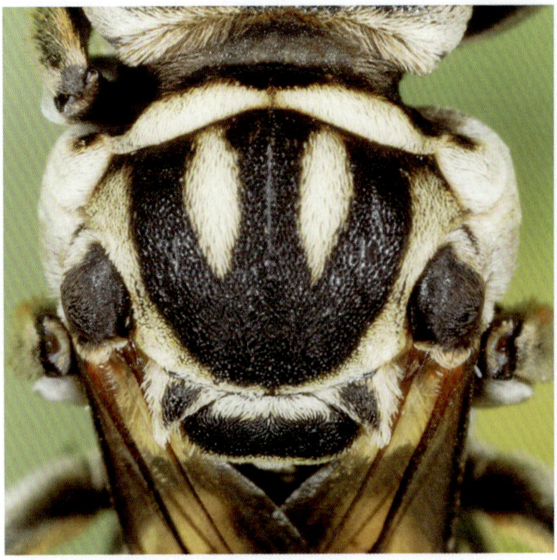

Thorax of a male *Triepeolus zacatecus*

ID tips: The unusually large size of this species makes it stand out in the field. Viewed more closely, (1) the thorax of *Triepeolus zacatecus* has long axillae that are sharply incurved at the tips; (2) the bee's pronotal collar is wide and protuberant; and (3) its mesepisternum (sides of the thorax) is densely covered with short, dense, pale hairs.

Triepeolus zacatecus is similar to the equally large species *T. grandis*. According to Rightmyer, *Triepeolus zacatecus* might be viewed as the black-legged form of *T. grandis*. The two differ only in slight respects: *Triepeolus grandis* usually has orange legs; the pale hair bands on its abdomen tend to be less yellow; the bands on T1 and T2 tend to be narrower; and the bee's axillae are usually triangular. Individual *Triepeolus grandis* specimens may have black legs and their axillae may be sharply incurved like those of *T. zacatecus*. The range of *Triepeolus grandis* is more northern, extending from central Mexico to Oklahoma, and it is more common in the United States.

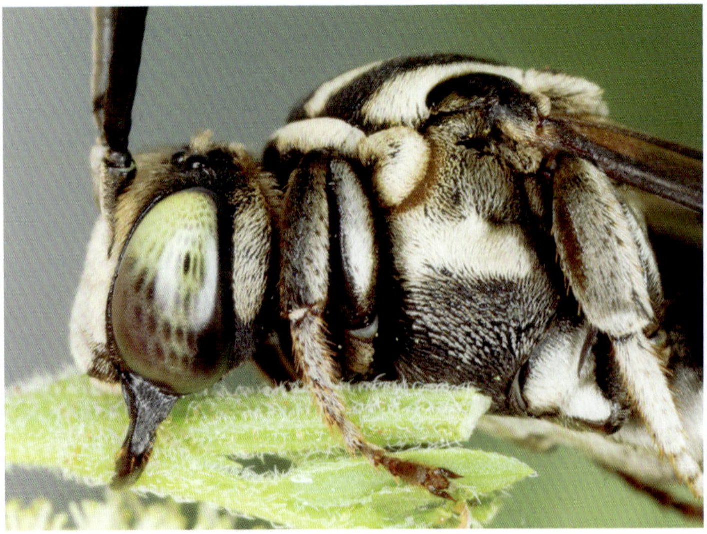

Lateral view of the thorax of a male *Triepeolus zacatecus*

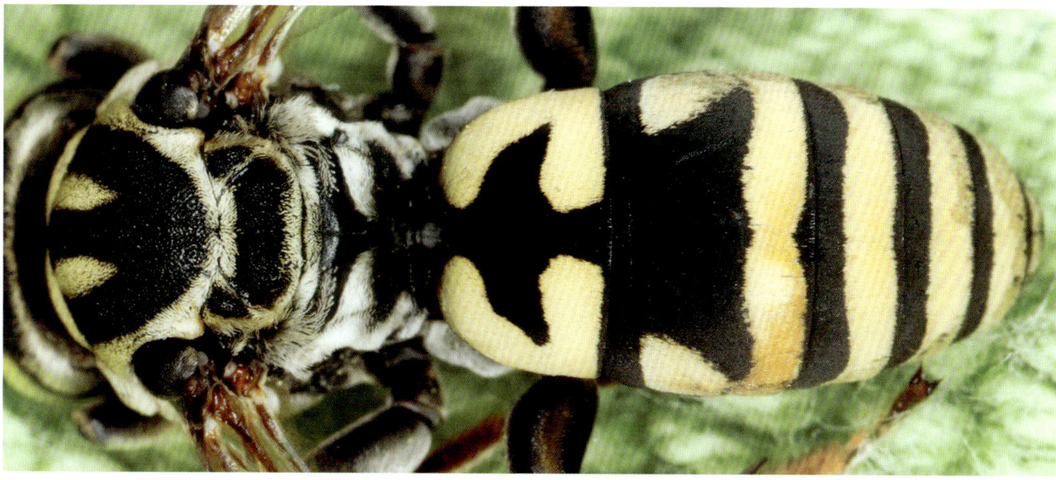

Dorsal view of the thorax, propodeum, and abdomen of a male *Triepeolus zacatecus*

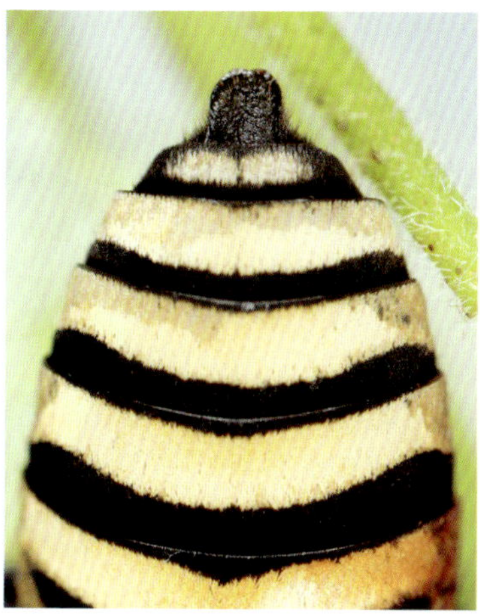

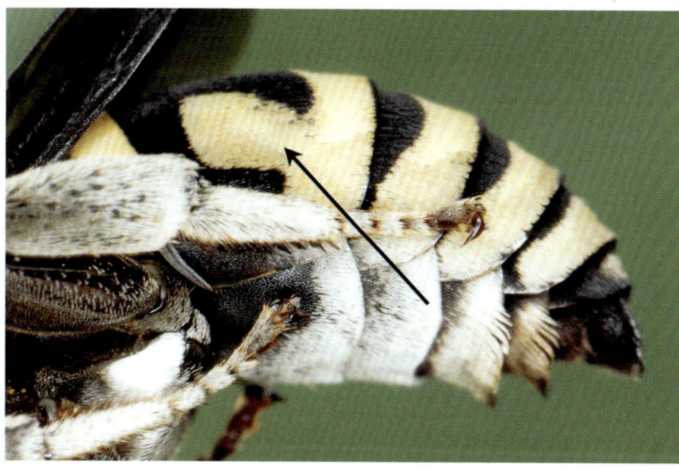

Left: The abdomen tip of a male *Triepeolus zacatecus. Right*: Lateral view of markings on T2.

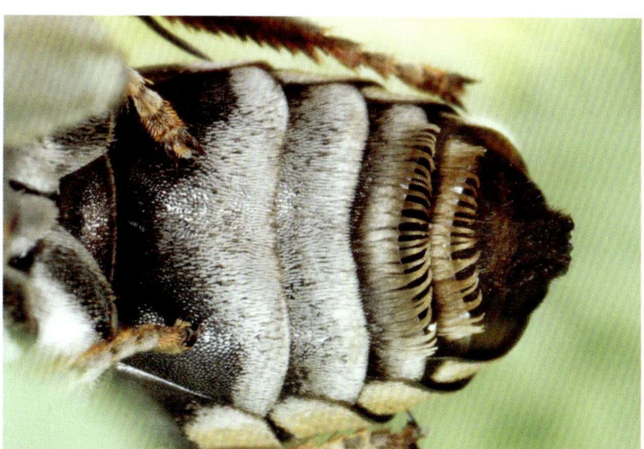

Sternum of a male *Triepeolus zacatecus*

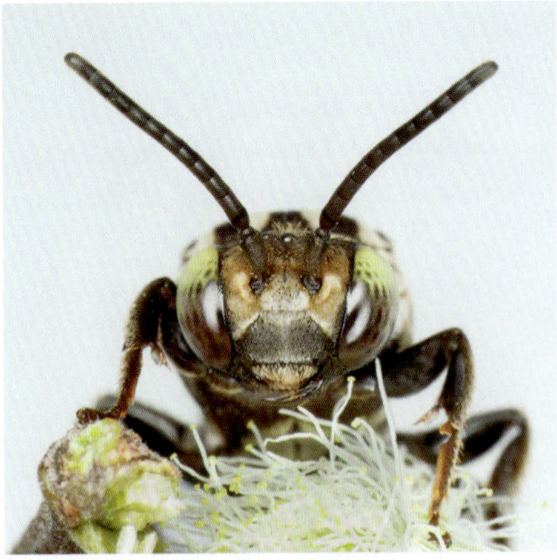

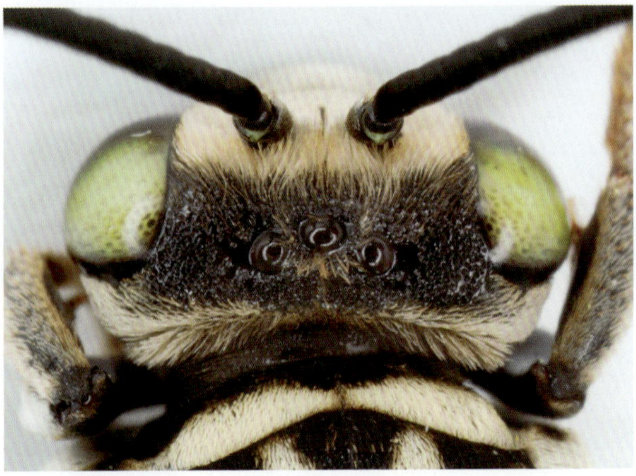

Face and vertex of a male *Triepeolus zacatecus*

Behavior: The host species of *Triepeolus zacatecus* remains unknown. The similarly large *Triepeolus grandis* is known to prey on large Diphaglossinae such as the Arizona feather-tongued bee (*Ptiloglossa arizonensis*), Jones's feather-tongued bee (*Ptiloglossa jonesi*), and Yarrow's fork-tongued bee (*Caupolicana yarrowi*). The latter is native to Tamaulipas and Nuevo León, which border Starr County. The Mexican feather-tongued bee (*Ptiloglossa mexicana*) and the glorious protoxaea (*Protoxaea gloriosa*) both occur in the immediate area in which the male *Triepeolus zacatecus* shown here was found.

Bumble Bees

Subfamily Apinae ◆ Tribe Bombini
Genus *Bombus*

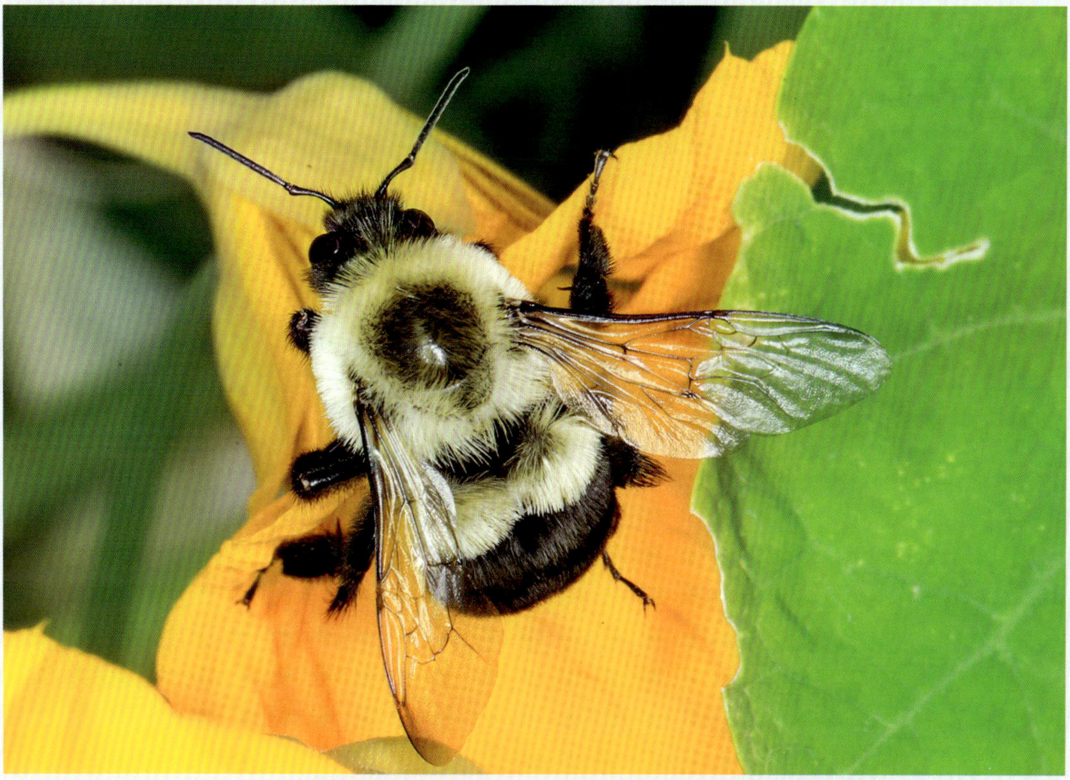

Bumble bees are loud, furry-looking bees, usually covered with black and yellow hairs.

Bumble bees are easily recognized by their plump, fuzzy, yellow-and-black- striped bodies and by their raucous buzzing when they zoom by. The genus name *Bombus* is Latin for "buzzing" or "rumble." Bumble bees have the distinction of being the only native bees in this guide that build colonies with a complex social structure. Like nonnative honey bees, bumble bees form colonies that each contain a queen, male drones, and female workers. Queens lay eggs; male drones mate with queens; and workers perform various tasks necessary to sustain the colony, such as incubating eggs, feeding larvae, collecting and storing pollen, and guarding the hive.

Bumble bee colonies are more modest in size and shorter lived than those of honey bees. While honey bee hives may contain tens of thousands of members and persist several years, North American bumble bee nests usually harbor a few hundred bees and last at most a year. Individual worker and male bees often live merely a few weeks. Only new queens continue on after each year's autumn, hibernating underground through the winter. Each spring, queen bumble bees emerge. Keeping themselves warm by vibrating their wing muscles, they make slow, brief flights, periodically landing to scout for potential nest sites. Unlike many of the solitary bees in this guide, bumble bees do not dig tunnels in earth or wood to build their own nests. Instead, queens tend to look for preexisting cavities, such as rodent burrows in the ground, or tree hollows. Bumble bee colonies can prove difficult to find because they are frequently subterranean and hidden under clumps of grass, leaf litter, and other plant debris.

Both bumble bees and honey bees use wax in nest construction. Honey bee nests consist of the hexagonal interlocking compartments most people visualize when they think of honeycombs. Most North American bumble bees, by contrast, create groups of oblong, rounded wax shells shaped like urns, which expand and divide as larvae grow inside them. The queen begins construction of a colony by making a "honey pot," usually from wax, and then filling it with nectar for her first offspring. Then she constructs her urn-shaped egg chambers and lays her eggs. Mated queens carry spermatozoa from the previous year in special sacs inside their abdomens called spermathecae, used to fertilize the eggs.

The first eggs that hatch produce female worker bumble bees, which continue to construct nest cells and food pots for the colony. While the queen continues to lay eggs, the workers tend and guard the nest and gather food for developing larvae. Toward the end of each summer, some of the eggs develop into potential queens. The females that hatch from these eggs fly off and leave the colony.

In the summer, bumble bee queens also produce eggs that hatch into male offspring. After maturing, the male bees leave the colony in search of mates. Males overnight outside the nest, seeking shelter under leaves and blossoms. On summer mornings, male bumble bees can often be found curled asleep inside flowers.

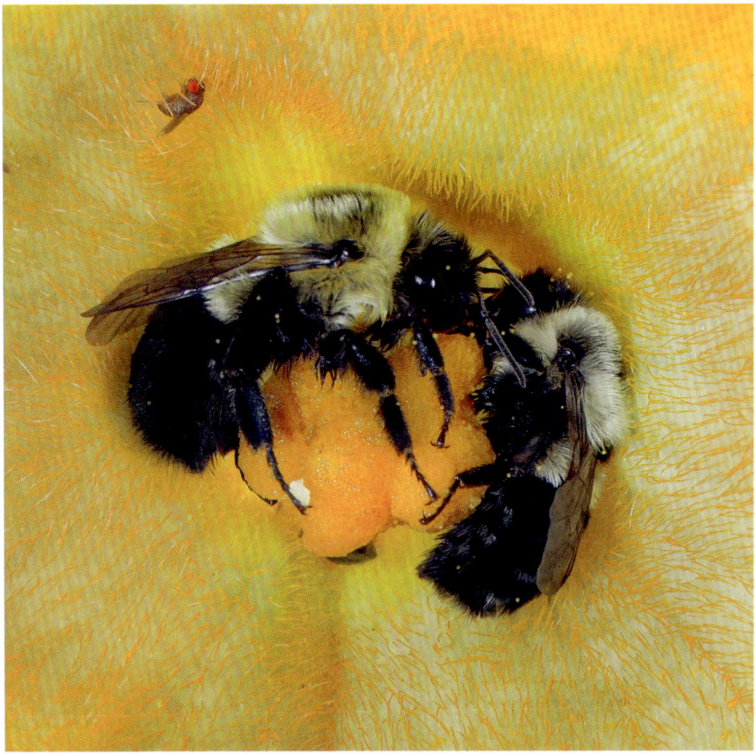

Male bumble bees sleeping inside a squash blossom in early morning

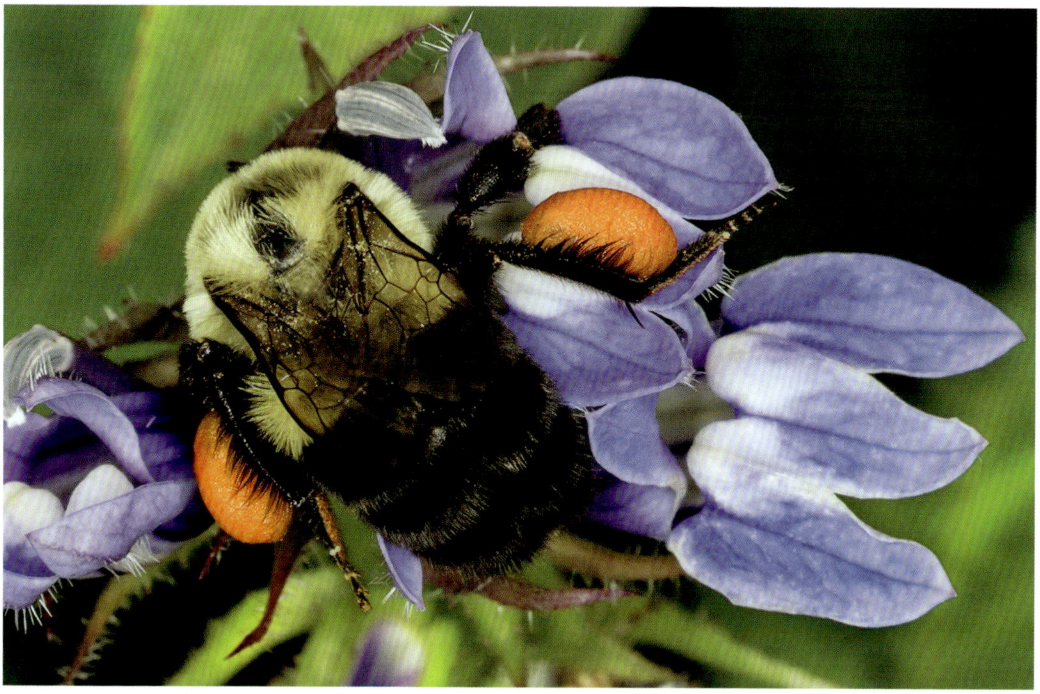

A female bumble bee carrying orange pollen in its corbiculae

Pollination by bumble bees: Female bumble bees carry pollen in rounded masses tucked into pollen baskets, called corbiculae, located on their hind legs and made up of fine, interwoven hairs. Bumble bees' hefty size allows them to carry relatively large pollen loads, making them efficient pollinators. Bumble bees are also capable of buzz pollination. These combined traits make them important pollinators of blueberries, cranberries, and tomato-family plants. Because honey bees are unable to practice buzz pollination, bumble bees are often instrumental in ensuring the pollination of such commercial crops.

Texas bumble bee species and bumble bee decline: At least nine bumble bee species have been documented in Texas: the American bumble bee (*Bombus pensylvanicus*), the black-and-gold bumble bee (*B. auricomus*), the brown-belted bumble bee (*B. griseocollis*), the common eastern bumble bee (*B. impatiens*), the yellow bumble bee (*B. fervidus*), the Sonoran bumble bee (*B. sonorus*), the southern plains bumble bee (*B. fraternus*), the variable cuckoo bumble bee (*B. variabilis*), and, sometimes straying into the western portions of the state, the Nevada bumble bee (*B. nevadensis*).

Many bumble bee species are currently in decline throughout the United States because of factors such as the overuse of pesticides, the loss of grasslands and other habitats to agriculture and urbanization, and bumble bees' exposure to disease communicated by commercial apiaries. Four of the nine bumble bee species found in Texas are threatened to some degree. The Texas Parks and Wildlife Department has classified the American, Sonoran, and variable cuckoo bumble bees in the category of "greatest conservation need" under the Texas Conservation Action Plan. The International Union for Conservation of Nature additionally has placed the yellow bumble bee on its Red List of vulnerable species.

Bumble bees favor cool climates and are more common in temperate regions. Most of the Texas bumble bee species that occur north of the Valley range far into the northeastern United States. Bumble bee populations are less diverse in southern Texas; it may be that the arid, subtropical habitats of the Lower Rio Grande Valley are too taxing for many of the already-threatened species that linger on farther north. Whatever the cause, bumble bee populations have dwindled in the Valley over the last 20 years.

During a three-year period of research between 2018 and 2021, this guide's author encountered a single *Bombus* species, the vulnerable American bumble bee (*Bombus pensylvanicus*). Anecdotal reports from regional park rangers suggest a current dearth of bumble bee nests in state parks and federal land tracts along the Lower Rio Grande border.

American bumble bee • *Bombus (Thoracobombus) pensylvanicus* (De Geer, 1773)

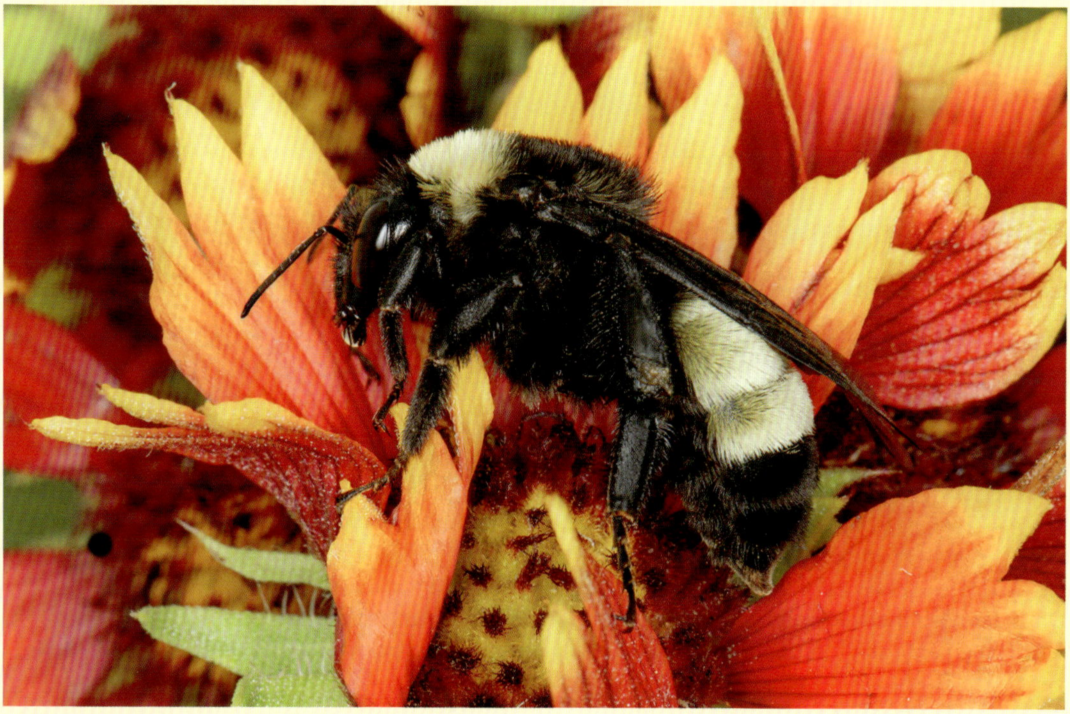

A female American bumble bee on blanketflower

Description: *Bombus (Thoracobombus) pensylvanicus* is a large bee with a black head covered entirely or mostly with black hairs; a thorax usually covered with dark hairs in front and black hairs behind; and an abdomen encircled with yellow and black hair bands. On females, the hair bands on the abdomen's front three segments are yellow. On males, yellow hairs usually cover the first four segments. The remainder of the abdomen is clothed mostly or entirely with black hairs. The legs and tegulae are black, and the wings dark with dark veins. **Size:** queens 22–26 mm (7/8–1 in.), female workers 13–19 mm (1/2–3/4 in.), males 16–22 mm (5/8–7/8 in.).

Range: The American bumble bee was first identified by Swedish entomologist Charles De Geer in 1773. For the next two centuries, this species ranged widely throughout the United States, northward into Canada, and southward as far as the Mexican state of Oaxaca. In the past 20 years, its range and numbers have decreased dramatically. A 2019 report by bumble bee experts MacPhail, Richardson, and Colla warned that *Bombus pensylvanicus* is now critically endangered.

Associated plants: The American bumble bee is a generalist pollinator that visits flora from a spectrum of plant families. Hundreds of *Bombus pensylvanicus* appeared at the National Butterfly Center in May 2021, feeding principally on silverleaf nightshade (*Solanum elaeagnifolium*) and lemon beebalm (*Monarda citriodora*). By November, the bees had spread throughout the grounds and were found on diverse flora, most notably crucita (*Chromolaena odorata*), esperanza (*Tecoma*

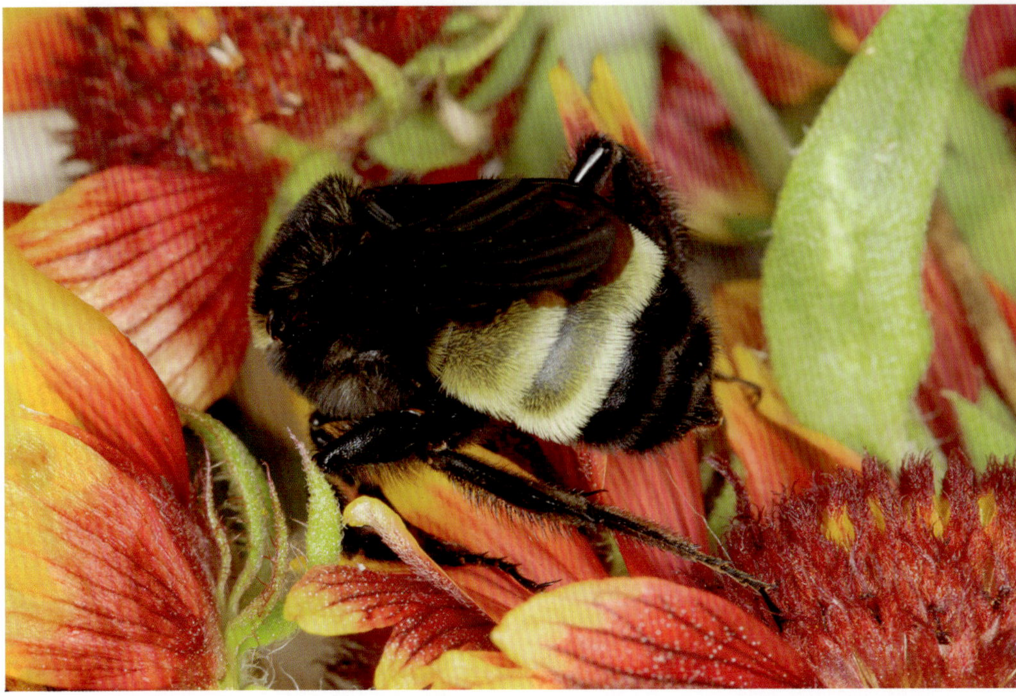

On all female American bumble bees, the fourth abdominal segment is always black. The fourth segment usually borders a yellow segment in front and a black segment behind.

stans), and tropical threefold (*Trixis inula*). In the same month, American bumble bees appeared at the nearby McAllen Nature Center, feeding on shrubby blue salvia (*Salvia ballotiflora*). In June and July of the same year, *Bombus pensylvanicus* appeared at the South Padre Island Birding and Nature Center (Cameron Co.), foraging on esperanza and cenizo (*Leucophyllum frutescens*).

ID tips: Bumble bee species are generally identified in the field by the hair-color patterns on their heads, thoraxes, and abdomens. Nonetheless, such qualities may vary considerably within a species, making identification tricky. The detailed traits of the American bumble bee are as follows:

Head: On all females (queens and workers), the hairs on the face and the top of the head are entirely black. On males, black hairs on the face and head may be intermixed with yellow hairs. Both male and female bees have black mandibles and antennae. *Thorax*: The hairs on the front third of the thorax are yellow on both males and females. The rear two-thirds are covered typically with black hairs or, less commonly, with black hairs intermingled with yellow hairs. On females, the hairs on the sides of the thorax are always completely black. *Abdomen*: As noted, the first three segments (T1–T3) are usually all yellow on females; on males, the first four (T1–T4) are usually all yellow. Less commonly, the first segment (T1) of either sex may be black, or black intermingled with yellow. On females, the hairs on the fourth abdominal segment (T4) are always black; and the fourth and fifth (T4–T5) are usually black. On males, the last two or three abdominal segments (T5–T7) are usually black, or black intermixed with yellow. Less commonly, on some males, the last segment (T7) may be entirely covered with orange hair, and the sixth segment (T6) may also be partly covered with orange hair.

American bumble bee behavior: American bumble bees nest most often in open grassland and farmland. Their tendency to nest aboveground, rather than in the safety of underground holes, exposes them to more likely encounters with predators. Perhaps as a consequence, *Bombus pensylvanicus* can be unusually aggressive for a bumble bee species.

In 1924, Phil Rau published a colorful article detailing *Bombus pensylvanicus* behavior. He observed that American bumble bees seemed to prefer nesting in shallow, bowl-shaped depressions in the dirt, hidden under long, dried, matted grass. Nonetheless, he discovered their colonies in other unique situations in the Missouri countryside: he stumbled on one colony nesting in an old paint bucket, and another in a sack filled with straw, lying on the floor of an abandoned clubhouse.

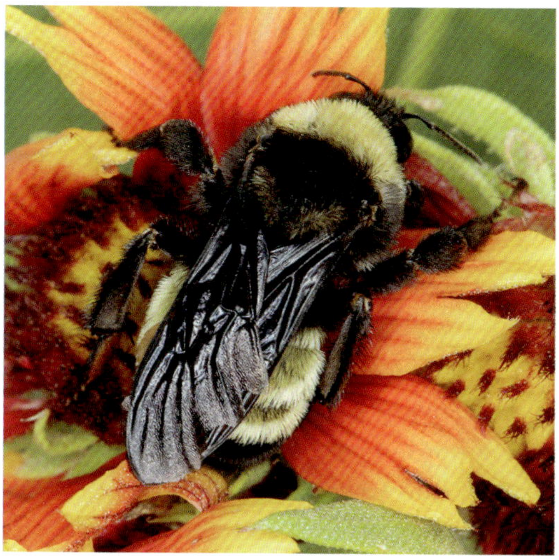
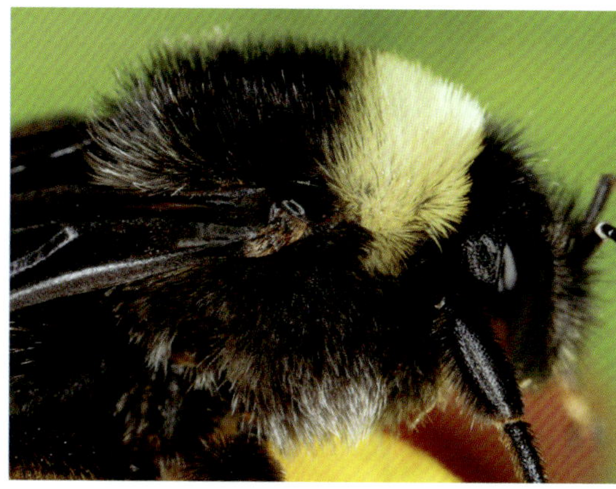

Female (*above left*) and male (*above right*) American bumble bees. The front third of the American bumble bee's thorax is yellow, and the back portion is black, or black with intermingled yellow hairs.

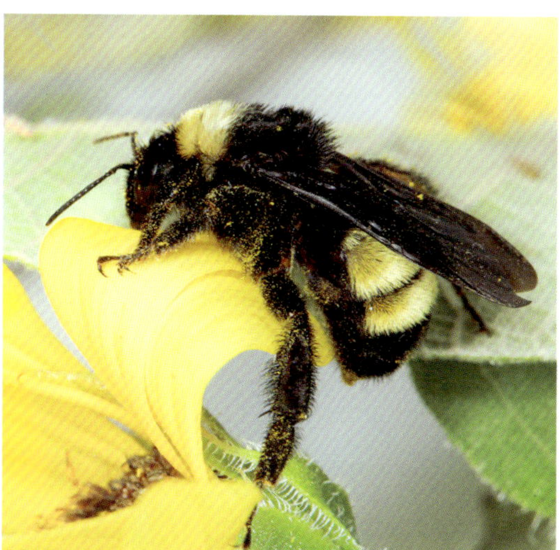
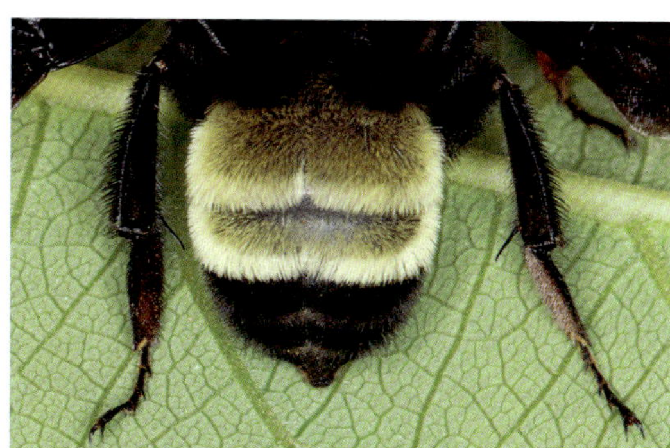

On the male (*left*) and female (*right*) bees shown here, the first abdominal segment is black, the second two are yellow, and the remainder are black. More commonly, the hairs on T1 are yellow.

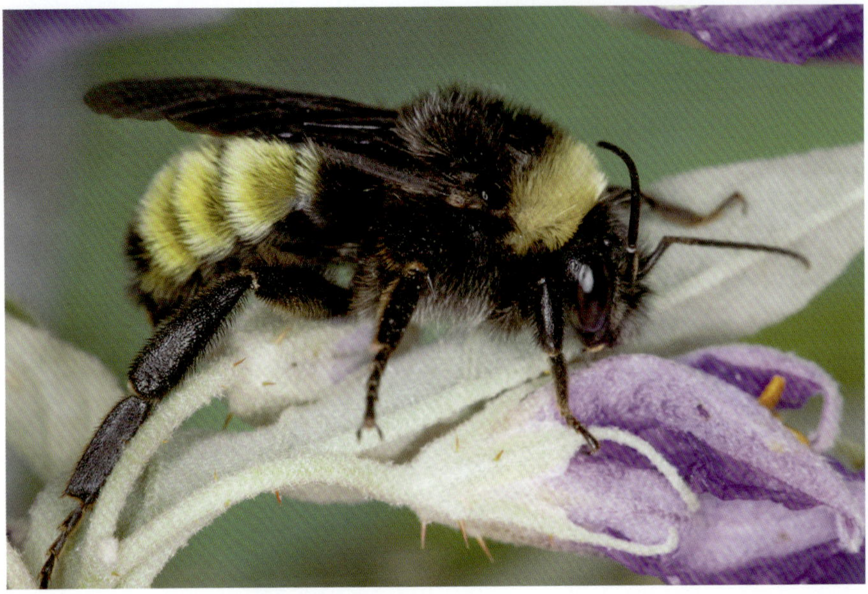

A male American bumble bee

He encountered an enormous colony in the ceiling of an old log house, and he happened upon an American bumble bee queen beginning a colony in the abandoned nest of a white-footed mouse, located in a hollow tree.

Inspired by his findings, Rau decided to establish some American bumble bee colonies on the roof of his house in Saint Louis. He anesthetized three groups of bees and transported them in flowerpots and cans to his city home. He was surprised by how well the bees survived the ether and the journey. He proffered the American bumble bees honey, part of which they ate; the rest they carried off and stored in wax honey pots constructed within Rau's flowerpots.

Rau placed one colonized flowerpot on the roof outside his third-story window. He covered the pot with glass in order to observe the colony better, leaving a small escape hole through which the bees could leave and enter the pot. When the first female bee exited into new surroundings, she spent a long period zigzagging around the area near the hive to orient herself and familiarize herself with the new landscape. She then flew in ever-widening paths around the roof and eventually disappeared for eight minutes—but then reappeared suddenly and zoomed directly back to the hive, laden with new pollen.

Thereafter, bees from that colony, and from a second colony that Rau placed on the same roof, were able to orient themselves quickly as well, and to begin collecting pollen provisions for their transplanted nests. Only one bee, which burst angrily and impetuously from the pot without checking its surroundings first, failed to return on its first day in its new setting. Sometime later, in response to a disturbance, a swarm of several bees angrily arose from one of the rooftop colonies and had trouble finding its way back, losing several members. "Thus mob psychology leads even these little creatures to reckless and disastrous action," Rau wrote.[13] Thereafter he dedicated himself to experimenting with the bees' homing abilities.

---

13. Rau (1924), p. 375.

Honey Bees

Subfamily Apinae ◆ Tribe Apini

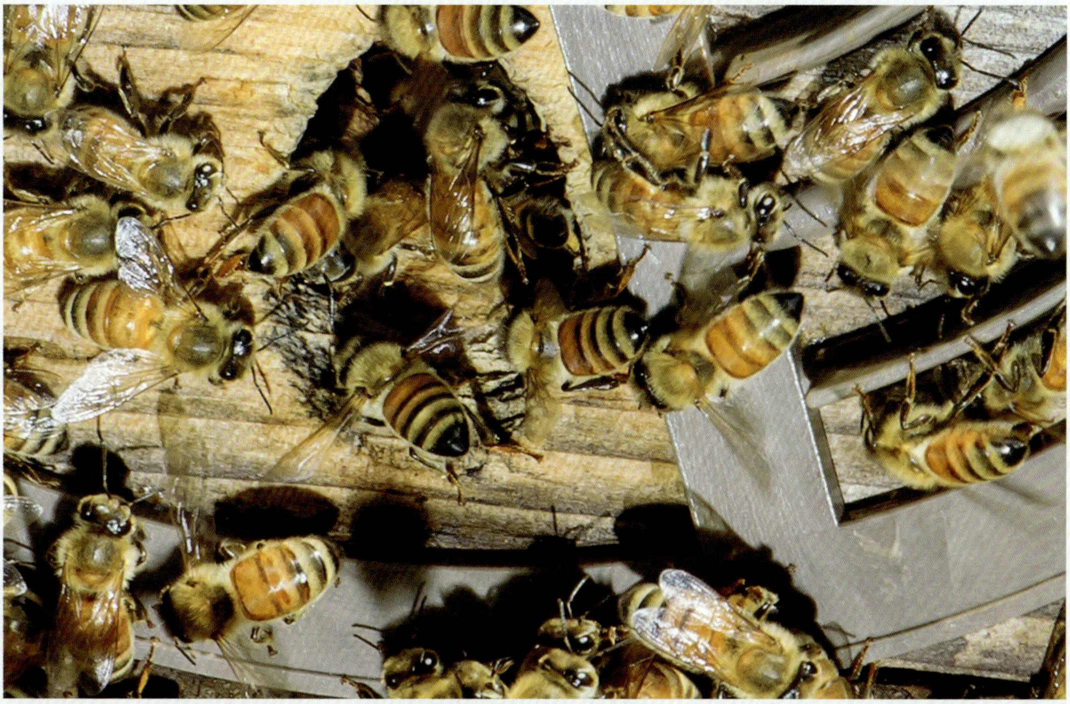

Worker honey bees entering a commercial hive

Honey bees are nonnative insects valued for their honey and their prowess as crop pollinators. Although there are more than 20,000 species of bees in the world, only a handful have proven capable of semidomestication for use as pollinators or honey producers. Of these, honey bees are the most important by far. Humans have kept honey bees for nearly 7,000 years.

Honey bees are particularly well suited for commercial management because their highly organized colonies concentrate large bee populations in a single nest—one hive can house as many as 80,000 bees. Honey bee society is divided into queens, which produce eggs and establish colonies; female workers, which collect pollen and honey, provision offspring, tend the nest, and guard the colonies; and male drones, which mate with new queens to produce future generations.

Other notable qualities make honey bees superior pollinators—sometimes to the detriment of native bee populations that cannot compete with them. Honey bees are able to range several miles in a single day when foraging for nectar and pollen. They are versatile generalist pollinators that thrive in diverse habitats, and they visit a seemingly endless array of flora from more than 50 plant families.

Honey bee societies are also adept at withstanding climate extremes. On hot days, honey bees collect outside the hive, fanning their wings to keep the colony cool. On chilly days, workers cluster around the queen to insulate her from the cold. During periods of extreme cold, workers may generate heat by *shivering*—they elevate their body temperatures by vibrating their flight muscles while keeping their wings still. On a day when temperatures are near freezing, a group of thousands of worker bees can raise the temperature of a hive to 95° Fahrenheit. Such collective actions designed to regulate hive temperature have allowed honey bees to thrive in a wide range of climates. They have proven capable of surviving on all continents except Antarctica.

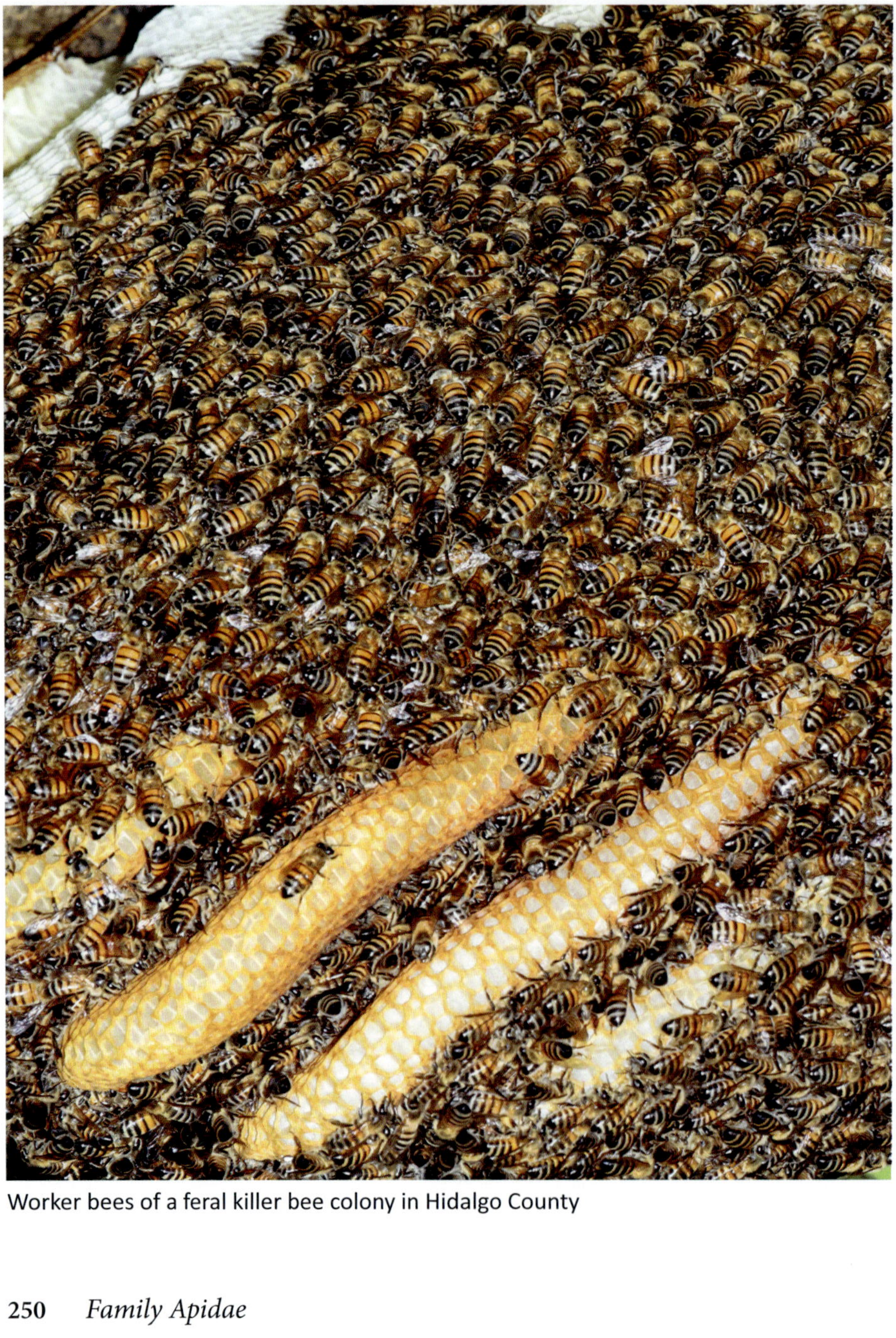

Worker bees of a feral killer bee colony in Hidalgo County

Western honey bee ◆ *Apis mellifera* Linnaeus, 1758

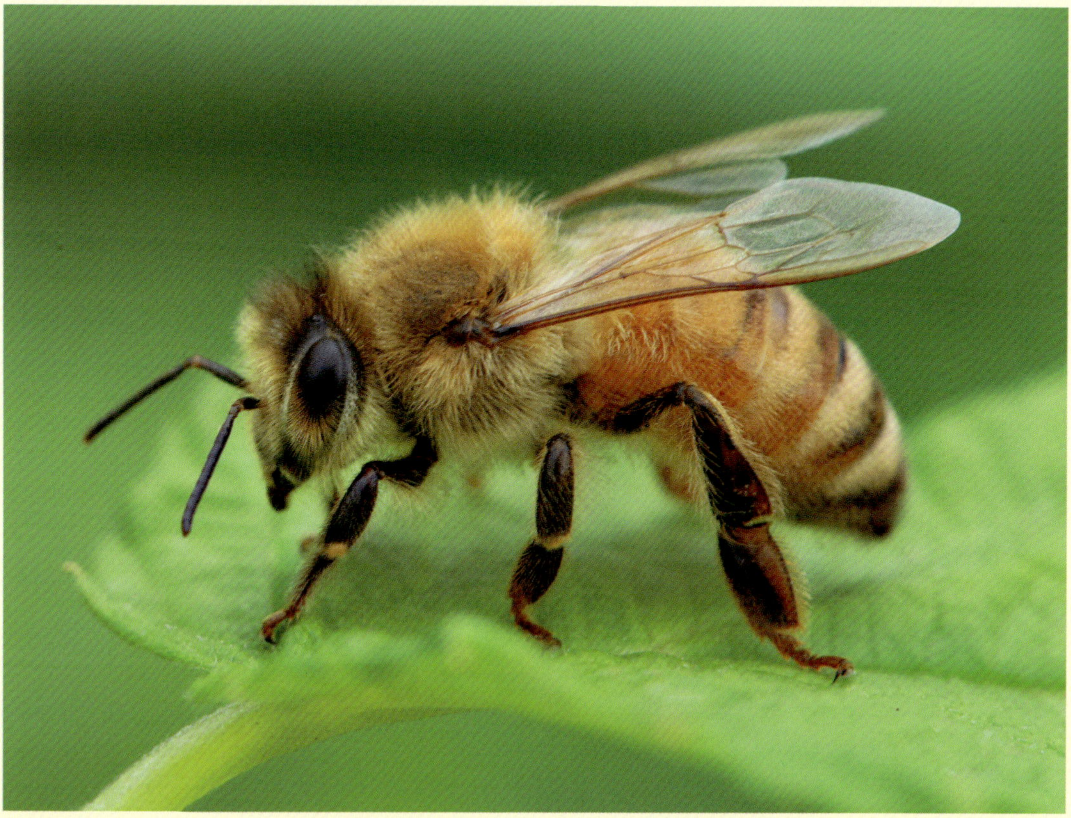

A honey bee worker. Photo credit Ross Eatman.

Description: *Apis mellifera* is a medium-sized bee, with a honey-colored or brown abdomen usually banded by dark stripes; transparent colorless wings; and a heart-shaped, furry-looking face—even the eyes of *Apis mellifera* are hairy. The most commonly seen honey bees are female workers, which transport pollen in corbiculae on their hind legs. Honey bees vary in color and size according to age, subspecies, and function (drone, worker, or queen). Male drones are larger than workers and have dark, bulbous eyes. Queens resemble workers but have much longer abdomens.

There are more than 20 subspecies of honey bees. One notorious honey bee variety prevalent in the Lower Rio Grande Valley is known as the "killer bee," or Africanized bee. Killer bees are not distinguishable by visual inspection from less aggressive regional varieties of the western honey bee. **Size:** queens typically 18–20 mm (7/10–4/5 in.), other females (workers) 12–16 mm (1/2–5/8 in.), males (drones) usually 14–16 mm (1/2–5/8 in.).

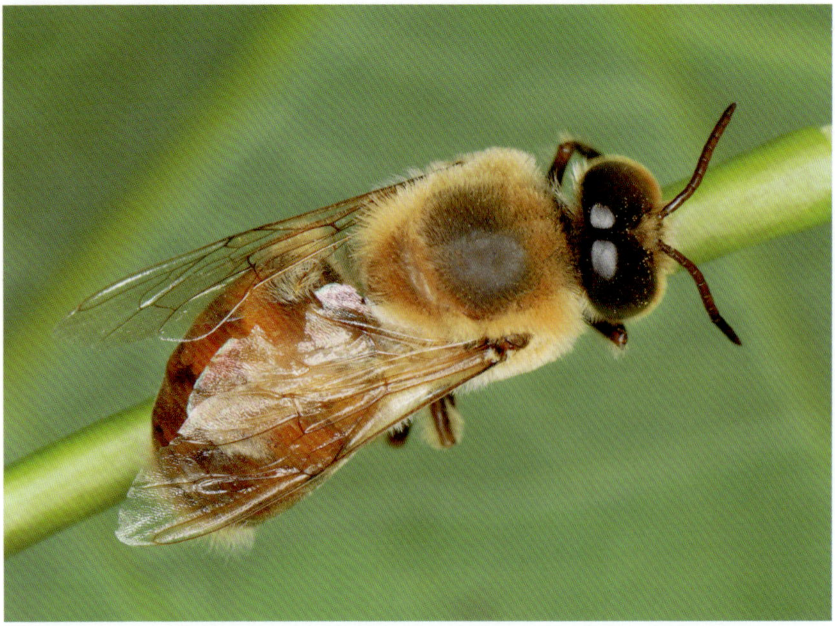

A honey bee drone: note the male bee's large, dark eyes.

Range: Recent genetic studies indicate that western honey bees originated in Asia about one million years ago and then traveled westward to Africa and Europe. Honey bees were first brought to the Americas in the 1600s by Spanish colonists, who utilized them for honey and beeswax production. The highly organized social life of honey bees allowed them to adapt to diverse environments in the New World. They spread throughout the Americas and are now naturalized in every part of the United States. As discussed below, the strain of honey bees regionally known as "killer bees" originated in Africa and arrived in Texas from South America during the 1990s.

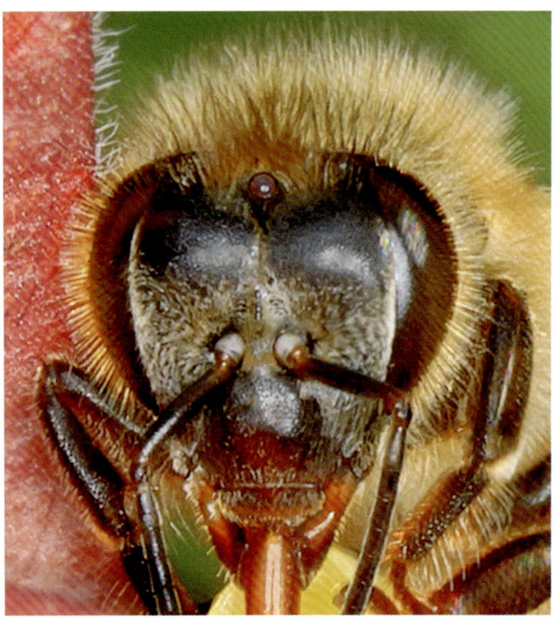

The faces of honey bees are hairy and heart shaped.

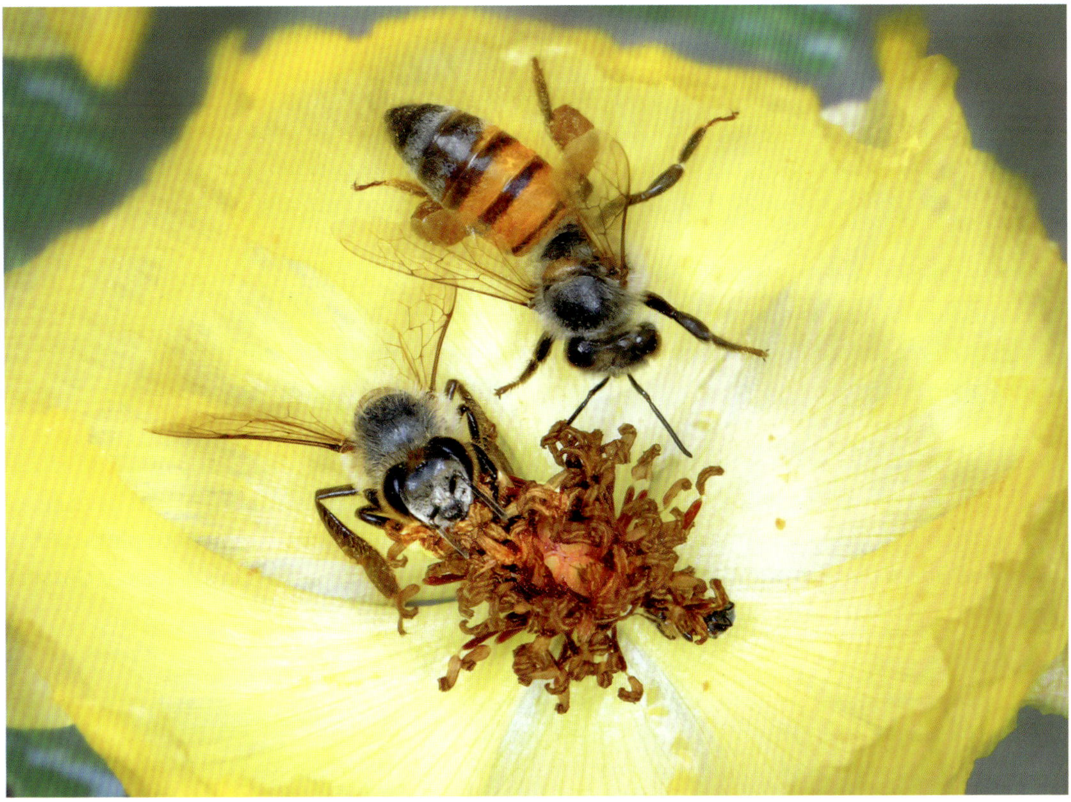

Honey bees gathering pollen from golden prickly poppy (*Argemone aenea*)

Plant associations: Honey bees are key pollinators of a diverse spectrum of crops, among them leafy vegetables, onions, berries, and fruit and nut trees. They are, in addition, ubiquitous pollinators of garden flowers, and they show remarkable adaptability to native flora of varied habitats.

In the Lower Rio Grande Valley, nonnative honey bees emerge in February to gather pollen, and they remain flying through late December. During spring in the Valley, honey bees visit early-blooming native flora such as guayacan (*Guaiacum angustifolium*), creosote bush (*Larrea tridentata*), and huisache (*Vachellia farnesiana*) before most native bees have emerged from their winter nests. Honey bees are also energetic pollinators of such native shrubs and trees as colima (*Zanthoxylum fagara*), coma (*Sideroxylon celastrinum*), Texas ebony (*Ebenopsis ebano*), and wild olive (*Cordia boissieri*). Honey bees are among a handful of bee species in the Valley that gather pollen from prickly poppies (*Argemone aenea*, *A. sanguinea*, and *A. mexicana*).

There are limits, however, to honey bees' versatility as pollinators. Honey bees do not gather pollen from many cacti native to the Valley. And, although they are standard commercial pollinators of orchards and truck vegetables, honey bees are incapable of buzz pollination. Thus, unlike many bumble bees and carpenter bees, honey bees are not efficient pollinators of crops that require vibratile pollination to produce fruit, among them blueberries, cranberries, and nightshade-family vegetables such as peppers, tomatoes, and eggplants.

Honey bee decline and vulnerability: In the last decade, honey bees have experienced a swift drop in numbers in both Europe and the United States. In 2006, American beekeepers began reporting steep annual decreases in honey bee populations. Honey bee decline has been attributed to the same factors that now imperil native bees—habitat loss and the overuse of pesticides—and to the spread of disease-carrying bacteria and pests such as hive beetles, and varroa and tracheal mites. Honey bee decline now threatens both large-scale agriculture and beekeeping industries in North America.

In the United States, some efforts to fortify honey bee populations have focused on cross-breeding various strains that are hardy and disease resistant. Tougher strains are frequently less docile. Bee breeding thus often requires a juggling of needs, in which hardiness may be balanced against aggressiveness.

The arrival of "killer bees" in the Lower Rio Grande Valley: In 1973, Michener penned an article titled "The Brazilian Honeybee" in which he noted, "Honey bees of European origin have never done particularly well in the tropics."[14] Although European honey bees exhibit both physical and behavioral adaptions that allow them to survive cold winters, they languish in sweltering climates and lack the general hardiness of African honey bees.

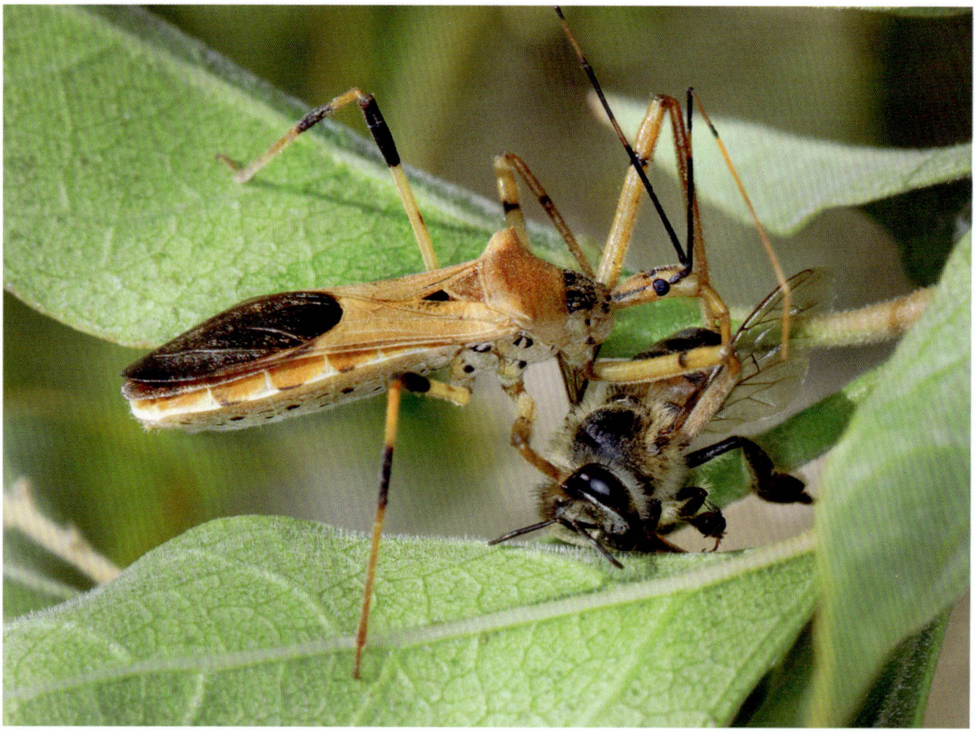

An assassin bug capturing a feral "killer bee" at Hidalgo County's Santa Ana National Wildlife Refuge

14. Michener (1973), p. 523.

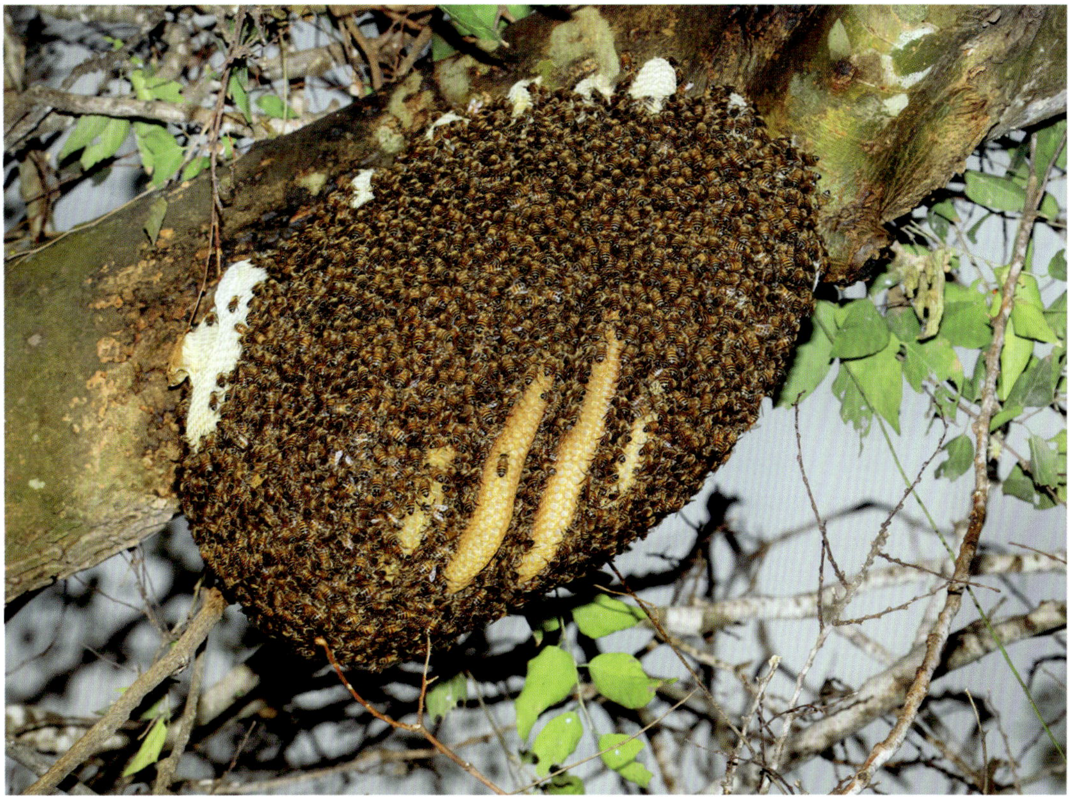

A feral hive of killer bees established in a tree in Hidalgo County

In the 1950s, beekeepers and farmers in tropical and subtropical South America faced a crisis: the European honey bees used as pollinators and honey producers in most of the world were faring poorly in hot, humid areas. Brazil's honey output was very low, and its lack of viable pollinators limited agricultural production.

In 1956, Brazilian geneticist Warwick Estevam Kerr traveled to Africa to search for honey bee subspecies that showed high productivity and survival rates in tropical climates. In Tanzania, Kerr found the heat-resistant, productive, and aggressive *Apis mellifera scutellata*, which he brought back to Brazil. Kerr set up quarantined hives for the African bees in Rio Claro, São Paulo, and commenced interbreeding their offspring with subspecies of European honey bees that flourished in warm Mediterranean areas. Kerr's aim was to produce a bee with the gentleness of the European strains and the toughness and productivity of the African bees.

As the story goes, in 1957, a visiting beekeeper overseeing Kerr's hives accidentally released from quarantine and into the wild 26 swarms headed by Tanzanian queen bees. These bees commingled and crossbred with European honey bees then extant in Brazil, producing "Africanized" honey bees—a hybrid that became known for the belligerent and sometimes dangerous behavior that earned it the popular name "killer bee."

Killer bees spread from Brazil into neighboring South American countries and then through Central America at a rate of about 200–300 miles per year. Several factors enabled Africanized bees to dominate existing European honey bee populations. These included Africanized bees' greater adaptability to hotter climates; their tendency to swarm and form new colonies with greater frequency than European honey bees; the ability of Africanized drones to outcompete European honey bee drones; and the faster development of Africanized honey bee offspring.

In October 1990, killer bees established their first documented colony in the United States—in Hidalgo, Texas, in the Lower Rio Grande Valley. Hidalgo has embraced its notoriety, pronouncing itself the "Killer Bee Capital of the World." It now has a statue of a giant killer bee on its town green, and a succession of Hidalgo hockey teams have called themselves the Killer Bees.

To date, killer bees are well established in the United States along the Mexican border from Texas to California. In the West, they have appeared as far north as Utah, and in the East, killer bee colonies have been discovered in Tennessee, Louisiana, Georgia, and Florida. Africanized bees' more northward expansion has been thwarted by their lack of adaptability to long winters. Among other limitations, killer bees are unable to warm themselves during the winter by shivering. Killer bees are also less able to withstand long periods of food shortage associated with winter climates.

Several factors make killer bee colonies more dangerous than those of European honey bees. Most notably, Africanized bees show a heightened defensiveness and willingness to attack when their nests are disturbed. Attacks by Africanized bees are also more prolonged, and killer bees chase their targets a longer distance than European honey bees typically do.

In 1973, Michener described an experiment in which a research team rapped on the outside of a killer bee hive and then dangled a piece of black leather at the nest opening. The Africanized bees, Michener wrote, "stung the leather 92 times in only about 52 seconds and followed our Committee member (protectively clothed!) carrying the leather for 1,044 meters, trying to sting the whole way."[15]

Africanized bees are also more likely to form large swarms when attacking. European honey bees swarm at intruders as well but tend to defend themselves in groups of 10 to 20 bees, while Africanized honey bees might respond in groups of several hundred. In the last decade, there have been a handful of reports of Africanized bees killing livestock and people on US soil.

Notably, individual Africanized honey bees are not dangerous to approach. A "killer" bee intent on collecting pollen from a flower far from its hive is unlikely to show aggression toward a human onlooker if left undisturbed. Wild honey bee colonies are potentially dangerous, however, anywhere in the Lower Rio Grande Valley.

15. Michener (1973), p. 525.

Family Colletidae

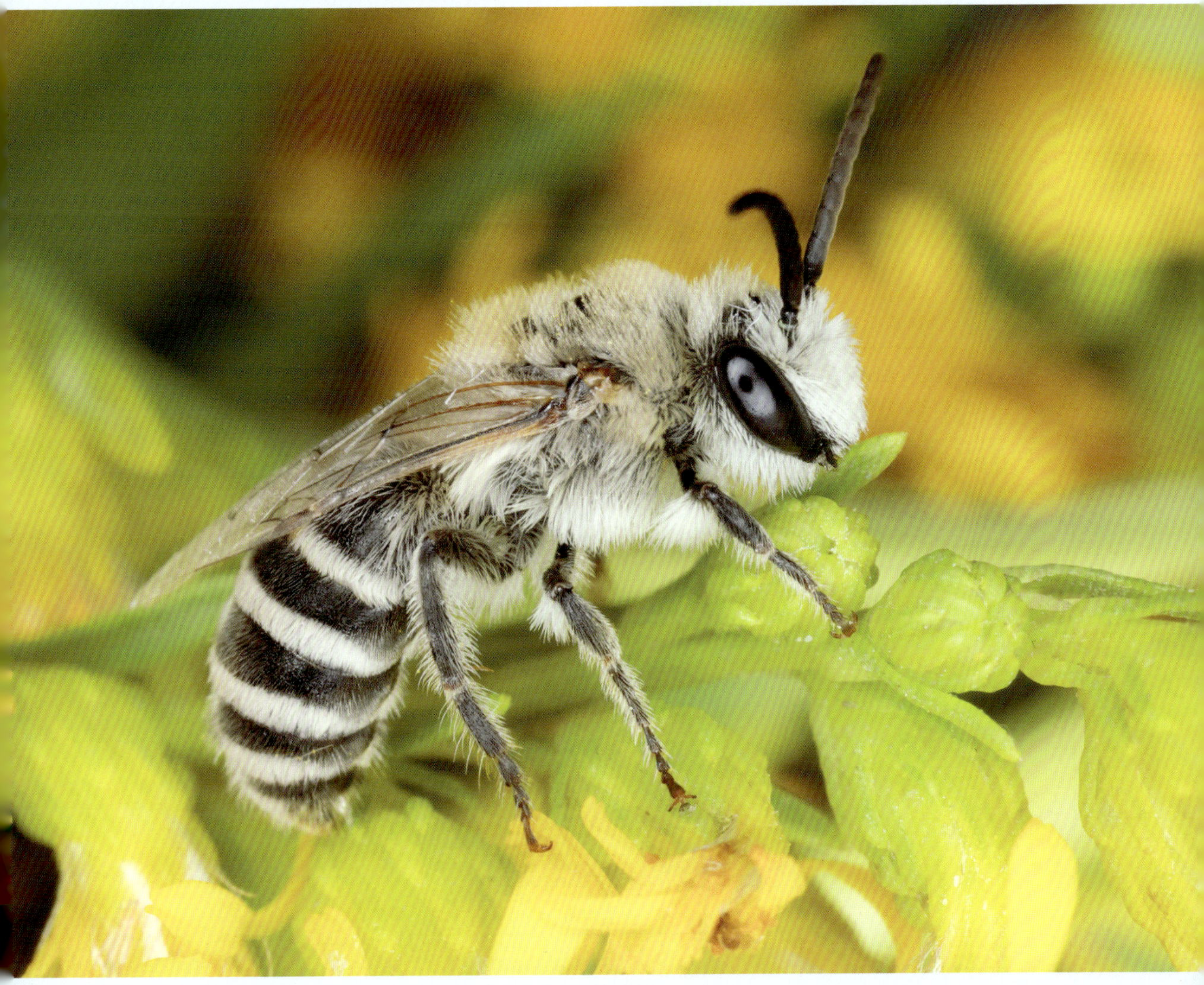

Feather-Tongued Bees
Cellophane Bees
Masked Bees

COLLETIDAE FAMILY GENERA
OF THE TEXAS LOWER RIO GRANDE VALLEY

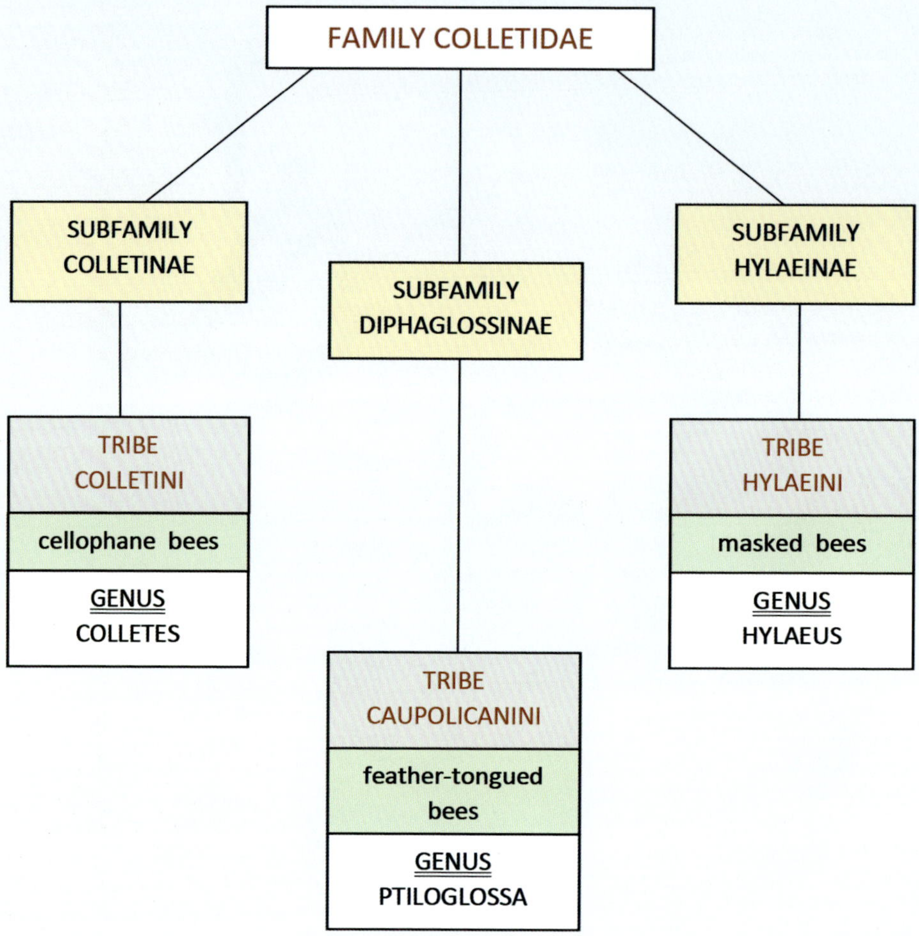

Feather-Tongued Bees

Subfamily Diphaglossinae ◆ Tribe Caupolicanini
Genus *Ptiloglossa*

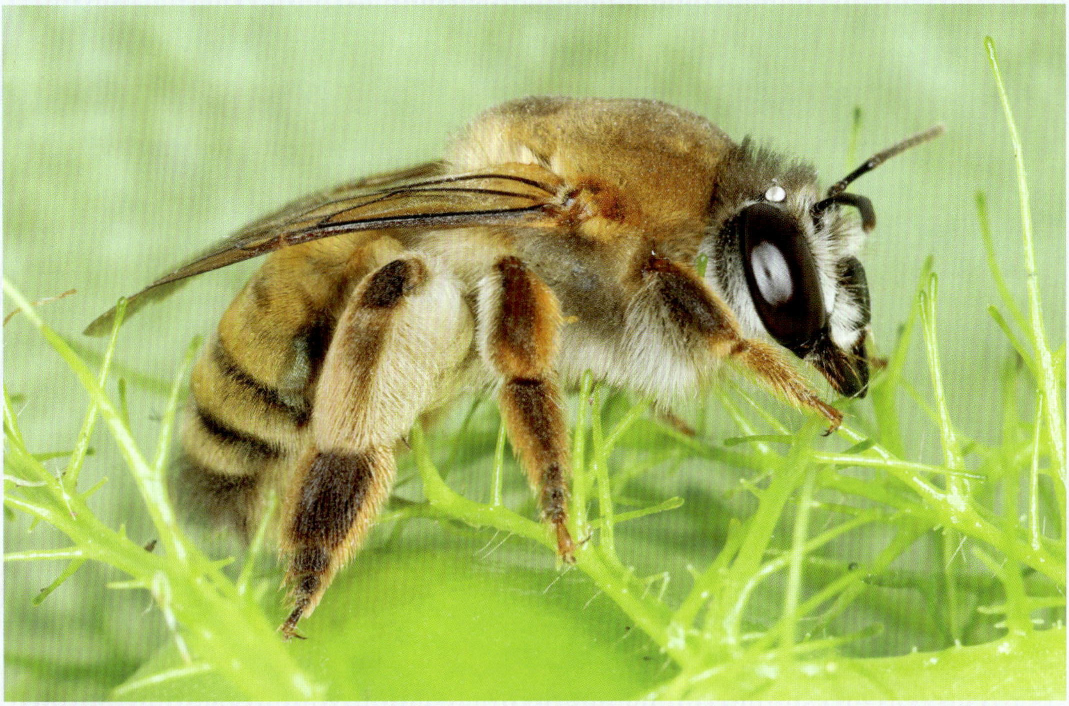

A female feather-tongued bee

Ptiloglossa are robustly built, hairy bees with enlarged ocelli adapted to navigating in low-light conditions. The bees avoid periods of intense sunlight and fly during the evening or early morning. Their tongues are forked and feathery—*ptiloglossa* is Greek for "feather-tongue."

Feather-tongued bees are found only in the New World and are generally tropical or subtropical. In the United States, they occur principally along the southern borders of Texas, New Mexico, and Arizona. In the past 100 years, three species have been recorded in the Valley—*Ptiloglossa arizonensis*, *P. hoplopoda*, and *P. mexicana*. All are buzz pollinators that target nightshade-family plants (Solanaceae). Feather-tongued bees are loud and large fliers, usually measuring 15–20 mm (3/5–4/5 in.).

Female *Ptiloglossa* carry pollen on their hind tibiae, and in corbiculae (pollen baskets) on their hind femurs. The corbiculae are formed by long, finely branched, interwoven scopal hairs well equipped to carry the fine-grained pollen of Solanaceae. *Ptiloglossa* also frequent other plants—among them passionflowers and pea-family flora such as *Cassia*, indigo, and sweetclover.

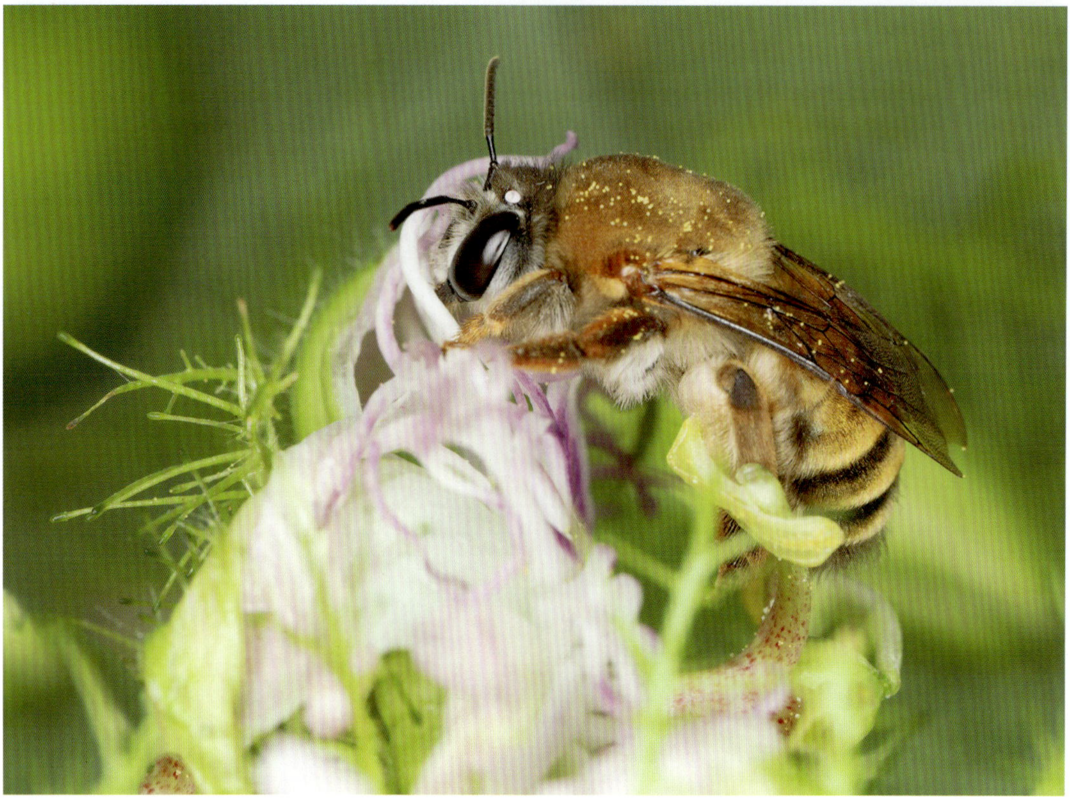

A female *Ptiloglossa mexicana*

Genus identification: All *Ptiloglossa* share distinctive traits that aid in their genus identification: (1) the forewing's stigma is shorter than the prestigma, and either slender or nearly absent; (2) the scutum is deeply grooved; (3) the face has a large malar space (the gap between the mandible base and the compound eye) that is at least one-third the length of the compound eye; and (4) on males, the third and fourth sternal segments (S3 and S4) overlap T4 and T5; where they overlap, the sternites are dull and covered with short, erect hairs.

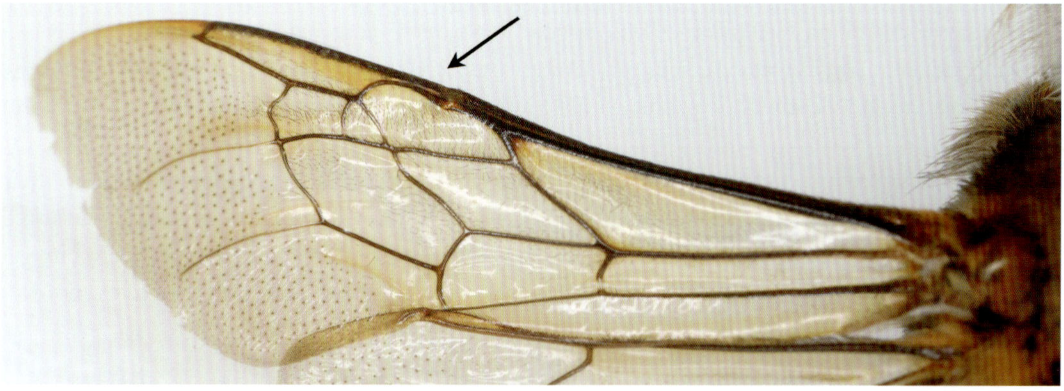

Ptiloglossa forewing: the stigma is short and narrower than the dark prestigma to its right.

Mexican feather-tongued bee ◆ *Ptiloglossa mexicana* (Cresson, 1878)

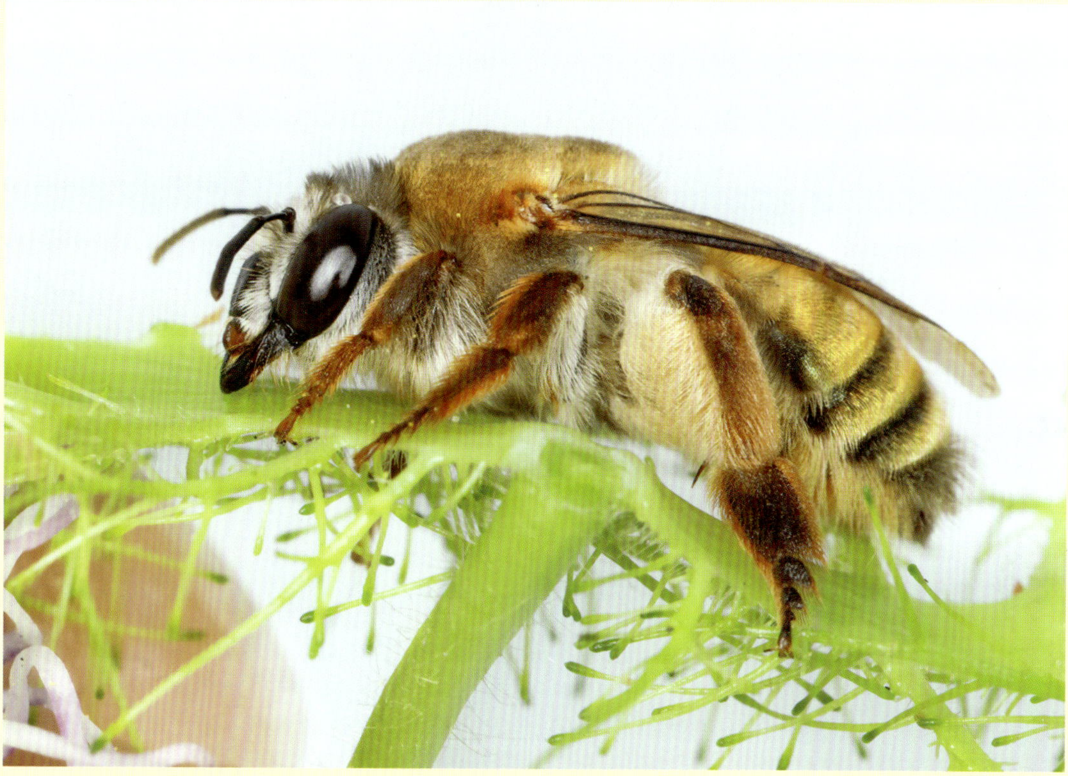

A female Mexican feather-tongued bee on cottonleaf passionflower

Description: *Ptiloglossa mexicana* is a hefty, furry-looking, gold bee with a dark abdomen girded by golden bands. Its legs are reddish gold, its tegulae are honey yellow and partly covered with golden hairs, and its wings are auburn with dark veins. Hairs on the Mexican feather-tongued bee's thorax and abdomen have a shimmery, chatoyant quality, with silky bluish-green reflections, so that under strong morning or evening light the bee appears radiant. The compound eyes are large and black. The female's enlarged ocelli are pink and appear luminous under direct light.

Feather-tongued bee species are separated in part by subtleties in the coloring of hairs on the vertex, thorax, and abdomen. Female Mexican feather-tongued bees have brown thorax hairs with dark tips, and black to brown hairs on the vertex. Males have golden thorax hairs with dark tips, and a mix of golden and dark brown hairs on the vertex. On both male and female *Ptiloglossa mexicana*, the first three abdominal segments (T1–T3) are covered with golden hairs. On females, T4 has some dark hairs, and T5–T6 are entirely covered with dark hairs. On males, T4 is covered with dark brown hairs and T5–T7 are clothed with white hairs; these white hairs intermingle with scattered dark hairs close to the tip of T7. **Size:** females 19 mm (3/4 in.), males 15 mm (3/5 in.).

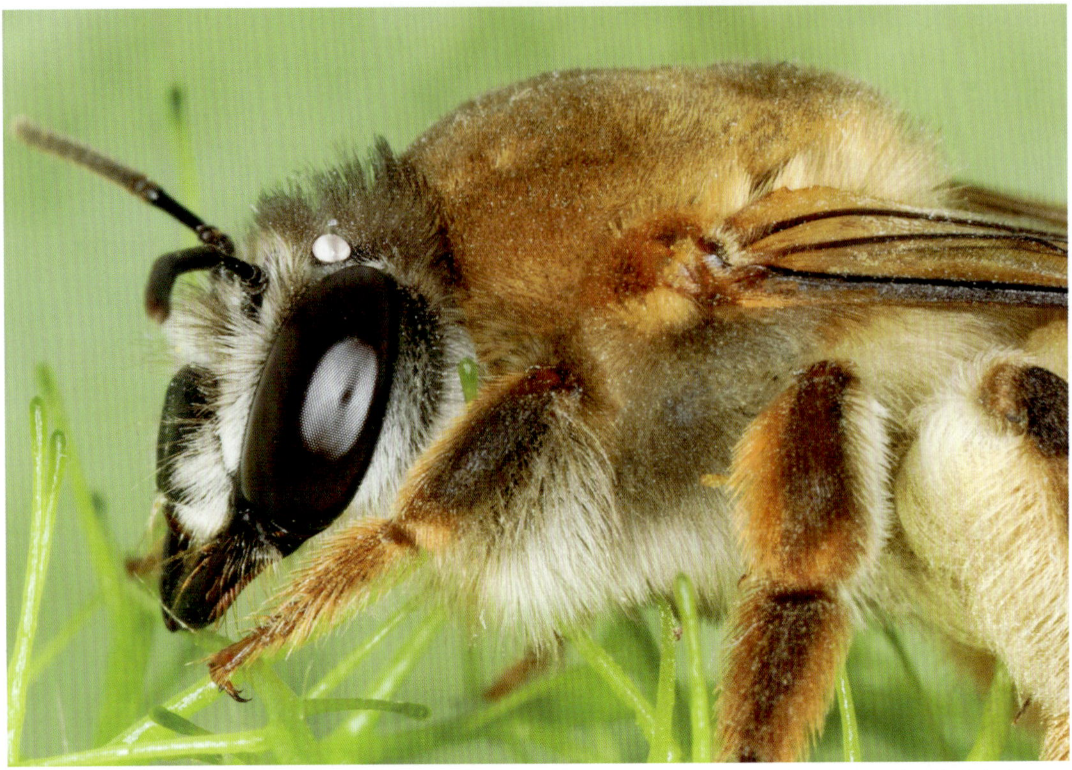

A female *Ptiloglossa mexicana*: note the luminous ocellus behind the compound eye, the dark hair on the top of the head, and the long malar space.

Range: The Mexican feather-tongued bee ranges from southern Texas to northern South America. In the United States, this species occurs almost exclusively in the Lower Rio Grande Valley. Documented sightings of *Ptiloglossa mexicana* in the Valley are relatively uncommon. In the last decade, most have occurred in Hidalgo County. Cresson first described this species in 1878 after examining a bee collected in Mexico by Sumichrast.

Associated plants: In Hidalgo County, *Ptiloglossa mexicana* commonly visits silverleaf nightshade (*Solanum elaeagnifolium*), Texas milkweed (*Asclepias texana*), and passionflower. The female and male bees shown here were found foraging in May and July 2021 on cottonleaf passionflower (*Passiflora foetida* var. *gossypiifolia*) at the estate of master gardener William Rich of McAllen.

ID tips: *Face*: The female Mexican feather-tongued bee's upper face is dark and pitted, while the lower part of its clypeus is burnished red and smooth. Its labrum is yellow and lobed, and its mandibles have two-pointed subapical teeth. The male's clypeus is protuberant and almost entirely honey yellow. *Legs*: *Ptiloglossa mexicana* females have a patch of dark hairs on the apical end of each hind femur. The outer spurs on the hind tibiae of males are broad at the base and curve downward, nearly parallel to the edge of the leg segment.

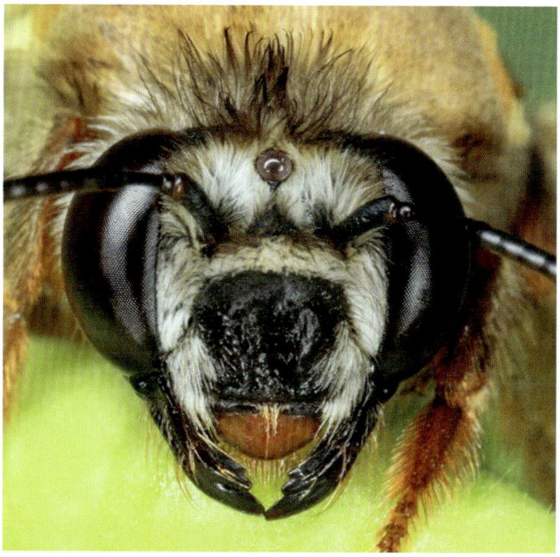

Face of a female *Ptiloglossa mexicana*

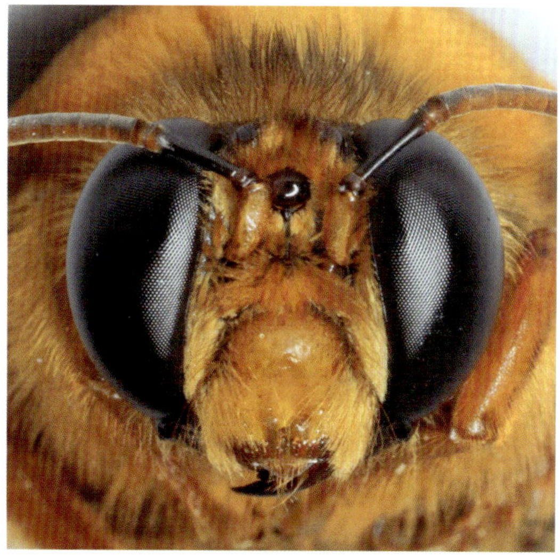

A male with a honey-yellow clypeus

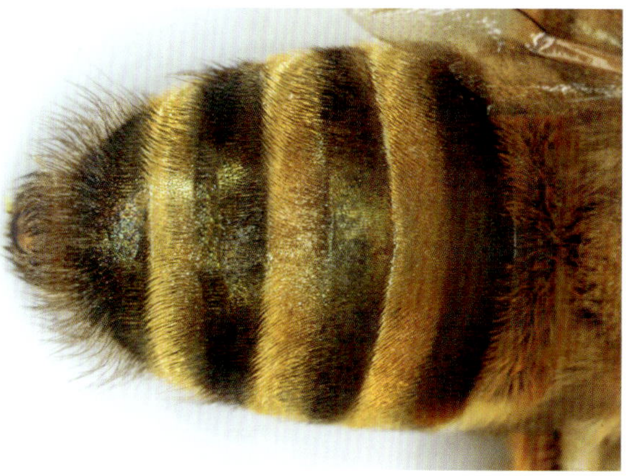

Abdomen of a female *Ptiloglossa mexicana*

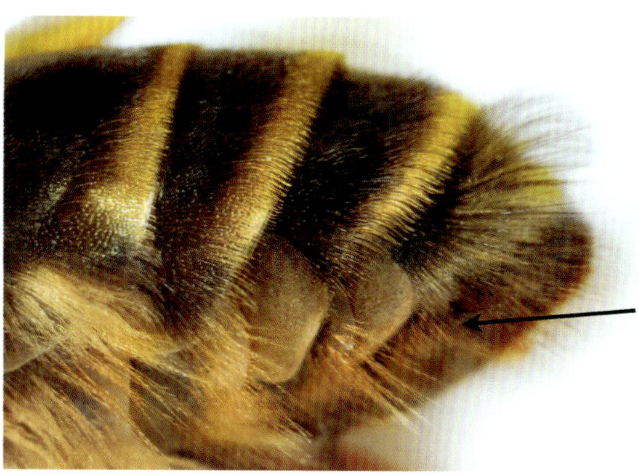

Abdomen of a male: the sternal plates of S3 and S4 overlap T4 and T5. This is a trait of all males of the genus *Ptiloglossa.*

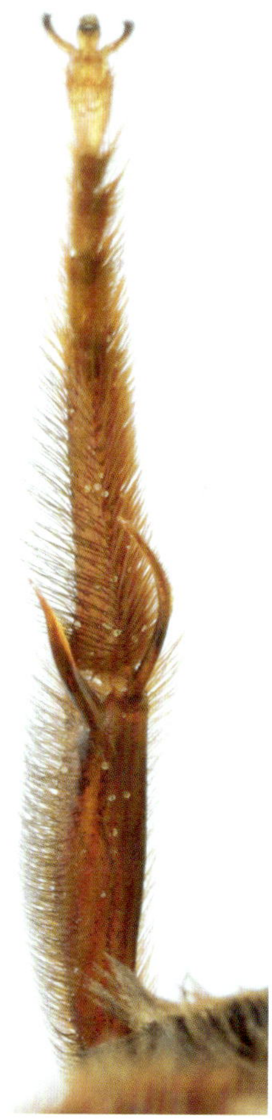

Hind-leg tibial spurs of a male *P. mexicana*

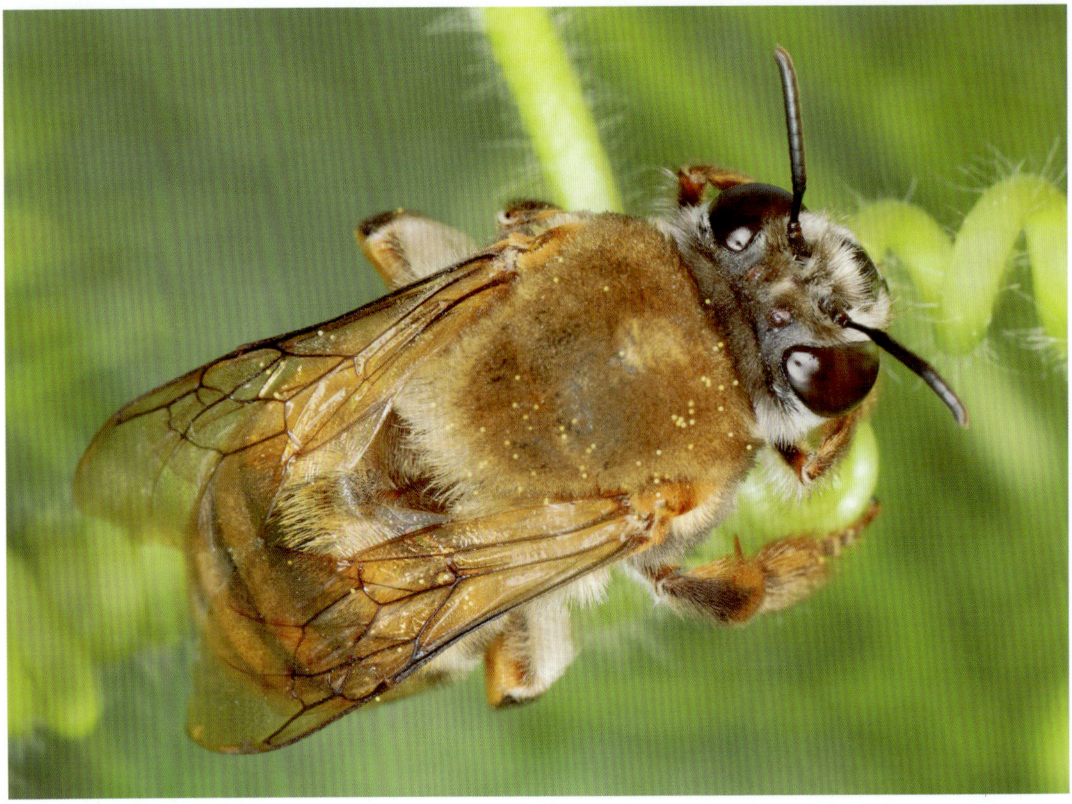

A female Mexican feather-tongued bee

Behavior: Sightings of *Ptiloglossa mexicana* in the Valley are uncommon in part because of the bees' unusual flight schedule. Mexican feather-tongued bees emerge at dawn to visit flowers. On hot days, they often retire by 9:00 a.m. They sometimes reappear very late in the afternoon: they can be found foraging on flowers just as the sun is beginning to set.

The nests of *Ptiloglossa mexicana* remain unknown. Bees of the genus *Ptiloglossa* are generally solitary (a trait of most Colletidae), although they often nest gregariously. In 1984, Jerome Rozen published an article about the nesting behavior of several *Ptiloglossa* species. He described the nests of two North American species, *Ptiloglossa arizonensis* and *P. jonesi*, found in Portal, Arizona. The nests of both had entrances surrounded by small dirt mounds and consisted of meandering burrows that terminated in lateral tunnels leading to egg chambers. In one nest excavated by Rozen, a female *Ptiloglossa arizonensis* had used her pygidial plate to smooth the walls of her egg cells and had lined the walls with a waterproof, plastic-like coating.

Cellophane Bees

Subfamily Colletinae ◆ Tribe Colletini
Genus *Colletes*

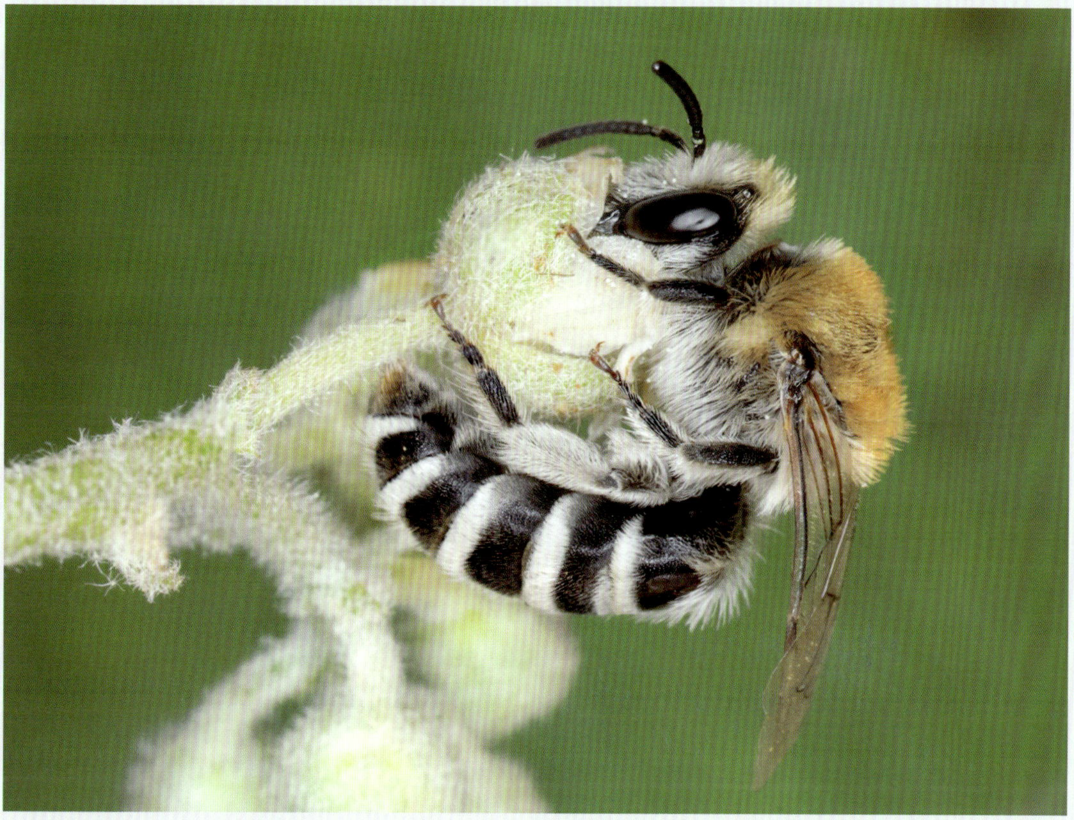

A cellophane bee drinking nectar from a flower

Colletes line the cells of their nests with a waterproof material made of macrocyclic lactones, which when dry resembles clear plastic. Accordingly, they are sometimes called "cellophane bees" or "polyester bees." *Colletes* are equipped with forked tongues, which they use to paint the plastic-like material onto their nest walls to keep the nests dry. The bees produce the material from a special abdominal organ called the Dufour's gland, named after French naturalist Léon Jean Marie Dufour, who first recorded his fascination with the bees' fabrication of *plastique* in 1835. Other Colletidae—such as the feather-tongued bee described in the previous section—similarly coat their egg cells with a waterproof substance produced by the Dufour's gland, although *Colletes* are unique in their ability to produce macrocyclic lactones.

Cellophane bees are about the size of honey bees (or slightly smaller or larger). They have black eyes and are generally hairy, with pale, brown, or rust-colored hairs on their heads and thoraxes. Their dark abdomens often have hair bands. Their faces are heart shaped—that is, they taper toward the mandibles and the eyes diverge upward. The heads of *Colletes* are usually

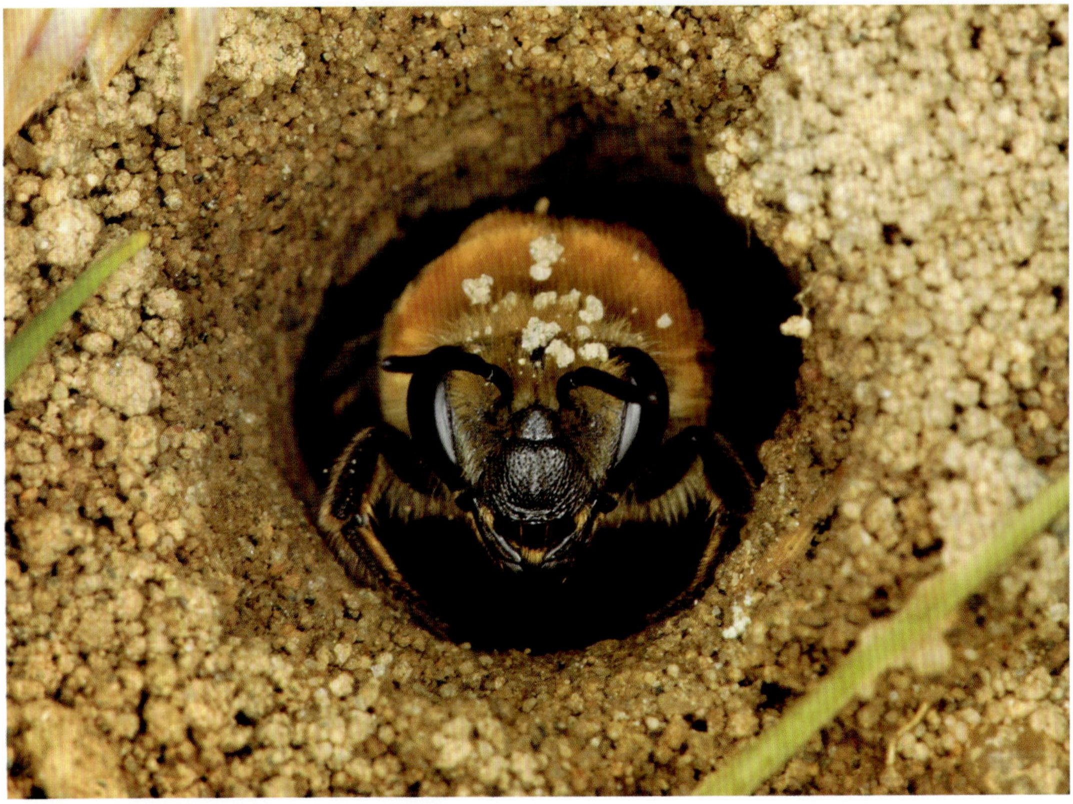

A female cellophane bee peering from its nest

covered with pale or light brown hairs, and their antennae are medium length. Female *Colletes* carry pollen on the upper parts of their hind legs.

The Lower Rio Grande Valley harbors many *Colletes* species that remain uncataloged. At least nine *Colletes* species have been identified in the Valley in the last 100 years. In addition to the five in this guide, these include the rarely collected *Colletes beamerorum*, *C. cercidii*, *C. punctipennis*, and *C. saritensis.*

Little has been written about the specific behavior of Lower Rio Grande Valley cellophane bees. *Colletes* are divided roughly into those that emerge in spring and those that emerge in fall. All five of the cellophane bee species in this guide—*Colletes birkmanni*, *C. intermixtus*, *C. mandibularis*, *C. texanus*, and *C. swenki*—emerge in spring. *Colletes birkmanni* and *C. texana* are thought to be bivoltine (producing two generations per year). Some evidence exists that *Colletes mandibularis* may be bivoltine in the Valley as well.

Female cellophane bees construct individual nests in the ground, excavating tunnels partitioned into series of egg chambers, which the bees provision with a pasty mixture of nectar and pollen. Despite their solitary status, *Colletes* tend to tunnel near one another. Groups of nesting cellophane bees sometimes number in the tens of thousands. *Colletes* nest aggregations should be left undisturbed; the bees are generally nonaggressive, and their nests usually last only a few weeks and serve to aerate the soil.

Birkmann's cellophane bee ◆ *Colletes birkmanni* Swenk, 1906

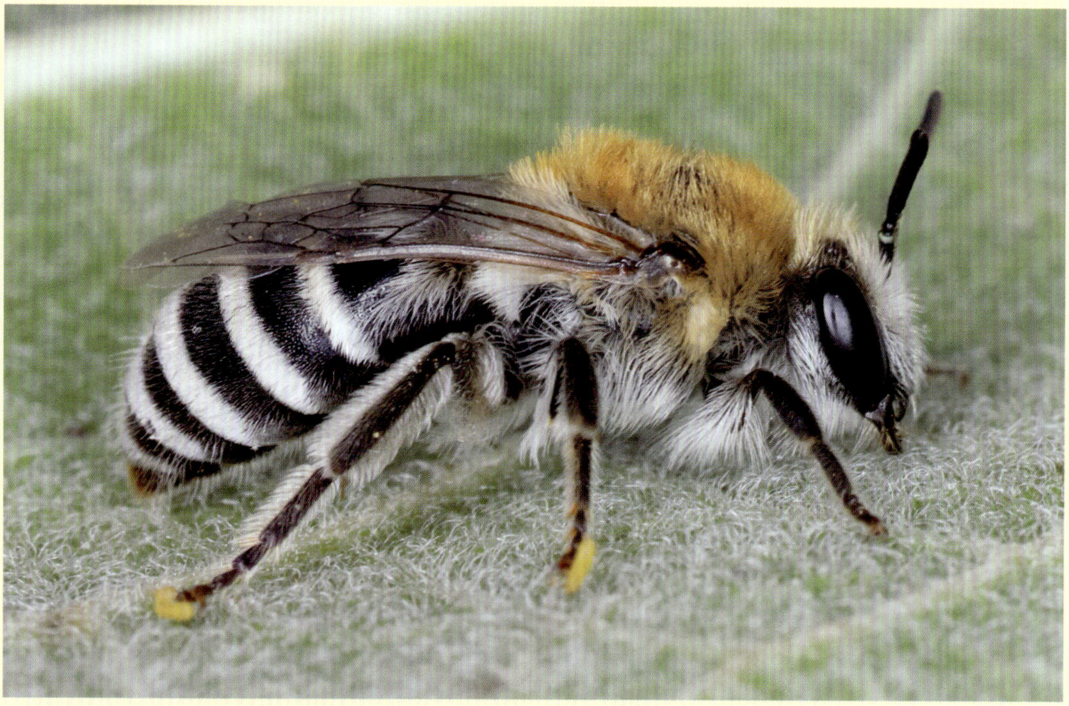

A female Birkmann's cellophane bee

Description: Birkmann's cellophane bee has a dark abdomen boldly striped with well-defined pale hair bands. The female's thorax is covered with rust-orange to pale orange hairs. On males, thorax hairs are predominantly white, sometimes with blond hairs on the scutum and scutellum. The head and underside of the thorax of both males and females are clothed in long, copious white hairs. The tegulae are brown, and the wings transparent brown with dark brown veins. **Size:** females 11–12 mm (2/5–1/2 in.), males 9–10 mm (1/3–2/5 in.).

Range: This is a common species in South Texas that flies from spring through December. It occurs west of the Mississippi and has been documented in Kansas, Oklahoma, Arizona, New Mexico, Colorado, and California. Birkmann's cellophane bee is found along the Texas border from Cameron County to Val Verde County.

Associated plants: *Colletes birkmanni* is a generalist pollinator often found feeding on composite flowers. The female and male bees shown here were foraging on seaside goldenrod (*Solidago sempervirens*) and woolly croton (*Croton capitatus*).

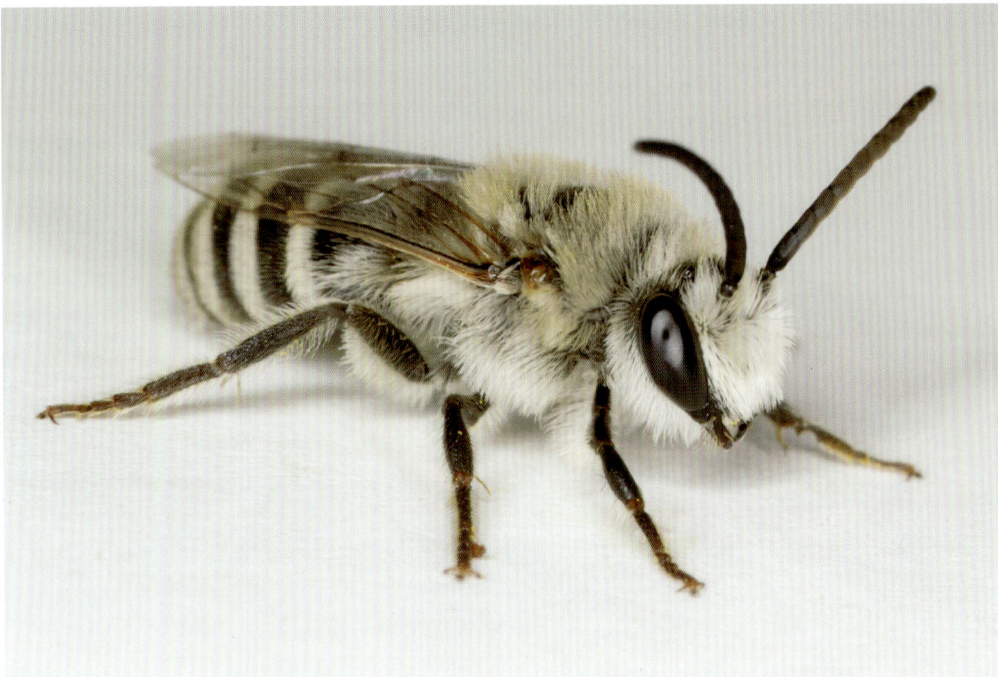

A male Birkmann's cellophane bee covered almost entirely with pale hairs

ID tips: Species identification of *Colletes* often hinges on such minute traits as the density of pits on the upper abdomen or the arrangement of hairs on the sternum. The male Birkmann's cellophane bee, for example, has specialized hairs on its sternum, visible with magnification. A distinctive trait of the female *Colletes birkmanni* is the presence of sparse pitting on T1 and T2. The rust-colored thorax hairs of the female *Colletes birkmanni* aid in field identifications of this bee. The male's vestiture of lush white hairs on its head and body is also striking and distinctive.

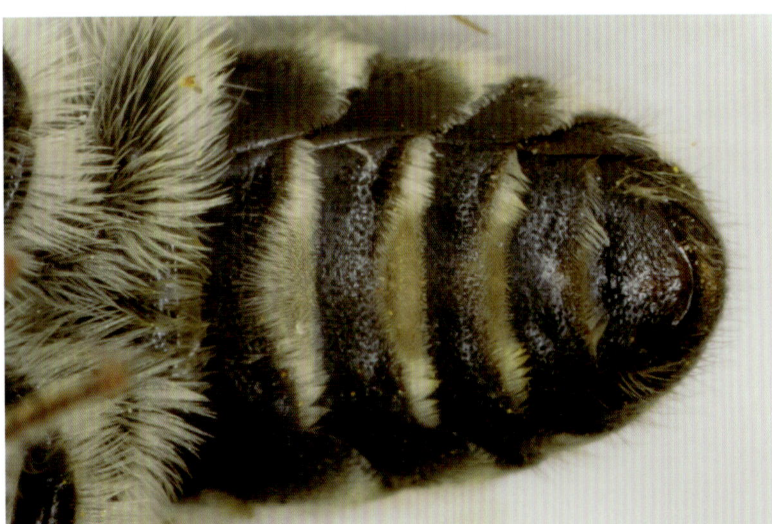

Short, specialized hairs on the sternum of a male *Colletes birkmanni*

Intermixed cellophane bee ◆ *Colletes intermixtus* Swenk, 1905

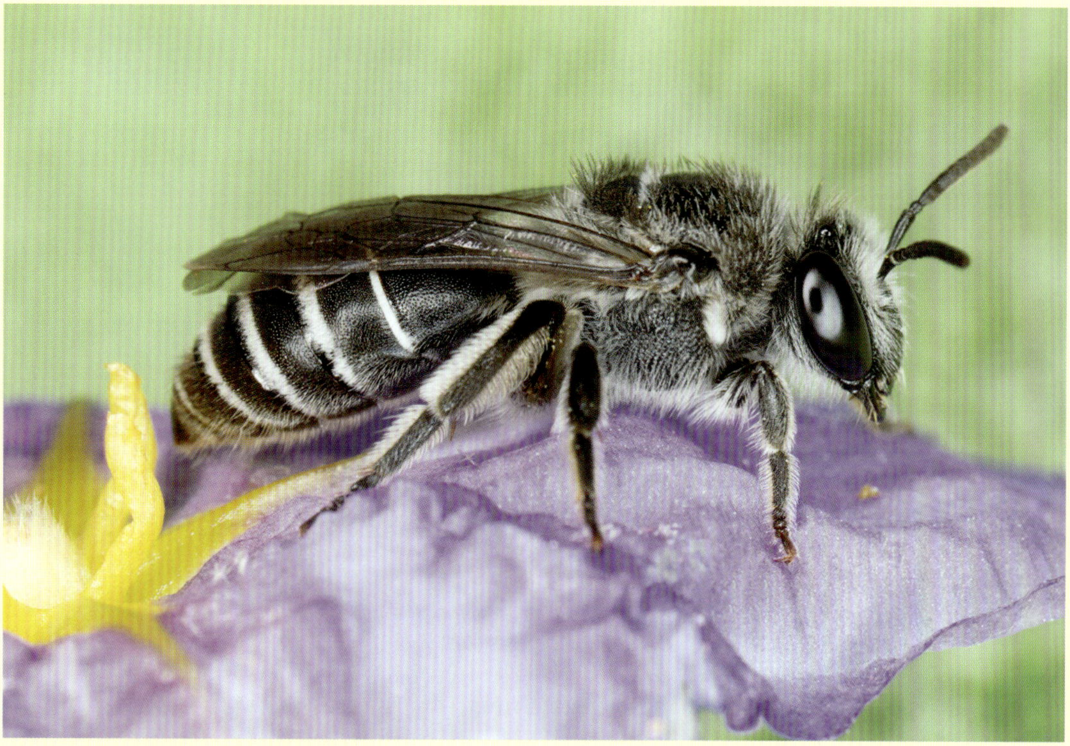

A female intermixed cellophane bee

Description: *Colletes intermixtus* is a black bee with a coarsely pitted, shiny thorax and narrow white bands on its abdomen. The bee's thorax and abdomen have a faint dark blue luster. The hairs covering the face and body are mostly white or pale gray. Dark hairs line the inner edges of the compound eyes. Scattered black hairs intermingle with light hairs on the scutum, and the hairs on the scutellum are entirely dark. The tegulae of *Colletes intermixtus* are dark brown, and the wings transparent brown with dark brown veins. The male's legs are relatively narrow. The scopal hairs on the female's hind legs are white. **Size:** females 10–12 mm (2/5–1/2 in.), males 9–10 mm (1/3–2/5 in.).

Range: This species ranges from the West Coast to East Texas, as far north as Dallas County. The female bee shown here appeared at Quinta Mazatlán (Hidalgo Co.) in early May 2021. *Colletes intermixtus* has been reported in Cameron County.

Associated plants: *Colletes intermixtus* is associated with nightshade-family plants and with flowers of the aster family, among them snakeweed, sunflower, and safflower. The female bee shown here was found on silverleaf nightshade (*Solanum elaeagnifolium*). There are early records of *Colletes intermixtus* feeding on a range of plants in the Valley, among them huisache (*Vachellia farnesiana*), whitebrush (*Aloysia gratissima*), and lemon beebalm (*Monarda citriodora*).

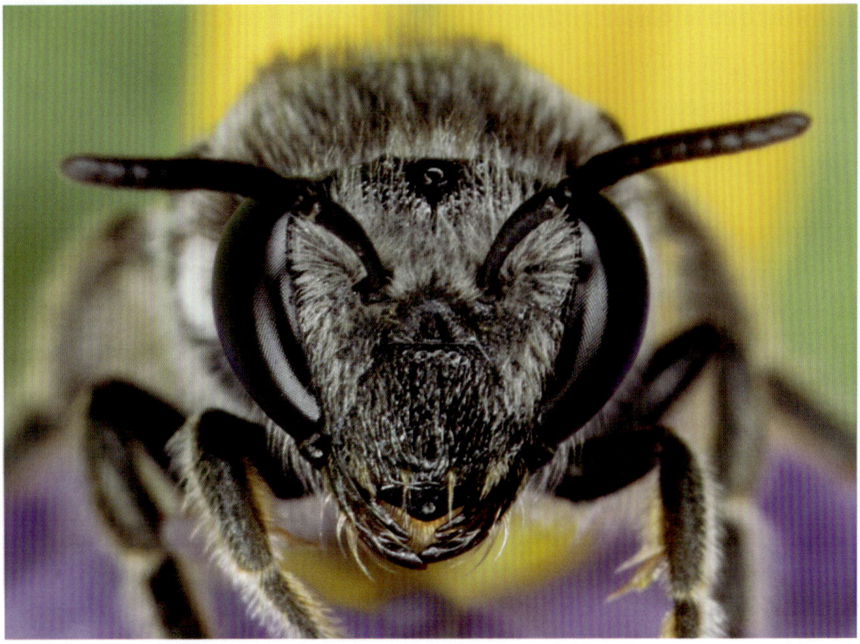

Face of a female *Colletes intermixtus*

ID tips: The nearly bald, coarsely pitted thorax of *Colletes intermixtus* aids in its identification. The face of both the male and female is also densely pitted, long, and black, with a short malar space. The clypeus has a longitudinal groove that runs from its base to its tip, a trait often hidden under facial hairs.

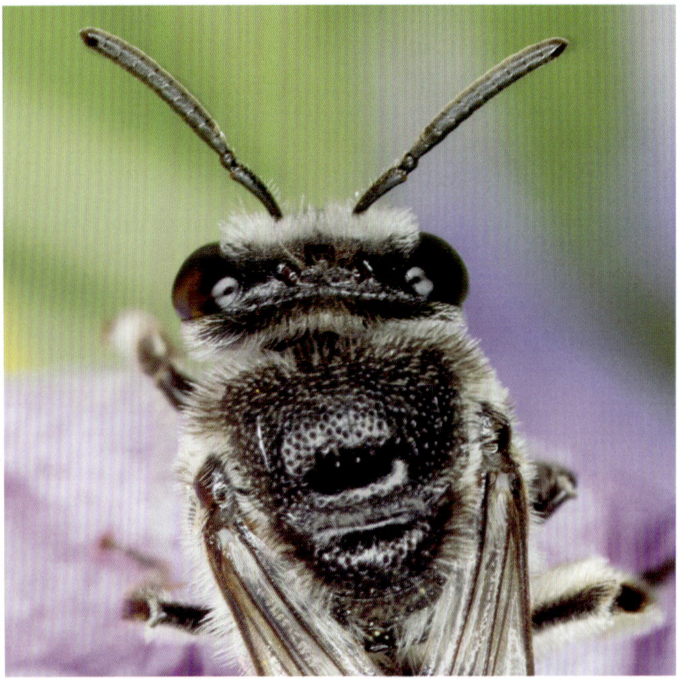

The pitted and shiny thorax of a female *Colletes intermixtus*

Texas cellophane bee ◆ *Colletes texanus* Cresson, 1872

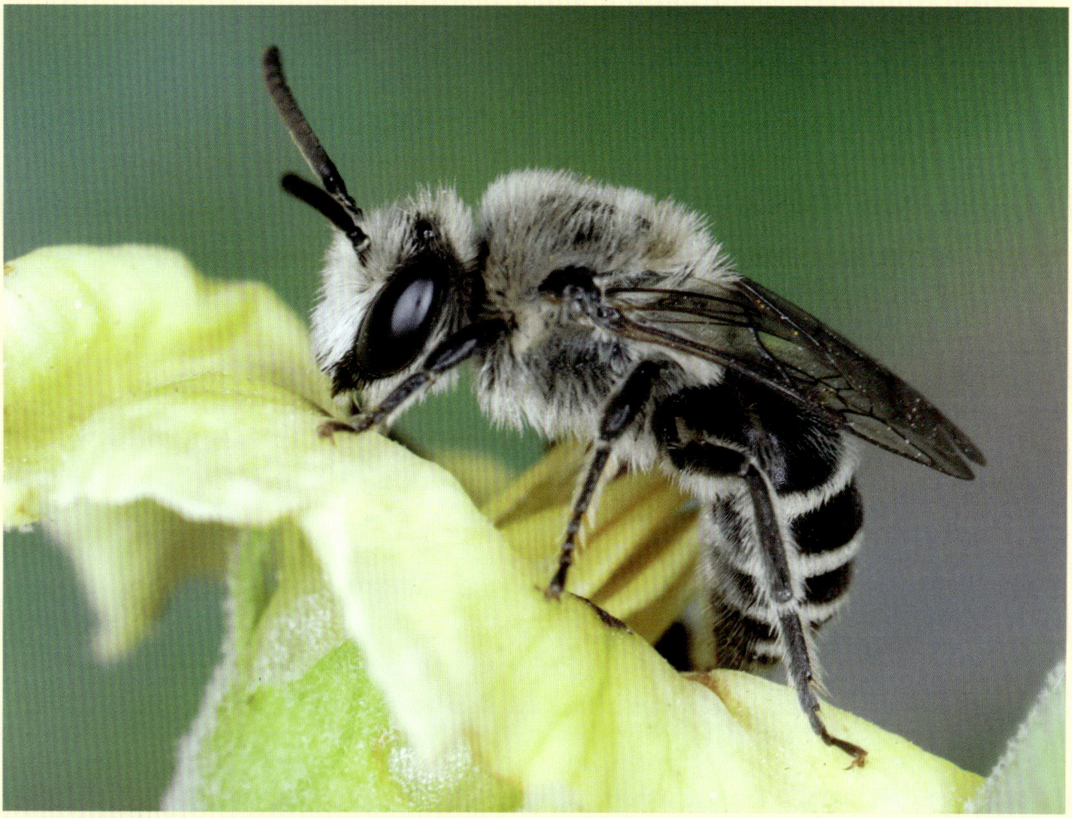

A male Texas cellophane bee on groundcherry

Description: *Colletes texanus* is a small, furry-looking, silvery-gray bee with dusky wings, black tegulae, and a black abdomen striped by bands of pale hairs. White hairs cover the bee's face, and intermixed white, gray, and dark hairs intermingle on the thorax. On females, long, mostly white hairs clothe the sides of the thorax. The hind legs of both males and females are narrow and covered in short white hairs. Females' hind femurs are covered with feathery white-and-gray scopal hairs. **Size:** females 9 mm (1/3 in.), males 7–8 mm (1/4–3/10 in.).

Range: *Colletes texanus* is a predominantly Texan species. Cresson named and described it in 1872 after examining a specimen found in Comal County, Texas. The Texas cellophane bee has been documented in both Cameron and Hidalgo Counties. This species appeared in early March 2020 at the National Butterfly Center (Hidalgo Co.).

Associated plants: The male Texas cellophane bee shown here was found feeding on downy groundcherry (*Physalis pubescens*). The Texas cellophane bee's proclivity for groundcherry was first noted more than a century ago: entomologist Myron Swenk reported finding Texas cellophane bees on *Physalis* in Dallas and San Angelo, Texas, between 1905 and 1908.

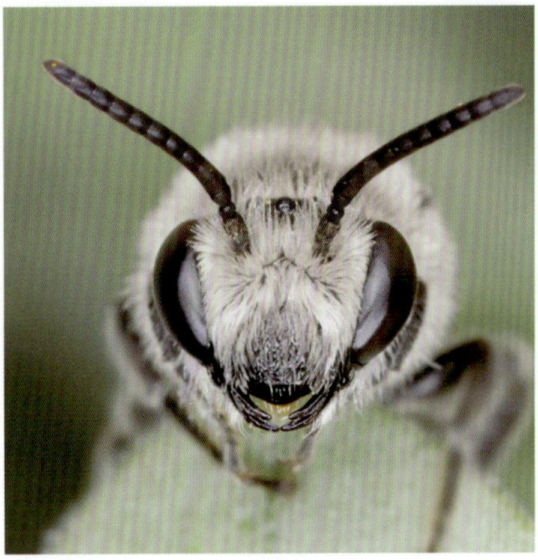

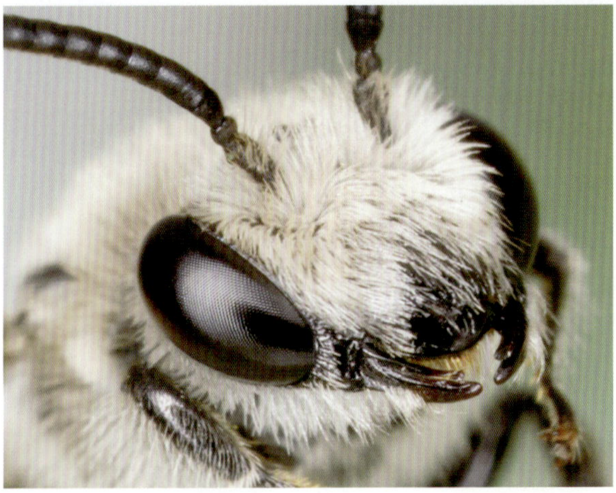

Face of a male Texas cellophane bee

ID tips: The relatively small size of *Colletes texanus*, and its association with groundcherry, aid somewhat in identifying the species. Minute traits help distinguish this bee from other regional *Colletes* that visit Solanaceae: (1) T1 of the male's abdomen is coarsely pitted, and T6 is covered with dark hairs; (2) the hair on the vertex of both males and females is pale or pale mixed with gray; and (3) the bees' faces are thickly covered with white hair. On males, the hair is particularly dense around the antennae and along the inner edges of the compound eyes, while on females, the hair is also white, but somewhat less dense and shorter.

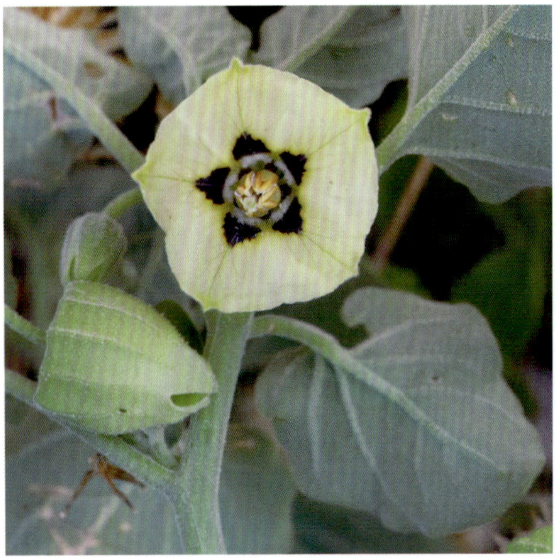

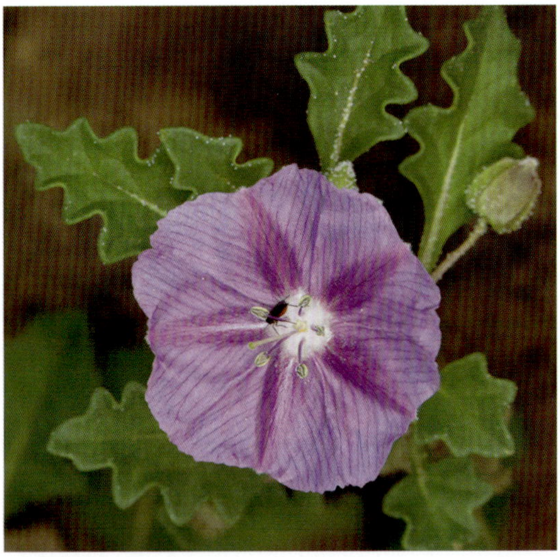

Downy groundcherry (*Physalis pubescens*) and purple groundcherry (*Quincula lobata*), pollen sources for *Colletes texanus* and *C. swenki*: both are members of the nightshade family (Solanaceae).

Swenk's cellophane bee ◆ *Colletes swenki* Stephen, 1954

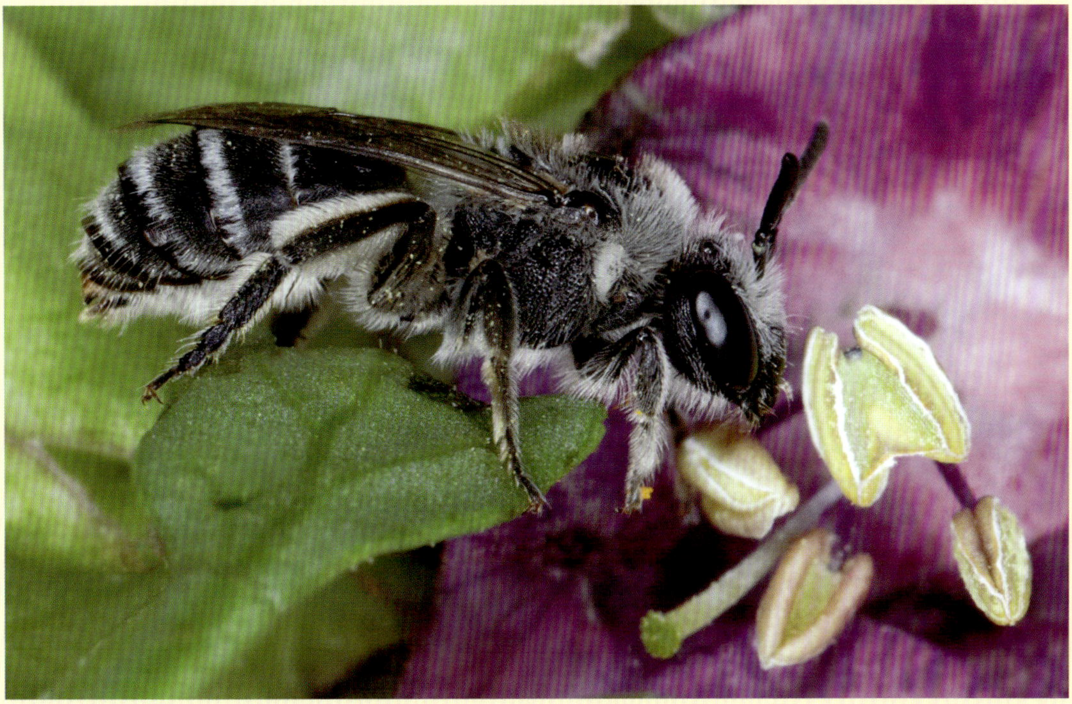

A female Swenk's cellophane bee

Description: *Colletes swenki* is a medium-sized bee with a fuzzy, grayish appearance. Its legs are black or dark brown, the tegulae dark brown, and the wings dusky with dark brown veins. Black hairs intermingle with white hairs on the thorax. The abdomen is banded by pale hairs. On females, the tergal hair bands continue onto the sternum. Females have feathery white scopal hairs on their hind femurs. **Size:** females 9.5–11 mm (3/8–2/5 in.), males 8.5 mm (1/3 in.).

Range: Swenk first described this species after examining specimens found in Texas in the 1950s. *Colletes swenki* is chiefly a Texas bee. It has been recorded in Bexar, Cameron, Hidalgo, Jim Wells, Kerr, Maverick, Shackelford, Travis, and Val Verde Counties. The female bee shown here appeared at the National Butterfly Center (Hidalgo Co.) in April 2019.

Associated plants: Swenk's cellophane bee is a noisy buzz pollinator. The bee featured here was visiting purple groundcherry (*Quincula lobata*). Swenk's cellophane bees are also known to visit prostrate ground cherry (*Chamaesaracha coniodes*), catclaw acacia (*Senegalia greggii*), littleleaf sumac (*Rhus microphylla*), honey mesquite (*Prosopis glandulosa*), Douglas's tickseed (*Coreopsis douglasii*), and horehound (*Marrubium vulgare*).

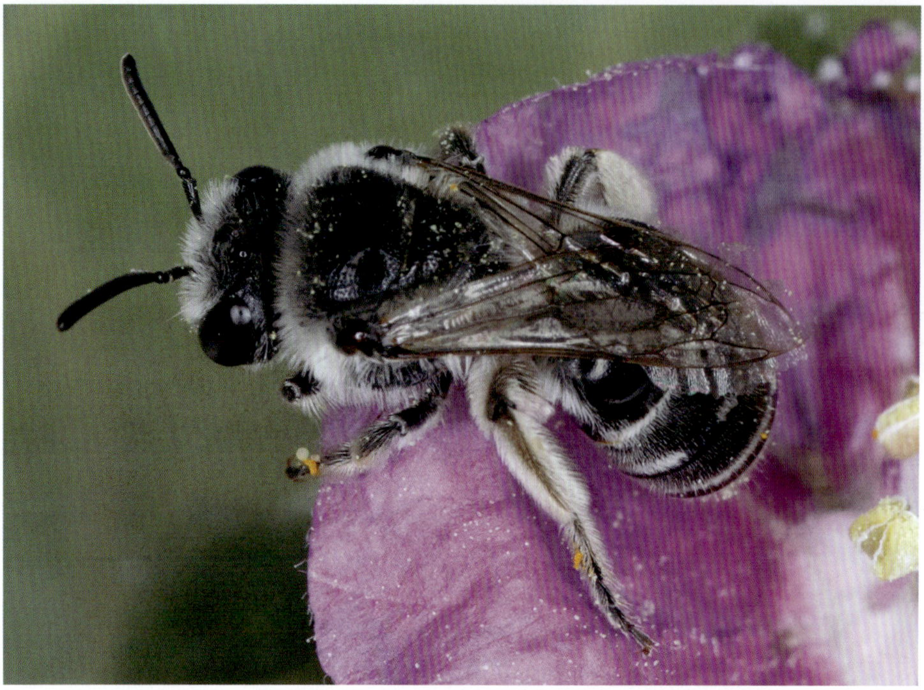

A female *Colletes swenki*, with feathery white scopal hairs on its hind legs

ID tips: *Colletes swenki* and *C. texanus* are similar in general appearance. The following traits may help differentiate them. *Males*: The male *Colletes swenki* is typically larger than the male *C. texanus*, and the first segment of the abdomen (T1) of the male *C. swenki* is banded with long, dense, white hairs, both medially and laterally. *Females*: The clypeus of the female *Colletes swenki* is black, bare of hair, and coarsely and conspicuously pitted. *Colletes swenki* females have unusually long hairs banding the sternum, and T1 is punctuated with coarse, sparse, well-separated pits. On *C. texanus* females, the pits on T1 are fine and dense.

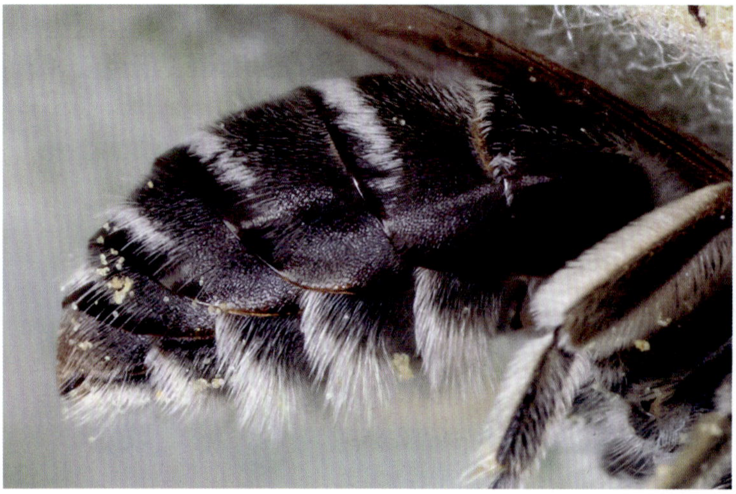

The abdomen of a female Swenk's cellophane bee: note the long white hairs on the sternum.

Pale gold cellophane bee ◆ *Colletes mandibularis* Smith, 1853

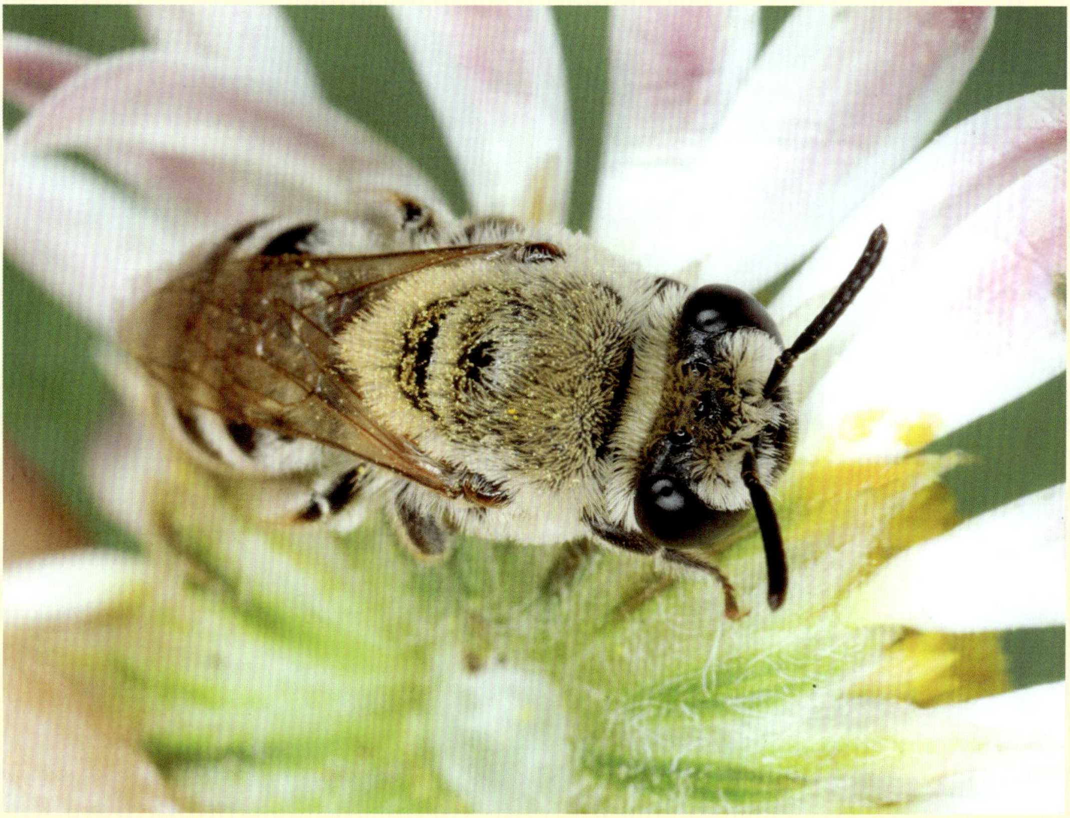

A female pale gold cellophane bee

Description: *Colletes mandibularis* is a small, broadly built bee with a head and thorax covered in ivory and pale gold hairs. Its abdomen is striped with beige or ivory hair bands. Its tegulae are dark reddish brown, and its wings glassy with reddish-brown veins. The legs are dark; males sometimes have yellowish tarsi. Females' scopal hairs are long and pale, and their hind-leg spurs yellowish or reddish. The basitarsi of both males and females are long and slender. **Size:** females 9 mm (1/3 in.), males 7–8 mm (1/4–3/10 in.).

Range: *Colletes mandibularis* has a far-flung range that extends from the Mexican border to Canada, as far west as Nevada, and as far east as Florida. Beamer and Michener recorded this species in Cameron County during the 1950s. *Colletes mandibularis* flies from late spring through November in the Valley and appears in Cameron, Hidalgo, and Starr Counties.

Associated plants: The pale gold cellophane bee is an aster-family specialist. The bees shown here were collecting pollen from Arkansas dozedaisy (*Aphanostephus skirrhobasis*), prairie broomweed (*Amphiachyris dracunculoides*), and cowpen daisy (*Verbesina encelioides*).

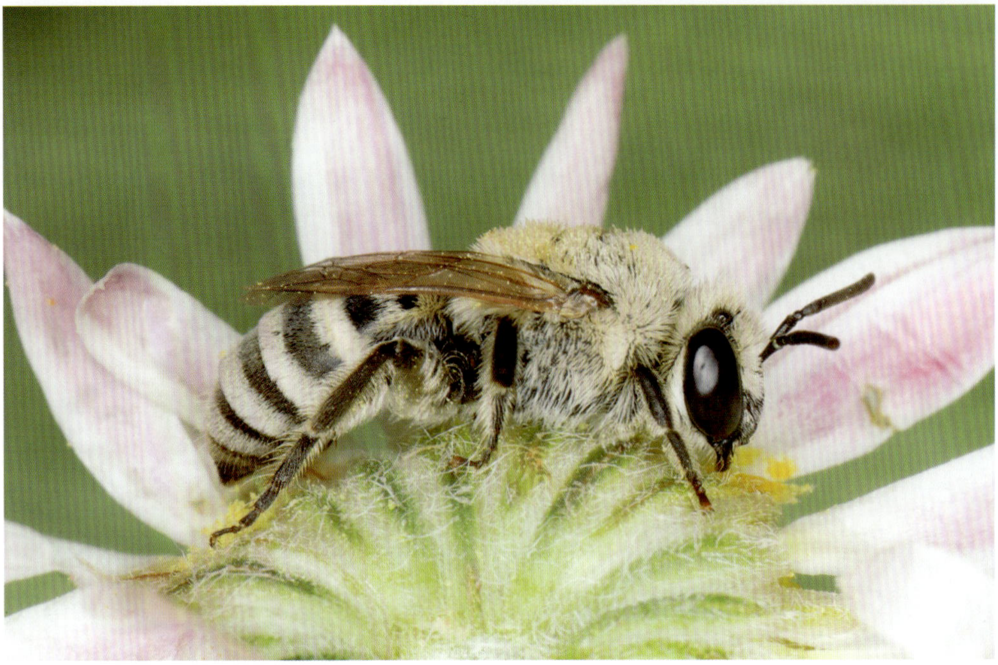

A female *Colletes mandibularis*

ID tips: The relatively small size of this cellophane bee and its general pale gold coloring make it distinctive. Other, more minute traits that aid in identification include the following: (1) on females, the third through fifth abdominal segments (T3–T5) are covered with wide bands of very short, dense, pale hairs; (2) the males' dark reddish-brown tegulae particularly aid in identifying them; and (3) the flagellar antennal segments of the male are long (at least 1.5 times as long as they are wide). *Colletes mandibularis* males might be confused with male *C. birkmanni*, which sometimes have blond hairs on the thorax. The latter, however, run larger and have specialized hairs on the sternum. *Colletes mandibularis* also has a broad rim on the metapleural prominence (at the rear of the thorax), a trait that *C. birkmanni* lacks.

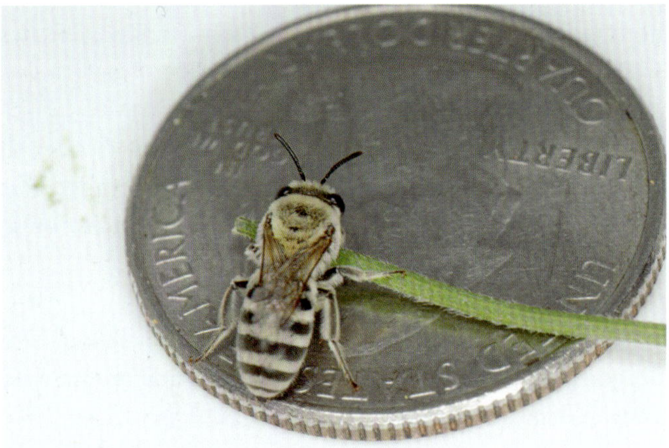

Colletes mandibularis, shown here on a quarter for scale, is small for a cellophane bee.

Masked Bees

Subfamily Hylaeinae ◆ Tribe Hylaeini
Genus *Hylaeus*

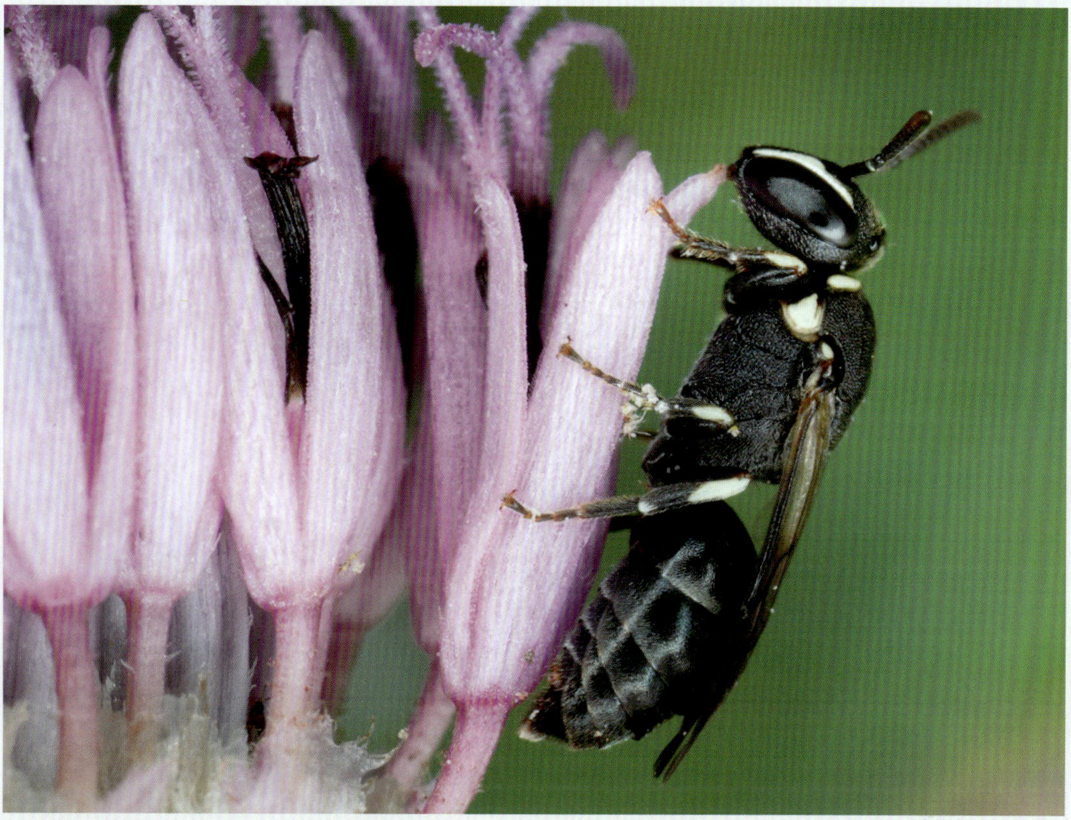

A female *Hylaeus affinis* on Texas palafox

Masked bees, sometimes called yellow-faced bees, make up the genus *Hylaeus*. They are found throughout the world, on every continent except Antarctica.

Masked bees are small and slender and, without the aid of magnification, easily mistaken for black ants. Typically, *Hylaeus* have dark bodies with pointed abdomens, pale "masks" on their faces, and black thoraxes and legs often adorned with pale markings. The abdomens are usually entirely black, a trait that helps distinguish them from many small black-and-yellow wasps. Some *Hylaeus* species of the Valley have red abdomens. The forewings of all *Hylaeus* have two submarginal cells; the first tends to be twice as long as the second. In most *Hylaeus* species, males are smaller than females.

Many *Hylaeus* build tunnel-like nests in dead plant stems. Others construct nests in preexisting cavities like nail holes; in burrows made by beetles or other bees in wood; or even in the cells of wasp nests. Because of their preference for plant-based nesting materials, bees of the genus *Hylaeus* (Latin for "of the woods") tend to abound near woodland habitats.

Unusual pollination practices of masked bees: Unlike most female bees, female *Hylaeus* do not have scopal hairs on their hind legs or other parts of their bodies. Instead, female masked bees carry pollen internally in stomach-like organs. Upon arrival at their nests, the bees regurgitate the pollen and store it for future offspring. Like female *Ptiloglossa* and *Colletes*, female masked bees secrete a plastic-like substance, used to coat the brood cells in which they deposit their eggs.

General identification information: There are more than a dozen species of *Hylaeus* in Texas. Most masked bees are 4–6 mm (1/6–1/4 in.). Differentiation among species is difficult with the naked eye because the bees are so small and nuances in their appearance minute. Species identification usually requires the aid of a macro lens or other magnifier.

Differences in facial masks are often the best means of separating types of *Hylaeus*. Masks may be ivory, pale yellow, or partly orange, and they vary in shape from one species to another. Facial markings also differ in male and female masked bees within a species: markings of males tend to cover a large area of the face, while on females, the markings are smaller and often restricted to two bands or shapes on either side of the face.

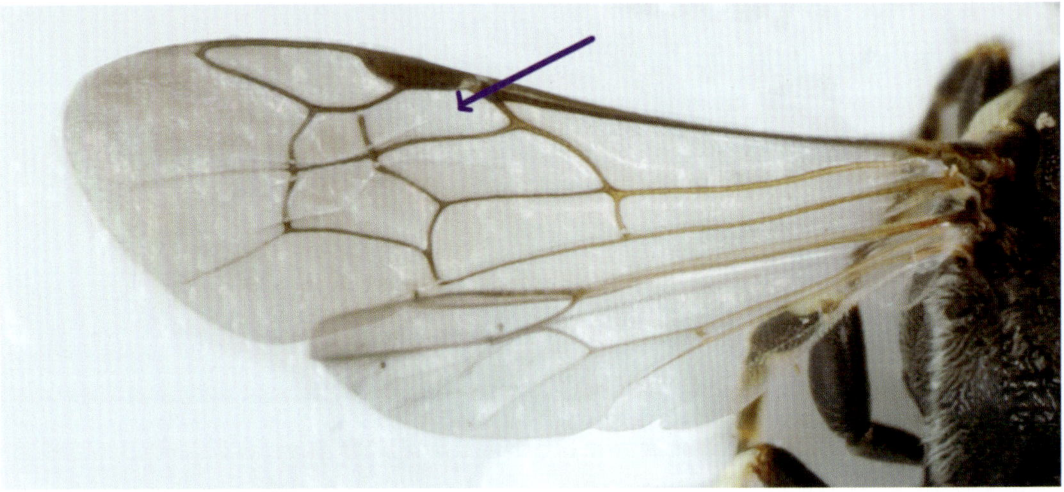

Forewing and hind wing of a female *Hylaeus affinis*: note that the first submarginal cell (black arrow) is roughly twice as long as the second submarginal cell to its left.

Panamanian masked bee ◆ *Hylaeus (Hylaeana) panamensis* Michener, 1954

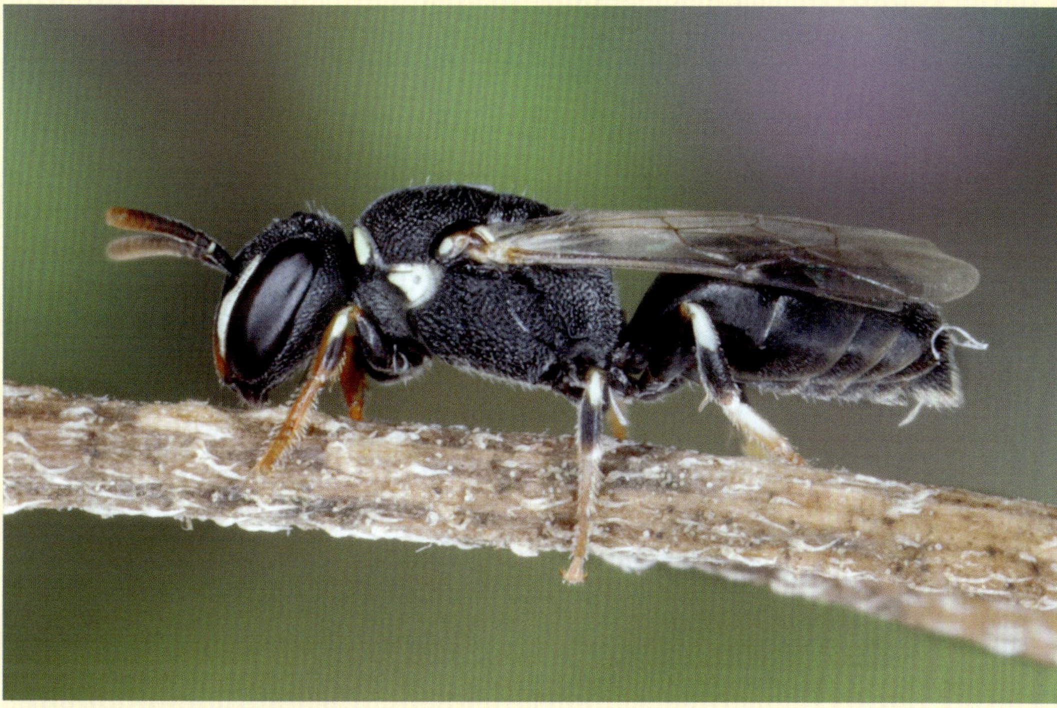

A female Panamanian masked bee

Description: *Hylaeus panamensis* belongs to the largely subtropical masked bee subgenus *Hylaeana*, represented by a single species in Texas—the Panamanian masked bee. Masked bees of this subgenus tend to run quite small (3.5–4.5 mm). *Hylaeus panamensis* is a dark, relatively hairless bee with a black to dark brown abdomen; ivory or pale yellow markings on its face, thorax, and legs; and dark brown mandibles and antennae. The female Panamanian masked bee has a striking partly orange clypeus. **Size:** females 4.5 mm (1/6–1/5 in.), males 4 mm (1/6 in.).

Range: In the United States, Panamanian masked bees have been sighted primarily along the Texas border in Hidalgo and Cameron Counties. At least one specimen has been recorded from Edwards County. The Panamanian masked bee has also been documented in southern Arizona. This species ranges southward from the border through Mexico and Central America as far as Panama.

Associated plants: The Panamanian masked bees shown here were found in November 2019 feeding on Texas palafox (*Palafoxia texana*) growing on sunlit trails near wooded areas of the National Butterfly Center and McAllen Nature Center (Hidalgo Co.). This small-flowered plant is native to Cameron, Hidalgo, and Willacy Counties. It blooms from spring through fall.

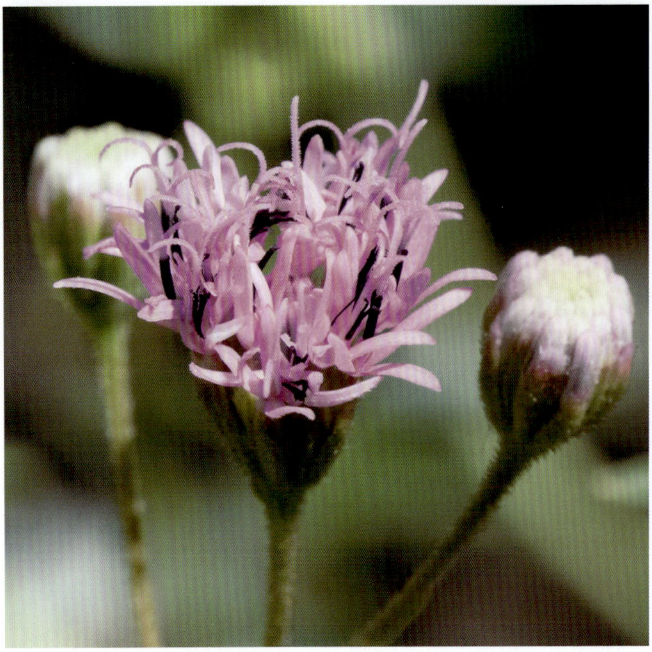

Texas palafox (*Palafoxia texana*), a food plant for *Hylaeus panamensis* in the Lower Rio Grande Valley

ID tips: *Hylaeus panamensis* is the only documented masked bee of the Lower Rio Grande Valley that has a partly orange clypeus. The female's "mask" consists of two ivory, roughly triangular patches that border the orange clypeal area and inside edges of the compound eyes. On males, the ivory mask is more extensive and covers most of the face below the antennae. Both males and females have various other pale markings. Both have a broken, curved, pale yellow stripe on the pronotal collar; pale yellow markings on the pronotal lobe and tegulae; and pale yellow spots on the femurs and tibiae. Females additionally have brownish-orange tarsi and orangish markings on all other leg segments.

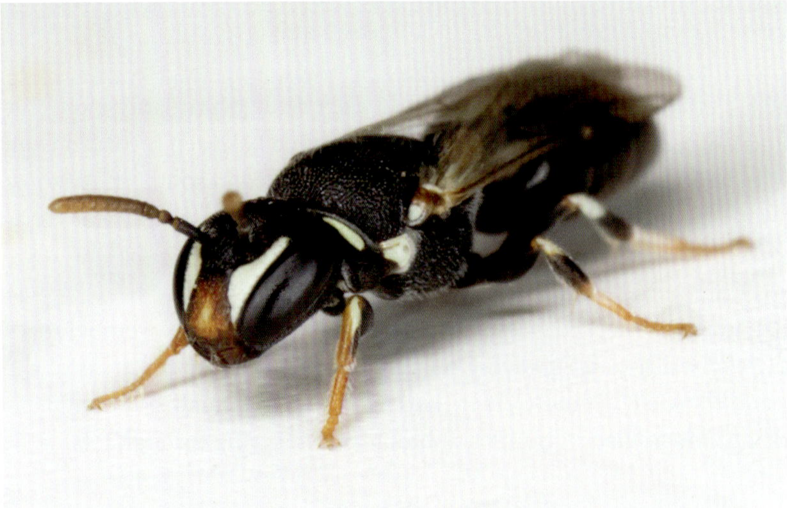

A female Panamanian masked bee

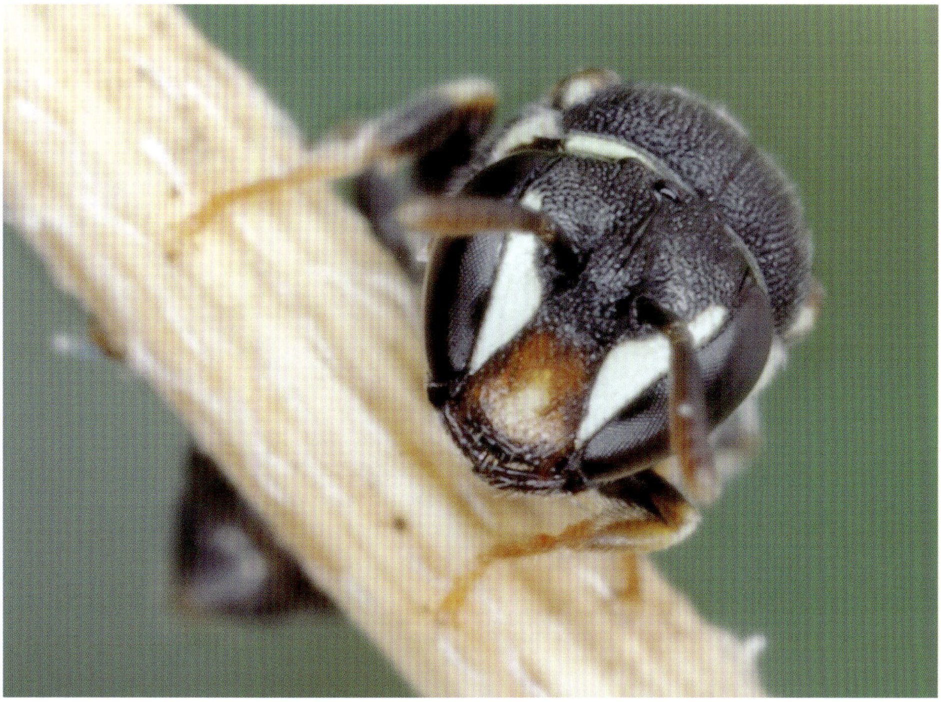

Face of a female *Hylaeus panamensis*: the clypeus is partly orange and flanked by pale triangular markings that extend well above the antennae.

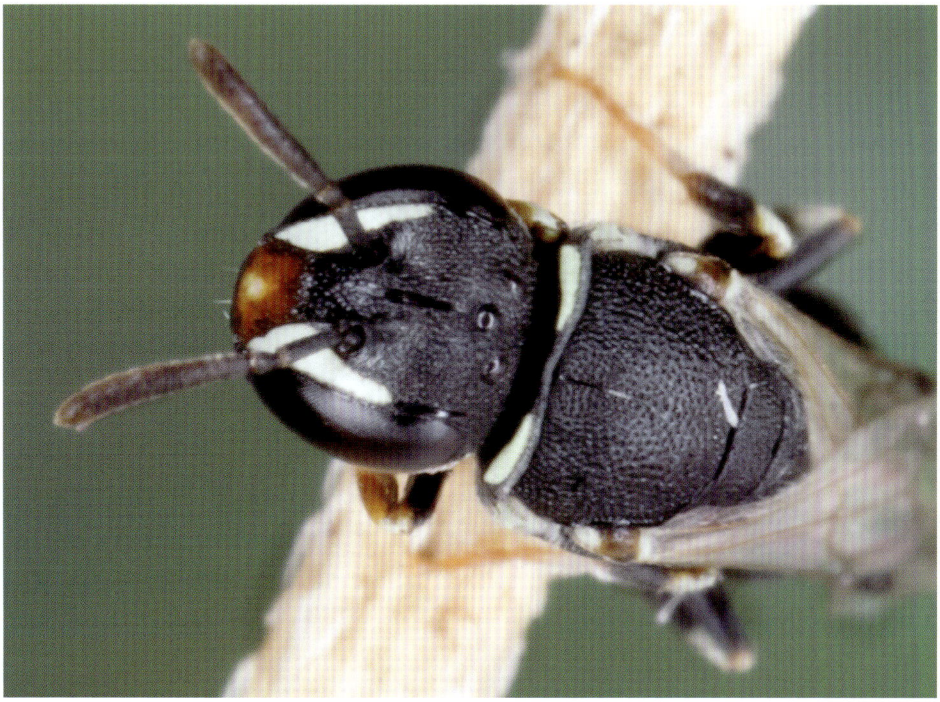

Dorsal view of a female *Hylaeus panamensis*: note the pitted vertex and thorax, and the broken yellow line on the pronotal collar.

Two notable traits of female Panamanian masked bees, visible only with strong magnification, aid in diagnosing the species: their thoraxes are coarsely and conspicuously pitted, and the first two segments of their abdomens (T1–T2) are smooth and unpitted.

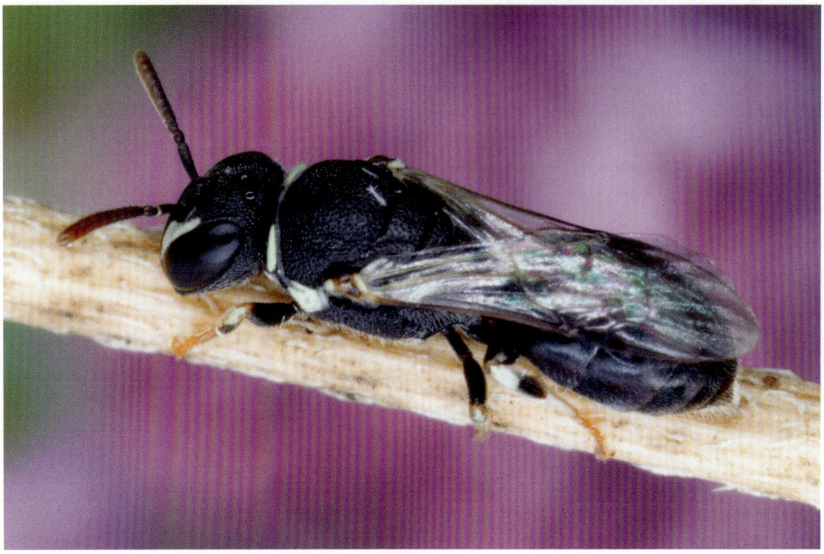

A female Panamanian masked bee

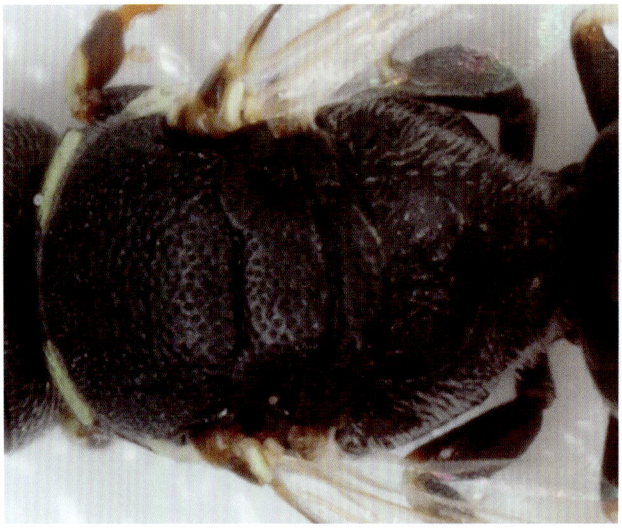

The pitted thorax of a female Panamanian masked bee

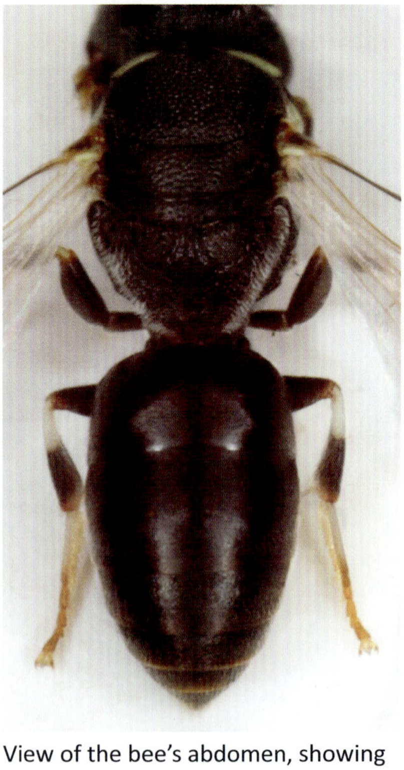

View of the bee's abdomen, showing the smooth, unpitted surfaces of T1 and T2

Eastern masked bee ◆ *Hylaeus (Prosopis) affinis* (Smith, 1853)

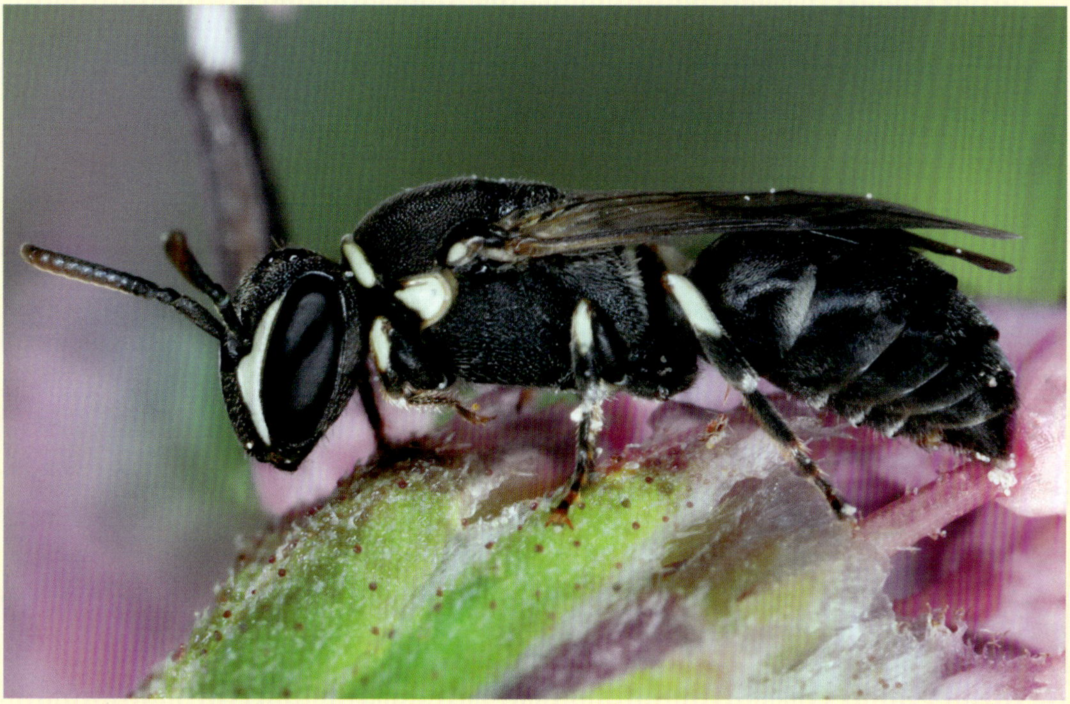

A female eastern masked bee

Description: The eastern masked bee belongs to the subgenus *Prosopis*, which is the dominant subgenus of Texas *Hylaeus*. This is a small black bee with ivory or pale yellow markings on its face, pronotal collar, pronotal lobes, tegulae, and legs. Males have pale masks and partly pale antennal scapes. Females have two roughly triangular pale patches on either side of the clypeus. The tegulae of males and females are reddish black with pale spots, and the wings are transparent reddish brown with dark brown veins. On females, the upper third or more of all of the tibiae, and the leg spurs, are pale. On males, pale areas appear on each lower front femur, parts of all tibiae, the upper tarsi, and leg spurs. **Size:** females 5–6.5 mm (1/5–1/4 in.), males 5–6 mm (1/5–1/4 in.).

Range: Masked bees of the subgenus *Prosopis* occur on all northern continents of the world. The eastern masked bee is widespread in North America and is found in the Pacific Northwest, throughout the eastern and central United States, and from British Columbia to eastern Canada. *Hylaeus affinis* appears from March through November in the Valley.

Associated plants: The bees shown here were found feeding side by side with Panamanian masked bees on native Texas palafox (*Palafoxia texana*) in November 2019 at the McAllen Nature Center (Hidalgo Co.). Eastern masked bees are known to feed on a remarkable range of plants as diverse as roses, asters, mints, sumacs, carrots, and allium.

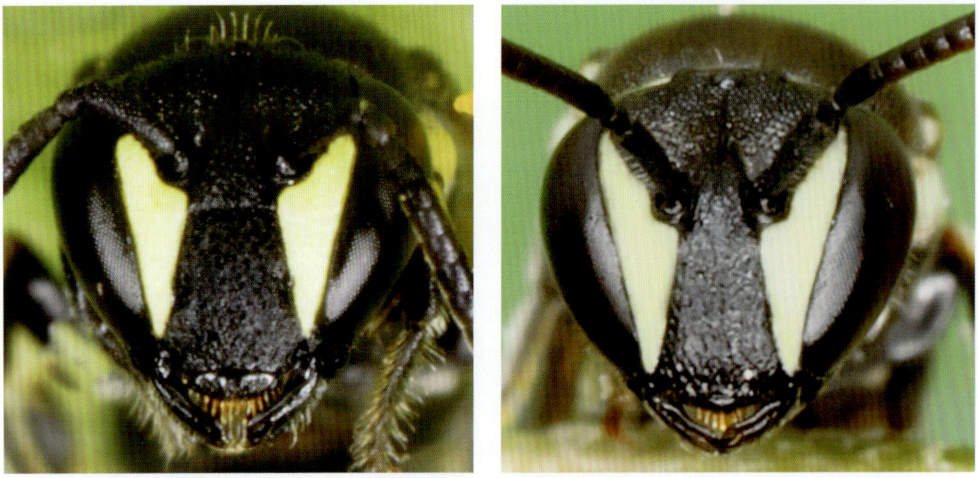

Faces of female *Hylaeus modestus* (*left*) and *H. affinis* (*right*)

ID tips—female bees: Female eastern masked bees (*Hylaeus affinis*) are difficult to distinguish from modest masked bees (*H. modestus*), described in the next section. Females of these two species are similar in size, and they have similar markings on their legs, pronotal lobes, and pronotal collars. They also have nearly identical markings on their faces: two roughly triangular shapes that border the clypeus and the inner orbits of the compound eyes. A minute trait is sometimes used to distinguish the two species: as shown above, the edges of the female's mask nearest the antennae are curved in *Hylaeus modestus* and relatively straight in *H. affinis.* A second way to differentiate female *Hylaeus affinis* and *H. modestus* is to examine the tegulae: female eastern masked bees have pale markings on the tegulae, and on the small tegula-like plates right behind them. Female *Hylaeus modestus* usually lack pale markings on these body parts. Distinctions between males of these two species are described in the next section.

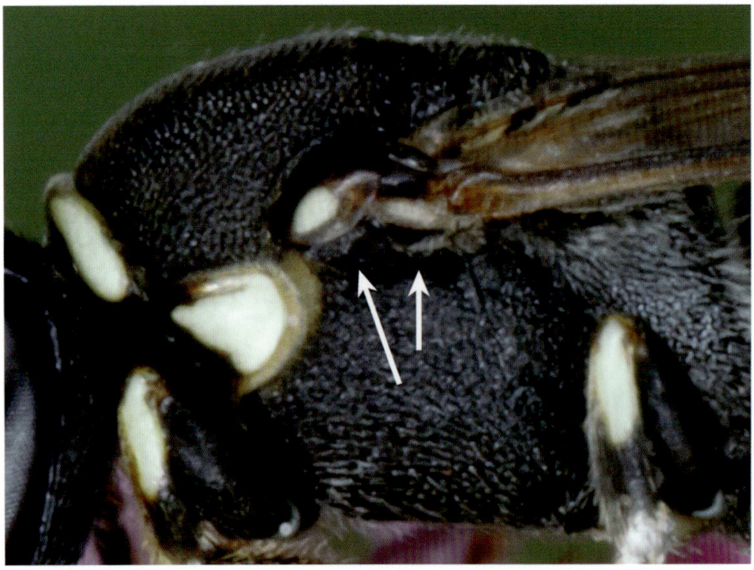

Close-up of a female eastern masked bee's thorax, showing pale markings on the tegula and on the small plate behind the tegula

Modest masked bee ◆ *Hylaeus (Prosopis) modestus* Say, 1837

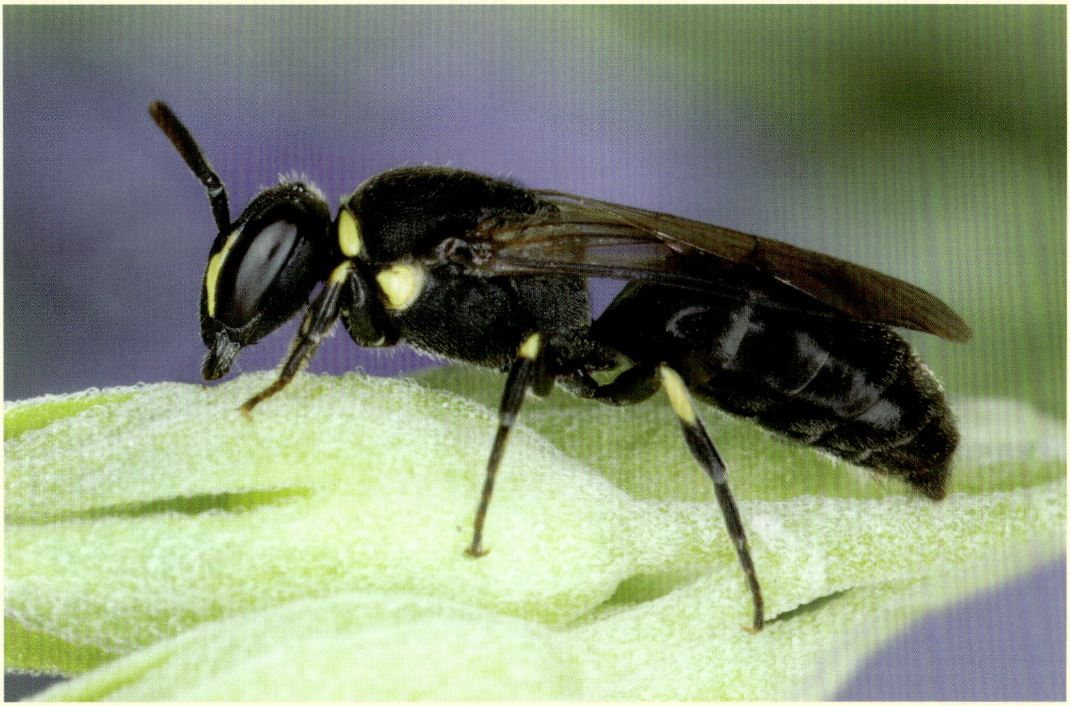

A female modest masked bee

Description: *Hylaeus modestus* belongs to the same subgenus (*Prosopis*) as the eastern masked bee, and, as noted in the previous entry, these two species closely resemble each other. Modest masked bees are black with ivory or light yellow markings on the face, pronotal lobes, legs, and (usually) along the pronotal collar. Males have extensive pale masks covering much of the face; females have two roughly triangular pale patches on either side of the clypeus. The bees' tegulae are black or dark brown and the wings transparent brown with blackish-brown veins. The legs of females have pale markings on the tibiae, and yellow spurs. The legs of males have pale areas on the tibiae, basitarsi, tarsi, and spurs. **Size:** females 5–7 mm (1/5–1/4 in.), males 4.5–7 mm (1/6–1/4 in.).

Range: The modest masked bee occurs throughout the United States from coast to coast, as far south as southern Texas and as far north as Canada and Alaska. This species appeared at the National Butterfly Center and the McAllen Nature Center (Hidalgo Co.) in spring of 2019.

Associated plants: Modest masked bees are extraordinary generalists that have been documented foraging on flowers from more than 25 different plant families. Both females and males show a preference in the Valley for small-flowered plants such as sticky florestina (*Florestina tripteris*) and small coastal germander (*Teucrium cubense*).

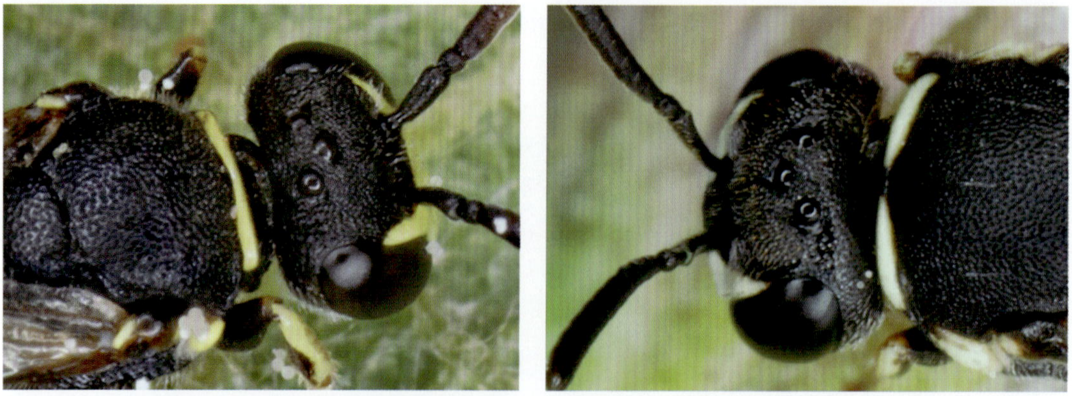

The vertex and thorax of the very similar male *Hylaeus modestus* (*left*) and *H. affinis* (*right*): on both species, a complete or narrowly interrupted ivory or pale yellow line usually rims the pronotal collar.

ID tips—male bees: Male *Hylaeus modestus* and *H. affinis* closely resemble each other: they have similar markings on their thoraxes and legs, and their facial masks are nearly identical. They can be distinguished, however, by their antennae: the antennal scapes of the male *Hylaeus modestus* are all black, while those of *H. affinis* have ivory or pale yellow markings on the front surfaces. Distinguishing traits of female *Hylaeus affinis* and *H. modestus* are described in the previous entry.

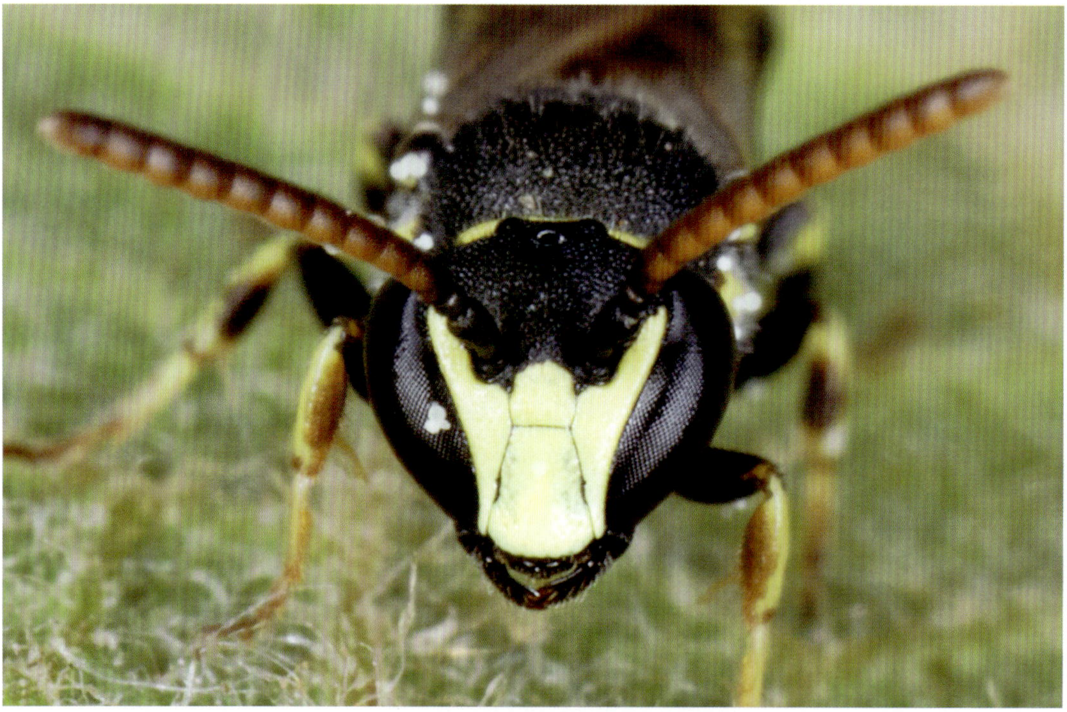

Face of a male modest masked bee: note the all-black antennal scapes.

Family Halictidae

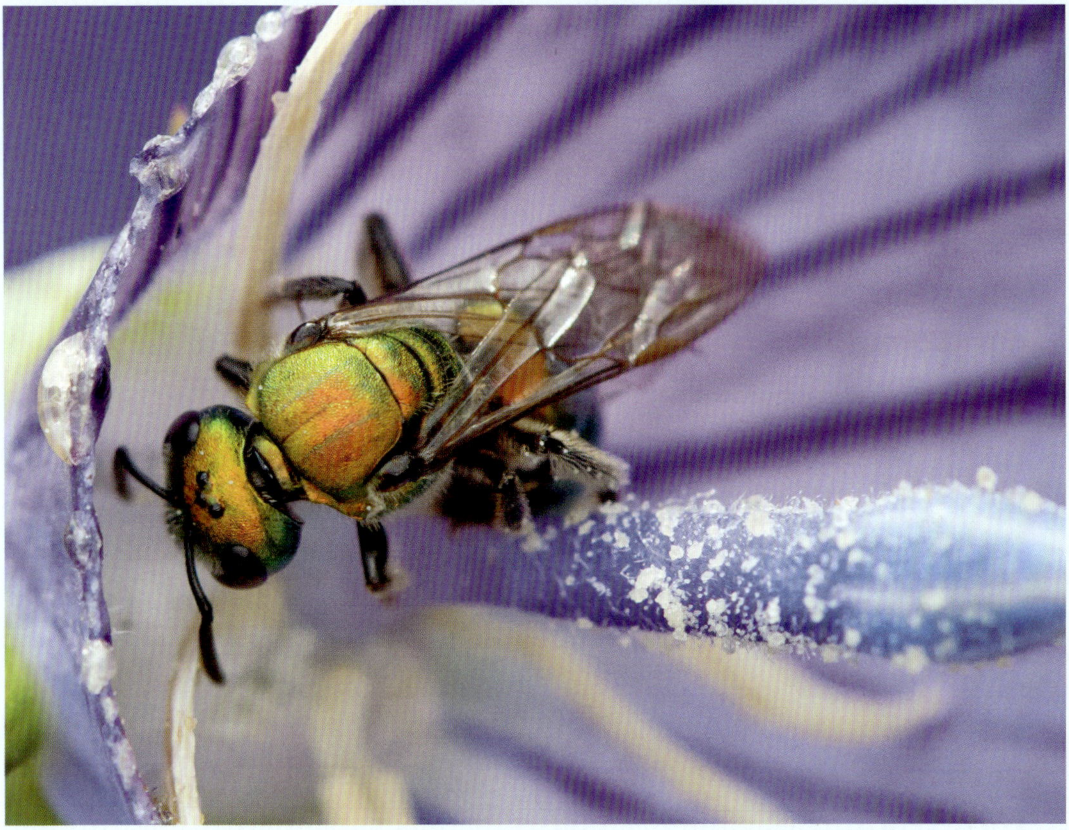

Green Sweat Bees
Agapostemon
Furrow Bees
Sphecodes Cuckoo Bees
Nomia and *Dieunomia*

HALICTIDAE FAMILY GENERA
OF THE TEXAS LOWER RIO GRANDE VALLEY

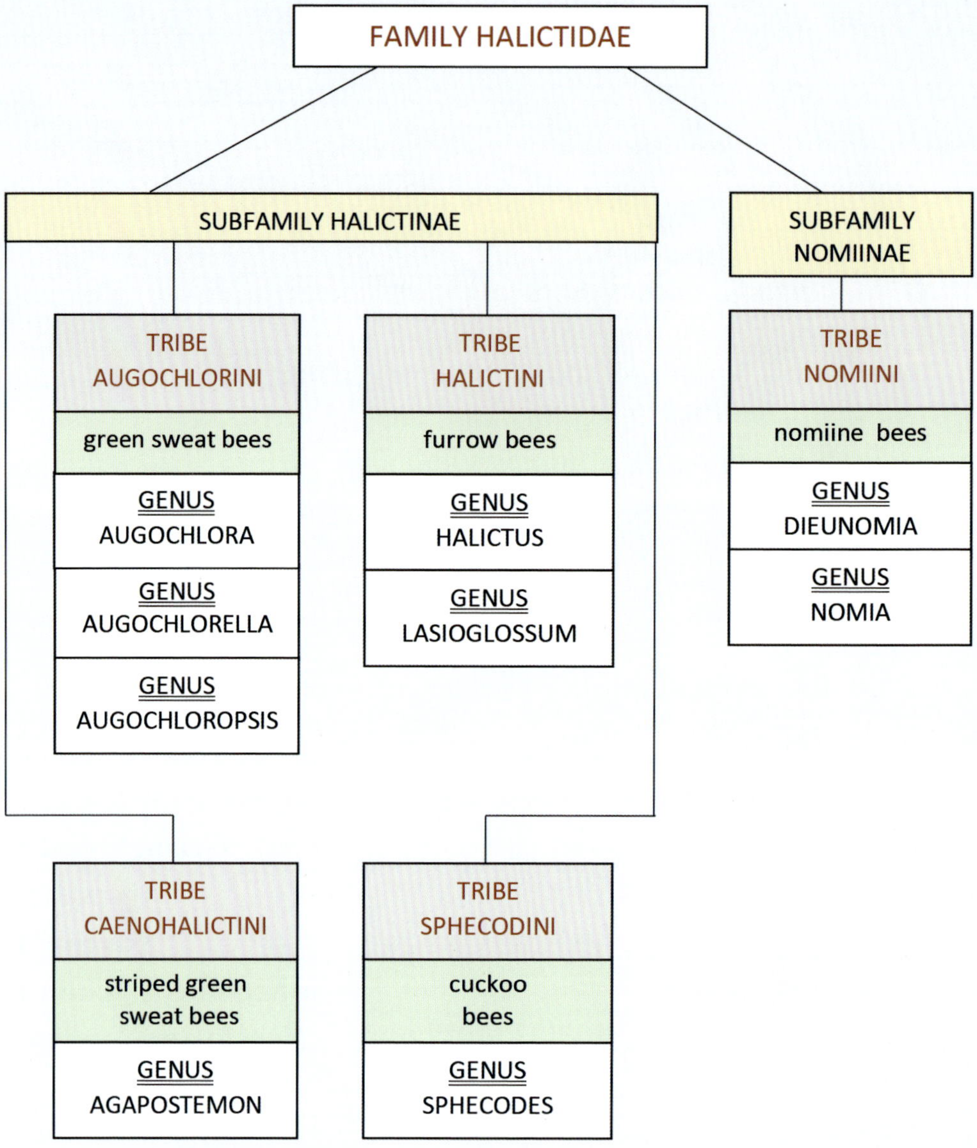

Green Sweat Bees

Subfamily Halictinae ◆ Tribe Augochlorini
Genera *Augochloropsis*, *Augochlora*, and *Augochlorella*

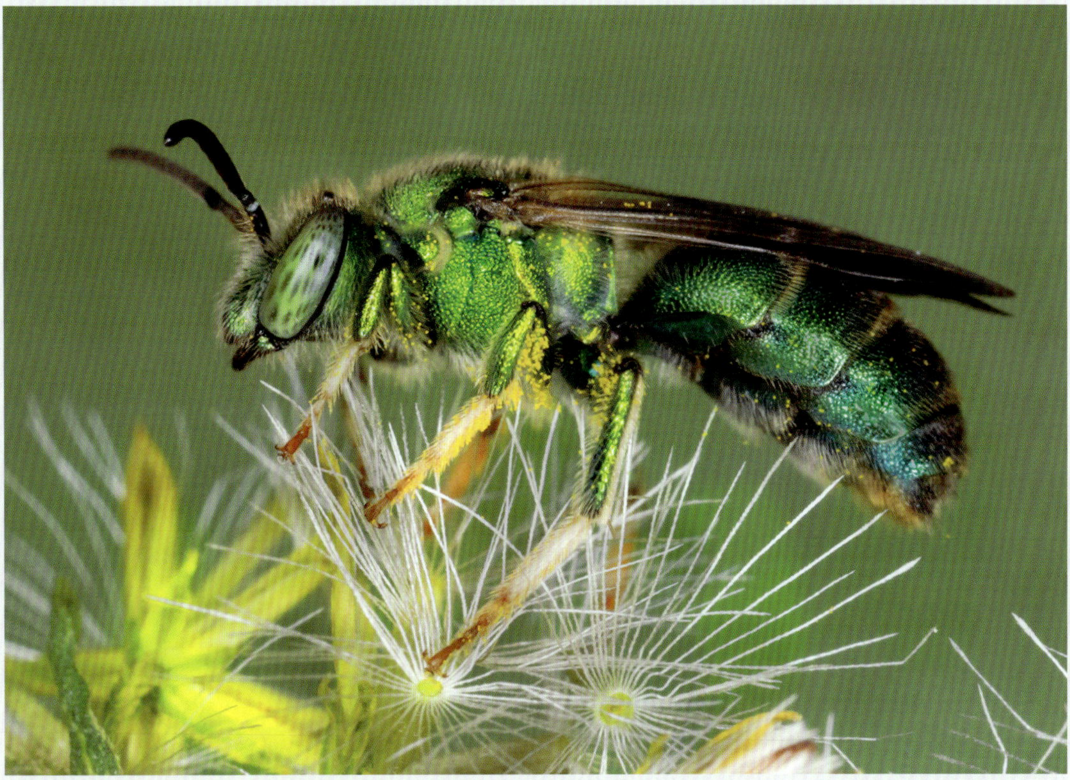

A male *Augochloropsis metallica*

Sweat bees belong to the subfamily Halictinae within the family Halictidae.[1] They are essential and ubiquitous pollinators of commercial crops, garden flowers, and native flora. Many Halictinae are too small, and their differences too nuanced, to identify in the field—typical Lower Rio Grande Valley sweat bees are 4–8 mm long (1/6–3/10 in.). A handful of them, however, such as green sweat bees of the tribe Augochlorini, striped green *Agapostemon*, and some dark-colored or metallic *Halictus* and *Lasioglossum*, can be recognized easily by the casual naturalist. Representatives of these groups are presented here.

1. The common name "sweat bee" is used somewhat vaguely and inconsistently in popular guidebooks and databases: sometimes it is meant to encompass the entire family Halictidae; sometimes to denote bees of the subfamily Halictinae; and sometimes to include only Augochlorini. In this book, the term "sweat bee" is used to refer to nonparasitic bees of the subfamily Halictinae.

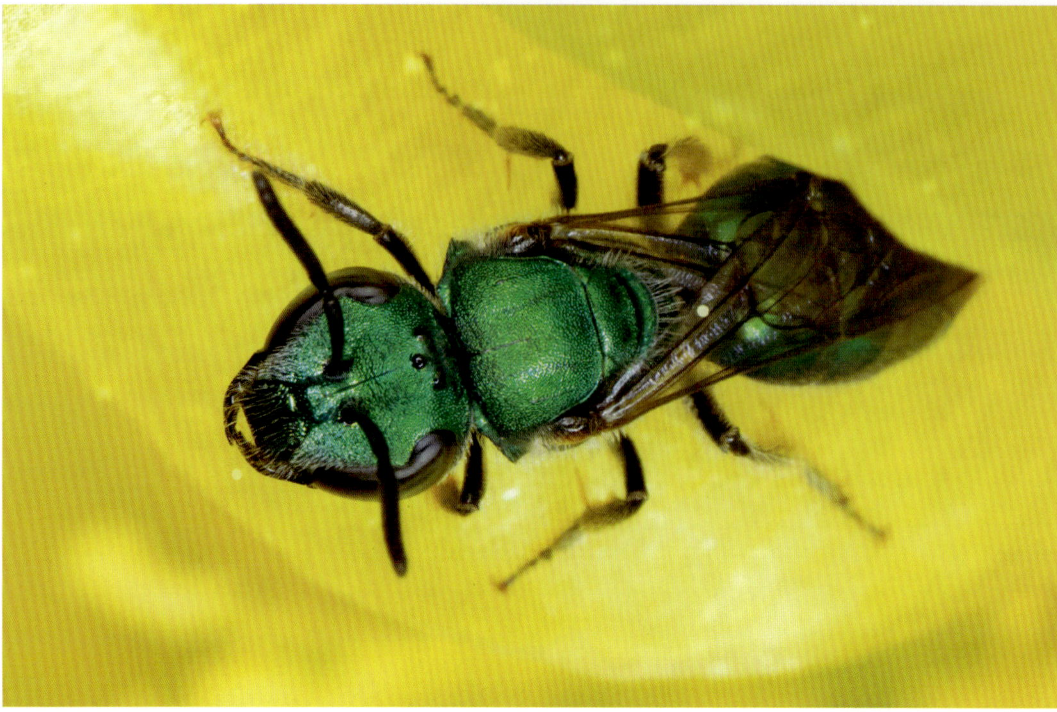

An Aztec sweat bee: the kidney-shaped eyes, notched just above the antennae; the flared ridge at the front of the thorax; and the intense green metallic coloring are all typical traits of Augochlorini.

The tribe Augochlorini: Augochlorini are brilliantly colored metallic bees, usually small to moderate sized, represented chiefly by three genera in the United States: *Augochloropsis*, *Augochlora*, and *Augochlorella*. Most North American Augochlorini are iridescent green, blue green, or golden green. All have compound eyes that are notched a third of the way down on the inner edges, giving them a kidney-shaped appearance. On all, the front of the thorax has a pronounced ridge. Augochlorini females carry pollen on scopal hairs on their hind legs.

Augochlorini are principally a tropical and subtropical bee tribe, most populous and diverse in South America. Nonetheless, a number of green sweat bee species range into North America, some as far north as Canada. Most North American Augochlorini are generalist pollinators. Most nest in soil, although some *Augochlora* nest in rotted logs. Augochlorini may be solitary or exhibit cooperative or even social behavior when nesting. Their nests are typically vertical tunnels that branch into lateral burrows housing egg chambers or leading to cell clusters.

The green bees of the tribe Augochlorini are sometimes confused with green *Agapostemon*, a genus belonging to the tribe Caenohalictini. Differences among these sweat bee genera are elucidated in the following entries.

ATTRIBUTES OF METALLIC GREEN BEE GENERA

AUGOCHLOROPSIS Tribe Augochlorini *Behavior*: Solitary or social ground nester.	Robustly built, usually (but not always) larger than *Augochlora* and *Augochlorella* and smaller than female *Agapostemon. Augochloropsis* are commonly metallic green or blue green but can be blue or purplish. *Distinguishing traits*: *Augochloropsis* have (1) metallic-green and kidney-shaped tegulae; (2) metallic-green upper legs; and (3) a pronounced pronotal ridge. (4) Females have a comb-shaped spur on each hind-leg tibia. (5) On males the rear edge of S4 has a median point. (6) The tip of the forewing marginal cell is blunt or roundly curved and bent away from the wing's edge. Most easily confused with *Agapostemon* females.
AUGOCHLORA Tribe Augochlorini *Behavior*: Nests in the ground or in rotted logs. Solitary, semisocial, or social.	Usually (but not always) larger than *Augochlorella*, and smaller than *Augochloropsis* and *Agapostemon*. Commonly metallic green, but may be blue green, blue, purplish, coppery pink, or black with metallic highlights. *Distinguishing traits*: (1) The tegulae are brown and oval. (2) The paraocular area (along the inner bottom edge of the compound eye) intrudes into the clypeus. (3) On males, the hind margin of S4 is smooth and rounded. (4) The tip of the forewing marginal cell is blunt or somewhat rounded. Most easily confused with *Augochlorella.*
AUGOCHLORELLA Tribe Augochlorini *Behavior*: Solitary to social ground nester.	Generally the smallest of the bee genera shown here. Usually metallic yellow green or blue green, but may be coppery, orangish, or coppery pink. *Distinguishing traits*: (1) The tegulae are oval and partly reddish or brown. (2) The vertex is narrow. (3) The bottom edge of the clypeus is beveled. (4) Males often have yellow facial markings. On males and females, the labrum is usually black or yellow. (5) The tip of the forewing marginal cell is pointed, distinguishing this genus from *Augochlora*, *Augochloropsis*, and *Agapostemon.* Most easily confused with *Augochlora.*
AGAPOSTEMON Tribe Caenohalictini *Behavior*: Solitary or communal ground nester.	Relatively large (although males within a species can vary greatly in size). Females are often entirely green with green faces, while males have striped abdomens and partly yellow faces. *Exceptions* include *Agapostemon melliventris* and *A. tyleri*, where both sexes are striped and have partly yellow faces. *Distinguishing traits*: (1) Striped abdomens distinguish males from all North American Augochlorini sweat bees. (2) All *Agapostemon* have a pronounced ridge on the propodeum. (3) *Agapostemon* females are most easily confused with *Augochloropsis*, but the tegulae of *Agapostemon* are brown, sometimes with yellow spots, rather than metallic green; and females' upper legs are brown rather than metallic.

Comparison of Forewings of *Augochlorella*, *Augochloropsis*, and *Augochlora*

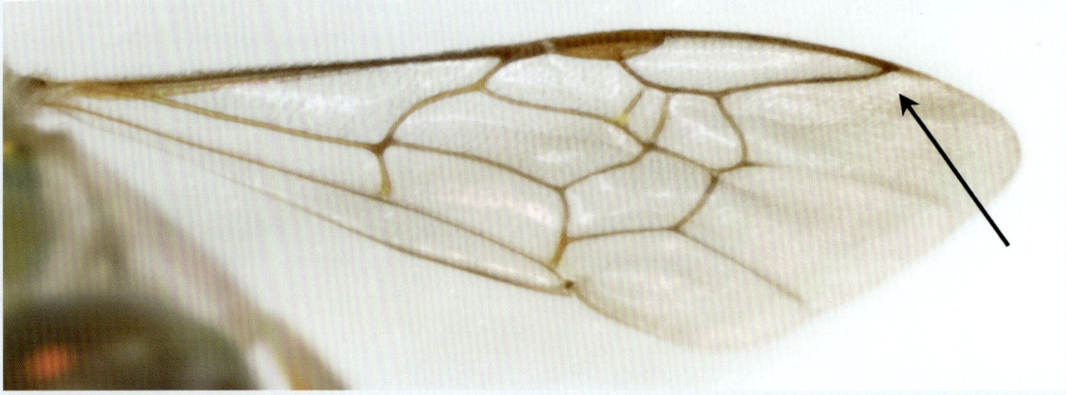

An *Augochlorella bracteata* forewing: the marginal cell's tip is pointed.

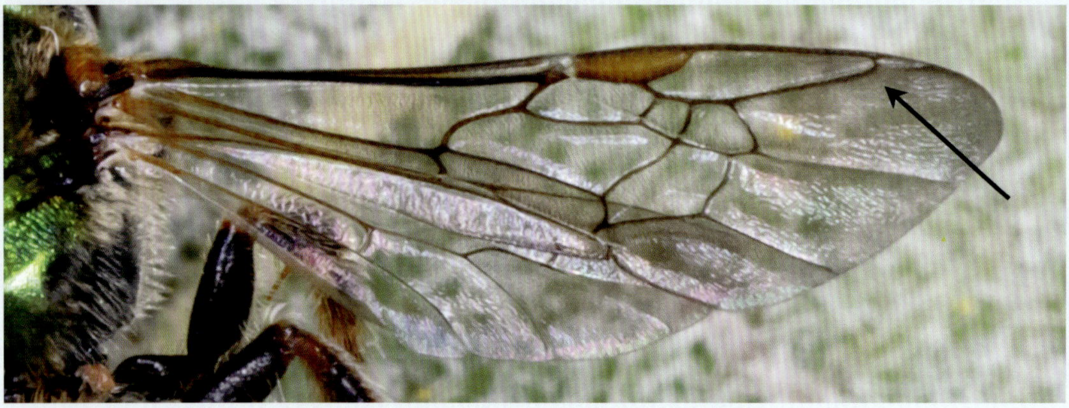

An *Augochlora aurifera* forewing: the marginal cell's tip is blunt or somewhat rounded.

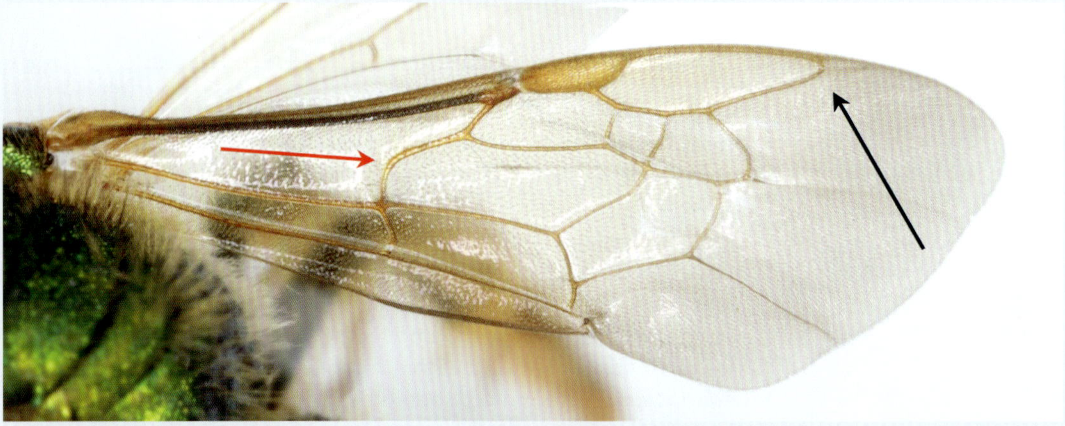

An *Augochloropsis metallica* forewing: the marginal cell's tip is blunt or somewhat rounded and bent away from the edge of the wing (black arrow). The curved basal vein (red arrow) is a characteristic of all bees of the subfamily Halictinae.

Metallic-green sweat bee ◆ *Augochloropsis (Paraugochloropsis) metallica* (Fabricius, 1793)

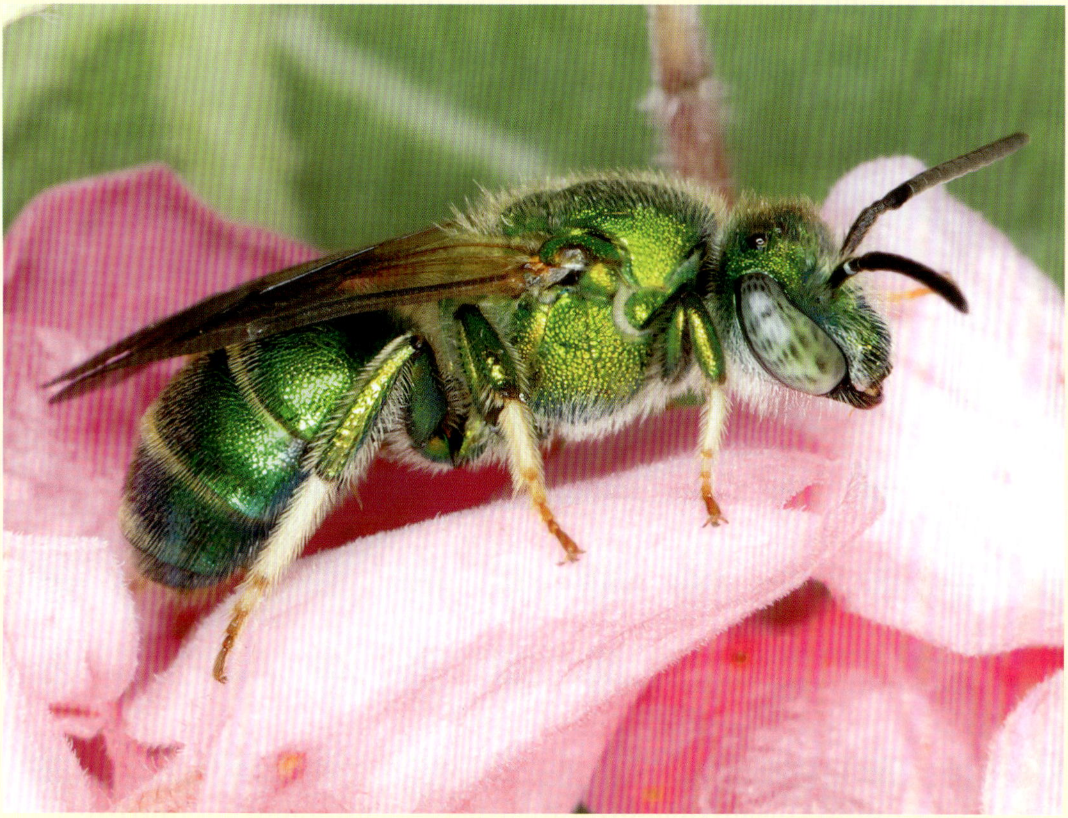

A male metallic-green sweat bee: the basitarsi are white and covered with fine white hairs.

Description: *Augochloropsis metallica* is a robust, beautifully iridescent sweat bee, usually green or blue green. Both males and females have kidney-shaped, iridescent green tegulae and metallic-green upper legs (tibiae and femurs). On female *Augochloropsis metallica*, the first two segments of the abdomen are rimmed with short, bristly hairs. Females' hind-leg basitarsi are metallic green, and the front and middle basitarsi brown. The basitarsi of males are white. Both males and females have reddish lower tarsi. The face of the female is iridescent green, with a narrow black strip on the bottom rim of the clypeus. The male's face and clypeus are also iridescent green, and its labrum is black. Both males and females have black mandibles, dark antennae, and green eyes. **Size:** females 9 mm (1/3 in.), males 8 mm (3/10 in.).

Range: *Augochloropsis metallica* is widespread in the eastern and central United States. This species is found as far north as Ontario and as far west as South Dakota and Arizona. It ranges southward from Texas through most of Mexico, and into Central America as far as Panama. *Augochloropsis metallica* is common in all three border counties of the Lower Rio Grande Valley.

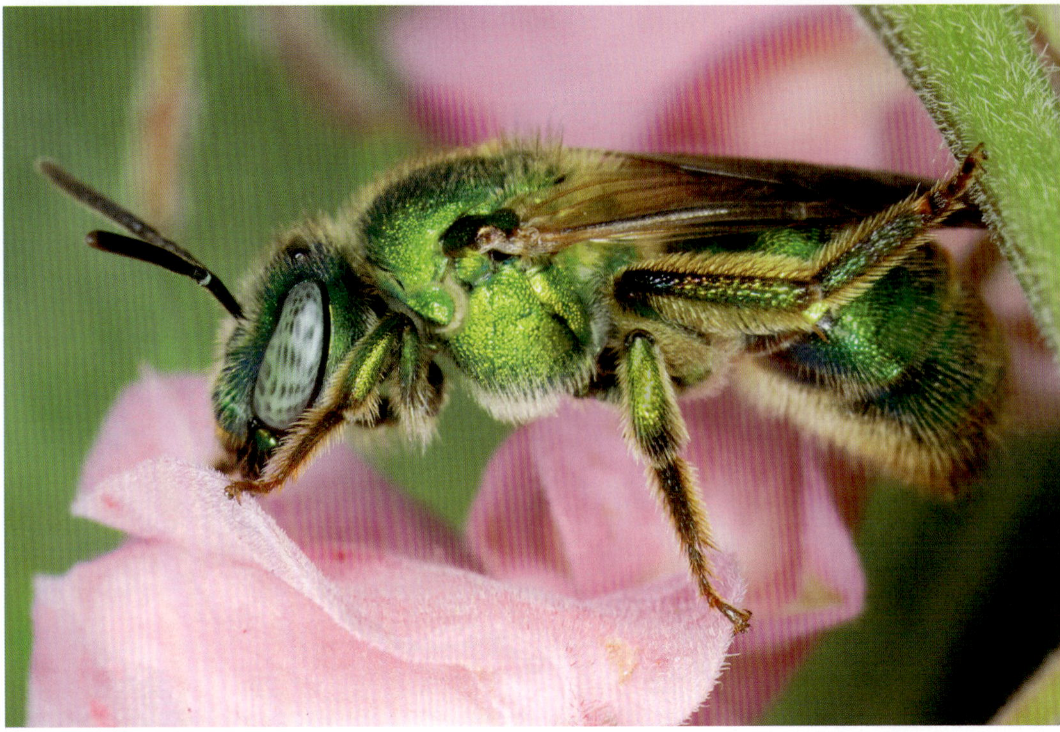

A female metallic-green sweat bee: note the metallic-green upper legs and the iridescent hind basitarsus. Pale and brownish hairs clothe the female's legs.

Associated plants: *Augochloropsis metallica* is a generalist pollinator that has been documented foraging on flora from more than 15 families. In the Valley, this species visits a variety of plants, among them bracted fanpetals (*Sida ciliaris*), hierba del marrano (*Symphyotrichum subulatum*), seaside goldenrod (*Solidago sempervirens*), shrubby blue salvia (*Salvia ballotiflora*), silverleaf nightshade (*Solanum elaeagnifolium*), and erect spiderling (*Boerhavia erecta*).

ID tips: In the field, *Augochloropsis metallica* is best recognized by its metallic-green upper legs and tegulae, defining traits of the genus *Augochloropsis*. The white legs of *Augochloropsis metallica* males also readily distinguish them from green *Agapostemon*. These traits are visible to the naked eye.

Other important traits, visible with magnification, include the following: (1) the thorax of both the female and male *Augochloropsis* has a pronounced pronotal ridge (located at the front of the scutum); (2) the inner hind-leg tibial spur of the female *Augochloropsis* is pectinate (comblike); (3) males lack a pygidial plate (a trait of all male Augochlorini); and (4) on all male *Augochloropsis*, the rim of the fourth segment of the sternum (S4) is jagged with a median point; this trait, however, is often obscured by hairs.

One other *Augochloropsis* species, *A. sumptuosa*, has been reported (uncommonly) in Willacy County. In that species, the projections on either side of the pronotal ridge tend to be more exaggerated and winglike, the two abdominal bands of hair on T1 and T2 are not well defined, and the surface of T1 has a distinctive tessellate pattern.

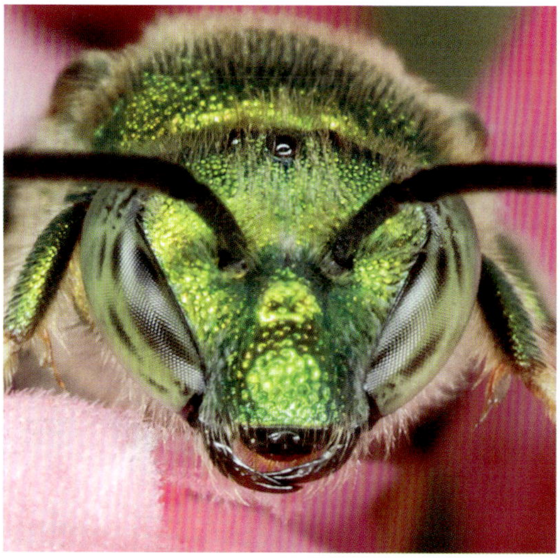

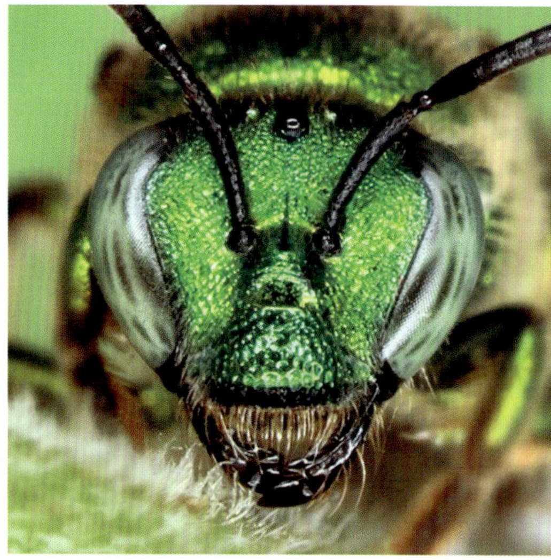

Faces of male (*left*) and female (*right*) *Augochloropsis metallica*. The clypei are mostly metallic green. The female's clypeus has a black strip on the bottom edge. The male's labrum is black.

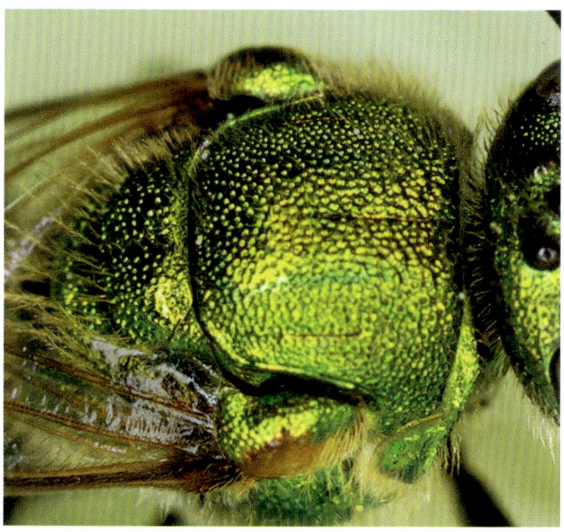

Close-up of the kidney-shaped tegulae of *Augochloropsis metallica.* The tegulae are predominantly metallic green. Note the wing-shaped projection at the front of the thorax.

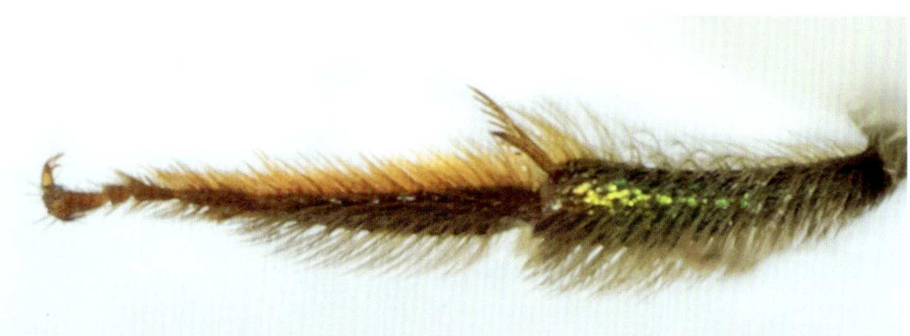

Hind leg of a female *Augochloropsis metallica*: the comblike inner hind tibial spur is a distinguishing trait of all female *Augochloropsis.*

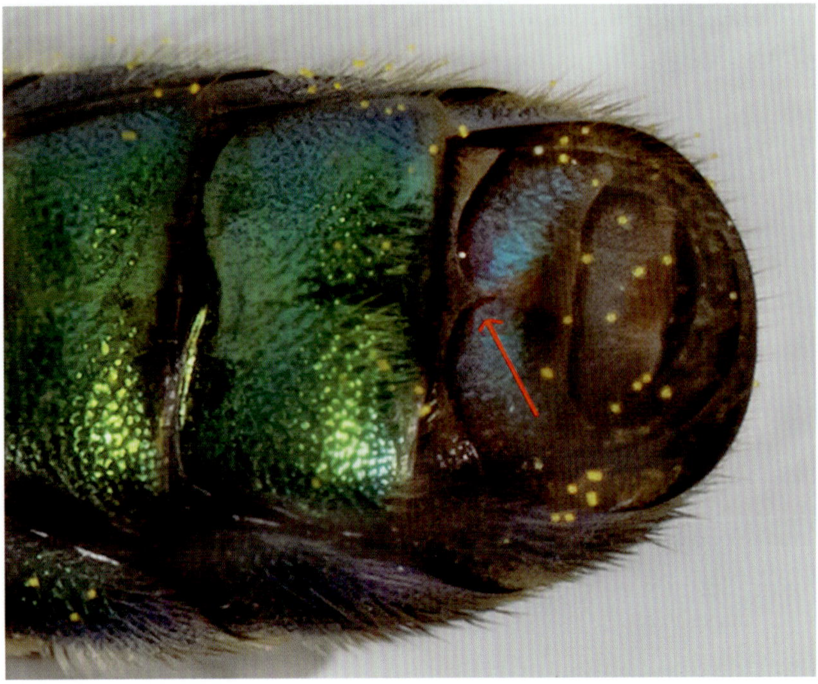

The sternum of a male *Augochloropsis metallica*: note the median point on the rim of S4. The lack of a pygidial plate is a characteristic of all male Augochlorini.

Behavior and additional information: *Augochloropsis metallica* was the first member of the tribe Augochlorini to be recognized. Danish zoologist John Christian Fabricius, a student of the great Swedish taxonomist Carl Linnaeus, first described the species in 1793.

Augochloropsis metallica nests in the soil, but its specific nesting habits remain uncertain. In 2017, entomologist Jason Gibbs wrote that although *Augochloropsis metallica* has been portrayed as both a solitary and a communal species, no published study has ever described its nesting behavior in detail. Studies of some other New World *Augochloropsis* species, however, have described them engaging in a spectrum of nesting patterns—solitary (with each bee constructing and provisioning its own nest), communal (sharing a common tunnel entrance), semisocial (dividing labor among sister bees), and even primitively eusocial (organizing labor via a multigenerational caste system).

In 1983, Charles C. Porter described *Augochloropsis metallica* he encountered in the Lower Rio Grande Valley. Although he did not investigate their nesting behavior, he chronicled their regional flight and foraging patterns. He found that *Augochloropsis metallica* in the Valley showed a preference for natural thornscrub and woodlands. The bees exhibited notable flexibility in their foraging, visiting flora ranging from low-growing herbs to the upper strata of trees. He also found evidence that Valley *Augochloropsis metallica* are multivoltine, noting that their populations peaked in early spring and in August, nearly disappeared during midsummer, and appeared in significant numbers from fall into winter.

Aztec sweat bee ◆ *Augochlora (Augochlora) azteca* (Vachal, 1911)

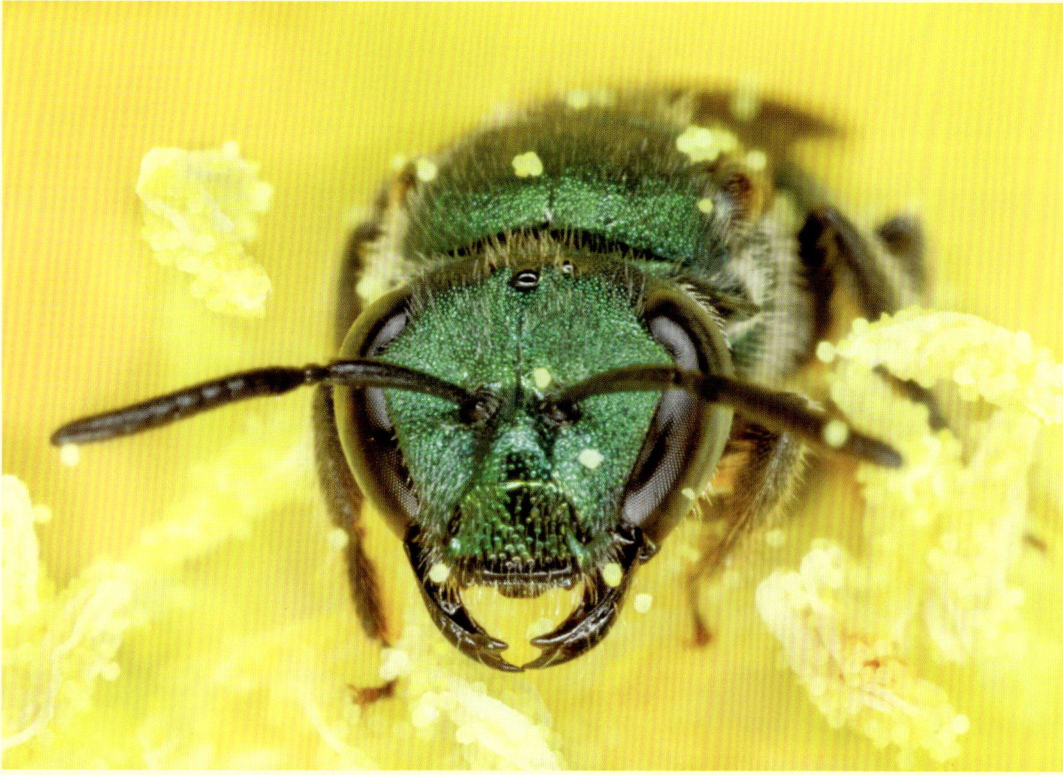

Face of a female *Augochlora azteca.* Photo credit Ross Eatman.

Description: *Augochlora azteca* is a slender bee with electric blue-green iridescence. Its tegulae are dark brown and oval, and its wings transparent brown with dark veins. The abdomen is mostly metallic blue green, with a narrow black hind margin on each of the first five segments (T1–T5); on females, T6 is entirely black. The female's legs are dark and covered with fine pale hairs; part of each front femur is green. The antennae, mandibles, and labrums of both males and females are dark colored and their clypei largely metallic, with a dark strip on the bottom rim. The female's mandibles have bidentate (two-toothed) tips. On males' antennae, the front surfaces of the flagella are reddish. **Size:** females and males 8 mm (3/10 in.).

Range: In the United States, this species appears only near the Texas-Mexico border. The bees shown here were found within half a mile of the Rio Grande. The range of the Aztec sweat bee extends from the border through Mexico, south to Veracruz. This species is rarer regionally than *Augochlora aurifera*, but *A. azteca* appears in all three border counties of the Valley from June through November.

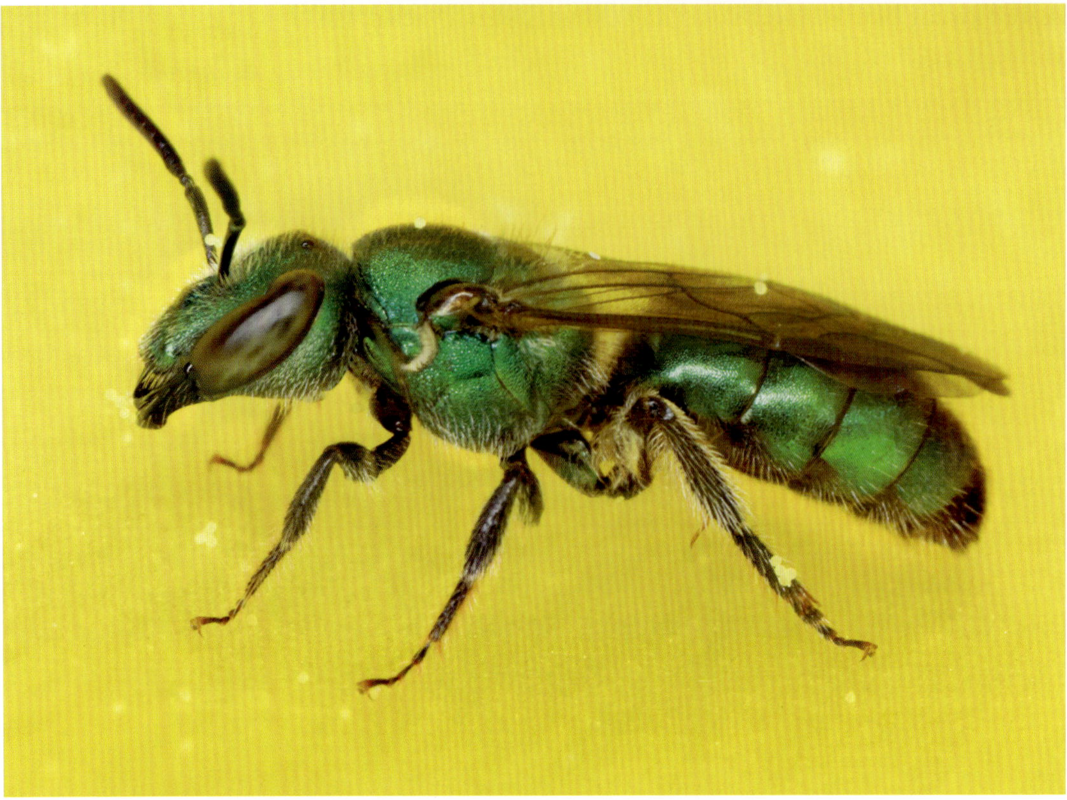

A female Aztec sweat bee. Photo credit Ross Eatman.

Associated plants: In the Valley, Aztec sweat bees are closely associated with alamo vine (*Merremia dissecta*), although they visit an array of other flora, including flame acanthus (*Anisacanthus quadrifidus*), cenizo (*Leucophyllum frutescens*), esperanza (*Tecoma stans*), Rio Grande abutilon (*Abutilon hypoleucum*), partridge pea (*Chamaecrista fasciculata*), teabush (*Melochia tomentosa*), and various cacti. The female bee shown here was one of several found feeding together on a Texas prickly pear cactus (*Opuntia engelmannii* var. *lindheimeri*).

ID tips: The oval brown tegulae of *Augochlora azteca* help differentiate them in the field from *Augochloropsis* sweat bees. All *Augochlora* have two traits, visible with the aid of a magnifying lens, that distinguish them from similar *Augochlorella*: the tip of the marginal cell of an *Augochlora* wing is blunt or somewhat rounded (rather than pointed); and the paraocular area of the face intrudes into the clypeus.

In the United States, there are a mere handful of *Augochlora* species, and only two subgenera of *Augochlora*. The Aztec sweat bee belongs to the subgenus *Augochlora*. It is nearly identical to the pure green sweat bee, *Augochlora (Augochlora) pura*—a common, widespread species found throughout much of the United States (including Texas). These two *Augochlora* species differ principally in color and geography—the pure green sweat bee lacks the intense blue-green coloration of *Augochlora azteca*, and it is unlikely to appear near the Texas border.

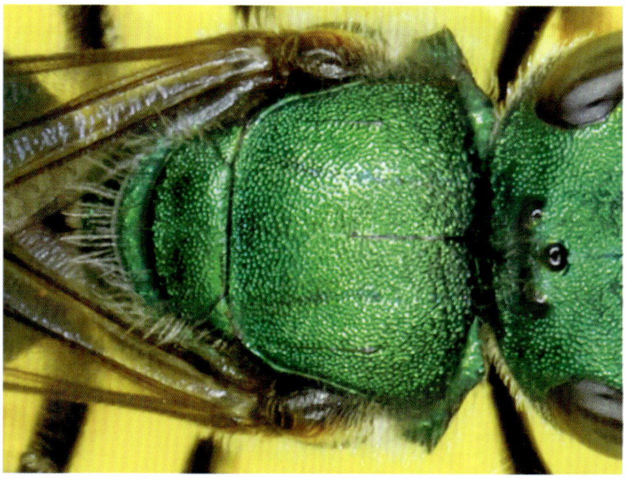

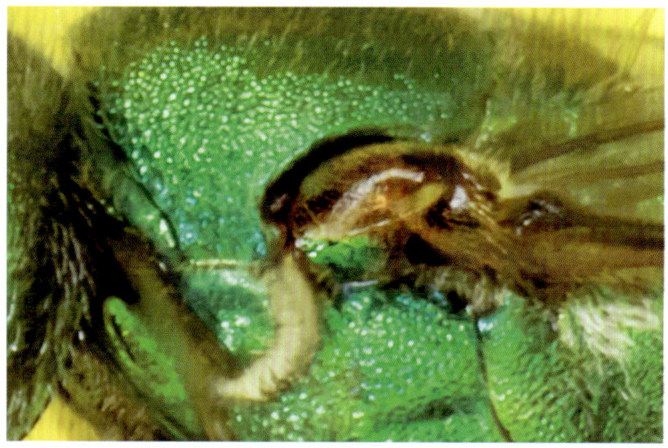

Close-ups of the thorax of an Aztec sweat bee. Note the rough, pitted, and wrinkled texture of the scutum and scutellum (*above*); the oval brown tegulae (*right*); and the radiating wrinkles of the propodeal disk, which extend all the way to its back rim (*below*).

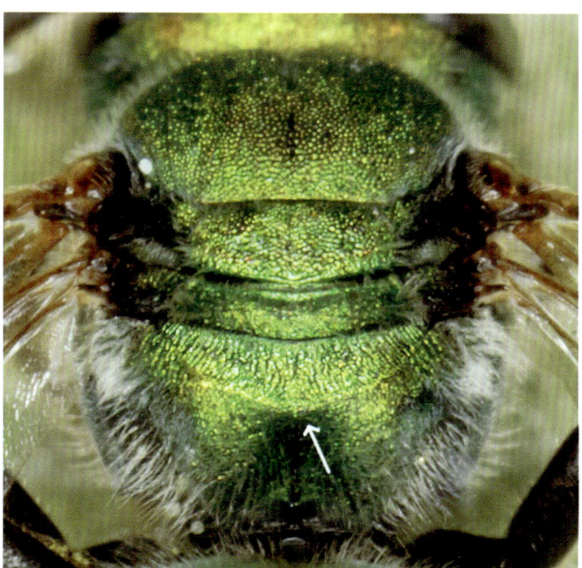

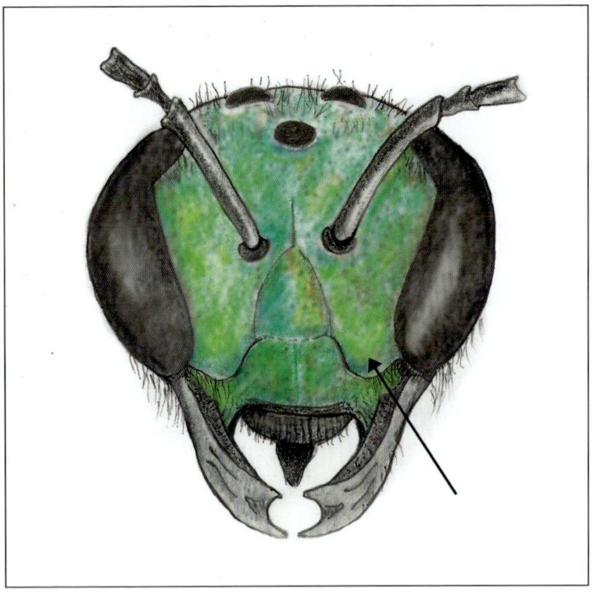

Augochlora face: the paraocular lobes intrude downward into the clypeus.

Behavior: The nesting habits of *Augochlora azteca* remain unstudied. The closely related *Augochlora pura*, however, has been written about extensively. *Augochlora pura* usually nest in rotting logs, although they have been found to dig nests in soil. The mandibles of all females of the subgenus *Augochlora (Augochlora)* have a distinct second tooth near each tip, which enables the bees to shape burrows in wood. As *Augochlora azteca* also possesses this trait, it is possible that the Lower Rio Grande Valley species similarly nests in rotted wood or other plant material.

In 1966, entomologist Karl Stockhammer conducted a comprehensive study of the nesting biology of *Augochlora pura*. He described females of this species as flexible and versatile nest builders. They often begin with preexisting insect burrows in old wood and then carve out their own architectural patterns. Their nests may be tunnel shaped or may take the form of corridors that are layered over each other to form comblike clusters. The bees typically line the walls of their burrows with a frame of wood particles, layered with a second interior wall and then coated with a waxlike substance. Stockhammer found the plasticity of the bees remarkable—they adapted nest-building strategies to specific environments, for example by dispensing with the second interior nest layer when the matrix in which they built furnished a sufficiently tough outer frame.

Augochlora pura sweat bees are remarkable in other aspects. They are capable of shifting between solitary and social behavior, depending on environmental conditions. They also produce multiple generations in a single year. Although the nesting behavior of *Augochlora azteca* is unknown, it is likely that this species is also multivoltine, given that it emerges in the Valley in early spring and remains through late fall. Whether *Augochlora azteca* sweat bees exhibit the nest-building flexibility characteristic of *A. pura* has yet to be tested.

Additional information: Cockerell recorded *Augochlora azteca* in 1923 in Victoria, Tamaulipas. The species had been described first by the French entomologist Joseph Vachal 12 years before. Vachal, a French lawyer and political figure with a passion for natural history, amassed an enormous Hymenoptera collection in the late 1800s and early 1900s by purchasing private specimens from insect dealers and collaborating with several French museums. He described 786 new Hymenoptera species during his lifetime. In the last year of his life, at the age of 72, Vachal described a Mexican specimen of *Augochlora azteca* as part of a 1911 publication in which he devised an influential classification system for the entire family Halictidae. In his will, Vachal left the *Augochlora azteca* specimen, along with his vast Hymenoptera collection, to the Paris museum of natural history.

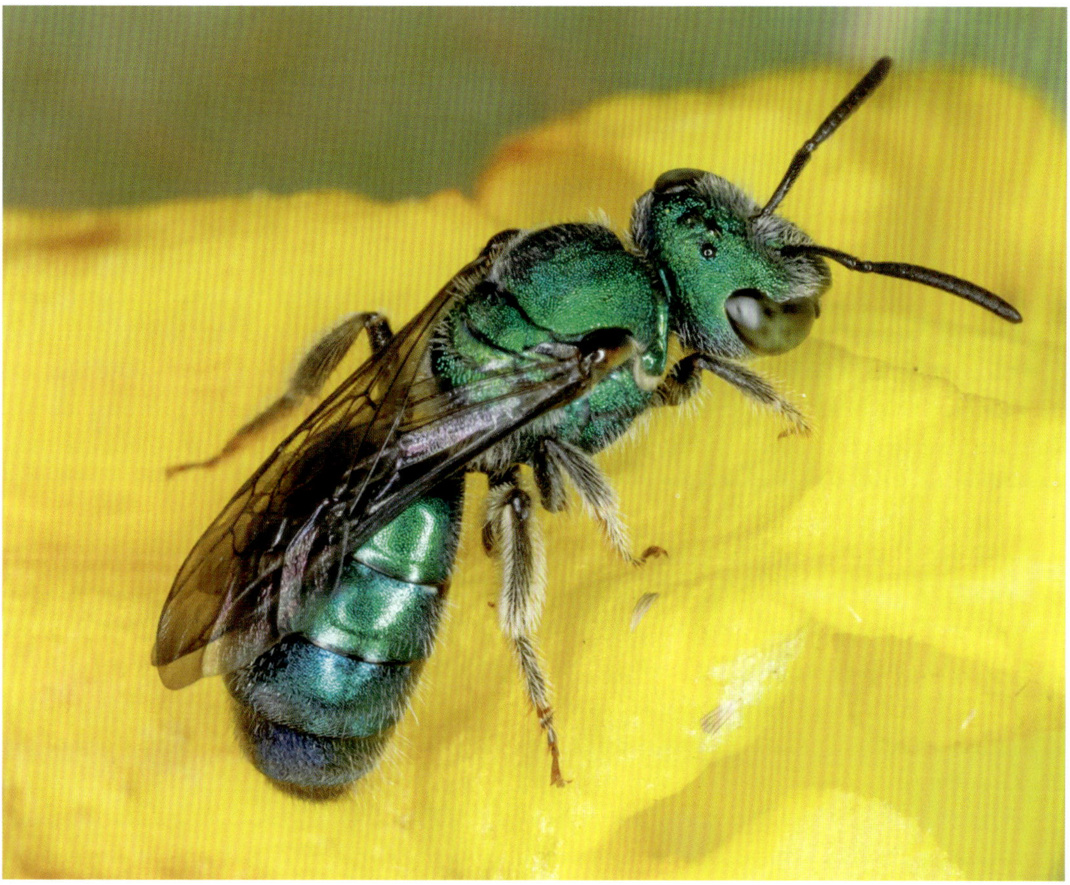

A female *Augochlora azteca*

Behavior of *Augochlora (Oxystoglossella)*

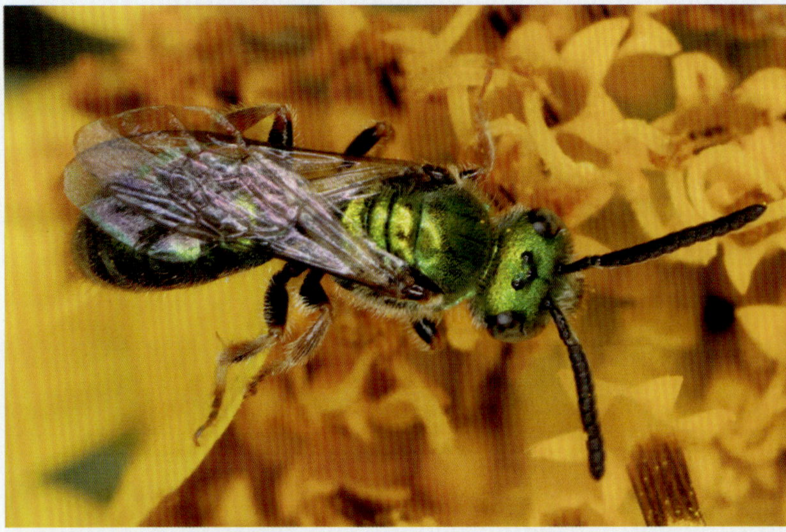

A male *Augochlora aurifera*

Behavior: *Augochlora aurifera* belongs to the subgenus *Oxystoglossella.* As noted above, female Aztec sweat bees of the subgenus *Augochlora (Augochlora)* have mandibles that terminate in two conspicuous teeth. *Augochlora (Oxystoglossella)*, however, lack this feature. Their mandibles appear to terminate in a single point; a second tooth is present but so small as to be inconspicuous. The bees' jaw structure affects their nesting habits: *Augochlora (Oxystoglossella)* do not use their mandibles to carve out nests in rotted wood—they nest in the ground.

No detailed reports exist regarding the behavior of *Augochlora aurifera.* In 1972, however, entomologists Eickwort and Eickwort studied the nesting habits of *Augochlora (Oxystoglossella)* in Costa Rica. Their work focused on *Augochlora (Oxystoglossella) nominata* nests constructed in hard clayey soil on a sloped embankment and dirt road bordered by shrubby vegetation between a sugarcane field and forest. This habitat is similar to Hidalgo County sites where *Augochlora aurifera* appears in large numbers, among them the National Butterfly Center, which is situated in a landscape that contains sugarcane fields, woodlands, an irrigation canal, and a bend of the Rio Grande.

The nests Eickwort and Eickwort studied showed that *Augochlora (Oxystoglossella)* are capable of primitively eusocial behavior involving a simplified caste system. Typically, among *Augochlora nominata*, a large foundress queen independently established, waterproofed, and provisioned a nest and laid eggs. The eggs of her first brood developed into males and small, infertile female workers that gathered pollen and guarded the nest. If the queen perished during this period, one of the workers would develop ovaries and become the new queen.

The queen thereafter laid eggs for a second brood in a newly constructed, deeper part of the nest. She remained within the nest during that brood's first weeks of development. When the second brood hatched, large fertile females emerged. Those that mated with males near the nest site became foundress queens the following year.

Golden-green sweat bee ◆ *Augochlora (Oxystoglossella) aurifera* Cockerell, 1897

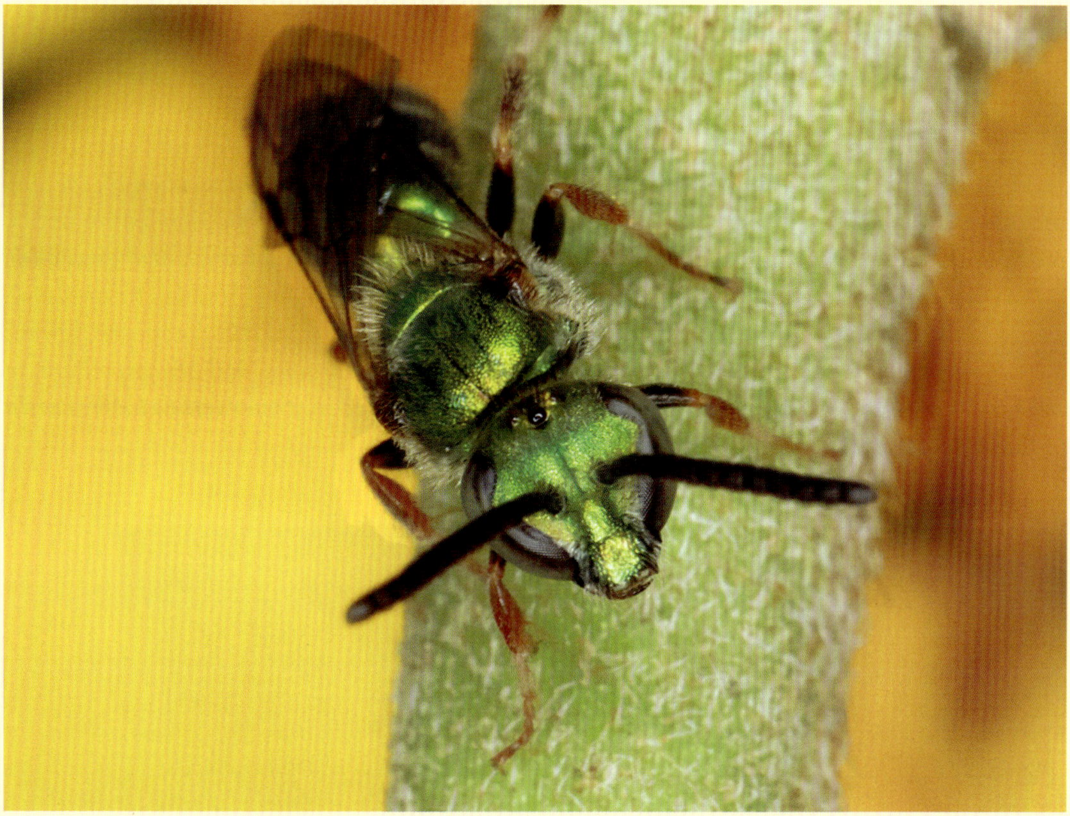

A male golden-green sweat bee

Description: *Augochlora (Oxystoglossella) aurifera* is a relatively small metallic sweat bee. The typical *Augochlora aurifera* is green with some golden highlights. The thorax and the first five segments (T1–T5) of the males' and females' abdomens are metallic green or golden green; on females T6 is black, and on males T6–T7 are black. The bees' tegulae are reddish brown, and their wings transparent brown with dark brown veins. The female's thorax is frequently patterned with an exquisite rainbow of blue, green, gold, and reddish swaths that run lengthwise along the scutum.

Both females and males have multicolored legs. The basal joints of the female's tibiae are red, and its tarsi reddish. The female's femurs, tibiae, and basitarsi are otherwise black and covered with white hairs. The male's tibiae are reddish brown, its femurs mostly black, its lower tarsi predominantly reddish brown, and its front basitarsi white. The trochanters of both females and males are metallic green. Both have reddish mandibles and metallic clypei. On females, the clypeus has a narrow black strip on the bottom rim; on males, the rim is pale. **Size:** females and males 5–6.5 mm (1/5–1/4 in.).

Range: *Augochlora aurifera* is predominantly Neotropical. Cockerell first described this species in 1897 in a brief article titled "On the Mexican Bees of the Genus *Augochlora*." He based his description on two female *Augochlora aurifera* found in San Rafael, Veracruz. In the United States, *Augochlora aurifera* occurs mainly in Texas and is found in all three border counties of the Lower Valley. Its range extends through Mexico and Guatemala to Costa Rica and Panama.

Associated plants: *Augochlora aurifera* is a generalist pollinator that appears on flowers of diverse families in the Valley, among them small coastal germander (*Teucrium cubense*), evening primrose (*Oenothera*), silverleaf and buffalobur nightshade (*Solanum elaeagnifolium* and *S. rostratum*), wavy-leaf nama (*Nama undulatum*), copper globemallow (*Sphaeralcea angustifolia*), spiked malvastrum (*Malvastrum americanum*), anglestem Indian mallow (*Abutilon trisulcatum*), cenizo (*Leucophyllum frutescens*), crucita (*Chromolaena odorata*), and various composite flowers. T. D. A. Cockerell found his San Rafael *Augochlora aurifera* on *Vernonia* and *Melampodium*. In 1917, Wilmatte Porter Cockerell discovered a female bee feeding on the flowers of native crown of thorns (*Koeberlinia spinosa*) in San Benito (Cameron Co.).

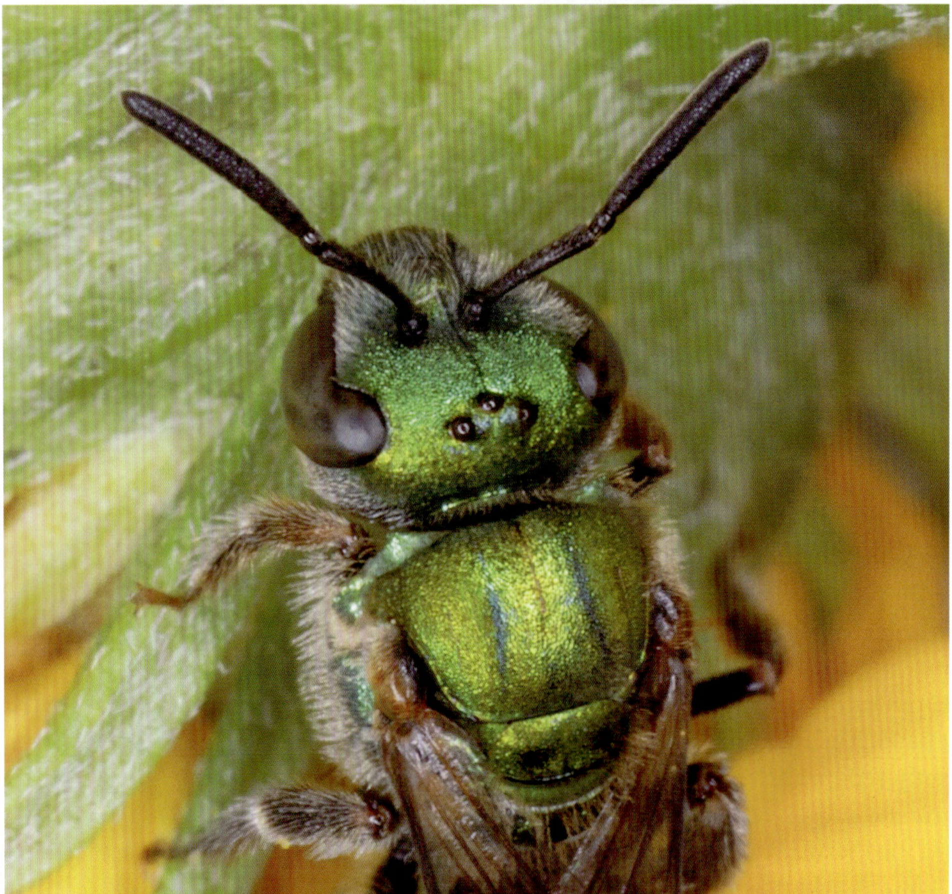

Close-up of the iridescent thorax of a female *Augochlora aurifera*

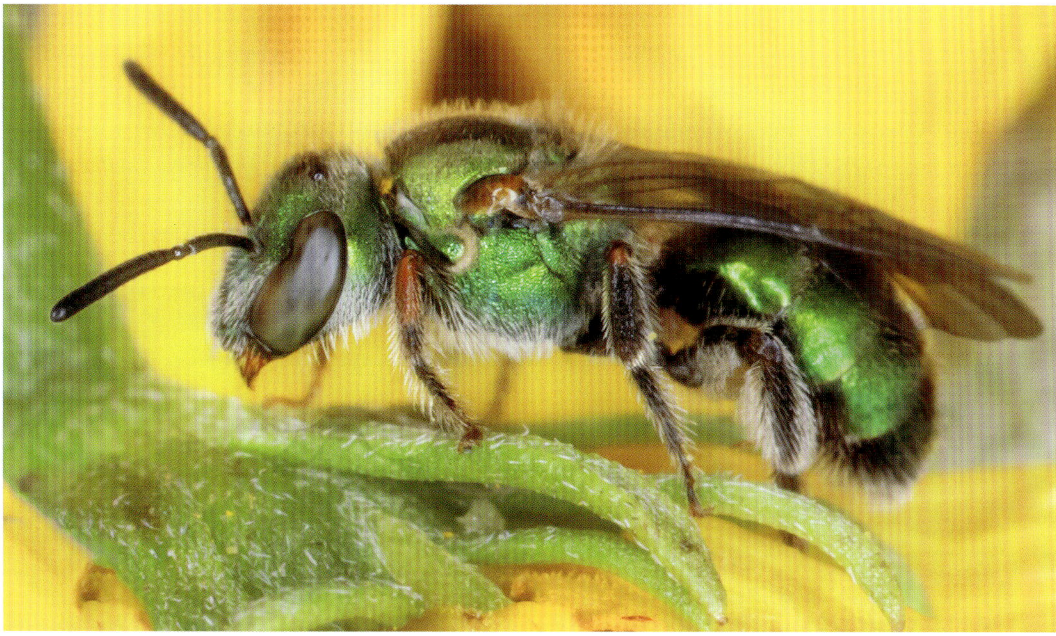

A female *Augochlora aurifera*: note the compact build, the red upper joints of the tibiae, the reddish mandibles, and the oval, reddish-brown tegula.

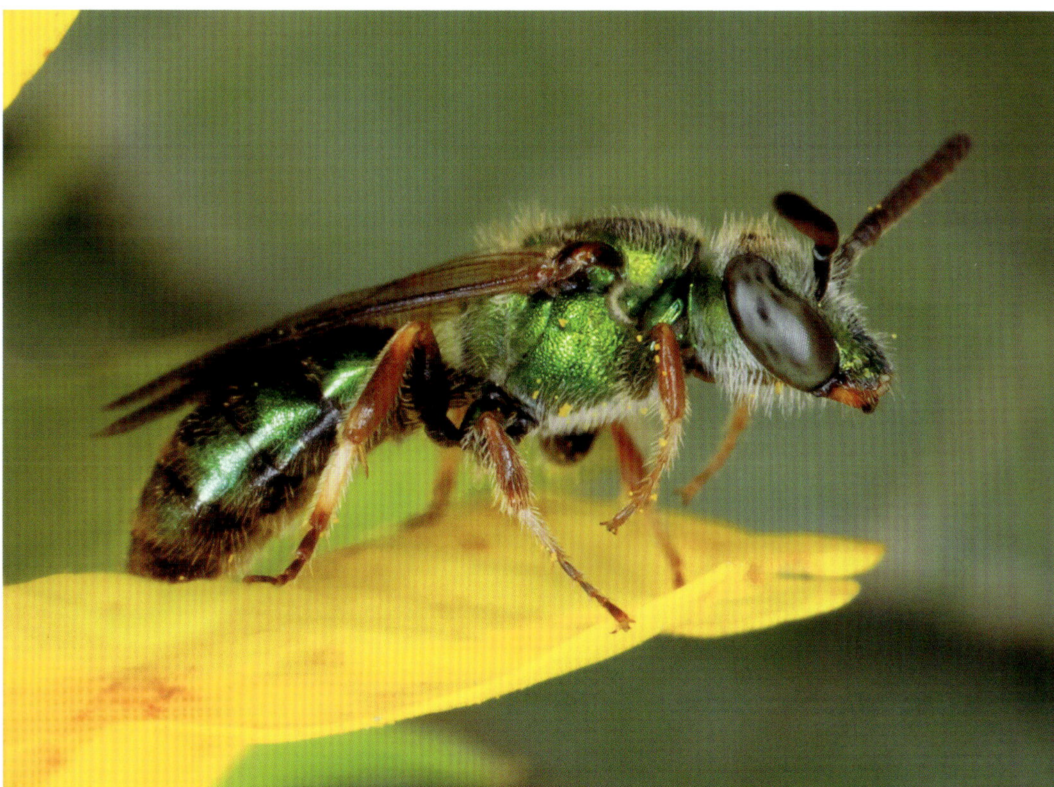

A male *Augochlora aurifera*: note the pale reddish mandibles and reddish-and-pale legs.

ID tips: The golden-green sweat bee is generally smaller and more common regionally than the Aztec sweat bee and lacks its intense blue-green coloration. The reddish mandibles of the female *Augochlora aurifera*, and the red spots on the upper joints of its tibiae, also readily distinguish it from the Aztec sweat bee. Minute details of the female's propodeal disk (the wrinkled area on the rear face of the thorax), shown here, also aid in distinguishing *Augochlora aurifera* from other regional *Augochlora*.

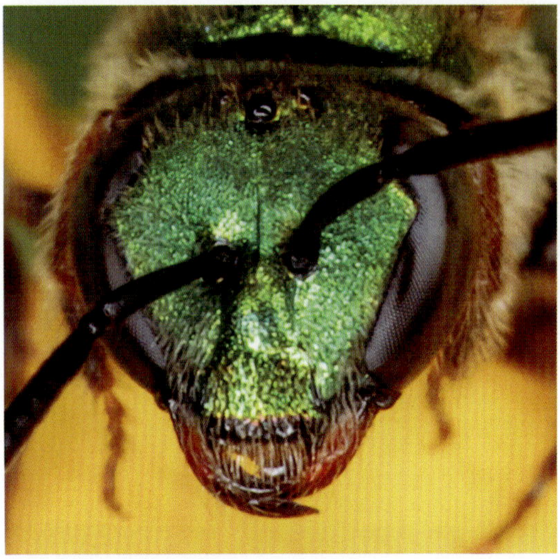

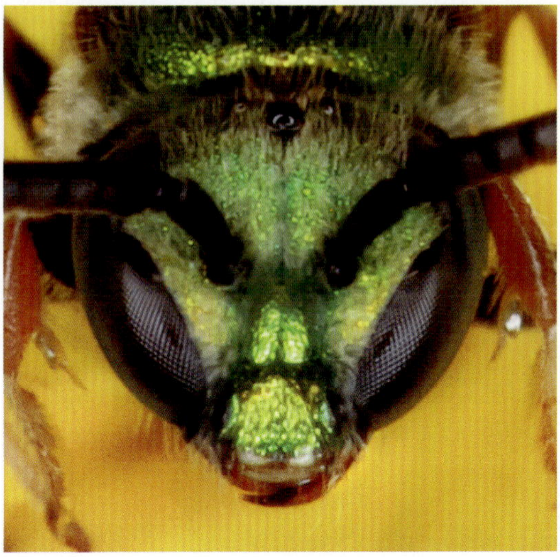

Faces of female (*left*) and male (*right*) *Augochlora aurifera*

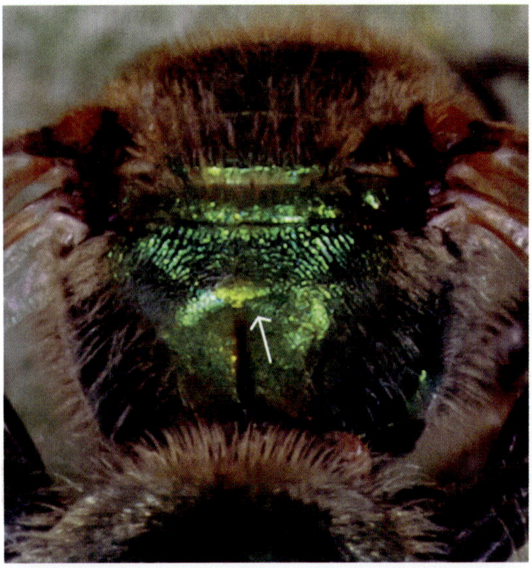

The propodeum of a female *Augochlora aurifera*: the wrinkles at the center of the propodeal disk do not reach its hind rim.

Gilded augochlorella ◆ *Augochlorella bracteata* Ordway, 1966

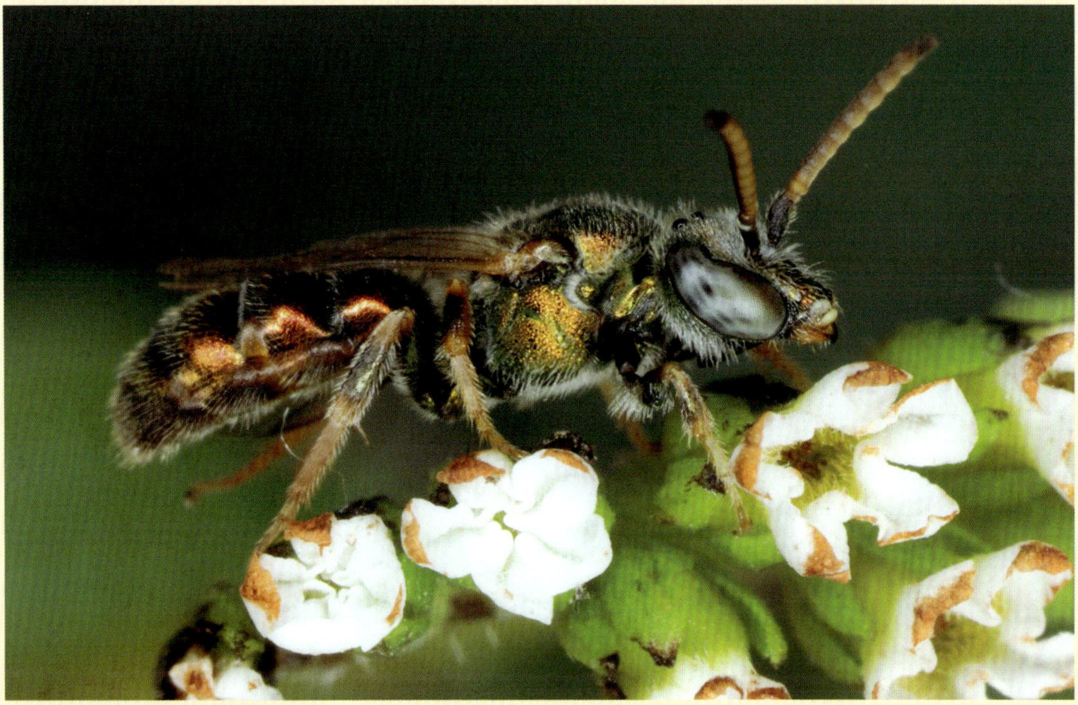

A male gilded augochlorella

Description: *Augochlorella* are small, slender bees with brilliant metallic coloring. They occur in North, Central, and South America. *Augochlorella* nest in the ground and, like *Augochlora*, range from solitary to primitively social in their nesting habits.

The gilded augochlorella is the smallest bee of the tribe Augochlorini in this guide. *Augochlorella bracteata* is typically green or yellow green, but some specimens are bright coppery pink. The thorax of *Augochlora bracteata* is smooth and shiny with minute, distinct pits. The wings are brown with brown veins, and the tegulae are oval and reddish on the hind portion and transparent yellow on the front portion. The legs are partly metallic green and partly brown, and they are covered with fine white hairs. The basitarsi and front tarsi of males are ivory.

The face of *Augochlora bracteata*, including most of the clypeus, is metallic. The bottom edge of the female's clypeus and its labrum are black. The bottom edge of the male's clypeus is ivory, with a reddish-metallic, pointed intrusion in the middle, and its labrum is also ivory. Males' antennae are black with golden-brown undersides, except on the scapes, pedicels, and last antennal segment (F11), which are entirely black. Both males and females have reddish mandibles that darken toward the tips. **Size:** females 5–7 mm (1/5–1/4 in.), males 4.5–6 mm (1/6–1/4 in.).

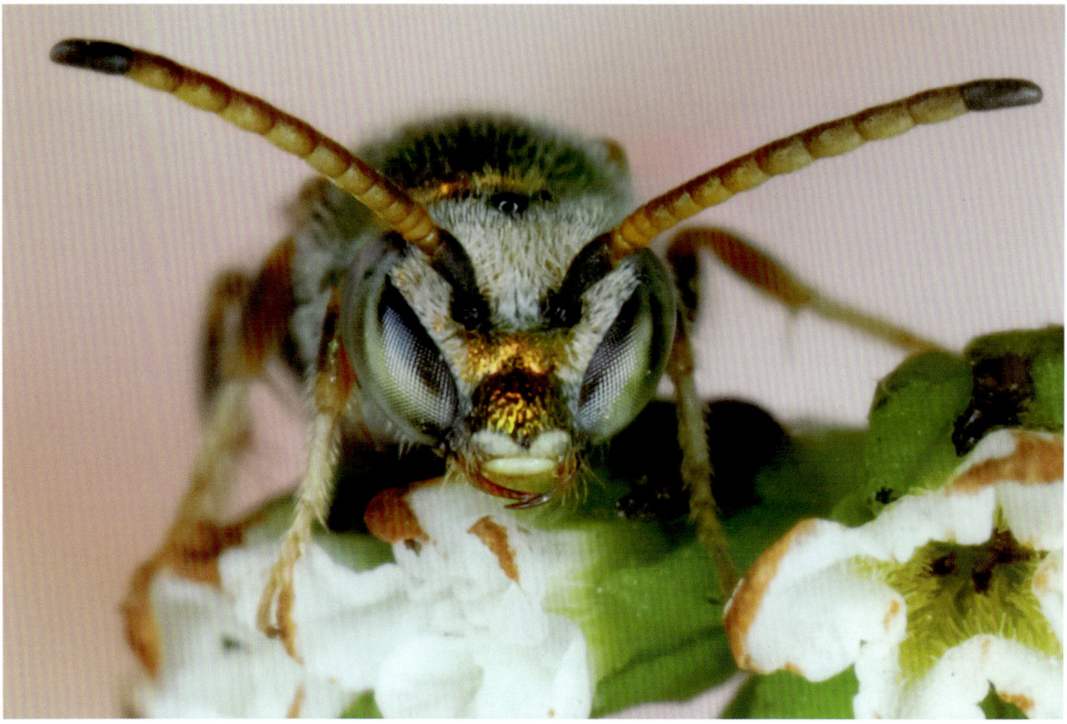

A male *Augochlorella bracteata*

Range: The distribution of *Augochlorella bracteata* is centered in Mexico. In the United States, this species occurs mostly in Texas and the Southwest. It is found in all three border counties of the Lower Rio Grande Valley. *Augochlorella bracteata* was first described in 1966 by Ellen Ordway. Her description was based on specimens collected in Cameron County during the early 1950s.

Associated plants: *Augochlorella bracteata* appears from March through November in the Valley. It shows a proclivity for small-flowered plants such as scorpion's tail (*Heliotropium angiospermum*) and prairie broomweed (*Amphiachyris dracunculoides*) but also feeds on other Valley flora, including bracted fanpetals (*Sida ciliaris*), coral vine (*Antigonon leptopus*), and composite flowers such as resinbush (*Viguiera stenoloba*) and hairy wedelia (*Wedelia acapulcensis* var. *hispida*).

ID tips—Augochlorella versus other sweat bee genera: Although typically smaller than *Agapostemon* and *Augochloropsis* sweat bees, *Augochlorella* can be hard to distinguish from small *Augochlora.* Traits that aid in separating these two genera require magnification and include the following: (1) the marginal cell of the *Augochlora* wing is pointed at the tip; (2) the typical *Augochlorella* has a narrow vertex (top of the head), and its clypeus is "beveled" (protrudes with a downward slant); the typical *Augochlora* clypeus, by contrast, is flat; (3) as shown here, male *Augochlorella* have distinctive faces, with compound eyes that converge toward the bottom, crowding the lower face and making it appear much narrower than the upper half; the intrusion of a metallic, reddish point into the pale rim of the clypeus is a common trait of male *Augochlorella*; and (4) on all male (and probably female) *Augochlorella*, F1 of the antennae is much shorter than F2.

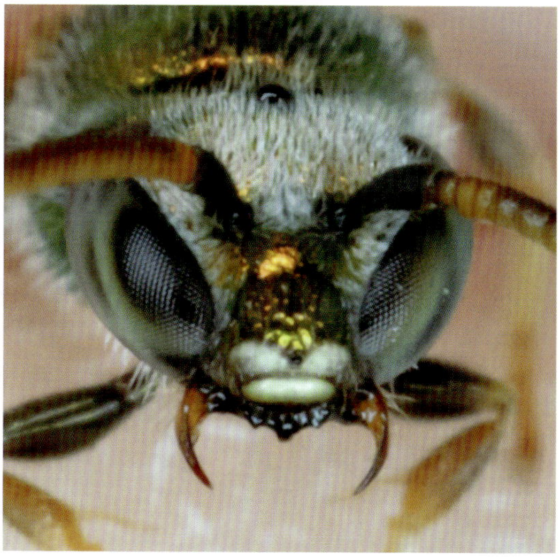

Face of a male *Augochlorella bracteata*: note the mostly pale bottom rim of the clypeus and the pale labrum.

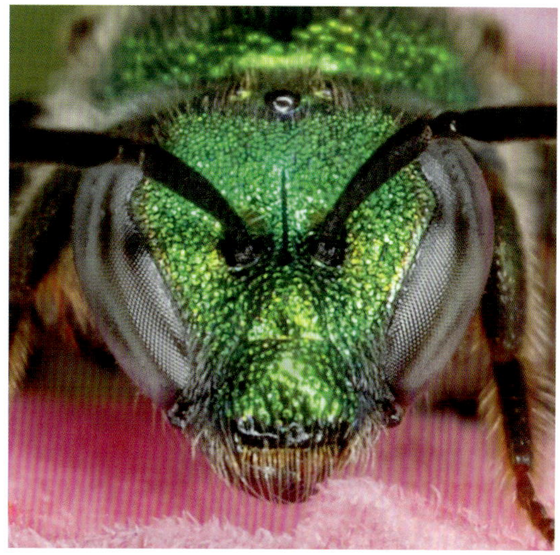

A female *Augochlorella bracteata*: the metallic-green clypeus has a narrow black strip on the bottom rim.

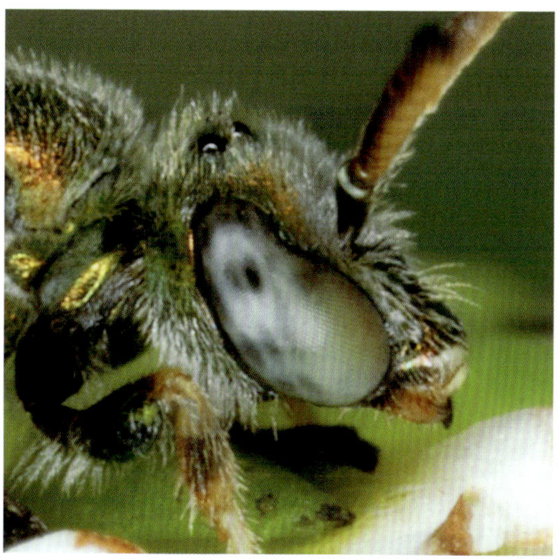

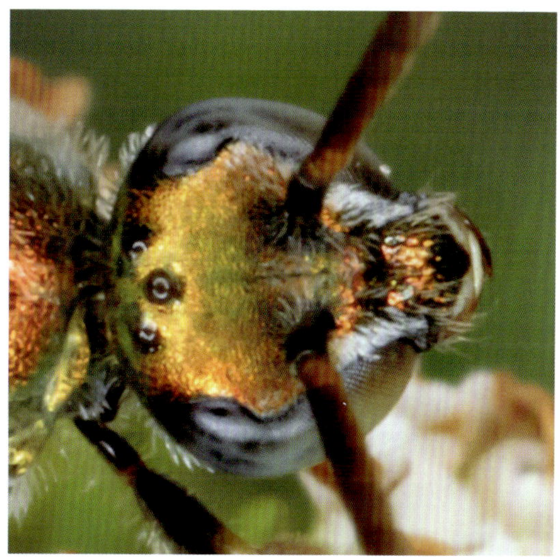

The clypeus of *Augochlorella bracteata* is protuberant and slants downward and forward. The vertex is narrow, a characteristic of all *Augochlorella*.

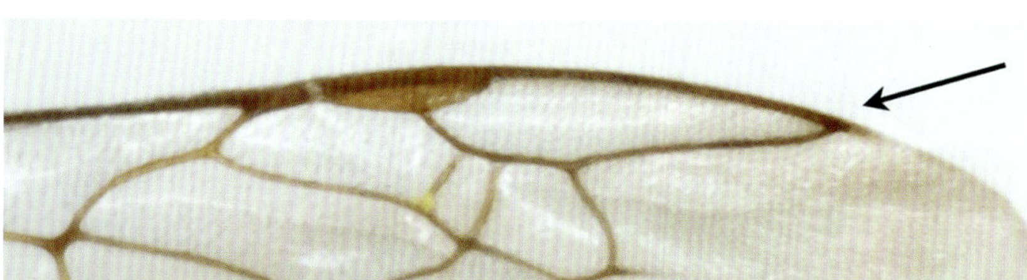

Close-up of the pointed marginal cell of an *Augochlorella bracteata* forewing

ID tips for Augochlorella bracteata: Seven species of *Augochlorella* are currently recognized in the United States. They are tricky to separate from one other and have proven difficult even for experts to define—over recent decades, entomologists have merged some species with others and even shifted them between *Augochlorella* and *Augochlora*. Species identification of *Augochlorella* frequently turns on "propodeal sculpturing"—that is, the general shape and qualities of the area on the rear of the thorax called the propodeal disk. On *Augochlorella bracteata* of both sexes, this area is semicircular with a smooth, curved hind edge. On other *Augochlorella*, this back rim may be bracket shaped and have one or multiple points in the middle.

Brazilian entomologist Beatriz Coelho, author of a comprehensive 2004 study of *Augochlorella*, noted that *A. bracteata* is the smallest and slenderest species of its genus. The 5–6 mm male *Augochlorella bracteata* can be differentiated from the common US species known as *A. aurata* in part by size. *Augochlorella aurata* is typically larger (6–8 mm). Male *Augochlorella* can also be differentiated by characteristics of the sternum. On male *Augochlorella bracteata*, the hind edge of the fourth segment of the sternum (S4) is deeply curved.

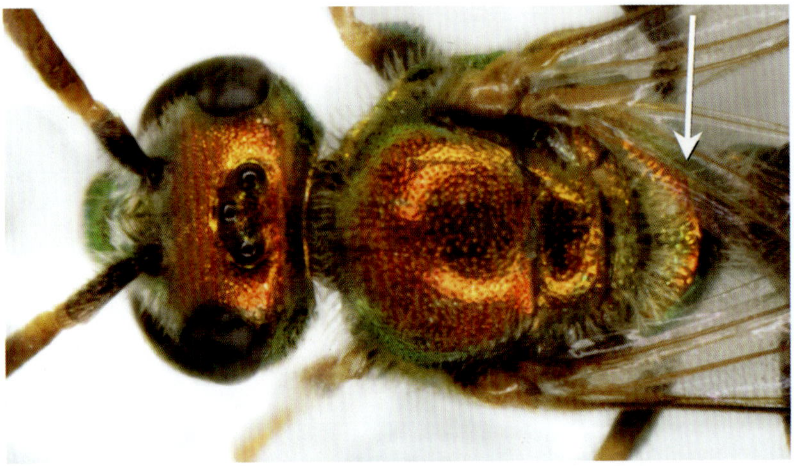

The thorax of *Augochlorella bracteata* is shiny and minutely pitted. Note the bright semicircle of the propodeal disk.

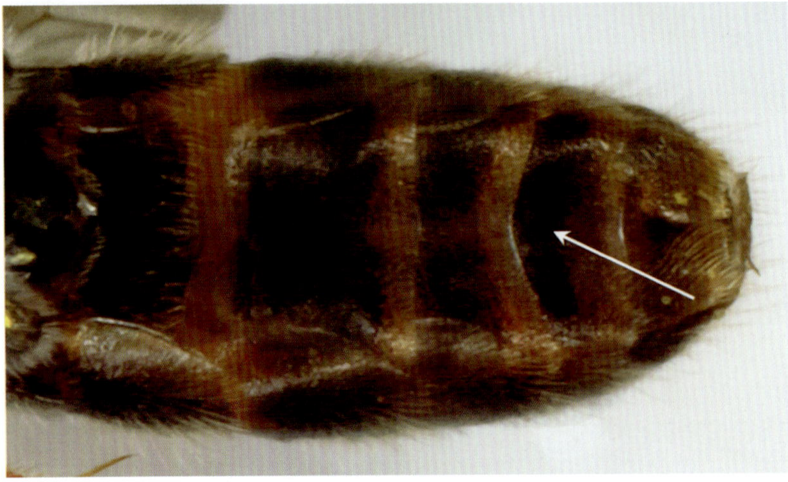

The sternum of a male *Augochlorella bracteata*: note the deeply curved hind edge of S4.

Striped Green Sweat Bees

Subfamily Halictinae ◆ Tribe Caenohalictini
Genus *Agapostemon*

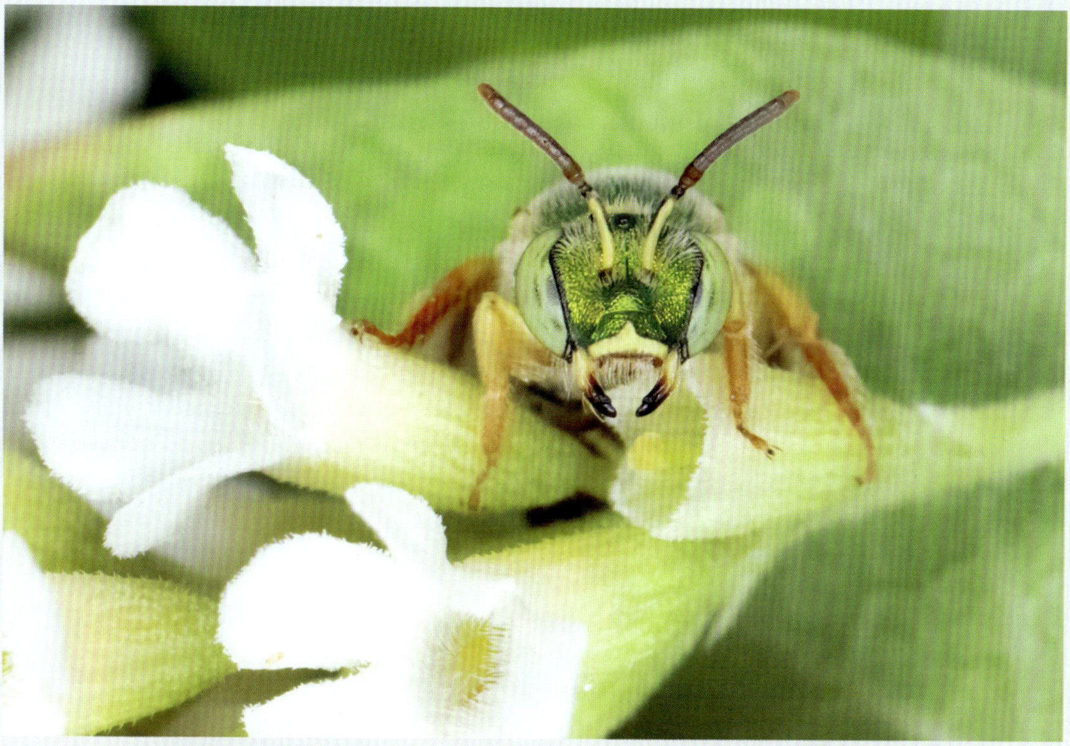

A female honey-tailed agapostemon

Agapostemon look like flying emeralds—they are intensely green, generalist bees that are instrumental in the pollination of wildflowers, gardens, and crops. To the naked eye, *Agapostemon* resemble sweat bees of the Augochlorini tribe discussed in the preceding entries. Nonetheless, *Agapostemon* tend to be larger than Augochlorini sweat bees, and differences in behavior, wing and leg structure, facial characteristics, and evolutionary history have led to the placement of the genus *Agapostemon* in a different tribe called the Caenohalictini. This tribe contains predominantly medium to large metallic bees. Unlike Augochlorini, which may be eusocial, Caenohalictini are typically solitary or communal.[2]

2. The genus *Agapostemon* is alternatively listed in popular and scholarly literature and databases as belonging to the tribe Halictini, in which case Caenohalictini is treated as a subtribe that includes *Agapostemon* and their kin. Both ways of treating *Agapostemon* are legitimate approaches to taxonomic classification; their discrepancies highlight the difficulties of categorizing this unique group of bees. An excellent and concise discussion of Caenohalictini and the other tribes of the subfamily Halictinae appears in Danforth et al. (2008).

Nearly all *Agapostemon*, whether male or female, have brilliant iridescent green or blue-green heads and thoraxes. The genus *Agapostemon* is sometimes referred to by the common name "striped green sweat bees" because males typically have boldly striped abdomens. Females are often entirely green, although some have striped abdomens as well. Radclyffe Roberts, author of a comprehensive 1973 treatment of the genus, wrote that *Agapostemon* can be identified by these distinctive traits: (1) all *Agapostemon* have a characteristic ridge on the propodeum (the back of the thorax); and (2) on male *Agapostemon*, the two basal tarsal segments of each hind leg are fused so that the bees appear to have four hind tarsal segments.

Range: *Agapostemon* are found only in the Western Hemisphere. They range from British Columbia to Paraguay and the Caribbean. This genus is relatively small, containing slightly more than 40 species.

Behavior: While the social habits of many *Agapostemon* species have been studied in detail, the nesting behavior of some species in the Lower Rio Grande Valley, such as *A. tyleri* and even the more common *A. melliventris*, remains largely unknown. Some, but not all, *Agapostemon* species nest close to one another in aggregations. They sometimes work cooperatively, sharing nest burrows whose entrances are watched by guard females. *Agapostemon*, however, do not construct complex hierarchical societies (comprising fertile foundress queens, males, and unfertilized females) like those of some *Augochlorella* and *Augochlora*.

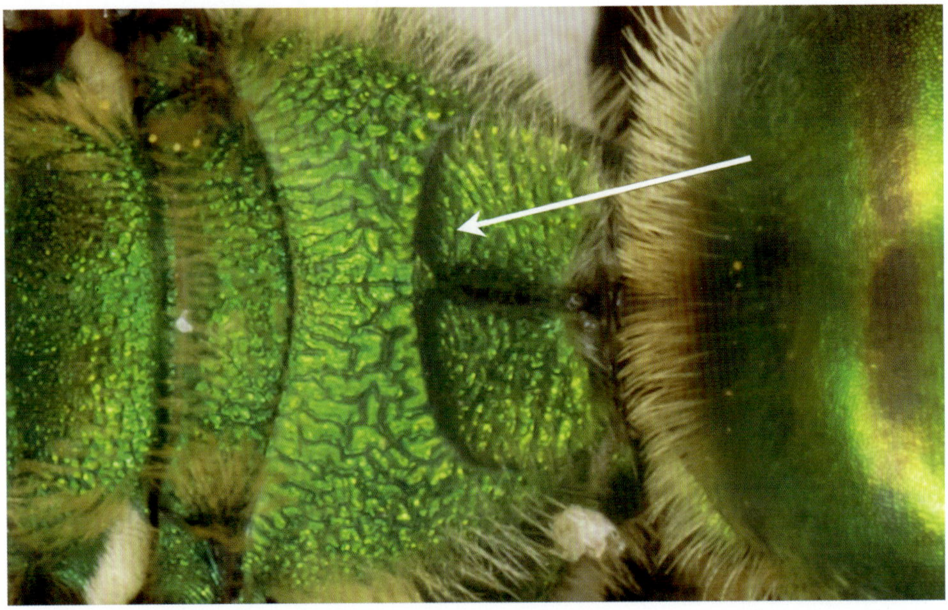

The propodeum on a female *Agapostemon*: the ridge rimming the rear of the propodeum is a distinguishing trait of all *Agapostemon*.

Texas agapostemon ◆ *Agapostemon (Agapostemon) texanus* Cresson, 1872

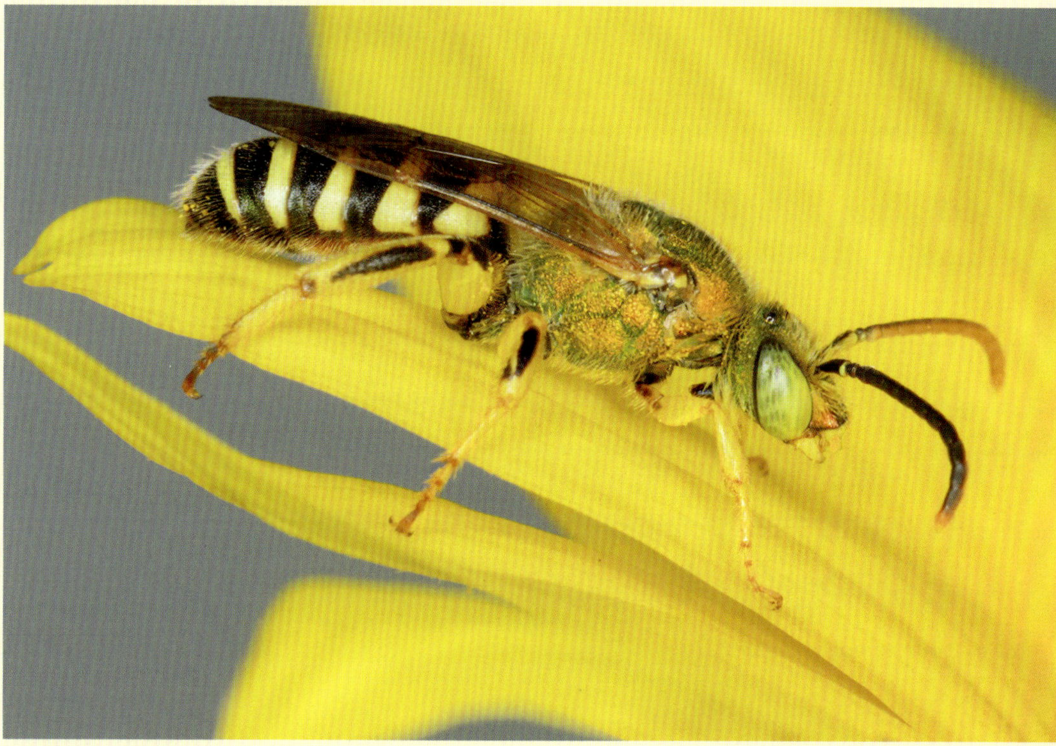

Agapostemon texanus on a sunflower

Description: The Texas agapostemon is a medium-sized iridescent bee with lustrous green eyes. Males and females differ considerably in appearance. The male *Agapostemon texanus* is slender, with a metallic-green or coppery-green head and thorax. Its abdomen is black and striped on the first through fifth segments (T1–T5) with yellow bands; T6–T7 are entirely black. The male's terga are faintly metallic near the base. The male's legs are predominantly yellow, with black spots on the femurs and tibiae; the femurs are wide. The male's tegulae, mandibles, clypeus, and antennal scapes are partly yellow. The front faces of its flagella are golden brown. The wings of both male and female *Agapostemon texanus* are brown with brown veins.

Female *Agapostemon texanus* are more robust than males and have entirely green heads and bodies. T1–T4 of females' abdomens are banded with white hairs; black hairs cover T6. Females' upper legs (coxae and trochanters) are partly metallic, and their lower legs are dark brown, except for small yellow spots on the tips of the front and middle tibiae. Females' lower legs are clothed mostly with white hairs; some dark hairs line the front surfaces of the hind-leg tibiae, and the hind-leg scopal hairs are largely pale. Females' antennae and labrums are black, and their mandibles are reddish at the base, darkening toward the tips. The female's clypeus is metallic green with a black lower rim. **Size:** females 11 mm (2/5 in.), males 8–10 mm (3/10–2/5 in.).

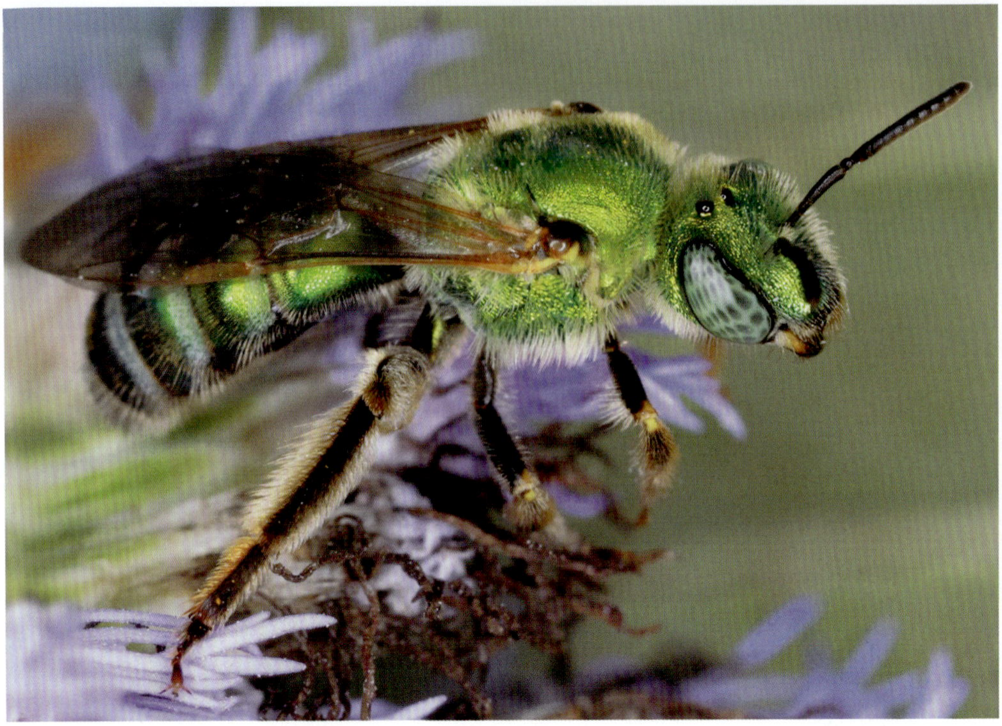

A female *Agapostemon texanus*

Range: The Texas agapostemon occurs throughout the United States and into Canada. It ranges southward from Texas through Mexico and Central America to Panama. *Agapostemon texanus* has been documented in both Hidalgo and Cameron Counties. It emerges in the Valley in midspring and remains visible through mid-September. This species, while common in Texas generally, is relatively uncommon in the Valley. It is most likely to appear in areas where sunflowers or ornamental composite flowers are cultivated rather than in settings dominated by native flora.

Associated plants: *Agapostemon texanus* is an instrumental pollinator of commercial crops, including apples, melons, beets, caneberries, and strawberries. The Texas agapostemon is known in particular for its prowess as a sunflower pollinator. In the Valley, it can sometimes be found on wild sunflowers (*Helianthus annuus*) and native composites such as cowpen daisy (*Verbesina encelioides*) and Mexican hat (*Ratibida columnifera*). In May 2021, *Agapostemon texanus* males and females appeared at Quinta Mazatlán (Hidalgo Co.), visiting *Salvia ballotiflora*. In mid-November 2018 and 2021, males and females were found foraging on crucita (*Chromolaena odorata*) at the National Butterfly Center (Hidalgo Co.).

ID tips: *Agapostemon* females are differentiated by such factors as the color of the clypeus, the color of the hind-leg hair, and the patterns of pitting and ridges on the thorax and propodeum. The female Texas agapostemon can be recognized in part by its black clypeus rim and the dark frontal hairs on its hind-leg tibiae. The female *Agapostemon texanus*, however, is nearly identical to *A. angelicus*, also found in Texas (although it has not been recorded in the Valley); females of both species have a distinctive pattern of alternating small and large pits on the scutum. Female *Agapostemon texanus* and *A. angelicus* are best differentiated by their association with more easily identified males. Male *Agapostemon* species are often told apart by the yellow-and-dark markings on their sterna, upper abdomens, and legs. The markings on the tibiae and femurs of the male *Agapostemon texanus*, as shown here, distinguish it from the very similar *A. angelicus*.

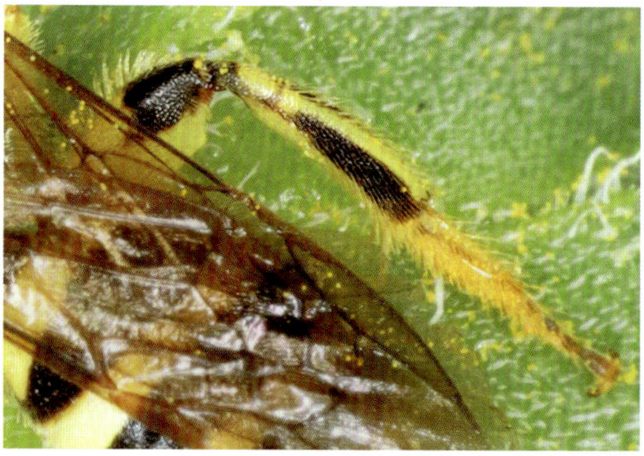
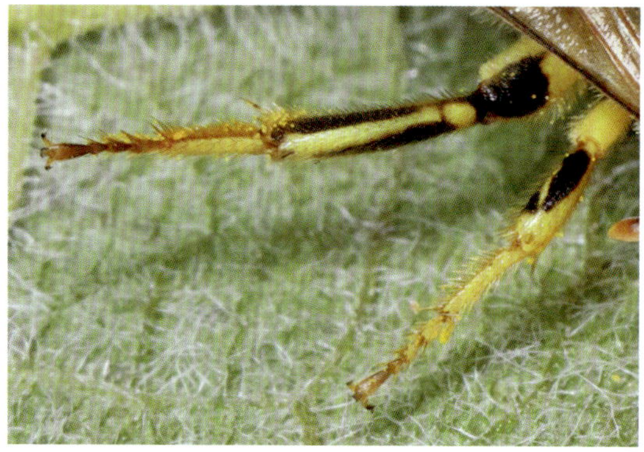

The apical interior hind femur of the male *Agapostemon texanus* is black. A black stripe also runs from the top to the bottom of the hind tibia, on both its inner and outer surfaces. The very similar male *Agapostemon angelicus* lacks a dark stripe on the outer surface of the tibia. Male *Agapostemon* typically have long tibiae.

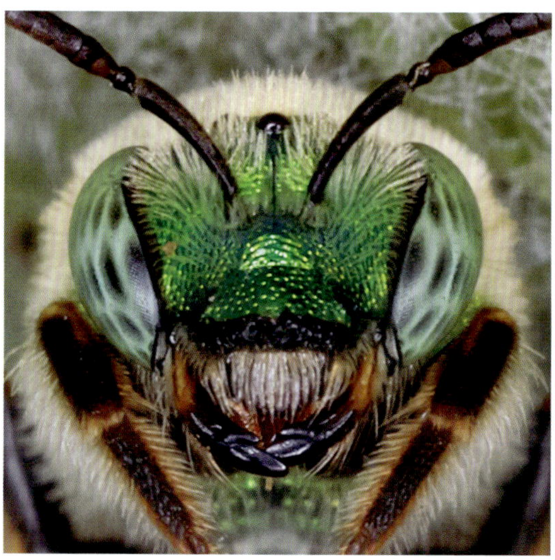
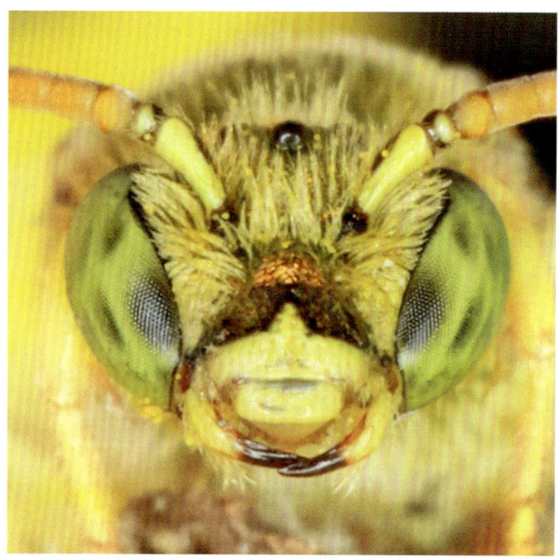

The ornate faces of female (*left*) and male (*right*) *Agapostemon texanus*: the female's clypeus is metallic green with a broad black strip on the bottom edge. Its mandibles are reddish at the base and otherwise black, and the antennae are black. The male's clypeus is coppery metallic at the top, with a black middle section and a broad yellow band with an upper median point on the apical rim. The male's labrum is yellow and its mandibles yellow with black tips.

Behavior: *Agapostemon texanus* is a solitary bee that builds and provisions its own nest without engaging in communal behavior. Roberts noted in 1973 that *Agapostemon texanus* females typically do not form large aggregations, unlike females of many other *Agapostemon* species. He believed that the two behavioral habits are connected—*Agapostemon* that avoid gathering in sizable aggregations also avoid cooperative behavior such as sharing nest entrances. He discovered, however, that female Texas agapostemon sometimes share underground cells known as hibernacula when overwintering.

The typical nest of *Agapostemon texanus* is a cratered cone with a nest entrance hole in the middle. The nest consists of a deep vertical burrow with multiple lateral burrows connecting to it, each containing one egg and sealed off with dirt. During the day, the female Texas agapostemon leaves the entrance hole open as she forages; at night, the bee blocks the hole with soil pushed up from the tunnel below—this practice is common to *Agapostemon* sweat bees generally. *Agapostemon texanus* nests can be difficult to discover. When founding nests, Texas agapostemon females are drawn to shady areas, and they tend to conceal their nest entrances under pebbles or leaves or within preexisting cracks and crevices.

Roberts described *Agapostemon texanus* as producing two seasonal generations each year. First, a fall-spring generation of fertilized females overwinters in the soil. These females emerge in early spring to gather pollen, provision their nests and lay eggs, and then die off before summer. Thereafter, a summer generation of female bees emerges, constructs nests, and lays unfertilized eggs that develop into males. At the end of summer, the final fall generation of bees emerges. Most are males, which sleep outside on flowers; during the day, they waylay females visiting the flowers and mate. Inseminated females then overwinter to begin the next year's cycle.

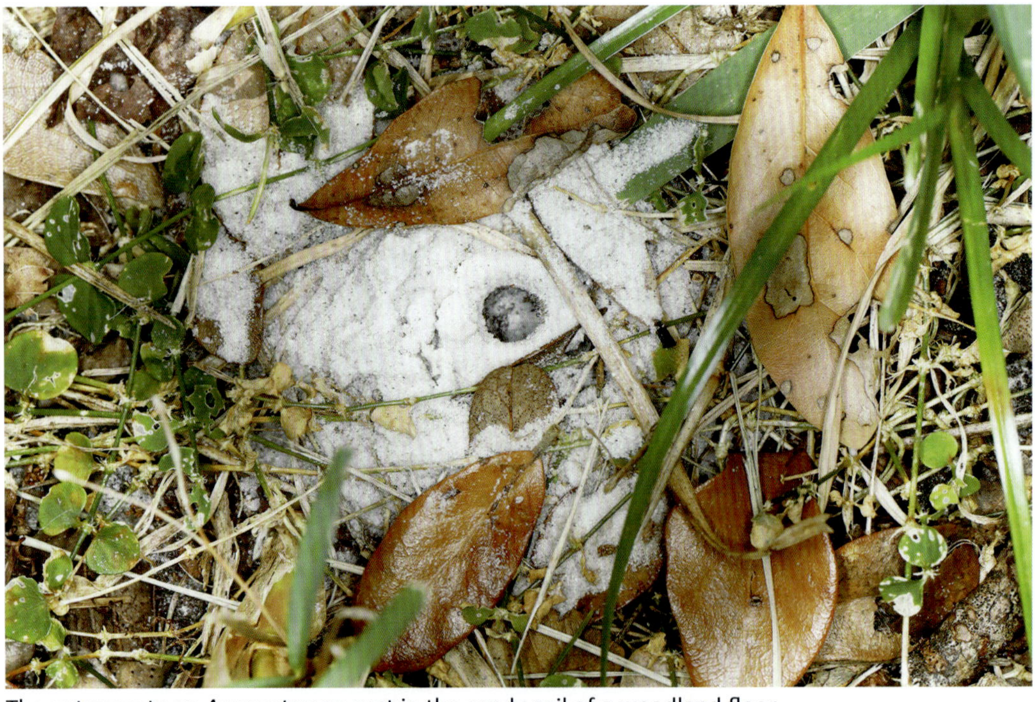

The entrance to an *Agapostemon* nest in the sandy soil of a woodland floor

Tyler's agapostemon ◆ *Agapostemon (Agapostemon) tyleri* Cockerell, 1917

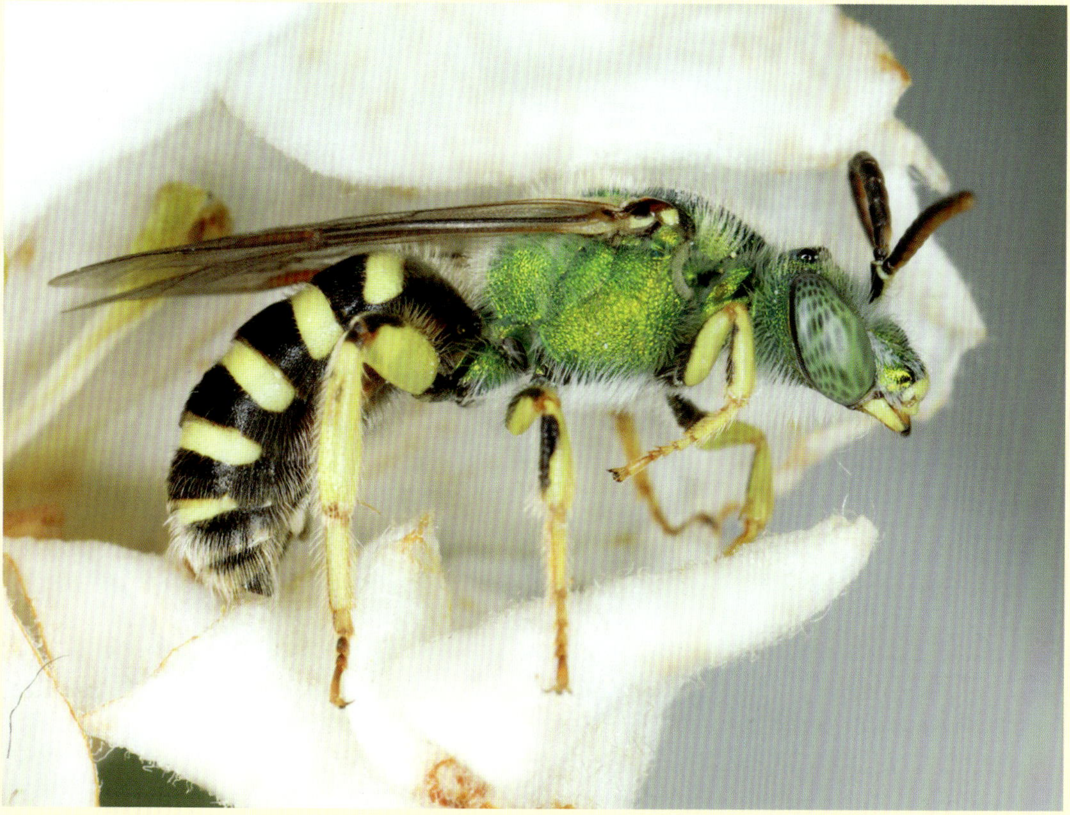

A male *Agapostemon tyleri* inside a wild olive blossom

Description: Both the male and female Tyler's agapostemon have iridescent green heads and thoraxes and banded black abdomens. On females, the abdominal bands are white and formed by short white hairs, and on males, they are pale yellow and part of the integument. Females' legs are dark and covered with pale hairs; males' legs are predominantly yellow, with some black markings on the upper segments.

The female's face has a wide black band with a rounded median point rimming the bottom edge of the clypeus, and dark mandibles with black tips. The male's clypeus is rimmed with a yellow band with a sharp median point; its labrum is also yellow, and its mandibles mostly yellow with black tips. Both male and female *Agapostemon tyleri* have pale (transparent to light brown) wings and dark tegulae. **Size:** females 9–10 mm (1/3–2/5 in.), males 8–9 mm (3/10–1/3 in.).

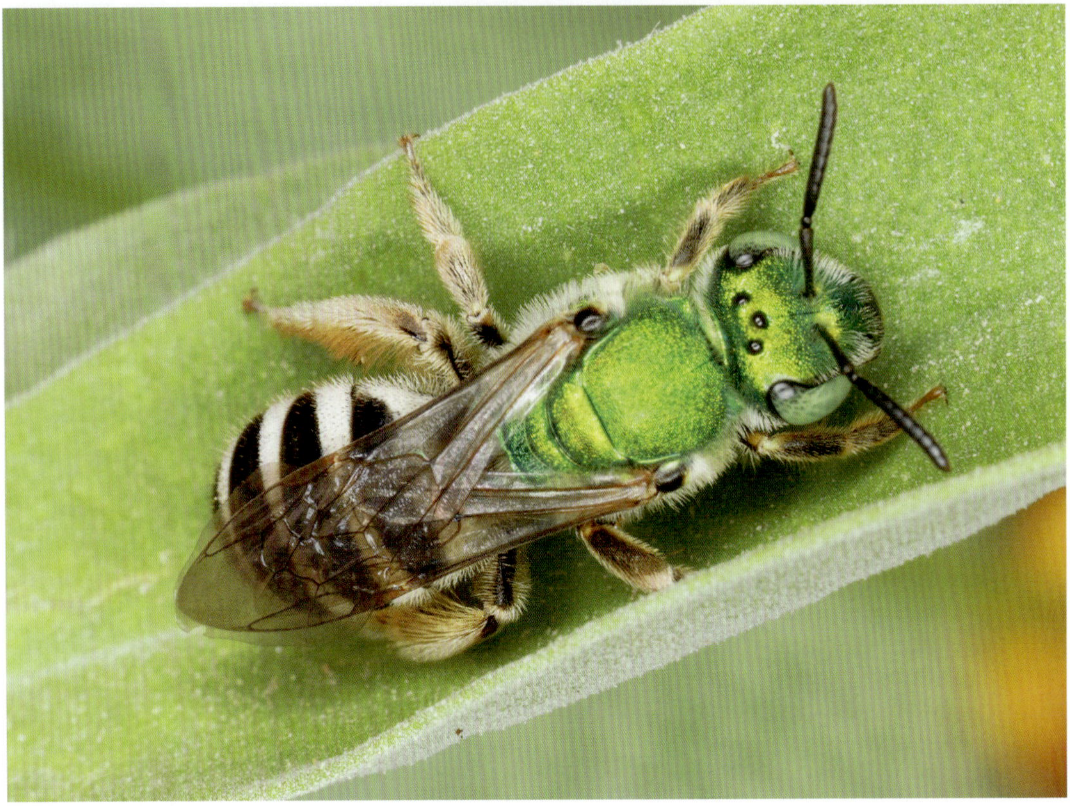

A female *Agapostemon tyleri*

Range: In the United States, *Agapostemon tyleri* is an uncommon species with a restricted range. It is found principally in deserts of southern Arizona, New Mexico, and West Texas, and in high-altitude regions of Mexico, where it ranges from the border as far south as Puebla. *Agapostemon tyleri* appeared in June 2021 at Campos Viejos Ranch (Starr Co.) in an area of Tamaulipan thornscrub with sandy soil.

Associated plants: Tyler's agapostemon is a generalist pollinator known to visit flowers as diverse as prickly pear, creosote bush, and Asteraceae. The female bees shown here were found collecting pollen from cenizo (*Leucophyllum frutescens*). Males were observed in large numbers drinking nectar from wild olive (*Cordia boissieri*). Several female bees were also discovered visiting candy barrel cactus (*Ferocactus wislizeni*).

ID tips: The male *Agapostemon tyleri* can be distinguished from all other regional *Agapostemon* by these two traits: its sternum is entirely black, without any yellow areas; and the first yellow band on its abdomen is interrupted. Traits that help differentiate the female *Agapostemon tyleri* from other South Texas *Agapostemon* include its black-and-white-striped abdomen, the lack of yellow on its clypeus, its size, its dark tegulae, and its light-colored, transparent wings.

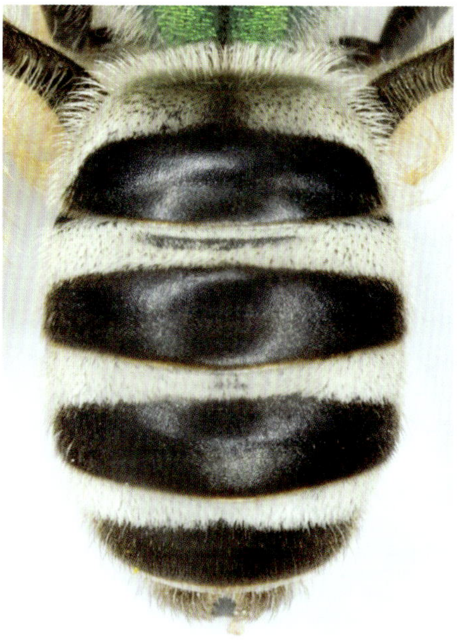

Abdomen of female *Agapostemon tyleri*

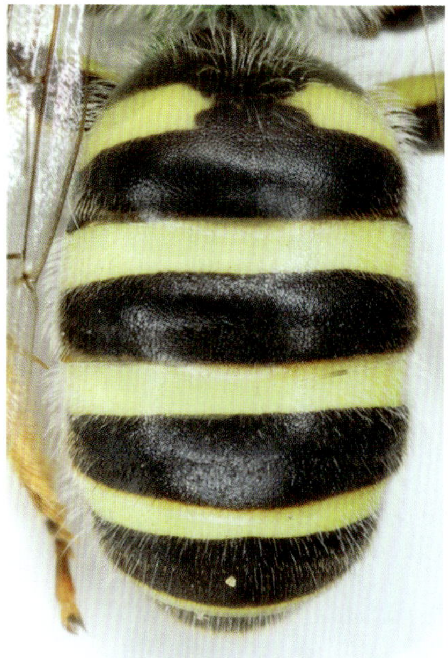

Abdomen of male *A. tyleri*

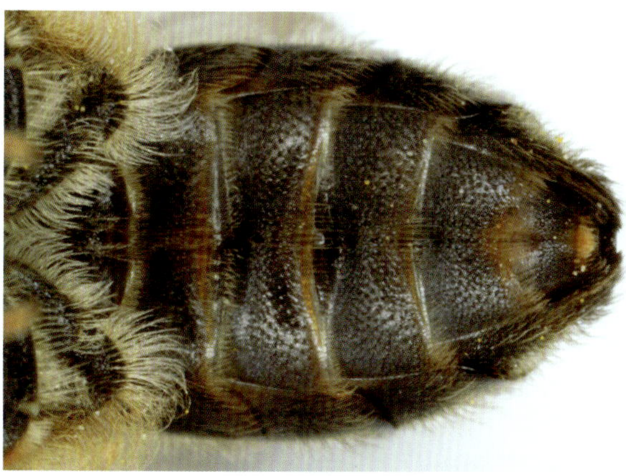

Sternum of male *Agapostemon tyleri*

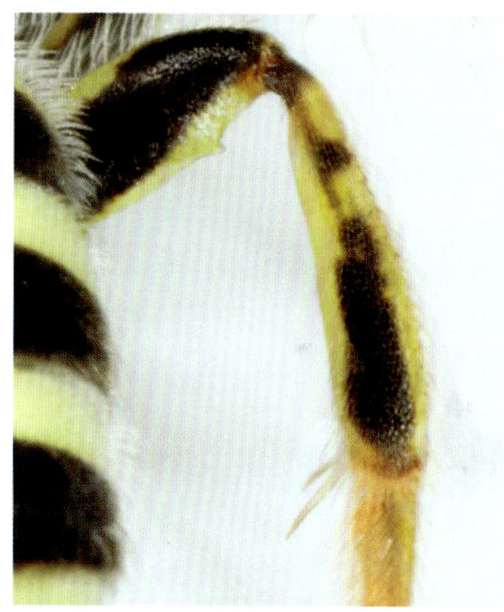

Hind femur and tibia of male *A. tyleri*

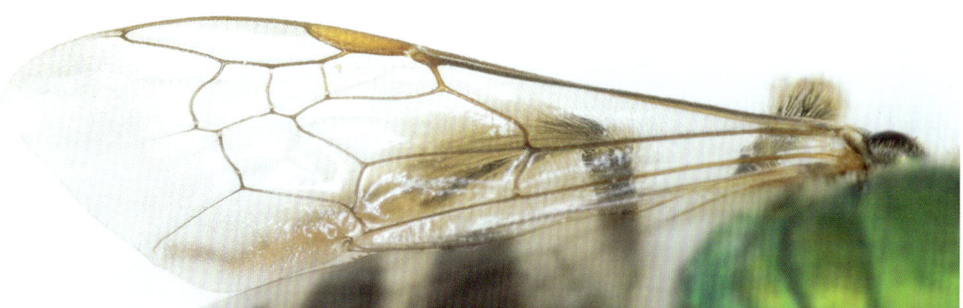

Forewing of a female *Agapostemon tyleri*

Behavior: In the Valley, male Tyler's agapostemon emerge in spring before females, often in large groups. In June 2021 at Campos Viejos Ranch, for example, hundreds of male *Agapostemon tyleri* were found together on a single wild olive tree (*Cordia boissieri*), a plant that, in the Valley, attracts relatively few bee species. Within days, females appeared individually on other nearby shrubs.

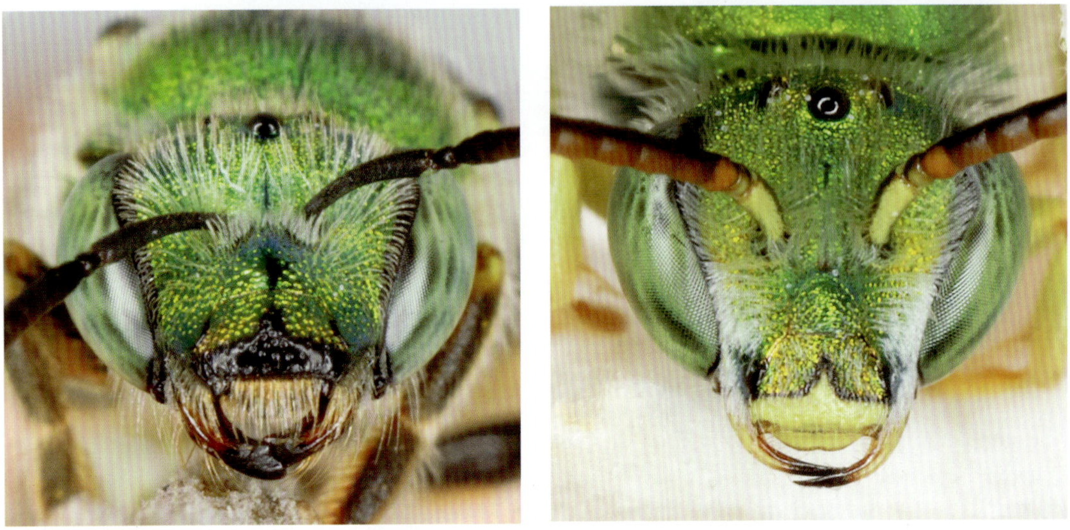

Faces of female (*left*) and male (*right*) *Agapostemon tyleri*

Honey-tailed agapostemon ◆ *Agapostemon (Agapostemon) melliventris* Cresson, 1874

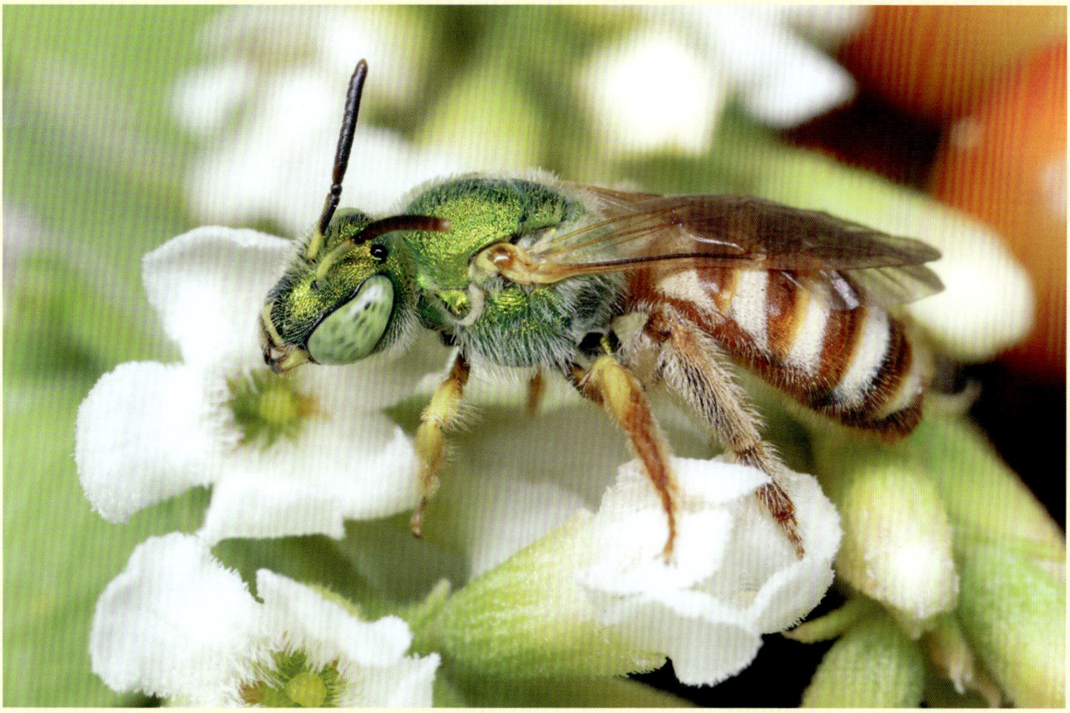

A female *Agapostemon melliventris*

Description: The male and female *Agapostemon melliventris* both have brilliant iridescent green heads and thoraxes and striped abdomens. The female's abdomen is usually honey colored (or occasionally black) and striped with bands of white hair. The male's abdomen is yellow banded with black and honey-colored bands, which are part of the integument of the abdomen and not formed by hairs.

Both the male and female honey-tailed agapostemon have yellow facial markings. The female's metallic-green clypeus is bordered on its bottom rim with a wide yellow strip and a narrow brown strip; its labrum is dark brown and obscured by long pale hairs. Its antennae are predominantly black, but the scapes are yellow on the front surfaces. Males have iridescent clypei bordered on the bottom edge by a yellow strip; yellow labrums; and antennae with yellow scapes, partly yellow pedicels, and flagella that are golden on the front surfaces and black on the hind surfaces.

The male's legs are mostly yellow. The hind legs of females are reddish brown; the middle legs' lower tibiae, basitarsi, and lower tarsi are brown. The female's front tibiae and basitarsi are yellow, and most of each middle tibia is also yellow. **Size:** females 8–9 mm (3/10–1/3 in.), males 7–8 mm (1/4–3/10 in.).

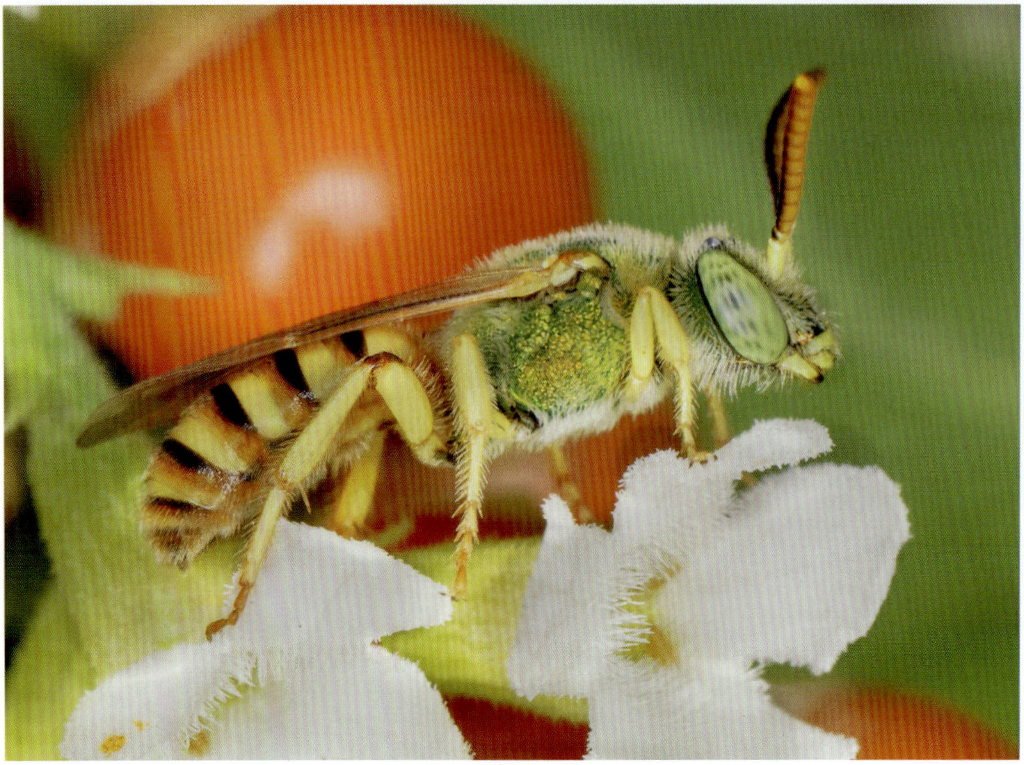

A male *Agapostemon melliventris*

Range: The distribution of *Agapostemon melliventris* is centered in the southwestern United States, Texas, and northern Mexico. This species, however, is found as far west as California, as far east as Kansas, and as far north as Montana, Idaho, and Oregon. It ranges southward from the Texas border to the Mexican state of Chiapas. It is the predominant *Agapostemon* species of the Lower Rio Grande Valley and appears in all three of its border counties.

Associated plants: The honey-tailed agapostemon is a pollinator generalist. It flies from May through November in the Valley, feeding on diverse native plants, among them Berlandier's fiddlewood (*Citharexylum berlandieri*), cenizo (*Leucophyllum frutescens*), crucita (*Chromolaena odorata*), guayacan (*Guaiacum angustifolium*), golden dewdrops (*Duranta erecta*), whitebrush (*Aloysia gratissima*), wild petunia (*Ruellia nudiflora* var. *runyonii*), seaside goldenrod (*Solidago sempervirens*), and the wild gourd known in the Valley as chilicote (*Cucurbita foetidissima*).

ID tips: The striped abdomen and partly yellow face of the female honey-tailed agapostemon readily identify it in the Valley. Males can usually be recognized by their honey-brown stripes. Two other minute traits set them apart from somewhat similar males elsewhere in Texas and in nearby southwestern states: the front face of the first abdominal segment (T1) of the male *Agapostemon melliventris* is yellow or honey colored rather than dark; and the hind leg is slender and almost entirely yellow—any dark spots on the hind legs are confined to the tip of the femur and base of the tibia.

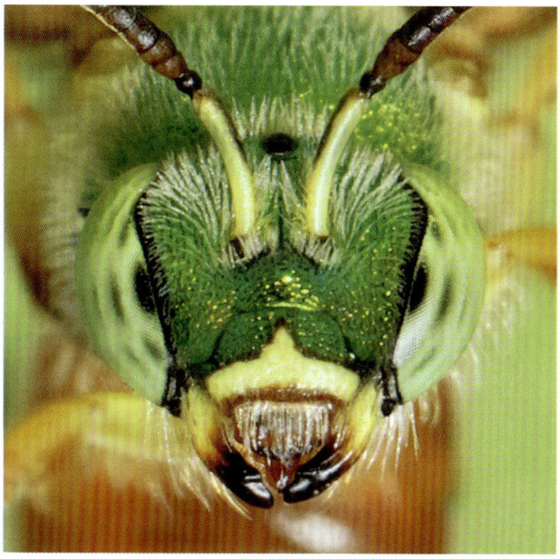
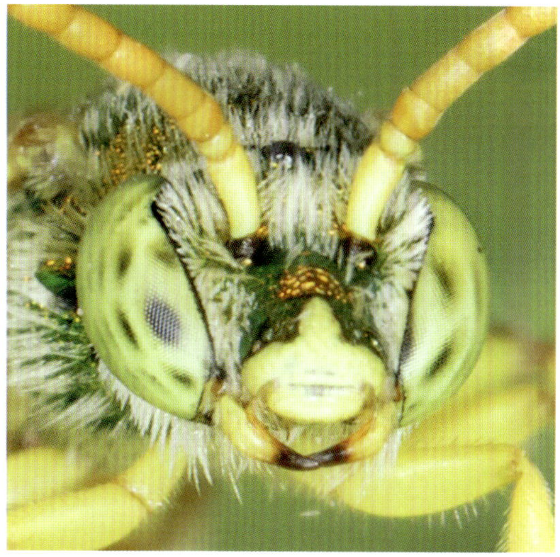

Faces of female (*left*) and male (*right*) *Agapostemon melliventris*

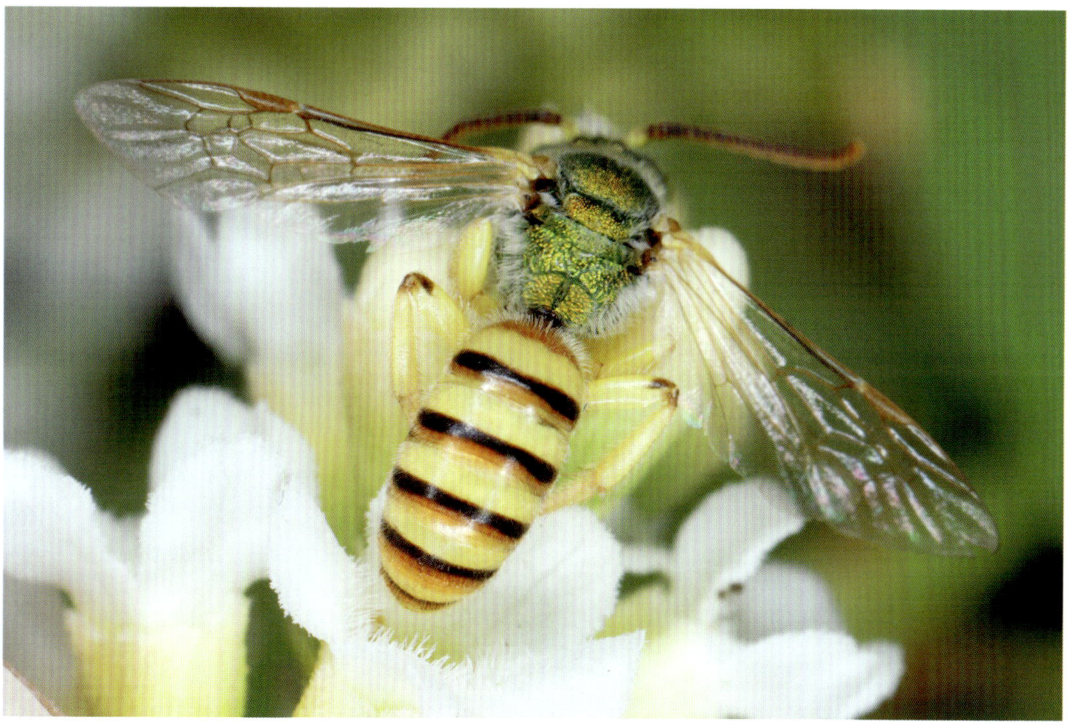

Dorsal view of a male honey-tailed agapostemon: note the slender, mostly yellow hind legs. The dark spots on the hind legs are confined to the junctures of the femurs and tibiae. The front face of the first segment of the abdomen is honey colored.

Behavior: Within the Lower Rio Grande Valley, *Agapostemon melliventris* emerges in early spring, vanishes in late summer, and reemerges in August, a pattern suggesting that the species is multivoltine (produces two or more generations per year). Linsley reported that *Agapostemon melliventris* females emerged before sunrise to pollinate flowers. In 1960, he observed female *Agapostemon melliventris* in the Valley gathering pollen from *Datura* so early at dawn that he required a flashlight to see them.

Furrow Bees

Subfamily Halictinae ◆ Tribe Halictini
Genera *Halictus* and *Lasioglossum*

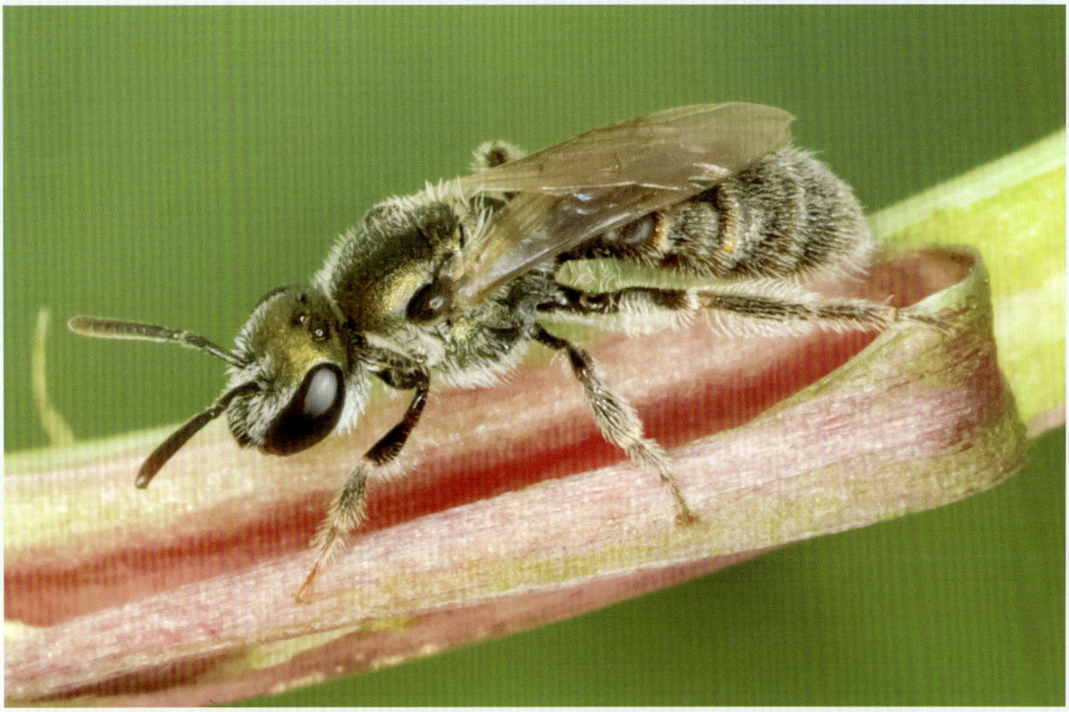

A female *Lasioglossum (Dialictus) coactum*

Halictus furrow bees are generalist pollinators represented by approximately 200 species worldwide. Ten *Halictus* species are found in the United States. All are dark brown or bronze with abdomens banded by pale hairs. *Halictus* females carry pollen on hind-leg scopal hairs. Males usually have some yellow markings on their legs and faces that aid in differentiating various *Halictus* species.[3]

Lasioglossum is a vast and varied genus that spans five subgenera and contains nearly 1,800 species worldwide and nearly 300 in the United States. Most *Lasioglossum* are tiny to small, dark brown, black, or darkly metallic bees with conspicuous or faint bands of pale hair on their abdomens. In the Lower Rio Grande Valley, the most commonly encountered *Lasioglossum* are metallic sweat bees in the subgenus *Dialictus*. Like *Halictus*, *Lasioglossum* carry pollen on hind-leg scopae.

3. Both *Halictus* and *Lasioglossum* are known by the common names "sweat bees" and "furrow bees." The term "furrow bee" has its origin in popular British and German entomological literature. The names derive from the fact that on bees of both genera, T5 is densely haired, with a bare, median, furrow-like structure.

In the United States, *Halictus* and *Lasioglossum* can sometimes be distinguished by the following two traits of the abdomens and wings: (1) on *Halictus*, the pale hair bands on the abdomens tend to hug the outer rim of each segment, while on *Lasioglossum*, the hair bands often (but not always) hug the base of each segment; and (2) on *Halictus*, the forewing veins are usually well defined, while on *Lasioglossum* the wing veins may be weakly defined, particularly on the outer edges.[4]

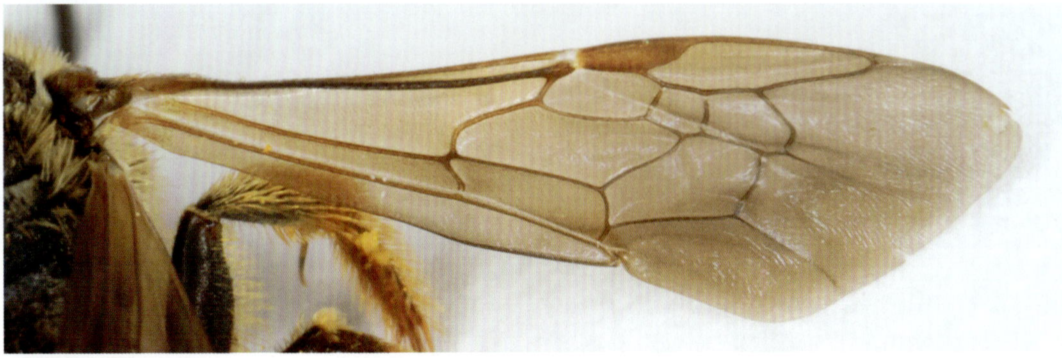

A *Halictus* forewing, with well-defined outer veins

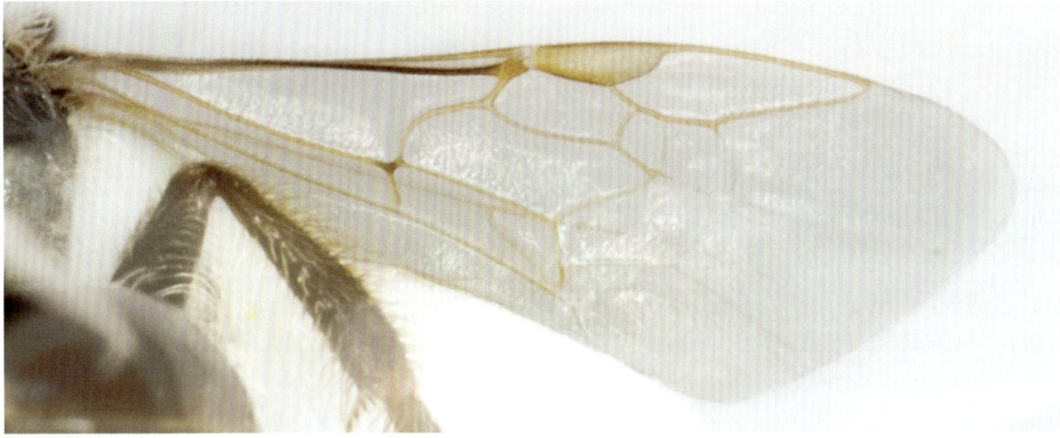

A *Lasioglossum* forewing, with weakly defined outer veins

4. These oft-cited distinctions, while useful, do not always hold true: some *Lasioglossum* have bands of hair on the rims of their terga; and some Lasioglossum—among them, many Lasioglossum (Dialictus)—lack conspicuous hair bands or have more visible wing venation than *Lasioglossum* sensu stricto.

Ligated furrow bee ◆ *Halictus (Odontolictus) ligatus* Say, 1837

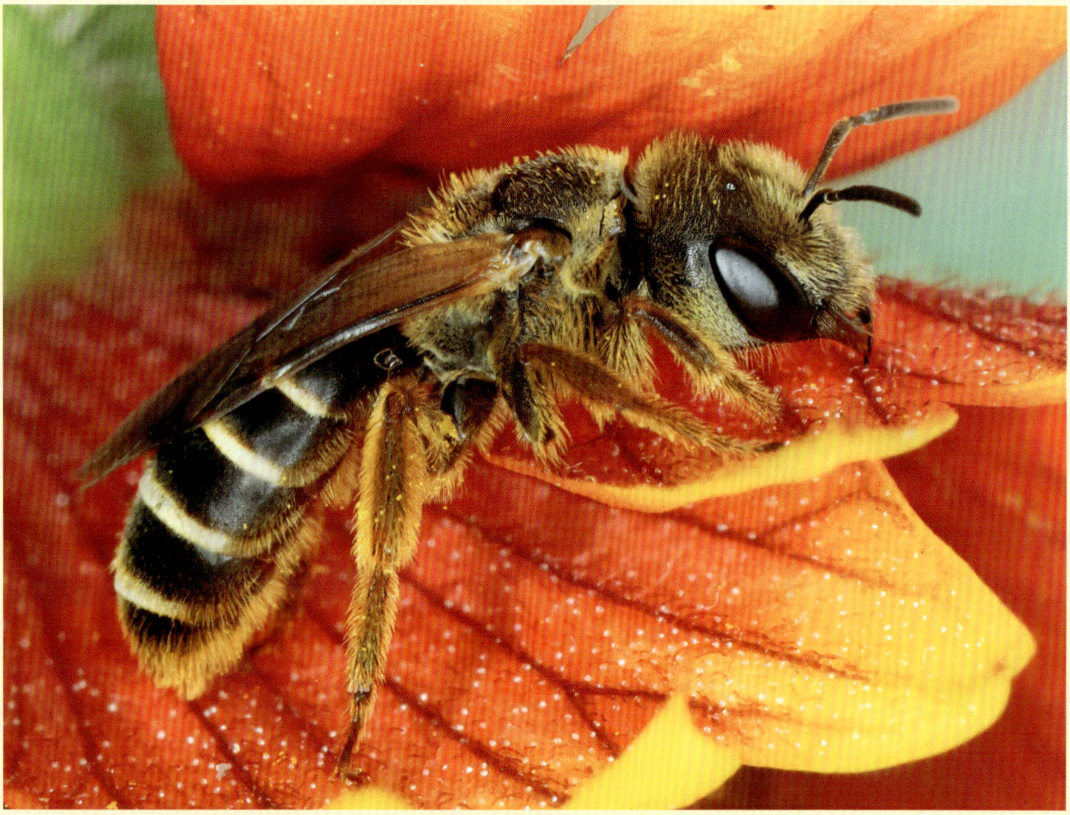

A female ligated furrow bee

Description: *Halictus (Odontolictus) ligatus* is a small, dark brown bee with broad, well-defined bands of pale hairs on its abdomen; dark wings; coppery tegulae; and dark eyes. Females differ somewhat from males. Females are robustly built, and their faces, mandibles, and antennae are dark brown. Their legs are brown and covered with fine pale and brown hairs. Male ligated furrow bees have narrower builds, and their clypei, mandibles, and legs are partly yellow. Their antennae are mostly dark, but golden brown on the front surfaces of the flagella. **Size:** females 8–10 mm (3/10–2/5 in.), males 7–9 mm (1/4–1/3 in.).

Range: Ligated furrow bees are among the most common native bees of North America—they appear seemingly everywhere, in gardens and wilderness, in urban and rural settings, and in both dry and humid habitats. They occur throughout most of the United States, from the West to the East Coast, as far south as the Texas border, and as far north as Arctic Canada. *Halictus ligatus* is found in all three border counties of the Lower Rio Grande Valley.

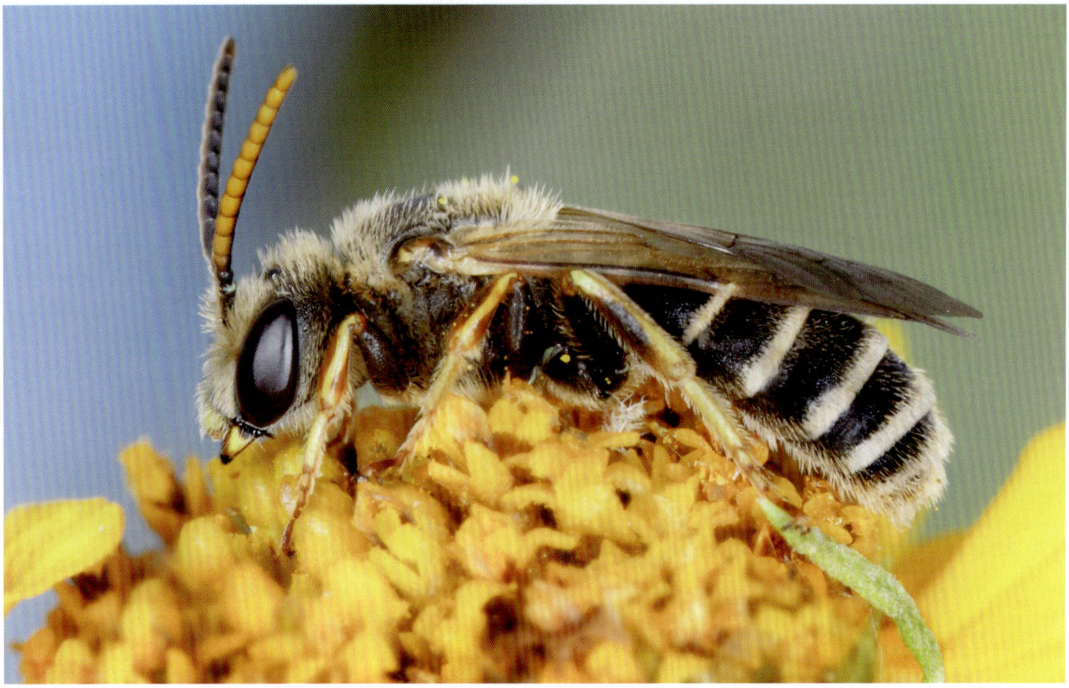

A male ligated furrow bee

Associated plants: Female *Halictus ligatus* are dependable pollinators of many crops (apples, asparagus, berries, melons, onions, and sunflowers). They visit cultivated and native flowers of more than 30 plant families.

In the Lower Rio Grande Valley, ligated furrow bees emerge in April to collect pollen from such native flora as creosote bush (*Larrea tridentata*) and honey mesquite (*Prosopis glandulosa*). They remain flying through November. From midspring to fall, they are most often encountered on regional aster-family plants, among them hoary blackfoot (*Melampodium cinereum*), blanketflower (*Gaillardia pulchella*), coat buttons (*Tridax procumbens*), common sunflower (*Helianthus annuus*), cowpen daisy (*Verbesina encelioides*), Mexican hat (*Ratibida columnifera*), plains dozedaisy (*Aphanostephus ramosissimus*), hierba del marrano (*Symphyotrichum subulatum*), resinbush (*Viguiera stenoloba*), romerillo (*Bidens alba*), and Texas thistle (*Cirsium texanum*).

ID tips: Individual female *Halictus ligatus* vary considerably in both general size and head size: larger females have disproportionately large heads, and thus two females found on the same plant may appear to be different species. All *Halictus ligatus* females, however, can be separated from other regional *Halictus* sweat bees by these traits: females' heads have wide cheeks, and their heads, thoraxes, and abdomens are a nonmetallic and nonreflective brown. Male *Halictus ligatus* can be distinguished from other Lower Rio Grande Valley sweat bees by their distinctly nonmetallic dark brown coloring, combined with the yellow markings on their clypei and mandibles.

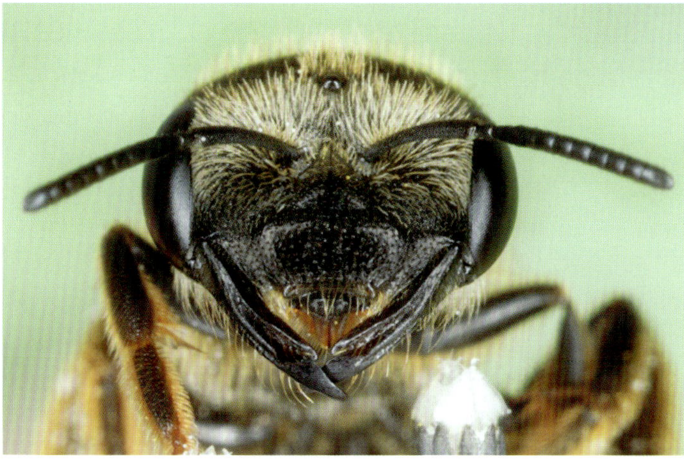

Face of a female ligated furrow bee: females have dark clypei, mandibles, labrums, and antennae.

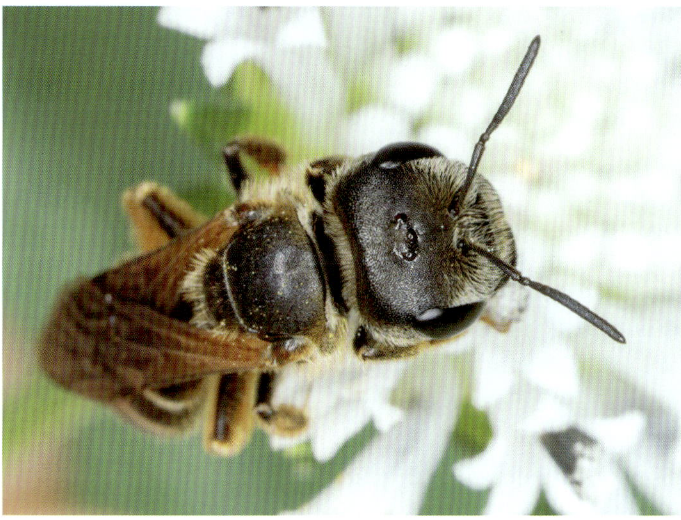

Dorsal view of a female ligated furrow bee: note the disproportionately large head.

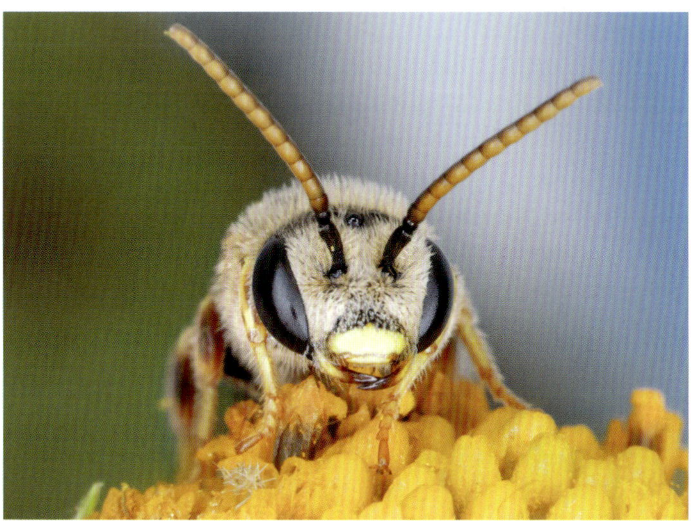

Face of a male ligated furrow bee: males have yellow markings on the clypei, mandibles, and labrums. The front surfaces of the male's flagella are yellowish brown, and the scapes and pedicels are dark colored.

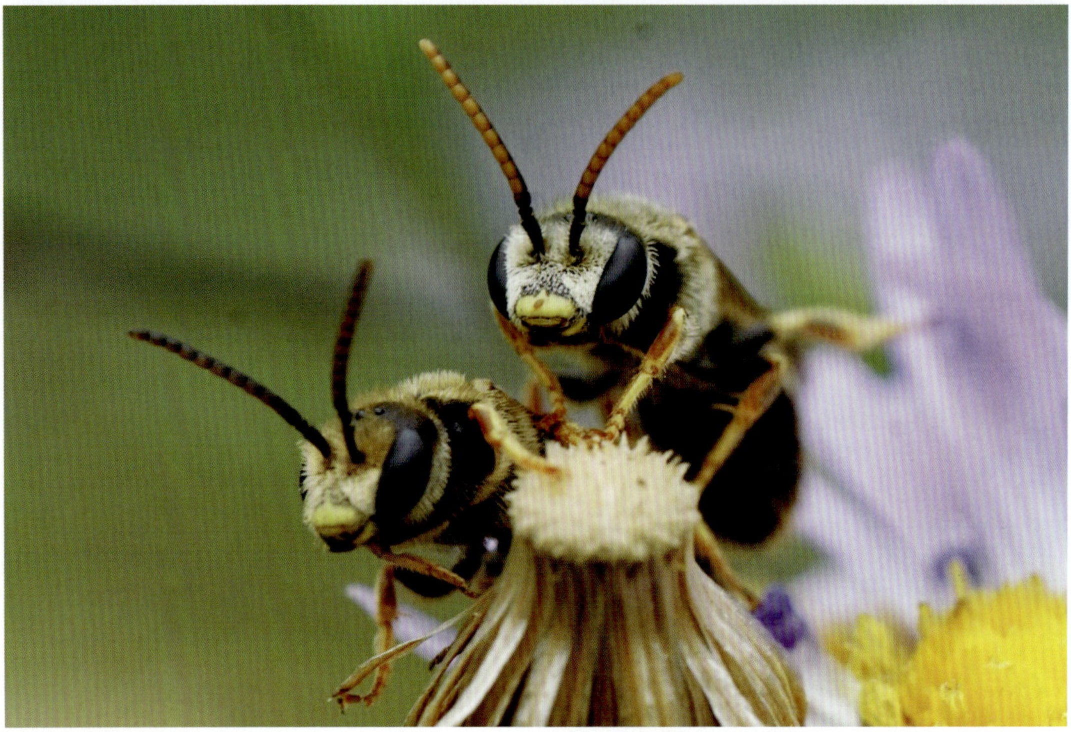

Two male *Halictus ligatus* resting on a flower stem. Photo credit Ernesto Herrera.

Behavior: Male *Halictus ligatus* sleep outside on plants, and in the Valley they can be found, from spring through fall, gathered in groups on aster-family flowers in early morning. Females engage in varied social activities and have been the subject of multiple studies.

As is true of all female bees of the genus *Halictus*, female ligated furrow bees nest in the ground, usually in loose soils. Females show remarkable behavioral flexibility, a trait that likely contributes to their ability to survive in vastly different climates and locations. Under harsh conditions, a *Halictus ligatus* female may function as a solitary nester. More commonly, *Halictus ligatus* females work together in primitively eusocial groups.

A typical social *Halictus ligatus* nest is established and organized as follows: Each spring, female foundress bees emerge to establish colonies. Working alone, or sometimes with other females, the foundress builds a ground nest and lays eggs. The eggs develop into smaller female worker bees, which gather pollen, provision the nest, and care for hatchlings. At its height, a *Halictus ligatus* colony may number as many as 200 bees. Over the course of a season, a foundress usually lays additional eggs that develop into male bees and fertile females. At the end of the season, males and infertile females die off, and fertile females overwinter in the ground to become the next year's foundresses.

Constricted metallic furrow bee ◆ *Lasioglossum (Dialictus) coactum* Cresson, 1872

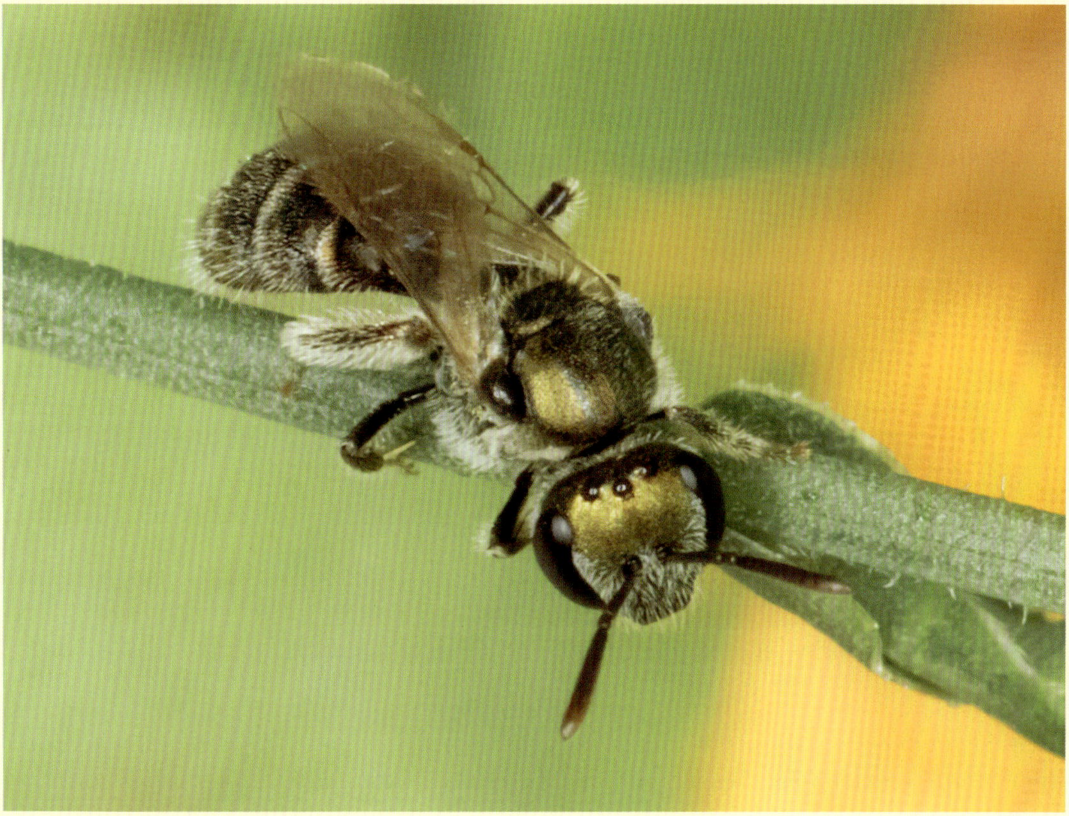

A female *Lasioglossum coactum*

Sweat bees of the subgenus *Lasioglossum (Dialictus)* are typically metallic bronze or bronze green, with dark abdomens banded by pale hairs. Some lack hair bands, and some have idiosyncratic traits that aid in their identification, such as blue-metallic iridescence or red abdomens. *Dialictus* is the largest North American subgenus of *Lasioglossum*, comprising around 200 bees in this country. There are roughly 60 species in Texas alone. In the Lower Rio Grande Valley, metallic *Dialictus* appear from April through November, visiting plants of diverse families.

Description: *Lasioglossum (Dialictus) coactum* is a small bee with a greenish-bronze head and thorax, and a dark brown abdomen that is finely pitted and roughly oval in shape. The first two abdominal segments (T1 and T2) are constricted apically. The legs are black. The tegulae are brown and the wings glassy and iridescent with honey-colored veins and stigmas. Pale hairs cover the thorax, abdomen, and legs. The head of *Lasioglossum coactum* is finely pitted and covered on the lower half with white hairs. The scutum is more coarsely pitted, and the propodeum has a wrinkled appearance. Both male and female *Lasioglossum coactum* have black mandibles with reddish tips, and antennae that are dark brown, with pale brown on the front surfaces of the flagella. **Size:** females and males 4.5–5 mm (1/6–1/5 in.).

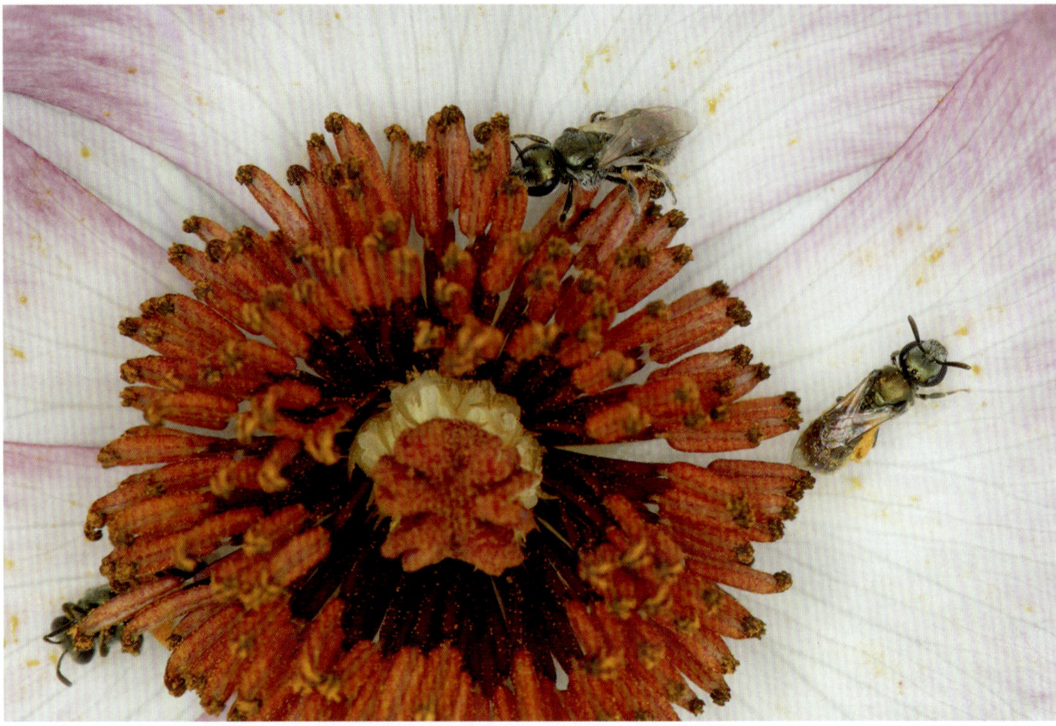

Lasioglossum (Dialictus) sweat bees collecting pollen from red prickly poppy

Range: *Lasioglossum coactum* is common in Cameron and Hidalgo Counties. It ranges generally from Central Texas through Tamaulipas. It has been found as far north as Kansas and Colorado. Cresson first identified this species in 1872 in "Hymenoptera Texana" after examining specimens collected in unspecified Texas locations.

Associated plants: The female bee shown here was found foraging in early November on wild poinsettia (*Euphorbia cyathophora*). In the Lower Rio Grande Valley, *Lasioglossum coactum* visits a gamut of flowers, among them crucita (*Chromolaena odorata*), shrubby blue salvia (*Salvia ballotiflora*), small-flower wrightwort (*Carlowrightia parviflora*), red prickly poppy (*Argemone sanguinea*), and composites such as camphor daisy (*Rayjacksonia phyllocephala*), cowpen daisy (*Verbesina encelioides*), hierba del marrano (*Symphyotrichum subulatum*), prairie broomweed (*Amphiachyris dracunculoides*), and resinbush (*Viguiera stenoloba*).

ID tips: Individual *Lasioglossum (Dialictus)* species can prove difficult to distinguish from one another, even with considerable magnification. Efforts to articulate the subtle differences among members of this subgenus have focused predominantly on species east of the Mississippi. Along the Texas-Mexico border, many *Lasioglossum (Dialictus)* species remain undescribed.

The first published record of *Lasioglossum (Dialictus)* sweat bees in the Lower Rio Grande Valley appears in Wilmatte Porter Cockerell's 1917 "Collecting Bees in Southern Texas." She discovered one new *Dialictus* species—*Lasioglossum rhodognathum*—and identified four others: *L. coactum*, *L. connexum*, *L. disparile*, and L. *semicaeruleum*. All were found in Cameron County, and all but *Lasioglossum disparile* were feeding on *Argemone* poppies. In the more than 100 years since Wilmatte Porter Cockerell's account, a handful of *Dialictus* species have been added to Lower Rio Grande Valley records. These include *Lasioglossum arenisaltans*, *L. hudsoniellum*, *L. hunteri*, and *L. illinoense*.

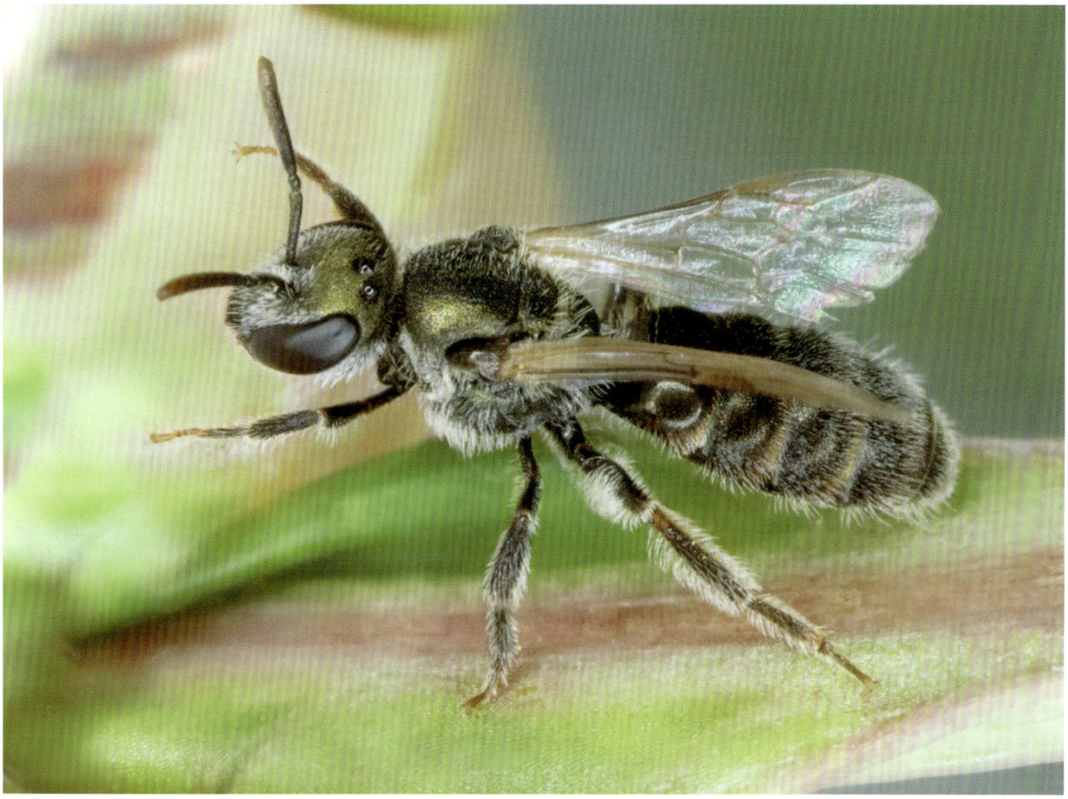

A female *Lasioglossum (Dialictus) coactum*

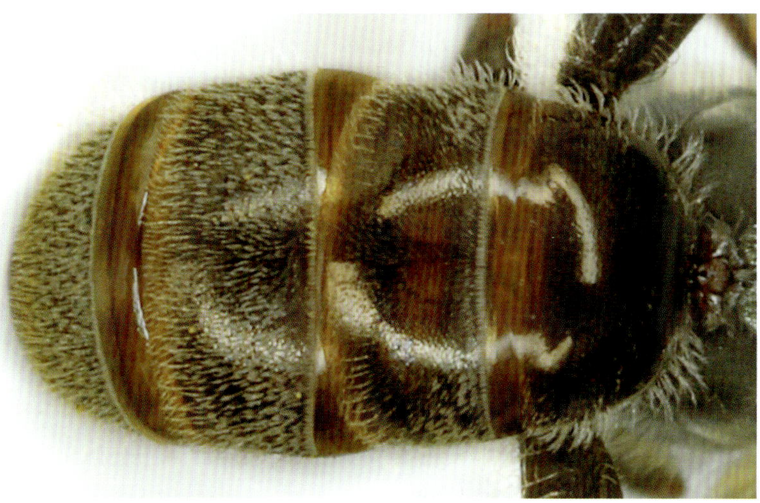

Abdomen of a female *Lasioglossum (Dialictus) coactum*

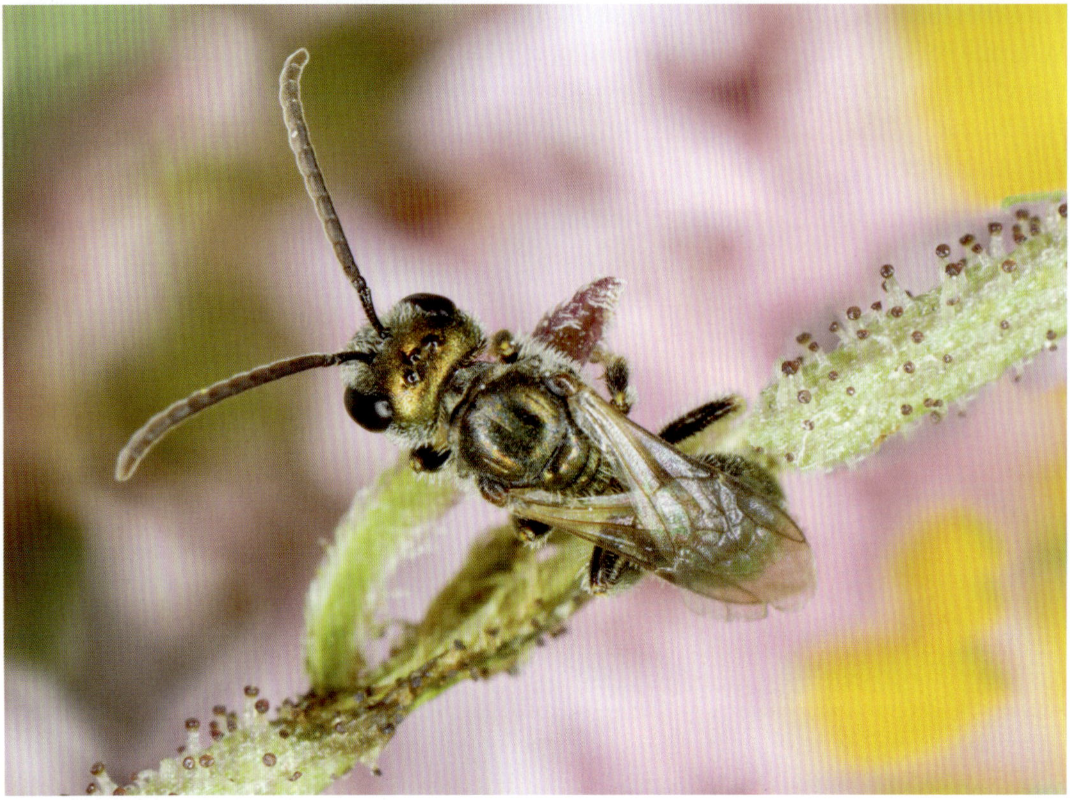

A male *Lasioglossum (Dialictus) coactum*

Identification of metallic *Lasioglossum (Dialictus)* species of the Valley often requires expert assistance, as the bees are small, and distinctions among them hairsplitting and elusive. *Lasioglossum coactum* is, however, fairly easy to identify: it can be recognized (under strong magnification) by its constricted T1 and T2; viewed from above or in profile, the segments appear as if they have been pulled in tightly with a belt. This species belongs to the *tegulare* group of *Lasioglossum (Dialictus)*, a complex of bees with modified tegulae that are relatively large, kidney shaped, and sometimes pitted. These attributes also aid in the bee's identification.

Seen only with the naked eye, *Lasioglossum coactum* is most easily confused in the Valley with *L. connexum*, *L. disparile*, and *L. rhodognathum*, which have similar bronze coloration of the head and thorax and a dark abdomen. *Lasioglossum semicaeruleum* is bluish metallic. *Lasioglossum arenisaltans*, *L. hudsoniellum*, and *L. hunteri* have predominantly red abdomens.

Behavior: *Lasioglossum (Dialictus)* sweat bees engage in a wide range of behavior, from solitary to primitively eusocial. Some, but not all, solitary varieties form large aggregations. The bees' nest structures range from simple to complex. Most *Lasioglossum (Dialictus)* nest in the ground; a minority nest in preexisting cavities created by other insects in wood.

Sphecodes Cuckoo Bees

Subfamily Halictinae ◆ Tribe Sphecodini
Genus *Sphecodes*

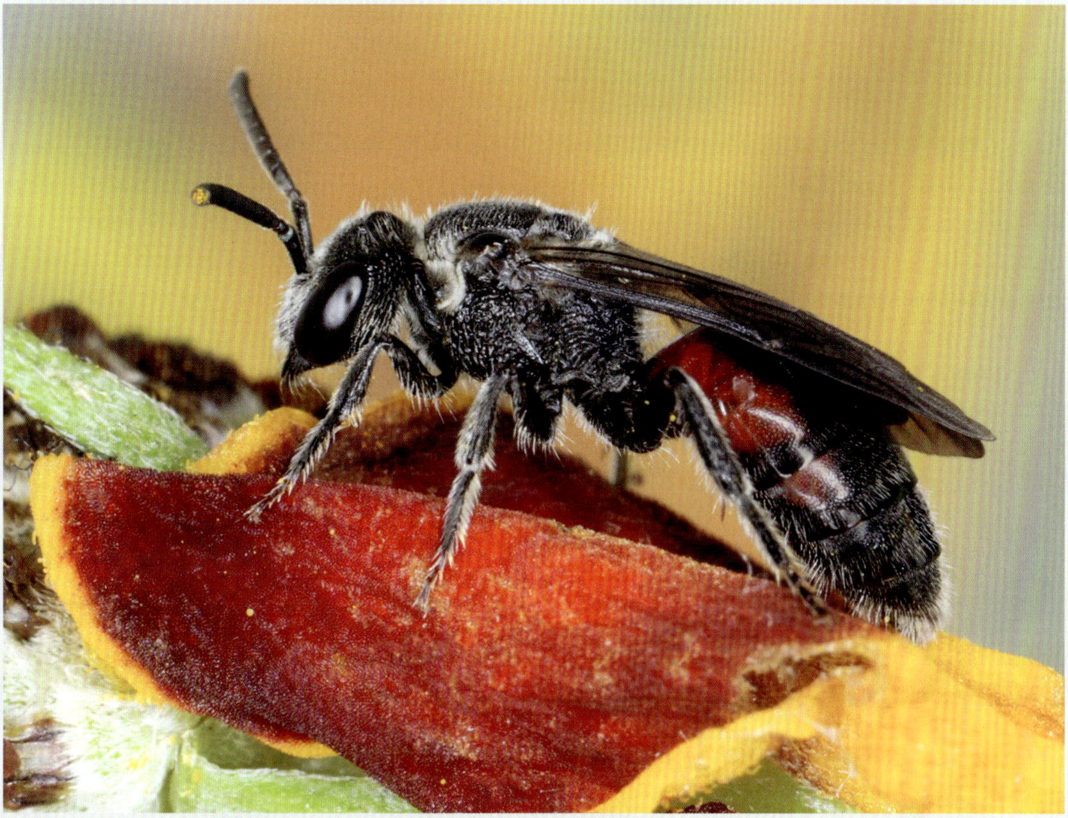

A female Hercules sphecodes

Sphecodes cuckoo bees, sometimes called blood bees, belong to the tribe Sphecodini of the sweat bee subfamily Halictinae.[5]

In the United States, *Sphecodes* are the only widespread cuckoo bees of this subfamily. They are cleptoparasites that usually invade the nests of *Halictus*, *Lasioglossum*, and Augochlorini. *Sphecodes* may parasitize the nests of other genera as well, among them *Colletes*, *Calliopsis*, *Perdita*, and *Andrena*. *Sphecodes* are known to target both solitary and social bees.

Sphecodes means "like a wasp"—like many of the cuckoo bees in this guide, *Sphecodes* tend to be slender and relatively hairless. In the United States, *Sphecodes* usually have black heads and black thoraxes, and their abdomens are either black or red, or both. Their bodies are often coarsely

5. The genus *Sphecodes* is alternatively listed in popular and scholarly literature and databases as belonging to the tribe Halictini, and the subtribe Sphecodina.

pitted. As is typical of Halictinae, the basal vein of a *Sphecodes* forewing is strongly arched. *Sphecodes* forewings have generally well-defined veins; most have three submarginal cells, although some have two.

Behavior: Most of the cuckoo bees in this guide lay eggs in their hosts' nests and depart; upon hatching, the cuckoo larvae outcompete or destroy the hosts' larvae. Adult *Sphecodes*, however, may attack host females or dispatch their eggs upon entering a nest. Occasionally, adult *Sphecodes* move into the nests of their hosts and cohabit with them.

Range: *Sphecodes* are found throughout the world. They inhabit Mexico, Canada, and every state of the United States. There are 72 recorded *Sphecodes* species in the United States and Canada, and at least a dozen have been documented in Texas. Nonetheless, *Sphecodes* are relatively uncommon in the Lower Rio Grande Valley. The Hercules sphecodes in the next entry is the single *Sphecodes* species encountered by the author in the Valley.

***Sphecodes* identification:** *Sphecodes* can be mistaken for small, dark furrow bees of the subfamily Halictinae. *Sphecodes*, however, differ from furrow bees in a few general respects: (1) *Sphecodes* lack the abdominal hair bands typical of many *Lasioglossum* and *Halictus* of the Valley; (2) as is true of other female cleptoparasitic bees, female *Sphecodes* lack scopal hairs on their legs because they do not collect pollen; they drink nectar and feed their young by pirating the provisions of other bees; and (3) *Sphecodes* males lack the pale or yellow markings on the clypeus typical of many male furrow bees.

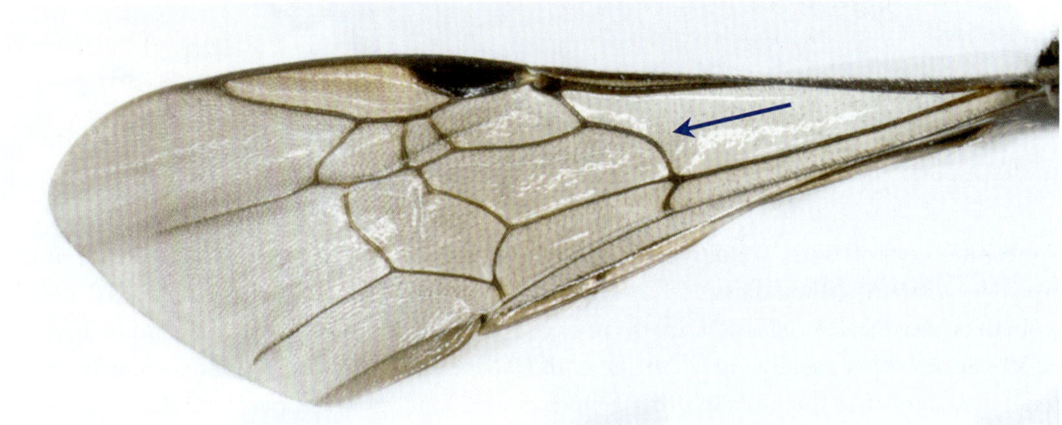

Forewing of a *Sphecodes*: the basal vein (blue arrow) is strongly arched.

Hercules sphecodes ◆ *Sphecodes heraclei heraclei* Robertson, 1897

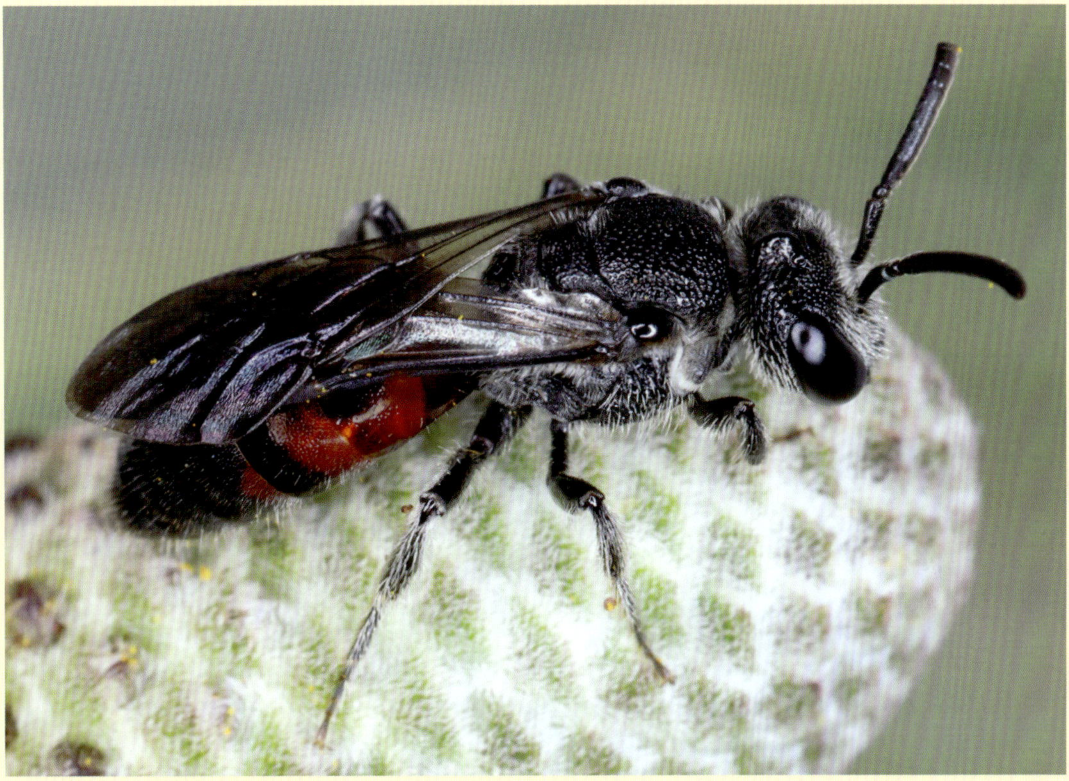

A female Hercules sphecodes

Description: The Hercules sphecodes has a heavily pitted black head and thorax. On females of the subspecies *Sphecodes heraclei heraclei* shown here, the first three segments of the abdomen (both the terga and sterna) are red or partly red, and the last three are black. Females' antennae are partly reddish, and their mandibles are reddish toward the tips, and otherwise black. On males, all parts of the head, thorax, and abdomen are entirely black. On both males and females, the vertex has a circular bump in the middle, right behind the hind ocelli. The bee's alternative common name "cyclops cuckoo bee" derives from this trait. The forewings of *Sphecodes heraclei heraclei* are somewhat dusky, with reddish-black veins and three submarginal cells. The tegulae are dark brown. The legs of both males and females are black. **Size:** females and males 7 mm (1/4 in.).

There are two recognized subspecies of the Hercules sphecodes: *Sphecodes heraclei heraclei* and *S. heraclei ignitus*, which occurs predominantly in Florida and Alabama. The female *Sphecodes heraclei ignitus* generally resembles *S. heraclei heraclei* but differs in having a mostly red thorax and a bright red clypeus. *Sphecodes heraclei ignitus* males have varying amounts of red on the thorax and abdomen.

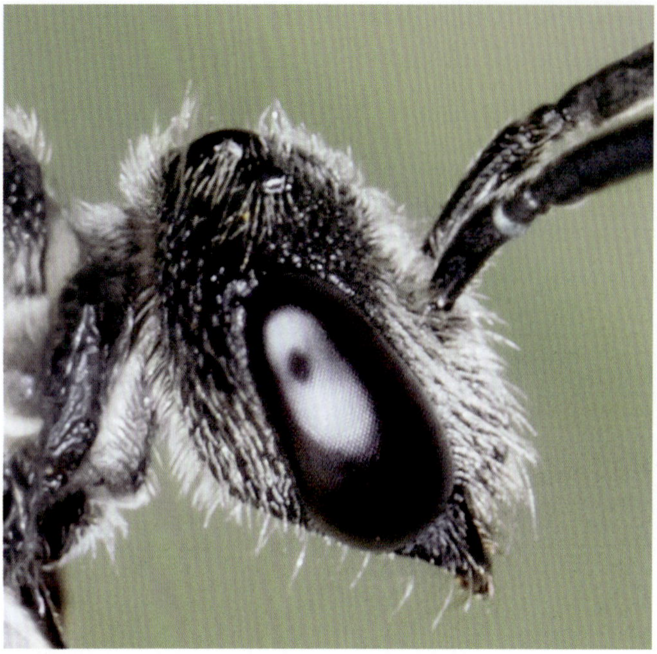

Close-up of the head of *Sphecodes heraclei heraclei,* showing the tubercle or "bump" behind the ocelli

Range: *Sphecodes heraclei heraclei* is found throughout the eastern United States from New England south to Alabama and Florida, and in the central United States as far west as Illinois, Kansas, Oklahoma, and Texas. It has been documented as far south as Chihuahua. *Sphecodes heraclei heraclei* occupies the Lower Rio Grande Valley from early spring through late fall and is most plentiful at cooler times of year. The bees shown here appeared at the National Butterfly Center (Hidalgo Co.) in November 2018, and at the McAllen Nature Center (Hidalgo Co.) in February 2020.

Associated plants: Hercules sphecodes drink nectar from plants as diverse as sumac, carrots, heather, strawberries, hydrangea, and aster-family flowers such as goldenrod and rudbeckia. In the Valley, they tend to appear on composites such as cowpen daisy (*Verbesina encelioides*), hierba del marrano (*Symphyotrichum subulatum*), Mexican hat (*Ratibida columnifera*), resinbush (*Viguiera stenoloba*), and Texas palafox (*Palafoxia texana*). *Sphecodes heraclei* parasitize the nests of halictids and tend to appear on flora where their hosts abound.

ID tips: The tubercle on the vertex is diagnostic of this species. It is nonetheless notable that *Sphecodes* identification in South Texas can be challenging: *Sphecodes* have yet to be systematically documented and cataloged in keys, and little information is available regarding *Sphecodes* of the Texas-Mexico border. Species identification often turns on minute traits such as the hairiness and sculpting of the propodeum and the first tergal segment of the abdomen.

Nomiine Bees

Subfamily Nomiinae, Tribe Nomiini
Genera *Nomia* and *Dieunomia*

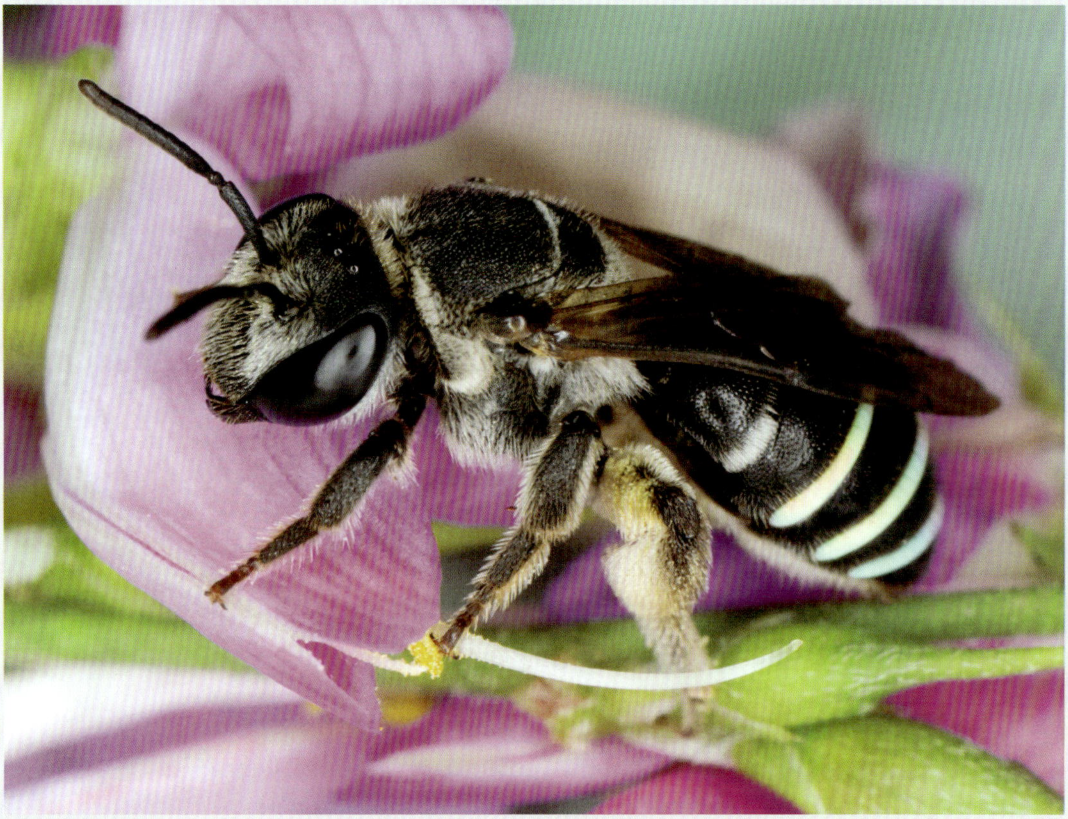

A female *Nomia*

Nomia* and *Dieunomia are two genera of ground-nesting bees in the subfamily Nomiinae. *Nomia* are found on five different continents and vary considerably in appearance. In the Western Hemisphere, they are represented by a mere handful of North American species. *Dieunomia* occur only in the New World. In the United States, *Dieunomia* appear principally in the Southwest, except for one species—*D. heteropoda*, a sunflower pollinator widespread throughout the country.

The typical Texas *Nomia* is around the size of a honey bee or slightly smaller. All *Nomia* in the United States are dark bees with abdomens girded by striking pearlescent bands. Such bands, rather than being formed by hairs, are part of the integument (exoskeleton), and their iridescence is the product of light interference caused by surface structures on the abdomen. Male *Nomia* in the United States have enlarged hind legs, used to grasp and subdue females during mating.

Dieunomia generally run larger than *Nomia* and frequently have dark wings and black or red abdomens banded by pale hairs. Both *Dieunomia* and *Nomia* have dark compound eyes that seem to meld with the clypeus, giving their faces an eerie appearance. Males of both genera have long antennae. *Dieunomia* are divided into two subgenera: *Dieunomia (Dieunomia)*, which can be as large as bumble bees (up to 23 mm or 9/10 in.); and *Dieunomia (Epinomia)*, which are relatively small.

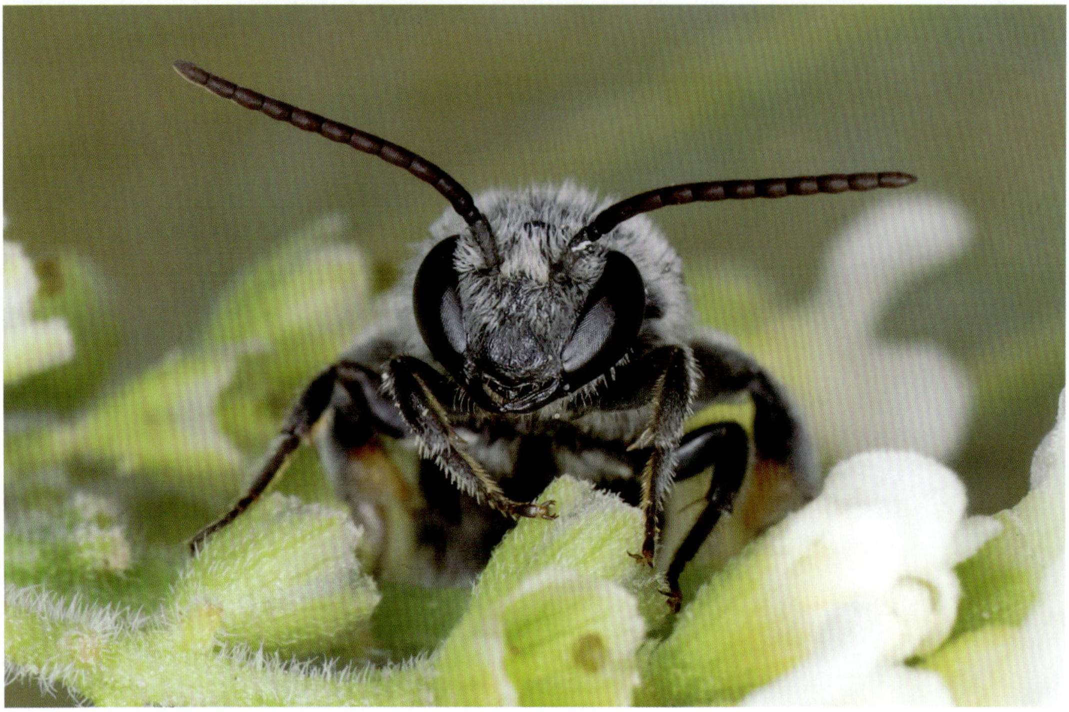

The face of a male four-banded nomia: note the way the compound eyes converge downward and touch the clypeus.

Four-banded nomia ◆ *Nomia (Acunomia) tetrazonata* Cockerell, 1910

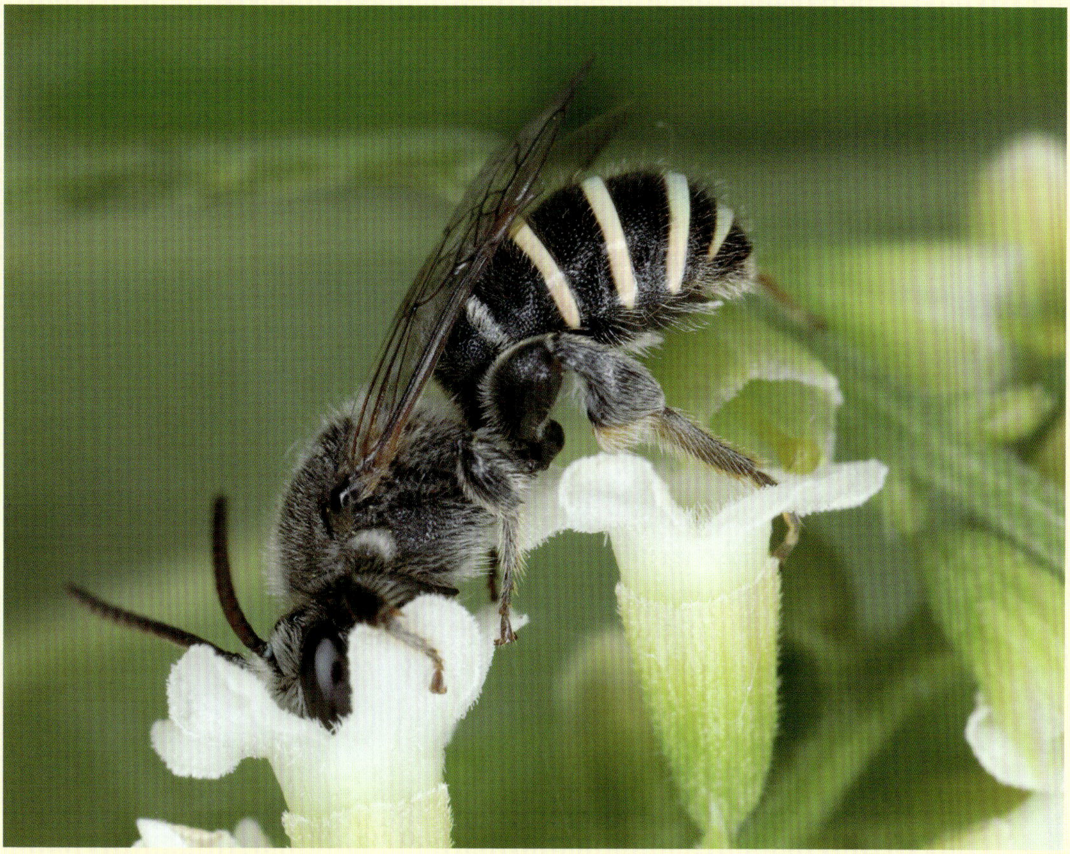

A male *Nomia tetrazonata*

Description: The four-banded nomia is also known as the pearly-banded bee. This *Nomia*'s abdominal bands iridesce in sunlight, appearing opalescent blue, yellow, or pale orange, depending on the angle of light hitting them. Only the male of this species has four opalescent bands (on T2–T5); the female has three (on T2–T4). On both males and females, T1 is rimmed with white hairs. The rest of the abdomen is dark brown. The tegulae, face, antennae, and mandibles are dark brown, and the eyes are black.

The males' forelegs and middle legs are dark brown, and their hind legs are mostly brown except for the tibiae. *Nomia tetrazonata* males have enlarged hind tibiae that are partly yellow brown and fringed with white and yellowish hairs. Females have brown legs without expanded parts; their hind-leg scopal hairs are long and pale. **Size:** females and males 9–10 mm (1/3–2/5 in.).

Range: *Nomia tetrazonata* is found in the western United States, as far north as Oregon and as far east as eastern Texas. It ranges into the northern half of Mexico. It was first identified in 1910 by Cockerell, who based his species description on four male bees collected in Arizona. This species appeared at La Puerta Tract and Campos Viejos Ranch (Starr Co.) in mid-May and mid-June 2021; at El Mesteño Ranch (northwest Hidalgo Co.) in May 2021; and at the National Butterfly Center (Hidalgo Co.) in late March 2019.

Associated plants: All North American *Nomia* are polylectic. *Nomia tetrazonata* has been documented visiting diverse flora, including, for example, salvias, mallows, legumes, asters, creosote, mesquite, and nightshade-family plants. Both the male and female bees featured here were found on Berlandier's fiddlewood (*Citharexylum berlandieri*). Males also appear in the Valley on a variety of native shrubs. During late spring 2021 at La Puerta Tract, groups of 100 or more male *Nomia tetrazonata* were found mobbing individual guayacan (*Guaiacum angustifolium*) and lotebush (*Ziziphus obtusifolia*) shrubs. Females appeared on the same plants about 10 days after the males, in smaller numbers.

ID tips: North American *Nomia* species are differentiated by such traits as the number and completeness of iridescent bands on the abdomens; the pitting on the faces, thoraxes, and terga; and traits of the males' hind legs and antennae. Males are often identified by the shape, size, and color of the enlarged areas on their hind-leg tibiae. On *Nomia tetrazonata*, a roughly triangular lobe appears above each hind-leg spur. The lobe's bottom edge is rounded on one end and pointed on the other. On male *Nomia tetrazonata*, pitting is equally coarse on T2 and T4.

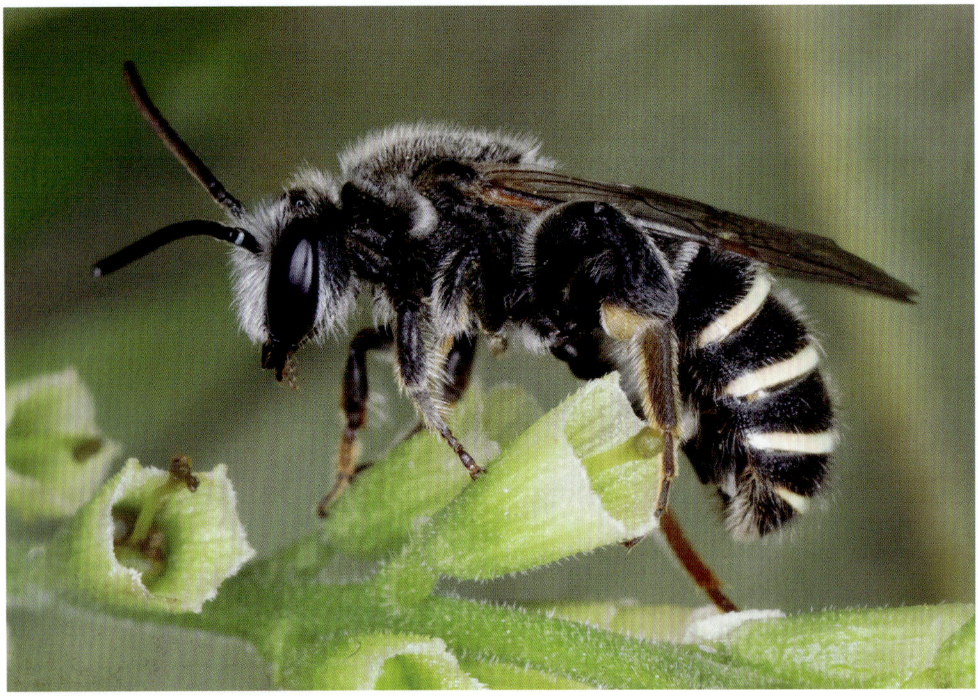

A male *Nomia tetrazonata*

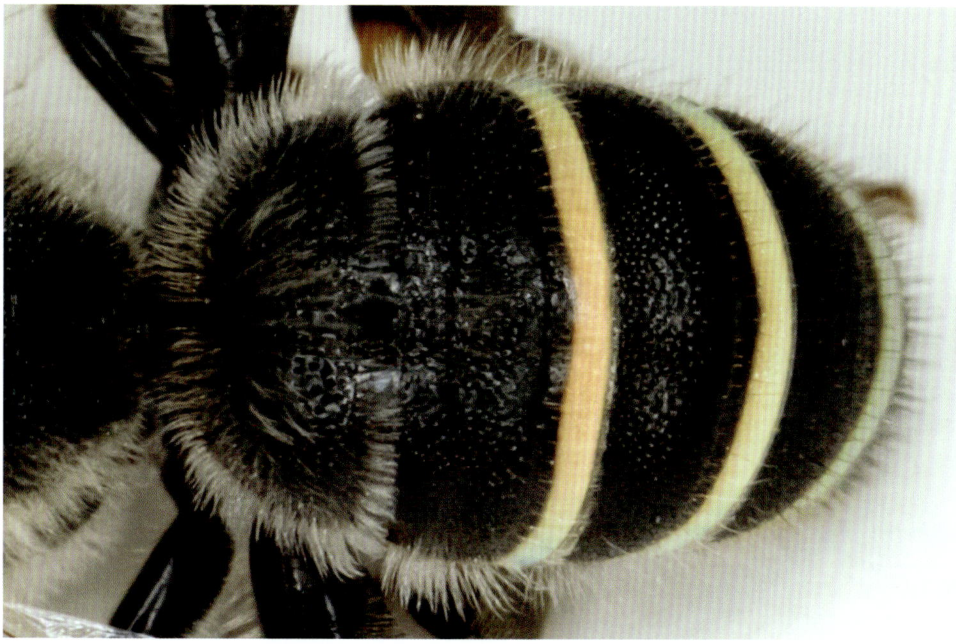

Close-up of the pitted abdomen of a male four-banded nomia: note that the first segment (T1) lacks a pearlescent band. Instead, its rear edge is rimmed by a faint interrupted strip of pale hairs.

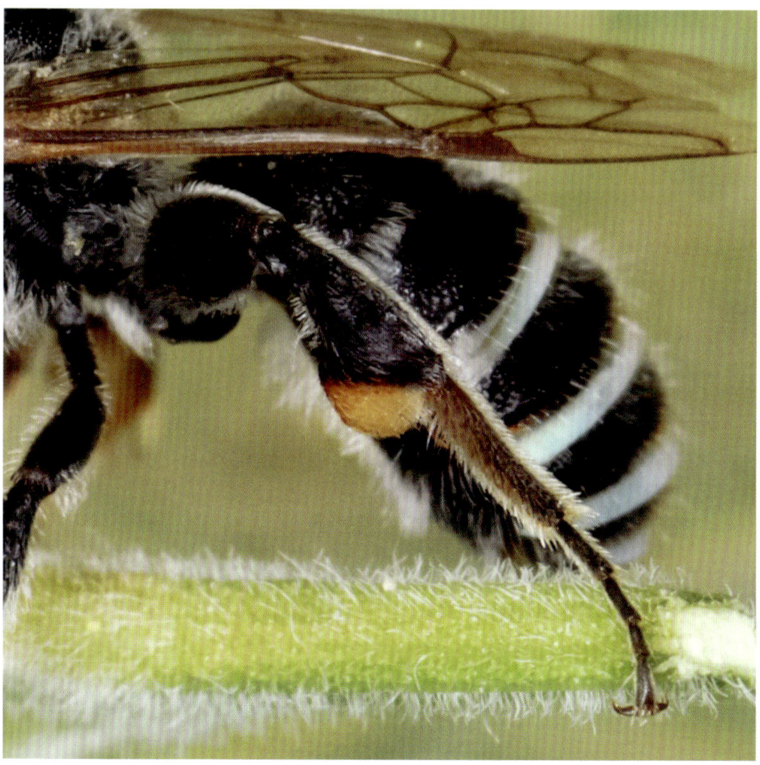

Close-up of the expanded tibia of a male four-banded nomia: the bottom of the expanded portion is light yellowish brown, and its front corner is rounded.

Behavior: Most *Nomia* are solitary and nest in the soil, usually in flat ground. *Nomia* females usually coat their egg cells with a waterproof material and construct turrets at their nest entrances. According to William T. Wcislo, author and coauthor of several studies of *Nomia tetrazonata*, four-banded nomia differ from the typical solitary bees of their genus. *Nomia tetrazonata* live communally, with up to 20 females sharing a single nest. Their nests lack turrets at the entrances and can be impressively deep, running as much as 50 cm (20 in.) beneath the soil. The cuckoo bee *Triepeolus verbesinae* has been documented parasitizing *Nomia tetrazonata* nests.

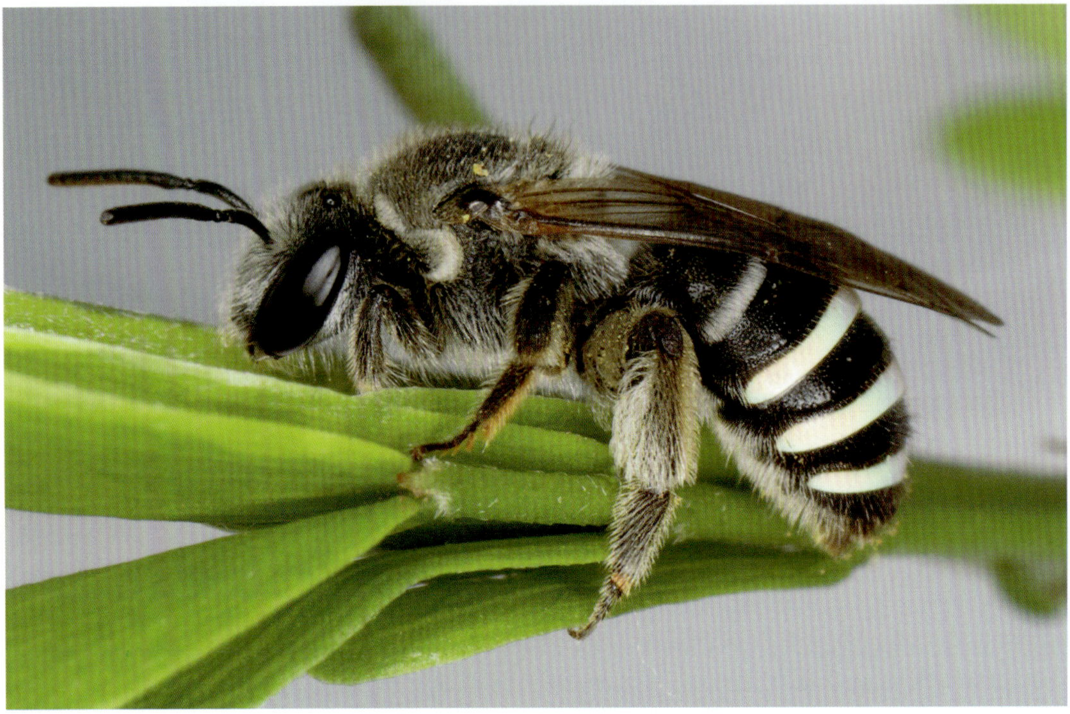

A female four-banded *Nomia*: females have three pearlescent bands on T2–T4.

Baker's dieunomia ◆ *Dieunomia (Epinomia) nevadensis bakeri* (Cockerell, 1898)

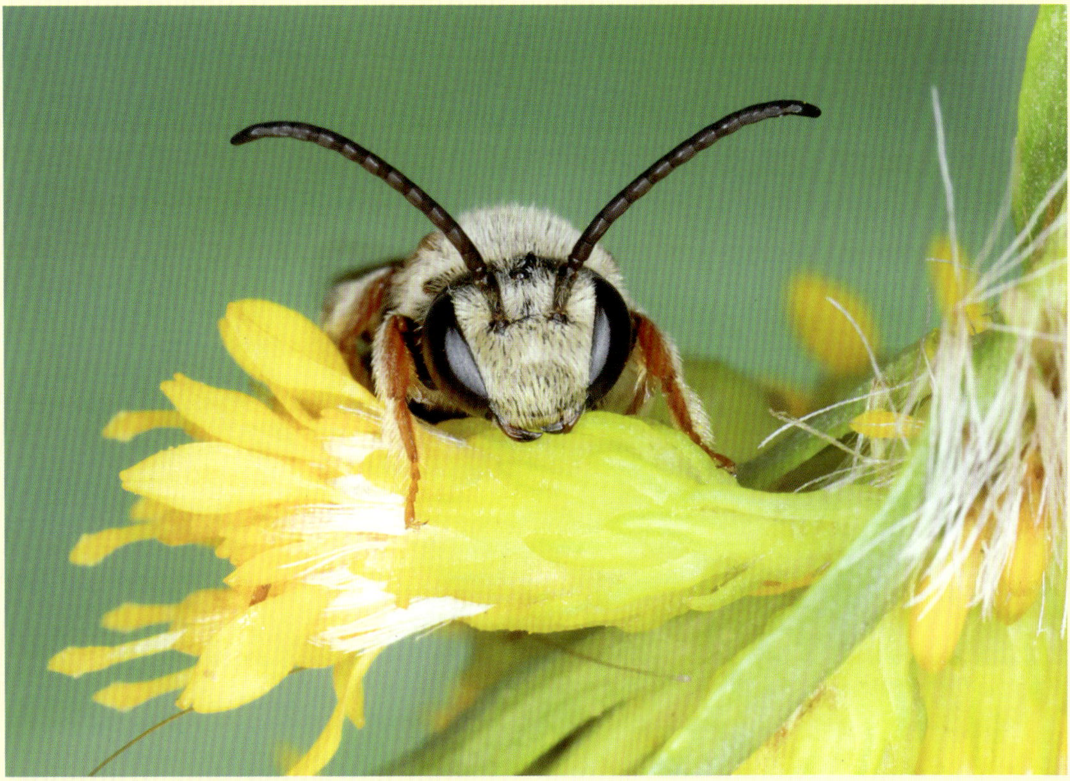

A male *Dieunomia nevadensis bakeri*

Description: *Dieunomia nevadensis bakeri* is a small black bee with red legs and tegulae, a dark abdomen banded by pale hairs, and smoky wings that darken toward the outer edges. The first segment of the abdomen is V shaped, and its front face concave and sometimes partly red. White hairs partly cover the top and sides of the thorax and clothe the legs. The female's hind-leg scopae are pale, fairly long, and dense. Males' faces are covered with thick white hair. Both males and females have red jaws with black tips. Females have mostly black antennae with red pedicels and lower scapes. Males' antennae are dark with reddish-black front surfaces. Males' hind tibiae are enlarged, with long apical spikes. **Size:** females and males 9–10 mm (1/3–2/5 in.).

Range: The female bee shown here appeared in June 2021 at Falcon State Park (Starr Co.). The male was found at the National Butterfly Center (Hidalgo Co.) in October 2019. *Dieunomia nevadensis bakeri* ranges eastward from Texas to Florida and North Carolina, westward as far as Colorado and Wyoming, and northward as far as Illinois. The subspecies *Dieunomia nevadensis arizonensis*, which has more extensive red coloration, occurs in an overlapping range in the southwestern and western United States and Mexico.

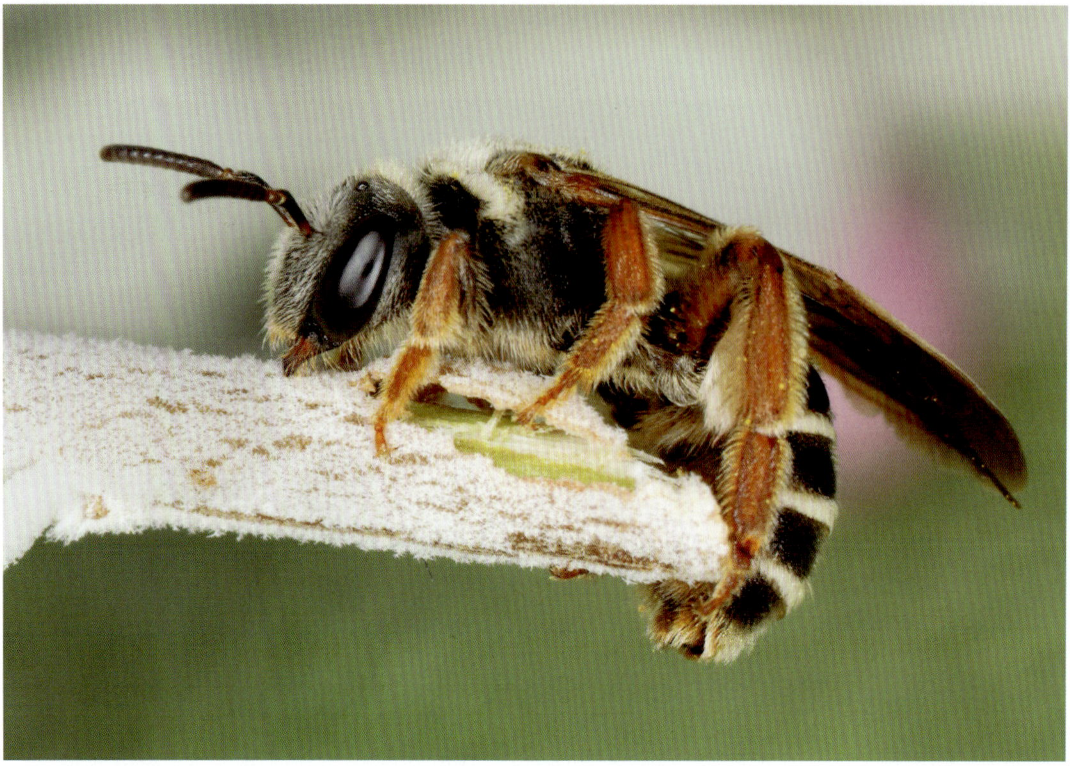

A female *Dieunomia nevadensis bakeri*, a member of the subgenus *Epinomia*

Associated plants: All North American *Dieunomia* are Asteraceae specialists. In the Valley, Baker's dieunomia tends to be found on small-flowered composites. The male bee shown here was nectaring on seaside goldenrod (*Solidago sempervirens*). The female was gathering pollen from hoary blackfoot (*Melampodium cinereum*).

ID tips: *Dieunomia nevadensis bakeri* can be differentiated from similar *Dieunomia* species in part by size—Baker's dieunomia are small even for bees of the subgenus *Epinomia*. The hind tibiae of Baker's dieunomia are also distinctive. The male's hind tibia has a bump in the middle and a spike on the lower (apical) end. The female's has an angled bump about halfway down the leg segment. In addition, like the bee shown here, Baker's dieunomia often has some red coloration, a trait somewhat singular for its genus.

Males of the two *Nomia* subgenera *Dieunomia* and *Epinomia* can be distinguished by examining the tips of the antennae—on those of the subgenus *Dieunomia*, the last antennal segment is flattened and broader at the tip than the base. In males of the subgenus *Epinomia*, such as Baker's dieunomia, the tips of the antennae are unmodified.

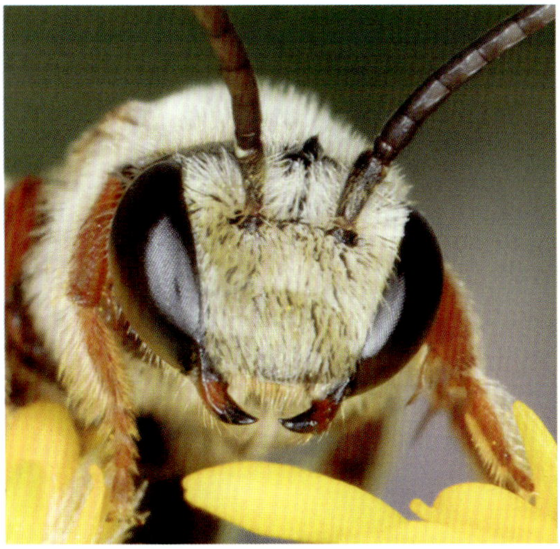

A male Baker's dieunomia: note the mandibles' inner teeth, and the thick patch of facial hair.

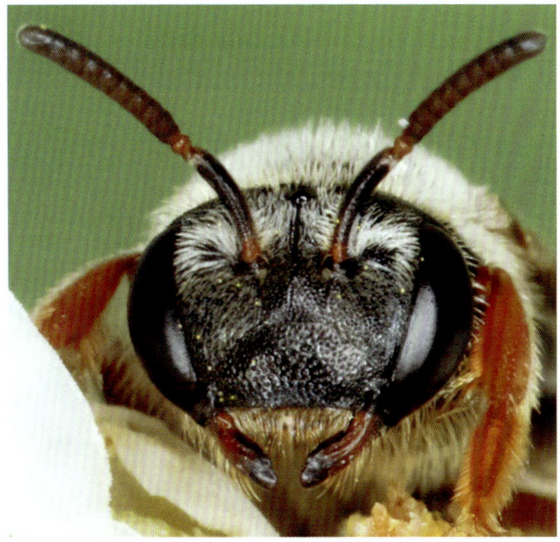

A female Baker's dieunomia: note the pitted bald face, reddish pedicels, and partly red scapes.

Behavior: All *Dieunomia* nest in the ground and all construct their own nests rather than using preexisting burrows or cavities. All are solitary, with each female provisioning her own nest, although females often settle near one another in aggregations.

Entomologist William B. Kerfoot first described the nests of *Dieunomia nevadensis bakeri.* During one afternoon in late July 1964, Kerfoot observed frenzied swarms of male and female *Dieunomia nevadensis* weaving over the dusty and sandy grounds of a Kansas schoolyard. The bees were mating with great intensity: males gathered over holes where females lay hidden, pouncing on any female that lifted her head above the earth and knocking low-flying females to the ground.

The 18 female *Dieunomia* of Kerfoot's schoolyard constructed 18 separate nests after erecting chimneylike tumuli over their nest holes. Each nest entrance consisted of a horizontal passage composed of cemented sand; whenever the female left her nest, she kicked sand over the entranceway. Upon return, she tunneled through the tumulus into the entrance.

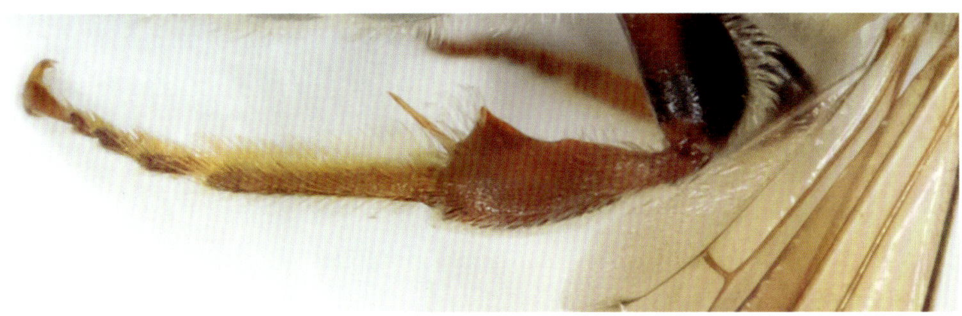

Hind leg of a male Baker's dieunomia: the tibia terminates apically in a spike and has a slight bump in the middle.

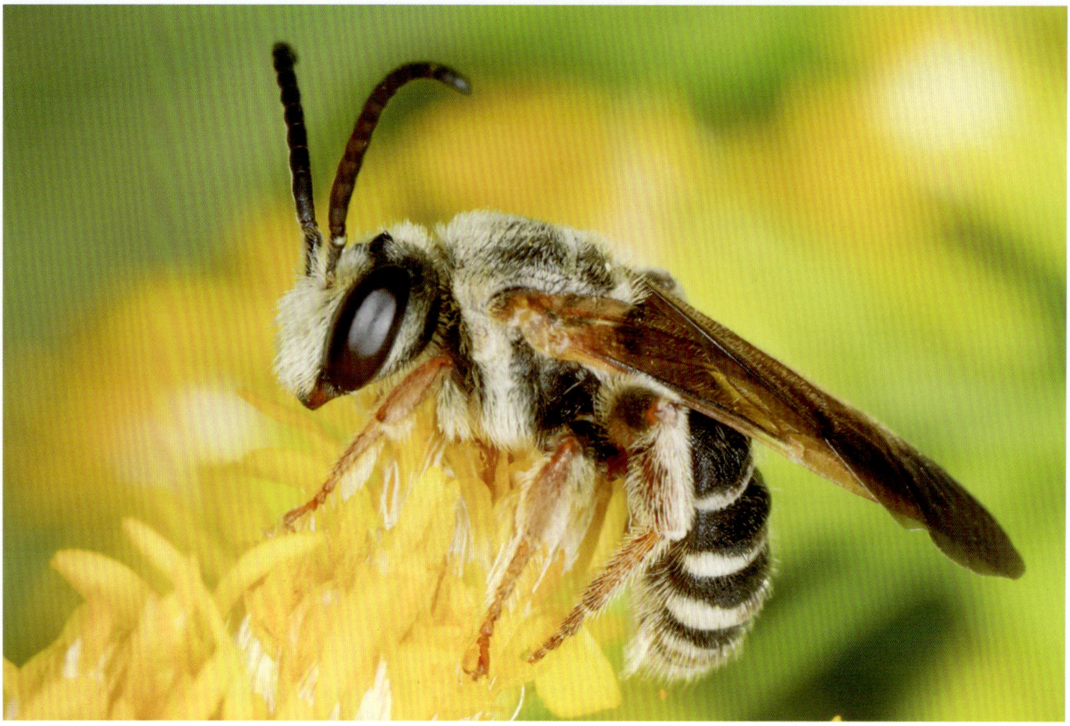
A male *Dieunomia nevadensis bakeri*

Typically, each female Baker's dieunomia dug a vertical shaft below the entrance that plunged to a surprising depth (45–72 cm, or 18–28 in.). Various horizontal passages ran from each main tunnel. The female bees excavated series of evenly spaced, narrow chambers below each shaft's floor. Each chamber hung vertically and tapered toward the neck, and each was lined with fine black soil or small grains of sand, and then coated with a waxy substance. The bees provisioned the cells with a mixture of pollen and nectar, laid eggs on top of the provisions, and then sealed the neck of each egg chamber by packing it with sand.

The *Nomia bakeri nevadensis* females Kerfoot observed formed their provisions into muffin shapes within their egg chambers: the wide rim of each muffin top touched the cell walls, ensuring that no other part of the provisions below made contact with outside surfaces. This ingenious architecture protected the provisions and the eggs perched on top of them from damp and mold.

Family Megachilidae

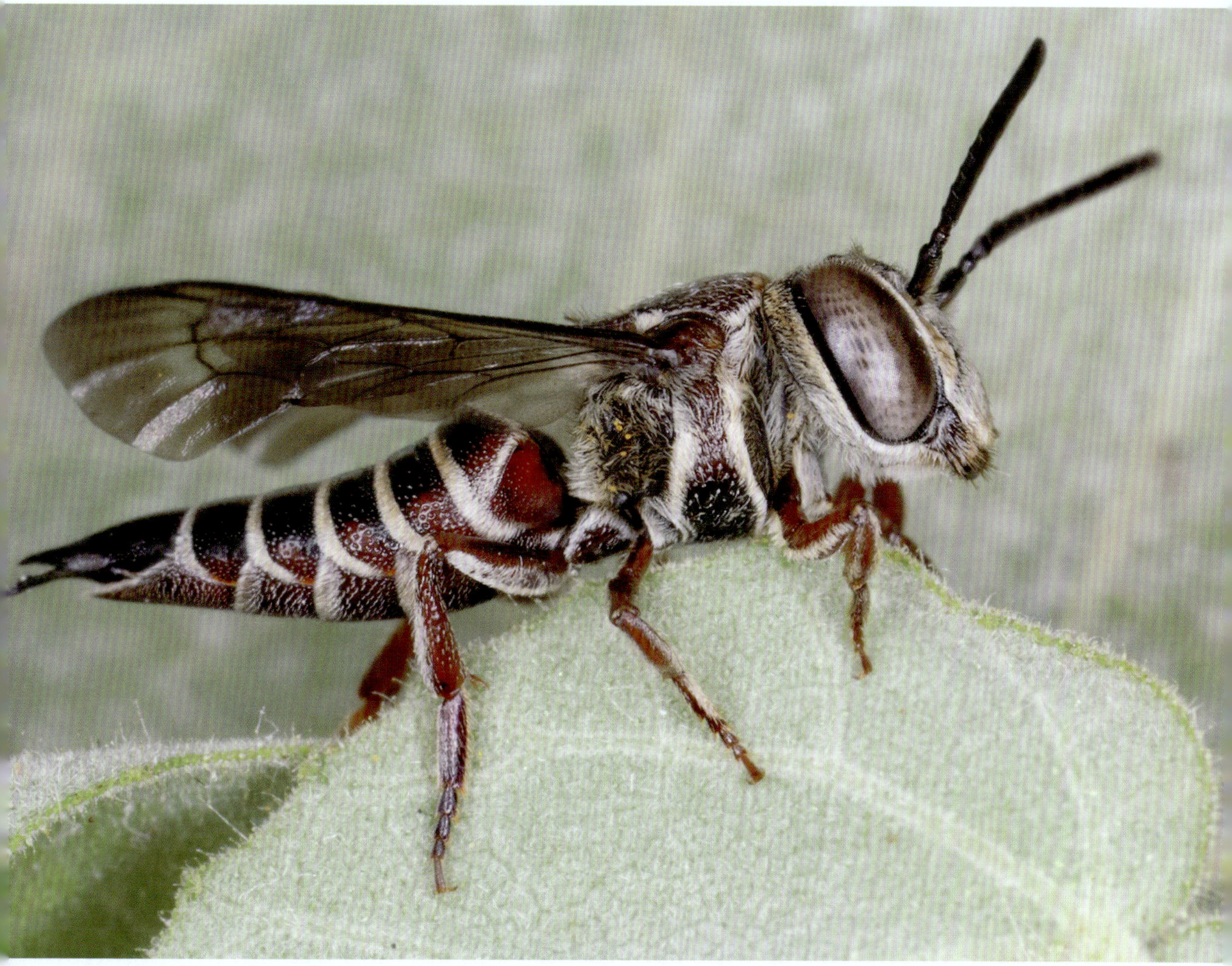

Cactus Woodborer Bees
Mason Bees
Resin Bees
Ashmeadiella
Pebble Bees
Stelis Cuckoo Bees
Leafcutter Bees
Cuckoo Leafcutter Bees

MEGACHILIDAE FAMILY GENERA
OF THE TEXAS LOWER RIO GRANDE VALLEY

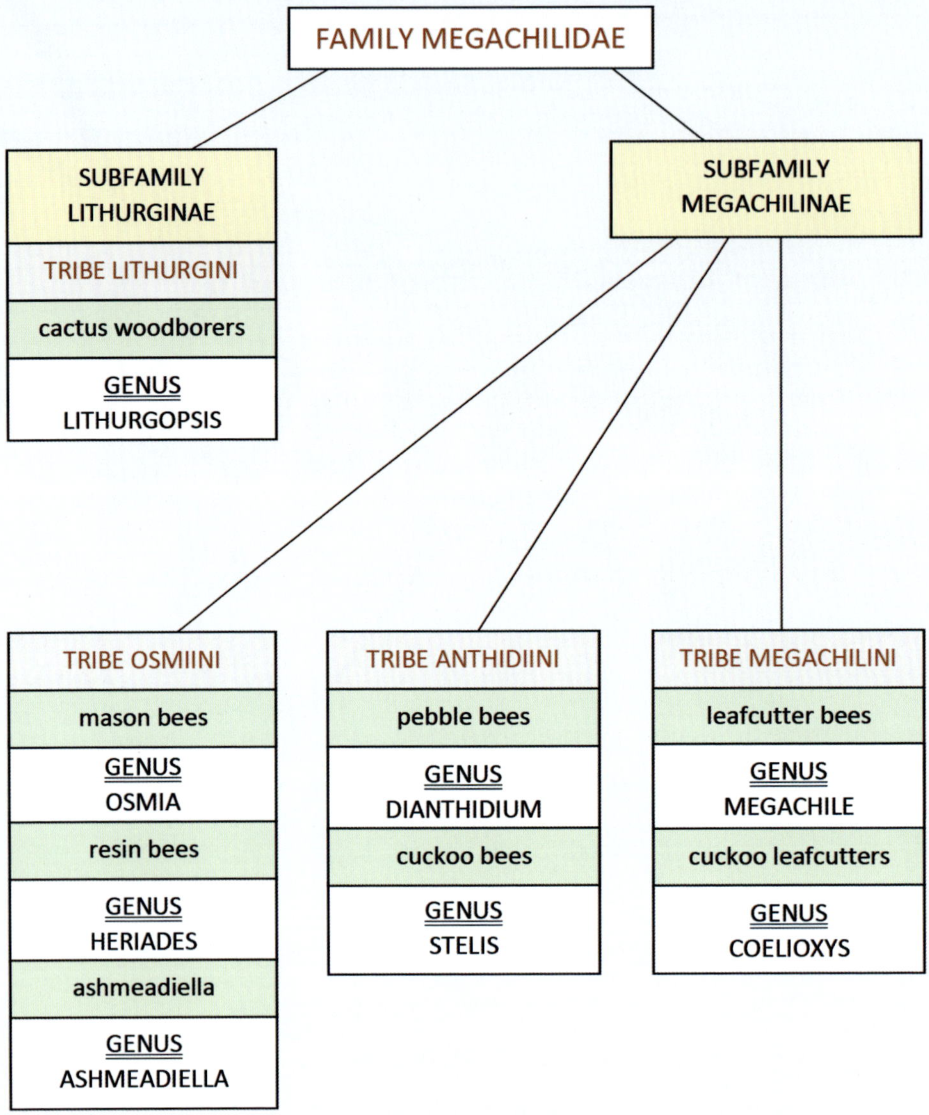

Cactus Woodborer Bees

Subfamily Lithurginae ◆ Tribe Lithurgini
Genus *Lithurgopsis*

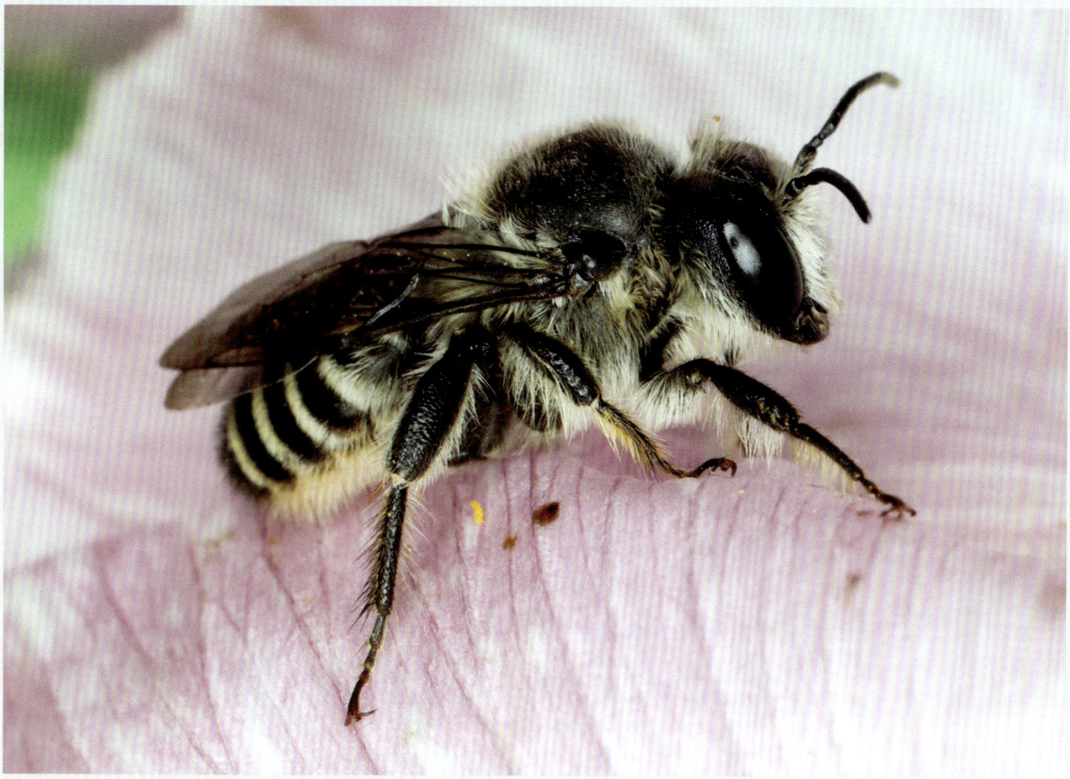

A male littoral cactus woodborer on a red prickly poppy *(Argemone sanguinea)*

In March 1917, Wilmatte Porter Cockerell took an adventurous road trip along the Texas border. She described the journey as an excursion through the Texas bush, in which she gratefully relied on the skillful driving of her companion Miss Mary Cowgill. Along the way, Cockerell stopped to document bees and other insects. She recorded her findings in her travelogue "Collecting Bees in Southern Texas."

Cockerell discovered several new species, among them the singular longhorn bee, *Eucera wilmattae*, described earlier in this guide. One of her other discoveries was a *Lithurgopsis* like the cactus woodborer bee shown here: she stumbled on a male of this unusual genus in the coastal town of Port Isabel, situated on the Gulf of Mexico at the eastern end of the Lower Rio Grande Valley.

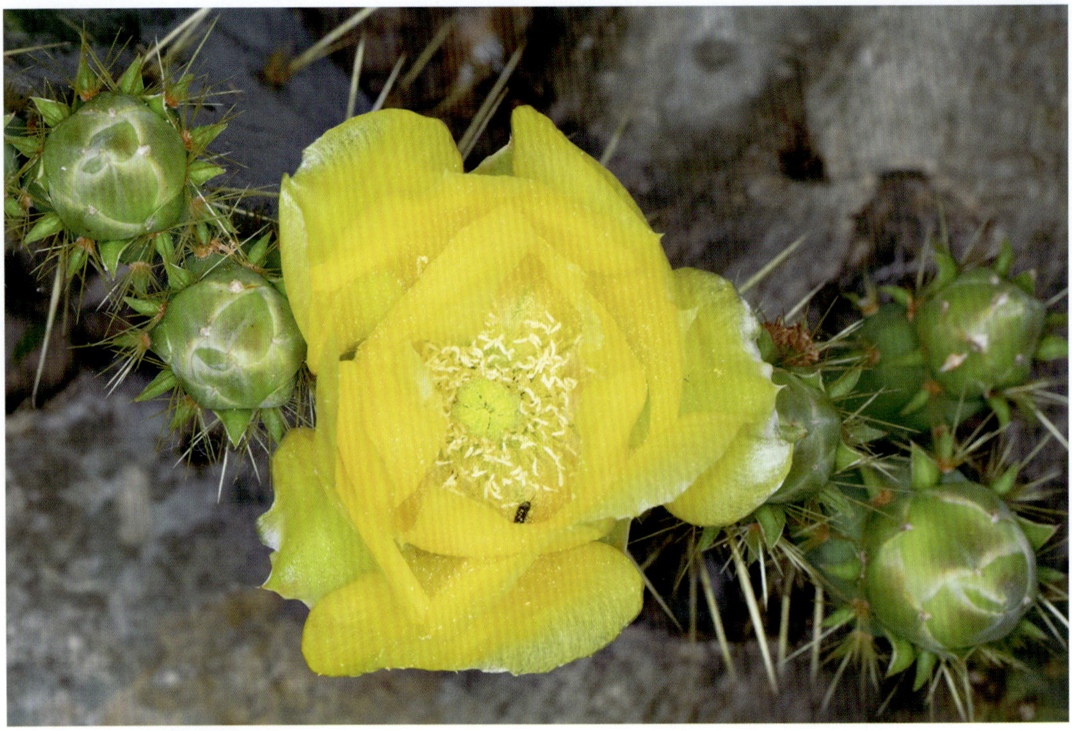

A Texas prickly pear blossom

T. D. A. Cockerell later described the Port Isabel bee as a new subspecies of the cactus woodborer known then as *Lithurgus apicalis*. He named the bee *Lithurgus apicalis littoralis*, noting that it might well deserve a species name of its own. This cactus woodborer is now considered a distinct species, *Lithurgopsis littoralis*.

Wilmatte Porter Cockerell's journal begins with a rapt description of prickly pear cactus and the many-colored prickly poppies that filled the landscape along the Lower Rio Grande Valley in March 1917. More than 100 years later, in March and April 2019 in Hidalgo County, male and female cactus woodborers were seen emerging from their holes, sheltering in prickly poppies, and visiting the flowers of Texas prickly pear for pollen and nectar.

The blossoms of Texas prickly pear (*Opuntia engelmannii* var. *lindheimeri*) produce exuberant amounts of pollen and are a source of nourishment for various bee genera of the Lower Rio Grande Valley. Female cactus woodborers use their long legs to propel themselves deeply into the sticky tangle of stamens at the centers of prickly pear cactus flowers. Upon exiting the blossoms, the bees often appear covered with pollen, as if they had been rolled in corn flour.

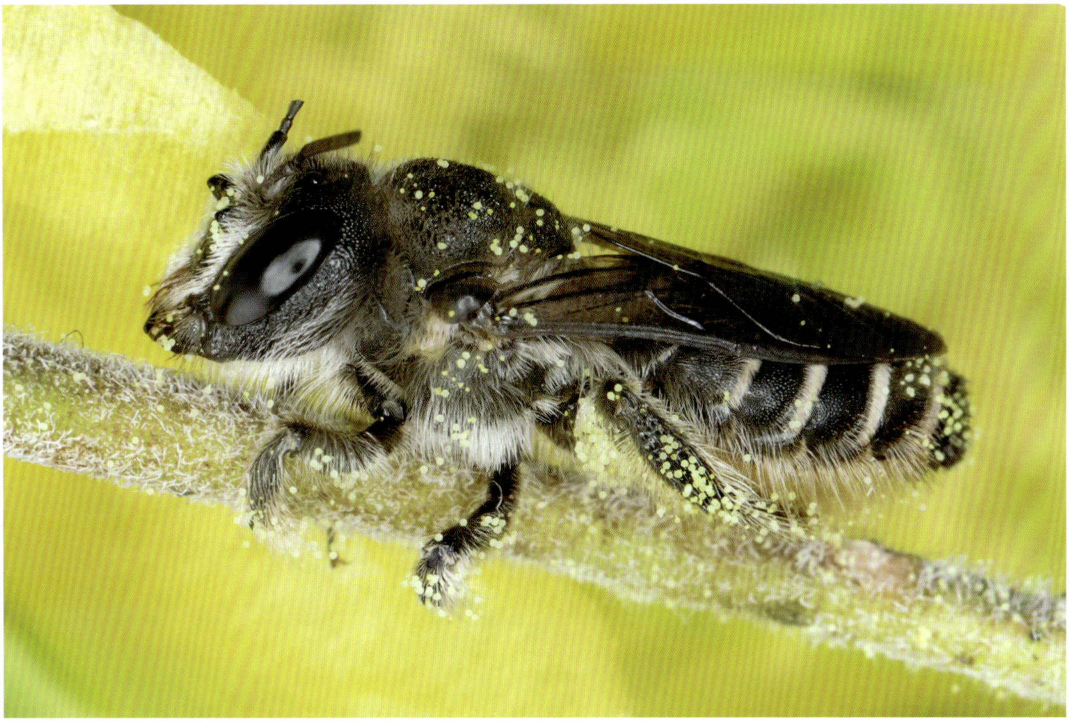

A female littoral cactus woodborer

Traits of the genus *Lithurgopsis*: *Lithurgopsis* belong to a different subfamily than all other Megachilidae in this guide—Lithurginae. In the United States and Mexico, this subfamily is represented by bees associated with cacti, known collectively as northern cactus woodborers.

Lithurgopsis look like large, fortified leafcutter bees. Cactus woodborers have sculpted dark heads and long, tapered abdomens that are banded with pale hairs. Like their leafcutter counterparts, female *Lithurgopsis* have forewings with two submarginal cells, and they carry pollen on long scopal hairs on their abdomens, rather than on their hind legs like most bees. Nonetheless, *Lithurgopsis* are distinctive—both males and females have tanklike builds, bulky heads, and long, sturdy legs. Their faces and legs are armored with spines, ridges, and bumps.

Additional information: Cactus woodborers are seldom featured in popular North American native bee guides. In the United States, they are represented by a mere six species (*Lithurgopsis apicalis*, *L. echinocacti*, *L. gibbosa*, *L. listrota*, *L. littoralis*, and *L. planifrons*). Except for *Lithurgopsis gibbosa*, these are found predominantly in arid habitats of the western and southwestern United States where cactus abounds.[1]

1. *Lithurgopsis* gibbosa has a more southeasterly distribution than other *Lithurgopsis* species in the United States. It is found from Texas to Florida, and northward into the Carolinas.

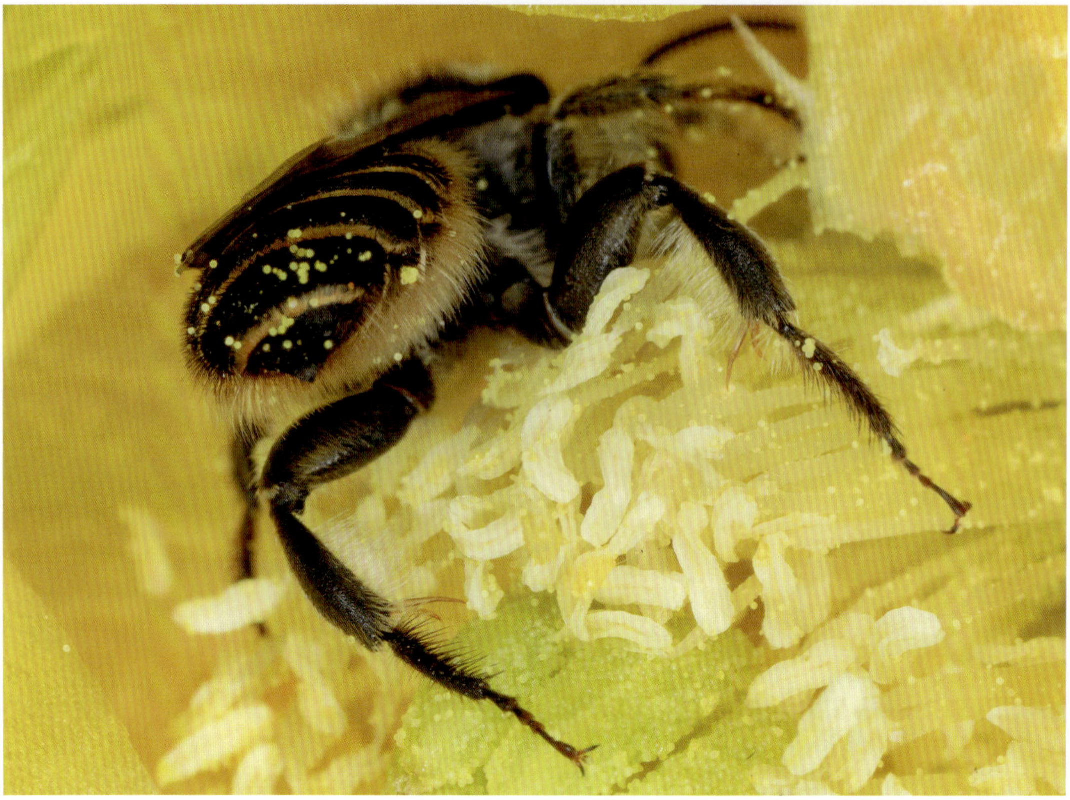

A female littoral cactus woodborer digging through the stamens of a prickly pear blossom

Lithurgopsis shares the subfamily Lithurginae with one other woodborer genus—*Lithurgus*, which is not native to North America. An exotic species of *Lithurgus* from Mediterranean Europe raised alarms in the 1970s when it arrived in Pennsylvania and New Jersey. Called *Lithurgus chrysurus*, the bee was said to be able to bore through the materials of human-made structures that ordinarily require axes, chisels, and hammers to penetrate, including new wood, compressed wood-fiber shingles, and even vinyl. This woodborer's potential for doing great damage to buildings and bridges spurred a flurry of studies, but ultimately the bee, a specialist on knapweed flowers, proved easy to contain.

Native *Lithurgopsis* lack the potential for destruction of their European cousin. In 2014, Rozen and Hall conducted a study on the nesting behavior of *Lithurgopsis apicalis*, a cactus woodborer found in much of the western United States. Rozen observed this species in Arizona building nests by chewing out tunnels in the dead and dying flower stalks of agave. He noted that despite the name "woodborer," *Lithurgopsis* of the United States are not equipped to bore through healthy wood. They require porous cellulose material for nesting, such as rotted logs or other partly decayed plant material.

Littoral cactus woodborer ◆ *Lithurgopsis littoralis* (Cockerell, 1917)

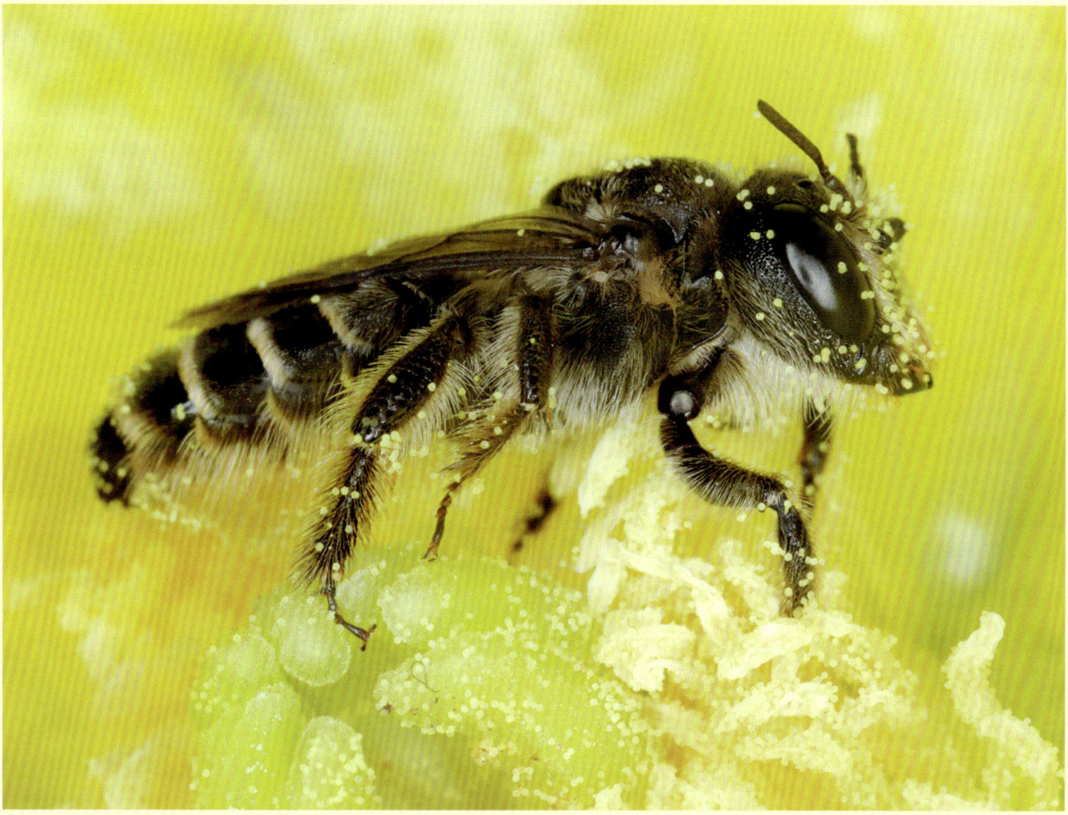

A female littoral cactus woodborer

Description: *Lithurgopsis littoralis* is a stalwart black bee with a hefty head and wide thorax; long, powerful legs; and a long abdomen banded with pale hairs. Both male and female littoral cactus woodborers have hornlike or toothlike structures on their faces, and spikes on their tibiae. Both have arolia between the tarsal claws of their front feet. The bees' tegulae are dark or reddish brown, and their forewings are a transparent smoky brown with dark veins. Their eyes are opaque black. Their mandibles and antennae are also predominantly black; the male's antennae are reddish toward the tip. **Size:** females 14–15 mm (1/2–3/5 in.), males 9–14 mm (1/3–3/5 in.). *Note*: Male *Lithurgopsis* often vary considerably in size within a single species. *Lithurgopsis* nests lack well-defined egg chambers; egg cell dimensions vary, and the size of a male hatchling tends to reflect the size of its individual egg chamber.

Range: *Lithurgopsis littoralis* occurs principally in South Texas and Mexico. The name *littoralis* (or "coastal") arose from Wilmatte Porter Cockerell's discovery of the first specimen in Port Isabel (Cameron Co.). Nonetheless, this species occurs inland in all three border counties of the Valley and into much of Texas. It has been documented as far north as Illinois.

Associated plants: *Lithurgopsis littoralis* is a cactus specialist commonly found in the Valley on Texas prickly pear (*Opuntia engelmannii* var. *lindheimeri*). Male *Lithurgopsis littoralis* first appear in the Valley during late March on red prickly poppies (*Argemone sanguinea*) and golden prickly poppies (*A. aenea*), right before the cactus bloom. Females emerge in the first weeks of April and occasionally gather pollen from prickly poppies.

This guide focused earlier on *Diadasia rinconis* males that gather on prickly poppies during spring in Hidalgo County. The poppies are sources of pollen but not of nectar, and thus male *Diadasia* seek them out not for food, but instead for shelter against wind and rain and as places to settle down at night. *Lithurgopsis* males engage in the same practice, using *Argemone* poppies as way stations while awaiting the cactus bloom and the emergence of *Lithurgopsis* females. After prickly pear cactus begins to bloom, males tend to abandon the poppies and shelter in cactus flowers.

In central Mexico, *Lithurgopsis littoralis* has been recorded on other *Opuntia* species, most notably twistspine prickly pear (*O. macrorhiza*), wheel cactus (*O. robusta*), and woolly joint prickly pear (*O. tomentosa*). The Texas apiarist H. B. Parks issued a short publication in 1930, titled "Notes on Texas Bees," in which he documented female *Lithurgopsis littoralis* collecting pollen in Bexar County from both tiger tongue prickly pear (*Opuntia ellisiana*) and twisted-rib cactus (*Thelocactus setispinus*).

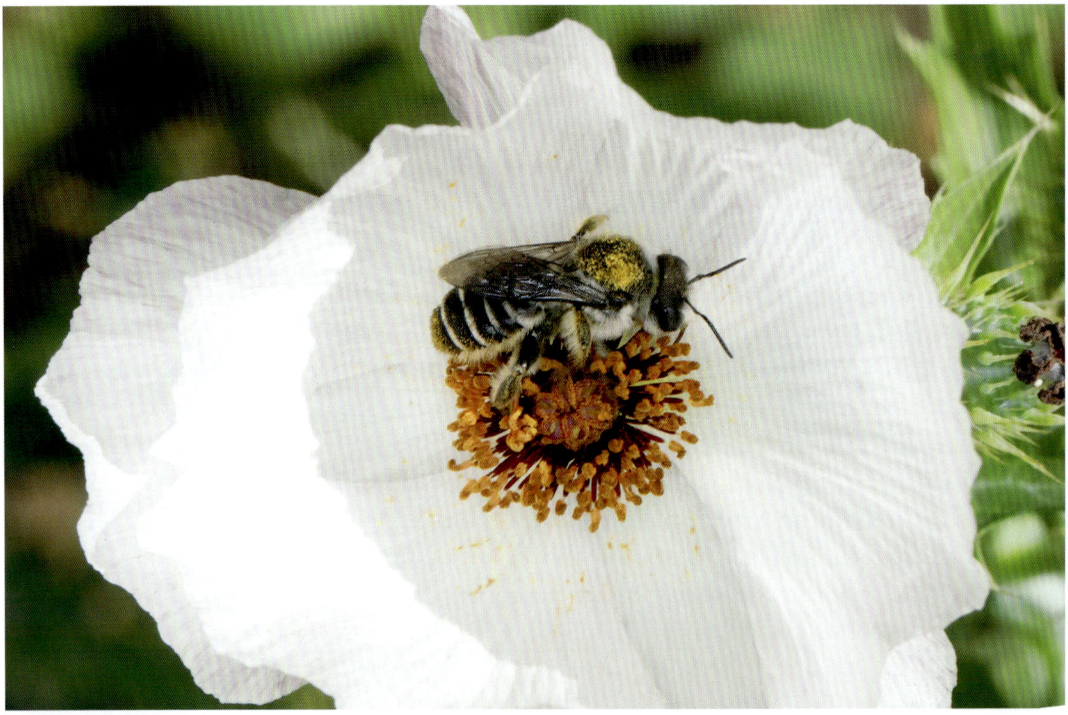

A female littoral cactus woodborer investigating a prickly poppy

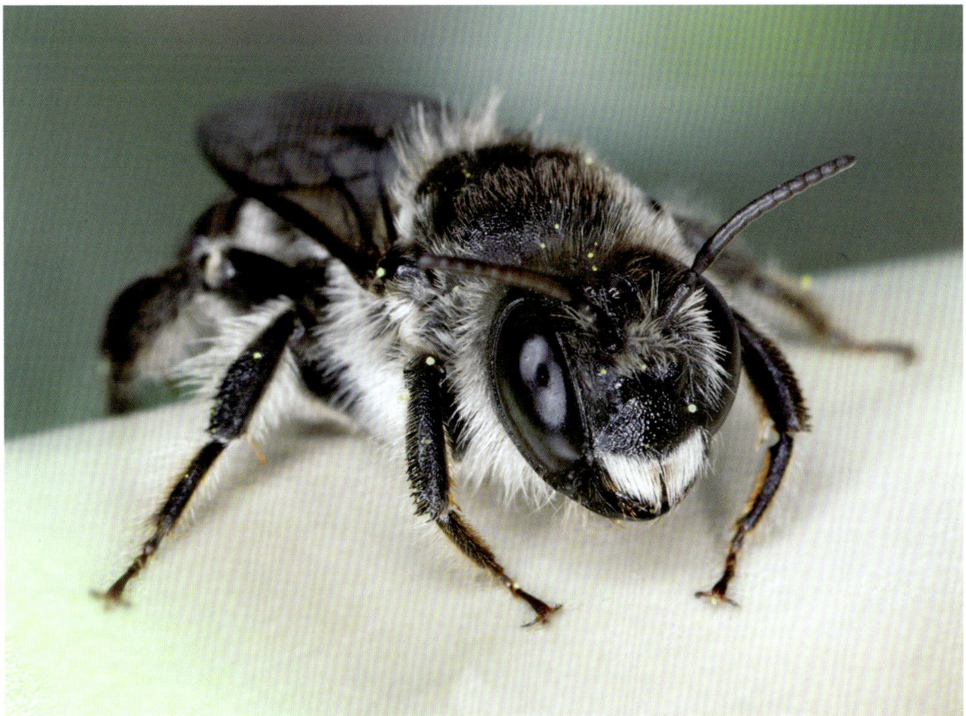

A male *Lithurgopsis littoralis*: note the tanklike build and spiked forelegs.

Littoral cactus woodborers occasionally forage on plants other than cacti. Parks observed male and female bees drinking nectar from mint-family plants such as spotted bee balm (*Monarda punctata*) and lemon beebalm (*M. citriodora*). In a 1983 review of North American *Lithurgopsis*, Roy Snelling noted unpublished records of littoral cactus woodborers visiting purple poppymallow (*Callirhoe involucrata*), Texas hogplum (*Colubrina texensis*), and Asteraceae. Wilmatte Porter Cockerell discovered her male littoral cactus woodborer on a yellow composite.

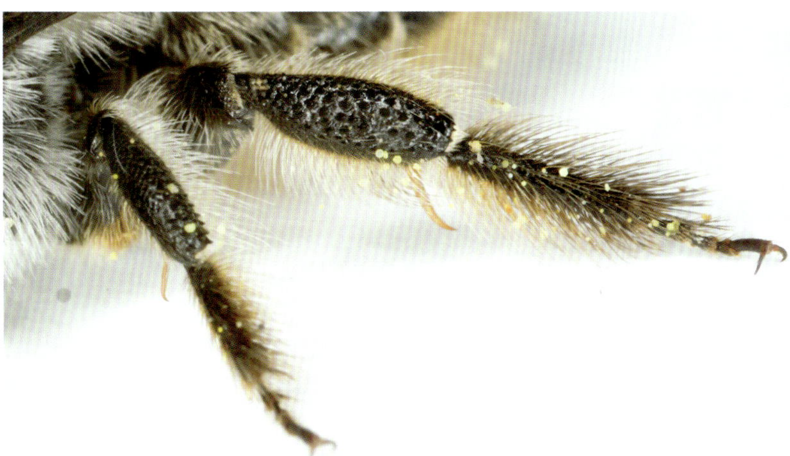

The middle and front legs of a female *Lithurgopsis littoralis*: spines rim the middle-leg tibia, and pits and bumps stud the hind-leg tibia.

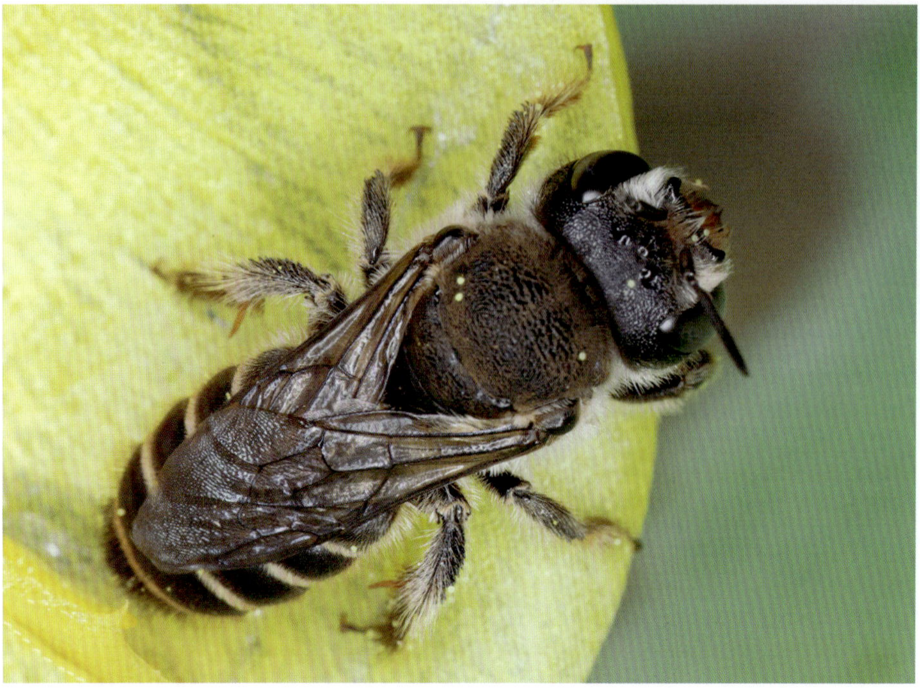

A female littoral cactus woodborer: note the hornlike projections on the face, the broad vertex, and the short, light brown hairs on the wide, faintly pitted and wrinkled thorax.

ID tips: General *Lithurgopsis* traits include the following: (1) the legs of *Lithurgopsis* have an armored appearance: large pits, bumps, and spikes stud the tibiae; (2) on both males and females, F1 of the antennae is shorter than F2, a distinguishing trait of the genus; and (3) because most leafcutter bees of the Valley have colorful (usually blue or green) eyes, the dark eyes of Valley *Lithurgopsis* are a useful, quick means of differentiating this genus in the field from leafcutters.

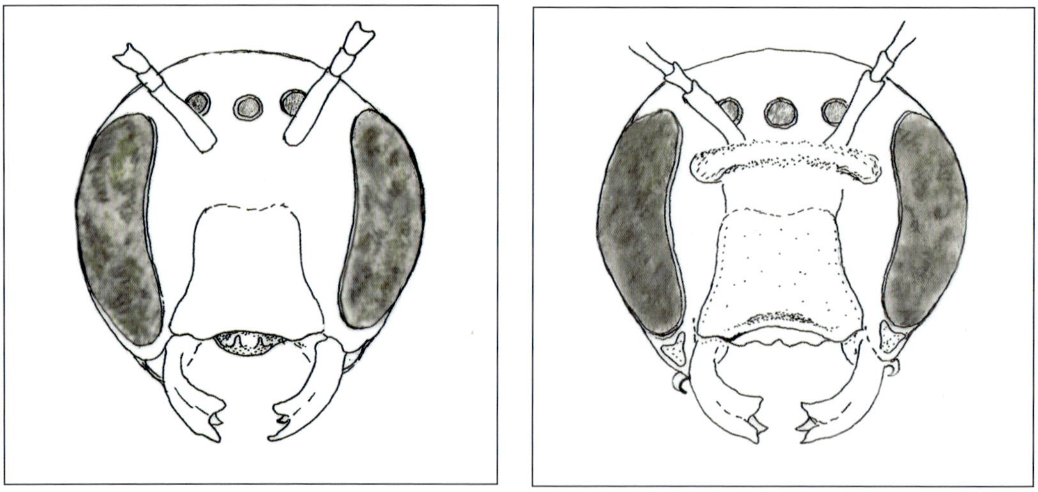

Faces of male (*left*) and female (*right*) *Lithurgopsis littoralis*

Other physical attributes of the littoral cactus woodborer bee: Descriptions of *Lithurgopsis littoralis* are far-flung and not widely available. Thus, the following details of this species' appearance are noted here.

Head: The male's face above the clypeus is protuberant and shiny, with scattered pits along the cheeks. Its labrum is armored with one or two toothlike spikes; this trait is diagnostic of the male *Lithurgopsis littoralis*. (The lower face, however, of both males and females is typically hidden by a dense fringe of long, straight, pale hairs, white in the middle and reddish black in the center.) The female littoral cactus woodborer has a thick, bowed ridge at the top of the clypeus, with two protuberances on either end that look like short bulbous horns, traits diagnostic of females of this species. The female's clypeus has a median indentation flanked by two smaller indentations. There are conspicuous depressions between the female's compound eye and the mandible base; thin ridges circle both the depressions and the eye.

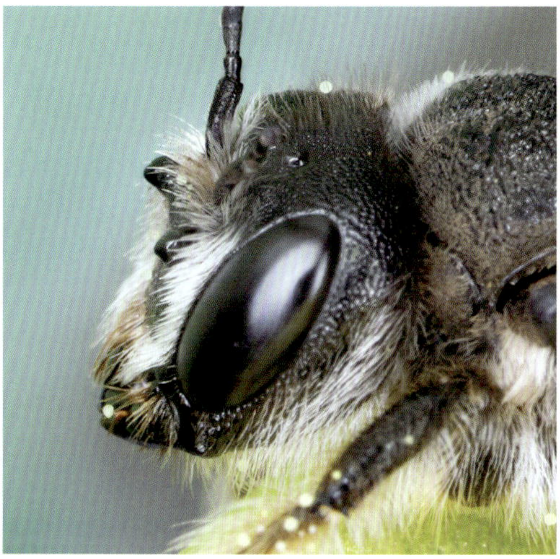

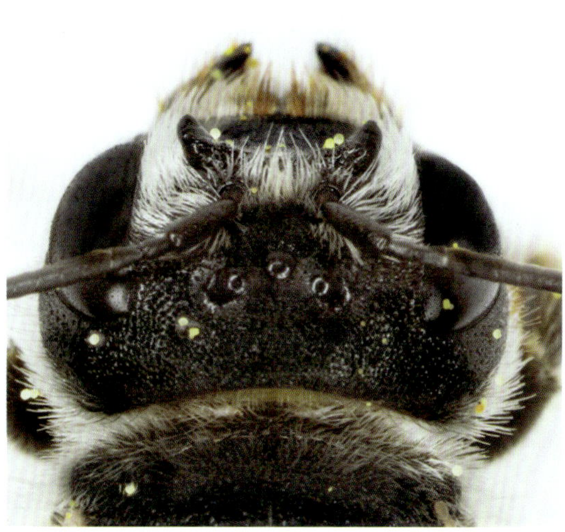

Hornlike projections on the face of a female *Lithurgopsis littoralis*

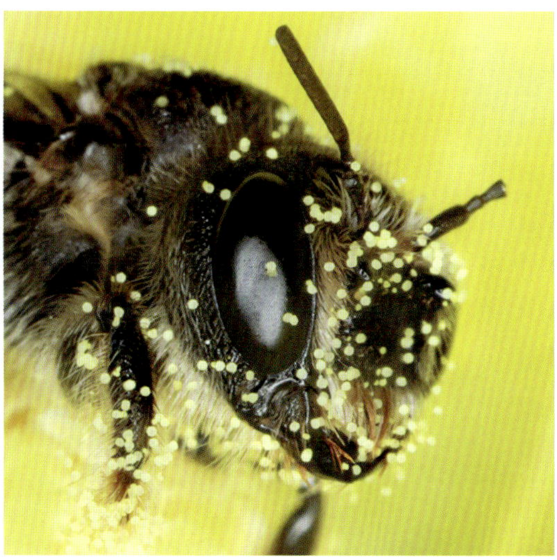

The depression between the eye and mandible of a female littoral cactus woodborer

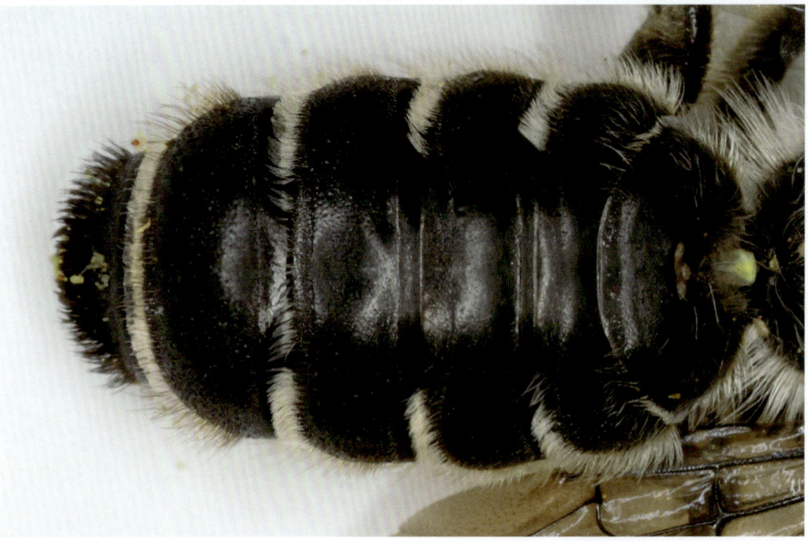

Dorsal view of the abdomen of a female *Lithurgopsis littoralis*

Thorax: The surface of the littoral cactus woodborer's thorax has a somewhat wrinkled and pitted appearance. Both males and females have very short, sparse, pale grayish-brown hairs on the top of the thorax (scutum and scutellum), and long, dense, white hairs on the sides of the thorax (mesepisternum).

Abdomen: The pale abdominal hair bands of the female *Lithurgopsis littoralis* appear on the first through fifth segments (T1–T5) and are interrupted in the middle, except on T5, where they extend sparsely across the top of the segment. Dark hairs intermingle with light hairs on T5, and dark hairs cover T6.

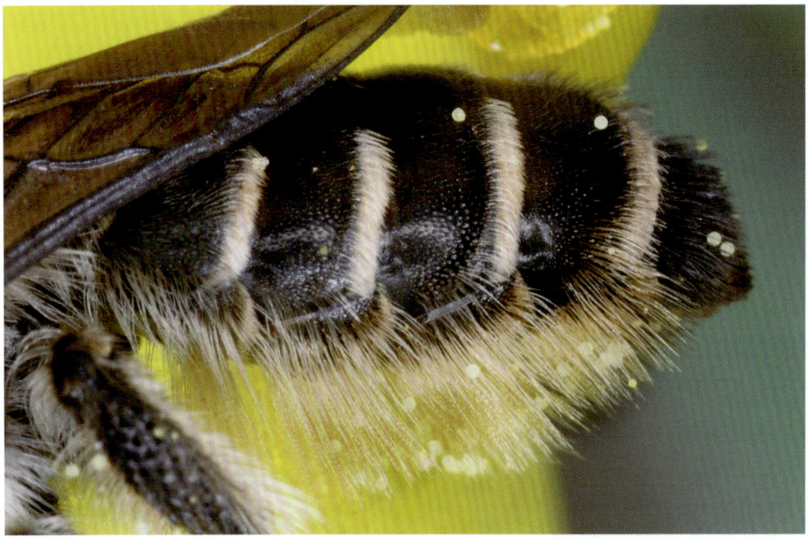

Scopal hairs of a female *Lithurgopsis littoralis*

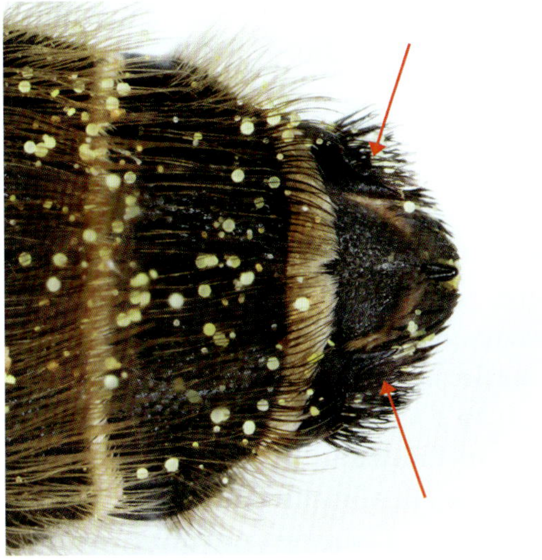

Tip of the sternum of a female *Lithurgopsis littoralis*: note the lateral teeth on S5.

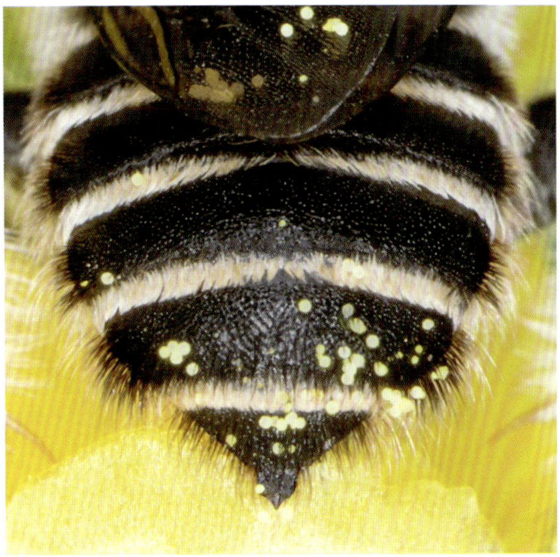

Toothed tip of the abdomen of a male *Lithurgopsis littoralis*: the pale ocher hair bands continue onto the sternum.

The female bee has long white scopal hairs on the first four segments (S1–S4) of its sternum, some dark hairs on S5, and all dark hairs on S6. The hind edge of the fifth sternal segment (S5) is armed with two lateral teeth. S6 has an apical spine.

On males, the abdominal hair bands continue onto the sternum. They are interrupted on T1–T3, almost complete on T4, and complete on T5–T6. The tip of the male's abdomen has a roughly rectangular protuberance with a point on the hind end.

Pollination and nesting behavior: Female *Lithurgopsis littoralis* observed in the Valley practice an unusual method of entering a cactus blossom: they dive headfirst into the flower anthers, facing the style, and then wriggle their legs and bodies, raking off pollen with their sternal plates and tibiae and combing it onto their scopal hairs. H. B. Parks noted similar behavior among intermingled female *Lithurgopsis littoralis* and *L. gibbosa* near San Antonio. Possibly, the spikes on the females' legs, and the lateral teeth on the hind edges of the females' fifth sternal segments, serve to dislodge the coarse pollen from cactus flowers.

Male *Lithurgopsis littoralis* tend to negotiate cactus blossoms differently. They rest upright on a cactus flower's pistil and then crawl along the inner edges of the petals to reach the nectary at the base of the blossom. After females emerge, male bees often nestle at the bottom of Texas prickly pear blooms, either to rest after nectaring or to await females.

In early morning during cactus blooms, male *Lithurgopsis littoralis* can be found sleeping in the depths of prickly pear blossoms. Mating tends to occur not in nest areas, as is common among ground-nesting bees, but instead on the *Opuntia* flowers themselves.

Lithurgopsis littoralis nest in preexisting holes in plant materials. In 1997, Mexican entomologist Olivia Yañez-Ordoñez conducted a comprehensive study of *Lithurgopsis littoralis* in the Pedregal de San Angel Ecological Reserve of Mexico City, where female littoral cactus woodborers gathered pollen from *Opuntia robusta* and *O. tomentosa.* In early spring, Yañez-Ordoñez laid trap nests made of dead agave stalks in prickly pear cactus beds in order to attract woodborer bees. Males emerged first in March to nectar on prickly pear, followed about two weeks later by females, which settled in the trap nests and commenced nest building.

The littoral cactus woodborer nests began as simple descending tunnels that evolved into complex but loosely defined structures in their final construction phase. One nest consisted of a main tunnel, 46 cm (18 in.) long and 34.8 cm (14 in.) deep, with a "booby-trap chamber" near the entrance filled with fibers from agave scapes, whose purpose Yañez-Ordoñez speculated might be to deter parasites or predators attempting to enter the nest. The tunnel then stretched onward through another area stuffed with more agave fibers. Beyond that protective bottleneck, the main tunnel branched laterally into shorter tunnels and then continued on until terminating in multiple branches.

The diameter of the nest corridors varied with the age of the nest tunnels and branches. In each branch, the female woodborer created a series of four egg chambers loosely separated from one another with agave fibers. She placed pollen provisions on top of each egg and then closed off the entrance by stuffing it with pollen, rather than sealing it by other, possibly more secure methods. Unlike many mason, resin, and leafcutter bees, the woodborer bee did not construct elaborate partitions or apply any waterproofing or smoothing materials to the egg chamber walls.

During the early phases of construction, the female woodborers of Pedregal de San Angel stopped their nest entrances with agave-fiber plugs that remained in place during the night. In the morning, the bees would forcibly expel the fibers in order to exit and begin gathering pollen. In later phases of construction, the plugs were no longer used. Eggs hatched within open-ended nests, the larvae spun protective cocoons, and the pupae entered diapause to overwinter for seven months until the next spring's cactus bloom.

Mason Bees

Subfamily Megachilinae ◆ Tribe Osmiini
Genus *Osmia*

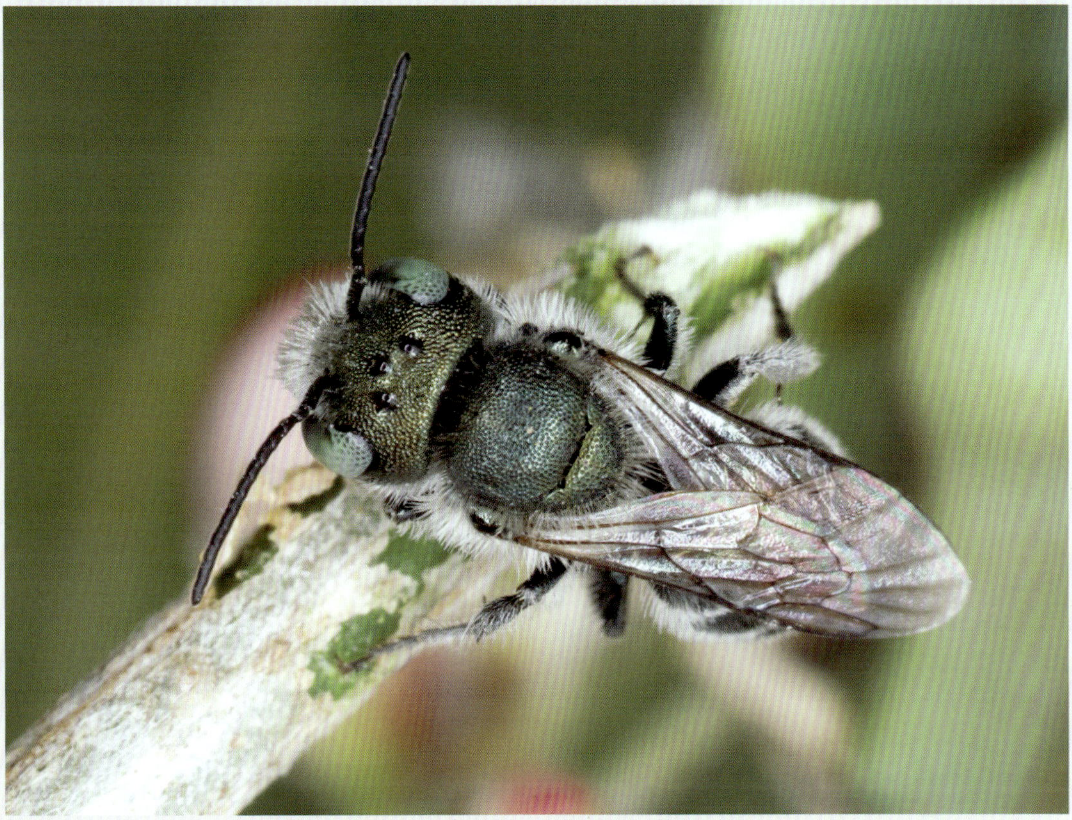

A male *Osmia subfasciata* mason bee

Osmia are solitary bees that nest in the hollows of dead pithy stems, in preexisting cavities in dead trees, stumps, and logs, or, less often, in soil or rock crevices. The name "mason bee" derives from *Osmia* bees' practice of constructing nests with plant parts and soil. Mason bees use masticated leaves, resin, mud, and grit to engage in "masonry." They employ such materials to build egg chamber partitions, seal nest entrances, and line nest walls with transported materials.

Traits of the genus *Osmia*: *Osmia* belong to the tribe Osmiini, a branch of the subfamily Megachilinae. *Osmia* are small to medium-sized bees with robust builds and relatively large heads. All Osmiini have forewings with two submarginal cells, and their feet sport arolia. Female *Osmia* carry pollen on scopal hairs located under their abdomens. The scopae may be white, yellowish, dark brown, black, or even electric orange. Most Texas *Osmia* are metallic blue green; some are deep blue. Their abdomens are often banded by narrow stripes of pale hairs. Their faces may be covered with dense mats of pale hair; some bees have short orange hairs on their mandibles.

There are at least 25 *Osmia* species in Texas. Species are distinguished by minute traits such as the grooves or pitting on the thoraxes and abdomens; differences in the structure of mandibles and face parts; the presence or absence of short orange hairs on the mandibles; the color and quality of females' sternal scopal hairs; and the shape of the tips of males' sterna.

Mason bees' importance as pollinators: In many parts of the United States, mason bees are bought and sold commercially for use as pollinators of apples and stone fruits. The blue orchard mason bee (*Osmia lignaria*), found in Texas north of the Lower Rio Grande Valley, is a widely used fruit and nut orchard pollinator. Many other Texas mason bees are essential pollinators of woodland, grassland, and desert plants, among them *Osmia ribifloris*, a pollinator of native bigberry manzanita (*Arctostaphylos glauca*); and *O. conjuncta*, *O. georgica*, *O. marilaunidii*, *O. pumila*, and *O. subfasciata*, generalist pollinators that appear on an impressive array of wildflowers and garden flowers.

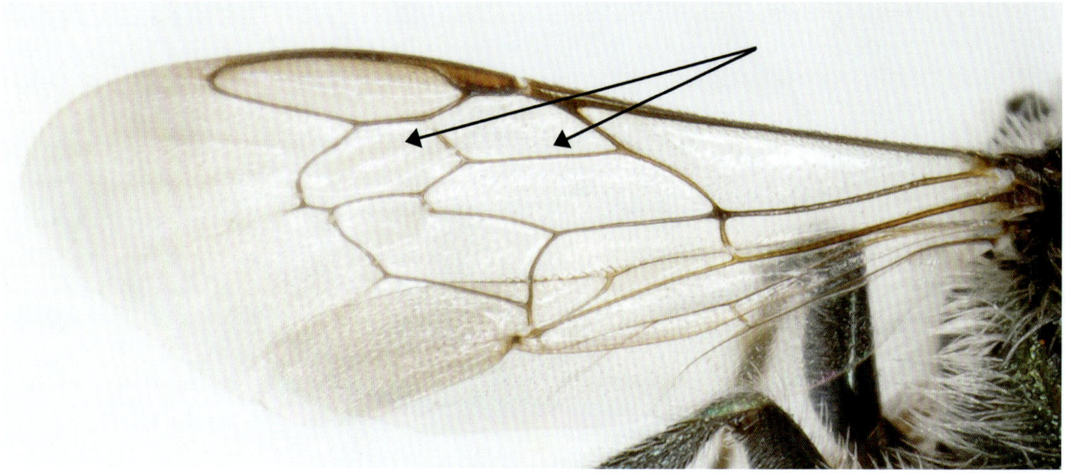

The forewings of *Osmia* have only two marginal cells.

Faintly banded mason bee ◆ *Osmia (Diceratosmia) subfasciata* Cresson, 1872

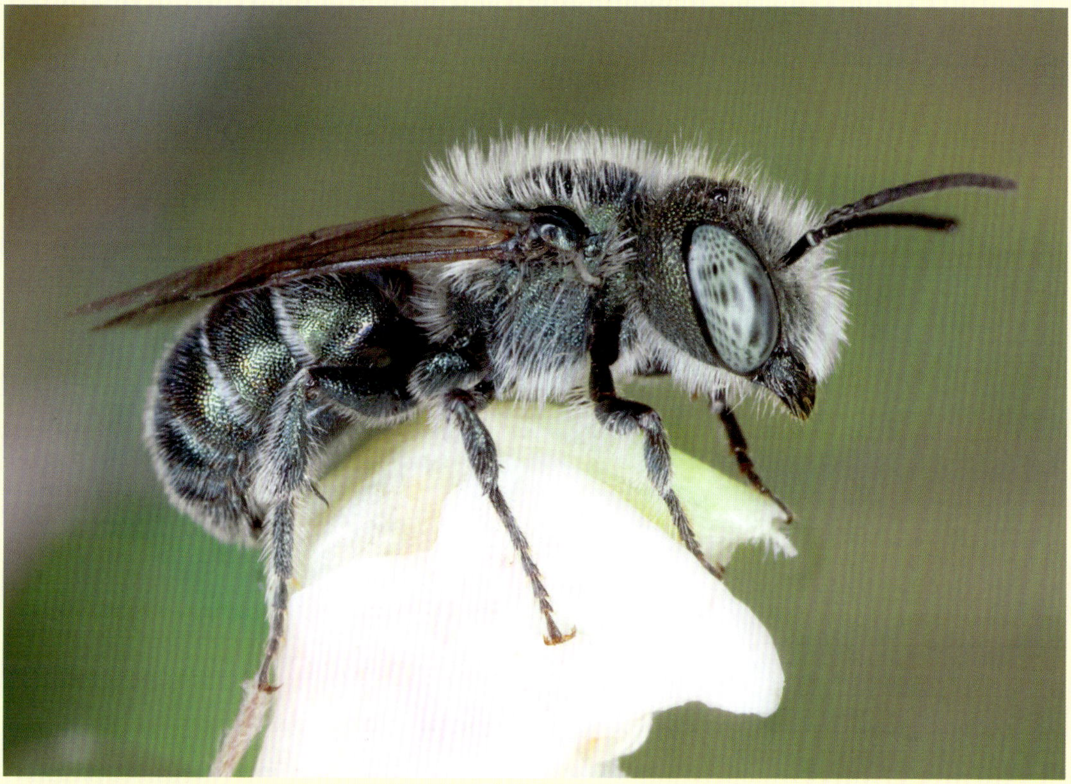

A male *Osmia subfasciata*

Description: *Osmia (Diceratosmia) subfasciata* is a small, metallic blue-green bee with green eyes. White hairs sprout from its face and thorax. *Fasciata* means "banded" in Latin, and *sub-* means "imperfectly" or "not quite." This name may derive from the attenuated and broken bands of white hairs girding the bee's abdomen.

Females have bright white scopal hairs on the underside of the abdomen. The tip of the male's abdomen is notched. The heads and bodies of both male and female *Osmia subfasciata* are pitted. Females have three-toothed mandibles, and males have two-toothed mandibles. Males have especially dense white hair covering their faces. **Size:** females 8 mm (3/10 in.), males 7 mm (1/4 in.).

Range: *Osmia* are common in the northern Western Hemisphere but absent in South America, and underrepresented in Mexico and along the Texas border. *Osmia* south of the border tend to be metallic blue-green bees in the subgenus *Diceratosmia*. *Osmia subfasciata* is native to Mexico and Texas; its range extends across the lower third of the United States, from southern California to Maryland. It is a common spring bee in Hidalgo and Starr Counties.

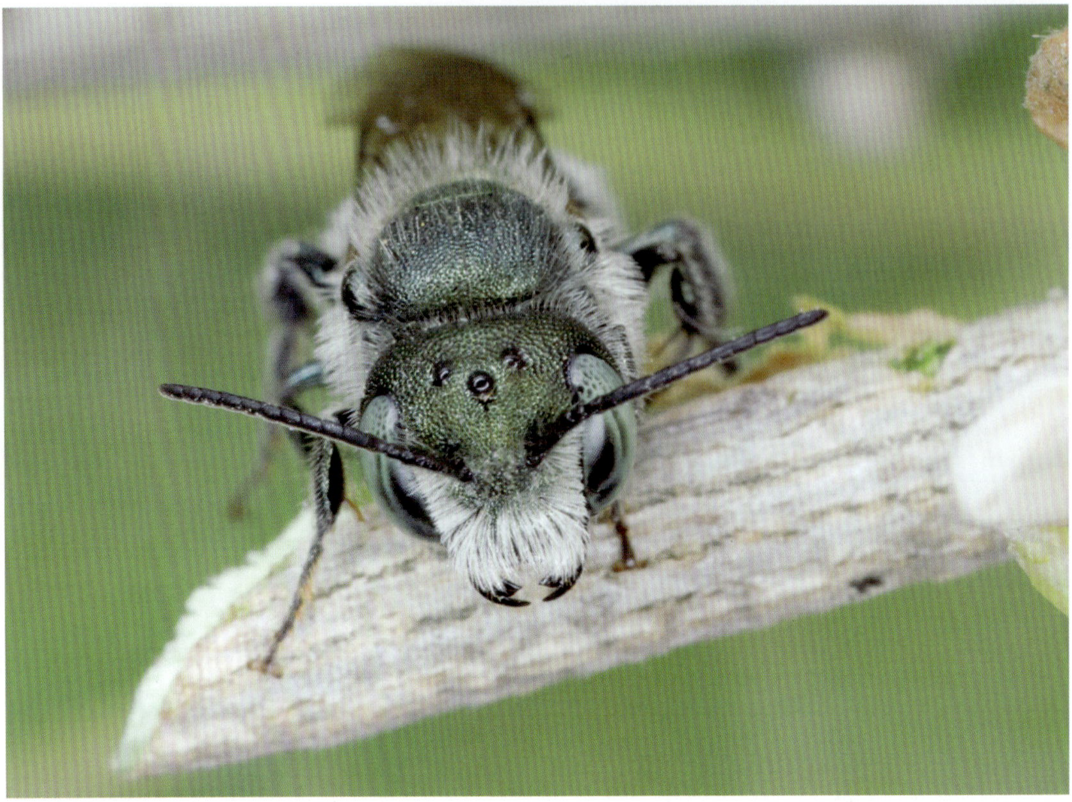

The face of the male *Osmia subfasciata* is covered with a thatch of long white hairs.

Associated plants: *Osmia subfasciata* are generalist pollinators. In the Lower Rio Grande Valley, females gather pollen from creosote bush (*Larrea tridentata*) and such pea-family plants as retama (*Parkinsonia aculeata*), honey mesquite (*Prosopis glandulosa*), and Louisiana vetch (*Vicia ludoviciana*). At the National Butterfly Center (Hidalgo Co.), both male and female *Osmia subfasciata* visit Texas babybonnets (*Coursetia axillaris*).

ID tips: Metallic blue-green mason bees of the subgenus *Dicerotosmia* can be difficult to identify in the field. The following traits, visible with magnification, aid in distinguishing *Osmia subfasciata*. *Females*: (1) On female bees, there are distinct bands of bright white hairs on the lateral edges of T1 and T2. These bands are interrupted on the top of each segment. (2) The three teeth on each side of the female's mandibles are evenly spaced. (3) The scopal hairs bell outward somewhat near the tip of the abdomen. On the very similar *Osmia (Diceratosmia) marilaunidii*, also native to the Valley, the scopal hairs taper toward the tip. *Males*: The tip of the male's abdomen (T7) is notched, and there are two prominent rounded lobes on either side of T6. On *Osmia marilaunidii*, there are acute, angular projections on the sides of T6 instead of lobes. All bees of the subgenus *Dicerotosmia* have elongate parapsidal lines. (Visible under magnification, these lines run lengthwise near each lateral edge of the scutum.)

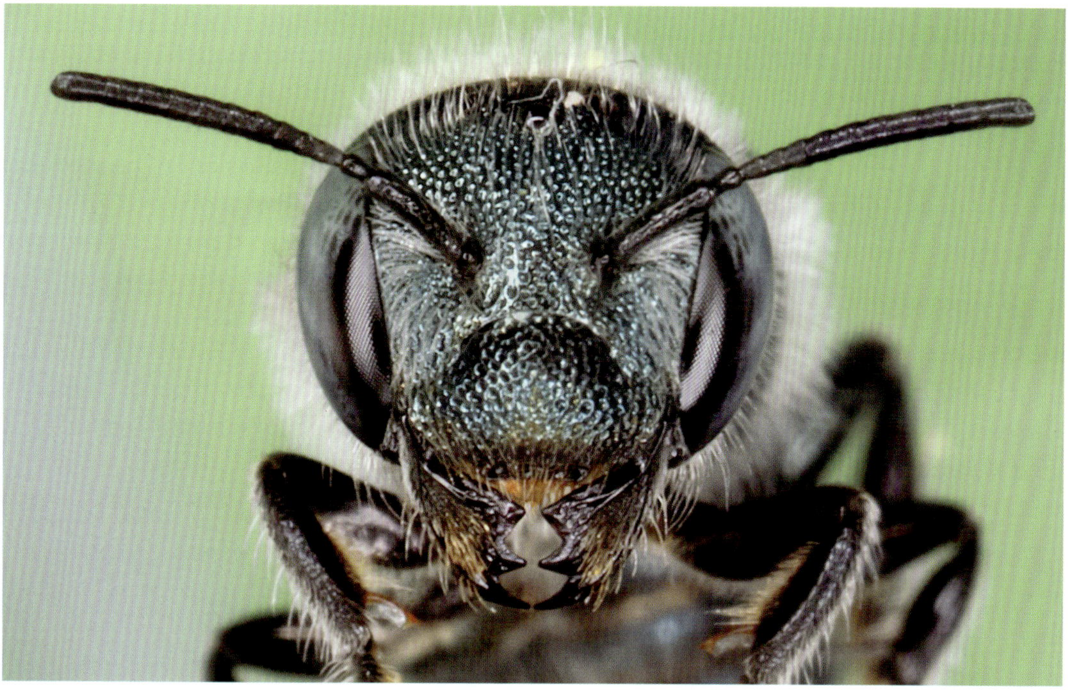

Face and mandibles of a female *Osmia subfasciata*: note the evenly spaced teeth on the mandibles.

Bee behavior: *Osmia subfasciata* are solitary bees that are not gregarious; their nests do not tend to appear in aggregations. A detailed 1992 study of this species by Neff and Simpson described *Osmia subfasciata* females in Central Texas transporting balls of masticated plant material to their nesting areas. Females would drop the balls to the ground and then chew and knead them while rolling them to incorporate sand or dirt into the mass. The bees then used this cement mixture to construct the walls and partitions of their nests.

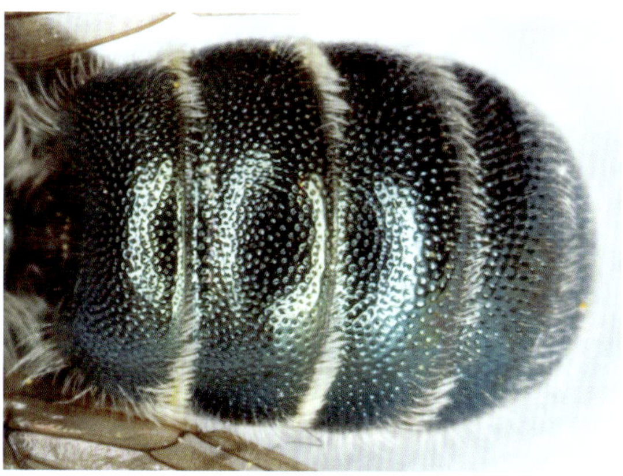

White hair bands line the rims of T1 and T2 on the female's abdomen. The hairs are dense and the bands well defined on the sides, but the bands are interrupted in the middle.

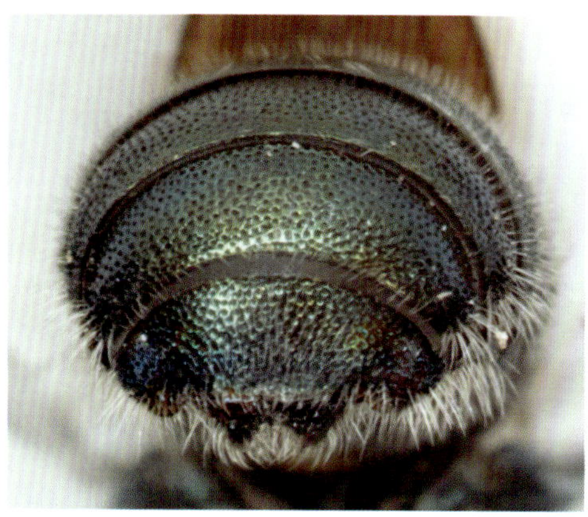

The tip of the male's abdomen (T7) is notched; on either side of T6 are rounded lobes.

Neff and Simpson reported that *Osmia subfasciata* females build their solitary nests in beetle burrows, abandoned snail shells, plant stems, and abandoned wasp nests. When using snail shells as nest cavities, the female bees deposit a single egg in a shell, provision it with food, and then enclose it within protective materials.

In the Lower Rio Grande Valley, *Osmia subfasciata* use the succulent flesh of prickly pear cactus as nest-building material. The female bee shown here was one of many observed in March 2019 in multiple locations in Hidalgo County, harvesting pieces of prickly pear (*Opuntia*) for nest construction. Individual female *Osmia subfasciata* were seen biting off mouthfuls of prickly pear flesh and then using their jaws to carry pieces of cacti to a wooded area. The mason bees made multiple trips daily but seemed to do no damage to the cactus: they focused on collecting soft materials gathered around the spines protruding from newly budding nopales (cactus pads). The viscous flesh of prickly pear cactus no doubt provides a good gluey base for mixing a natural cement useful in nest construction.

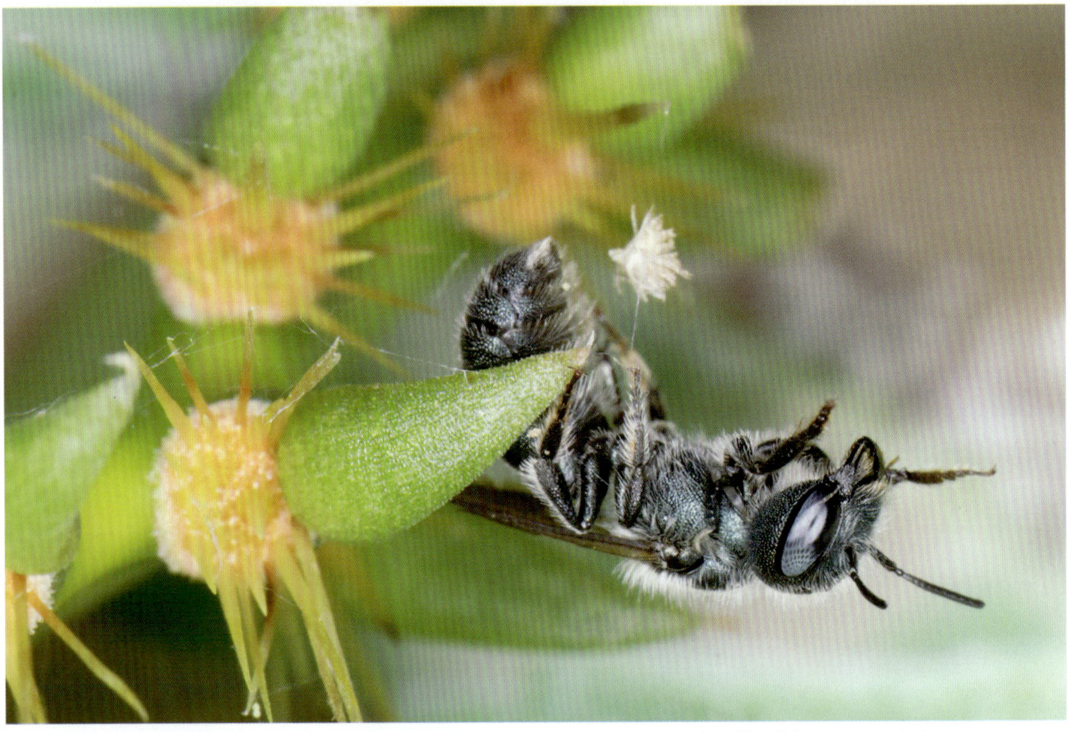

A female *Osmia subfasciata* navigating a prickly pear cactus in search of building materials

Steel-blue mason bee ◆ *Osmia (Helicosmia) chalybea* Smith, 1853

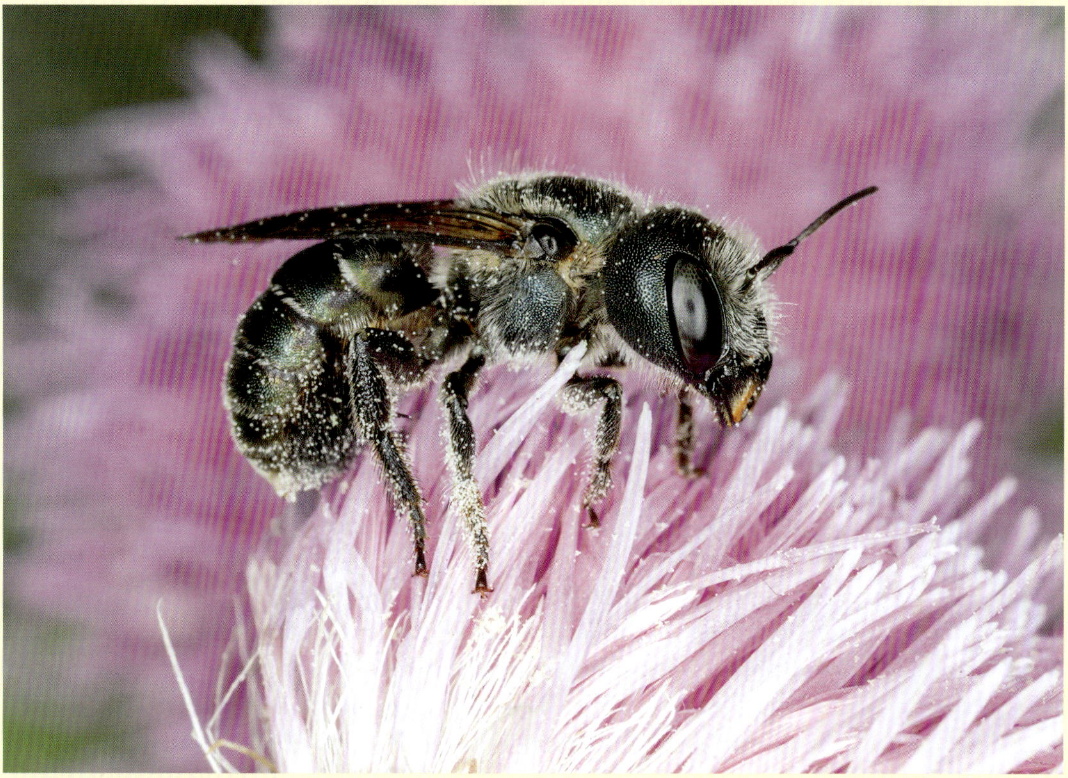

A female *Osmia chalybea* on Texas thistle

Description: *Osmia (Helicosmia) chalybea* is easily recognized by its large size, steel-blue metallic coloring, and association with thistles. Both male and female bees have very dark wings. Their heads and bodies are covered with sparse pale hairs. Females' scopal hairs are black. Females also have distinctive mandibles with short orange hairs near the tips, and depressed areas near the base of each mandible—traits of the subgenus *Helicosmia*. The clypeus of the female *Osmia chalybea* has two toothlike projections in the middle. The male has a tuft of dark hair on T4, and its abdomen is notched on T6. **Size:** females 14–15 mm (1/2–3/5 in.), males 10–11 mm (2/5 in.).

Range: *Osmia chalybea* is a widespread species that ranges along the Eastern Seaboard of the United States from New York to Florida, and across the Gulf States. It is rarely found west of East Texas, although a sighting of the species has been reported in Utah. The bees shown here appeared at the National Butterfly Center (Hidalgo Co.) in late April 2021, feeding on Texas thistle (*Cirsium texanum*).

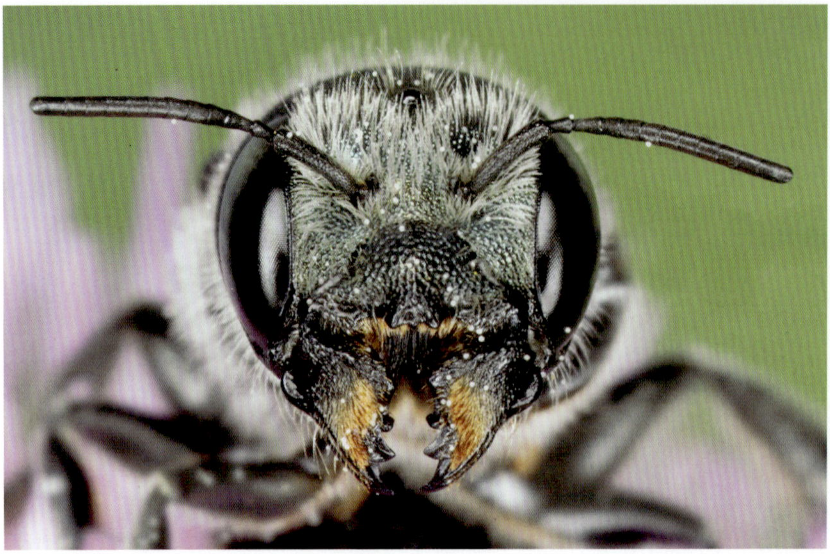

Face of a female *Osmia chalybea*

Associated plants: *Osmia chalybea* usually occurs in association with thistles; in some areas, it also pollinates lilacs and blueberries.

ID tips: The steel-blue mason bee is similar to one other species, sometimes found in South Texas, that shares the subgenus *Helicosmia*—the Texas mason bee (*Osmia texana*). These traits aid in separating the two: (1) *Osmia chalybea* tends to emerge in spring and feeds on a narrow range of plants (usually thistles), while *O. texana* emerges in summer and forages on plants of diverse families; (2) the wings of *Osmia texana* are not as dark, and its ocelli are relatively small; (3) the hairs on the hind tarsi of the female *Osmia chalybea* are black, while on female *O. texana* they are brown; and (4) *Osmia texana* runs somewhat smaller, with females around 12 mm and males around 9 mm.

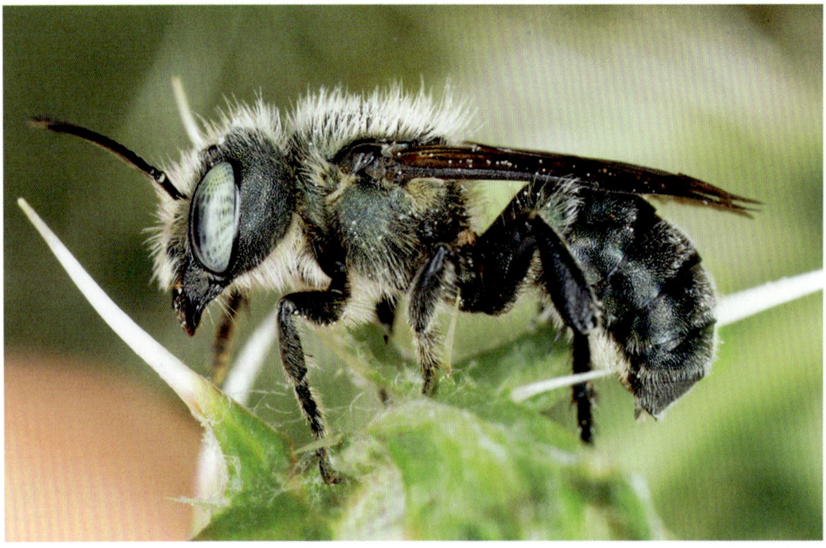

A male *Osmia chalybea*

Resin Bees

Subfamily Megachilinae ◆ Tribe Osmiini
Genus *Heriades*

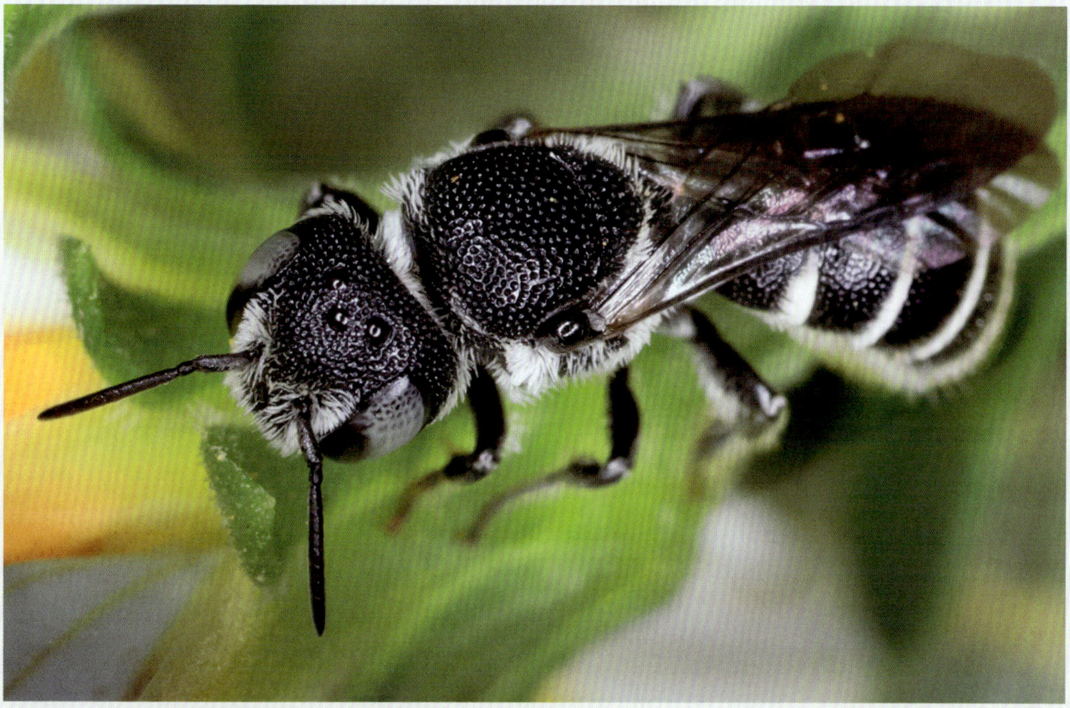

A female variegated resin bee

Heriades are known by the common name "resin bees." These native bees nest in preexisting holes in wood and hollow stems, and they use plant resin to plug entrances to their egg chambers. *Heriades* are specially equipped with mandibles designed to scrape resin from plants. Like the *Osmia* mason bees in the previous section, *Heriades* belong to the tribe Osmiini in the family Megachilidae. Thus, females carry pollen on scopal hairs under their abdomens, the bees' forewings have two submarginal cells, and their feet have arolia.

Heriades of several subgenera are found throughout the world, but a single subgenus, *Neotrypetes*, occurs in North America, ranging from Canada to Panama. Approximately 10 *Heriades* species in this subgenus inhabit the United States, at least seven of which are found in Texas.

Traits of the genus *Heriades*: *Heriades* are small, black, slender, parallel-sided bees with heavily pitted heads and bodies. According to Michener, defining traits of *Heriades* include the following: (1) the front face of the abdomen is concave and (2) bounded by a distinct ridge; and (3) the abdomens of male *Heriades* curl under, the tips nearly touching the front segments, hiding most of the sternum from view. The last segment of the male's abdomen (T7) is concealed by the sixth segment (T6).

North American *Heriades* of the subgenus *Neotrypetes* are black with pale hair bands. Coarse indentations pit their heads, thoraxes, and abdomens. Females' sternal scopal hairs are usually white and often long and conspicuous even to the naked eye, despite the bees' small size. An additional trait helpful in identifying North American *Heriades* is that each segment of the abdomen bells out slightly in the middle and is constricted at the base and rim, as shown here.

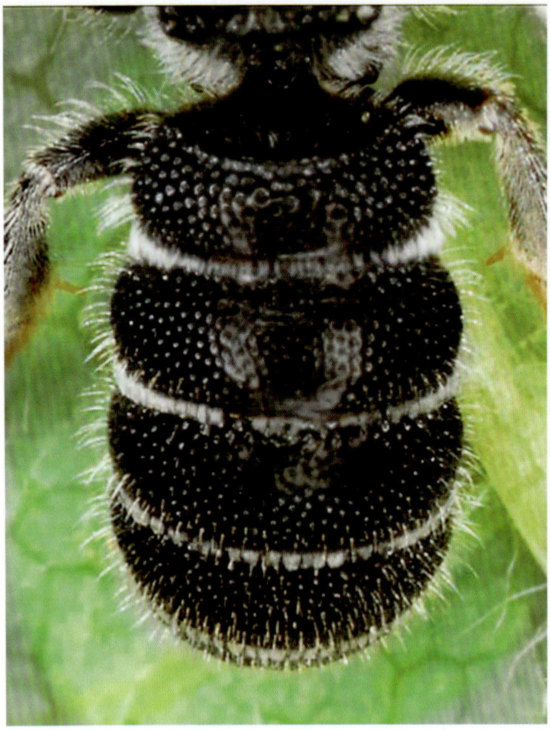

Abdomen of a female *Heriades* resin bee: note how each segment of the abdomen expands slightly in the middle and is pinched on either end.

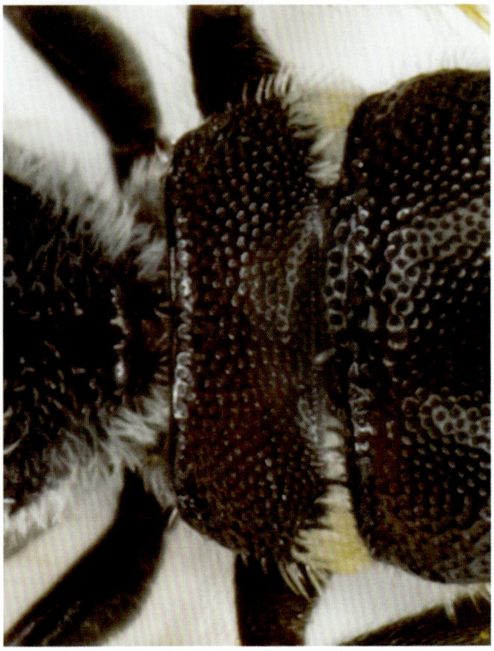

Close-up showing the ridge on the front of the first tergal segment

Variegated resin bee ◆ *Heriades (Neotrypetes) variolosa* (Cresson, 1872)

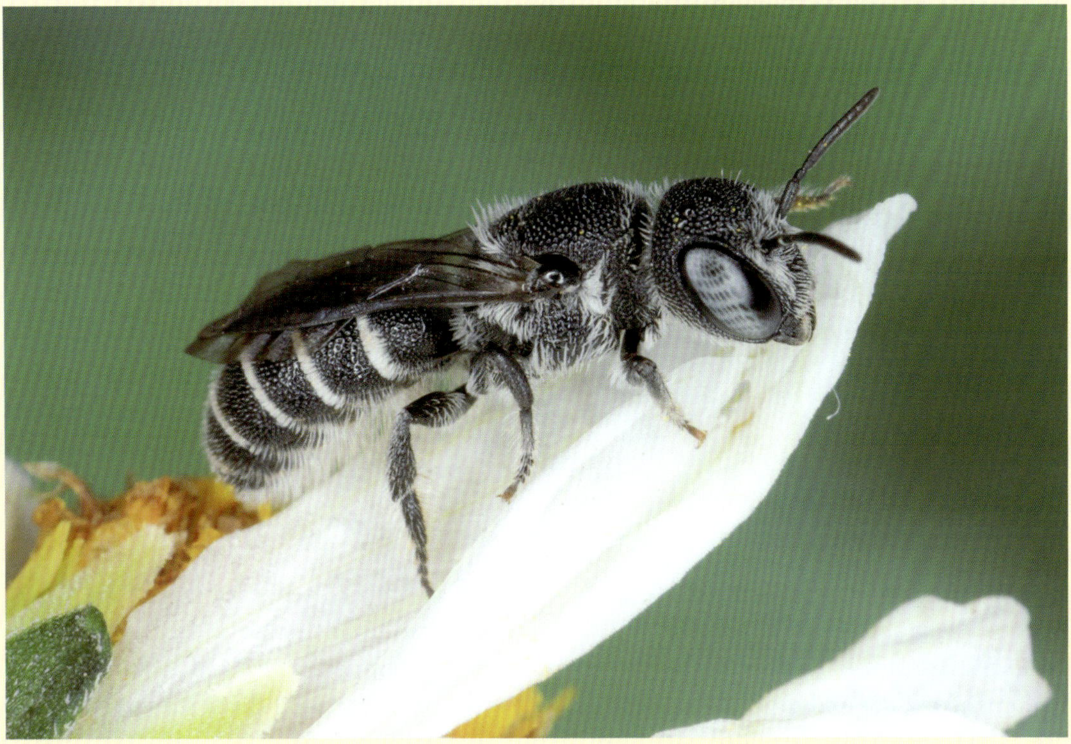

A female variegated resin bee

Description: *Heriades (Neotrypetes) variolosa* is a small black bee with a heavily pitted head and body, and an abdomen girded by narrow bands of white hairs. The tegulae, antennae, and mandibles are black. There are woolly patches of white hairs in front of the tegulae. Females have white scopal hairs on the sternum. On females, all segments of the abdomen are banded; on males, only the first four segments (T1–T4) are banded. The legs of both males and females are mostly black, with pale yellow leg spurs. The front femurs and tibiae of males may be reddish. As is typical of *Heriades* generally, females have three-toothed mandibles, and males have bidentate mandibles. The clypeus of the male is obscured by a patch of long white hairs. Females have blue-gray eyes; males have green eyes. **Size:** females 6–7 mm (1/4 in.), males 5 mm (1/5 in.).

Range: This is a widespread species. *Heriades variolosa* is found throughout Mexico and most of the United States east of California. It ranges as far north as British Columbia, Alberta, and Ontario. It is relatively common in all three border counties of the Valley.

Associated plants: *Heriades* tend to be generalist pollinators. In the Valley, *Heriades variolosa* is commonly found foraging on small-flowered plants such as small coastal germander (*Teucrium cubense*), hierba del marrano (*Symphyotrichum subulatum*), resinbush (*Viguiera stenoloba*), Texas palafox (*Palafoxia texana*), and Texas crownbeard (*Verbesina microptera*).

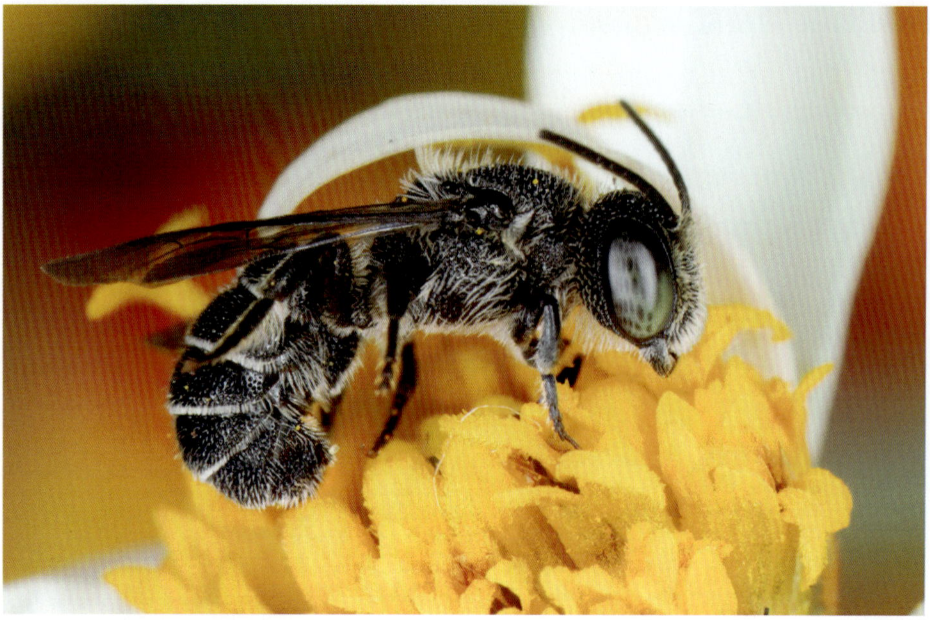

A male variegated resin bee

ID tips: Helpful traits for distinguishing *Heriades variolosa* from similar bees of the subgenus *Neotrypetes* are the following: (1) the pits on the first and second segment (T1–T2) of the female's abdomen are finer and more closely spaced than those on T3; and (2) the first segment of the male's sternum (S1) is elongated with a point on the rear rim.

Additional note on subspecies: A subspecies of *Heriades variolosa* known as the purple variegated resin bee (*H. variolosa purpurascens*) is native to the Mexican border state of Tamaulipas. This subspecies ranges as far south as Yucatán and sometimes strays into South Texas. A female of this subspecies was first identified by Cockerell in 1931. He characterized it as a variation, rather than a subspecies, noting that it differed from the typical *Heriades variolosa* in only one aspect: the female bee has purplish eyes.

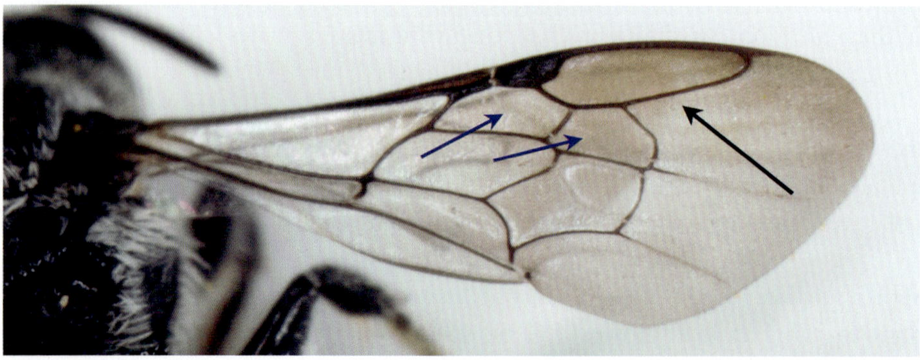

The veins and stigma of the *Heriades variolosa* forewing are black, and the marginal cell (black arrow) is dusky. The forewing has two submarginal cells (blue arrows).

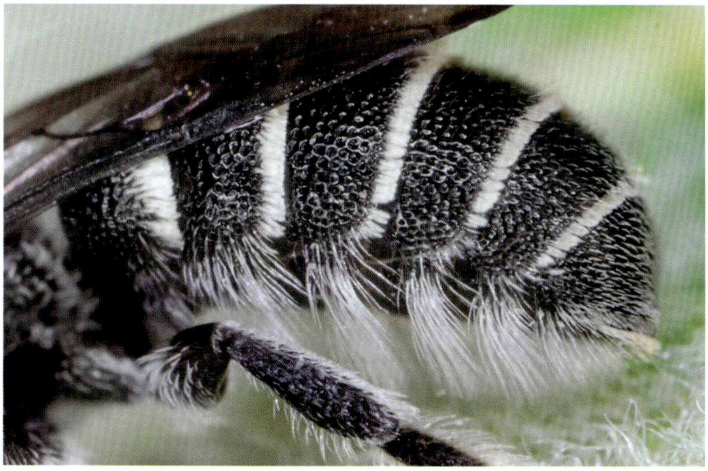

Abdomen of a female *Heriades variolosa*: note the long white scopal hairs and well-defined hair bands.

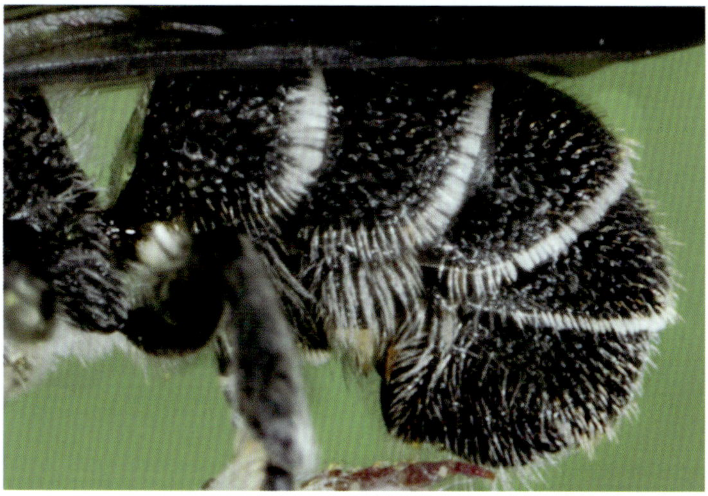

Abdomen of a male *Heriades variolosa*: note the way the abdomen tucks under, hiding the last segment (T7). This is a defining trait of male *Heriades*.

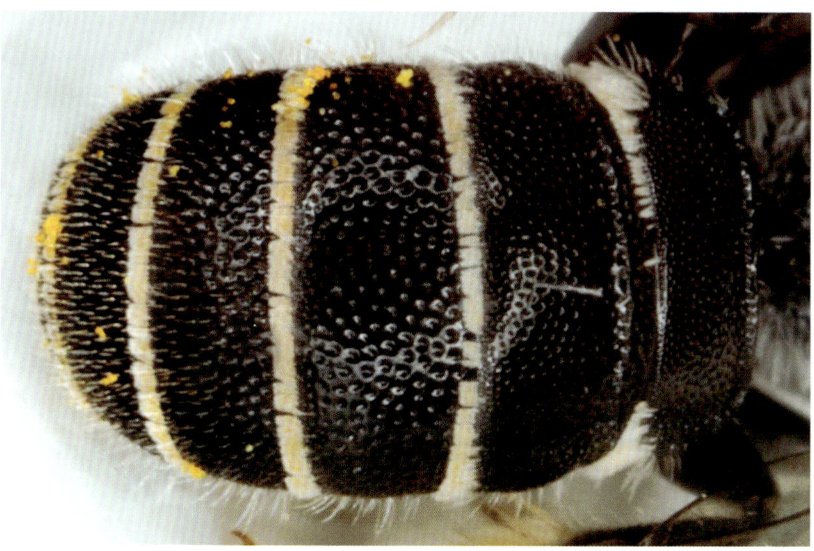

Abdomen of a female *Heriades variolosa*: the pits on T1 and T2 are finer and more closely spaced than those on T3.

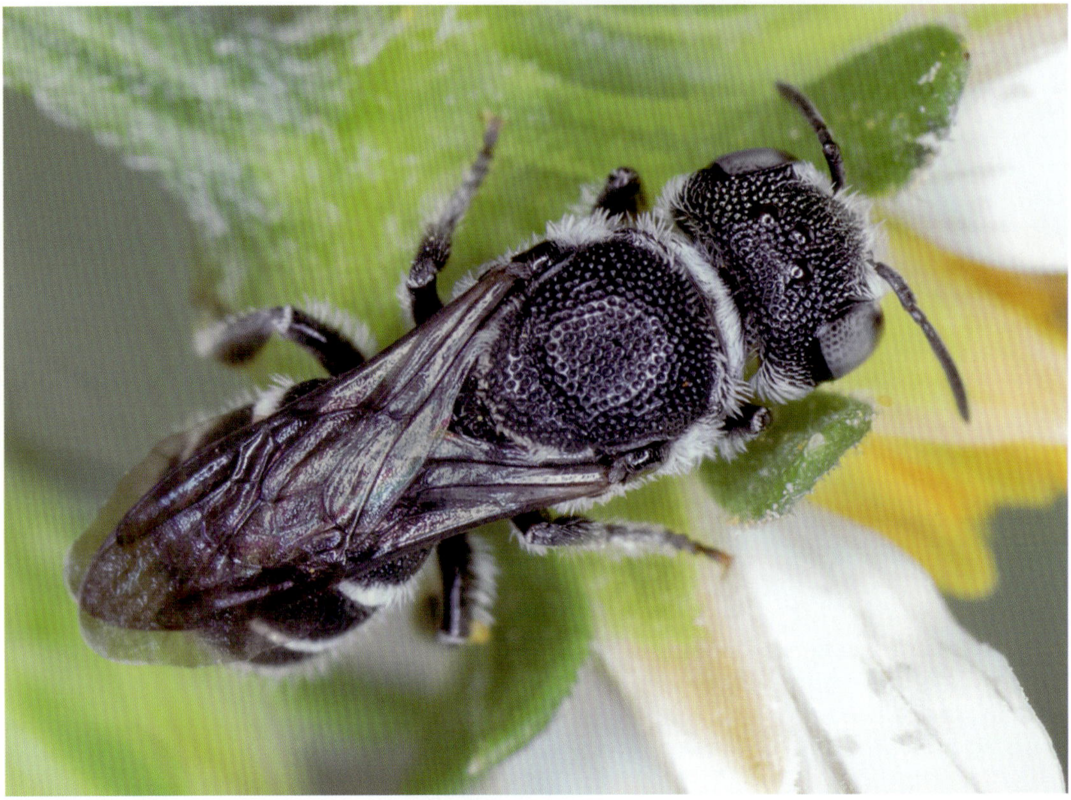

The head and thorax of the variegated resin bee are heavily pitted.

Bee behavior: Although humans like to claim that as *Homo faber*, we are unique in our ability to wield tools, the construction skills of the small *Heriades variolosa* might give us pause. The nest of a variegated resin bee was first described by Fischer in 1955. The nest, found in a dead sumac twig in Minnesota, was assembled in a tunnel that measured a mere 7.5 cm (3 in.) in length. The nest consisted of a series of five chambers arranged in linear fashion to conform to the interior of the twig, while ingeniously allowing for overhead air space within the tunnel. The chambers were divided ornately with double-layered partitions, or "caps," each made of an inner layer of resin and an outer layer of fine sand and soil. Each partition was reinforced by a support made of material resembling crinkled cellophane. This complex structure served to protect and waterproof the chambers, each of which contained an egg and provisions.

Ashmeadiella

Subfamily Megachilinae ♦ Tribe Osmiini
Genus *Ashmeadiella*

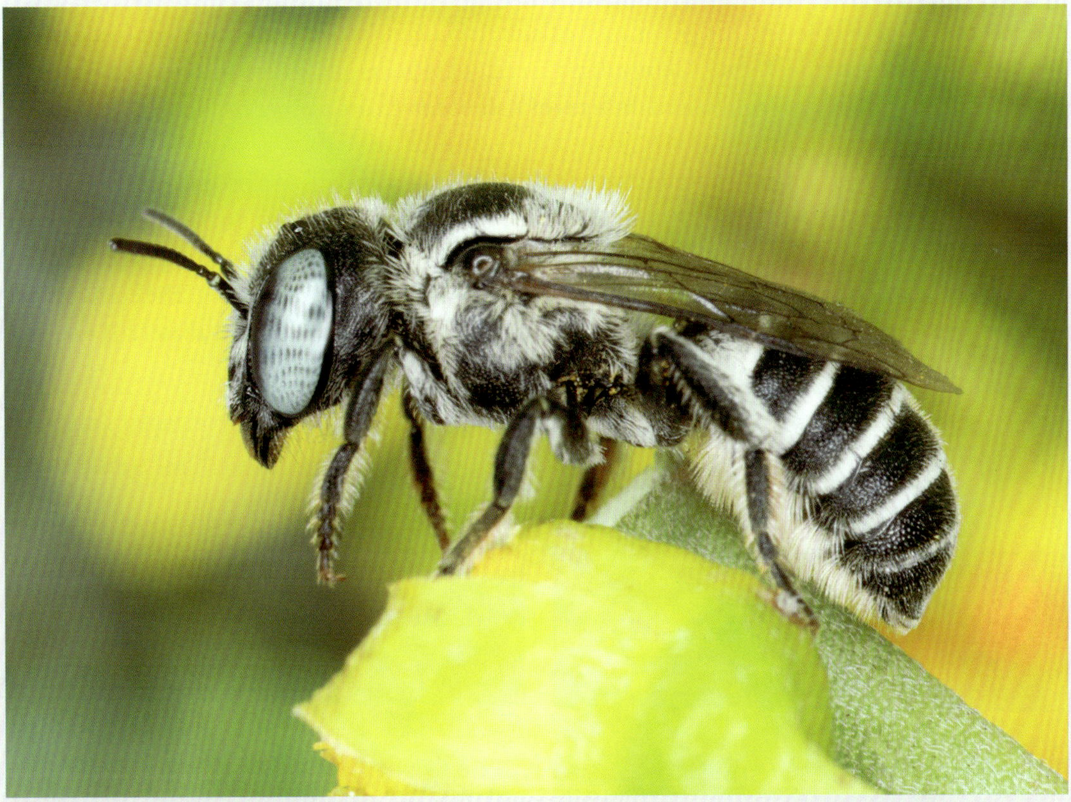

A female *Ashmeadiella*

The genus *Ashmeadiella* of the tribe Osmiini comprises small to medium-small bees with robust builds; nonmetallic, usually black heads and bodies; and abdomens banded with pale hairs. Some species have eye-catching red or partly red abdomens. As with other members of the family Megachilidae, the forewings of *Ashmeadiella* have two submarginal cells, and females transport pollen on scopal hairs under the sternum. The tarsal claws of *Ashmeadiella* possess arolia. The front surface of the first segment of an *Ashmeadiella* abdomen is widely concave and bounded with a ridge.

Ashmeadiella are found only in North America and Central America. In the United States, *Ashmeadiella* occur principally in western desert and semiarid areas. Two species are native to the Valley, and both are found in dry habitats where cactus and other xeric plants abound.

Characteristics of *Ashmeadiella*: *Ashmeadiella* can be separated from similar bee genera by the following two traits: (1) the tip of the male's abdomen (T6) usually has four prongs; and (2) on both males and females, the mesepisternum (side of the thorax) is divided by a weak ridge: the area in front of the ridge is smooth and shiny, and the area behind it is pitted. The second of these traits requires strong magnification to isolate, and it may be obscured under hairs. The distinctive four-pronged abdomen of the male, however, is readily recognizable even to the naked eye and is the best means of diagnosing the genus in the field. *Ashmeadiella* are sometimes confused in the field with *Heriades*, but in Texas, *Ashmeadiella* females are generally larger and stouter; *Heriades* males lack the abdominal prongs; and the eyes of *Ashmeadiella* are a more brilliant blue or green.

Behavior: *Ashmeadiella* tend to fly in the spring and early summer, with some exceptions. All *Ashmeadiella* are solitary cavity nesters. They nest in a variety of places, including preexisting holes in wood, dead or decaying cactus, beetle burrows, shells, or (less often) the ground. *Ashmeadiella* nests usually take the form of a tunnel divided at the end into egg cells. The cells contain one or several eggs and are separated by partitions made of chewed-up plant materials and sap.

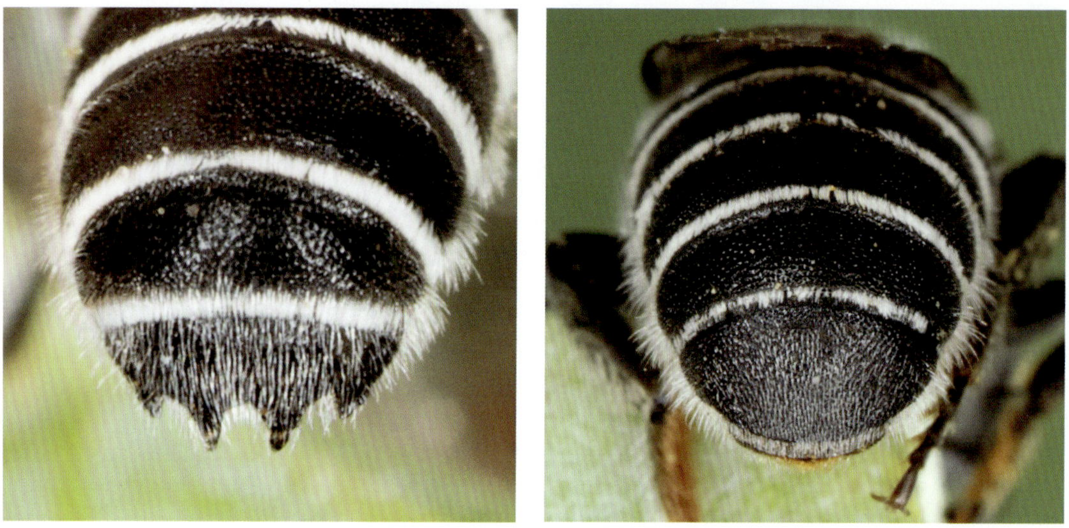

Rear views of the abdomens of a male (*left*) and female (*right*) *Ashmeadiella*

Greater ashmeadiella ◆ *Ashmeadiella (Ashmeadiella) maxima* Michener, 1936

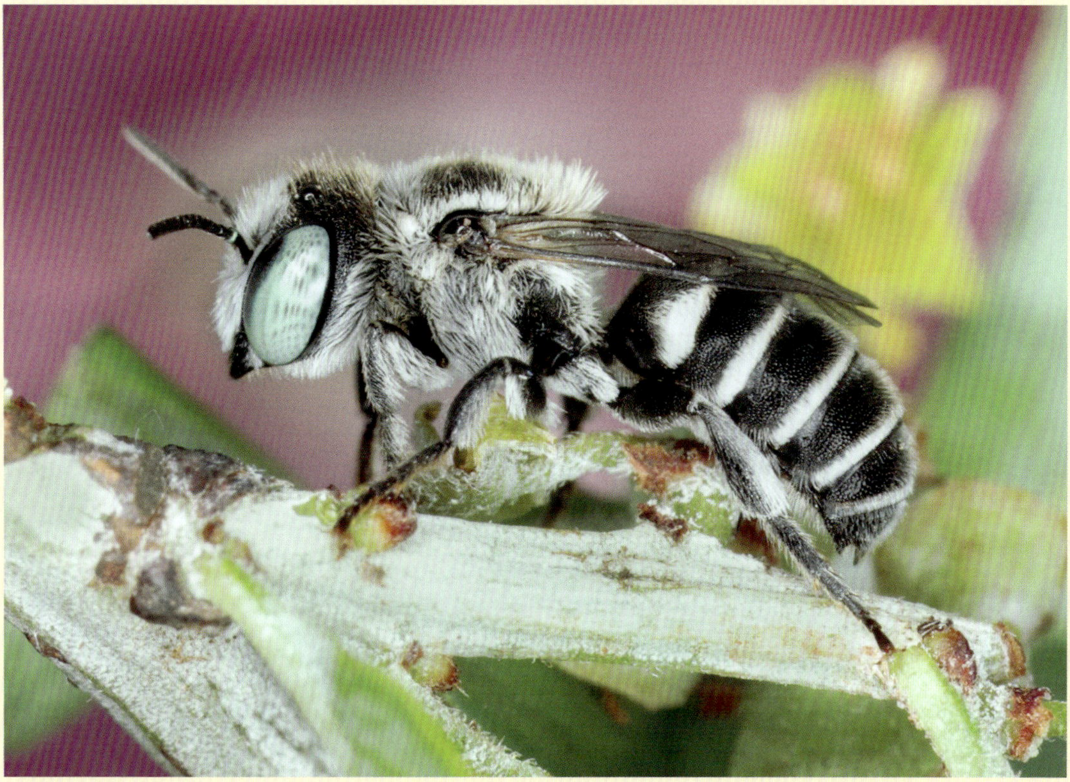

A male *Ashmeadiella maxima*

Description: *Ashmeadiella maxima* is a black bee with a robust build; an abdomen striped by pale hairs; a head and thorax covered with white hairs; black legs clothed with pale hairs; and a broad face with green eyes. The bee's tegulae are brownish black, and its wings transparent light brown with dark brown veins. Females have white scopal hairs on the sternum. On both females and males, the abdominal hair bands appear on T1–T5. On males, T6 bears four long, narrow prongs. **Size:** females 7–9 mm (1/4–1/3 in.), males 8 mm (3/10 in.).

Range: In the United States, *Ashmeadiella maxima* is found principally along southern border areas of Arizona, New Mexico, and Texas. This species appeared at the Pixie Preserve (Hidalgo Co.) in early May 2021.

Associated plants: The male bees shown here were feeding on the small-flowered native shrub known as lotebush (*Ziziphus obtusifolia*). *Ashmeadiella maxima* has also been documented visiting prickly pear (*Opuntia*), sage (*Salvia*), and Gregg's prairie clover (*Dalea greggii*).

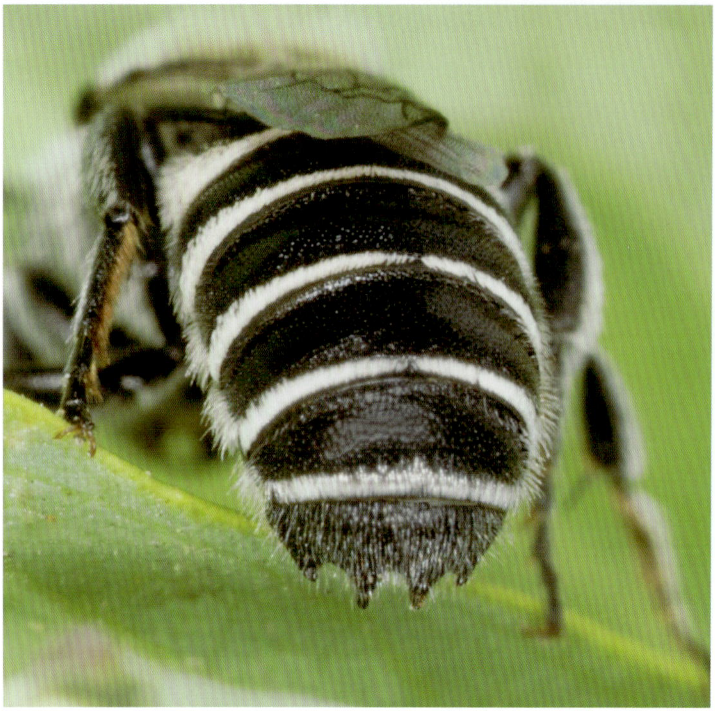

The abdomen tip of a male *Ashmeadiella maxima*

ID tips: As its name implies, *Ashmeadiella maxima* can be recognized by its size—it is large for its genus. The black tarsi, the absence of any red body color, and the dense white hairs on the male's face also help distinguish this species from other *Ashmeadiella.* Male *Ashmeadiella* species are separated in part by characteristics of the prongs on the abdomens, and by the positioning of the ocelli on the vertex. On *Ashmeadiella maxima* males, the four prongs at the end of the abdomen (T6) are slender and well separated (by gaps more than twice the width of each prong), and the depression between the middle prongs is deeper than a semicircle. The male's front ocellus is at a midpoint between the bases of the antennae and the hind edge of the vertex, and its two outer hind ocelli are widely separated.

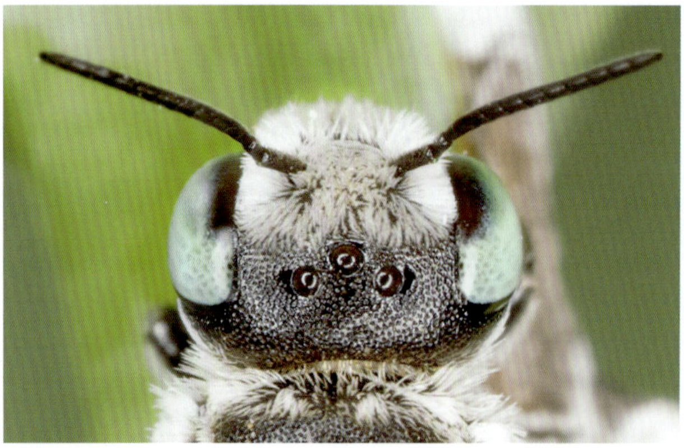

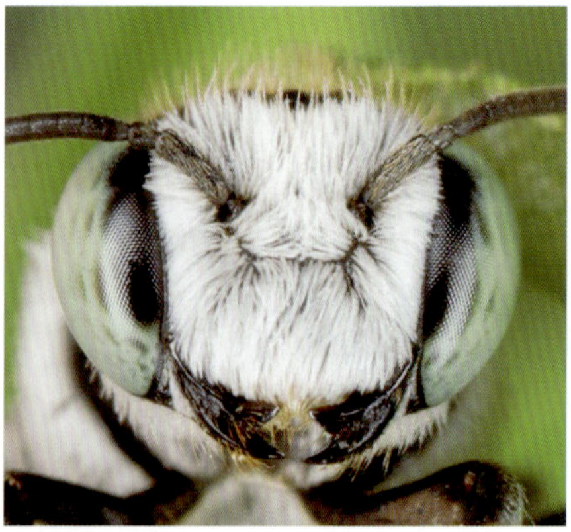

The vertex and densely haired face of a male *Ashmeadiella maxima*

Cactus ashmeadiella ◆ *Ashmeadiella (Ashmeadiella) cactorum* (Cockerell, 1897)

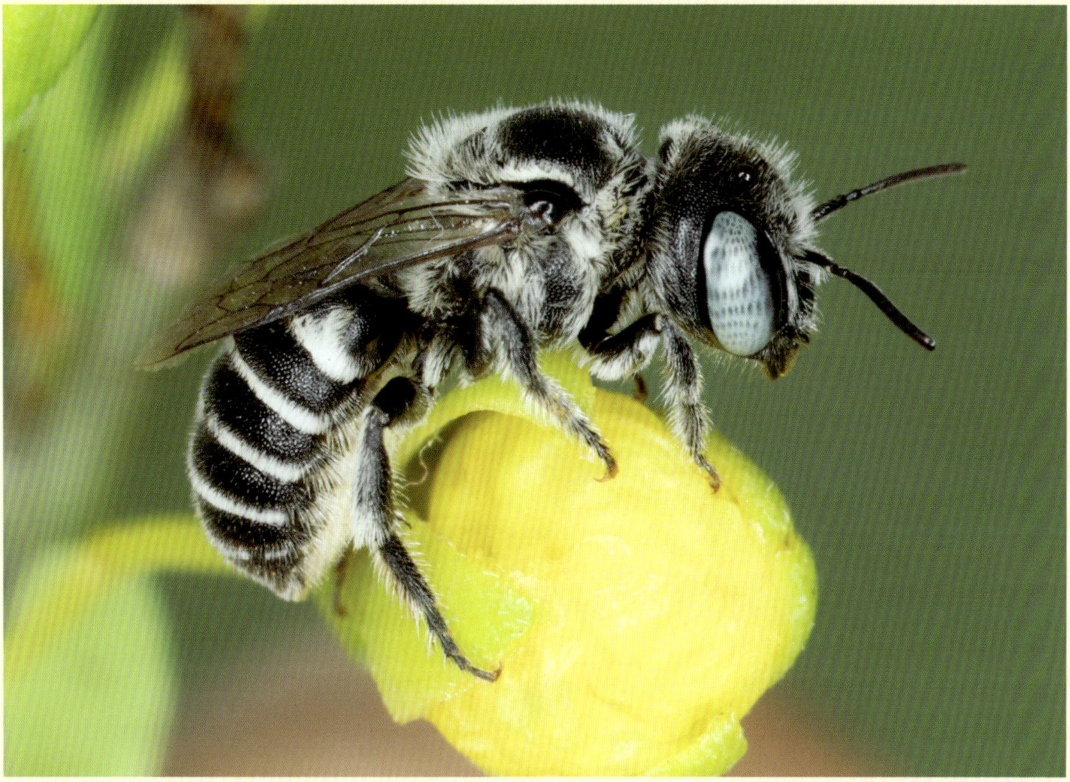

A female *Ashmeadiella cactorum* on Texas paloverde

Description: *Ashmeadiella cactorum* is a moderately small black bee with a coarsely punctate body; an abdomen striped by pale hair bands on T1–T5; a thorax and face sparsely covered with white hairs; black legs; and blue eyes. The tegulae are black and the wings are glassy brown with dark brown veins. Females have yellowish-white scopal hairs under their abdomens. On females, the hair band on T5 is sparse in the middle and less well defined than the bands on T1–T4. On males, the hind margins of the abdominal segments are brownish, and T6 bears four long, narrow prongs. **Size:** females 6–8 mm (1/4–3/10 in.), males 6 mm (1/4 in.).

Range: *Ashmeadiella cactorum* is found in the southwestern and western United States and throughout Mexico. It occurs in both Starr and Hidalgo Counties.

Associated plants: Despite its name, *Ashmeadiella cactorum* is not a cactus specialist. It is a generalist pollinator that visits plants from a broad range of families. The female bees shown here were found gathering pollen from Texas paloverde (*Parkinsonia texana*) at the La Puerta Tract (Starr Co.) in early May 2021; and from Arkansas dozedaisy (*Aphanostephus skirrhobasis*) and bladdermallow (*Herissantia crispa*) at El Mesteño Ranch (northwest Hidalgo Co.) in late June 2021.

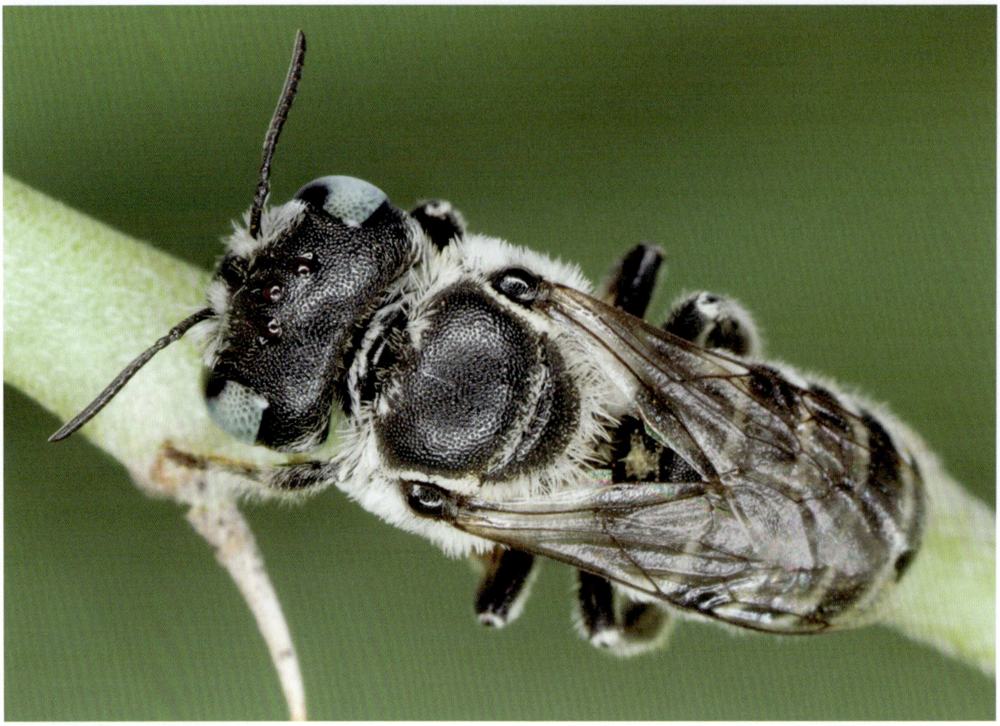

A female *Ashmeadiella cactorum*

ID tips: *Ashmeadiella cactorum* is smaller than *A. maxima* and less densely haired, and lacks its broad face. The face of the female *Ashmeadiella cactorum* is distinctive: the tip of its clypeus is finely punctate and its bottom edge slightly concave. The female also has sparse orange hairs on the outer mandibles. All females of the subgenus *Ashmeadiella* have three-toothed mandibles.

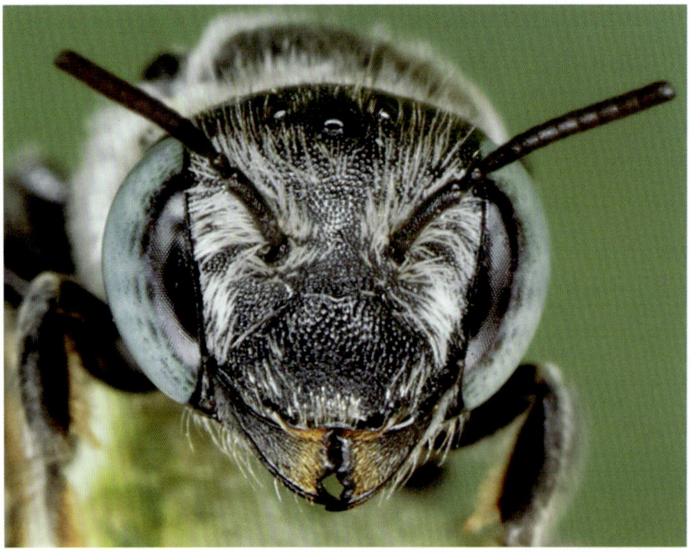

Face of a female *Ashmeadiella cactorum*

Pebble Bees

Subfamily Megachilinae ◆ Tribe Anthidiini
Genus *Dianthidium*

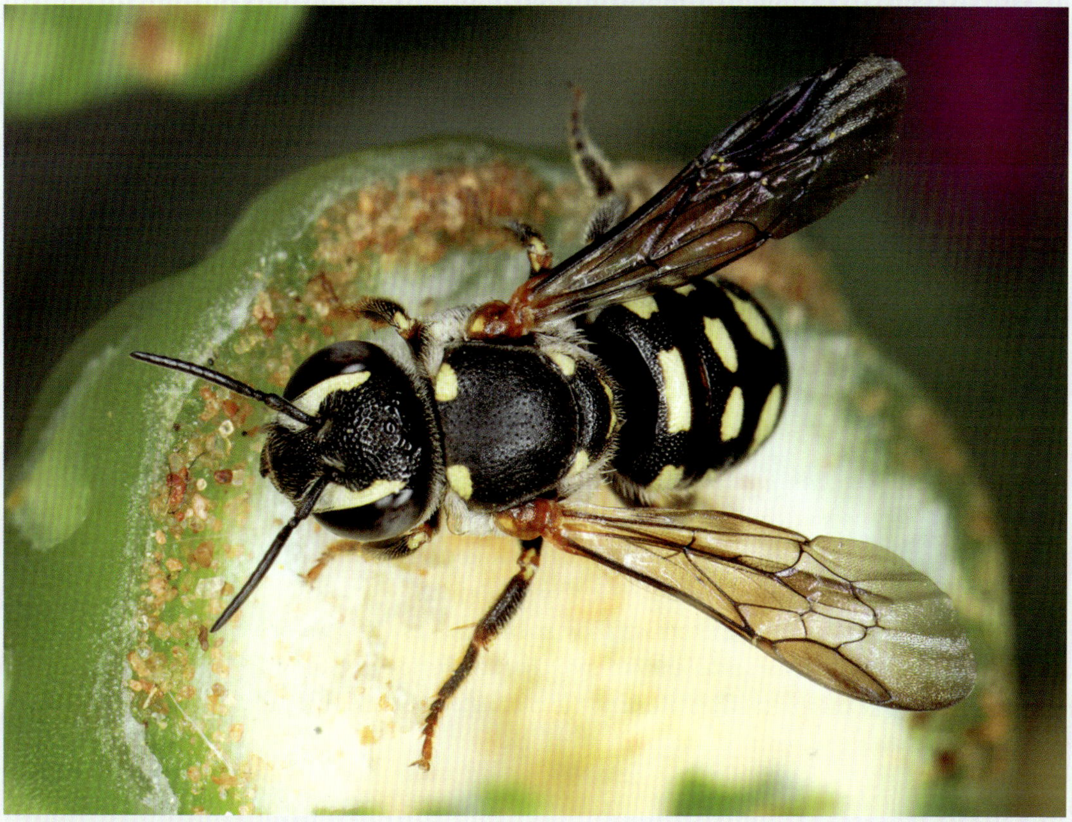

A female *Dianthidium discors*

Dianthidium is one of more than 30 genera in the Anthidiini tribe. Known by the common name "pebble bees," *Dianthidium* mix plant resin with materials such as soil, mud, pebbles, or plant parts to construct nests. *Dianthidium* are solitary, but sometimes their nests are conjoined or constructed near one another.

Dianthidium usually have bold yellow markings, and their bodies tend to be relatively hairless: they can be mistaken for robust wasps. Within a given species, the markings on males and females may vary considerably. *Dianthidium* carry pollen on scopal hairs under their abdomens, their forewings have two submarginal cells, and their feet have arolia.

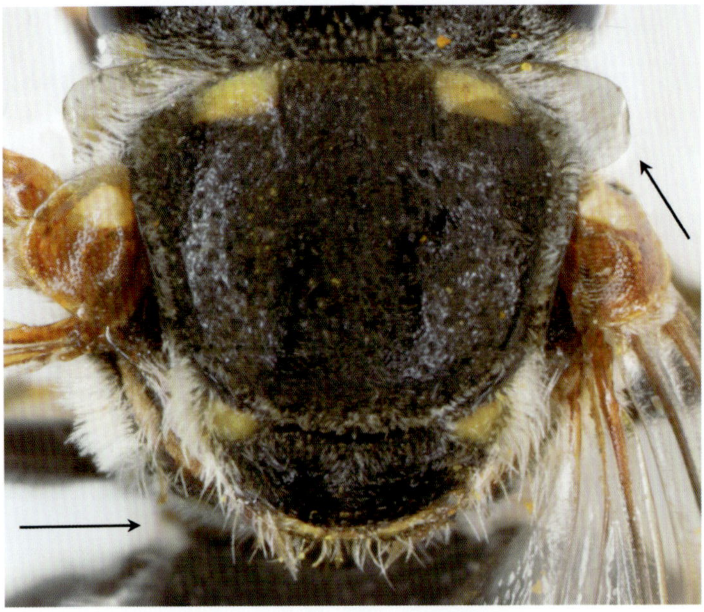

Pronotal lobes and scutellum of a male *Dianthidium discors*

Traits of the genus *Dianthidium*: Identifying characteristics of the genus *Dianthidium* include the following: (1) as shown here, the rear margin of the scutellum is semicircular and convex; (2) the pronotal lobes are expanded and have thin and translucent edges; and (3) the male has a prominent spine on each hind coxa, visible from the side.

Dianthidium species are differentiated by such traits as the pale or yellow patterns on the face, body, and legs; the bee's size; and minute characteristics such as qualities of the mandibles, the structure of T7 on males, the nature of the pitting on the thorax and face, and the presence, location, and quality of hairs on the body and legs.

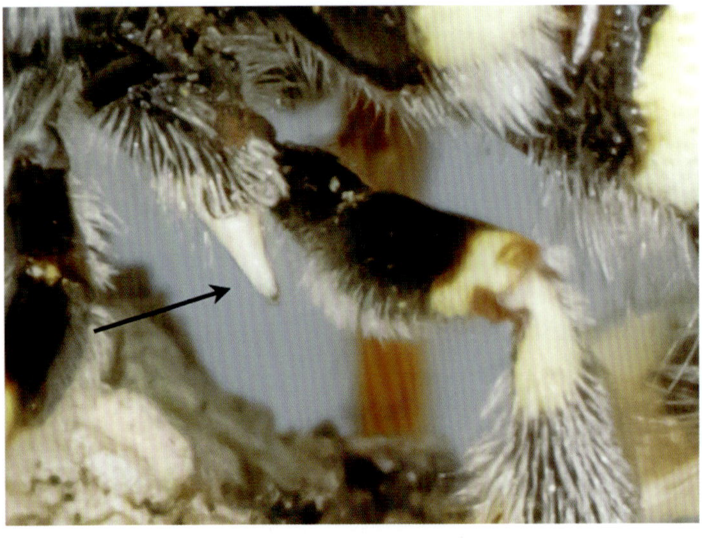

Coxal spine of a male *Dianthidium discors*

Discordant pebble bee ◆ *Dianthidium (Dianthidium) discors* Timberlake, 1948

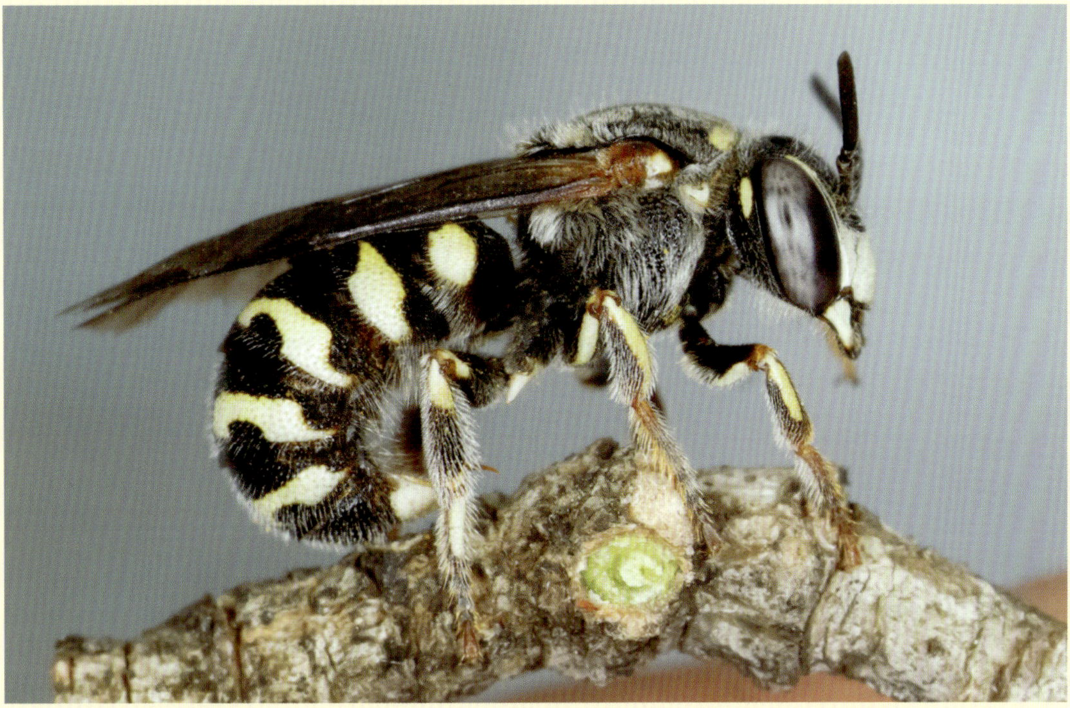

A male *Dianthidium discors*

Description: *Dianthidium (Dianthidium) discors* is a striking medium-sized bee with a hefty build. Yellow markings adorn the bee's face, mandibles, cheeks, upper legs, tegulae, pronotal collar, pronotal lobes, and abdomen. The head and body are principally black, and heavily pitted. The lower legs are black and reddish. The wings are smoky black, and the tegulae reddish with yellow spots near the front rims. The yellow abdominal patterns of females and males differ. A series of oblong yellow spots bedeck the female's first five terga (T1–T5). Males have jagged, curved yellow bands on T1–T5; the male's T7 is entirely yellow and shaped like a bow tie. The female's sternal scopal hairs are white. **Size:** females 8 mm (3/10 in.), males 9–10 mm (1/3–2/5 in.).

Range: *Dianthidium* are found in North and Central America, from Canada through Costa Rica. In the United States, they occur predominantly in the West. *Dianthidium discors* is found principally in Arizona and West Texas. Nonetheless, the male and female bees shown here appeared in southern Starr County, at La Puerta Tract and Campos Viejos Ranch in early May and mid-June 2021. Both locations were dominated by Tamaulipan thornscrub and sandy soils. Timberlake first described this species in 1942 after examining a female bee discovered at an elevation of 4,200 feet in Sierra Blanca, Texas.

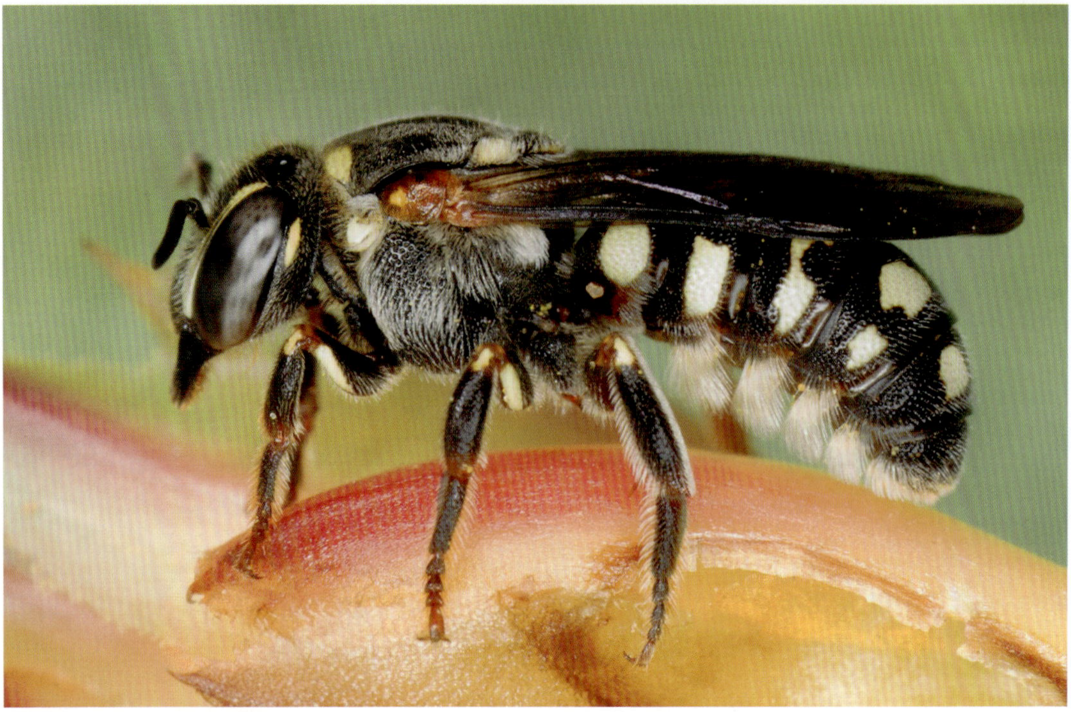

A female *Dianthidium discors*

Associated plants: *Dianthidium discors* is associated with creosote bush (*Larrea tridentata*). The male bee shown here was one of many males visiting native guayacan (*Guaiacum angustifolium*) in early May 2021 at La Puerta Tract. This plant, like *Larrea*, is a member of the family Zygophyllaceae. The female bee shown here was observed with dozens of other females and males visiting candy barrel cactus at Campos Viejos (*Ferocactus wislizeni*).

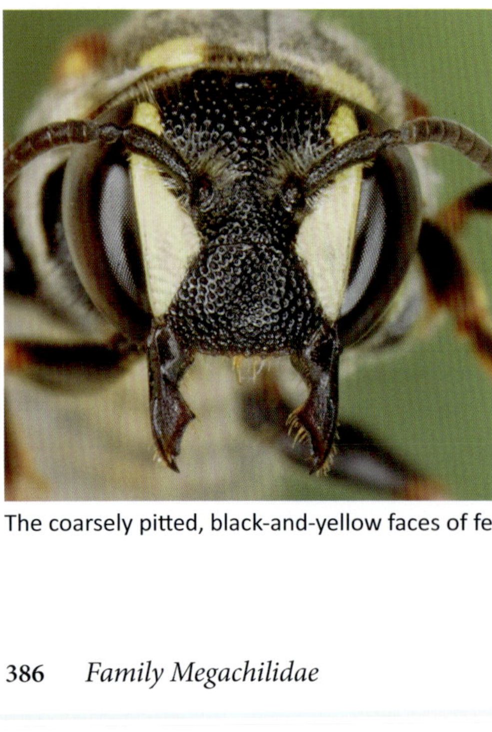

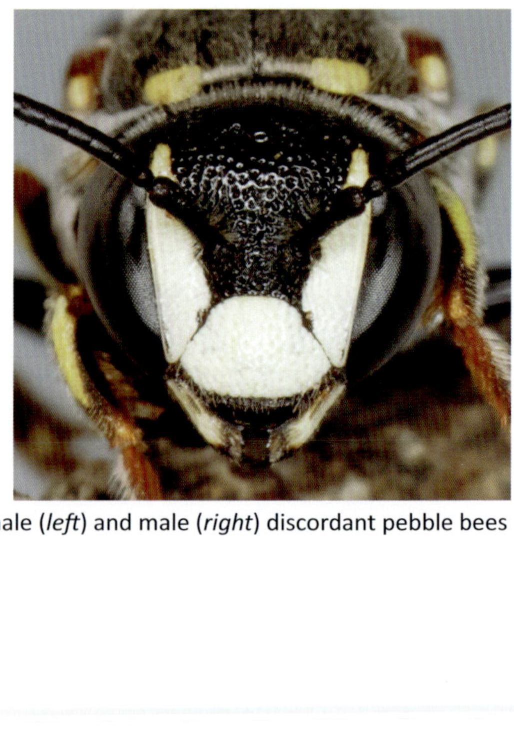

The coarsely pitted, black-and-yellow faces of female (*left*) and male (*right*) discordant pebble bees

ID tips: The yellow patterns on the abdomens of male and female *Dianthidium discors*, shown here, differ enough to make the bees appear at first glance to belong to separate species. Both patterns, however, are distinctive and aid in identifying *Dianthidium discors*. Other features that distinguish this species include the following: (1) the female's clypeus is entirely dark; (2) males have yellow markings covering most of the clypeus, labrum, and mandibles; and (3) the females' and males' thoraxes are largely black, with yellow spots on the front edges of the scutum and scutellum, and a narrow yellow medial strip lining the scutellum's back edge. Timberlake wrote when describing the species in 1948 that the female *Dianthidium discors* can be distinguished from all other *Dianthidium* by two minute traits: its frons is coarsely pitted, and its scutum is doubly punctate with alternating large and small pits.

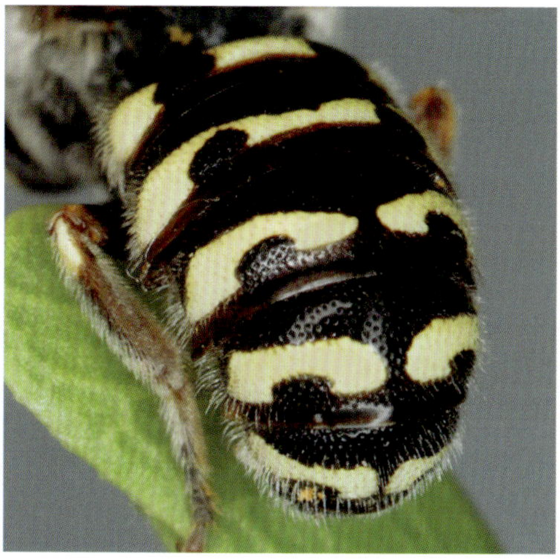

Markings on the abdomen of a male *Dianthidium discors*

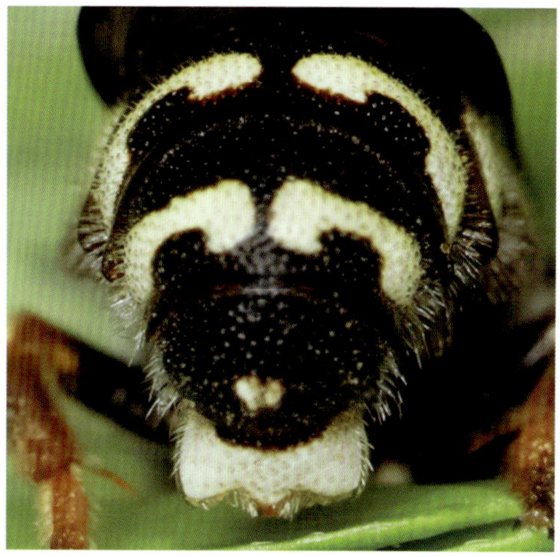

The male's distinctive yellow T7

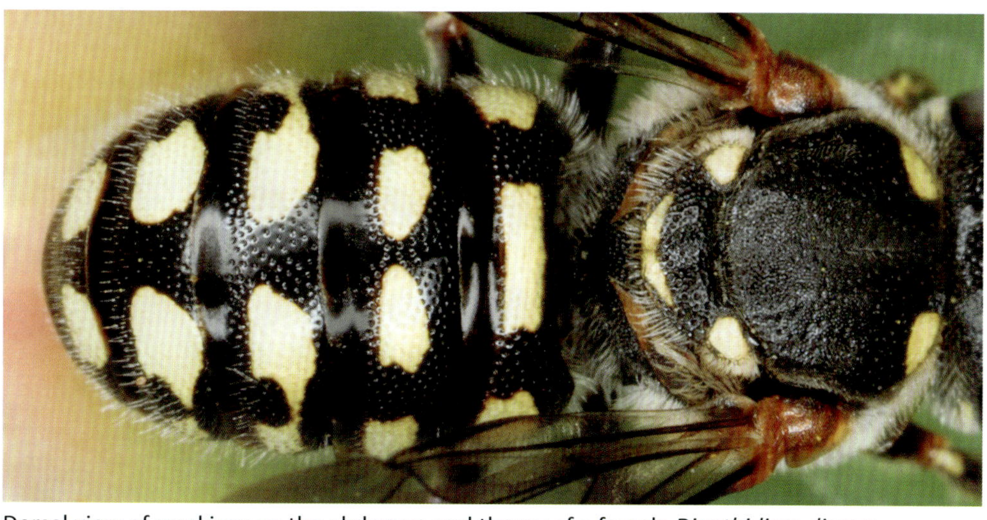

Dorsal view of markings on the abdomen and thorax of a female *Dianthidium discors*

Behavior: In mid-June 2021 at the Campos Viejos Ranch, the author observed numerous male and female bees engaging in a mating ritual consisting of the following behavior: Female *Dianthidium discors* would land on candy barrel cacti and, without stopping to collect pollen or drink nectar, would search for shallow holes in the stalks of the cactus blossoms or buds. The females then buried their heads in the holes while pointing their exposed abdomens upward, displaying their bold yellow-and-black patterns. This behavior attracted male bees, which would descend and mate with the females directly on the cactus flowers.

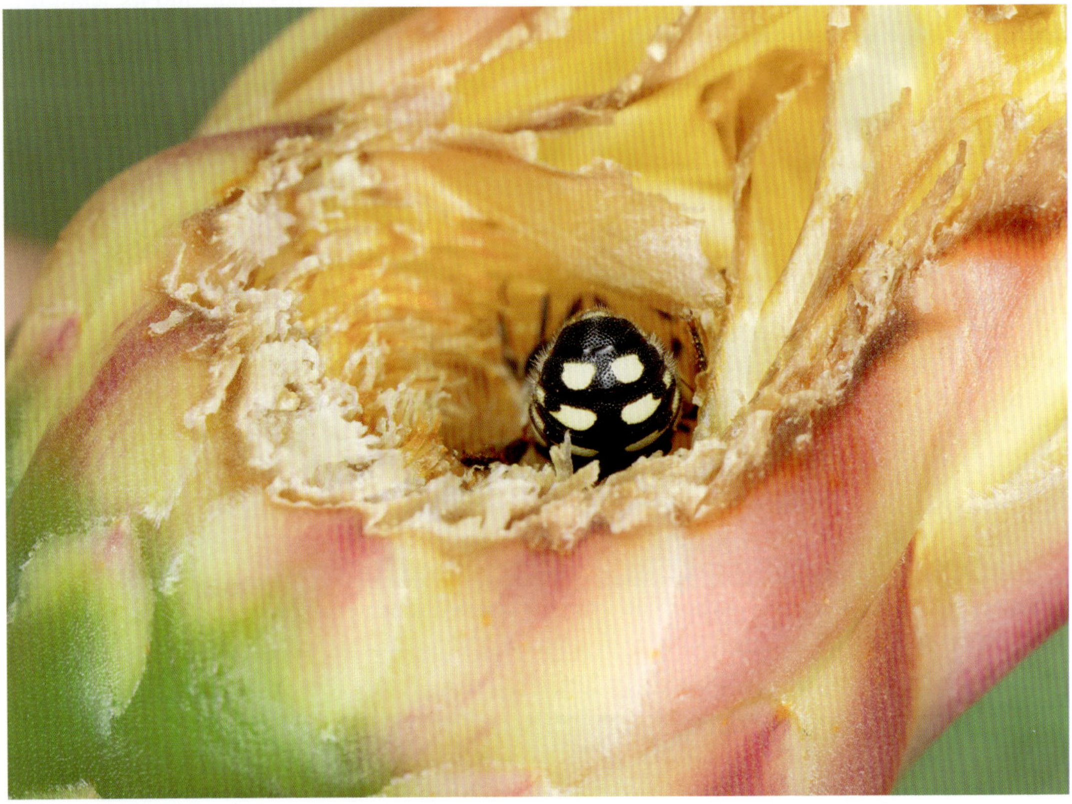

A female *Dianthidium discors* flashing its spotted abdomen, its head buried in a hole bored into a candy barrel cactus blossom

Stelis Cuckoo Bees

Subfamily Megachilinae ◆ Tribe Anthidiini
Genus *Stelis*

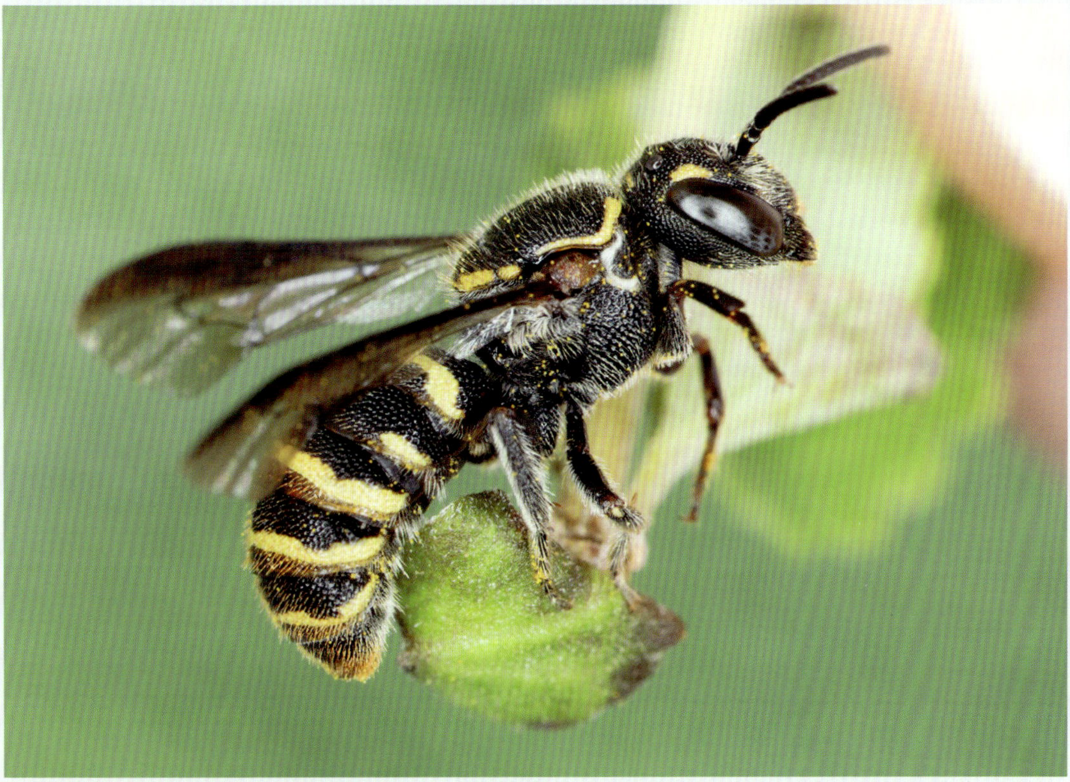

A female *Stelis costalis* cuckoo bee

Stelis, like *Dianthidium* in the preceding entry, belong to the tribe Anthidiini. *Stelis* are a genus of cuckoo bee that typically parasitizes the nests of other Megachilinae. Stelis cuckoo bees occur throughout the Northern Hemisphere, in North and Central America as well as Eurasia. In the United States, *Stelis* range widely from California to the East Coast, and from southern Canada to the Mexican border. From there, *Stelis* range through Mexico and Central America as far as Costa Rica. Stelis cuckoo bees are not found in South America.

Traits of the genus *Stelis*: Stelis cuckoo bees are small to moderate in size (4–14 mm), and usually black or a dark metallic blue or green. Their bodies tend to be heavily pitted, and they often have yellow or white patterns on their abdomens; some have reddish-orange abdominal markings, and some have colorful markings on the face and thorax as well as the abdomen. *Stelis* share traits typical of other Anthidiini: they have forewings with two submarginal cells and arolia on their feet—female *Stelis*, however, lack scopal hairs under their abdomens because, like other cuckoo bees, they do not gather pollen but instead rely on the pollen stores of their hosts.

***Stelis* behavior:** In the Western Hemisphere, *Stelis* parasitize the nests of bees belonging to the tribes Osmiini, Megachilini, and even their own tribe, Anthidiini. Many North American *Stelis* target resin bees of the subgenus *Megachile (Chelostomoides)*.

Most *Stelis* enter their hosts' nests while the hosts are still building their egg chambers, before the cells are sealed. The cuckoo larvae that later hatch use their mandibles to destroy the hosts' eggs or larvae and then devour pollen stores left behind by the hosts. Bees of the subgenus *Dolichostelis*, however, such as the ribbed stelis in the next entry, operate differently. *Stelis (Dolichostelis)* are "closed-cell parasites"; that is, they break into their hosts' nests after the egg chambers are sealed. The cuckoo bees then dispatch the hosts' eggs and replace them with *Stelis* eggs.

In 1987, at a nest site in Logan, Utah, entomologist F. D. Parker and his colleagues reported observing a female *Stelis (Dolichostelis) rudbeckiarum* chew through an entrance plug on a resin bee nest. As it worked, the cuckoo deposited "glistening droplets from the tip of its abdomen . . . the substance, acting as a solvent, enabled the *Stelis* to soften and mold the resin."[2]

The cuckoo removed a portion of the resin plug, slept overnight near the nest entrance, and upon awakening the following morning, pulled the resin back into place, smoothed it over, and flew away. Resin bee hosts defend themselves from such behavior by reopening cells and removing the pollen provisions in them, or by resealing cells with a fortified mixture of resin and leaf pulp.

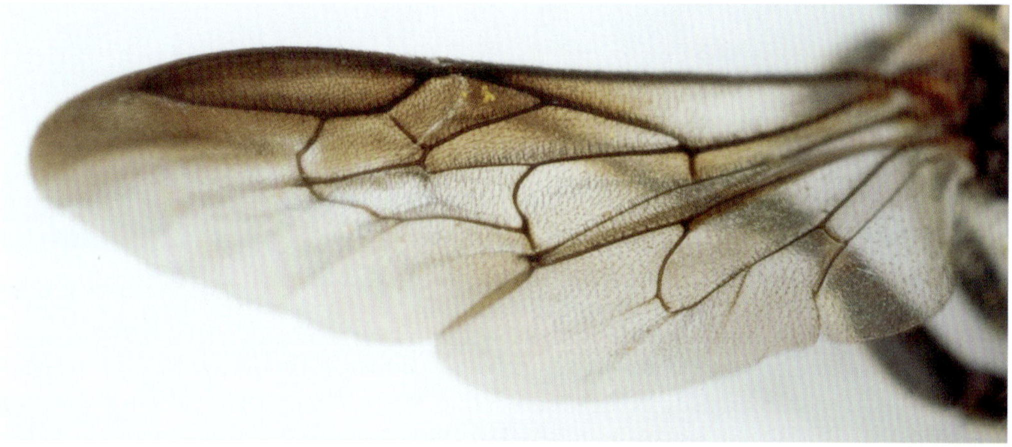

The forewing and hind wing of *Stelis costalis*

2. Parker et al. (1987), pp. 176–177.

Ribbed stelis cuckoo bee ◆ *Stelis (Dolichostelis) costalis* Cresson, 1872

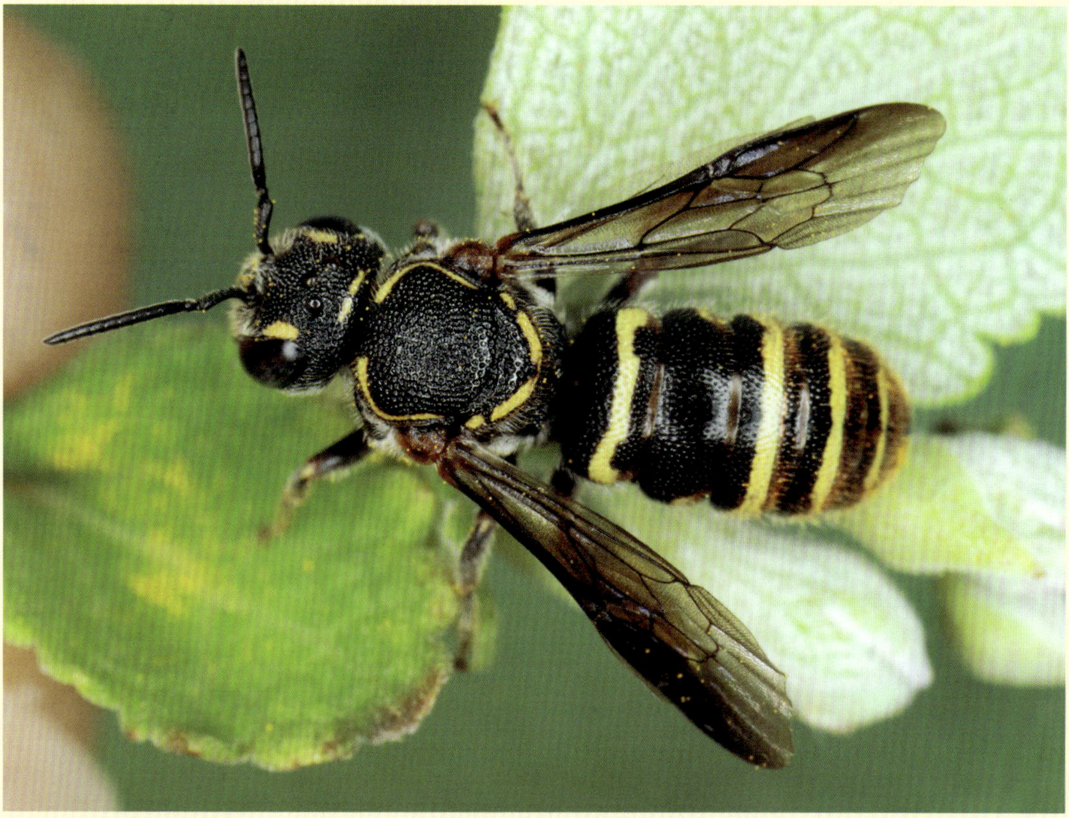

A female *Stelis costalis*

Description: *Stelis costalis* has a dark, elongate, coarsely pitted head and body, embellished with yellow and reddish markings. The bee's legs may be all black, black and dark red, or entirely dark red. The wings are transparent brown, with dark brown veins and brownish-black outer edges, sometimes with bluish reflections. The tegulae are reddish brown to honey yellow. The thorax (scutellum and scutum) of *Stelis costalis* is outlined by a narrow, nearly continuously curved yellow border. The abdomen is banded by several yellow stripes (T1 and T3–T5 on females, and T1 and T3–T6 on males), and it has two lateral spots on T2.

Both male and female *Stelis costalis* have predominantly dark faces, dark antennae, and dark blue-gray eyes. Oblong yellow markings hug the inner rims of both the females' and males' upper compound eyes; a yellow line also rims the back of the vertex of both males and females. The male has two additional yellow spots on the bottom edge of the clypeus. Sparse white hairs partly cover the face, vertex, thorax, legs, and sternum, and erect black hairs sprout from the center of T6. **Size:** females 9 mm (1/3 in.), males 8 mm (3/10 in.).

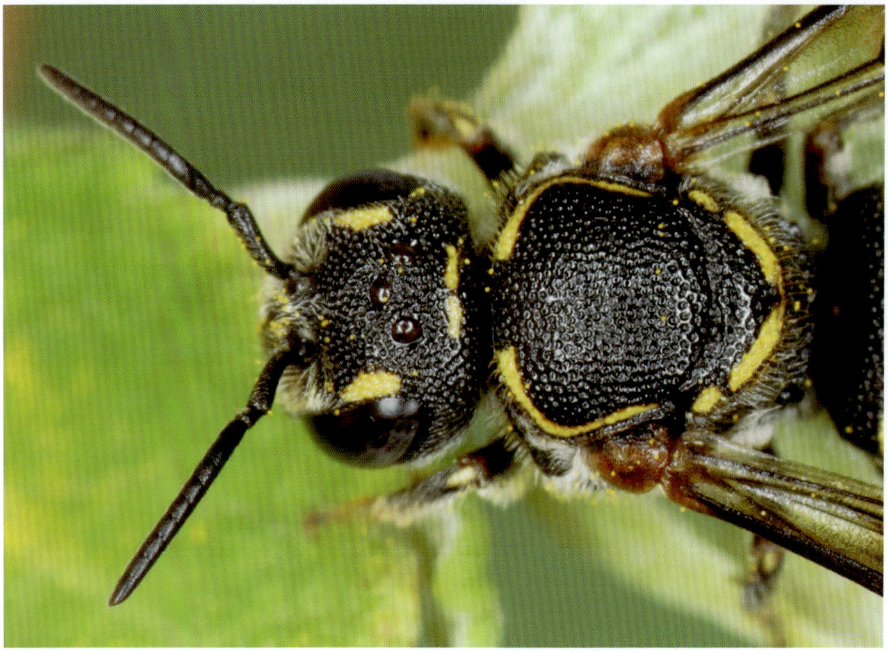

Dorsal view of the head and thorax of a female *Stelis costalis*

Range: The female bee shown here appeared in mid-June 2021 in Harlingen (Cameron Co.). *Stelis costalis* was first identified by Cresson in his 1872 "Hymenoptera Texana," based on two bees found in Bosque and Comal Counties in Texas. In the United States, the distribution of this species is centered in South Texas. Its population reaches into western Texas, and as far north as Dallas County. *Stelis costalis* occurs throughout Mexico, from Baja California and the Texas border south to Yucatán. There are isolated records of *Stelis costalis* in North Florida and Illinois.

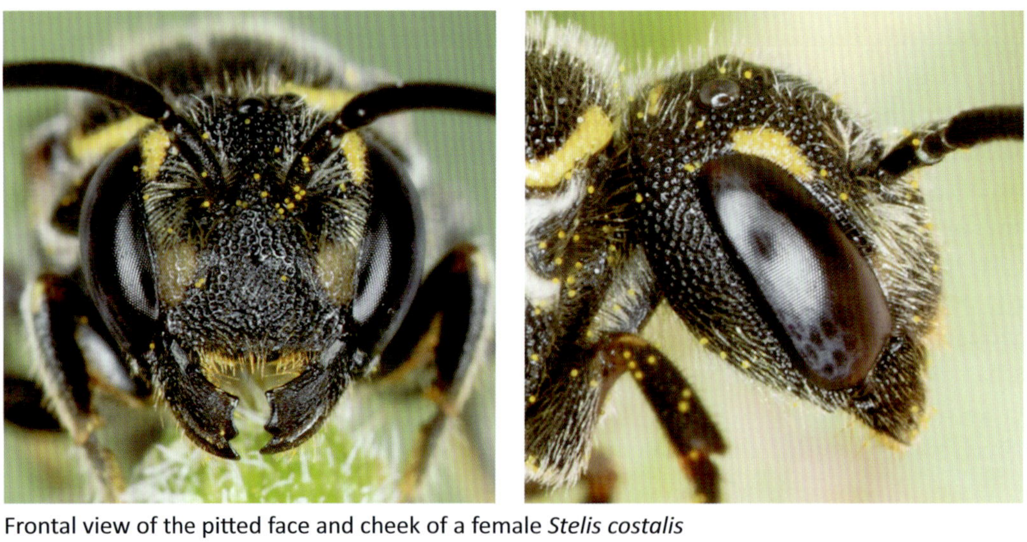

Frontal view of the pitted face and cheek of a female *Stelis costalis*

Associated plants: *Stelis costalis* visits a range of plants. The bee featured here was discovered nectaring on the yellow composite hairy wedelia (*Wedelia acapulcensis* var. *hispida*) in a garden frequented by many *Megachile (Chelostomoides) exilis*. In the Lower Rio Grande Valley and Mexico, *Stelis costalis* has also been documented visiting coral vine (*Antigonon leptopus*), a pink-flowered plant native to Mexico and grown as an ornamental throughout the Valley.

ID tips: *Stelis costalis* is very similar to, and often confused with, the more widespread *Stelis (Dolichostelis) louisae*. The color patterns on the two species are nearly identical. According to Neff, the two species are best separated by minute inspection of the density of pitting on the gena (cheek): in *S. costalis*, pits on the gena nearly touch one another, while on *S. louisae* there are larger, shiny interspaces between the pits. The geographic ranges of these two species in the United States just barely overlap and differ widely: while *Stelis costalis* has a restricted range centered in southern Texas, *S. louisae* is widespread throughout the eastern half of the United States, from East Texas and Nebraska to the Atlantic coast, as far north as Maine and as far south as South Florida.

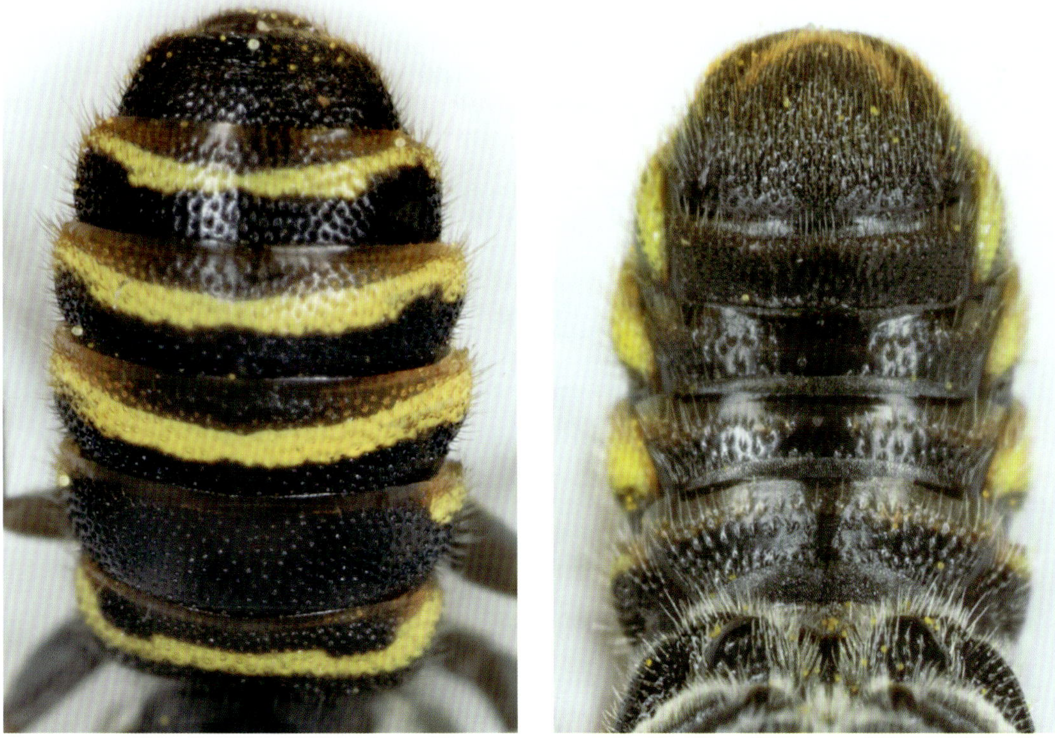

Dorsal view of a female's abdomen (*left*) and sternum (*right*)

MANDIBLE AND LEG CHARACTERISTICS OF VALLEY LEAFCUTTERS

Megachile (Argyropile) parallela	*Females*: 4 teeth on mandibles. *Males*: 4 teeth; forelegs black and unmodified; middle and hind legs black.
Megachile (Chelostomoides) exilis	*Females*: 4 teeth without cutting edges. *Males*: 3 teeth; front basitarsi red and dilated; middle and hind legs black.
Megachile (Chelostomoides) prosopidis	*Females*: 5 teeth without cutting edges. *Males*: 3 teeth; front basitarsi red and dilated; middle and hind legs red and black.
Megachile (Chelostomoides) texensis	*Females*: mandible slender with a tridentate tip. *Males*: mandible with a tridentate tip; legs dark brown with reddish lower tarsi; forelegs not modified.
Megachile (Litomegachile) brevis	*Females*: 4 teeth. *Males*: 3 teeth; forelegs black and unmodified; middle and hind legs black.
Megachile (Megachile) montivaga	*Females*: 5 teeth without cutting edges. *Males*: 3 evenly spaced teeth; forelegs dark and unmodified; middle and hind legs black.
Megachile (Neochelynia) chichimeca	*Females*: 4 teeth. *Males*: 3 teeth, with a jagged middle tooth; forelegs dark and unmodified; middle and hind legs black; lower tarsi of all legs reddish.
Megachile (Pseudocentron) sidalceae	*Females*: 4 teeth. *Males*: 4 teeth; foreleg basitarsi yellow and dilated; middle and hind legs black with pale tarsi.
Megachile (Sayapis) frugalis	*Females*: 4 teeth. *Males*: 3 teeth; forelegs dark and unmodified; middle and hind legs black.
Megachile (Sayapis) inimica inimica	*Females*: 4 teeth. *Males*: 3 teeth; all legs mostly red; foreleg basitarsi dilated, with pale areas on outer surfaces.
Megachile (Sayapis) policaris	*Females*: 4 teeth. *Males*: 3 teeth; forelegs orange and black with pale lower legs; foreleg basitarsi greatly dilated, thumb shaped; middle legs black and orange; hind legs black.
Megachile (Sayapis) zaptlana	*Females*: 4 teeth. *Males*: 3 teeth; forelegs orange and black with yellow-green, dilated basitarsi; middle and hind legs black.
Megachile (Tylomegachile) cf. *toluca*	*Male*: 3 teeth; forelegs red and unmodified; middle and hind legs red.

Leafcutter Bees

Subfamily Megachilinae ◆ Tribe Megachilini
Genus *Megachile*

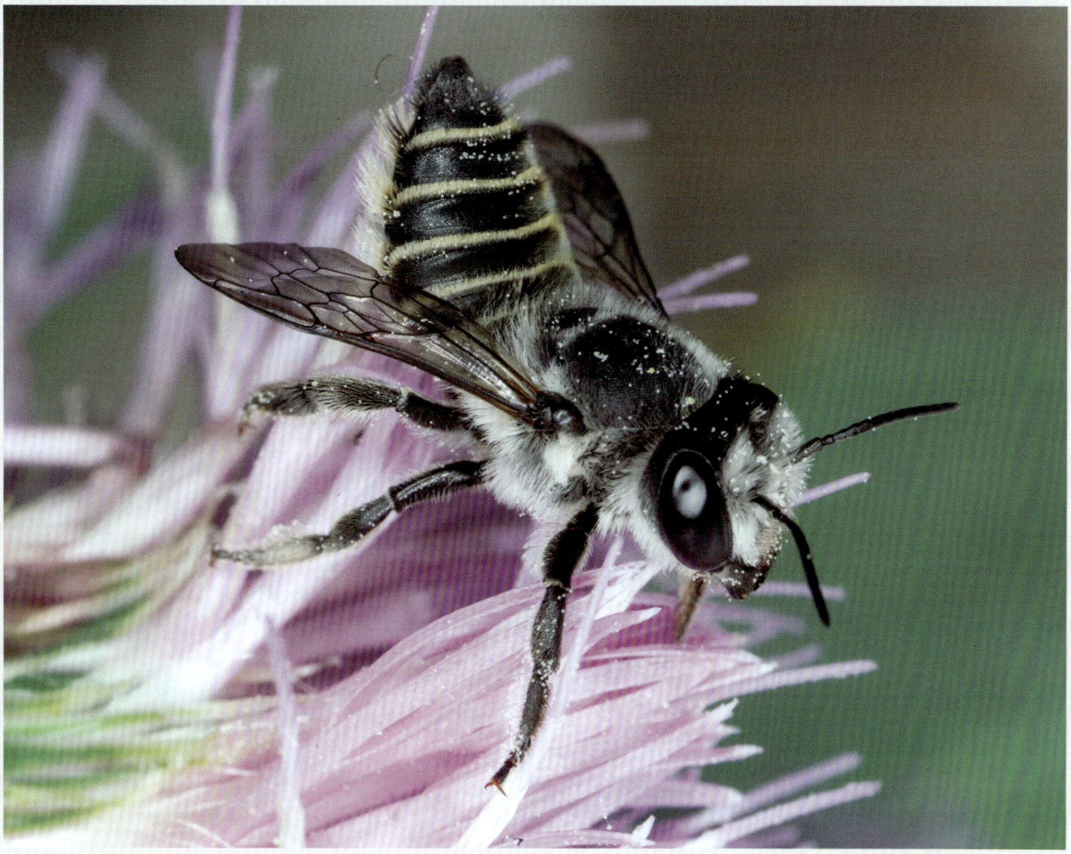

A female *Megachile (Litomegachile)* lifting its abdomen in a pose typical of leafcutters

Most female *Megachile* use their sharp-edged mandibles to trim leaves and other plant parts; the common name "leafcutter" derives from this practice. The bees use such plant matter to plug nest tunnels and line nest walls. Leafcutters construct their solitary nests in a variety of environments, including rotted logs, dead twigs, insect holes in wood, and crevices in human-made structures such as brick walls and masonry. Most leafcutters are cavity nesters, but some species, such as the parallel leafcutter (*Megachile parallela*), nest in the ground. Leafcutter nests vary in size and complexity according to species. Many leafcutter nests are tunnel shaped, consisting of an entry chamber followed by a series of egg chambers provisioned with pollen and nectar.

Most female leafcutters have large, toothed mandibles, usually equipped with cutting blades between the teeth. Those with more formidable mandibles cut leaves, while those with simpler ones may trim only delicate materials such as flower petals. The pieces leafcutters snip from plants tend to be nearly uniform in shape—oval for constructing the walls of egg cells and circular for

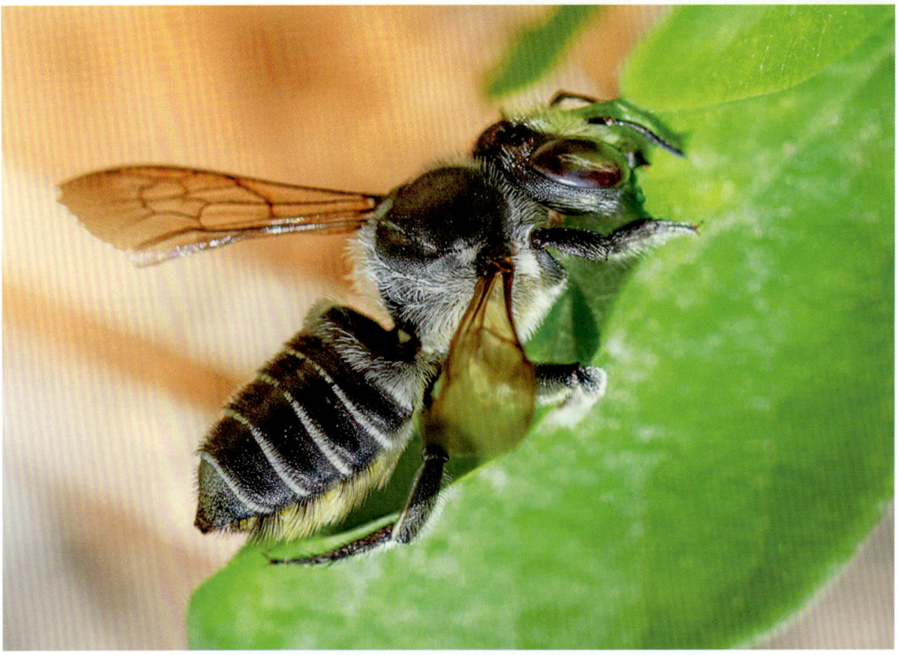

A female leafcutter bee trimming a leaf. Photo credit Laura Zurro.

covering cell openings. Various *Megachile* species use other materials to build or buttress their nests, including chewed-up leaf pulp, whole leaflets from small-leafed plants, tatters of flower petals, plant resins, or small grains of soil or sand.

Traits of the genus *Megachile*: Leafcutters typically have wide, somewhat flattened abdomens that taper abruptly at the ends, and broad, sculpted-looking heads. Like other Megachilidae, leafcutters have forewings with two submarginal cells, and females carry pollen on scopal hairs on the sternum. (Unlike Osmiini and most Anthidiini, however, leafcutters do not have arolia.) Male leafcutters tend to be smaller than females and to have hairier faces. In some species, males' forelegs may be brightly colored and greatly enlarged. Males also often have long spines on their foreleg coxae. These features are used to blindfold and grasp females during mating.

The Lower Rio Grande Valley has a diverse leafcutter population. Thirteen species of leafcutters, spanning eight subgenera, are featured in this guide. Some, such as *Megachile chichimeca* and *M. zaptlana*, are rarely found in the United States outside the Valley. Others, such as *Megachile brevis* and *M. inimica*, are found throughout the country.

Female bees of individual leafcutter species are identified by such traits as the number and arrangement of teeth and cutting blades on their mandibles; the color of their scopal hairs; the qualities of hair bands on their abdomens; and the shape of their abdomens' tips (concave, flat, or convex; with or without ridges, hairs, or an upturned rim). Male leafcutters are distinguished in part by traits of their mandibles and forelegs; the contours of their abdomens' tips (whether they are notched, jagged, toothed, or smooth); and whether their sternal segments are exposed or tucked under and hidden.

Common little leafcutter ◆ *Megachile (Litomegachile) brevis* Say, 1837

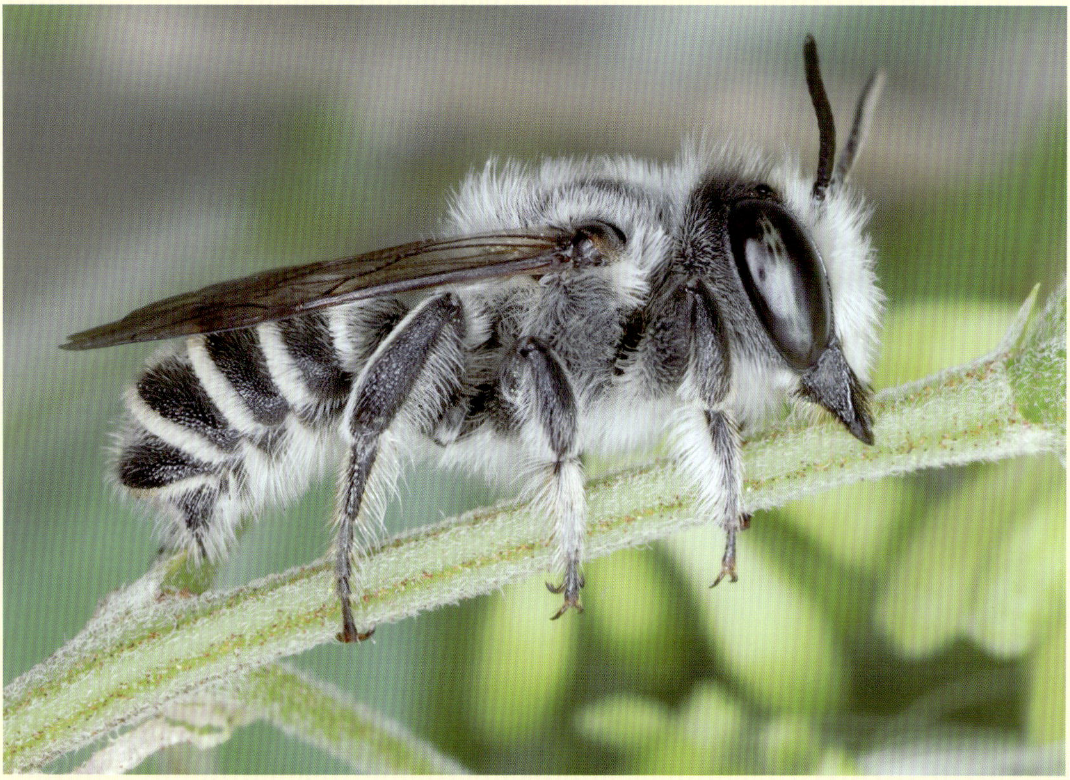

A male common little leafcutter

Description: Leafcutters of the subgenus *Litomegachile* are medium-sized bees with broad bodies, dark heads, dark thoraxes partly covered with white hairs, and dark abdomens with five bands of pale hair (T1–T5). Female *Megachile brevis* have mostly white scopal hairs on the first five segments of the sternum (S1–S5), and some dark hairs on S6. On males, the tergal hair bands continue onto the sternum. Both males and females have dark legs, tegulae, and antennae. Males' faces are densely covered with white hairs, particularly along the inner edges of the compound eyes. Females have four-toothed mandibles, with cutting edges between the second and third, and the third and fourth teeth. Males have three-toothed mandibles. Both males and females have greenish eyes. **Size:** females 9–13 mm (1/3–1/2 in.), males 7–9 mm (1/4–1/3 in.).

Range: *Megachile brevis* is among the most common leafcutter species in North America; it is found from coast to coast, and from southern Canada to the Mexican state of Oaxaca. This species appears in the Valley from April through November.

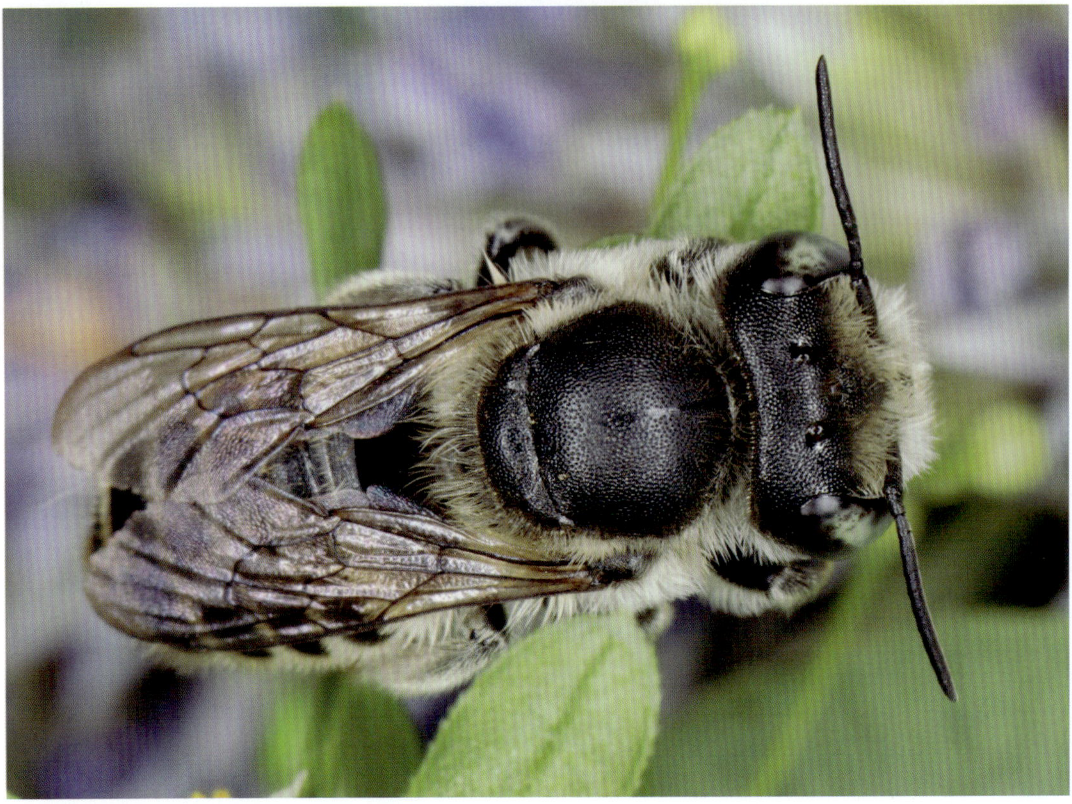

A female common little leafcutter

Associated plants: During spring in the Valley, common little leafcutters can be seen mobbing the blossoms of native pea-family plants such as tenaza (*Havardia pallens*), guajillo (*Senegalia berlandieri*), and blackbrush acacia (*Vachellia rigidula*). *Megachile brevis* is known for its prowess as a pollinator of alfalfa, a pea-family crop. In a 1953 study of the species, Michener wrote that a single common little leafcutter female pollinated an average of 15 alfalfa flowers per minute and visited 1,620 flowers to provision a single egg chamber. He estimated that one female *Megachile brevis* might pollinate 32,400 alfalfa flowers in a lifetime.

Despite its predilection for pea-family flowers, *Megachile brevis* is a generalist pollinator that forages on flowers from diverse families, including carrots, morning glories, mints, mallows, verbenas, varied Asteraceae, and even cacti. Michener believed that common little leafcutters tend to gather pollen from white, blue, purple, and pink flowers of various families but rarely collect from yellow composites such as sunflowers.

ID tips: Traits that aid in identifying *Megachile brevis* include the following. *Females*: T6 of the abdomen is concave, with a rim on the tip. The sternal hairs on S6 are mostly pale, with a fringe of dark hairs on the apical (hind) edge. The hind basitarsi are also quite broad, as wide as the tibiae above them. *Males*: T6 is notched and covered with short, flattened, pale hairs. On either side of the notch are two median teeth that entomologist Charles Roberts likened to the cogs of a circular saw.

Behavior: *Megachile brevis* is so nimble and fast in flight that it can be difficult for the human eye to see. In 1953, Michener described *Megachile brevis* as "a bee that flies very rapidly with a distinct high pitched buzz audible on calm days for as much as ten or twelve feet. It is usually more easily found by means of this sound than by sight." He clocked a female common little leafcutter flying 150 feet in 15 seconds, from her nest to a copse where she cut leaves, and he noted that *Megachile brevis* leafcutters had been observed in winds "strong enough to roll a bee over and over on the ground if it happened to attempt to land in a gust."[3]

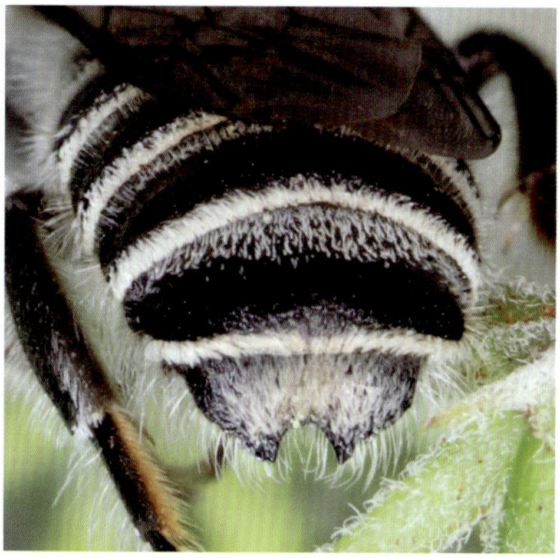

T6 of the male's abdomen has a distinct notch and is covered with short, pale, appressed hairs.

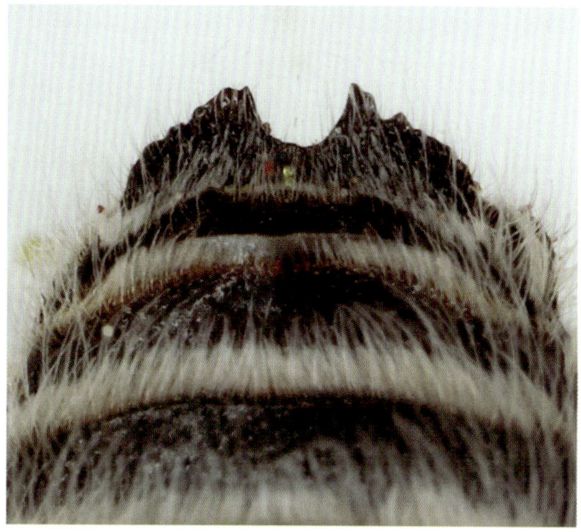

Cog-like teeth flank the notch on T6.

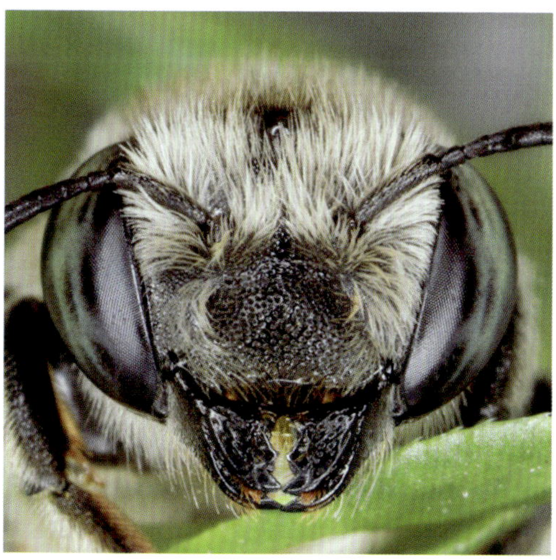

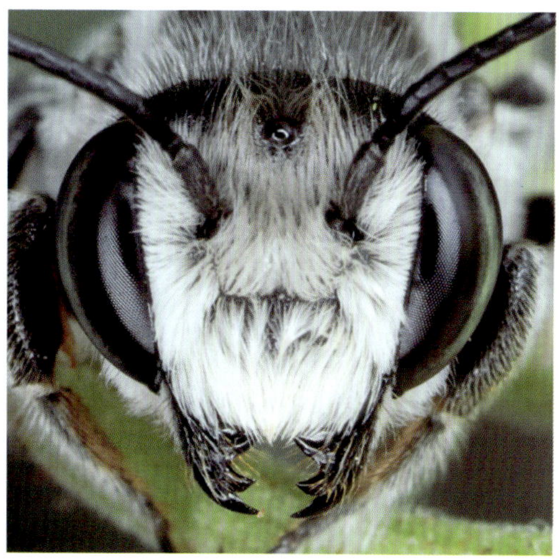

The four-toothed mandibles of a female *Megachile brevis* (*left*) and the three-toothed mandibles of a male (*right*)

3. Michener (1953), pp. 1689–1690.

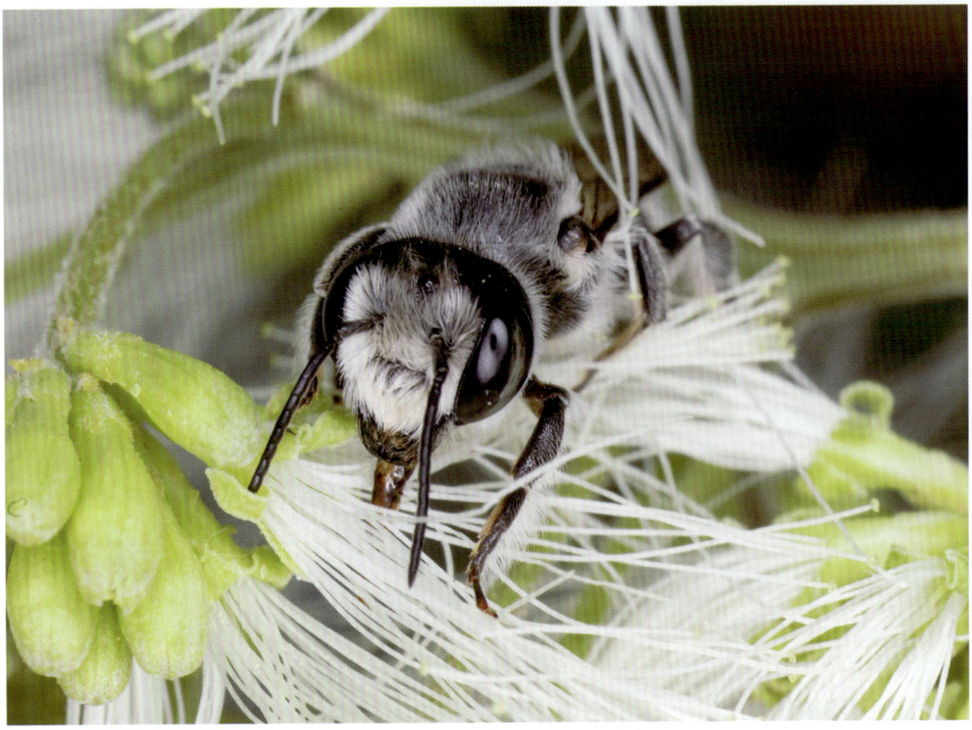

A male common little leafcutter sipping nectar from tenaza blossoms

Michener observed that *Megachile brevis* nests in preexisting cavities "almost anywhere"—in hollow plant stems and grass stalks, rolled-up leaves, or bits of wood.[4] Common little leafcutters are also known to nest in cavities in the ground. While many solitary bees place all their eggs in one nest, *Megachile brevis* females make multiple nests containing one to several eggs each, sometimes located miles from one another. At night, female bees vanish into their nests, while males often seek overnight refuge in small, random shelters such as curled-up leaves.

Megachile brevis females are notably meticulous nest builders. They select firm materials such as rose leaves to strengthen cell walls, and more flexible leaves like those of waxberry for padding, varying the shapes of cuttings depending on their function. A notable feature of *Megachile brevis* behavior is the bees' use of flowers in nest construction. Although *Megachile brevis* leafcutters rely on leaves as primary building materials, Michener described them using petals from flowers such as roses, violet lespedeza, evening primrose, and partridge pea to fill empty nest space; layering circular petal cuttings alternately with leaves to fashion caps for egg cells; and using elongate petal trimmings to line the insides of egg-cell cups.

Common little leafcutters produce a succession of generations each year, so that, except in spring, both sexes and all life stages of the bee, including larvae and pupae, can be found at one time. Male *Megachile brevis* emerge at the same time as the females and live as long as they do, a phenomenon Michener pronounced unusual among solitary bees.

4. Michener (1953), p. 1676.

Chichimec leafcutter ◆ *Megachile (Neochelynia) chichimeca* Cresson, 1878

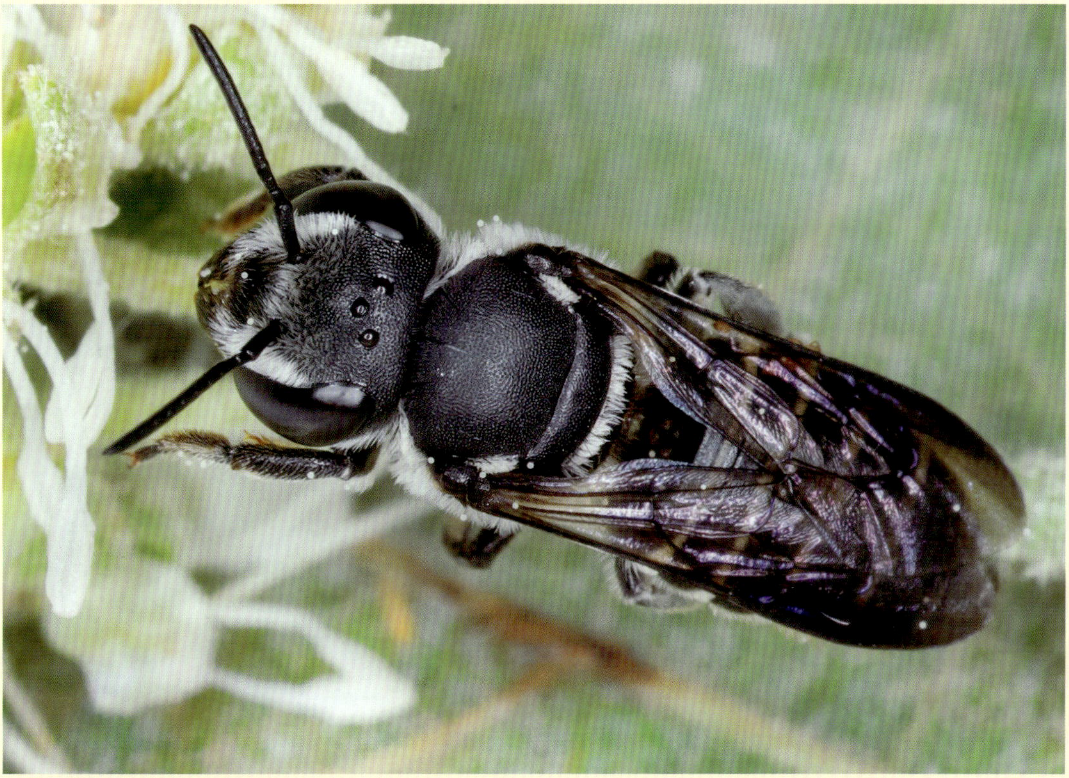

A female Chichimec leafcutter

Description: Leafcutters of the Neotropical subgenus *Neochelynia* are small, often slender bees with forewings that are dark on the outer edges. Males have elongate, tapering abdomens and distinctive three-toothed mandibles on which each middle tooth has two toothlike points.

Megachile (Neochelynia) chichimeca is a black bee with prominent tufts of white hairs behind the tegulae, white hairs rimming the hind edge of the scutellum, and uninterrupted bands of pale hairs on T1–T5. The tegulae are black with reddish outer margins. The legs are mostly black: on males, the lower tarsi are reddish, and on females, rust-colored hairs line the inner hind basitarsi. Females' scopal hairs are white toward the front, partly rust colored on T5, and mostly rust colored on T6, with some black hairs toward the tip. The pale tergal hair bands of the male continue onto the sternum (S1–S5), and T6 of the male's abdomen has a semicircular notch. Females have four-toothed mandibles with subtle, incomplete cutting edges between the second and third, and between the third and fourth teeth. Both males and females have black antennae and greenish eyes. **Size:** females 10 mm (2/5 in.), males 7–8 mm (1/4–3/10 in.).

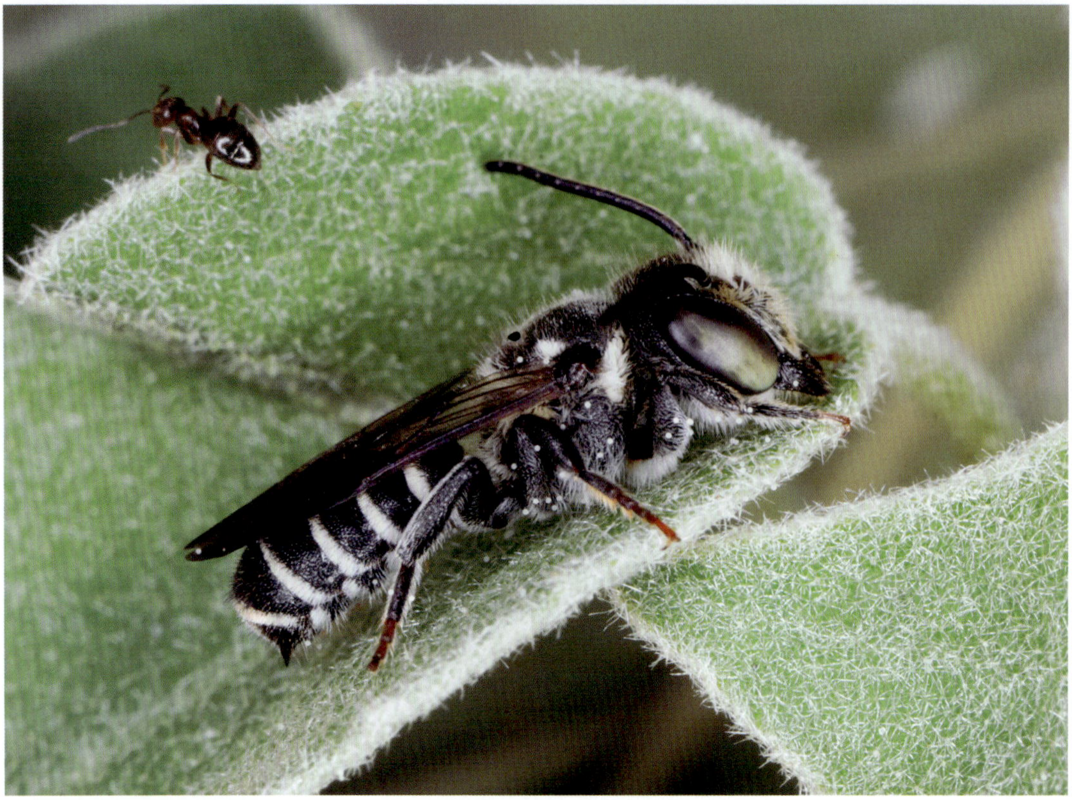

A male Chichimec leafcutter

Range: *Megachile chichimeca* is native to Mexico and was first cataloged in 1878 as a Mexican species. Nonetheless, it has appeared in southern Texas for more than a century and is a common leafcutter in the Lower Rio Grande Valley. Wilmatte Porter Cockerell recorded a Chichimec leafcutter in Cameron County in her 1917 "Collecting Bees in Southern Texas." The Chichimec leafcutter occurs as far south as Brazil and Guyana.

Associated plants: *Megachile chichimeca* is a generalist pollinator that has been documented south of the border feeding on an array of plants, among them flora in the dogbane, borage, spurge, and pea families. Chichimec leafcutters emerge in May in the Valley and remain visible through November. They are common, conspicuous pollinators of native Texas ebony (*Ebenopsis ebano*), whitebrush (*Aloysia gratissima*), low croton (*Croton humilis*), shrubby blue salvia (*Salvia ballotiflora*), and regional aster-family plants such as tropical threefold (*Trixis inula*) and resinbush (*Viguiera stenoloba*).

ID tips: In the field, *Megachile chichimeca* is best recognized by the distinctive white hairs behind the tegulae and scutellum, the reddish hairs on the tip of the female's sternum (S6), and the dark-edged forewings with dusky marginal cells. Under magnification, the distinctive middle teeth of the male's mandibles aid in diagnosing the species. The female's clypeus has a protuberance in the middle of its bottom edge.

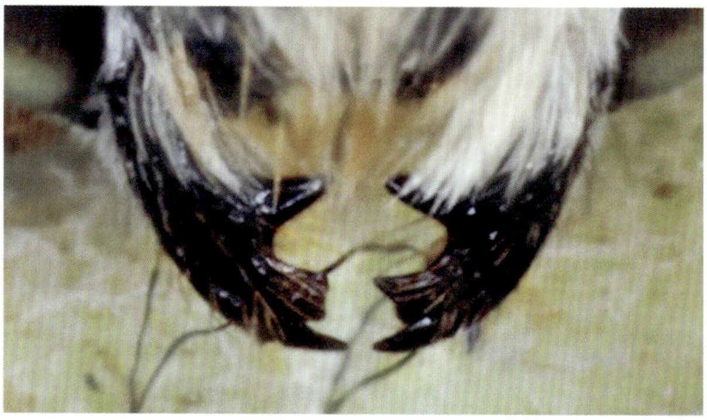

The mandibles of a male *Megachile chichimeca*: the middle tooth on each half of the mandibles has a jagged edge, with two corners that make it look like two separate teeth.

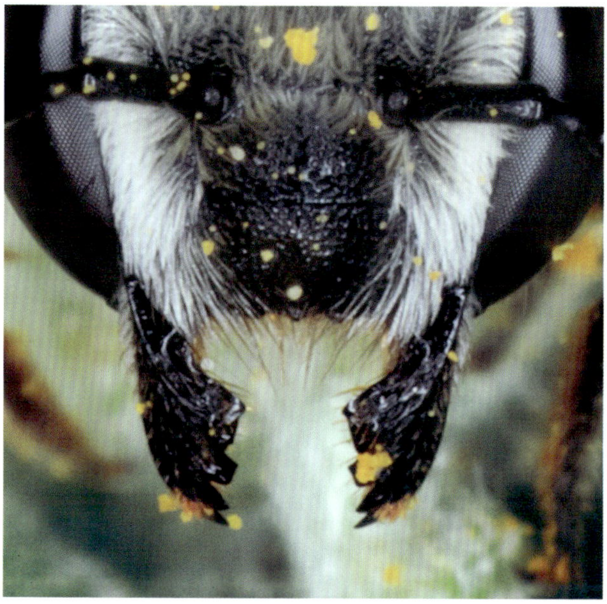

Lower face of a female Chichimec leafcutter: note the median bump on the bottom edge of the clypeus. Female bees have three-toothed mandibles.

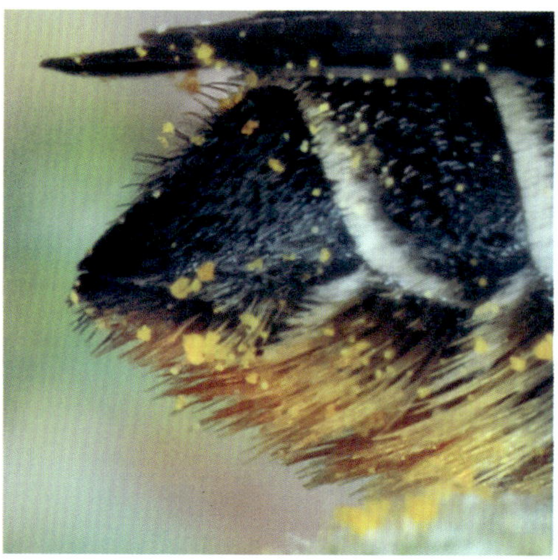

Reddish scopal hairs on T5 and T6 of a female Chichimec leafcutter's abdomen

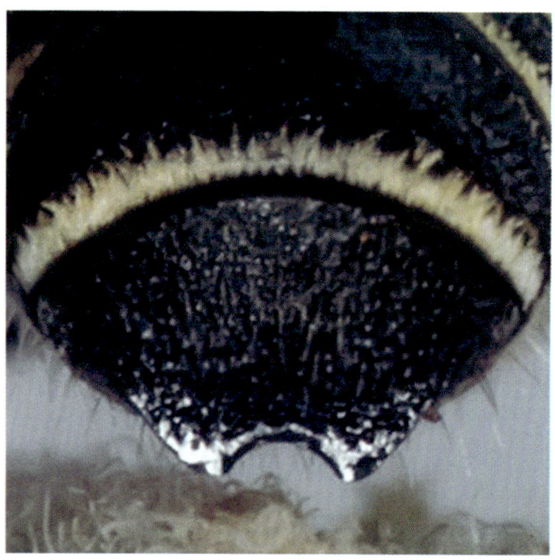

The semicircular notch at the tip (T6) of a male's abdomen

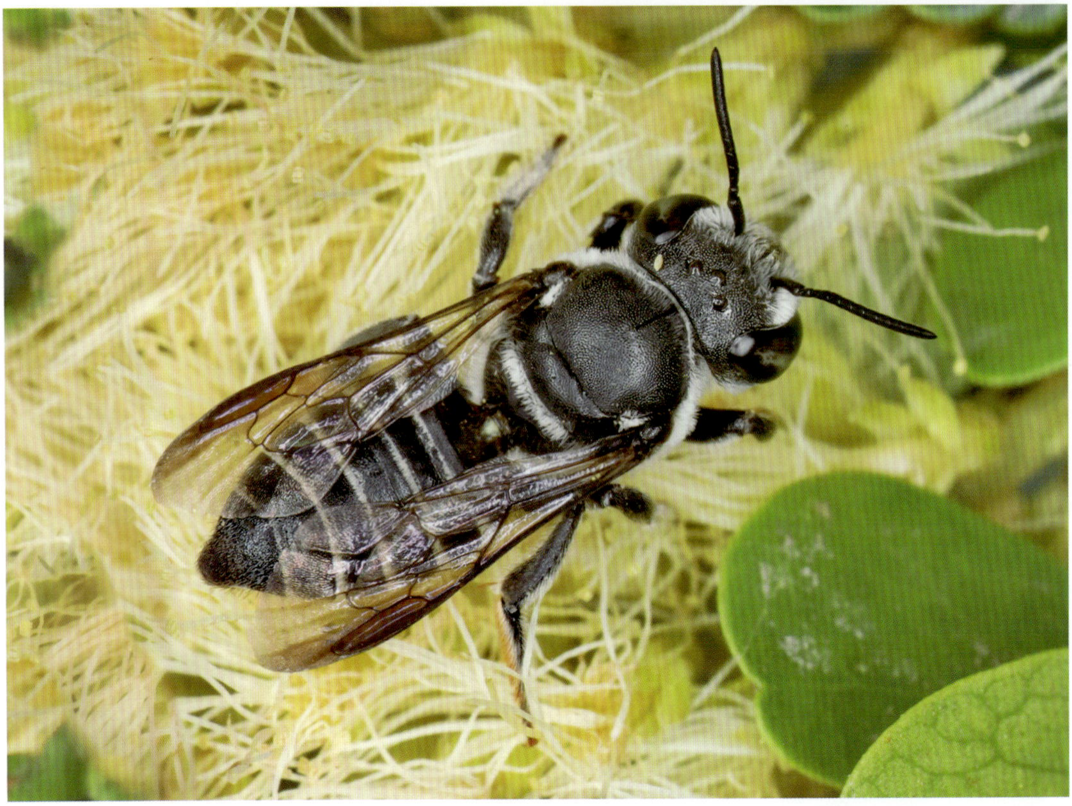

A female *Megachile chichimeca* on Texas ebony: note the forewings' dark outer edges and marginal cells

Behavior: Very little published research about leafcutters of the subgenus *Neochelynia* is available. Entomologists Matos, Silva, and Teodoro conducted a 2016 study of *Megachile (Neochelynia) brethesi* in an agricultural area of the Brazilian state of Maranhão. They observed that this species built nests in traps fashioned out of hollow bamboo canes strapped together. In 2004, Brazilian entomologist Elder Morato successfully used trap nests to attract *Megachile (Neochelynia) paulista* during research for his PhD thesis. These studies suggest that *Neochelynia* leafcutters are cavity nesters rather than ground nesters.

Toluca leafcutter ◆ *Megachile (Tylomegachile)* cf. *toluca* Cresson, 1878

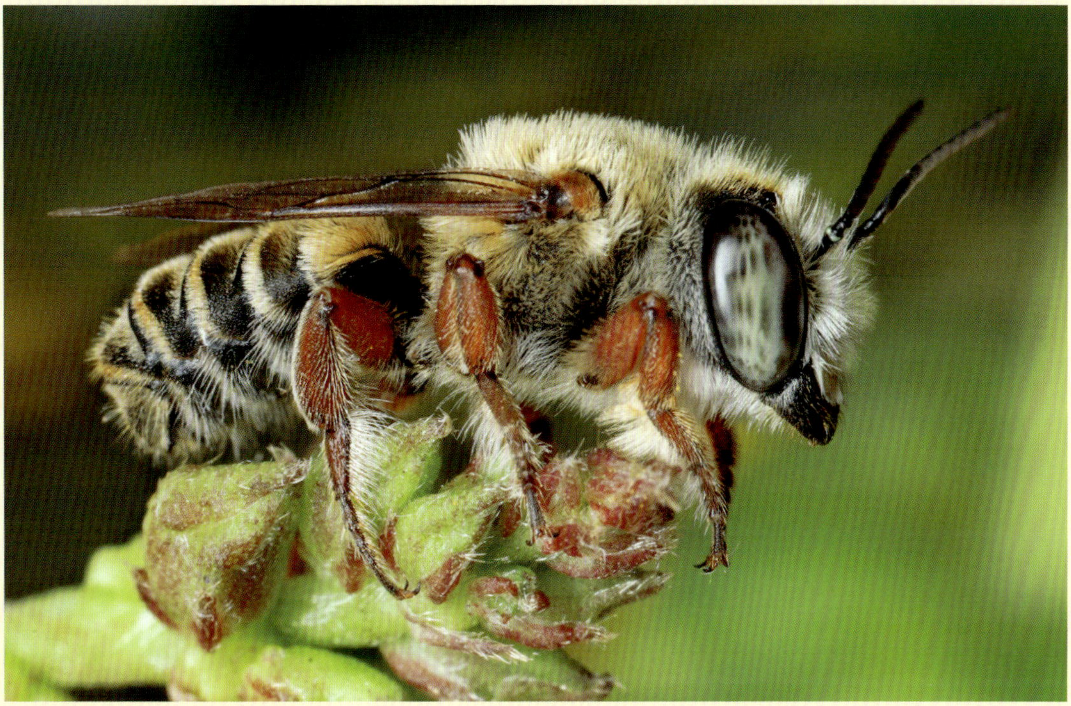

A male *Megachile* cf. *toluca*

Description: This male red-legged leafcutter bee, tentatively identified as a Toluca leafcutter, or *Megachile (Tylomegachile)* cf. *toluca*, appeared at the National Butterfly Center in September 2018. This marked the first sighting of any member of the subgenus *Tylomegachile* in the United States. The male leafcutter was identified through the collaboration of John L. Neff, John S. Ascher, and Hadel Go.

The male leafcutter shown here is a medium-small bee with a robust build, red legs, and reddish tegulae. White hairs cover its face and head, and copious yellowish hairs cloak its thorax and ring its abdomen. The mandibles and antennae are black, the eyes are green, and the tip of the abdomen, which curls under slightly, bears two conspicuous teeth. **Size:** male 10 mm (2/5 in.).

Range: *Tylomegachile* leafcutters are Neotropical, generally ranging from Argentina to the Mexican border. The first published record of *Megachile toluca*, furnished in 1878 by Cresson, describes a male bee found in Mexico. In the past 50 years, this species has been documented in Mexico, Guatemala, and the Guanacaste province of Costa Rica. Its most recent previously recorded sighting occurred in 2010, in Mexico.

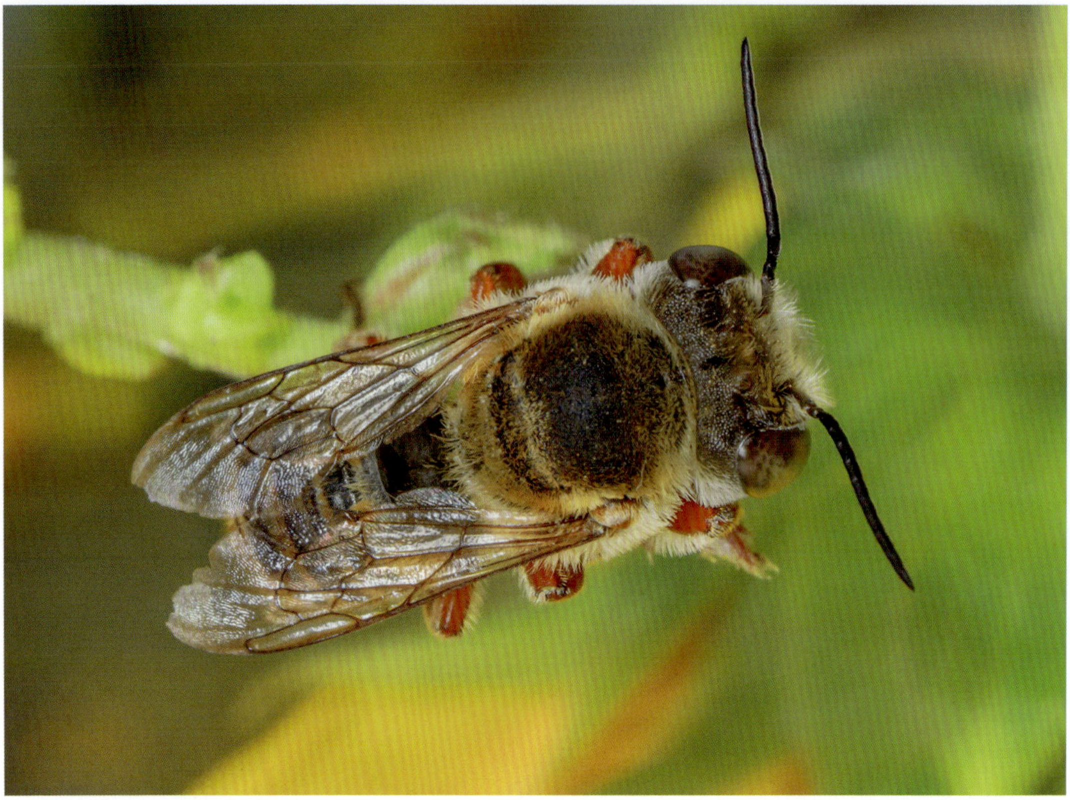

Dorsal view of a male *Megachile* cf. *toluca*

Associated plants: The male bee shown here was feeding on the yellow flowers of Texas snout bean (*Rhynchosia senna* var. *texana*) growing on the banks of an irrigation canal. Male *Megachile toluca* has been recorded in Costa Rica on the subtropical and tropical flowering plants *Andira inermis*, *Casearia aculeata*, *Casearia nitida*, *Cordia inermis*, *Cordia toqueve*, *Crotalaria incana*, *Croton* sp., *Machaerium biovulatum*, *Muntingia calabura*, *Pterocarpus rohrii*, and *Securidaca sylvestris*. Female bees have been found foraging on *Dalbergia retusa* and *Bursera tomentosa*.

ID tips: The combination of this bee's red legs, the exuberant yellowish hairs on its thorax, the yellowish bands on its abdomen, and the two large teeth on T6 make it distinctive and unlikely to be confused with other leafcutters of the Valley. Because *Megachile toluca* remains partly undescribed, a more comprehensive description of *Megachile* cf. *toluca* follows here.

Head: The male *Megachile* cf. *toluca* featured here has a head that is wider than its thorax, and a broad vertex. (The lateral ocelli are slightly closer to one another than to the back of the head.) The face, antennae, and mandibles are black. The mandibles are three toothed, with evenly spaced teeth. The antennae are long. F1 is slightly longer than the pedicel, and slightly longer than it is wide; F2 is nearly twice the length of F1; F2–F10 are the same length; and F11 is slightly longer than F2. The face, cheeks, and vertex are covered with dense off-white hairs that completely hide the clypeus, form a thicket over the ocelli, and make a curtain of hair under the cheeks and compound eyes. There are some intermingled dark hairs on the vertex.

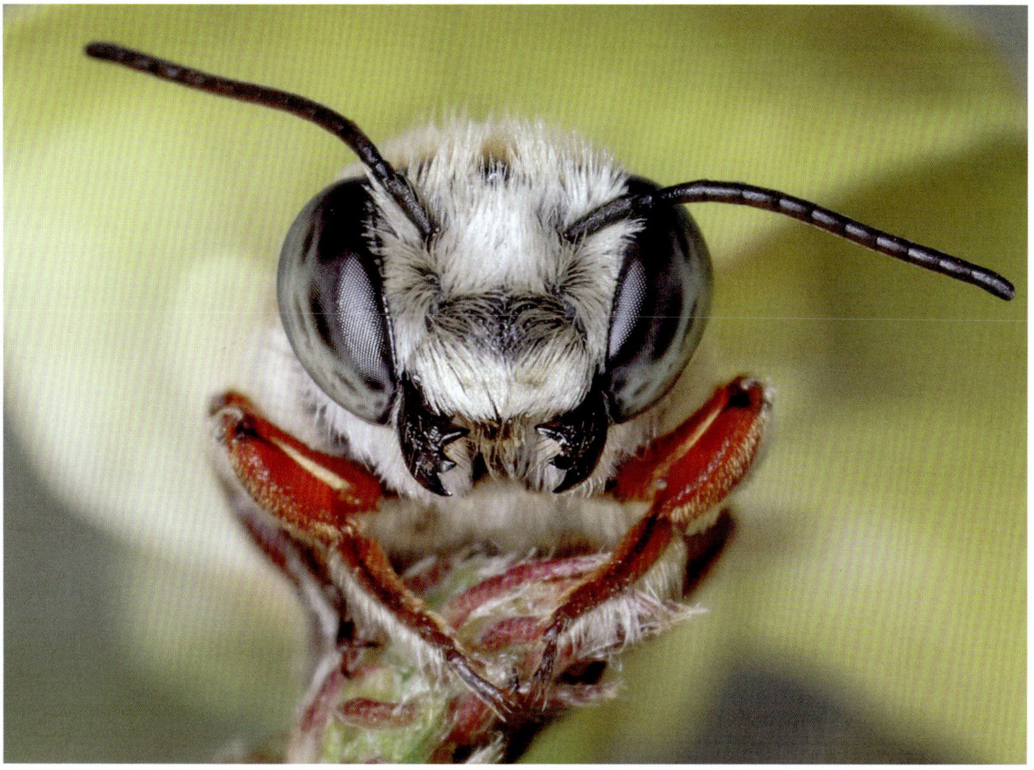

Dense pale hairs cover the face of the male leafcutter. The male bee has three-toothed mandibles.

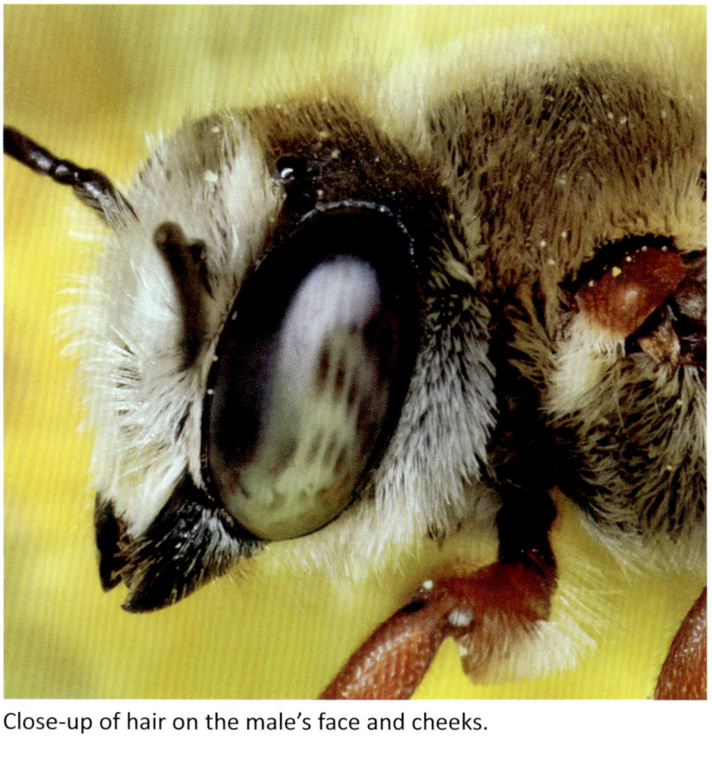

Close-up of hair on the male's face and cheeks.

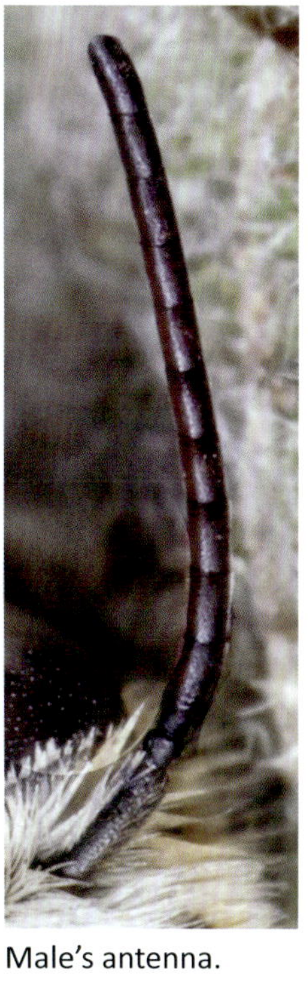

Male's antenna.

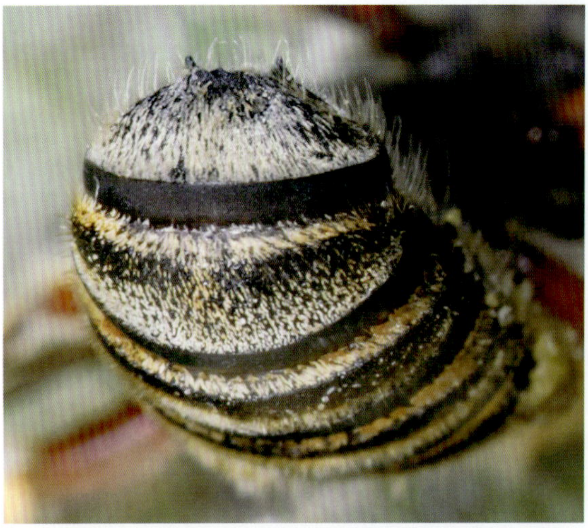

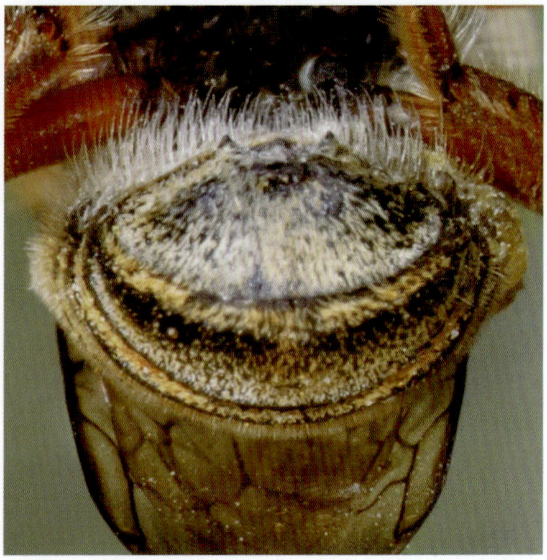

The rim of T6 of the male *Megachile* cf. *toluca* has a small point flanked by two larger median teeth. T6 is partly covered with pale hairs, some long and some appressed.

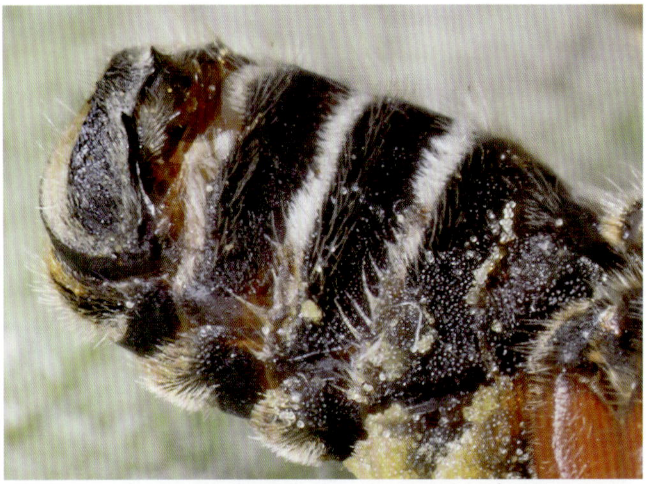

The pale hair bands on T2–T4 continue onto the sternum.

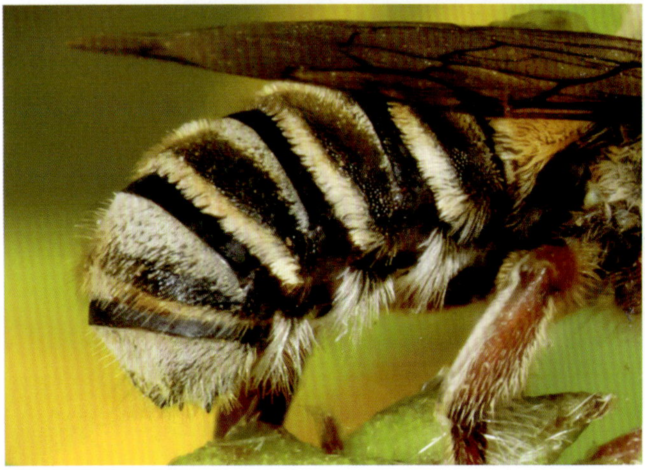

The apical edges of T1–T5 are girded by black, hairless, parallel-sided strips. Pale hairs band T1–T4, and long pale hairs appear on the sides of T2–T4.

Thorax, legs, and wings: Long, pale gold hairs cover the scutum. These hairs continue more sparsely onto the sides of the thorax. Whiter hairs form patches under the tegulae and fringe the lower edges of the mesepisternum. The upper and lower legs are mostly red, with some black on the upper hind legs (hind coxae and trochanters). Each front coxa sports a black spine. Long white hairs fringe the basitarsi and lower tibiae. The hind-leg spurs are reddish yellow. The tegulae are reddish, and the wings are glassy, with brown veins and vaguely darkened outer hind edges.

Abdomen: A wide band of long, pale gold hairs lines the base of T1. On T2–T4, short pale hairs edge the base of each segment, and long, pale gold hairs band each segment's apical rim. Short, appressed, pale yellow hairs edge the bases of T5 and T6, but most of these segments are covered with long yellowish hair. T6 has a small median tooth flanked by two larger submedian teeth on its outer rim, a trait common to leafcutters of the subgenus *Tylomegachile*.

Behavior: The nesting habits of Toluca leafcutters are unknown, and scarce published information is available regarding the behavior of members of the subgenus *Tylomegachile*. In conducting research for his graduate thesis in 2004, Morato successfully used trap nests to attract *Megachile (Tylomegachile) orba*. This report suggests that at least some *Megachile (Tylomegachile)* leafcutters are cavity nesters.

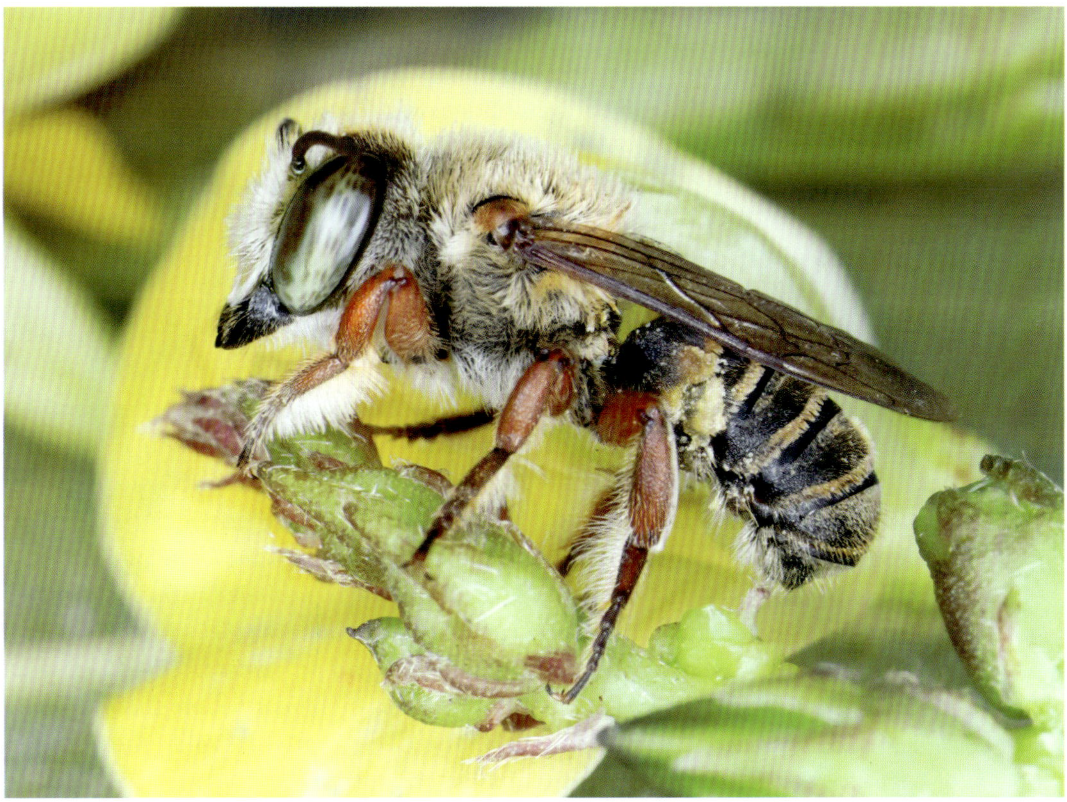

Lateral view of a male *Megachile* cf. *toluca*

Lower Rio Grande Valley Leafcutter Names

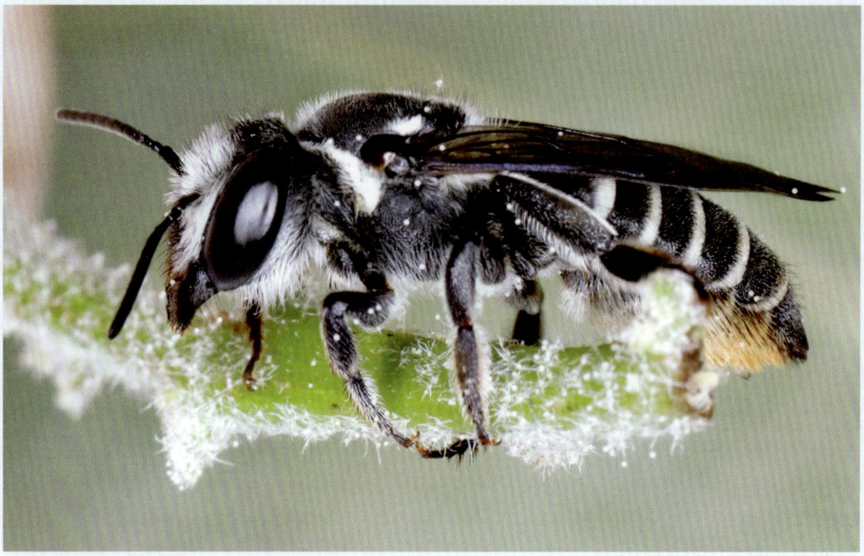

A female *Megachile chichimeca*

Three leafcutter species in this guide—*Megachile chichimeca*, *M. toluca*, and *M. zaptlana*, were first described by Cresson in an 1878 article titled "Descriptions of New North American Hymenoptera in the Collection of the American Entomological Society." In this work, Cresson cataloged 38 species of leafcutters, nearly half of which came from Mexico and were unknown in published annals of entomology. Most of the Mexican specimens—including *Megachile chichimeca*, *M. toluca*, and *M. zaptlana*—had been collected by the Swiss entomologist Adrien Louis Jean François Sumichrast during expeditions in Mexico earlier in the nineteenth century.

More than 20 years before, Sumichrast had traveled to Mexico on an insect-collecting trip with fellow Swiss entomologist Louis Frédéric de Saussure. After a year, Saussure returned to Switzerland, citing the difficulties of travel and political unrest generated by the Mexican-American War. Sumichrast stayed behind, married a Mexican woman, and remained in Mexico until his death in 1882. For nearly 30 years, he journeyed through the states of Veracruz, Puebla, México, Oaxaca, and Chiapas, collecting insects. In 1878, when Cresson undertook the task of cataloging new species, 55 of his bee specimens, spanning more than a dozen genera, came from Sumichrast.

Cresson liked to christen Mexican bees after ancient Mesoamerican civilizations and Aztec place names. Thus, the leafcutters he cataloged in 1878 bear such names as *Megachile chichimeca*, *M. azteca*, *M. montezuma*, *M. toluca*, *M. tuxtla*, *M. zapoteca*, and *M. zaptlana*. The Chichimec leafcutter in the previous entry derives its name from a seminomadic people that occupied north-central Mexico at the time of the Conquest. The name *zaptlana* evokes the Mexican kingdom of Zapotlán, the site of several pre-Columbian civilizations before it was overtaken by conquistadors in the 1500s. *Toluca* is the name of the highland valley where the ancient Aztec ruins of Calixtlahuaca are situated.

Zaptlan leafcutter ◆ *Megachile (Sayapis) zaptlana* Cresson, 1878

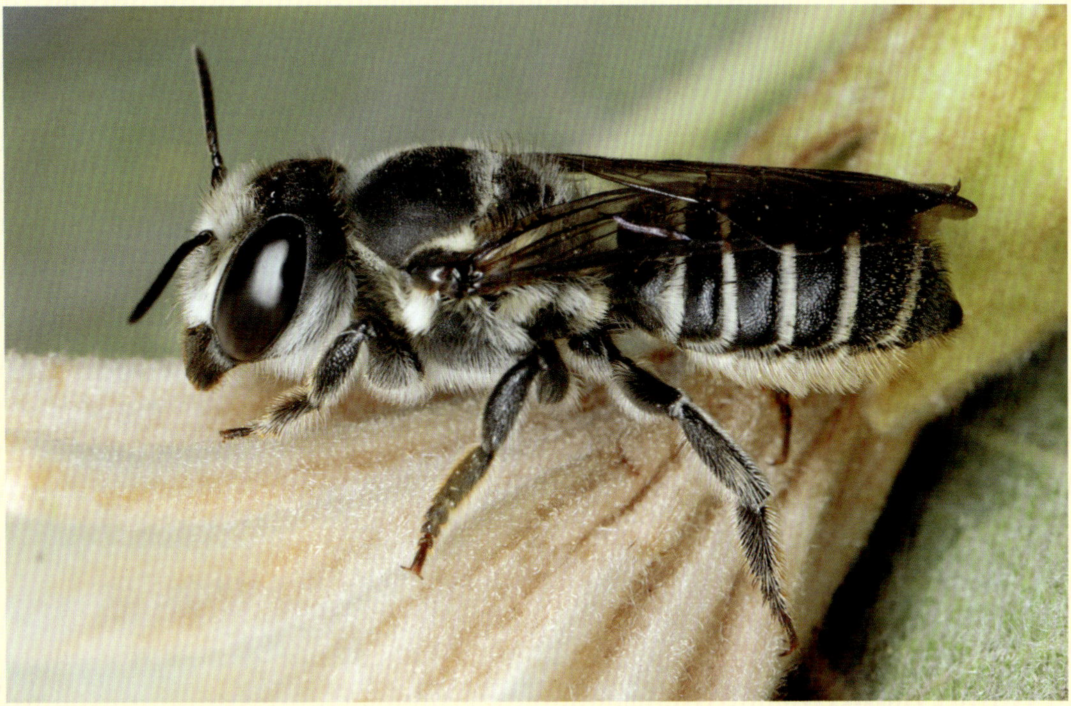

A female Zaptlan leafcutter

Description: Leafcutters of the subgenus *Sayapis* are relatively large *Megachile* and have long, parallel-sided bodies. The sixth sternal segment of females of this subgenus (S6) is covered, at least in part, with erect black hairs. Males have four exposed sternal segments. (That is, the last three segments of the sternum are not visible when the bee is turned over.) This trait helps separate them from the narrow, parallel-sided leafcutter subgenus *Chelostomoides*, whose members have only three exposed sternites. The mandibles of female *Megachile (Sayapis)* are four toothed. Males' mandibles are three toothed.

Megachile (Sayapis) zaptlana is a large leafcutter with a narrow black abdomen ringed with five bands of white hair. The bee's tegulae are dark brown, and its wings brownish with dark brown veins and dusky outer and hind edges. The antennae are dark and slender. The scopal hairs of the female Zaptlan leafcutter are white except on the last segment (S6), which is covered with longish light-and-black hairs. The mandibles of the female *Megachile zaptlana* have beveled edges (rather than sharp cutting edges) between the second and third teeth.

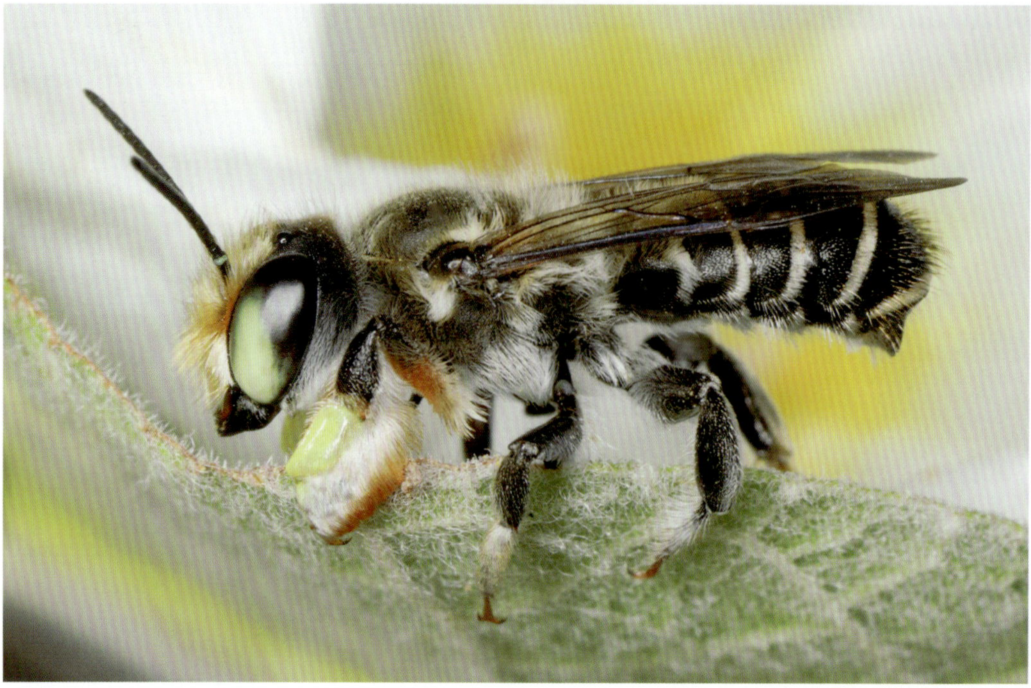

A male Zaptlan leafcutter

Male and female *Megachile zaptlana* show pronounced sexual dimorphism: the female's legs are largely dark, while the male has striking forelegs with enlarged lower segments that are pale greenish yellow and fringed with white and rust-colored hairs. Exuberant yellowish hairs cover parts of the male's face. The female's face, cheeks, and clypeus are covered with white hairs. Females have black eyes, and males' eyes are light to dark green. **Size:** females 11–14 mm (2/5–3/5 in.), males 9–12 mm (1/3–1/2 in.).

Range: Zaptlan leafcutters have been found in Mexico, in Central America as far south as Panama, in Jamaica, and in Paraguay. In the United States, this species appears principally in southern Texas. *Megachile zaptlana* is a common leafcutter species in all three border counties of the Valley.

Associated plants: Zaptlan leafcutters have been documented feeding on a range of plants, among them aster-family flowers, spurges, and pea-family flowers. In the Valley, Zaptlan leafcutters can be found foraging on such plants as whitebrush (*Aloysia gratissima*), honey mesquite (*Prosopis glandulosa*), bladdermallow (*Herissantia crispa*), low croton (*Croton humilis*), and Mexican hat (*Ratibida columnifera*).

ID tips: *Females*: Traits that help distinguish the female *Megachile zaptlana* from others of its subgenus include the following: (1) the vertex is broad and relatively hairless; (2) although white hairs cover most of the face, there are some black hairs between the antennae and on the vertex; (3) white hairs rim the rear, front, and lateral edges of the scutum; and (4) intermingled dark-and-light hairs cover the scutellum, and its back rim is edged with long, sparse white hairs.

Males: The front basitarsi of *Megachile zaptlana* males are greenish yellow, boat shaped, and carved with hollows. The tips of the front tibiae are greenish yellow, and the tarsi are bright yellow and ivory. The front femurs are partly orange, and the front tibiae mostly black. The male also has coarse dark hairs on T5, and the tip of the abdomen has a toothed edge.

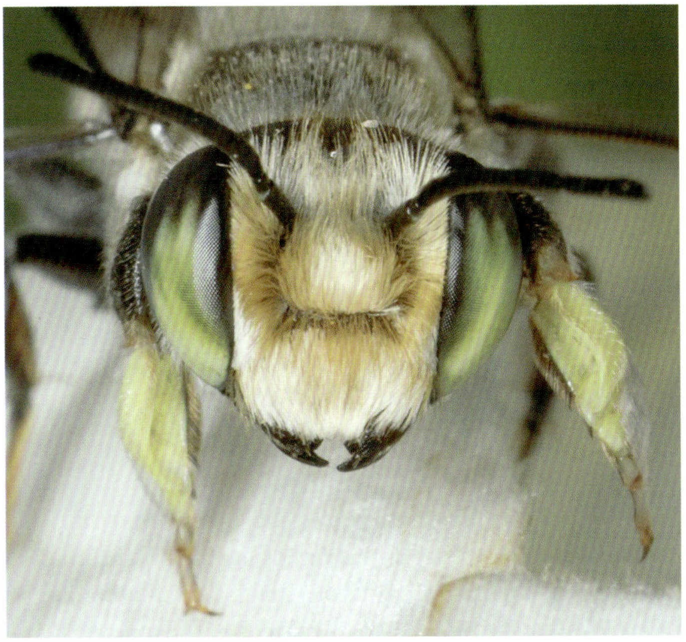
Face of a male Zaptlan leafcutter

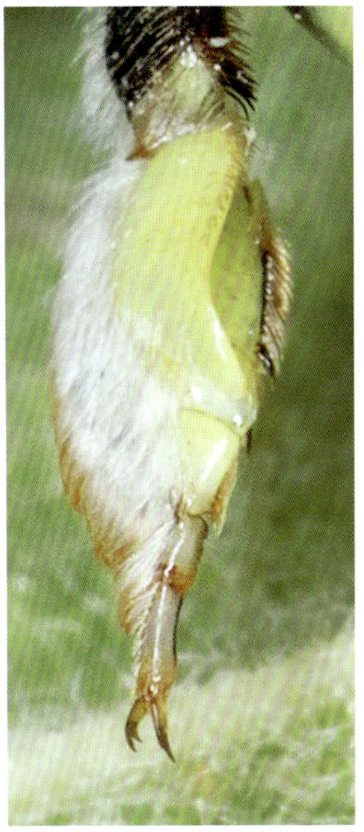
Foreleg basitarsus of a male Zaptlan leafcutter

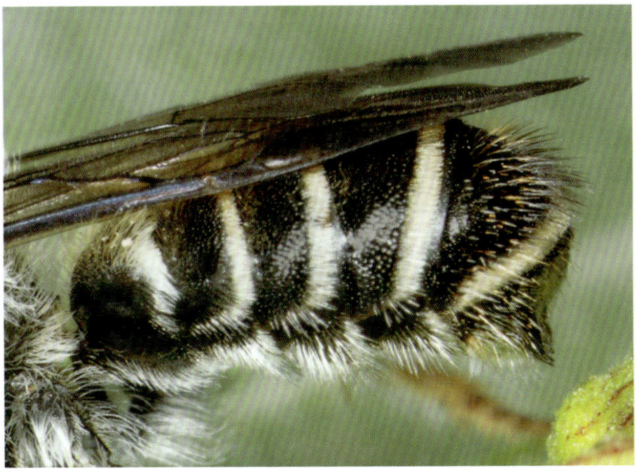
Abdomen of a male *Megachile zaptlana*

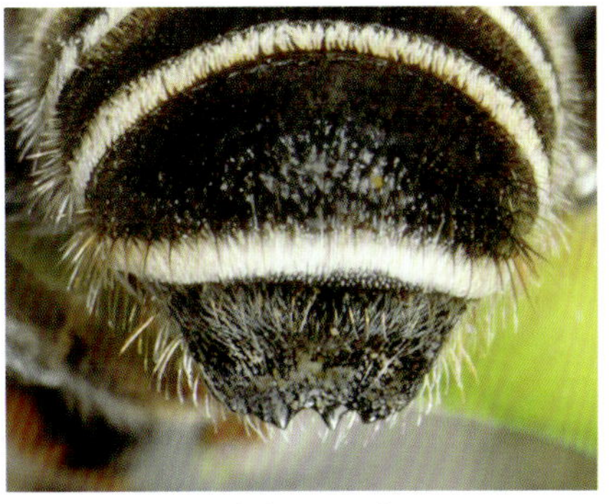
Toothed edge of the tip of T6

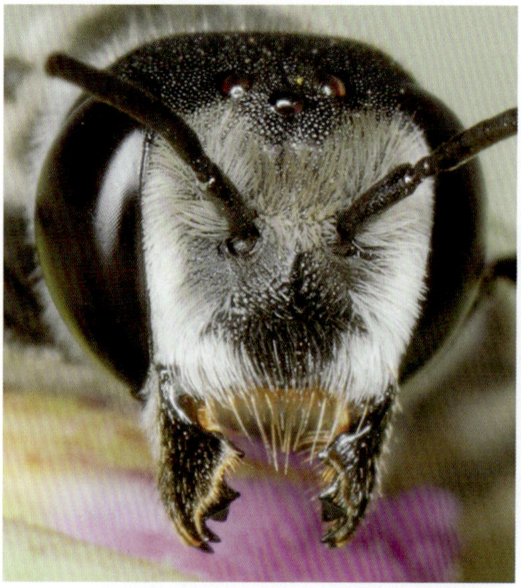

The mandibles of the female *Megachile zaptlana* are four toothed, with beveled edges between the second and third teeth.

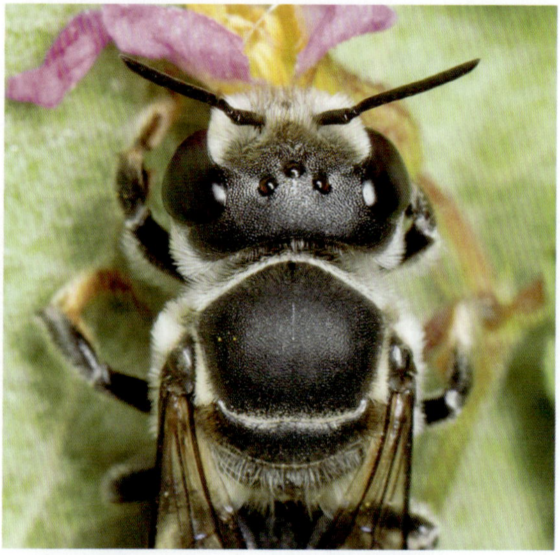

The finely pitted vertex and thorax of the female: note the short white hairs rimming the scutum.

Behavior: Leafcutters of the subgenus *Sayapis* nest in preexisting holes in plant matter or human-made structures. They do not line their nests with leaves. They do, however, plug their nest entrances with leaf cuttings (or a mix of materials) to protect them from parasites. In the 1980s, entomologist Anthony Raw conducted a study of *Megachile zaptlana* in Jamaica. He discovered Zaptlan leafcutters nesting in abandoned beetle burrows in fence posts—in one post, Raw found 129 *Megachile zaptlana* nests. Each nest was modest in size, typically consisting of a tunnel partitioned into five or six egg chambers, with an empty vestibular chamber at the entrance.

Additional information: When Cresson first described *Megachile zaptlana* in 1878, he had viewed only one Zaptlan leafcutter, a female collected by Sumichrast in Mexico. In 1913, 36 years after Cresson's first description of a female *Megachile zaptlana*, the Russian-born entomologist Curt Schrottky discovered a male leafcutter in Paraguay with a banded abdomen and enlarged forelegs, which he christened *Megachile melochiae.* He had found what would later prove to be a male *Megachile zaptlana*, roughly 7,000 miles from where Sumichrast had collected the female bee examined by Cresson. Perhaps because of the vast geographic range between the two discoveries, or because of the remarkable sexual dimorphism of the species, decades would pass before the two bees were united under the name *Megachile zaptlana.*

Policicaris leafcutter ◆ *Megachile (Sayapis) policaris* Say, 1831

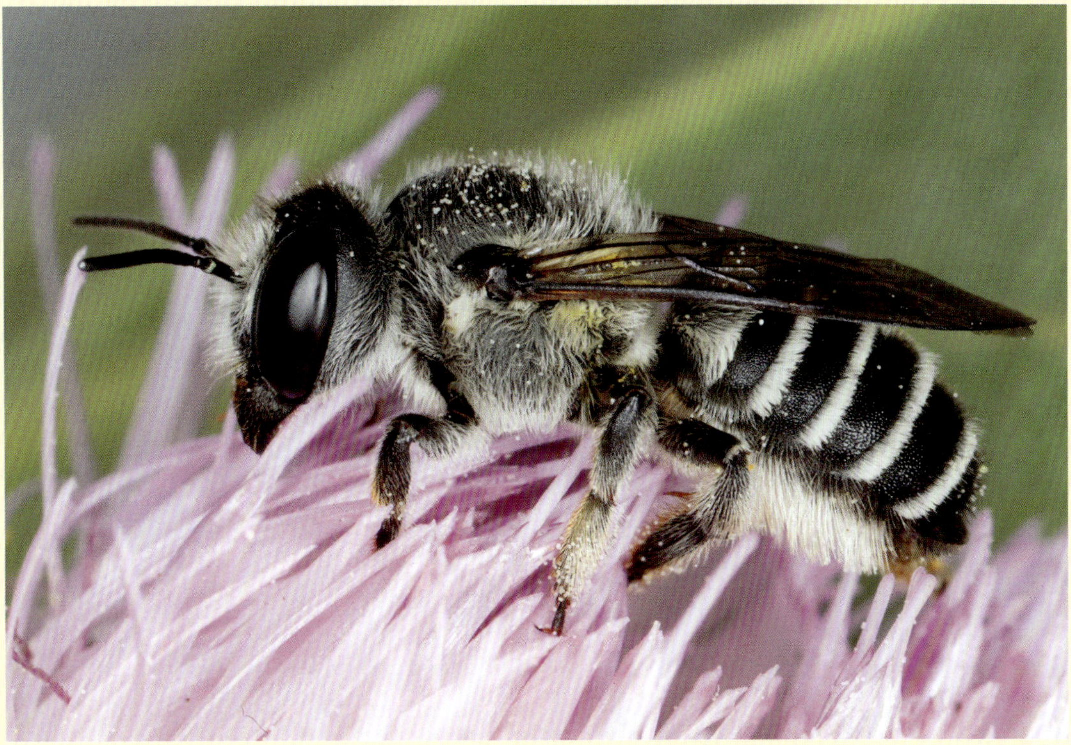

A female *Megachile policaris* on Texas thistle *Cirsium texanum*

Description: Male and female *Megachile (Sayapis) policaris* have the long, narrow, parallel-sided abdomens characteristic of the subgenus *Sayapis*. Their heads and bodies are black and their abdomens encircled by five white hair bands. Females have white scopal hairs on S1–S5 and dark hairs on S6; four-toothed mandibles with a beveled edge between the second and third teeth; and black eyes. Males have forelegs that are orange, ivory, and yellow, with each basitarsus expanded into a wide thumb shape. (The name *policaris* derives from the Latin *pollicis*, meaning "thumb.") Males' forelegs are fringed with long white-and-orange hairs. Their middle legs are also partly expanded, with partly orange tibiae and basitarsi. T6 of the male's abdomen is notched. On females, T6 is concave and has a ridged, upturned tip. Both males' and females' tegulae are dark and their wings a glassy brown, with brown veins. Males have three-toothed mandibles and green eyes. **Size:** females 14 mm (1/2–3/5 in.), males 13 mm (1/2 in.).

Range: *Megachile policaris* is widespread through most of the United States excluding the Northeast: it is found as far north as Ohio and Montana, and as far south as Florida. It ranges throughout Texas and southward from the Lower Rio Grande Valley to Costa Rica.

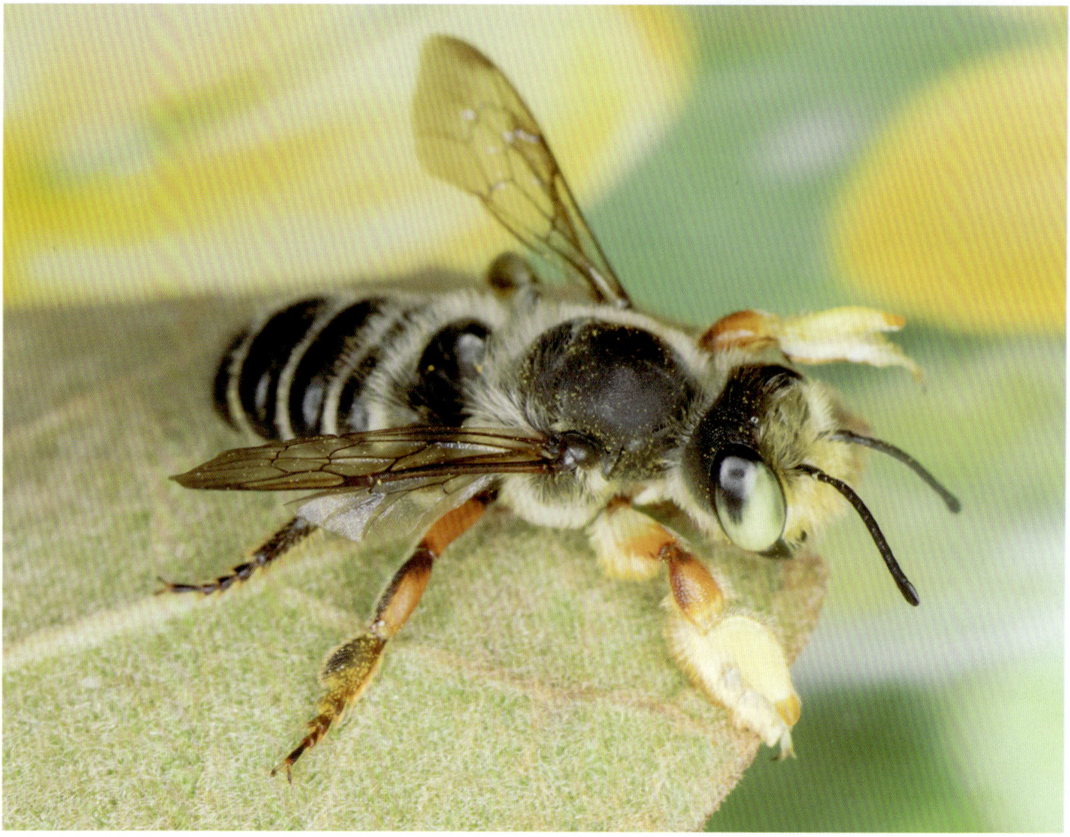

A male *Megachile policaris*

Associated plants: *Megachile policaris* flies in the Valley from March through November. It is a broad generalist pollinator but is commonly associated with Asteraceae. In the Valley, it visits such plants as creosote bush (*Larrea tridentata*) and catclaw acacia (*Senegalia greggii*), as well as aster-family flowers such as resinbush (*Viguiera stenoloba*), Texas thistle (*Cirsium texanum*), and coat buttons (*Tridax procumbens*).

ID tips: *Females*: In the Valley, *Megachile policaris* females most resemble other long-bodied female leafcutters of the subgenus *Sayapis*. *Megachile policaris* females can be distinguished in the field from female *M. zaptlana* because the latter have a distinct line of bright white hairs between the scutum and scutellum. The legs of *Megachile inimica inimica* females are typically red in the Valley, and thus that species is also readily distinguished from the black-legged female *M. policaris*. Subtle diagnostic differences in the clypei of female *Megachile policaris*, *M. inimica inimica*, and *M. frugalis* are discussed in the next sections.

Males: The male *Megachile policaris* is most likely to be confused in the Valley with *M. zaptlana*. The two species can be distinguished by differences in their abdomens and legs: (1) T6 of the male *Megachile policaris* abdomen is notched, rather than having a jagged, toothed edge like that of the male *M. zaptlana*; (2) *Megachile policaris* males have extensive orange markings on their middle legs, while those of the male *M. zaptlana* are predominantly black; (3) the middle-leg basitarsus of the male *M. policaris* also has a prominent expanded protuberance, wider than the tibia; and (4) the size, sculpting, and color of the forelegs of *Megachile policaris* distinguish it from all other males of the subgenus *Sayapis* in the Valley. The "thumbs" on the male's forelegs extend as far as the third tarsal segment below them and are pale yellow with orange tips. The inner hind surface of each foreleg basitarsus behind the "thumb" has an elongated black dot midway down the segment. Long foreleg hairs, white with rust-colored tips, appear on the front trochanters, femurs, tibiae, and basitarsi. The foreleg femurs and tibiae are also partly orange and pale. The front tarsi are ivory.

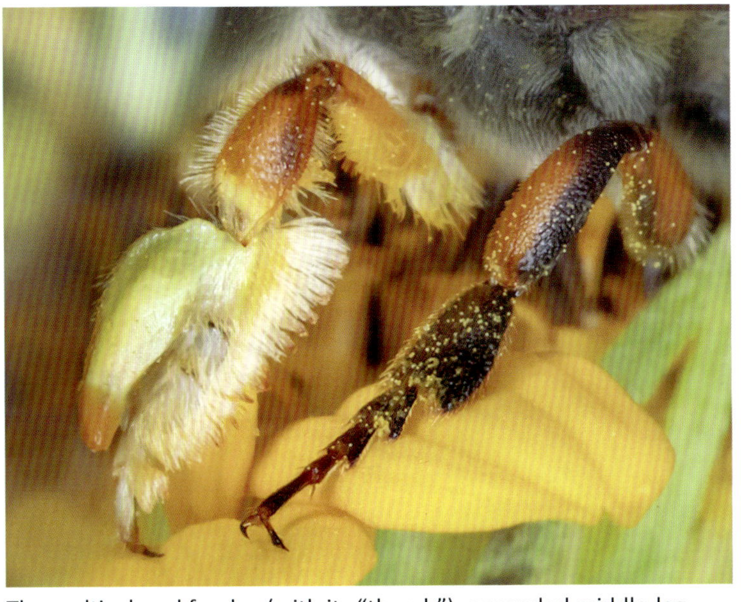

The multicolored foreleg (with its "thumb"); expanded middle-leg basitarsus; and partly orange middle-leg tibia of a male *Megachile policaris*

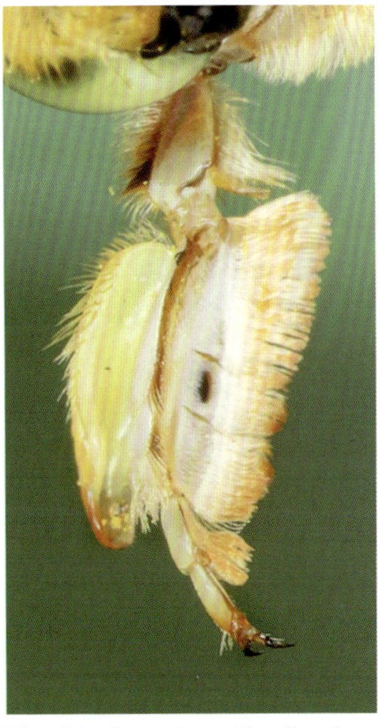

The black spot on the foreleg basitarsus

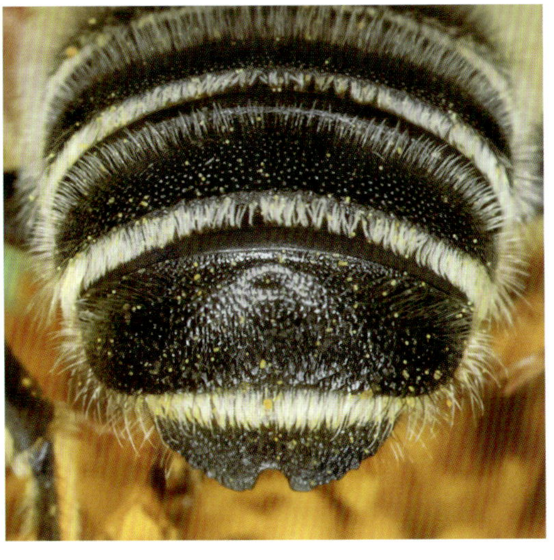

The abdomen of the male *Megachile policaris* is notched at the tip (T6).

Behavior: *Megachile policaris* nests in preexisting cavities in rotten logs, twigs, other plant matter, or the ground. In the late 1960s, entomologist Karl Krombein set trap nests for *Megachile policaris* in both Arizona and Florida and discovered that in both areas this species was multivoltine (producing several generations per year). Krombein found *Megachile policaris* unusual. He wrote: "This species differs markedly from other known megachilid bees and, in fact, from almost all aculeate Hymenoptera in that several larvae develop amicably in a single large brood cell with no apparent cannibalism."[5] It has since been determined that *Megachile policaris* females do divide their brood cells: they form discrete masses of pollen around individual eggs and sometimes construct partitions between them made of leaflets and leaf pulp. Nonetheless, female bees occasionally omit building partitions between cells, so that when their larvae achieve the fifth instar, they are able to eat their way through the whole length of the nest.

In the nests Krombein observed, female *Megachile policaris* placed a thin layer of gummy leaf pulp at the entrances of their tunnels before beginning to collect and store pollen and nectar. After provisioning their nests and laying eggs, the leafcutters sealed the entrances with plugs as thick as 25 mm (1 in.), made of alternating layers of leaf pulp, leaflets, and in some cases slices of cut leaves. In Arizona, the leaflets were collected in part from mesquite (*Prosopis juliflora*), catclaw acacia (*Senegalia greggii*), and kidneywood (*Eysenhardtia* sp.), all plants found in the Valley.

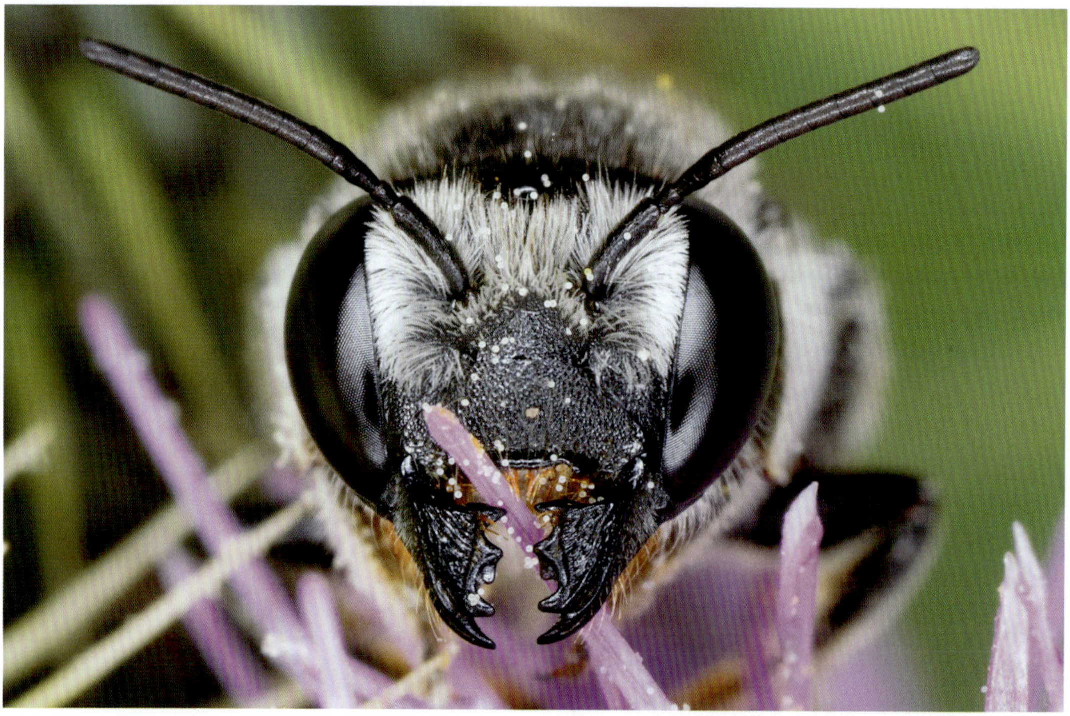

Face of a female *Megachile policaris*: the mandibles are four toothed, with a beveled edge between the second and third teeth. Photo credit Ross Eatman.

5. Krombein (1967), p. 334.

Hostile leafcutter ◆ *Megachile (Sayapis) inimica inimica* Cresson, 1872

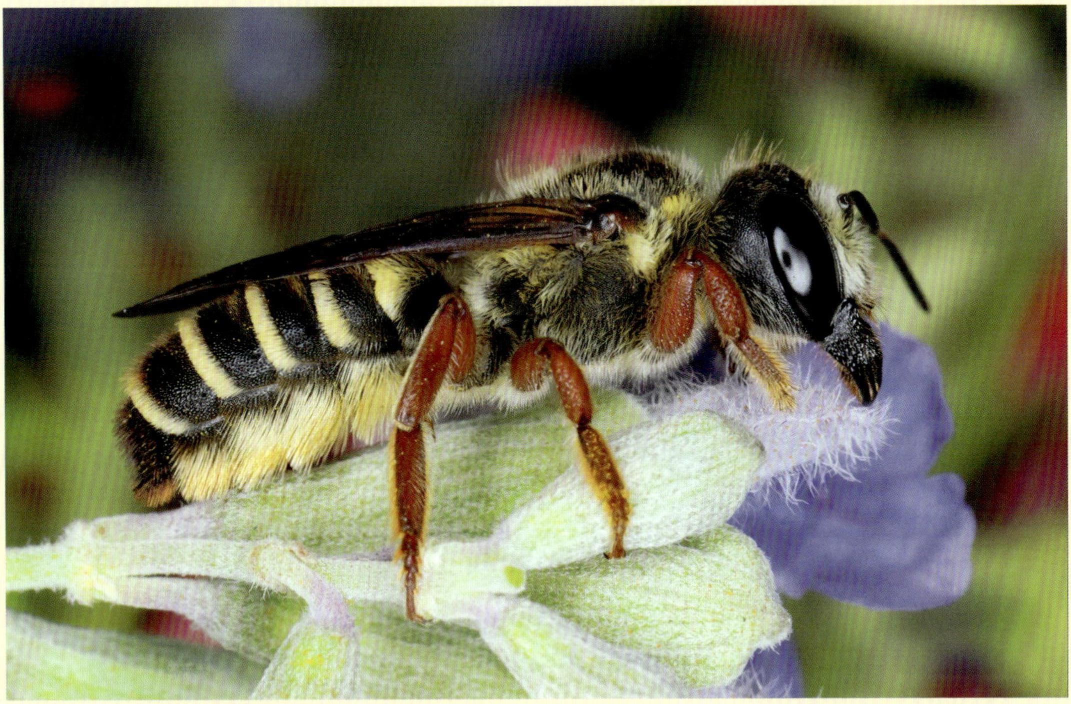

A female *Megachile inimica inimica*

Description: The female *Megachile inimica inimica* is a large bee with a black head and body, red legs, five bands of yellow hair on its abdomen, mostly yellow scopal hairs, and yellowish hairs on its face and thorax. It has the narrow, parallel-sided body and dark scopal hairs on S6 typical of the subgenus *Sayapis*. The female's vertex is broad, and its mandibles large and four toothed, with a well-defined beveled edge between the second and third teeth. Its clypeus has a distinctive point in the middle. The wings of both male and female *Megachile inimica inimica* are brown, darkening toward the tips, with dark brown veins. Their tegulae, antennae, and mandibles are dark. Males have mostly red legs. The male's foreleg basitarsi are enlarged, with white outer surfaces and fringes of pale hairs. Males' abdomens have five bands of pale hairs and are notched on T6. Males have three-toothed mandibles. **Size:** females 13–16 mm (1/2–5/8 in.), males 11–14 mm (2/5–3/5 in.).

Range: The hostile leafcutter subspecies *Megachile inimica inimica* ranges from Florida to Texas, and southward through Mexico to Guatemala. A more northerly subspecies, *Megachile inimica sayi* (which has black legs, and white rather than yellow hairs), has a broader range, extending as far north as New York, as far west as California, and as far south as Florida and Mexico. Female *Megachile inimica inimica* appeared at the National Butterfly Center (Hidalgo Co.) in April 2019, and at La Puerta Tract (Starr Co.) in mid-May 2021.

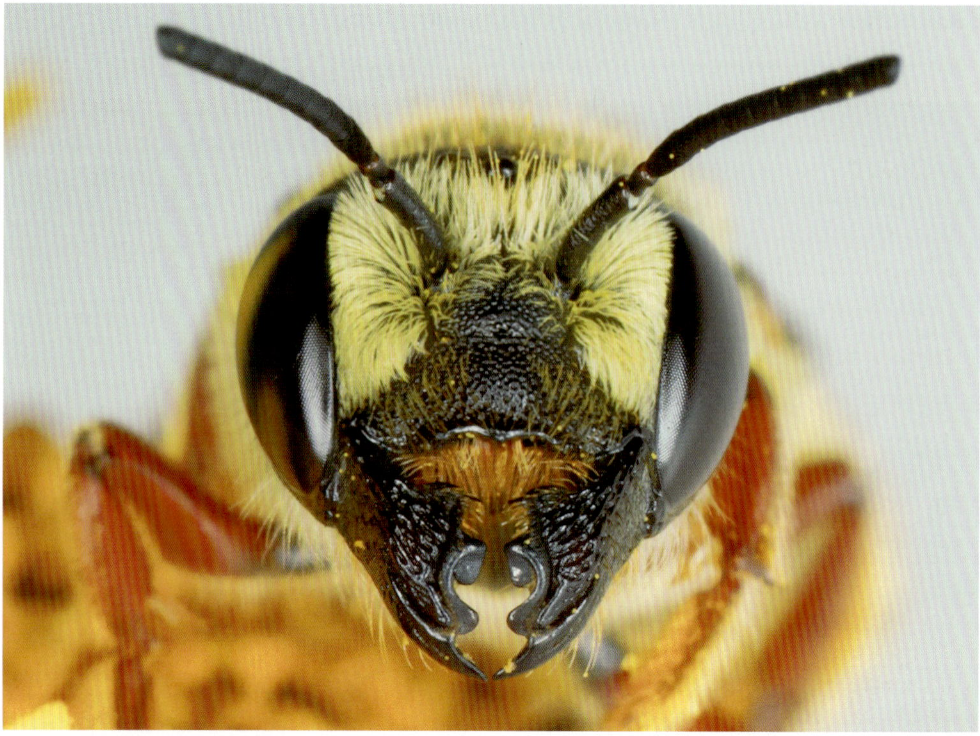

Face of a female *Megachile inimica inimica*, showing its large four-toothed mandibles with well-defined beveled edges between the second and third teeth. Note the median point on the rim of the clypeus.

Associated plants: *Megachile inimica inimica* females tend to gather pollen from Asteraceae. The female bees shown here were found collecting pollen from resinbush (*Viguiera stenoloba*) and nectaring on guajillo (*Senegalia berlandieri*).

ID tips: The combination of this bee's size, long and narrow body, and red legs distinguishes it from all other leafcutters of the Lower Rio Grande Valley. The female's clypeus, with its median point, is also a useful diagnostic trait that separates *Megachile inimica inimica* from similar *Megachile (Sayapis)* females. The dilated pale areas of the male's basitarsi are relatively small; they are narrower than those of *Megachile policaris* and *M. zaptlana* and do not extend as far as the third tarsal segment below them.

Behavior: Krombein described nests of *Megachile inimica inimica* as consisting of unlined tunnels with caps formed of one or two circular leaf cuttings attached to a tier of agglutinated sand, or of sand mixed with debris. The nests he observed were plugged with multiple circular leaf cuttings, sand, or a combination of these. Krombein also examined nests belonging to *Megachile inimica sayi*, in which the leafcutters had intermixed fine and coarse small pebbles and gummy leaf pulp in their construction materials. The nests sometimes incorporated empty entrance areas that buffered the tunnel portion containing brood cells from the nest's plugged entrance.

Frugal leafcutter ◆ *Megachile (Sayapis) frugalis* Cresson, 1872

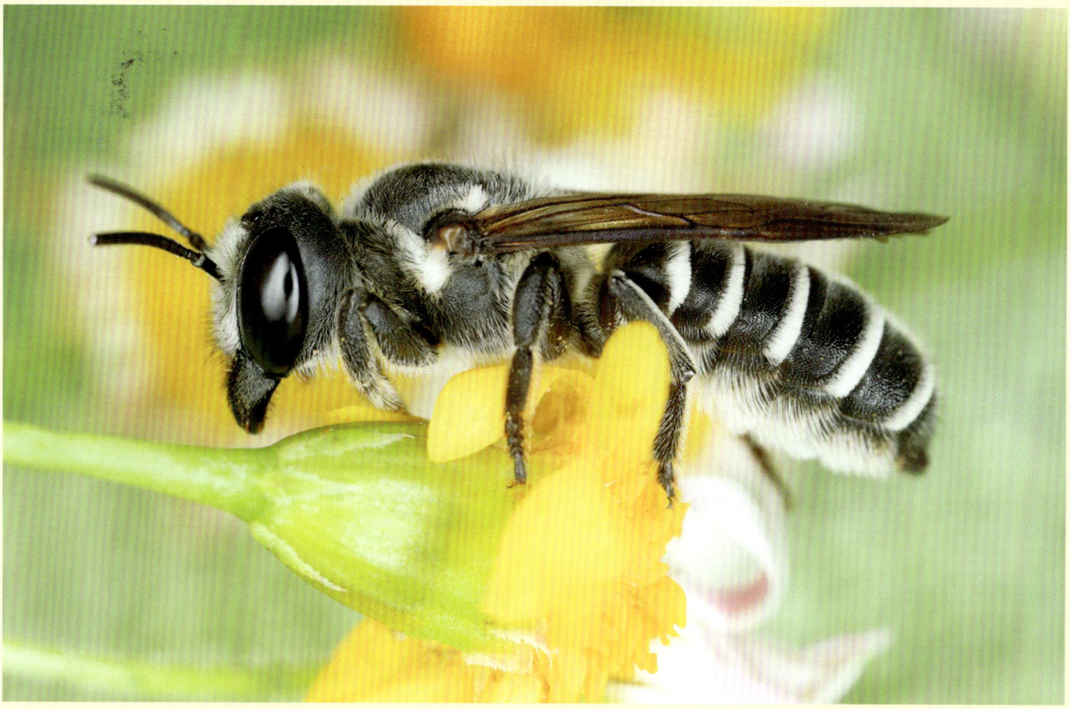

A female *Megachile frugalis*

Description: The frugal leafcutter has a long, narrow body banded on T1–T5 with white hairs. White hairs partially cover the head, thorax, and legs. Both males and females have black legs, reddish-brown tegulae, and brown-veined, smoky wings with dark edges and dark marginal cells. The female's scopal hairs are dark on S6, but otherwise pale. The tip of the female's abdomen, viewed in profile, is convex with a narrow, upturned rim. The forelegs of males are unmodified and fringed with longish white hairs. The tip of the male's abdomen (T6) has a medial notch flanked on either side by indentations. Both males and females have dark eyes and dark antennae. Females' mandibles are four toothed, with beveled edges between the second and third teeth. Males' mandibles are three toothed. **Size:** females 12–13 mm (1/2 in.), males 10–11 mm (2/5 in.).

Range: Frugal leafcutters range from the West to the East Coast of the United States, as far north as New England on the Eastern Seaboard, and as far south as Florida. *Megachile frugalis* also ranges from the Texas border through Mexico, and into Central America as far south as Costa Rica. It is nonetheless uncommon in the Lower Rio Grande Valley: other members of the subgenus *Sayapis*, particularly *Megachile zaptlana* and *M. policaris*, are far more abundant. The male and female bees shown here appeared at El Mesteño Ranch (northwest Hidalgo Co.) in early June 2021.

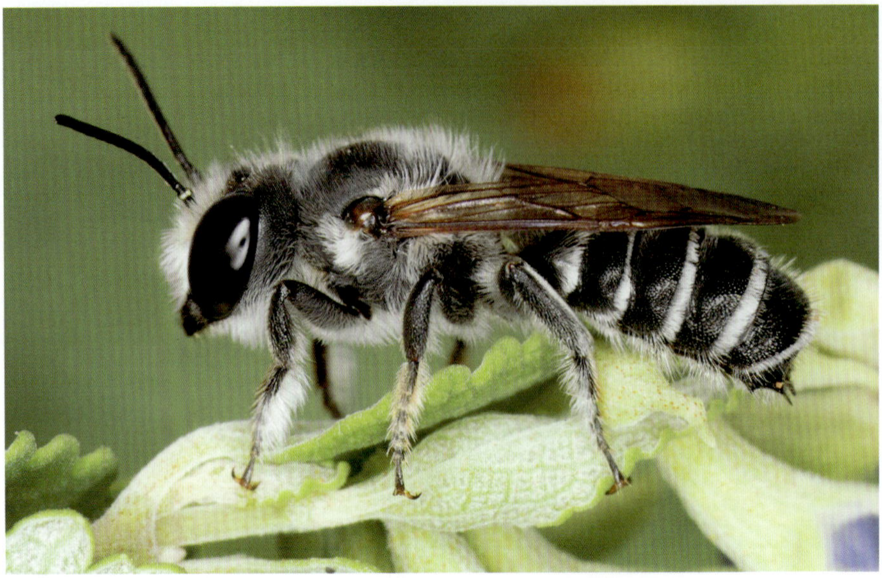

A male frugal leafcutter

Associated plants: Frugal leafcutters are broad generalist pollinators. The female bee at El Mesteño Ranch was gathering pollen from Arkansas dozedaisy (*Aphanostephus skirrhobasis*). The male bee was nectaring on shrubby blue salvia (*Salvia ballotiflora*).

ID tips: Both male and female *Megachile frugalis* have distinctive black hairs on the clypeus, most easily seen in profile. Males, in part because of their unmodified forelegs, are superficially similar in appearance to *Megachile (Litomegachile)*, but male frugal leafcutters have proportionately longer, narrower bodies, and on male *M. frugalis*, T6 is not covered by the short white hairs typical of *Litomegachile* in the Valley.

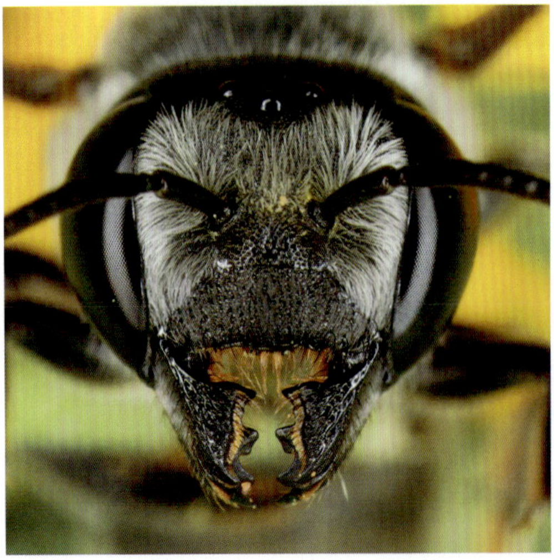

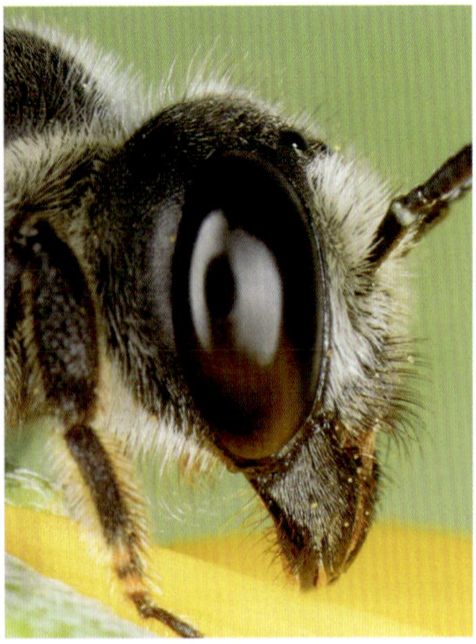

Face and profile of a female frugal leafcutter: note the black hairs on the clypeus.

Silver-tailed petal cutter ◆ *Megachile (Megachile) montivaga* Cresson, 1878

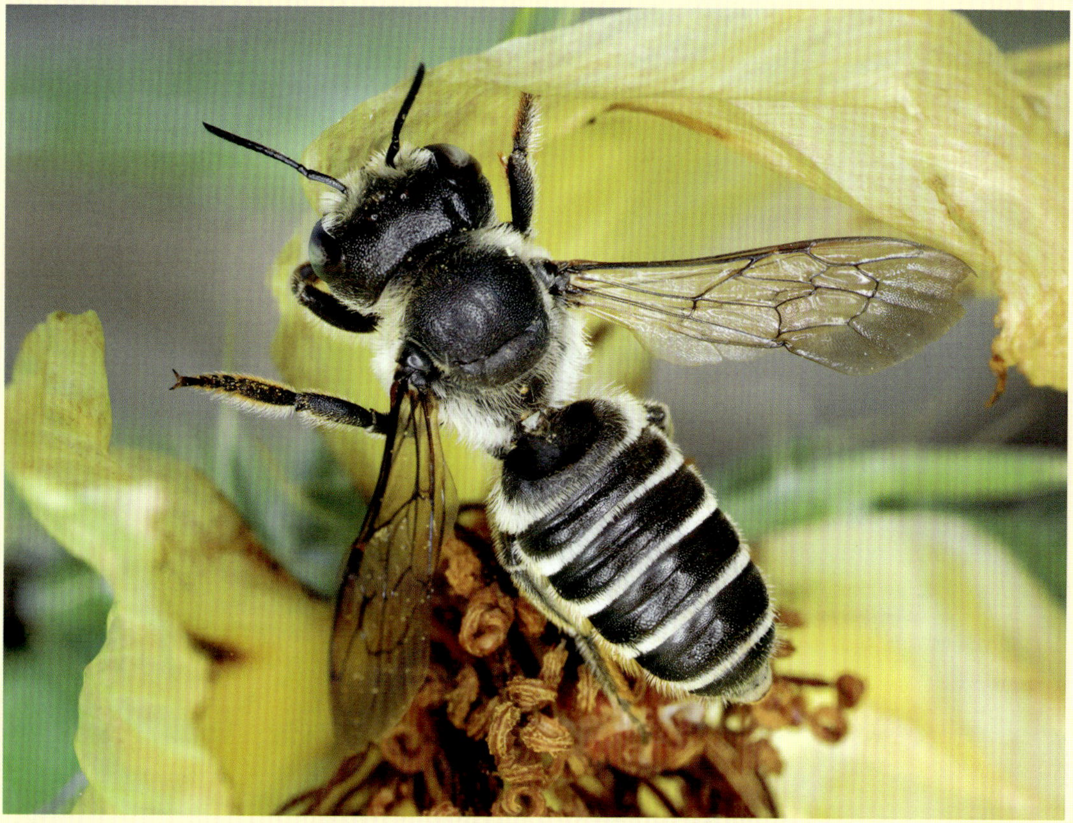

A female *Megachile montivaga* unfurling a prickly poppy before trimming its petal

Description: *Megachile montivaga* is an uncommon leafcutter whose Latin name translates to "mountain wanderer." This graceful bee belongs to the subgenus *Megachile.* Leafcutters of this subgenus vary fairly widely in physical appearance and behavior. Both male and female silver-tailed petal cutters are black bees with slender abdomens that are deeply grooved and banded with pale hairs. The scopal hairs of females are entirely pale, often a yellow white. The legs of both females and males are black, their tegulae dark, and their wings faintly brown with dark brown veins. Dense white hairs cover the bees' faces and cheeks and rim the sides and rear of the thorax. The mandibles of female *Megachile montivaga* are five toothed but lack cutting edges. Males' mandibles are three toothed, with the teeth evenly spaced. The bees' eyes are dark green. **Size:** females 11–13 mm (2/5–1/2 in.), males 9–11 mm (1/3–2/5 in.).

Range: *Megachile montivaga* is found throughout the United States, northward into Canada, and southward to the Mexican state of Guerrero. The female bees shown here appeared at the National Butterfly Center (Hidalgo Co.) in July 2019 and April 2021.

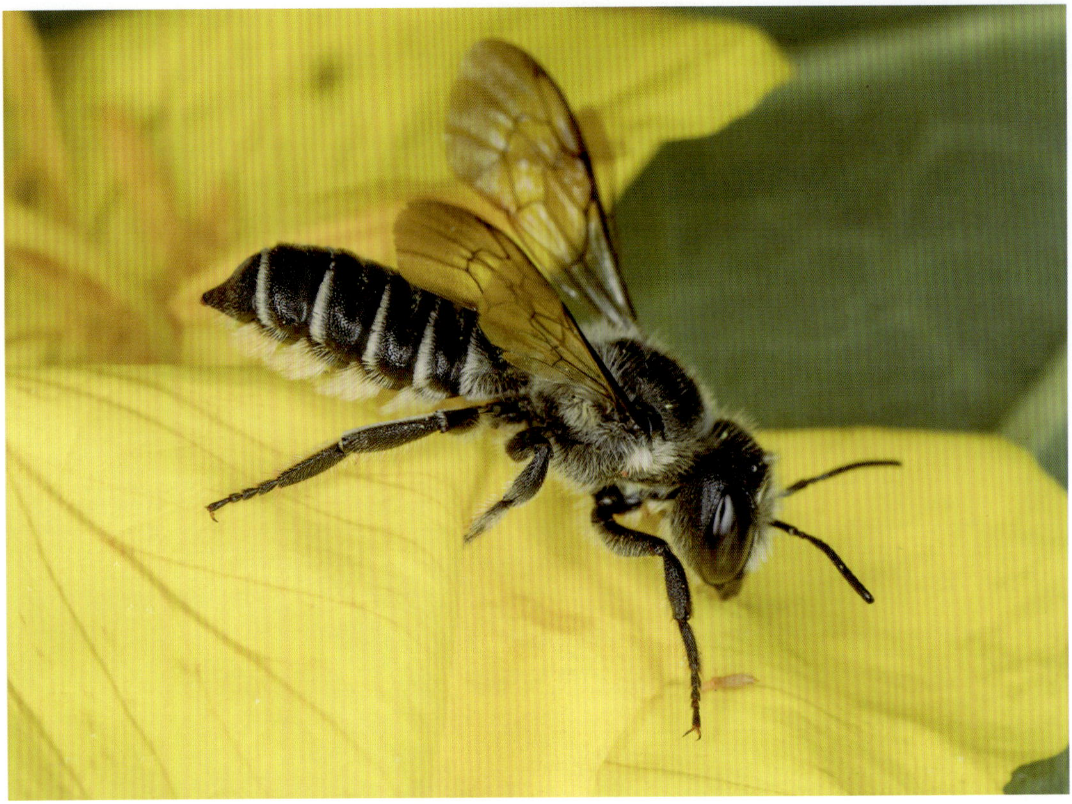

Lateral view of a female *Megachile montivaga*

Associated plants: Silver-tailed petal cutters are generalist pollinators that have been documented foraging on diverse flora, including mallows, salvias, pea-family plants, roses, chokecherry, milkweed, and prickly pear cactus. Female *Megachile montivaga* often choose the hollow stems of dead, large-stalked flowers such as sumac, mullein, elderberry, and sunflowers as nest locations.

Female *Megachile montivaga* do not usually trim leaves from plants for use as nesting materials. Instead, the bees line their nest walls with flower petals. They tend to use the broad, diaphanous petals of clarkia and other flowers of the evening primrose family (Onagraceae), and *Megachile montivaga* populations are sometimes linked to the availability of such flora. Female *Megachile montivaga*, however, can be ecumenical in choosing flora for nest building: phlox, *Cassia*, asters, and possibly *Sphaeralcea* have been found lining their egg chambers. The female bees at the National Butterfly Center were found trimming the large, delicate petals of golden prickly poppy (*Argemone aenea*), as well as pink evening primrose (*Oenothera speciosa*).

ID tips: Distinguishing traits of the female *Megachile montivaga* include its entirely pale scopae, its five-toothed mandibles, and the appearance of S6, which is concave in profile and bare of hairs near the rim. On male bees, T6 is protuberant and its hind edge is toothed. Males' forelegs are unmodified. T7 of the male's abdomen is visible from above, and S1–S4 of the sternum are exposed.

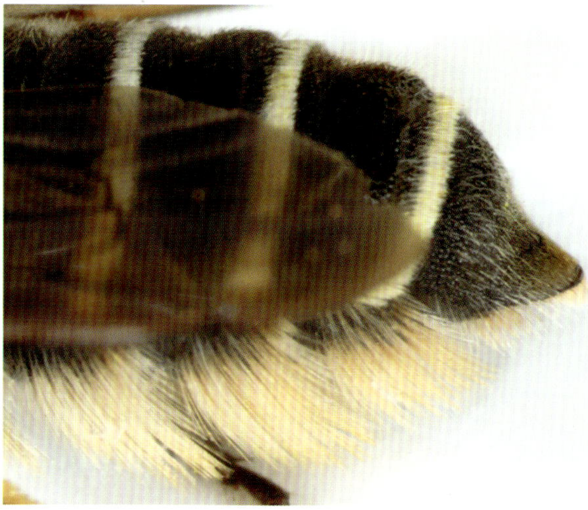

The tip of the abdomen (T6) of a female *Megachile montivaga* is concave in profile. The scopal hairs are entirely pale (white or light yellow).

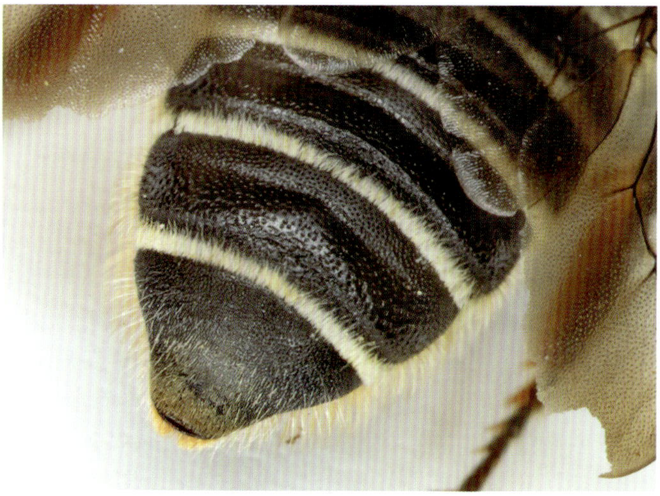

Dorsal view of the tip of a female's abdomen. T6 narrows abruptly, and T5 is narrower than T4.

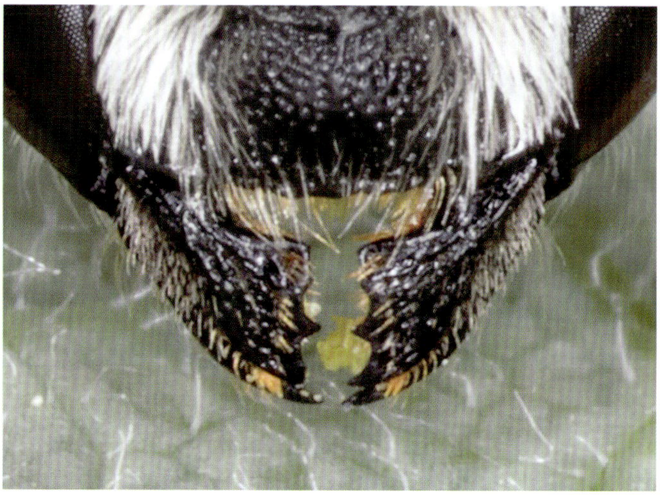

The mandibles of the female *Megachile montivaga* lack cutting edges and have five teeth on each side, which are sometimes mistaken for four teeth because the fourth tooth (second from top) is quite small.

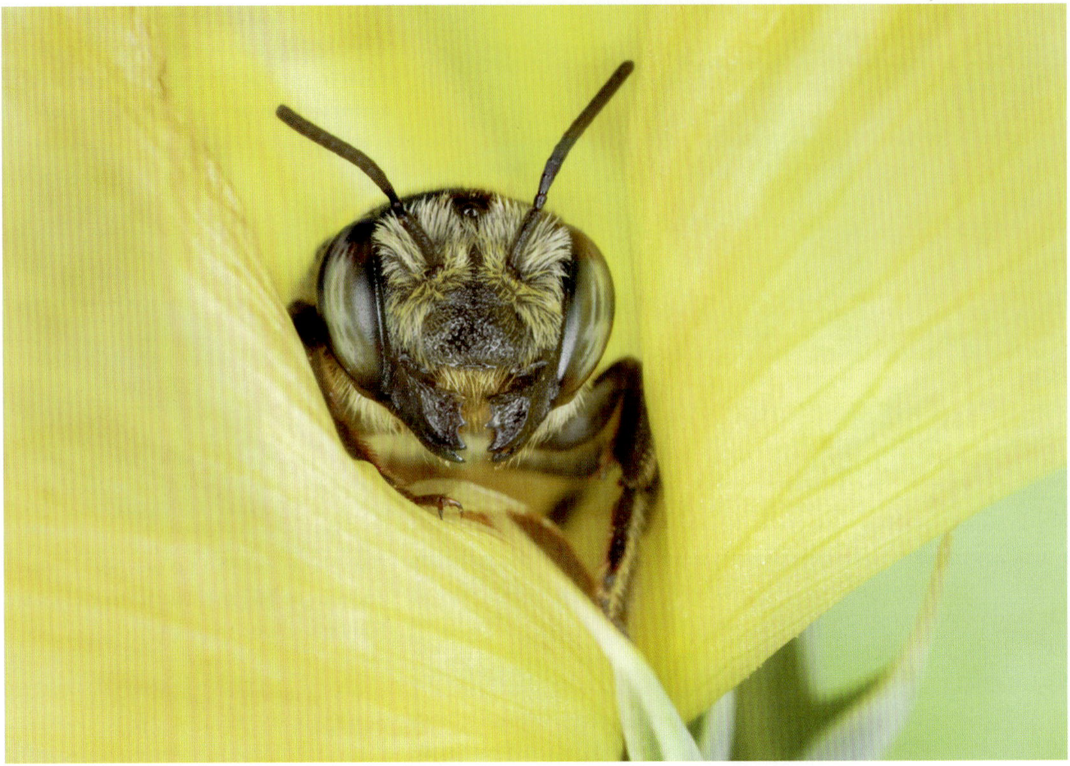

A female silver-tailed petal cutter peering from a golden prickly poppy

Petal-cutting and nesting behavior: According to a 2013 study by entomologists Sheffield, Ratti, Packer, and Griswold, silver-tailed petal cutters' use of flowers rather than leaves in nest construction is unique among leafcutters of the *Megachile* subgenus. A handful of Texas leafcutters of other subgenera also trim flower petals: the evening primrose leafcutter (*Megachile oenotherae*) and Chomsky's leafcutter (*Megachile chomskyi*), both members of the subgenus *Megachiloides*, are petal cutters. As noted above, *Megachile (Litomegachile) brevis* is known to incorporate petals as lining and padding materials in nests constructed largely of leaves.

Megachile montivaga nests both in cavities in the ground and in dead plant matter. A remarkable trait of this species is its occasional tendency to build nests in live flora. Living plants are often well armored with such features as spines, hard outer coatings, and chemical defenses, and thus cavity-nesting bees customarily avail themselves of brown, rotted, overwintered, and otherwise dead plant materials. Nonetheless, in 2015, entomologists Orr, Portman, and Griswold discovered *Megachile montivaga* building nests in live thistles (previously damaged by stem-boring insects) in the Hualapai Mountains of Arizona. Orr and colleagues reported that an earlier entomological team, Parker and Bohart, had also observed silver-tailed petal cutters nesting in live elderberry stems in 1966.

Although silver-tailed petal cutters prefer using flower petals for nest construction, Michener found a *Megachile montivaga* nest in Los Angeles County in 1953 whose egg-cell walls were lined in part with the red autumn leaves of poison oak. The nest had been constructed in a green weedy stem.

Sidalcea leafcutter ◆ *Megachile (Pseudocentron) sidalceae* Cockerell, 1897

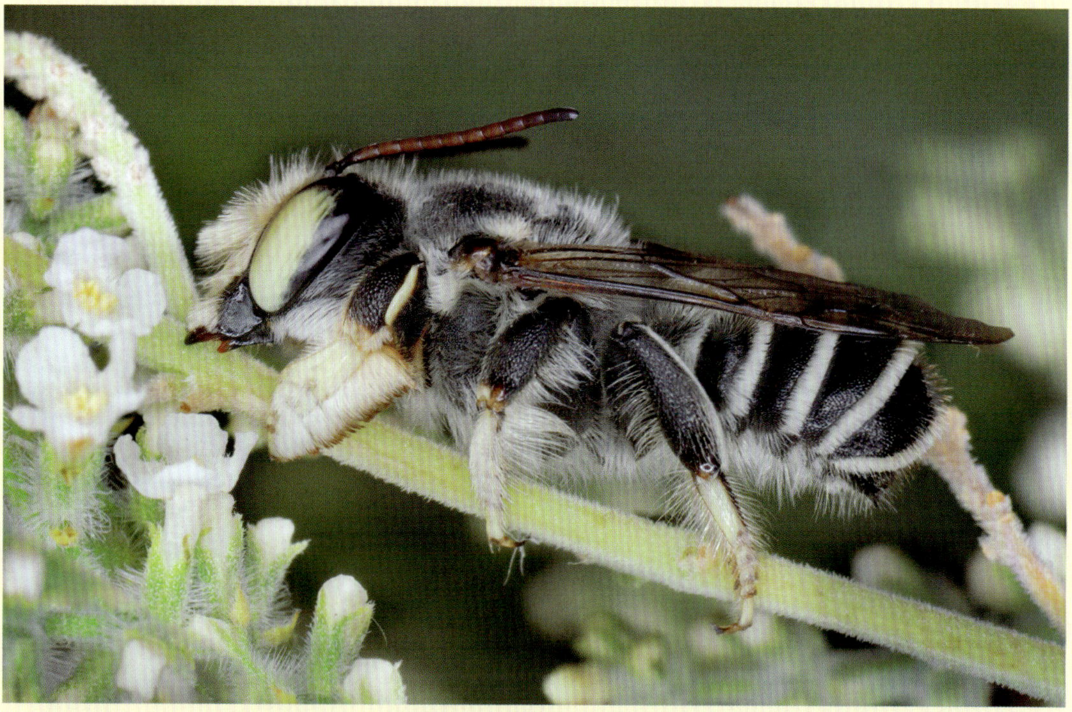

A male *Megachile sidalceae* on whitebrush: note the yellow stripe on the front tibia.

Description: *Megachile (Pseudocentron) sidalceae* is a handsome black bee with bands of bright white hair on its abdomen. A strip of white hairs clearly marks the border between this leafcutter's scutum and scutellum, white hairs line the back of its scutellum, and distinctive tufts of white hairs appear behind each tegula. The tegulae are yellowish or reddish with a black mark, and the wings are brown veined and glassy. The scopae of female *Megachile sidalceae* are white on S1–S5 and dark on the base of S6.

Males have a conspicuous yellow stripe on the rear of each foreleg tibia, a distinguishing trait of the subgenus *Pseudocentron*. The forelegs of male *Megachile sidalceae* also have dilated, pale yellow basitarsi. All of the male's tarsi, on all legs, are predominantly pale yellow. Copious yellowish-white hairs cover the male's face, and long white hairs appear on its legs and lower body. The tip of the male's abdomen (T6) is notched and rimmed with jagged teeth. Its tergal hair bands continue onto the sternum. Females of the subgenus *Pseudocentron* have four-toothed mandibles, with a cutting edge between the third and fourth tooth, and an incomplete edge between the second and third tooth. Females sometimes have a small median tubercle on the lower edge of the clypeus. Males also have four-toothed mandibles, sometimes equipped with subapical teeth. **Size:** females 13–15 mm (1/2–3/5 in.), males 11–14 mm (2/5–3/5 in.).

Range: The leafcutter subgenus *Pseudocentron* is mostly Neotropical—it is the largest leafcutter subgenus of South America, containing at least 64 species. In the United States, it is represented by two species, *Megachile sidalceae* and *M. pruina.* (The latter has been documented in Galveston County, Texas, but not in the Lower Rio Grande Valley.) In the United States, *Megachile sidalceae* is found mostly in the West and Southwest. It occurs in Texas, northward to Kansas, and westward to southern California. It appears in both Hidalgo and Starr Counties and ranges southward from the border through Mexico at least as far as Oaxaca.

Associated plants: *Megachile sidalceae* is a generalist pollinator that visits plants from diverse families, among them redbuds, mints, mallows, amaranths, and asters. In the Valley, this species can be found from midspring through midsummer foraging on such flora as whitebrush (*Aloysia gratissima*), Texas kidneywood (*Eysenhardtia texana*), retama (*Parkinsonia aculeata*), Texas palafox (*Palafoxia texana*), and resinbush (*Viguiera stenoloba*). It has been documented elsewhere feeding on mesquite (*Prosopis juliflora*) and creosote bush (*Larrea tridentata*).

ID tips: (1) On all female *Megachile (Pseudocentron)*, the sixth abdominal segment (T6) looks like a blunted triangle from above. On female *Megachile sidalceae*, S6 has dark hairs at the base but is otherwise sparsely haired, with a bare lip. (2) In addition to other distinctive leg traits shown here, all male *Megachile (Pseudocentron)* lack tibial spurs on their middle legs.

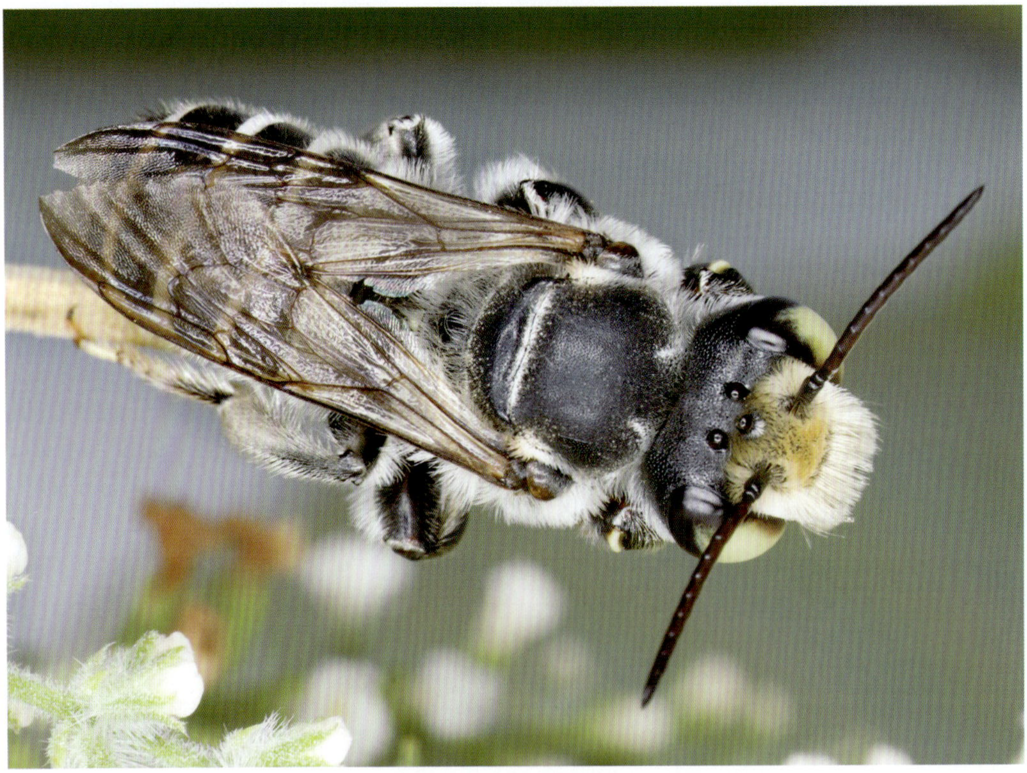

Dorsal view of a male *Megachile sidalceae*: note the tufts of white hair behind the tegulae and the bright line of white hair on the thorax, between the scutum and scutellum.

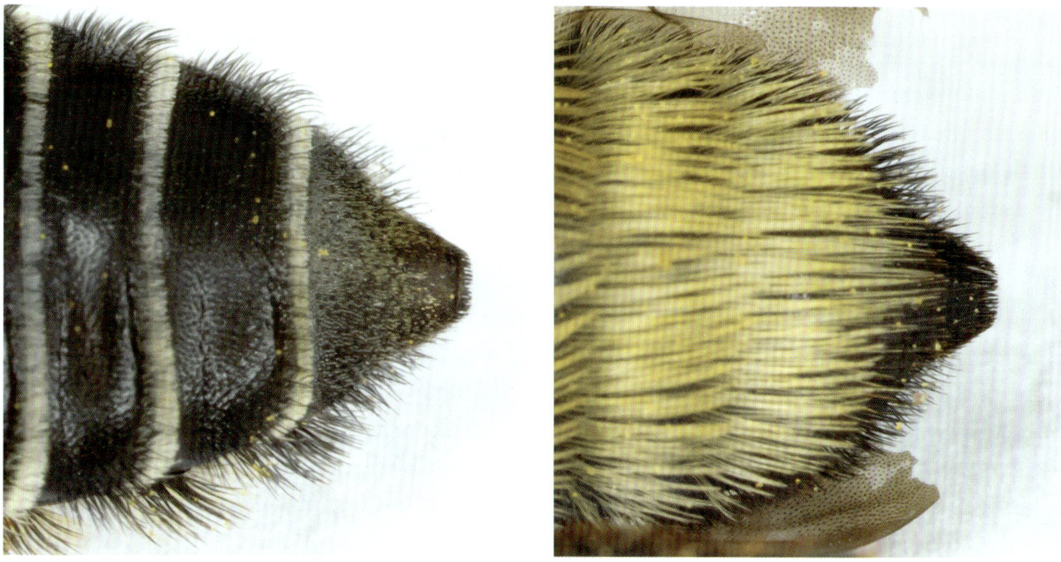

The tip of the abdomen of a female *Megachile (Pseudocentron)*: T4–T6 (*left*) and S4–S6 (*right*).

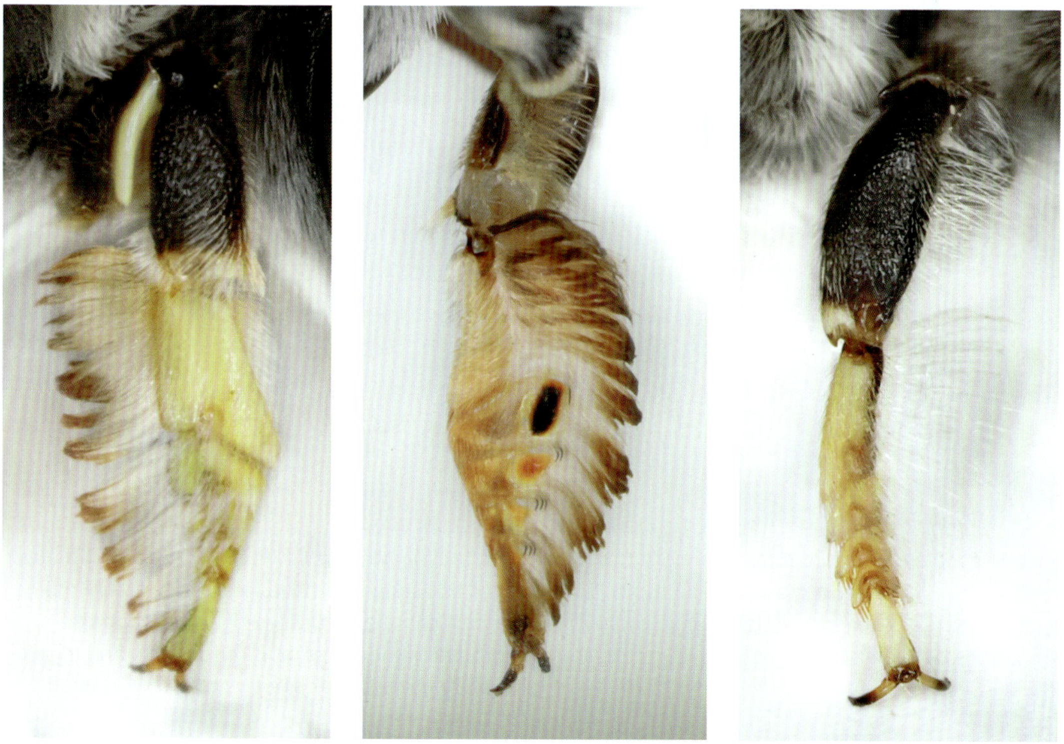

The outer foreleg basitarus (*left*); inner foreleg basitarsus (*center*); and middle leg (*right*) of a male *Megachile sidalceae*. The front femur bears a conspicuous yellow stripe. The front basitarsus is greatly dilated, yellow, and fringed behind with long white hairs with brown tips. The second joint of each front tarsus has a dark spot on the inner face. The middle leg lacks a tibial spur; instead, it has a short protuberance (known as a pseudocentron).

Behavior: Little has been published about the behavior and nesting biology of leafcutters of the subgenus *Pseudocentron.* Available literature indicates that they may nest in cavities in the ground or in plant materials. In 1865, Cresson described a Cuban species, *Megachile poeyi*, which nested under a pile of rocks and constructed its nest from leaves. A 2014 study by Landry, Elliott, and Vitale described a closely related Bahamian species, *Megachile (Pseudocentron) alleni*, nesting in cavities in rocks and asphalt and collecting leaves for nest building from plants of the aster and pea families.

In 2012, entomologist Juan Torretta and colleagues studied the South American *Megachile (Pseudocentron) gomphrenoides* on the northeastern coast of Argentina, in an area of the pampas characterized by pastureland, sunflower fields, and seminatural grassland. This species, an oligolectic pollinator with a preference for sunflowers and other Asteraceae, proved to be bivoltine. It nested in hollow bamboo canes supplied by the researchers as trap nests. The leafcutters' nests were tunnel shaped, with entrances sealed with circular leaves: some had empty vestibules buffering egg chambers from the nest entrances. Circular cuttings from black locust, *Trifolium*, and *Chenopodium* were used to partition egg chambers, while oval cuttings were used to line them.

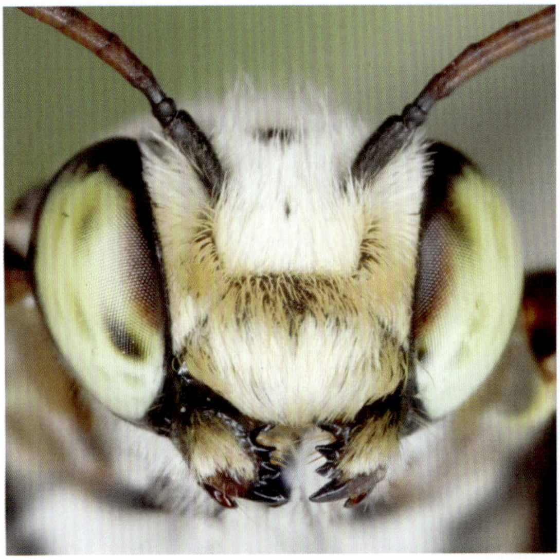

Face of a male *Megachile sidalceae*: note the four-toothed mandibles, and the subapical tooth. The face and part of each mandible are covered with long, dense, yellowish-white hairs.

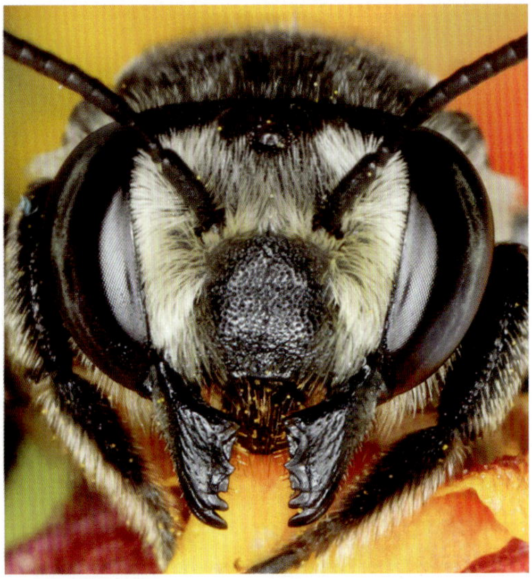

A female *Megachile (Pseudocentron)*: females of this subgenus have four-toothed mandibles, with complete cutting edges between the fourth (innermost) and third teeth, and incomplete cutting edges between the second and third teeth.

Parallel leafcutter ◆ *Megachile (Argyropyle) parallela* Smith, 1853

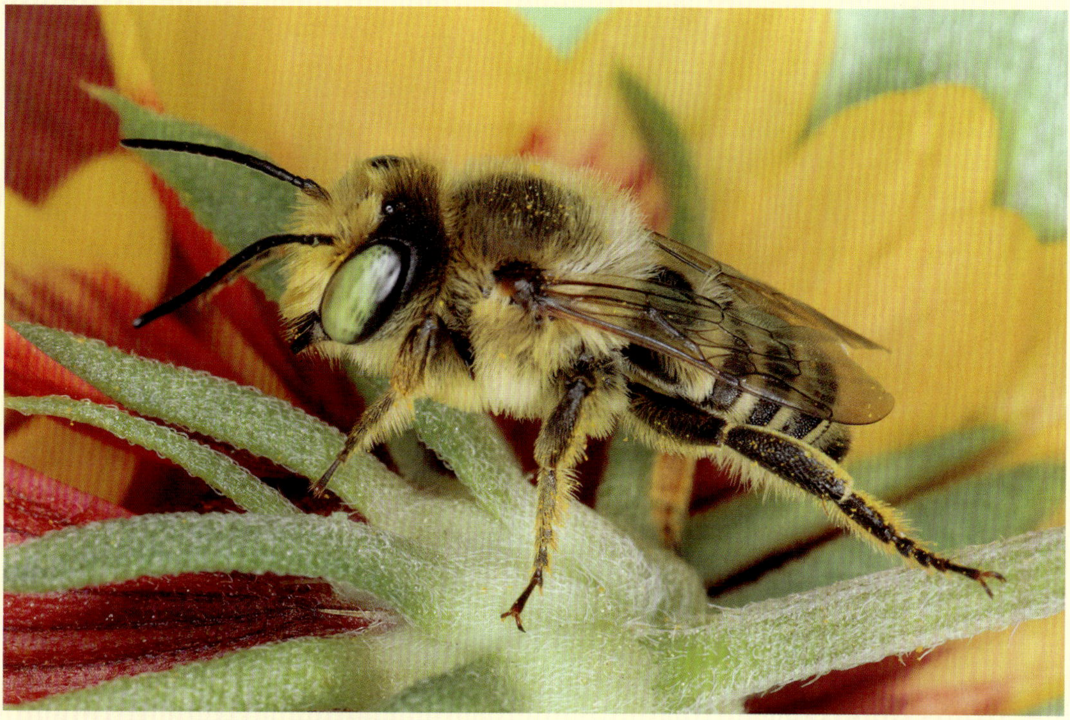

A male parallel leafcutter on blanketflower

Description: The male *Megachile (Argyropyle) parallela* has an overall yellowish-brown appearance that makes it stand out in the field. Golden-brown hairs clothe the thorax and band the abdomen, and long yellowish hair covers the face. On females, the tergal bands are white and crisply defined. The female's facial hair is white, and the hairs on its thorax white or pale yellow. The heads, bodies, legs, antennae, and mandibles of both males and females are black. Their tegulae are dark, and their wings are glassy brown with brownish-black veins.

Both males and females have four-toothed mandibles and green eyes. Females' mandibles have a long cutting edge between the third and fourth teeth and a shorter one between the second and third. **Size:** females 13–14 mm (1/2–3/5), males 12–13 mm (1/2 in.).

Range: The parallel leafcutter is widespread throughout the United States and Mexico and as far north as British Columbia. It has existed in the Lower Rio Grande Valley for more than 100 years: in 1917, Wilmatte Cockerell found a female parallel leafcutter feeding on a composite flower in Port Isabel in Cameron County. This species appeared at the National Butterfly Center (Hidalgo Co.) in late March 2019 and late June 2021.

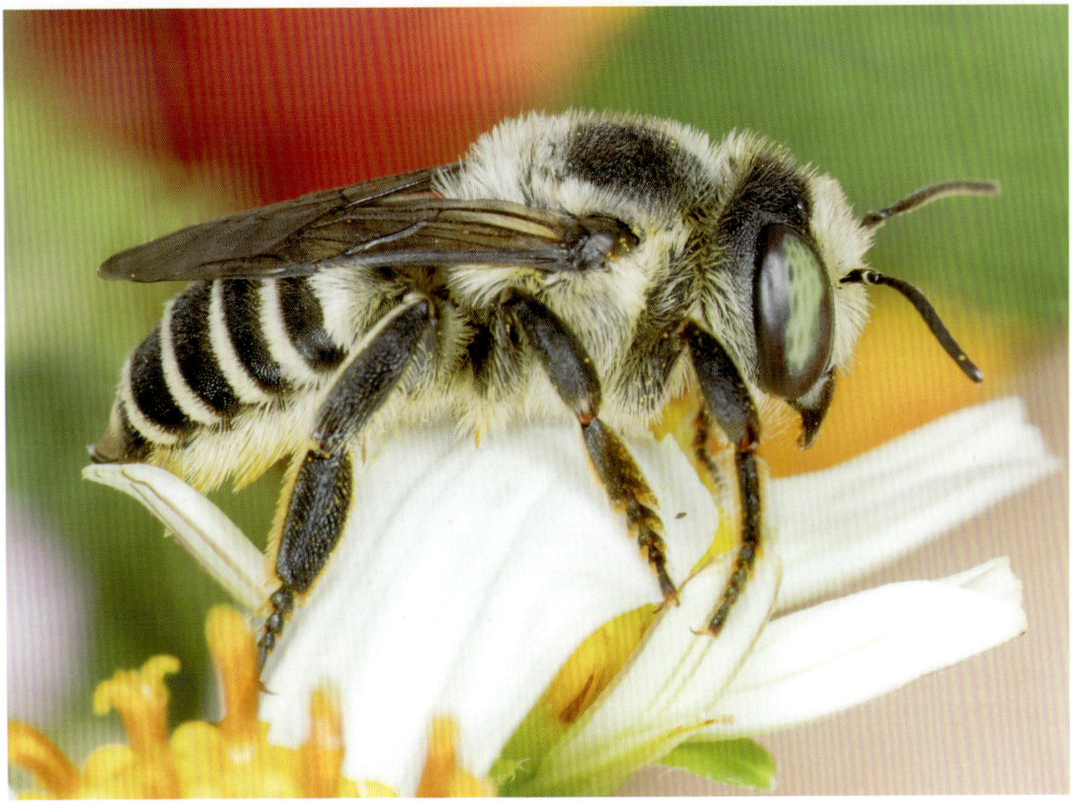

A female parallel leafcutter

Associated plants: Parallel leafcutters are generalist pollinators that feed on flora of many families, among them mints, pea-family plants, mustard, spurge, mallows, and verbenas. In 1923, USDA plant physiologist Thomas Kearney observed parallel leafcutters pollinating cotton flowers and extolled the quantity of pollen they carried as "remarkable." Parallel leafcutters nonetheless show a preference for composites. In the Lower Rio Grande Valley, they visit flora such as cowpen daisy (*Verbesina encelioides*), blanketflower (*Gaillardia pulchella*), and romerillo (*Bidens alba*).

ID tips: Female parallel leafcutters are easily recognized by the tips of their abdomens: the upturned rim of the last segment of the sternum (S6) extends beyond T6. The lower margin of the female's clypeus is straight, with a distinct tubercle in the middle. Males can be identified by their combination of four-toothed mandibles, unmodified legs, and double rows of two teeth on the tip of the abdomen.

In Texas, this species is somewhat different from parallel leafcutters in other parts of the United States. Elsewhere, the typical female parallel leafcutter has black hairs along the sides of the second through fifth abdominal segments (T2–T5), and dark hairs and a bare patch on S6. The female Texas parallel leafcutter may lack these features.

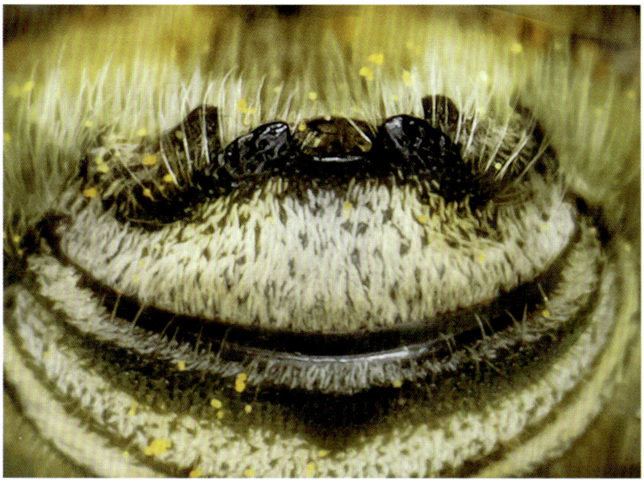

T6 of the male parallel leafcutter's abdomen has two sets of two teeth.

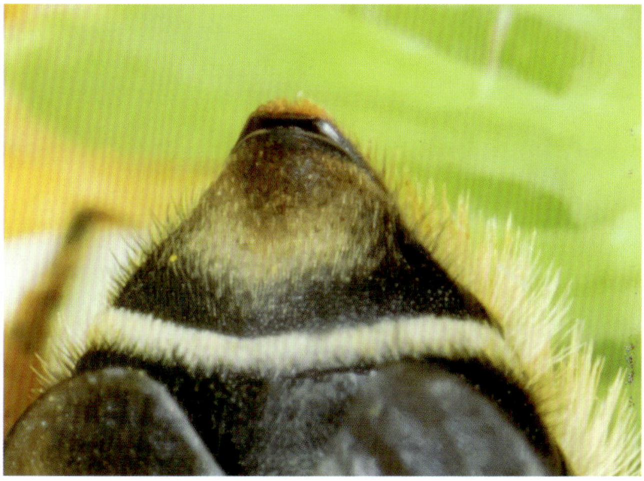

The tip of the female parallel leafcutter's abdomen is distinctive: the last sternal segment is upturned and protrudes beyond the rim of T6.

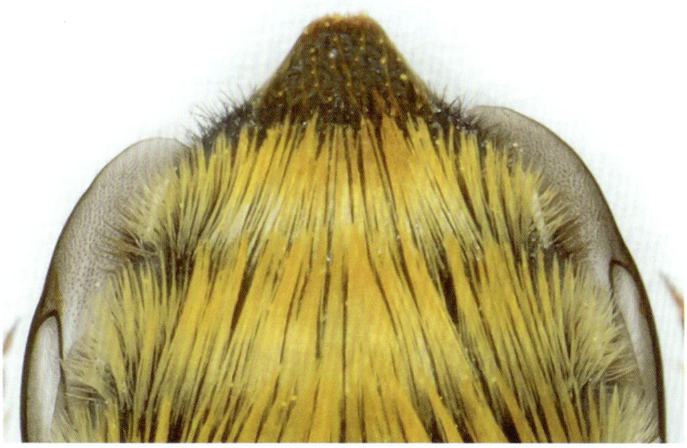

The sternum of a female parallel leafcutter: pale yellowish hairs cover most of the sternum (S1–S5), but S6 is usually bare or partly bare. On some Texas parallel leafcutters, there is pale hair on S6.

Behavior: Parallel leafcutters are known to nest in the ground, but little has been written about the behavior of this species. In 1951, entomologist Roland Fischer described the ground nest of a female parallel leafcutter that he stumbled on in a fallow field in Leavenworth, Kansas. The nest's entrance was partly concealed by a clump of earth. A two-inch burrow extended from the entrance at a 45-degree angle and terminated in a single egg cell that ran parallel to the ground. The egg chamber was constructed of 15 entire intermingled leaves, rather than leaf pieces. Fischer identified the leaves as those of white clover and Van Houtt spirea. Fischer remarked that the use of whole leaves snipped off at the base was remarkable behavior in a leafcutter, and efficient, since it spared the bees the laborious task of cutting disks out of the leaves.

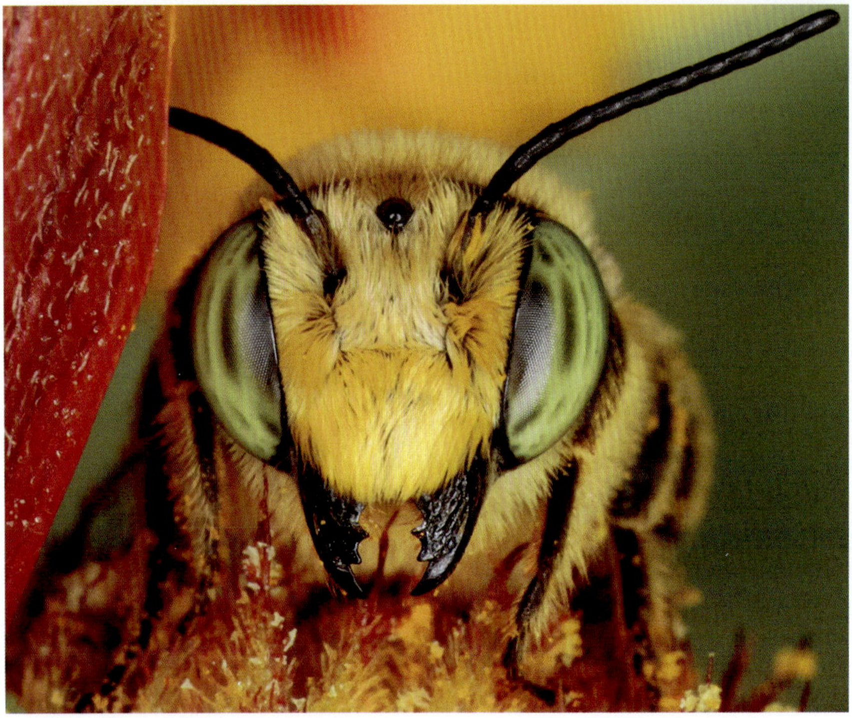

The face of the male parallel leafcutter is covered with copious yellow hairs. Males have four-toothed mandibles.

Slender resin bee ◆ *Megachile (Chelostomoides) exilis* Cresson, 1872

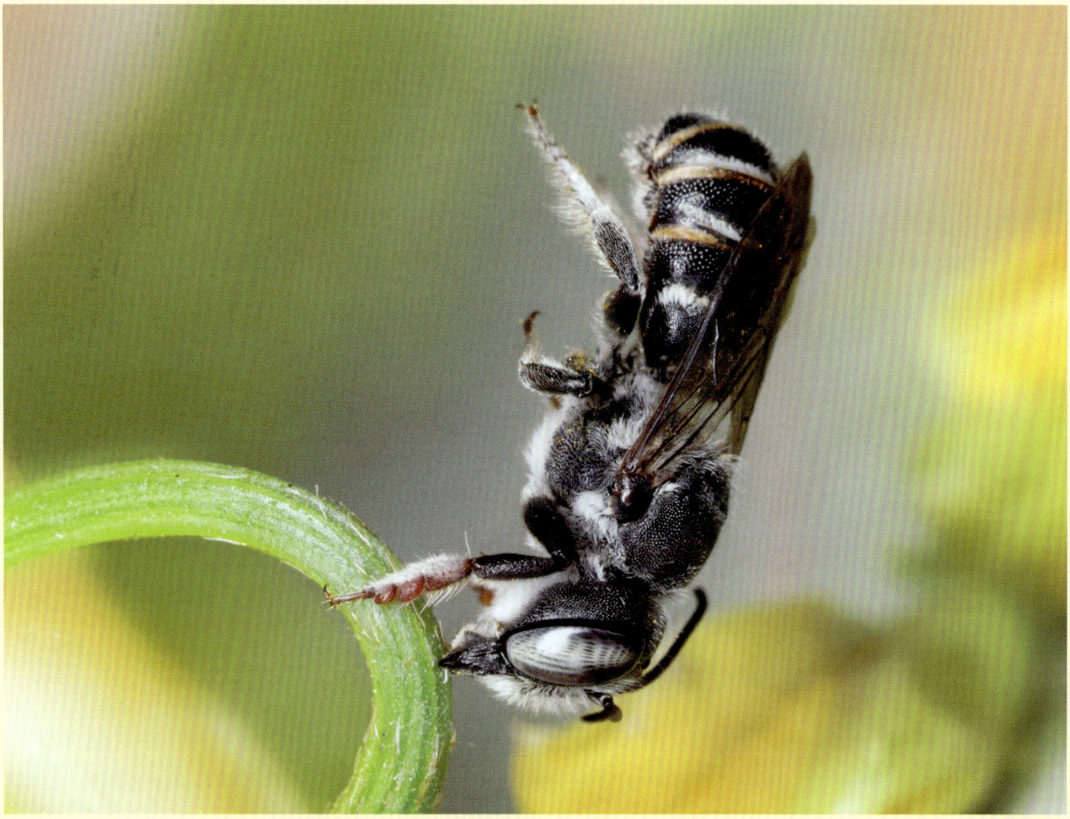

A male *Megachile exilis* standing upside down by clamping a pea vine in its jaws

Resin bees: Leafcutters of the subgenus *Chelostomoides* are sometimes referred to by the common name "resin bees." Like other *Megachile*, they have large, toothed mandibles. *Megachile (Chelostomoides)*, however, lack cutting edges on their mandibles and do not trim leaves or petals from plants. Instead, they use plant resins and saps to construct nests. They have long, roughly cylindrical bodies, rather than the somewhat flattened abdomens more typical of leafcutters. The abdomens of male *Megachile (Chelostomoides)* curl under toward the tips so that they have only three exposed sternal segments, rather than the four typical of most *Megachile*.

Description: Both female and male *Megachile (Chelostomoides) exilis* are black, narrow-bodied, petite bees with dark, conspicuously pitted and grooved abdomens banded by white hairs. Their thoraxes are pitted and relatively hairless, and their heads are partly covered with white hairs. Their tegulae are black or (on males) dark red, and their wings are faintly smoky with brown veins. Females' legs are black. Males have striking forelegs: their front basitarsi are red, expanded and flattened, and rimmed with cottony white hairs behind and long white hairs in front. Males' legs are otherwise black except for reddish lower tarsi and front basitarsi. Males lack coxal spines.

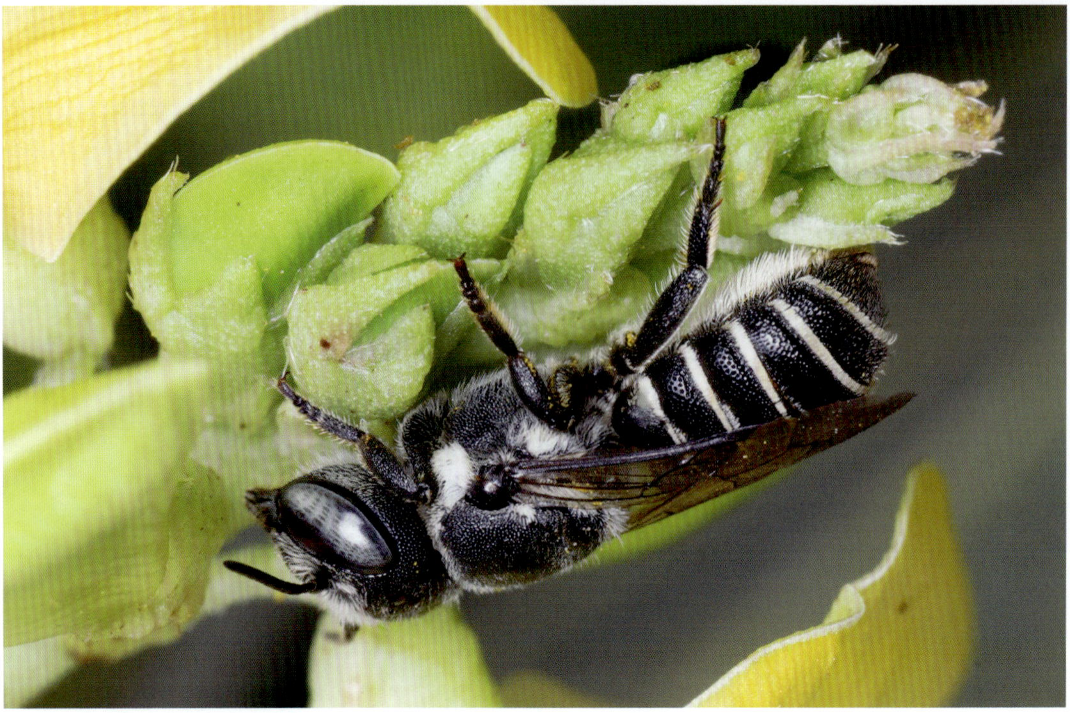

A female *Megachile exilis* on snout bean

The white abdominal hair bands of *Megachile exilis* appear on T1–T5 on females and T2–T6 on males. Females have some dark hairs on T5 and T6, and their scopal hairs are white except for some dark hairs on the tip of S6. On males, the apical rims of T2–T4 are a translucent amber, and T6 has a wide, shallow notch flanked by two teeth. Female slender resin bees have four-toothed mandibles, and males have three-toothed mandibles. Females have black antennae; males' antennae are mostly black, but reddish on the undersides. Females have gray-blue eyes, and males greenish eyes. The ocelli of both females and males are faintly translucent and reddish amber. **Size:** females 9–10 mm (1/3–2/5 in.), males 7–8 mm (1/4–3/10 in.).

Range: *Chelostomoides* is a Western Hemisphere subgenus, most heavily represented in desert areas of Mexico and the southwestern United States. It ranges southward into Central America and northern South America. *Megachile exilis* is found mostly in Texas and Arizona. It is relatively common in Hidalgo and Cameron Counties.

Associated plants: This species emerges in June and remains through September in the Valley, where it is usually found on pea-family plants such as mesquite (*Prosopis glandulosa*), low prairie clover (*Dalea scandens*), vetch, and snout bean. The bees shown here formed part of a large 2019 resin bee population visiting Texas snout bean (*Rhynchosia senna* var. *texana*), growing along an irrigation canal near the National Butterfly Center (Hidalgo Co.). Slender resin bees are known to visit diverse plants, among them milkweeds, asters, evening primrose, sumac, blueberry, and salvia.

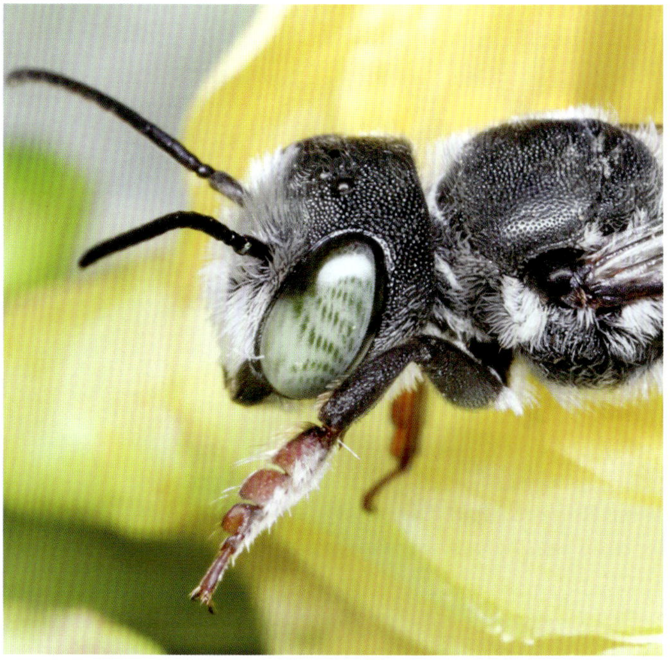

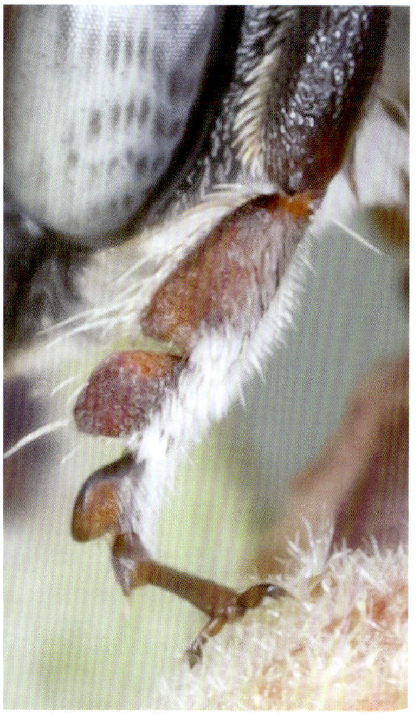

The front basitarsus of a male *Megachile exilis*

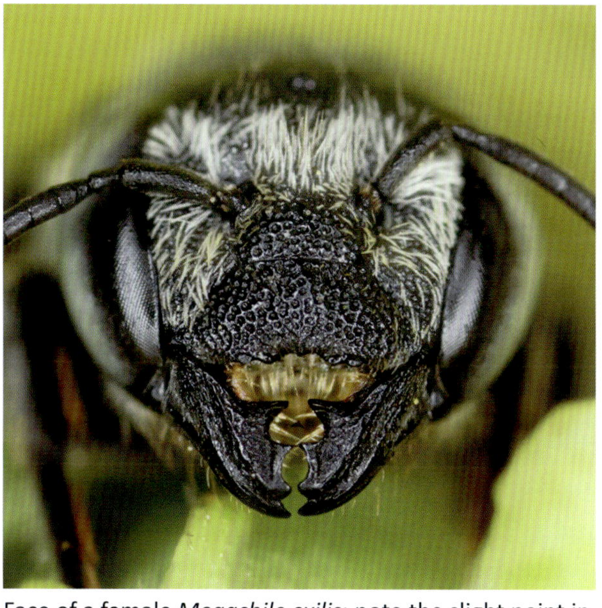

Face of a female *Megachile exilis*: note the slight point in the recessed middle portion of the bottom edge of the clypeus.

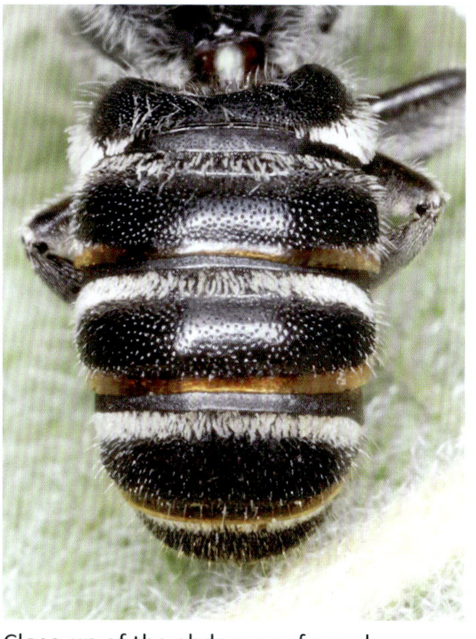

Close-up of the abdomen of a male *Megachile exilis*, showing the translucent amber bands on the tergal rims

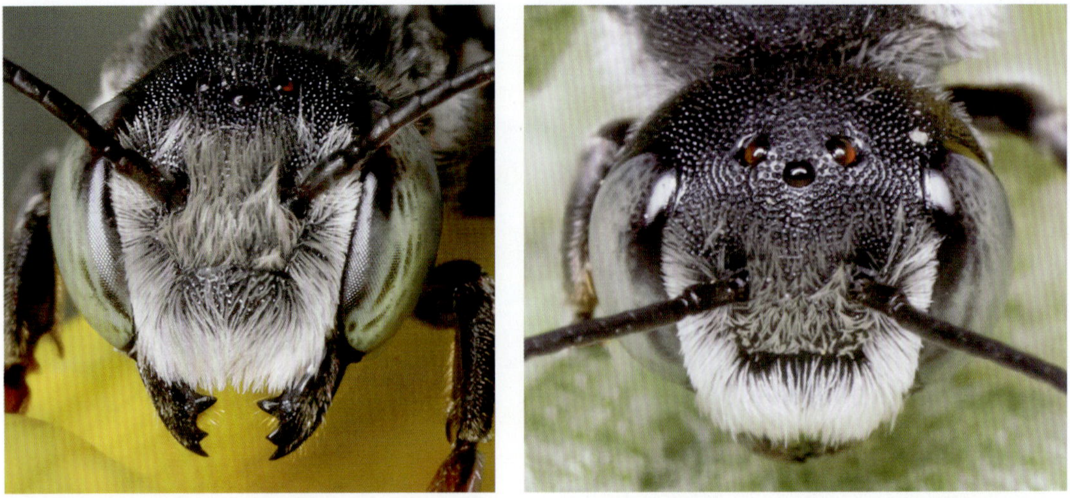

Face of a male *Megachile exilis*: the male has three-toothed mandibles, and dense white hair covering its clypeus. Both female and male slender resin bees have amber-colored ocelli.

ID tips: *Females*: The female *Megachile exilis* is most easily identified by a small median tooth set in a shallow, narrow indentation on its otherwise relatively straight clypeus. *Males*: The combination of the males' small size, expanded red forelegs, and amber tergal bands makes them fairly easy to recognize in the field. Minute traits of the male's forelegs (shown here) distinguish *Megachile exilis* of the Valley both from similar resin bee species and a variant in the eastern United States, *Megachile exilis parexilis*. On the typical male *Megachile exilis*, the basitarsus is wider than the tibia. It is divided into three expanded red parts: the second part is especially broad, belling out into a roughly rectangular shape. The third part is not as wide and narrows toward the tip. On male *Megachile exilis parexilis*, the third part is nearly as wide as the second and does not narrow toward the tip. This single detail diagnoses this variant. Females of the *Megachile exilis parexilis* variant are identical to the typical *M. exilis*.

Resin bee behavior: The ability to use their mandibles to cut leaves and petals for nest building is a salient characteristic of most North American *Megachile*. The inclusion of resin bees in this genus is thus somewhat problematic. Not too long ago, resin bees of the subgenus *Chelostomoides* occupied their own genus, *Chalicodoma*, a worldwide group of bees with parallel-sided abdomens, whose females had mandibles that lacked cutting blades.

Megachile (Chelostomoides) resin bees possess a distinctive versatility and resourcefulness in their methods of nest construction. Females fashion nest linings and plugs by mixing resin and saps with mud, dirt, sand, small pebbles, bits of wood, or vegetation. They commonly construct their nests—tunnel-shaped series of egg chambers provisioned with pollen—in preexisting burrows carved in wood or twigs by beetles or other insects. *Megachile (Chelostomoides)* sometimes settle in cavities in human-made structures, such as nail holes or crevices in brick and adobe walls.

Texan resin bee ◆ *Megachile (Chelostomoides) texensis* Mitchell, 1956

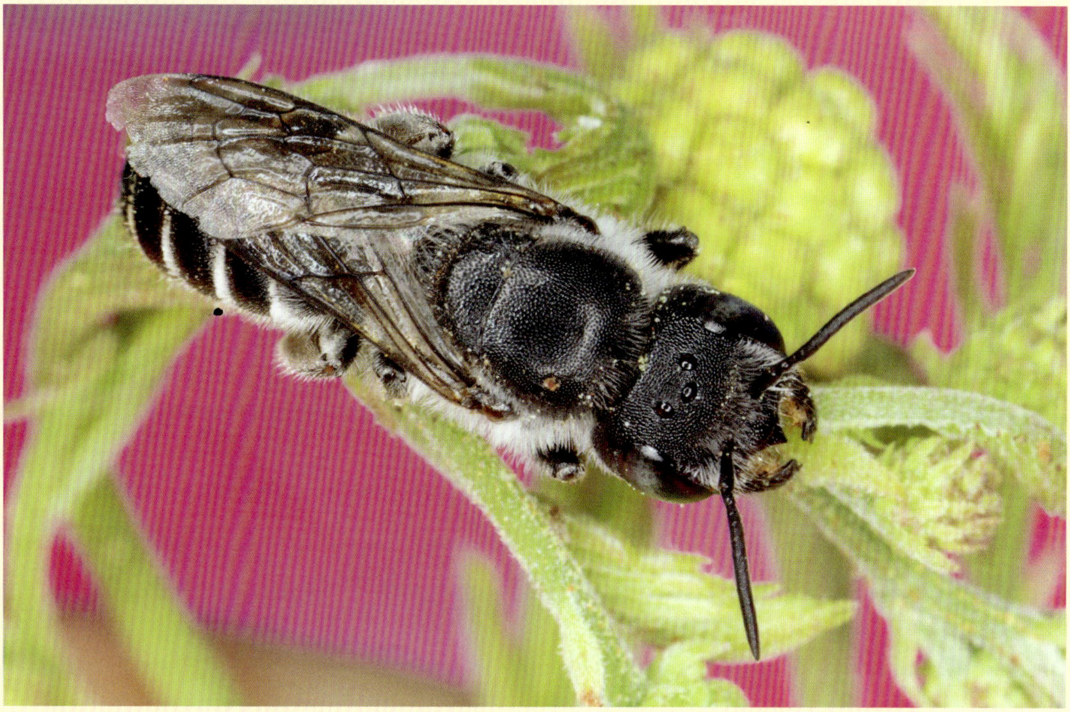

A female *Megachile texensis*

Description: The female *Megachile texensis* is an imposing bee with prominent long mandibles and a conical horn in the center of its face. The female also has two tubercles on the bottom edge of the clypeus, and, true to its subgenus *Chelostomoides*, lacks cutting edges on its mandibles. Both female and male *Megachile texensis* have black heads and predominantly black bodies banded by pale hairs. The female's scopal hairs are a faint yellowish white except on the sixth segment, where they are dark. On the male's abdomen, T6, S2, and S3 are reddish. The male's legs are mostly black, with reddish lower tarsi. The undersides of the flagella of the male's antennae are also reddish. On both males and females, the tips of the mandibles are three toothed. **Size:** females 9–10.5 mm (1/3–2/5 in.), males 8–10.5 mm (3/10–2/5 in.).

Range: This species was first described in 1956 by Mitchell after he examined a single female found in Cameron County. The Texan resin bee, despite its name, is a subtropical species that ranges from Central America through Mexico. In the United States, it appears principally and uncommonly in the Lower Rio Grande Valley. The female bee shown here was discovered in late April 2021 in the Teniente Tract, on the border of Hidalgo and Willacy Counties.

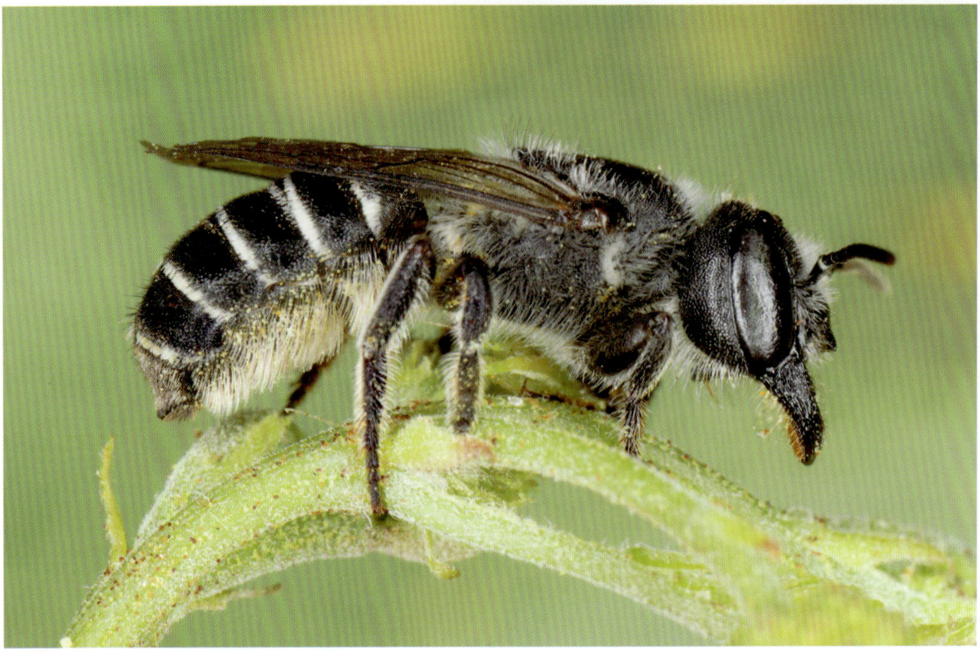

A female *Megachile texensis*: note the hornlike protrusion on the face

Associated plants: Texan resin bees are associated with pea-family plants. The bee shown here was found in a gas pipeline accessway devoid of wildflowers, but bordered on all sides by a scrub woodland in which various pea-family trees and shrubs were in bloom—these included guajillo (*Senegalia berlandieri*), honey mesquite (*Prosopis glandulosa*), and huisachillo (*Vachellia tortuosa*). The soil of this habitat was sandy and strewn with wood debris.

ID tips: The distinctive jaws of this species, and the hornlike protrusion on the female's face, make it readily recognizable.

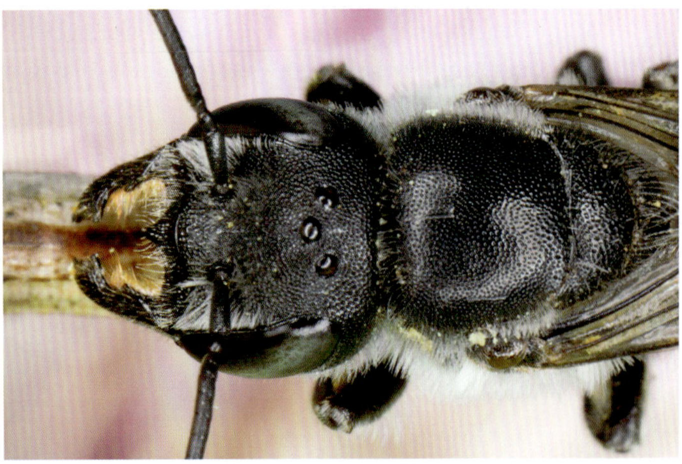

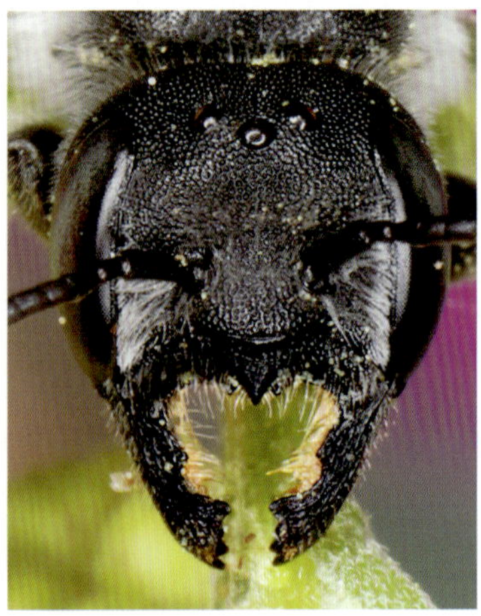

Close-up of the head of a female *Megachile texensis*

Mesquite resin bee ◆ *Megachile (Chelostomoides) prosopidis* Cockerell, 1900

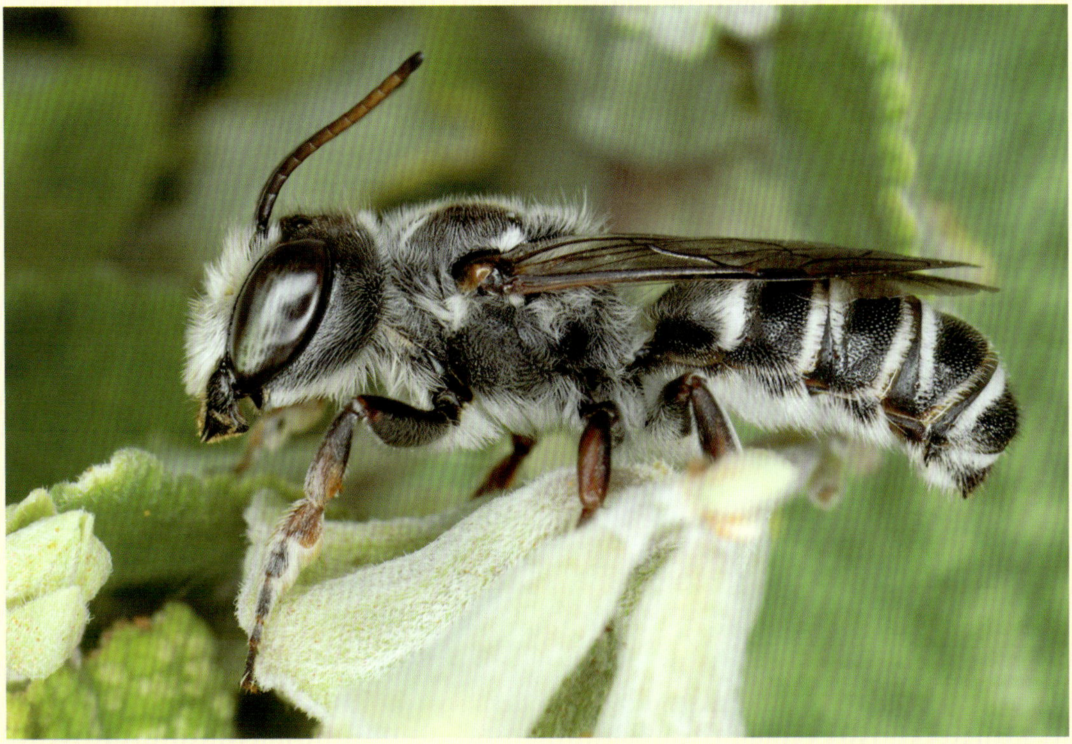

A male *Megachile prosopidis*

Description: The mesquite resin bee resembles *Megachile exilis* but runs somewhat larger. *Megachile prosopidis* is a black, narrow-bodied bee with an abdomen banded by pale hairs; dark, partly reddish tegulae; and glassy, brown-veined wings that are smoky near the outer edges. The rims of the terga are a translucent amber. Both males and females have two distinctive swatches of white hair slightly in front of and below the tegulae. The female's scopal hairs are white except on S6, which is covered with black hairs. Dense white hairs cover the face, cheeks, and undersides of the male's thorax. Its sternum is banded with rows of long white hairs. The tip of the male's abdomen (T6) has a deep, rounded notch. Male *Megachile prosopidis* have red, expanded foreleg basitarsi, and partly red middle and hind legs.

There are two tufts of white hair on either side of the female's face, the labrum is covered with a brush of brown hairs, and the clypeus bears a conspicuous rounded median tooth on the bottom edge. Female *Megachile prosopidis* have five-toothed mandibles, and males have three-toothed mandibles. The antennae of both males and females are black, with reddish front surfaces on the flagella. **Size:** females 12–13 mm (1/2 in.), males 10–11 mm (2/5 in.)

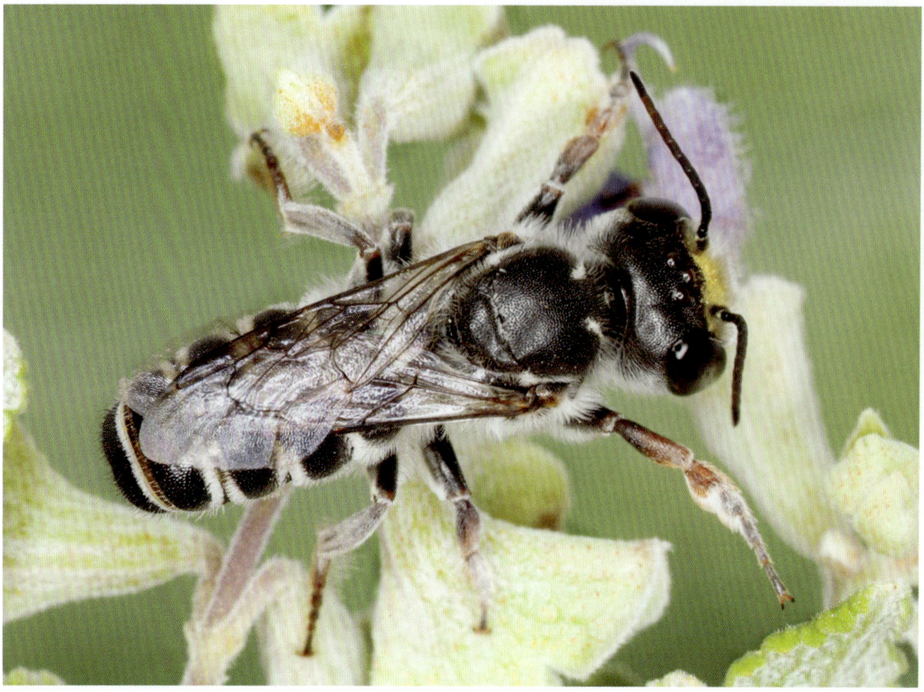

Dorsal view of a male *Megachile prosopidis*

Range: *Megachile prosopidis* ranges along the Mexican border from Texas to southern California. It is found as far north as Utah. This species also occurs in the northern half of Mexico, as far south as Zacatecas and the tip of Baja California. The male bee shown here appeared in an area covered by Tamaulipan thornscrub at Falcon State Park (Starr Co.) in early June 2021.

Associated plants: *Megachile prosopidis* is considered an oligolectic pollinator of plants of the family Fabaceae, particularly leguminous trees. Cockerell first described and named this species after examining female bees found feeding in part on honey mesquite (*Prosopis glandulosa*). In the Lower Rio Grande Valley, mesquite resin bees can be found pollinating both honey mesquite (*Prosopis glandulosa*) and retama (*Parkinsonia aculeata*) in late spring and summer. Despite their oligolectic status, both male and female *Megachile prosopidis* have been recorded visiting plants of other families. Some of Cockerell's female specimens were collected on desert willow (*Chilopsis linearis*) and *Phacelia*. The male bee shown here from Falcon State Park was one of many nectaring on shrubby blue salvia (*Salvia ballotiflora*).

ID tips: The primarily red coloration of the middle and hind legs of the male *Megachile prosopidis* sets it apart from other male *Megachile (Chelostomoides)* of the Valley. The middle and hind legs of *Megachile exilis* and *M. texensis* are black. Females can be distinguished from similar resin bees of the Valley by their five-toothed mandibles.

Behavior: In 2004, entomologist Elizabeth Armbrust conducted a study of *Megachile prosopidis* and other members of the resin bee subgenus *Chelostomoides* in arid desert and mountain habitats in the Tucson region. Armbrust found that in those areas, *Megachile prosopidis* appeared to be bivoltine, emerging from mid to late spring and then again in late July. Some of the *Megachile prosopidis* females Armbrust studied formed multilayered caps for their nests, which sometimes incorporated a layer of loose fill made up of bark, stones, leaves, flowers, insect parts, or other litter. Armbrust observed female *Megachile prosopidis* harvesting petals for nest construction from Mexican holdback (*Caesalpinia mexicana*).

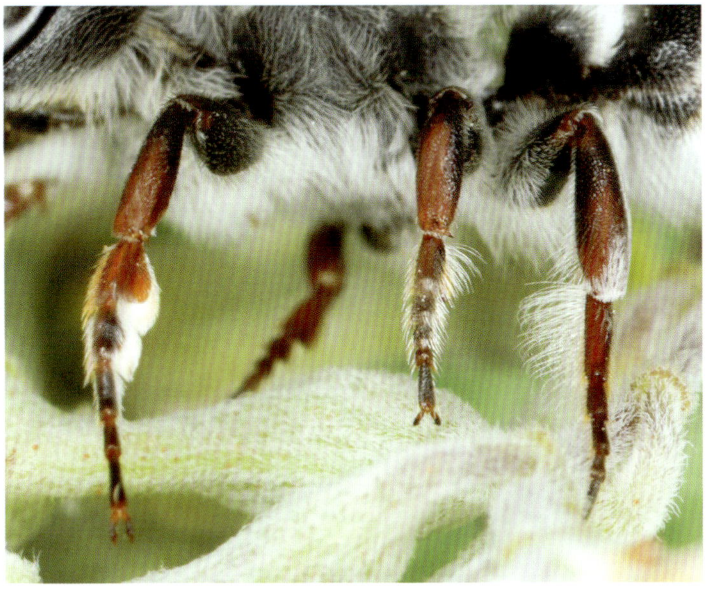

Legs of a male *Megachile prosopidis*

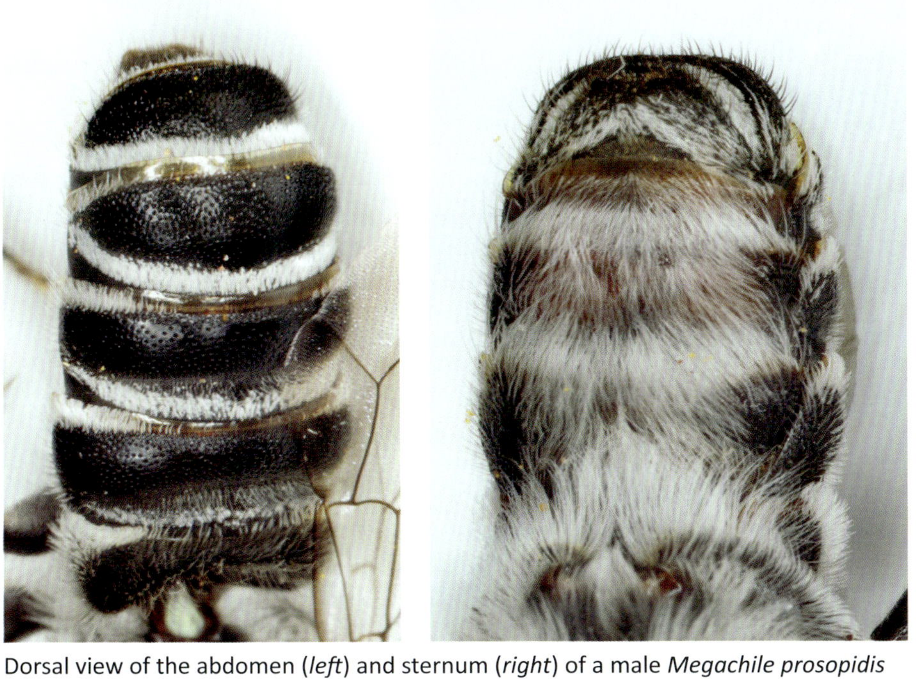

Dorsal view of the abdomen (*left*) and sternum (*right*) of a male *Megachile prosopidis*

NESTING BEHAVIOR OF LOWER RIO GRANDE VALLEY LEAFCUTTERS

Megachile (Argyropile) parallela	Ground nester. *Nest materials*: whole leaves from plants such as *Trifolium repens* and *Spirea vanhouttei.*
Megachile (Chelostomoides) exilis *Megachile (Chelostomoides) prosopidis* *Megachile (Chelostomoides) texensis*	Nest in cavities made by beetles and other arthropods in wood or twigs; crevices in brick and adobe walls; or nail holes in wood. *Nest materials*: plant sap and resin, pebbles, sand, soil, or bits of insects, wood, or plant matter
Megachile (Litomegachile) brevis	Nests in preexisting cavities in hollow plant stems, grass stalks, rolled-up leaves, or pieces of wood. May occupy cavities in ground but does not excavate soil. *Nest materials*: leaves and petals from such plants as waxberry, roses, evening primrose, and partridge pea.
Megachile (Megachile) montivaga	Usually nests in dead stems of pithy plants or other plant matter. May nest in preexisting ground cavities but does not excavate soil. *Nest materials*: usually flower petals, such as poppy and prickly poppy, clarkia, evening primrose, phlox, *Cassia*, asters, and possibly globemallow.
Megachile (Neochelynia) chichimeca	Species behavior unknown. Other members of the subgenus are probable cavity nesters.
Megachile (Pseudocentron) sidalceae	Species behavior unknown. Other members of the subgenus nest in cavities in the ground or plant matter and use leaf cuttings as nest materials.
Megachile (Sayapis) frugalis	Observed by the Minnesota Bee Atlas project nesting in preexisting holes in bee blocks. *Nest materials*: cut leaves, chewed plant matter, and soil particles.
Megachile (Sayapis) inimica inimica	Nests in preexisting cavities. *Nest materials*: leaf cuttings, leaf pulp, sand, small pebbles, or sand mixed with debris.
Megachile (Sayapis) policaris	Nests in preexisting cavities. *Nest materials*: tiny, whole leaflets, leaf parts, and gummy leaf pulp. Whole leaflets are collected from plants such as mesquite and kidneywood.
Megachile (Sayapis) zaptlana	Nests in preexisting cavities in plant matter or human-made structures. Found in abandoned beetle burrows in fence posts. *Nest materials*: leaf cuttings.
Megachile (Tylomegachile) cf. *toluca*	Unknown. Other members of the subgenus *Tylomegachile* are probable cavity nesters.

Cuckoo Leafcutter Bees

Subfamily Megachilinae ◆ Tribe Megachilini
Genus *Coelioxys*

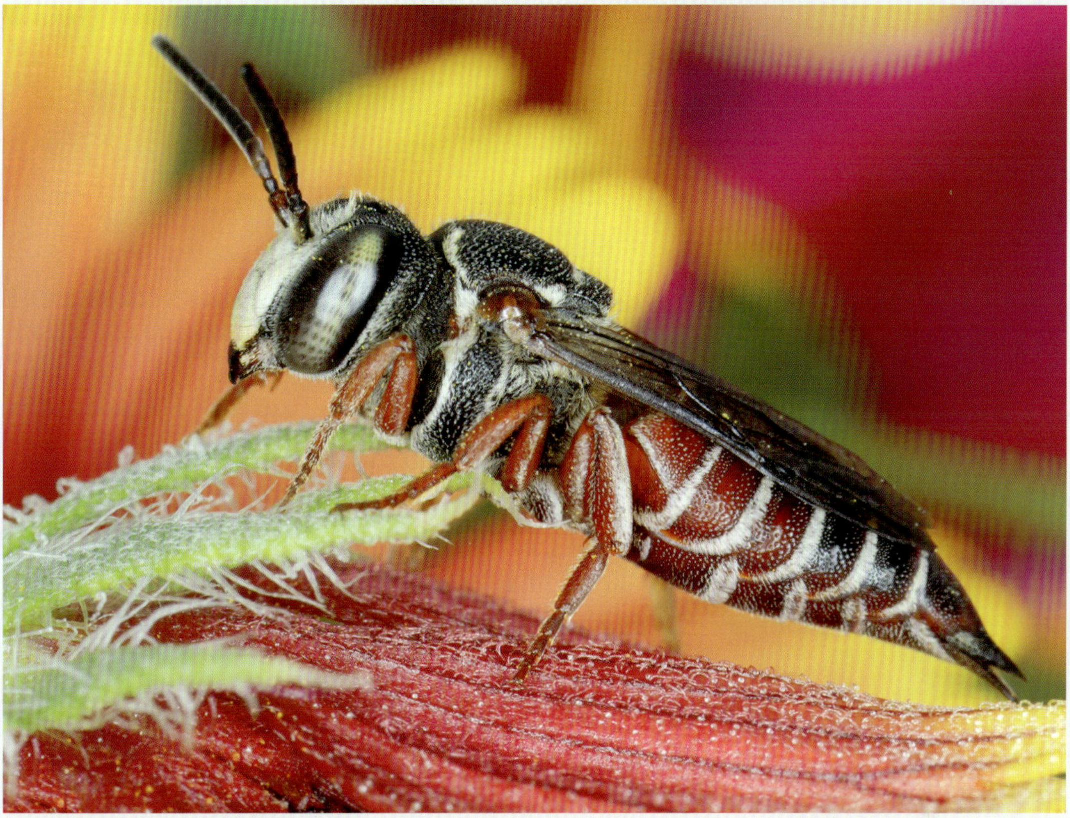

A female cuckoo leafcutter

Coelioxys are cleptoparasitic bees that target the nests of leafcutters and belong, like them, to the tribe Megachilini. Female *Coelioxys* have long, tapered, flexible abdomens with spade-like tips (the Greek *coelioxys* means "sharp belly"). This feature allows the bees to tuck their eggs into host provisions or the linings of cell walls. Male cuckoo leafcutters have abdomens armed with multipronged tips.

Traits of the genus *Coelioxys*: All *Coelioxys* share a notable trait: short, straight hairs cover their compound eyes. The *Coelioxys* scutellum is also distinctive: it is sculpted and well defined, and flanked by conspicuous toothlike axillae. The cuckoo leafcutter's long abdomen is often grooved longitudinally, coarsely pitted, and banded by somewhat sparse pale hairs. The forewings of *Coelioxys* have two submarginal cells, and a wide marginal cell with a rounded tip. Like other female cuckoo bees featured in this guide, *Coelioxys* females do not collect pollen and thus do not have scopal hairs.

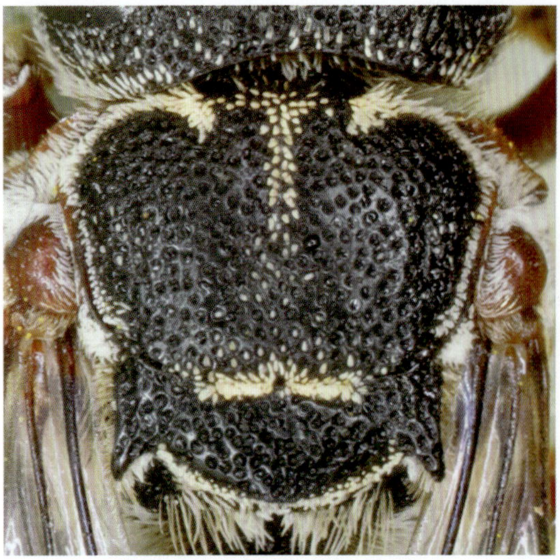

The coarsely pitted thorax of *Coelioxys* has large, toothlike axillae on the rear of the scutellum.

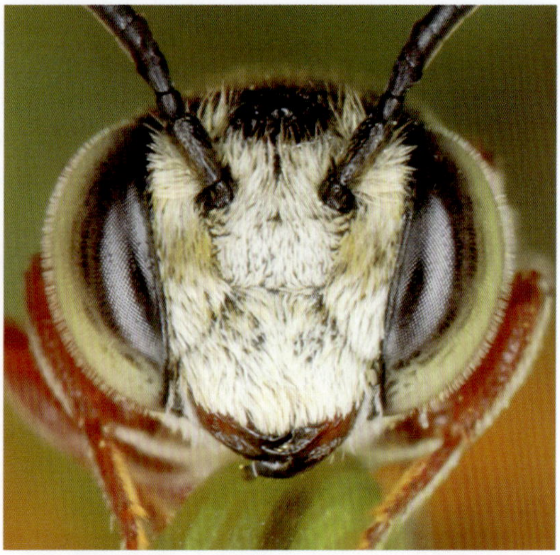

The compound eyes of *Coelioxys* are covered with short hairs.

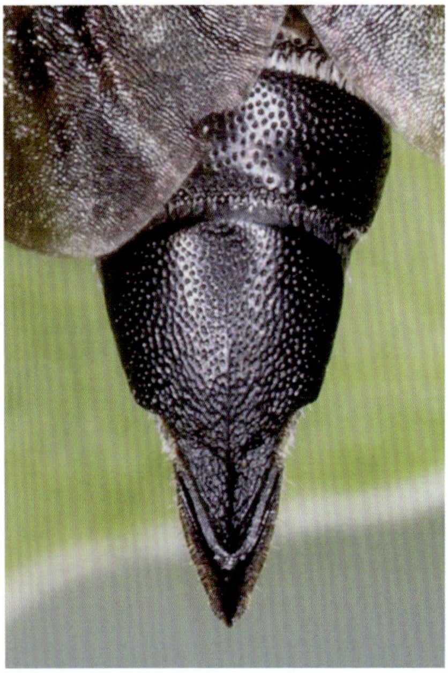

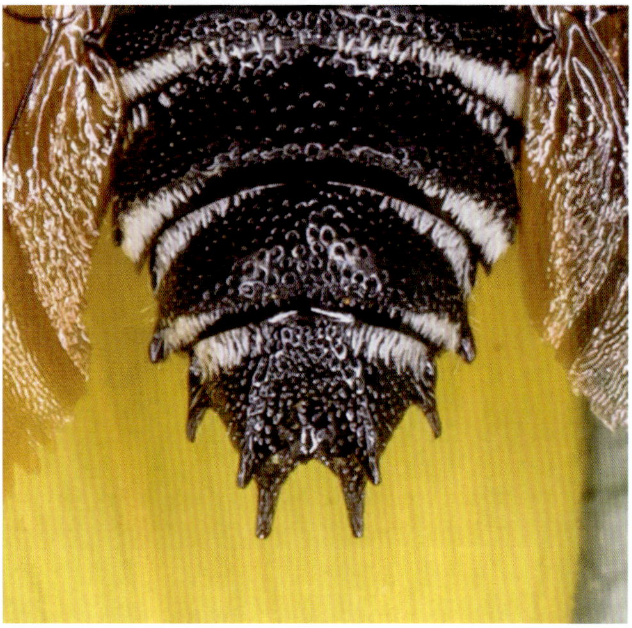

The tip of the abdomen of a female *Coelioxys* (*left*) is spade shaped, while the male's abdomen (*right*) has a multipronged tip.

***Coelioxys* behavior:** Both male and female cuckoo leafcutters overnight outside. They can often be found during the morning sleeping together in small groups on plants. *Coelioxys* have a distinctive way of resting: they tend to sleep upside down on stems, their hind legs and wings held tightly against their bodies, their antennae pointing forward, and their jaws gripping the plant.

Coelioxys females typically enter open leafcutter nest chambers to deposit eggs while the hosts are away foraging for food. As noted above, many leafcutters in the Lower Rio Grande Valley build nests comprising a linear series of egg chambers within a tunnel. They provision each cell separately. *Coelioxys* can often be observed stationed outside the holes of such nests, waiting for the leafcutters inside to depart between provisioning trips.

Coelioxys behavior seems to reveal an uncanny strategic intelligence. Rather than simply entering a host's nest to lay its own eggs all at once, a female *Coelioxys* will loiter near the nest, waiting for the leafcutter to build and begin provisioning a single chamber. When the leafcutter departs to gather new provisions, the cuckoo slips into the nest and lays one egg in the chamber before the host returns to seal it. The *Coelioxys* then scrambles out of the nest and again perches a few feet from the entrance to repeat the process. Thus, as the unsuspecting leafcutter constructs her nest cells one by one, the cuckoo shadows her and steals into each cell in sequence. This guarantees that each *Coelioxys* offspring will hatch in its own chamber and awake to its own pile of provisions.

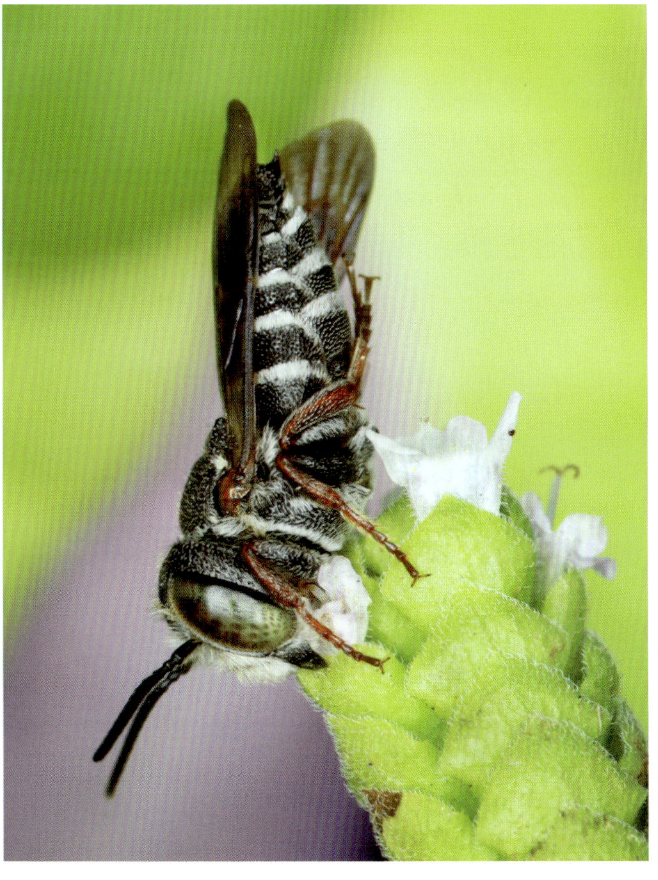

A male *Coelioxys* resting upside down on a flower stem

When a *Coelioxys* egg hatches, it passes through various developmental stages, or instars, within the leafcutter nest. During the second instar, the *Coelioxys* has large mandibles, which it uses to destroy leafcutter eggs or larvae. After dispatching its host's offspring and devouring food stores, the cuckoo leafcutter larva spins a cocoon within the nest cell. Upon emerging, it leaves the leafcutter nest permanently.

Distribution and species differentiation: *Coelioxys* species are found worldwide. There are more than 50 species in the United States and nearly 30 in Texas. At least five different subgenera are found in the Lower Rio Grande Valley. Some, like the Aztec and Totonac cuckoo leafcutters, are rare in the United States and occur only along its southern border with Mexico.

Most North American *Coelioxys* have predominantly black heads, legs, and bodies. In hotter parts of the United States, *Coelioxys* are more colorful. In the Lower Rio Grande Valley, they often have red or partly red legs, tegulae, sterna, and terga. *Coelioxys* species are distinguished by such features as body and leg color; differences in the sculpting of the scutellum and its axillae; characteristics of the grooves traversing the abdomen; and details of the spade-shaped or multipronged tips of their abdomens.

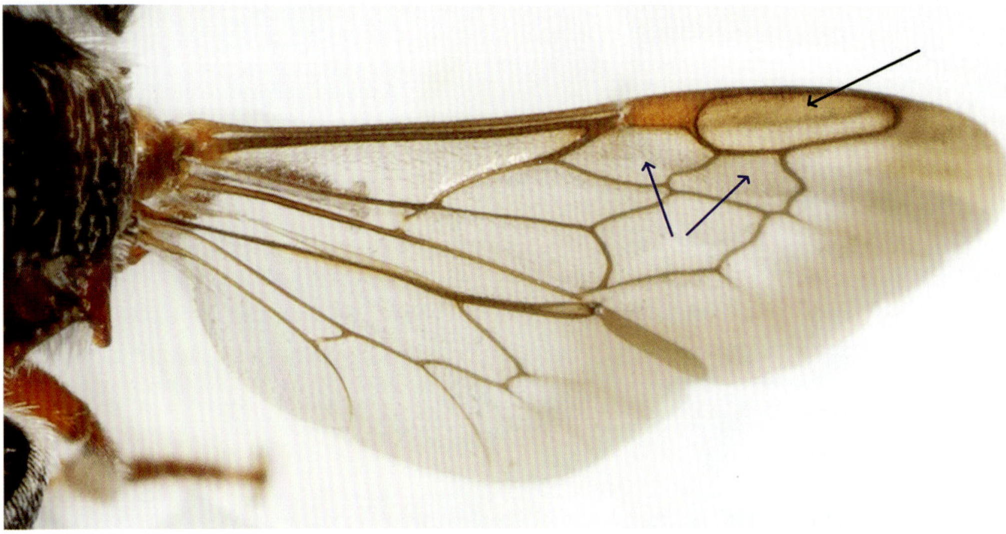

Forewing and hind wing of *Coelioxys*: the forewing has two submarginal cells (blue arrows), and the marginal cell is rounded at the tip (black arrow).

Aztec cuckoo leafcutter ◆ *Coelioxys (Leuraspidia) aztecus* Cresson, 1878

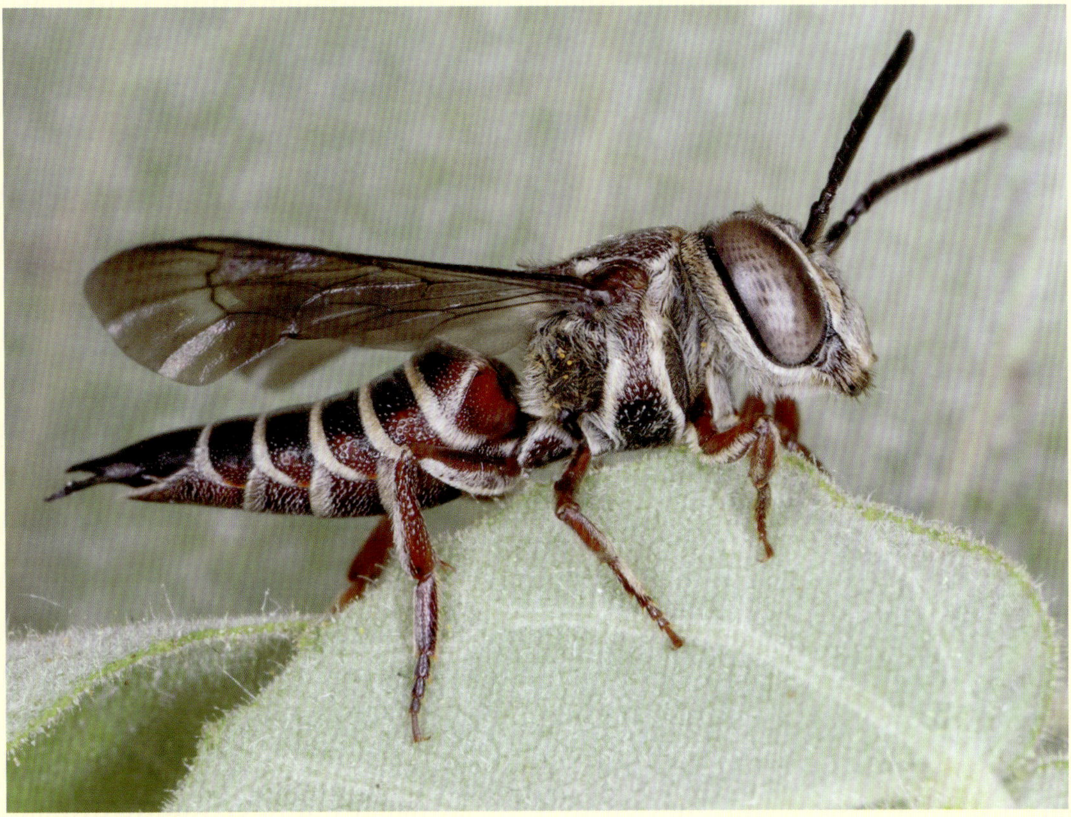

A rarely seen female Aztec cuckoo leafcutter

Description: *Coelioxys (Leuraspidia) aztecus* is a striking cuckoo leafcutter with a red-and-black thorax; an abdomen that is black above and red on the sternum; red legs and tegulae; a face covered with dense white hairs; and dark wings. Its mandibles and antennae are black and its eyes liver colored. The subgenus name *Leuraspidia* means "smooth shield" and derives from the appearance of the scutellum, which is shiny and relatively unpitted. *Coelioxys* of this subgenus share these characteristics: the rear of the scutellum protrudes in a rounded tip and is ridged on the hind rim; interrupted grooves on the abdomen traverse T2 and T3; and the face just below the ocelli is flat and pitted. **Size:** females 10 mm (2/5 in.), males 7 mm (1/4 in.).

Range: *Coelioxys aztecus* is uncommon in the United States and has rarely been photographed. It appears infrequently in Hidalgo and Cameron Counties. The bees shown here appeared at the National Butterfly Center in November and December 2021 and November 2018. This species ranges from the Texas border through Panama. All other cuckoo leafcutters of the small subgenus *Leuraspidia* are found exclusively in Mexico and Central and South America.

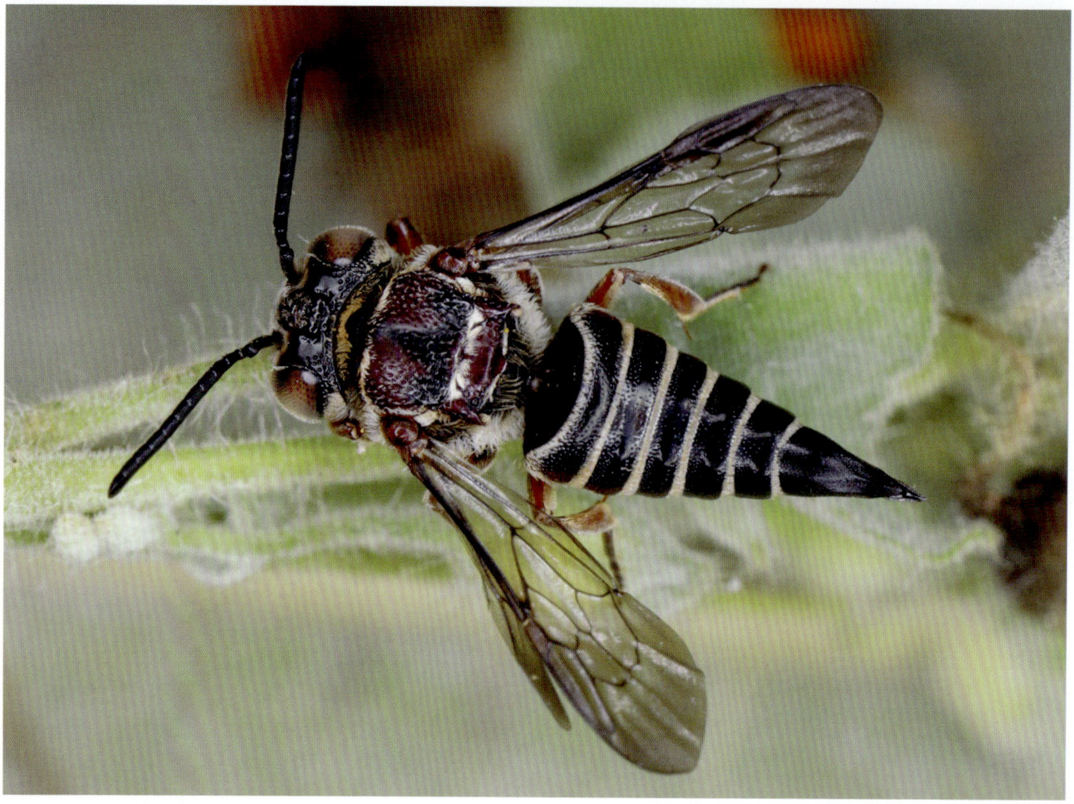

Dorsal view of a female Aztec cuckoo leafcutter

Associated plants: The female Aztec cuckoo leafcutters shown here were found nectaring on Rio Grande abutilon (*Abutilon hypoleucum*). The males were foraging on mallows, shrubby blue salvia (*Salvia ballotiflora*), resinbush (*Viguiera stenoloba*), and romerillo (*Bidens alba*).

ID tips: The shiny surface of the Aztec cuckoo leafcutter's scutellum, combined with the red coloring of the thorax (on both the scutum and most of the scutellum), sets this bee apart from all other Texas cuckoo leafcutter species.

Notable attributes of *Coelioxys aztecus*, not already listed above, include the following: (1) the scutum is swollen in the middle; (2) the tip of the female's abdomen is indented laterally, and the point of S6 extends beyond that of T6; (3) the male's abdomen has conspicuous teeth on either side of T5; and (4) the area around the middle ocellus of both females and males is flat rather than swollen. Cockerell noted in 1905, after examining a Mexican specimen, that the female's clypeus is deeply notched, with a W-shaped apical (lower) rim, a trait often obscured by dense facial hairs.

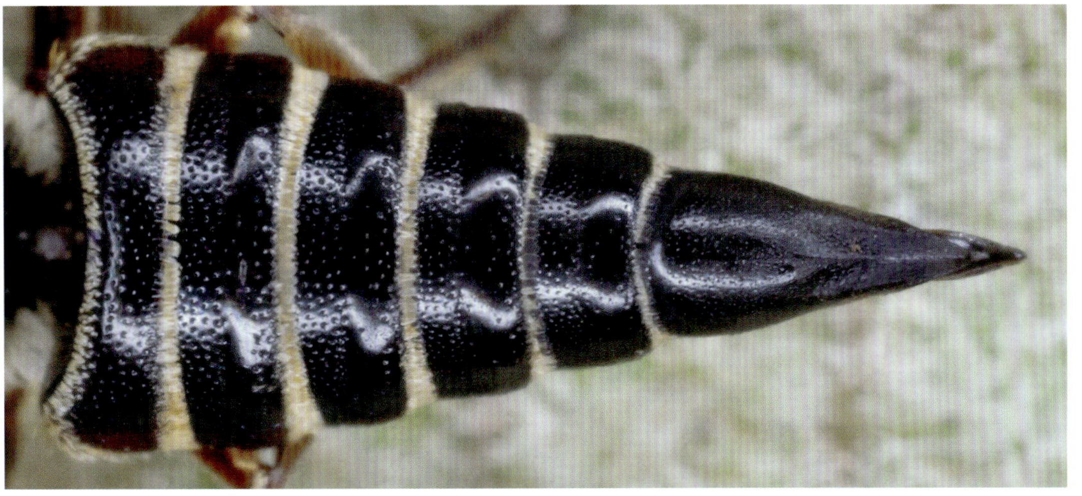

Abdomen of a female *Coelioxys aztecus*: S6 protrudes farther than T6 above it.

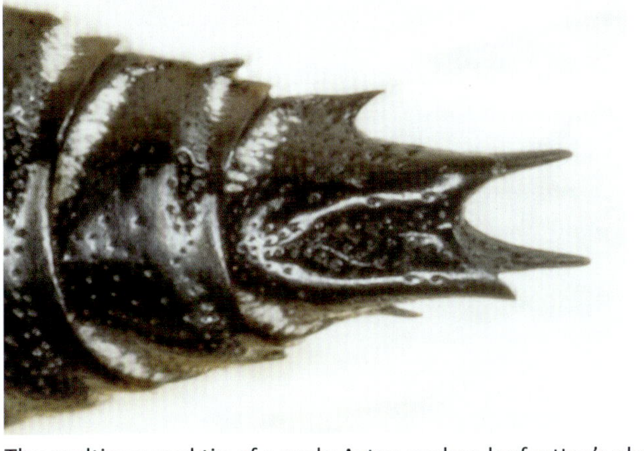

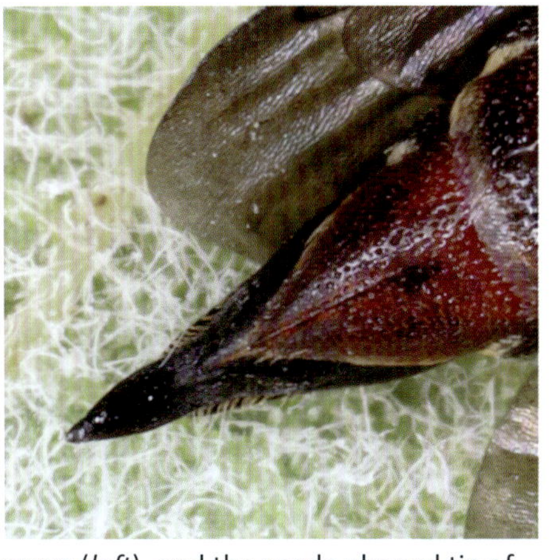

The multipronged tip of a male Aztec cuckoo leafcutter's abdomen (*left*), and the spade-shaped tip of a female's abdomen (*right*). T5 of the male has teeth on either side, and S6 of the female is indented on either side, with a raised median ridge.

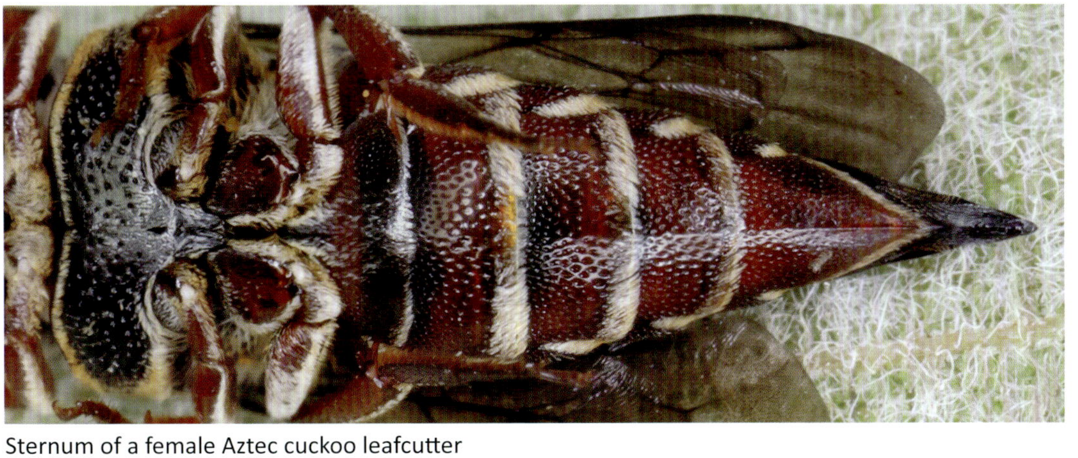

Sternum of a female Aztec cuckoo leafcutter

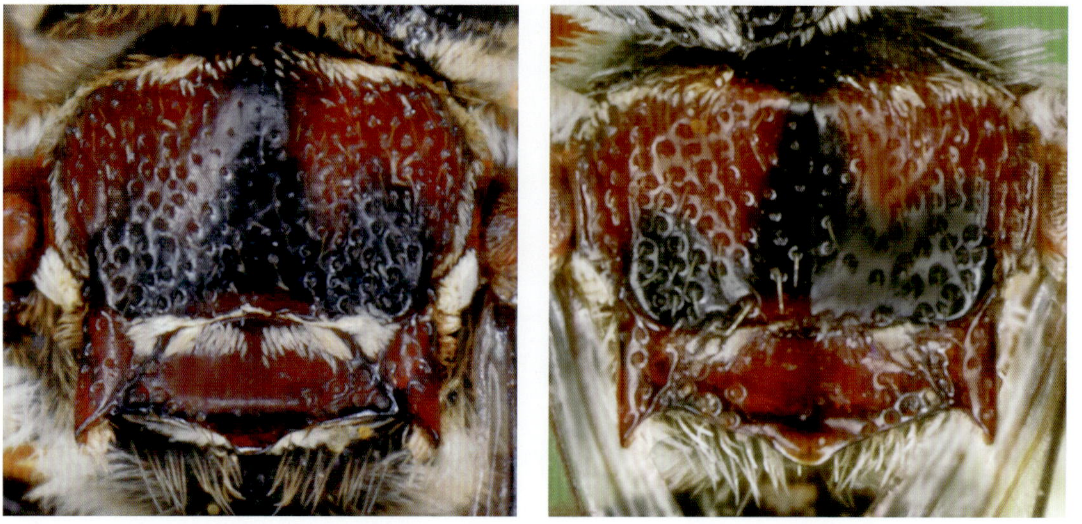

Thoraxes of female (*left*) and male (*right*) Aztec cuckoo leafcutters. The scutellum is swollen in the middle, and its hind rim is narrowly ridged.

Behavior: There is little published information on the behavior of the Aztec cuckoo leafcutter. Its host species is unknown. During late autumn when *Coelioxys aztecus* appears at the National Butterfly Center, the following leafcutter species have been documented in the immediate area: *Megachile chichimeca*, *M. exilis*, *M.* cf. *toluca*, and *M. zaptlana*.

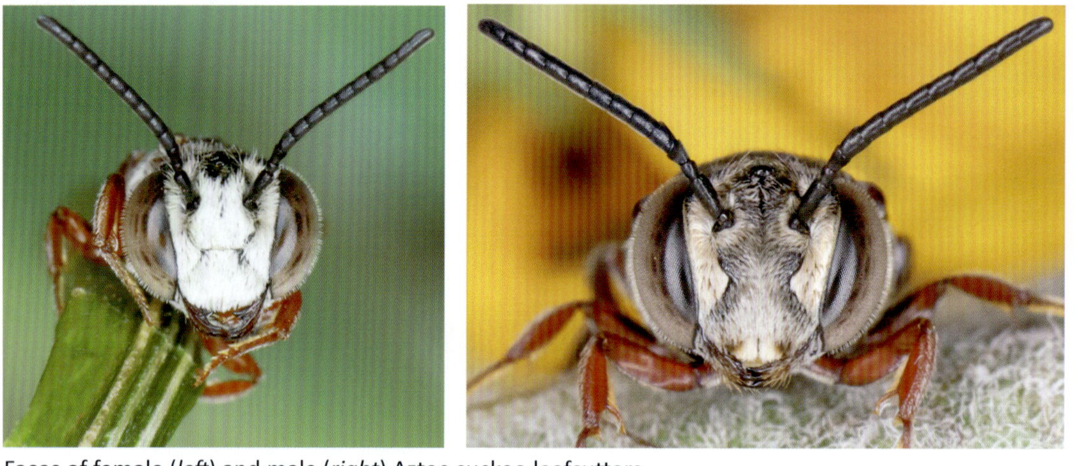

Faces of female (*left*) and male (*right*) Aztec cuckoo leafcutters

Totonac cuckoo leafcutter ◆ *Coelioxys (Glyptocoelioxys) totonacus* Cresson, 1878

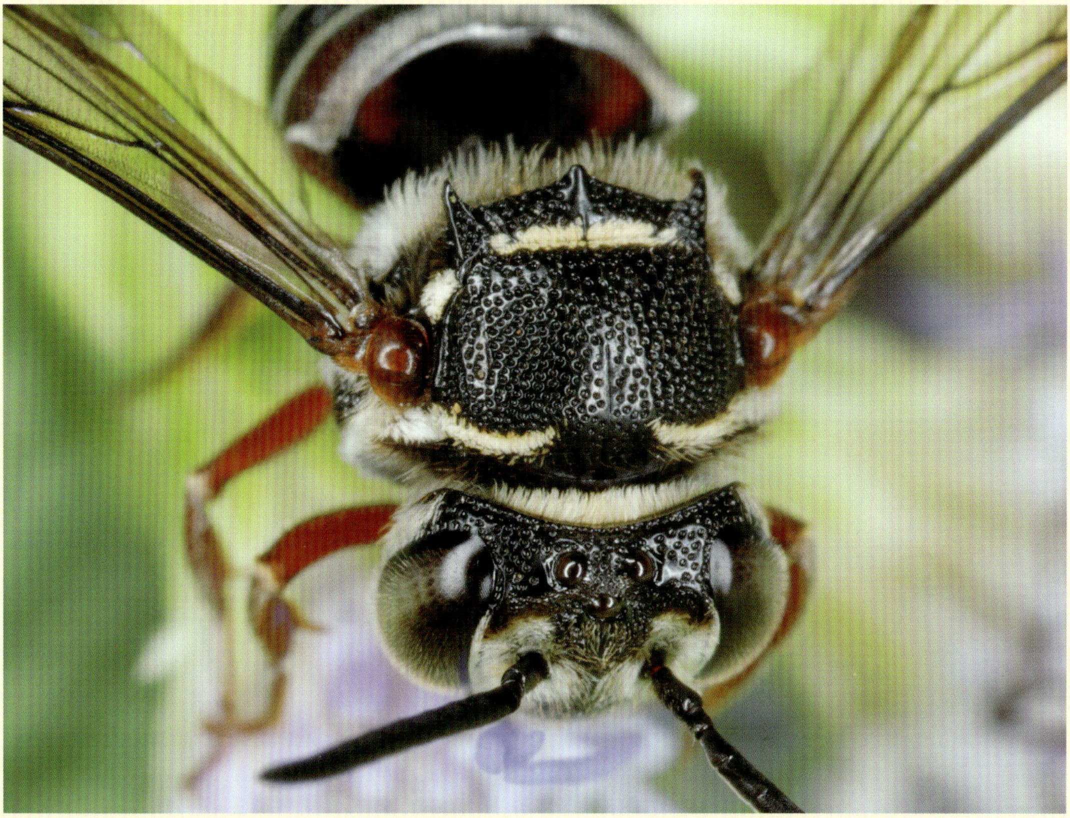

A female *Coelioxys totonacus*

Description: *Coelioxys (Glyptocoelioxys) totonacus* can be recognized by the three prominent spines protruding from the back of its scutellum. The head and body are predominantly black with some red coloration. The legs and tegulae of the Totonac leafcutter are red. Its wings are a transparent brown. On females, T1 is partly red on the front face and sides; T2 is red laterally; and sternal segments S1–S5 are black at the base and otherwise dark red. The female's head, scutum, and scutellum are coarsely pitted. The female's face is covered with a distinctive pattern of dense white hairs; pale hairs cover most of the male's face.

On both males and females, the rear of the vertex has a prominent ridge. The mandibles are dark red with black tips, the antennae black, and the eyes dark green. On females, pale hairs form two curved bands on the front of the scutum; broken interrupted grooves traverse T2 and T3; and T6 is notched on either side and has a median ridge on the apical half. Males' abdomens have a four-pronged tip. **Size:** females 11 mm (2/5 in.), males 8.5 mm (1/3 in.).

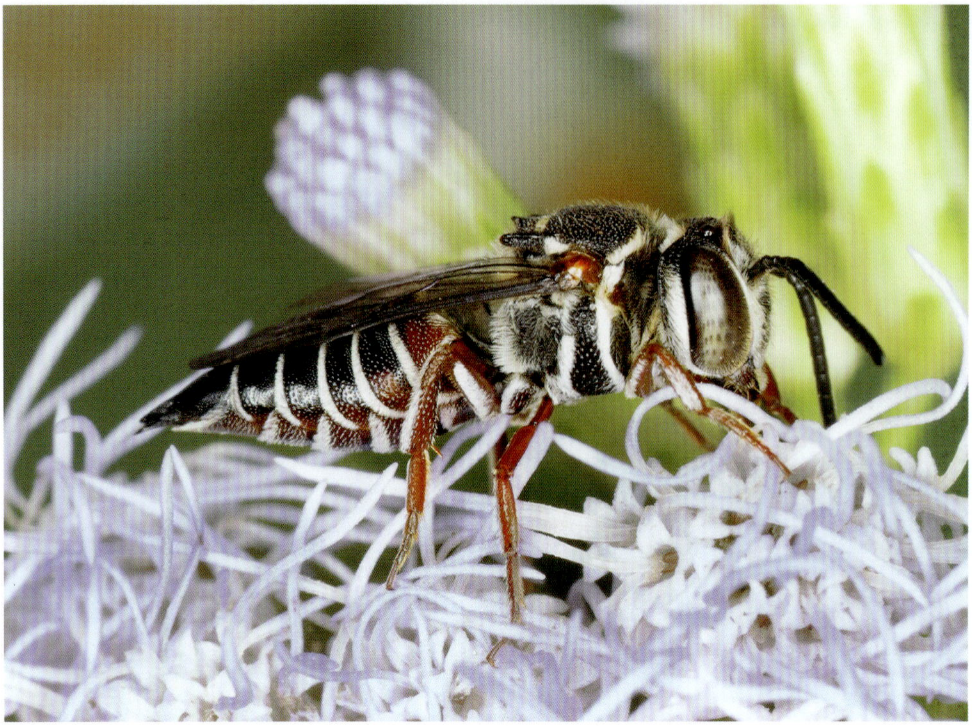

A female *Coelioxys totonacus*

Range: The female bee shown here is the first recorded example of this species in the United States. It appeared at Quinta Mazatlán (Hidalgo Co.) in early November 2021. The subgenus *Glyptocoelioxys* is primarily subtropical. *Coelioxys totonacus* occurs from northern Mexico (Chihuahua and Tamaulipas) through Central America and into South America, as far south as the Brazilian state of Minas Gerais. The Totonac cuckoo leafcutter was first identified in 1878 by Cresson, who based his description on a specimen collected by Sumichrast in Mexico. "Totonac" is a reference to a fifteenth-century indigenous Mexican civilization based in Mexico.

Associated plants: The female bee shown here was found feeding on crucita (*Chromolaena odorata*). In Mexico, this species has been recorded on wild olive (*Cordia boissieri*), pignut (*Hyptis suaveolens*), and Spanish cedar (*Cedrela odorata*).

ID tips: Only one other species of subgenus *Glyptocoelioxys* is found in the United States—the German cuckoo leafcutter (*Coelioxys germanus*). That species occurs mainly along the Eastern Seaboard, although it has been found as far west as Kansas. It is less extensively red than *Coelioxys totonacus*—in particular, T1 and T2 of *C. germanus* are black (although its legs and S1 may be red). Females of the subgenus *Glyptocoelioxys* tend to have interrupted grooves on T2 and T3 and the notched T6 noted above. Both *Coelioxys totonacus* and *C. germanus* have the diagnostic three spines on the scutellum.

Additional information. The host of *Coelioxys totonacus* is unknown.

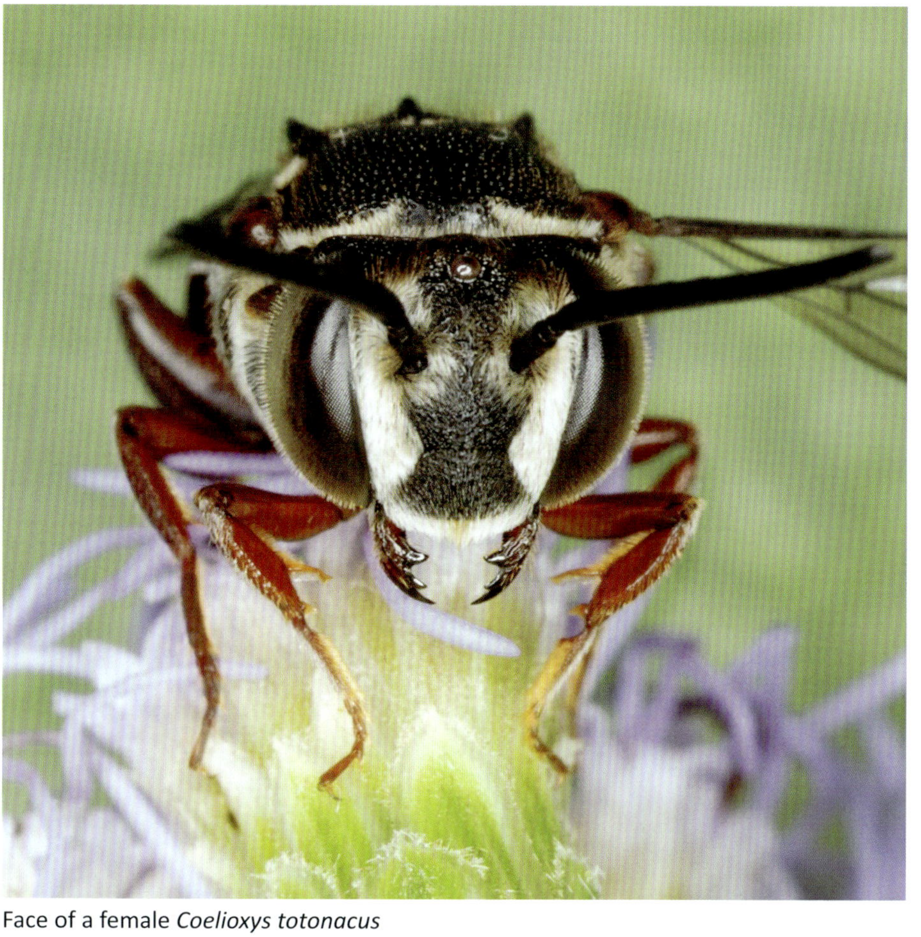

Face of a female *Coelioxys totonacus*

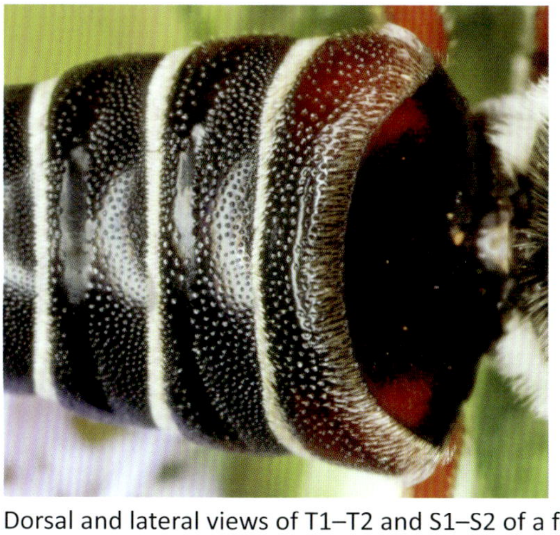

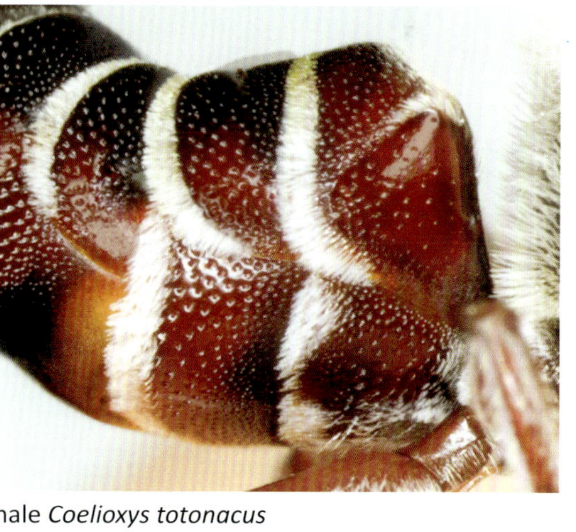

Dorsal and lateral views of T1–T2 and S1–S2 of a female *Coelioxys totonacus*

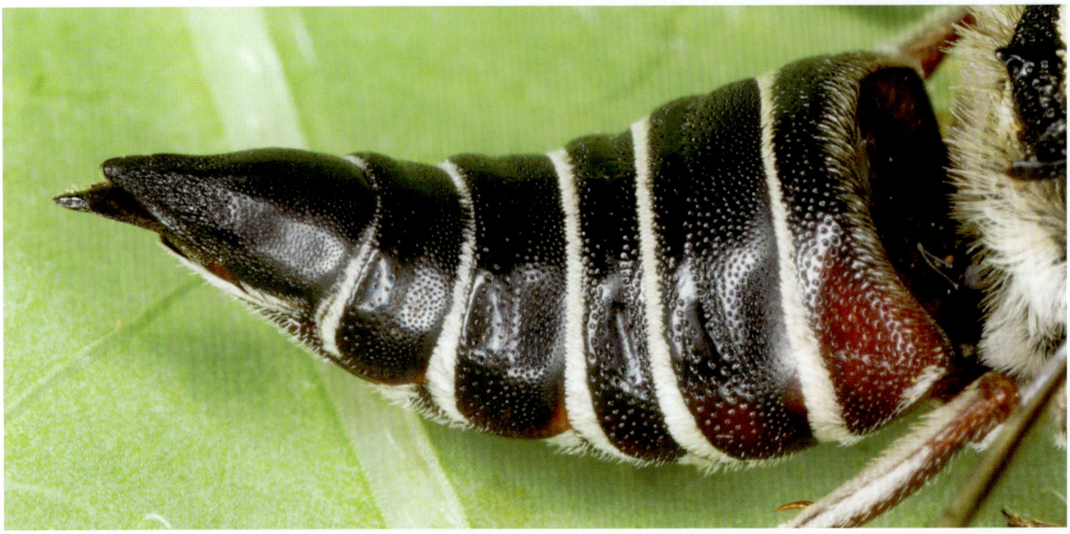

Lateral view of the abdomen of a female *Coelioxys totonacus*

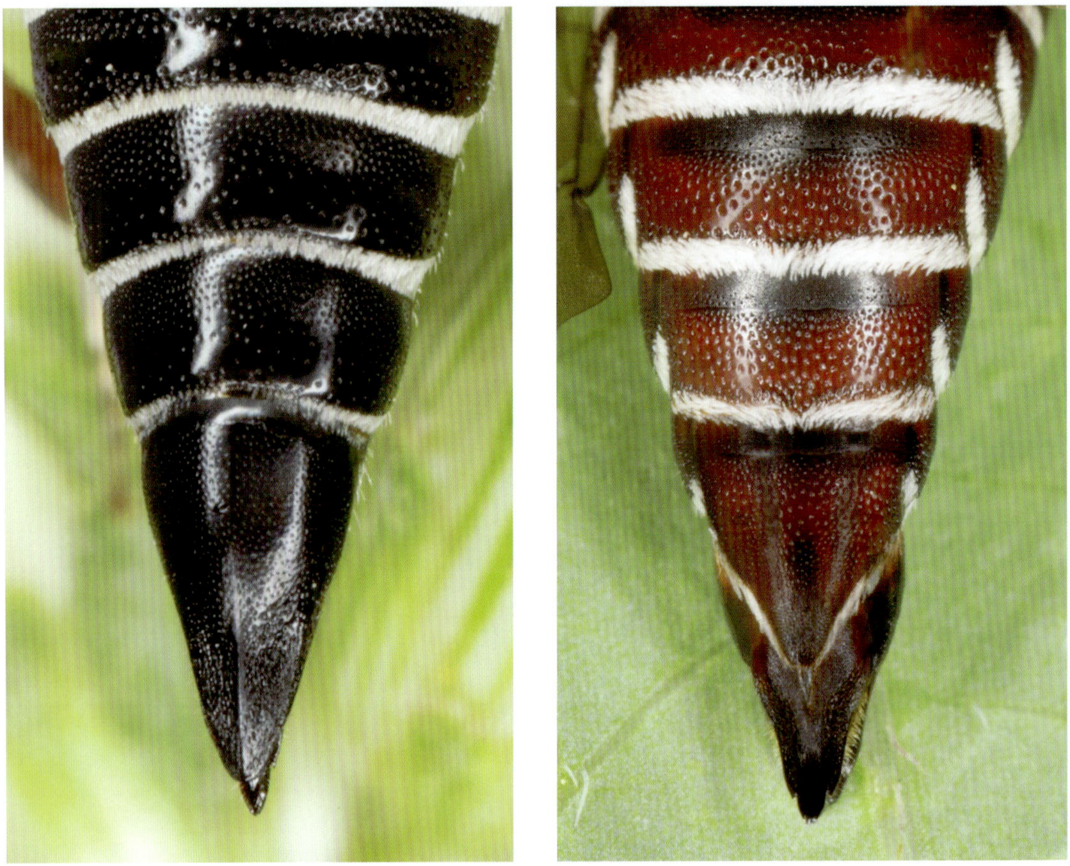

Dorsal and ventral views of the tip of the abdomen of a female *Coelioxys totonacus*: T6 has a median ridge and is indented on each side.

Scaly cuckoo leafcutter ◆ *Coelioxys (Xerocoelioxys) editus* Cockerell, 1872

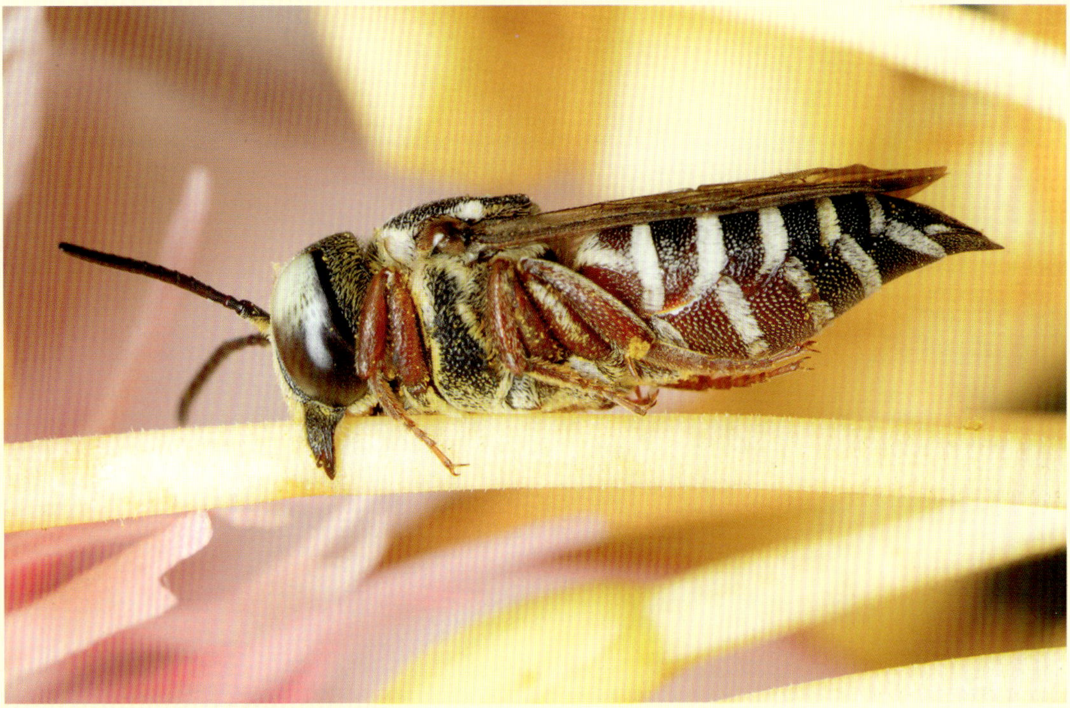

A female scaly cuckoo leafcutter. Photo credit Ross Eatman.

Description: *Coelioxys editus* is a black bee with red legs and tegulae, and a red-and-black abdomen traversed with bands of white hairs. Its head and thorax are coarsely pitted, and its face is covered with white-and-yellow hairs. Its mandibles are black and robust, its antennae black, and its eyes green. On both males and females, the first segment of the upper abdomen (T1) is red, and the rest of the upper abdomen is black; sometimes the bases of the black segments are reddish. On females, the first three sternal segments (S1–S3) are red, and the rest are black. On males, the first four sternal segments are red (S1–S4), and the rest black. The tegulae of males and females are reddish brown. **Size:** females 14 mm (1/2–3/5 in.), males 8.5–10.5 mm (1/3–2/5 in.).

Range: *Coelioxys editus* appeared at the National Butterfly Center (Hidalgo Co.) in September 2018. This species emerges in April in Texas and remains visible through October. *Coelioxys editus* occurs in the central and western United States, Florida, British Columbia, and Alberta.

Associated plants: The scaly cuckoo leafcutters shown here from the NBC were found feeding on Rio Grande abutilon (*Abutilon hypoleucum*), romerillo (*Bidens alba*), Mexican hat (*Ratibida columnifera*), and common sunflower (*Helianthus annuus*). Elsewhere, *Coelioxys editus* has been documented foraging on plants of the aster, pea, mint, and tamarisk families.

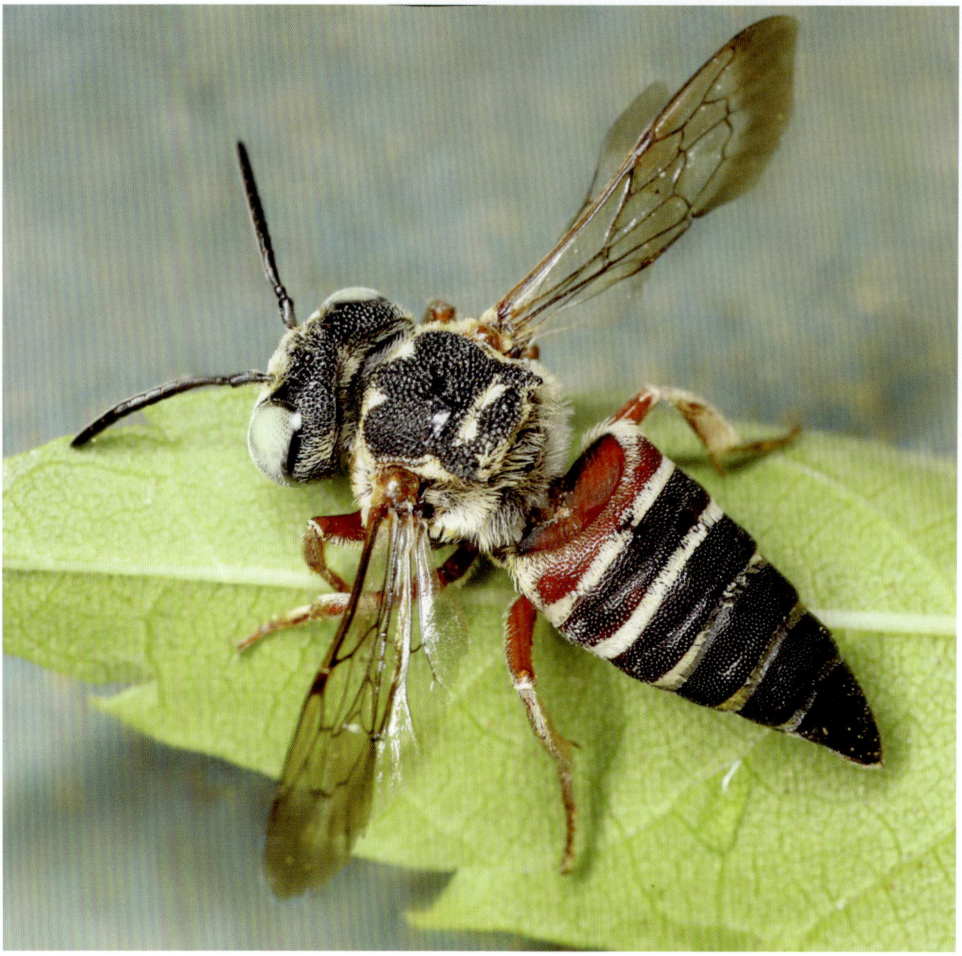

A female scaly cuckoo leafcutter: note the red T1 and the grooves on T2 and T3. Photo credit Ross Eatman.

ID tips: The red legs and coloration of the abdomen, combined with the following distinctive traits, aid in identifying *Coelioxys editus*: (1) the abdomens of both males and females have deep, uninterrupted creases or grooves that extend across the second and third segments (T2–T3); (2) the thoraxes are closely pitted; (3) in both sexes, coarse pitting on the axillae gives their edges an uneven appearance, and each axilla has a slight ridge on the hind edge; (4) the scutum of the male has squamose (embedded) hairs on the middle of the front rim; (5) males have a brush of longish pale hairs along the hind bottom rim of the compound eye; (6) the tip of the female's abdomen (S6) is indented on either side, while the tip of the male's abdomen has four prongs, and its T5 has two lateral teeth; and (7) males and females have triangular patches of white hair at the front of the scutum, a trait that aids in quick field identification of the species.

Behavior: The host species of *Coelioxys editus* is unknown. During the period when *Coelioxys editus* appeared at the National Butterfly Center, these leafcutter species were documented: *Megachile chichimeca*, *M. exilis*, *M.* cf. *toluca*, and *M. zaptlana*.

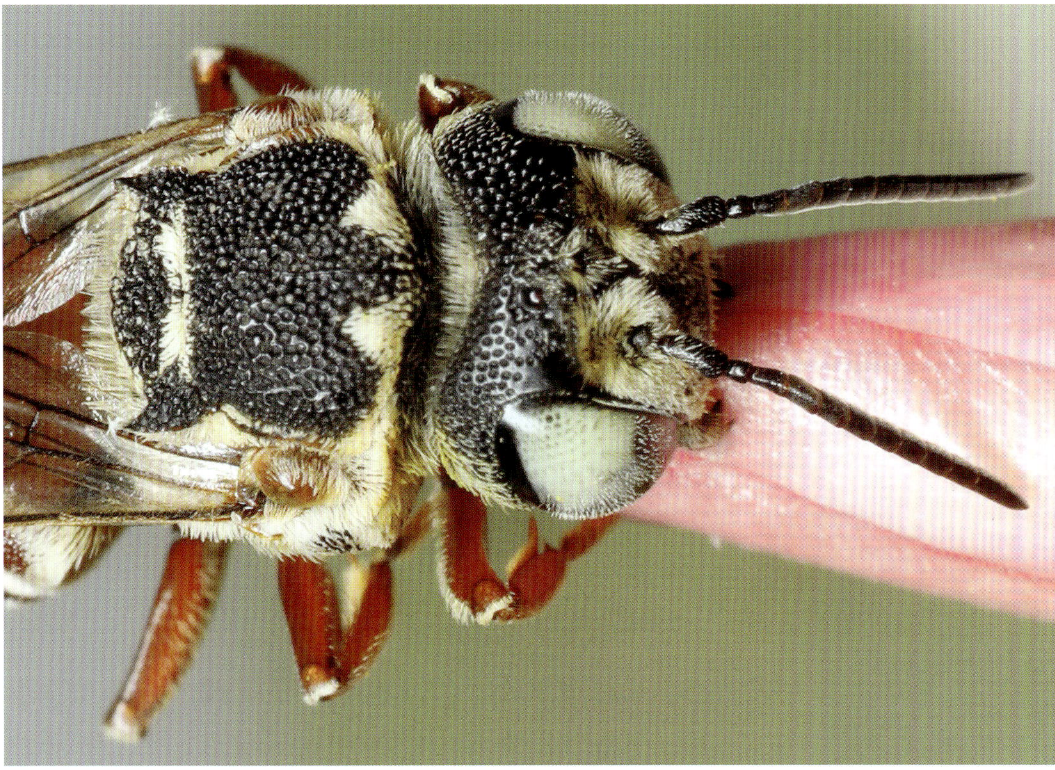

Scutum of a female scaly cuckoo leafcutter: the scutum is densely pitted. Note the triangular hair patches on the front of the thorax.

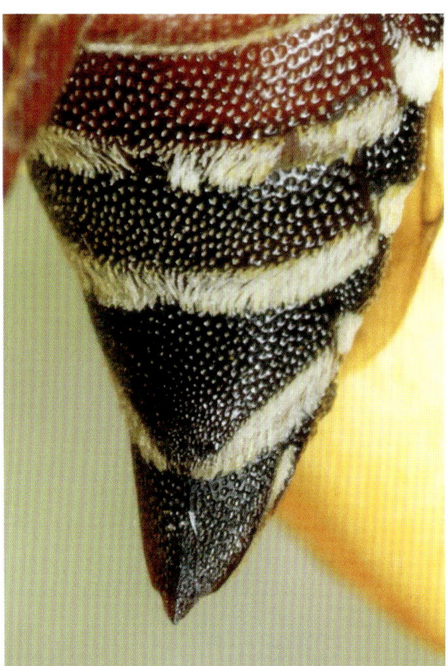

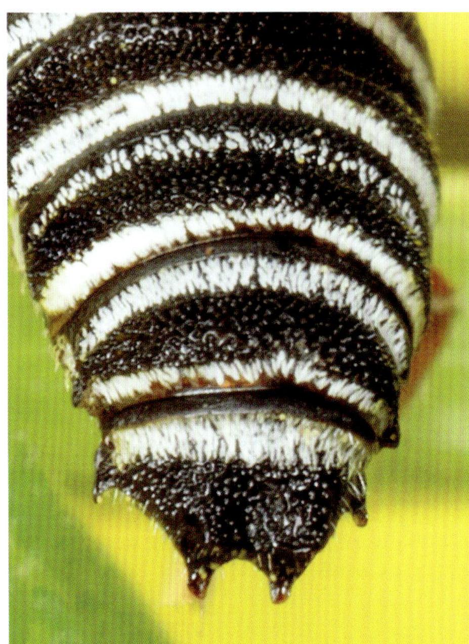

Left: Sternum of a female scaly cuckoo leafcutter: T6 is notched. *Right*: Abdomen of a male bee: T6 has four prongs, and there are short teeth on either side of T5.

Comparison of Abdomens of *Coelioxys editus* and *Coelioxys slossoni arenicola*

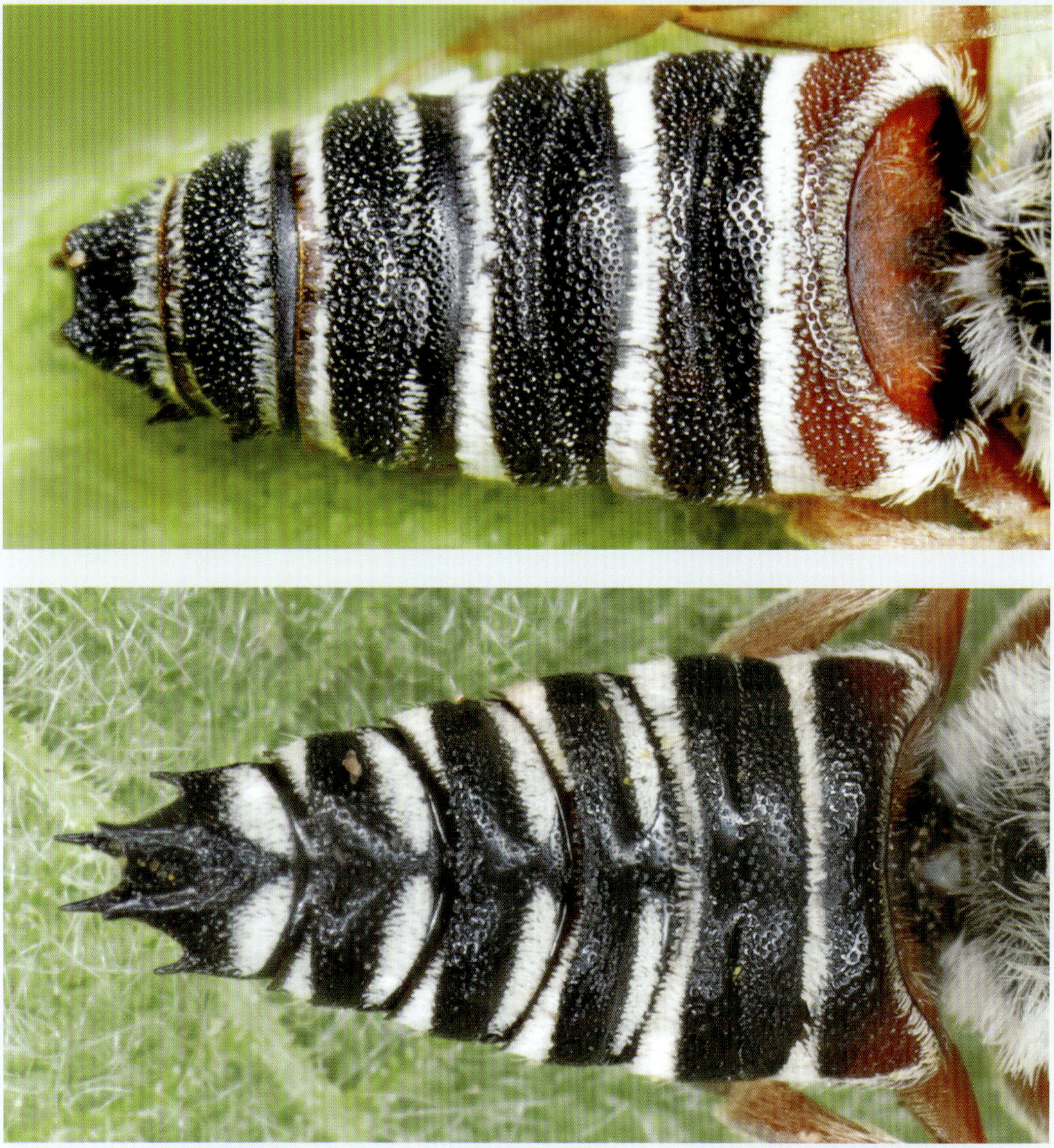

Dorsal views of the abdomens of male *Coelioxys editus* (*top*) and *C. slossoni arenicola* (*bottom*): both bees have predominantly black abdomens banded by white hairs, with red or reddish coloration on T1. Note, however, that the grooves on T2 and T3 of *Coelioxys editus* are continuous; on *C. slossoni arenicola*, the grooves are interrupted. The intricate pattern formed by the hair bands on the male *Coelioxys slossoni arenicola* is also distinctive.

Sand-dwelling cuckoo leafcutter ◆ *Coelioxys (Neocoelioxys) slossoni arenicola* Crawford, 1914

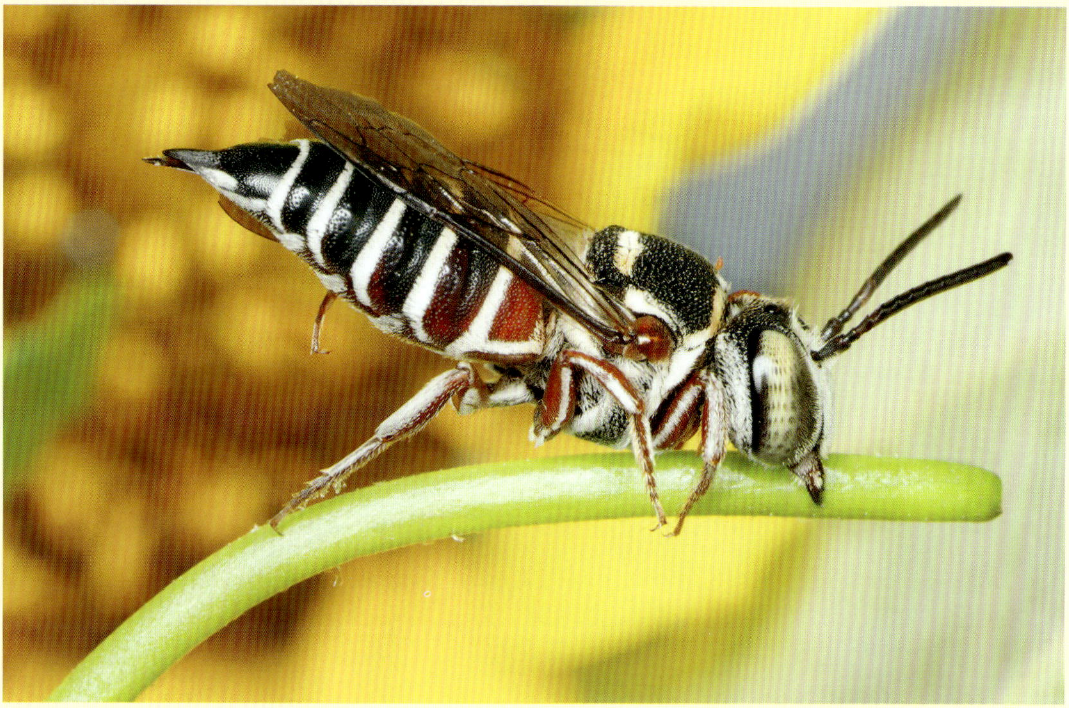

A female *Coelioxys slossoni arenicola*

Description: *Coelioxys slossoni arenicola* is a black bee with red legs and tegulae; a black, coarsely pitted thorax; a black abdomen with a red first tergal segment and pale tergal hair bands; and brown wings with dusky margins. Viewed from above, the abdominal hair bands form an arresting pattern of complete and partial stripes. The sternum is entirely red except for the last segment. The male's abdomen terminates in six prongs, with four on the tip, and two on either side of the sixth segment (S6). The tip of the female's sternum is notched on either side. In both sexes, the face is covered with white hairs. The antennae are mostly black, sometimes with partly reddish lower segments; the mandibles are partly red; and the eyes are green. **Size:** females 12–15 mm (1/2–3/5 in.), males 12 mm (1/2 in.).

Range: *Coelioxys slossoni arenicola* is common in all three border counties of the Lower Rio Grande Valley. It was first described in 1914 by James Chamberlain Crawford, a curator at the United States National Museum. He based his description in part on a female bee found in Brownsville (Cameron Co.). *Coelioxys slossoni arenicola* ranges from Mexico into Texas, and eastward to Georgia and the Carolinas. A second subspecies, *Coelioxys slossoni slossoni*, is restricted to Florida: on this subspecies, a larger portion of the upper abdomen is red, and the hair patterns on the thorax are less distinct.

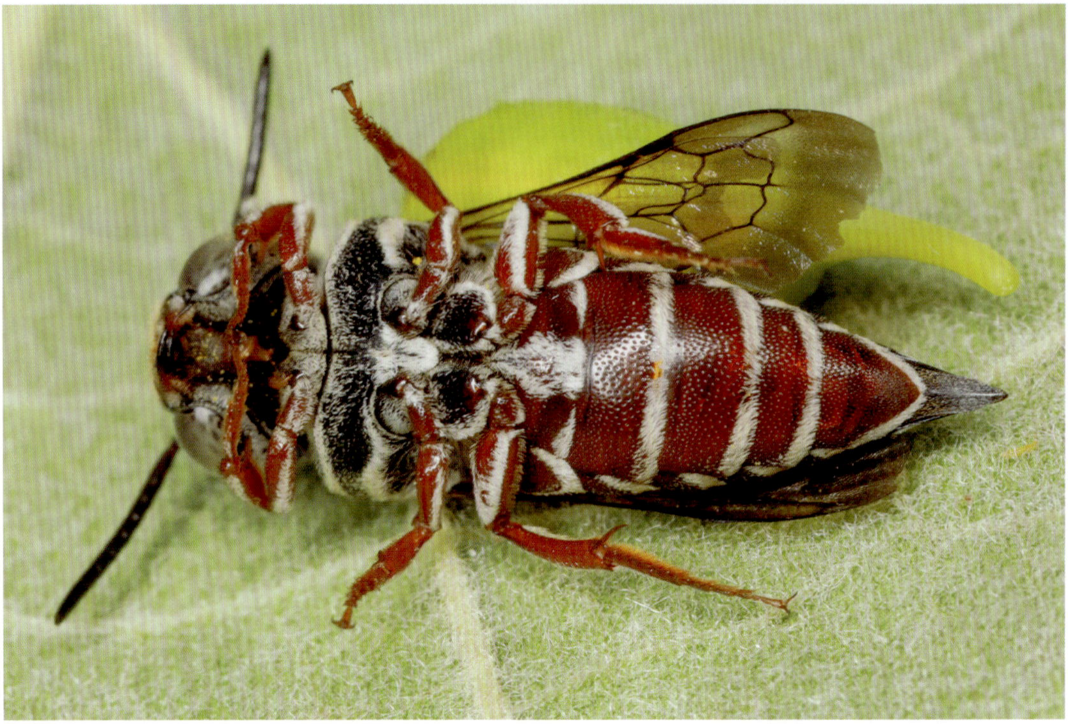

A female *Coelioxys slossoni arenicola*

Associated plants: In the Valley, *Coelioxys slossoni arenicola* visits an array of flora, including verbena-family plants such as Berlandier's fiddlewood (*Citharexylum berlandieri*), whitebrush (*Aloysia gratissima*), brushland lantana (*Lantana achyranthifolia*), and Texas frogfruit (*Phyla nodiflora*); aster-family flowers such as sunflowers (*Helianthus*), resinbush (*Viguiera stenoloba*), and hairy wedelia (*Wedelia acapulcensis* var. *hispida*); pea-family plants such as guajillo (*Senegalia berlandieri*); and Euphorbiaceae such as low croton (*Croton humilis*).

ID tips: In the Valley, *Coelioxys slossoni arenicola* is most likely to be confused with the scaly cuckoo leafcutter (*C. editus*). The latter, however, lacks the distinctive pattern formed by yellowish thorax hairs typical of *Coelioxys slossoni arenicola*, a trait visible in the field. The scaly cuckoo leafcutter's sternum also has a larger black portion, and its red portions are less brilliantly colored. Other differences between the abdomens of male *Coelioxys editus* and *C. slossoni arenicola* are shown preceding this entry.

Cuckoo leafcutters of the subgenus *Neocoelioxys* share the following characteristics: (1) the grooves on T2 and T3 are incomplete; (2) the area underneath the middle ocellus is swollen; (3) on females, S6 is notched on either side; and (4) the back rim of the scutellum is smoothly curved, without a median point or ridge, and it has two shallow, semicircular depressions beside each axilla. All of these traits help distinguish *Coelioxys slossoni arenicola* from other Valley cuckoo leafcutters in this guide.

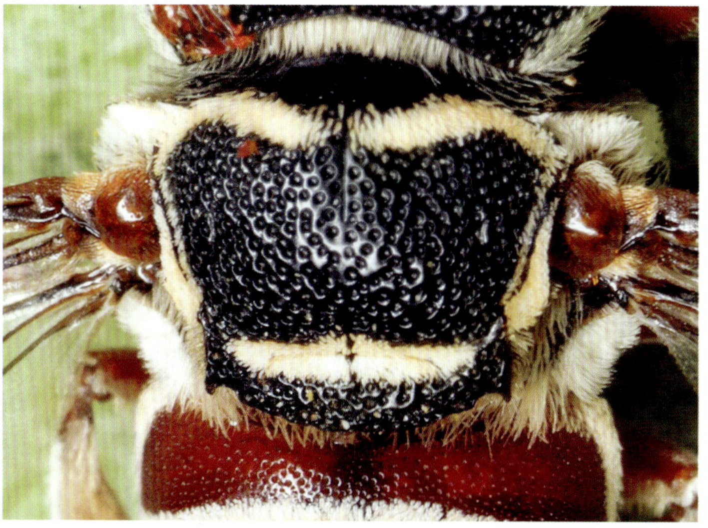

Close-up of the thorax of a female *Coelioxys slossoni arenicola*: note the distinctive patterns of yellow hairs framing the scutum, and the semicircular indentation beside each axilla.

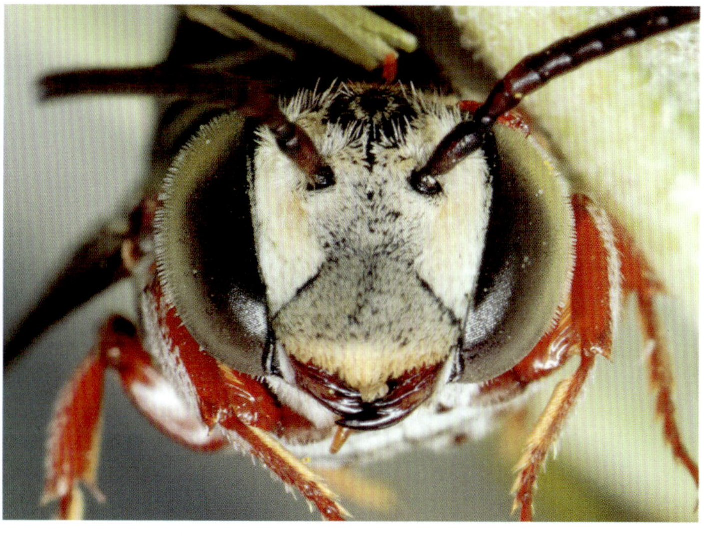

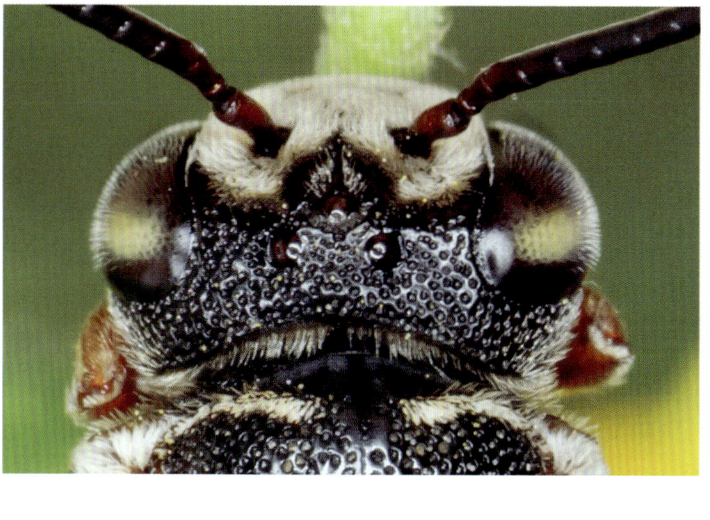

Frontal and dorsal views of the head of a female *Coelioxys slossoni arenicola*: white hairs cover the face. The mandibles, scaped, pedicels, F1, and F2 are partly reddish. Note the swollen area in front of the middle ocellus.

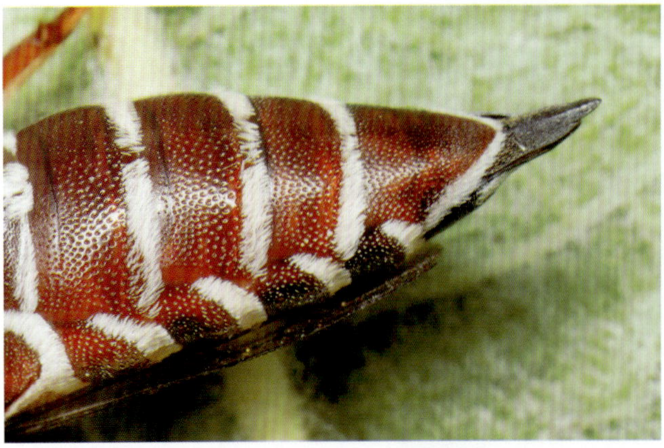

Sternum of a female *Coelioxys slossoni arenicola*

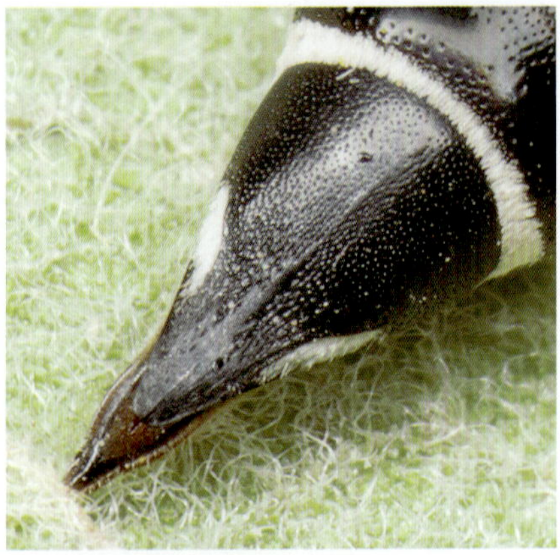

Dorsal view of the tip of a female's abdomen: S6 is indented on either side, and it protrudes farther than T6.

Behavior: *Coelioxys slossoni arenicola* is a documented cleptoparasite of the carpenter-mimic leafcutter bee, *Megachile (Melanosarus) xylocopoides*. Neither this leafcutter species nor any other member of its subgenus occurs in the Lower Rio Grande Valley. Some *Coelioxys* have been known to show flexibility, however, in selecting hosts.

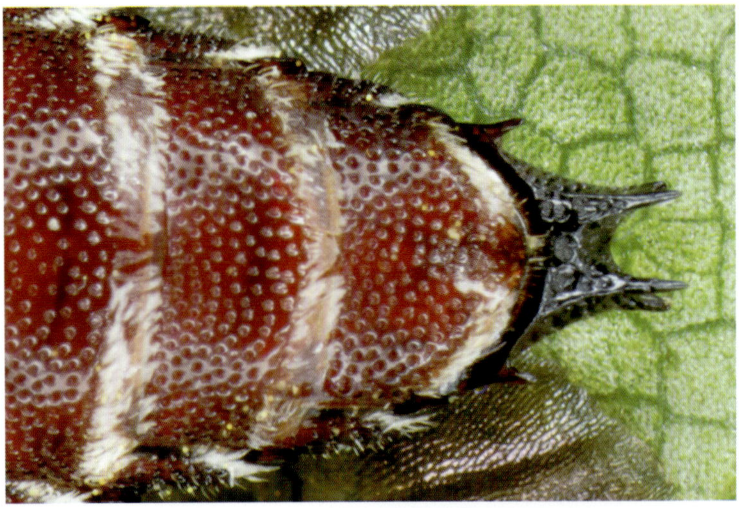

Tip of the sternum of a male *Coelioxys slossoni arenicola*

Texas cuckoo leafcutter ◆ *Coelioxys (Synocoelioxys) texanus* Cresson, 1872

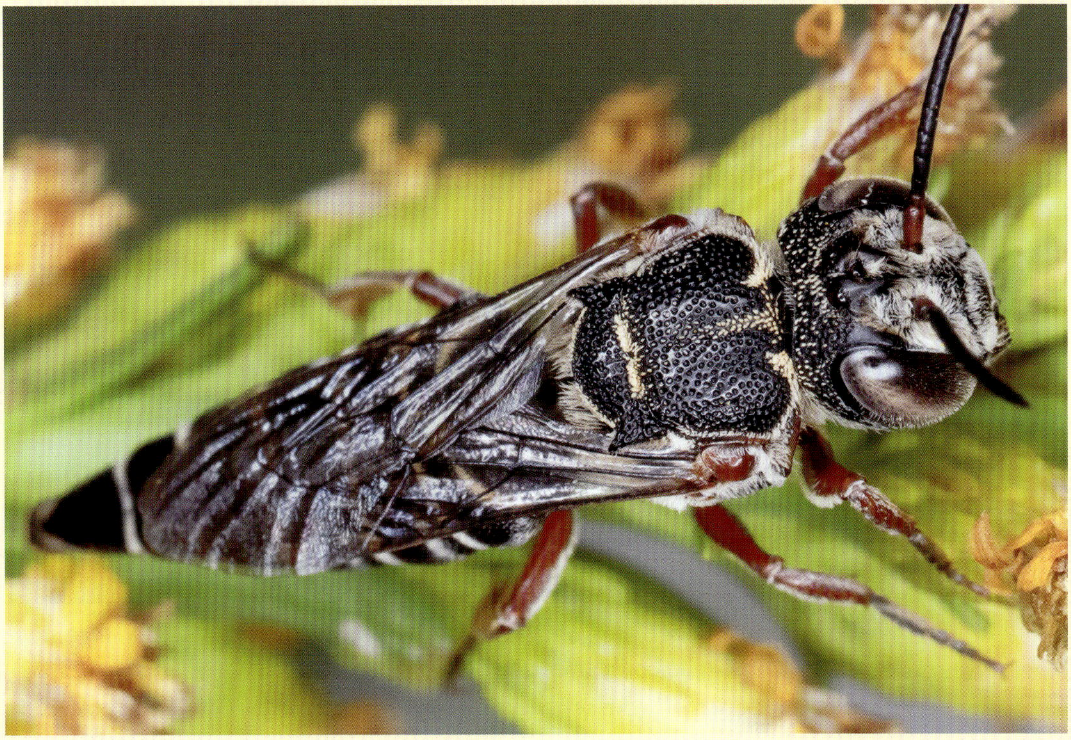

A female Texas cuckoo leafcutter

Description: Texas cuckoo leafcutters are black bees with coarsely pitted black thoraxes; abdomens banded by white hairs; red legs; red or yellowish-red tegulae; and faces covered by dense white hair. All tergal segments of the abdomens are black. The sternum of the female is dark red on the first five segments (S1–S5). The sternum of the male is dark red on S1–S3. Both males and females have black mandibles. Females have liver-colored eyes, and males' eyes are greenish brown. The bees' antennal scapes and pedicels are usually entirely red. **Size:** females 13 mm (1/2 in.), males 11–12 mm (2/5–1/2 in.).

Range: Cuckoo leafcutters of the subgenus *Synocoelioxys* are found throughout the United States, and in Mexico and the Caribbean. *Coelioxys texanus* occurs predominantly in the southwestern and western United States, but its population extends through the central part of the country as far north as Minnesota, and it is also found in Florida. This species ranges southward from the Texas border through Mexico and Central America to Costa Rica. It is common in all three border counties of the Valley. According to James R. Baker, author of a comprehensive 1975 study of North American *Coelioxys*, Texas cuckoo leafcutters are adaptable to a range of climates and geographies. They have been found at sea level and at altitudes as high as 9,000 feet, in habitats as diverse as mangrove swamps, humid floodplain forests, and deserts.

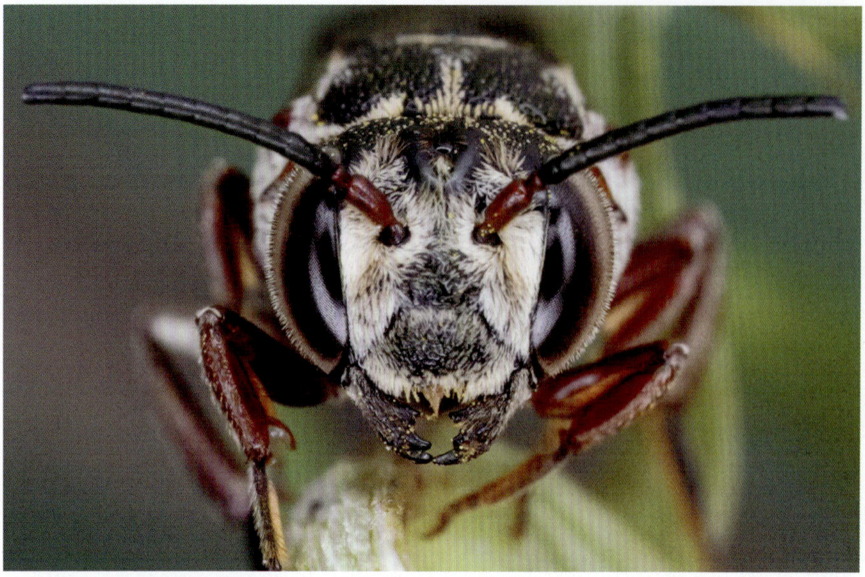

Face of a female Texas cuckoo leafcutter

Associated plants: In the Valley, this species forages on varied flora, including sticky florestina (*Florestina tripteris*), shrubby blue salvia (*Salvia ballotiflora*), seaside goldenrod (*Solidago sempervirens*), resinbush (*Viguiera stenoloba*), and hairy wedelia (*Wedelia acapulcensis* var. *hispida*). *Coelioxys texanus* has been documented elsewhere nectaring on a wide gamut of plants, among them *Hilaria*, *Bouteloua* grasses, mountain mahogany, pepperweed, buckwheat, prickly pear cactus, tarwort, creosote bush, oak, juniper, various composite flowers, spurges, and pea-family plants such as *Dalea greggii*, *Parkinsonia*, *Hocksackia*, *Mimosa*, and *Indigofera*.

ID tips: The red antennal scapes and pedicels of typical Texas cuckoo leafcutters distinguish them from other regional *Coelioxys*. Some individual *Coelioxys texanus*, however, have entirely black antennae. Where this is the case, the species is still fairly easy to differentiate from other cuckoo leafcutters of the Valley. Female and male *Coelioxys texanus* can be recognized by their black upper abdomens and red legs, combined with the following traits: (1) on both males and females, the coarsely pitted scutellum has a rounded back edge, and it lacks a median point on the hind rim or a median line; (2) the axillae are well defined and do not extend past the rim of the scutellum; (3) on females, the tip of S6 is shaped like a half oval, has smoothly curved sides, and extends well beyond T6; (4) on males as well as females, the grooves on T2 and T3 are widely interrupted; and (5) on males, the margin of T6 has a distinctive scalloped edge.

Behavior: Cuckoo leafcutters of the subgenus *Synocoelioxys* tend to be cleptoparasites of leafcutters of the subgenus *Sayapis*. The host of the Texas cuckoo leafcutter is *Megachile (Sayapis) policaris*. In 1960–61, Karl Krombein reared *Coelioxys texanus* in Portal, Arizona, from trap nests occupied and provisioned by *Megachile policaris*. Krombein noted that the *Coelioxys texanus* offspring pupated in April and May within their hosts' nests and tended to emerge four to five weeks later.

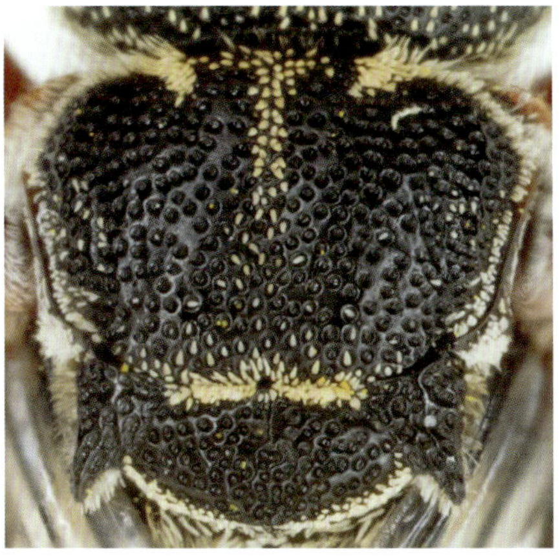

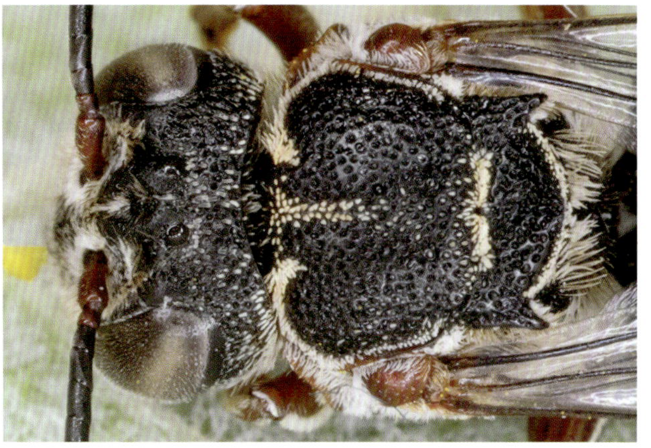

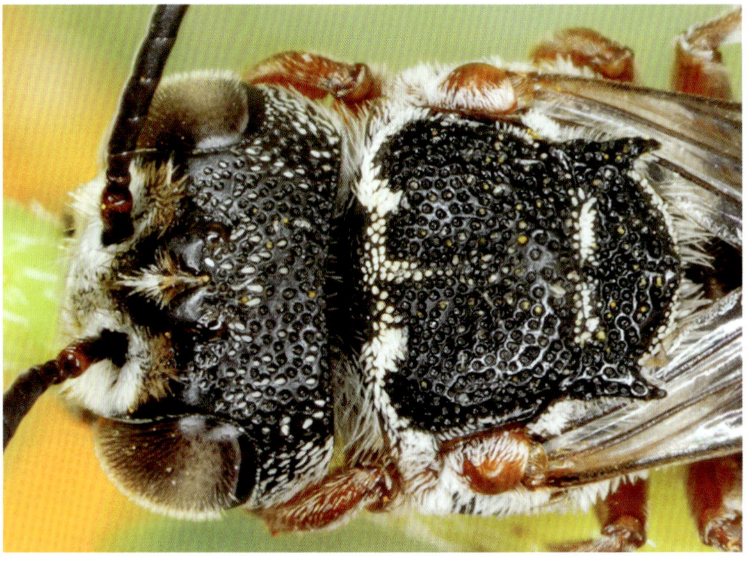

The scutum and scutellum of the female (*above*) and male (*left*) Texas cuckoo leafcutter are coarsely pitted. The scutellum has a rounded hind edge and lacks a median ridge or point on the rear rim. The toothlike axillae do not reach past the rear rim. Note also the densely pitted vertex: the pits behind the ocelli are distinct and well separated.

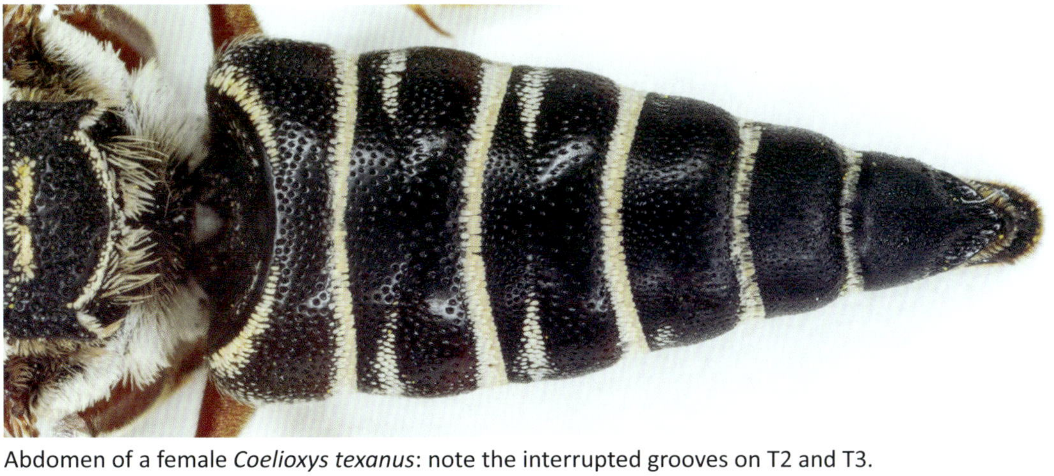

Abdomen of a female *Coelioxys texanus*: note the interrupted grooves on T2 and T3.

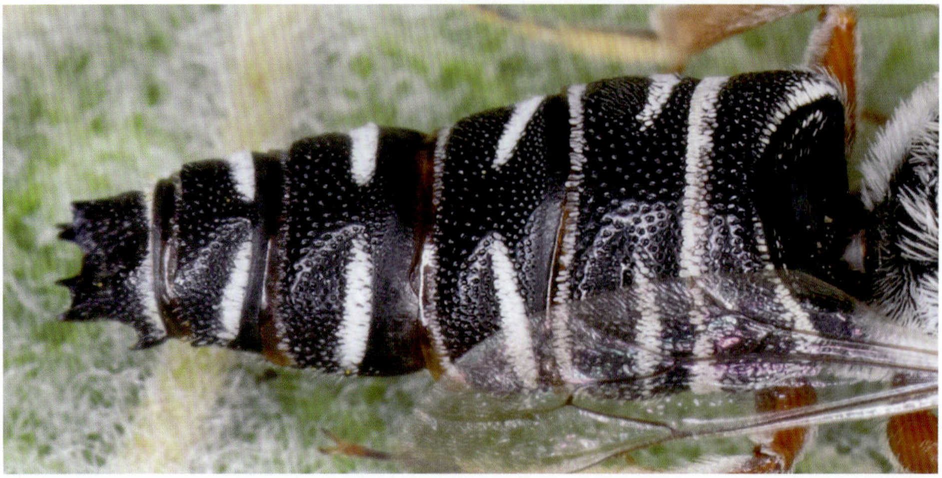

Abdomen of a male *Coelioxys texanus*

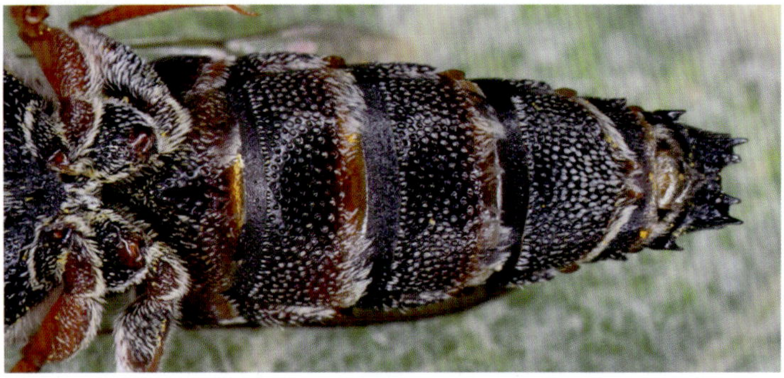

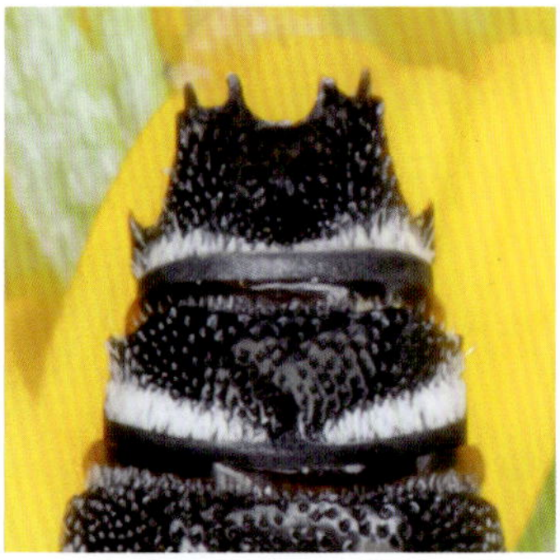

The sternum (*left*) and a dorsal view of the abdominal tip (*right*) of a male *Coelioxys texanus*: S5 and S6 have lateral teeth, and S7 has four prongs. The hind rim of T6 is scalloped.

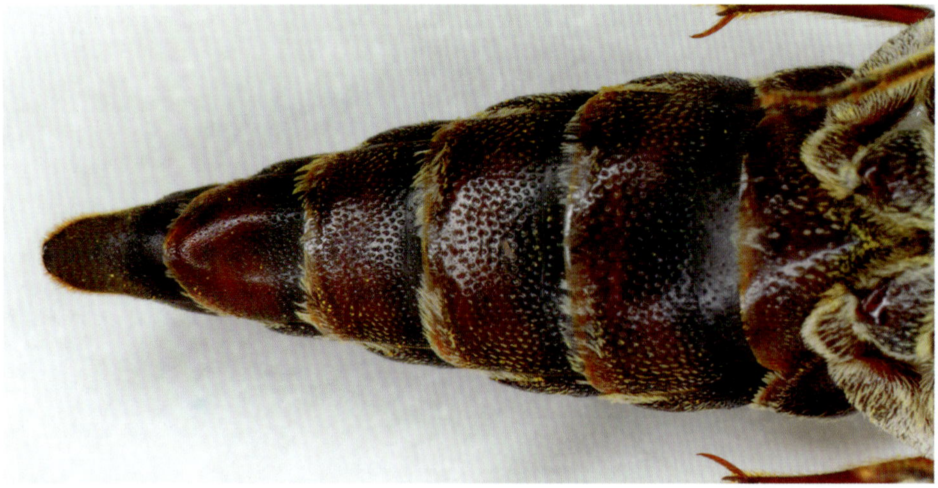

Sternum of a female *Coelioxys texanus*: S5 has a broadly rounded hind rim. The tip of S6 is gently rounded in the form of a half oval.

Appendix

List of Lower Rio Grande Valley Bee Species

With Rio Grande Valley Locations and Flora

Research for this book was conducted from September 2018 to November 2021 and focused on 25 sites in the Texas Lower Rio Grande Valley (LRGV). The sites visited included the following:

Cameron County: Bahia Grande Unit of Laguna Atascosa National Wildlife Refuge (NWR) (Laguna Vista/Brownsville); Hugh Ramsey Nature Park (Harlingen); Laguna Atascosa National Wildlife Refuge, northern sector (Los Fresnos); Las Palomas Wildlife Management Area—Ebony Unit (San Benito); Resaca de la Palma State Park (SP) (Brownsville); Sabal Palm Sanctuary (Brownsville); South Padre Island Birding and Nature Center (BNC) (South Padre Island).

Hidalgo County: Bentsen–Rio Grande Valley State Park (SP) (Mission); El Mesteño Ranch and Arboretum (Puerto Rico); Estero Llano Grande State Park (SP) (Weslaco); Hub of Prosperity (Edinburg); La Sal del Rey Tract of the LRGV National Wildlife Refuge (northeast of Edinburg); McAllen Nature Center (MNC) (McAllen); Mission Hike and Bike (Mission); National Butterfly Center (NBC) (Mission); Oleander Acres (Mission); Pixie Preserve (Mission); Quinta Mazatlán (McAllen); Santa Ana National Wildlife Refuge (NWR) (Alamo); Teniente Tract of the LRGV National Wildlife Refuge (11 miles west of Raymondville); garden of master gardener William Rich (McAllen).

Starr County: Campos Viejos Ranch (north of Rio Grande City); Falcon State Park (SP) (Roma); La Puerta Tract of the LRGV National Wildlife Refuge (Rio Grande City); Salineño Tract of the LRGV National Wildlife Refuge (Salineño).

Identification of Specimens: Dr. Karen Wright of Texas A&M University identified *Svastra sabinensis. Megachile* cf. *toluca* was identified through the collaboration of John L. Neff, director of the Central Texas Melittological Institute (CTMI); Dr. John Ascher of the National University of Singapore; and Hadel Go of the American Museum of Natural History; this specimen is now housed in the Entomology Collection of the University of Texas at Austin. John L. Neff identified all other species; vouchered specimens are in the collection of the CTMI.

FAMILY ANDRENIDAE

Andrena (Larandrena) miserabilis Cresson, 1872 (miserable mining bee). **Sites: Hidalgo Co.** Mission Hike and Bike (*2021*) April 26 (♀); (*2019*) April 14 (♀). **Flora visited:** Nesting in vicinity of blooming *Prosopis glandulosa*.

***Andrena (Opandrena)* sp.** (undescribed species). **Sites: Starr Co.** Falcon SP (*2021*) May 26 (♀♂), May 29 (♀), June 3 (♀). **Flora visited:** *Leucophyllum frutescens*.

Calliopsis (Calliopsis) hondurasica Cockerell, 1949 (Honduran calliopsis). **Sites: Cameron Co.** Hugh Ramsey (*2021*) June 11 (♀♂). Laguna Atascosa NWR (*2021*) Nov. 9 (♀♂). **Hidalgo Co.** MNC (*2021*) April 25 (♀), April 27 (♀). Quinta Mazatlán (*2021*) May 13 (♀♂). **Flora visited:** *Aphanostephus ramosissimus, Nama hispidum, Phyla nodiflora, Teucrium cubense.*

Mesoxaea (Mesoxaea) texana (Friese, 1898) (Texas mesoxaea). **Sites: Hidalgo Co.** El Mesteño (*2021*) May 20 (♂), May 21 (♂), May 23 (♀), June 2 (♂), June 16 (♂). **Starr Co.** Falcon SP (*2021*) May 29 (♂). **Flora visited:** *Aloysia gratissima, Salvia ballotiflora.*

Perdita (Cockerellia) scopata Timberlake, 1953 (scopate fairy bee). **Sites: Hidalgo Co.** El Mesteño (*2021*) May 21 (♀♂), June 2 (♀). Hub of Prosperity (*2021*) May 10 (♀). MNC (*2020*) Feb. 28 (♀♂). **Starr Co.** Campos Viejos (*2021*) June 17 (♀♂). **Flora visited:** *Helianthus annuus, Ratibida columnifera, Verbesina encelioides.*

***Perdita (Hexaperdita)* sp.,** possibly *Perdita foveata brachycephala* Timberlake, 1956. **Sites: Starr Co.** Campos Viejos (*2021*) June 17 (♀♂). Falcon SP (*2021*) May 26 (♀), June 6 (♀). **Flora visited:** *Leucophyllum frutescens, Melampodium cinereum, Nama hispidum, Tecoma stans.*

Perdita (Hexaperdita) ignota Cockerell, 1896 (ignored fairy bee). **Sites: Cameron Co.** Bahia Grande (*2021*) Nov. 18 (♀). South Padre Island BNC (*2021*) Nov. 17 (♀♂). **Hidalgo Co.** El Mesteño (*2021*) May 21 (♀♂), June 2 (♀). Hub of Prosperity (*2021*) May 10 (♀♂). MNC (*2021*) April 27 (♀♂), April 28 (♀♂), May 22 (♀♂). **Flora visited:** *Aphanostephus ramosissimus, Rayjacksonia phyllocephala, Symphyotrichum* sp., *Tridax procumbens, Verbesina encelioides.*

Perdita (Perdita) missionis Timberlake, 1958 (Mission fairy bee). **Sites: Hidalgo Co.** MNC (*2021*) April 24 (♀♂), (*2019*) March 18 (♂). **Flora visited:** *Nama hispidum, Nama undulatum.*

Protandrena (Protandrena) bancrofti Dunning, 1897 (Bancroft's protandrena). **Sites: Hidalgo Co.** El Mesteño (*2021*) June 2 (♀), June 16 (♀). **Flora visited:** *Sphaeralcea angustifolia.*

Protandrena (Pseudopanurgus) texana (Timberlake, 1976) (Texas protandrena). **Sites: Starr Co.** Falcon SP (*2021*) May 26 (♀), June 3 (♀), June 12 (♀). **Flora visited:** *Aloysia macrostachya, Palafoxia texana, Wedelia acapulcensis* var. *hispida.*

Protandrena (Pterosarus) ornatipes (Cresson, 1872) (fancy-footed protandrena). **Sites: Cameron Co.** Bahia Grande (*2021*) Nov. 9 (♀♂), Nov. 16 (♀♂), Nov. 18 (♀). **Flora visited:** *Rayjacksonia phyllocephala.*

Protoxaea gloriosa (Fox, 1893) (glorious protoxaea). **Sites: Hidalgo Co.** El Mesteño (*2021*) May 20 (♂), May 21 (♂), May 24 (♀), June 2 (♂). **Starr Co.** Falcon SP (*2021*) May 26 (♂), May 29 (♂), June 3 (♀). **Flora visited:** *Aloysia gratissima, Salvia ballotiflora.* Also found in jaws of lynx spider (♀), and flying near *Saliva ballotiflora*, aggressively defending its territory (♀♂).

FAMILY APIDAE

Ancyloscelis apiformis (Fabricius, 1793) (apiform crooked-legged bee). **Sites: Hidalgo Co.** NBC (*2021*) April 25 (♀♂), May 31 (♀♂), June 27 (♂), Nov. 5 (♀), Nov. 10 (♀♂); (*2020*) March 4; (*2018*) Sept. 7 (♂), Sept. 9 (♀), Sept. 11 (♂). **Starr Co.** Campos Viejos (*2021*) June 21 (♂). **Flora visited:** *Cirsium texanum, Merremia dissecta, Tecoma stans.*

Anthophora (Anthophoroides) californica Cresson, 1869 (California digger bee). **Sites: Hidalgo Co.** Bentsen–Rio Grande Valley SP (*2020*) Feb. 27 (♂). El Mesteño (*2021*) May 24 (♀). NBC (*2021*) Nov. 5 (♀♂); (*2019*) April 8 (♀), April 29 (♀); (*2018*) Sept. 18 (♂). Quinta Mazatlán (*2021*) May 4 (♂), Nov. 6 (♂). **Starr Co.** Campos Viejos (*2021*) June 17 (♀♂). Falcon SP (*2021*) May 29 (♀♂), June 3 (♀♂). La Puerta (*2021*) May 6 (♀♂), May 9 (♀♂). **Flora visited:** *Citharexylum berlandieri, Duranta erecta, Guaiacum angustifolium, Parkinsonia texana, Salvia ballotiflora.*

Anthophora (Mystacanthophora) capistrata Cresson, 1878 (masked salvia digger bee). **Sites: Hidalgo Co.** NBC (*2021*) Nov. 10 (♀♂); (*2019*) Oct. 22 (♂); (*2018*) Sept. 7 (♀). Quinta Mazatlán (*2021*) Nov. 6 (♀♂). **Flora visited:** *Duranta erecta, Salvia ballotiflora.*

Anthophorula (Anthophorula) compactula Cockerell, 1897 (compact anthophorula). **Sites: Hidalgo Co.** El Mesteño (*2021*) May 20 (♂), June 27 (♂). Mission Hike and Bike (*2019*) April 11 (♀♂). MNC (*2021*) April 24 (♀), May 22 (♀). NBC (*2019*) April 5 (♂), April 14 (♀). Pixie Preserve (*2021*) May 18 (♂). **Starr Co.** Campos Viejos (*2021*) June 17 (♀). Falcon SP (*2021*) June 12 (♀♂). La Puerta (*2021*) May 7 (♂). **Flora visited:** *Aloysia gratissima, Argemone sanguinea, Echinocereus enneacanthus, Melampodium cinereum, Nama undulatum, Parkinsonia texana, Opuntia engelmannii* var. *lindheimeri, Sida ciliaris, Thelocactus setispinus, Thymophylla pentachaeta.*

Apis mellifera Linnaeus, 1758 (western honey bee) (♀ = worker; ♂ = drone). **Sites: Cameron Co.** Bahia Grande (*2021*) Nov. 9 (mobbing *Colima* and *Coma* shrubs) (♀), Nov. 16 (♀). Hugh Ramsey (*2021*) June 6 (♀), June 9 (♀). Laguna Atascosa NWR (*2021*) Nov. 7 (♀) (♂ on ground near colony). Resaca de la Palma SP (*2021*) Nov. 23 (♀). South Padre Island BNC (*2021*) June 15 (♀). **Hidalgo Co.** Bentsen–Rio Grande Valley SP (*2021*) April 9 (♀); (*2020*) Feb. 27 (♀). El Mesteño (*2021*) May 2 (♀), May 20 (♀). MNC (*2021*) April 24 (♀). NBC (*2021*) April 24 (♀),

Nov. 5 (♀), Nov. 24 (♀); (*2018*) Nov. 9 (♀♂ hive). Quinta Mazatlán (*2021*) May 4 (♀), May 28 (♀), Nov. 5 (♀). Santa Ana NWR (*2021*) April 9 (♀). Teniente (*2021*) Nov. 21 (♀). **Starr Co.** Campos Viejos (*2021*) June 17 (♀), June 21 (♀). Falcon SP (*2021*) May 29 (♀). Salineño (*2021*) May 26 (♀). **Flora visited:** *Abutilon hypoleucum, Agave americana, Aloysia gratissima, Antigonon leptopus, Argemone aenea, Argemone mexicana, Argemone sanguinea, Bidens alba, Chromolaena odorata, Cirsium texanum, Cordia boissieri, Ebenopsis ebano, Guaiacum angustifolium, Helianthus annuus, Isocoma drummondii, Karwinskia humboldtiana, Larrea tridentata, Leucophyllum frutescens, Opuntia engelmannii* var. *lindheimeri, Phyla nodiflora, Prosopis glandulosa, Ratibida columnifera, Salvia ballotiflora, Sideroxylon celastrinum, Solidago sempervirens, Symphyotrichum subulatum, Trixis inula, Vachellia farnesiana, Verbesina encelioides, Zanthoxylum fagara, Zornia bracteata.*

Bombus (Thoracobombus) pensylvanicus (De Geer, 1773) (American bumble bee). **Sites: Cameron Co.** South Padre Island BNC (*2021*) June 5 (♀). **Hidalgo Co.** MNC (*2021*) May 20 (♀). NBC (*2021*) May 31 (♀♂), Nov. 5 (♀♂); (*2019*) Oct. 22 (♂); (*2018*) Sept. 7 (♀). **Flora visited:** *Chromolaena odorata, Duranta erecta, Leucophyllum frutescens, Monarda citriodora, Salvia ballotiflora, Solanum elaeagnifolium, Tecoma stans, Trixis inula, Viguiera stenoloba.*

Centris (Heterocentris / Hemisiella) nitida Smith, 1874 (shining oil-digger bee). **Sites: Hidalgo Co.** MNC (*2018*) Nov. 17 (♂). NBC (*2018*) Nov. 20 (♀). **Flora visited:** *Chromolaena odorata, Malpighia emarginata, Malpighia glabra.*

Centris (Paracentris) atripes Mocsáry, 1899 (black-legged oil-digger bee). **Sites: Cameron Co.** Resaca de la Palma SP (*2021*) Nov. 23 (♀). **Hidalgo Co.** Estero Llano Grande SP (*2021*) June 16 (♀). MNC (*2021*) May 22 (♀). NBC (*2021*) April 24 (♀), May 2 (♀); (*2019*) April 8 (♀), April 11 (♀), May 6 (♀), Sept. 8 (♀), Oct. 20 (♀), Oct. 28 (♀). Oleander Acres (*2021*) May 31 (♀). Quinta Mazatlán (*2021*) May 4 (♀♂), May 13 (♂). **Starr Co.** Campos Viejos (*2021*) June 21 (♂). Falcon SP (*2021*) May 29 (♂), June 25 (♂). La Puerta (*2021*) May 6 (♀), May 7 (♀), May 9 (♀). **Flora visited:** *Arctostaphylos glauca, Duranta erecta, Galphimia glauca* (nonnative), *Malpighia emarginata, Malpighia glabra, Mascagnia macroptera* (nonnative), *Rhynchosia senna* var. *texana, Salvia ballotiflora, Tecoma stans.*

Ceratina (Calloceratina) cobaltina Cresson, 1878 (cobalt small carpenter bee). **Sites: Hidalgo Co.** NBC (*2018*) Sept. 18 (♀). Oleander Acres (*2021*) May 31 (♀). Quinta Mazatlán (*2021*) May 13 (♀). **Flora visited:** *Cirsium texanum, Merremia dissecta, Opuntia engelmannii* var. *lindheimeri.*

Ceratina (Zadontomerus) texana Daly, 1973 (Texas small carpenter bee). **Sites: Cameron Co.** Hugh Ramsey (*2021*) June 21 (♀), Nov. 11 (♀♂). **Hidalgo Co.** El Mesteño (*2021*) May 21 (♀). MNC (*2021*) April 29 (♂). NBC (*2019*) Nov. 5 (♀♂). Oleander Acres (*2021*) May 31 (♀). Quinta Mazatlán (*2021*) Nov. 6 (♂). **Starr Co.** Campos Viejos (*2021*) June 17 (♀). Falcon SP (*2021*) May 26 (♂). Salineño (*2021*) May 26 (♀). **Flora visited:** *Argemone aenea, Cucumis melo* (nonnative), *Echinocereus enneacanthus, Melampodium cinereum, Nama undulatum, Palafoxia texana, Wedelia acapulcensis* var. *hispida.*

Diadasia (Coquillettapis) diminuta (Cresson, 1878) (globemallow bee). **Sites: Hidalgo Co.** El Mesteño (*2021*) May 20 (♂). MNC (*2021*) April 28 (♀). NBC (*2019*) April 27 (♀), May 3 (♂). **Flora visited:** *Cylindropuntia leptocaulis, Malvastrum americanum, Melochia tomentosa, Vicia* spp.

Diadasia (Coquillettapis) rinconis Cockerell, 1897 (Rincon chimney bee). **Sites: Hidalgo Co.** El Mesteño (*2021*) May 2 (♀), May 20 (♂), May 21 (♂), May 24 (♂). La Sal del Rey (*2021*) April 26 (♀♂). Mission Hike and Bike (*2021*) April 26 (♀♂); (*2019*) April 14 (♀♂). MNC (*2021*) April 24 (♂), May 22 (♂); (*2019*) March 24 (♂). NBC (*2021*) April 25 (♂), May 2 (♂), May 7 (♂); (*2019*) March 25 (♂), March 27 (♂), March 29 (♂), April 14 (♀♂). Pixie Preserve (*2021*) May 1 (♂), May 18 (♂); (*2019*) May 18 (♀♂). Quinta Mazatlán (*2021*) May 13 (♂). Santa Ana NWR (*2021*) April 9 (♀♂). **Starr Co.** Campos Viejos (*2021*) June 17 (♀♂), June 21 (♀♂). Falcon SP (*2021*) May 26 (♂), May 29 (♂), June 12 (♀♂), June 25 (♀♂). La Puerta (*2021*) May 6 (♂). Salineño (*2021*) May 26 (♂). **Flora visited:** *Argemone aenea, Argemone sanguinea, Astrophytum asterias, Cirsium texanum, Cordia boissieri, Cylindropuntia leptocaulis, Ferocactus wislizeni, Opuntia engelmannii* var. *lindheimeri, Parkinsonia texana, Prosopis juliflora, Tecoma stans.*

Diadasia (Dasiapis) ochracea (Cockerell, 1903) (ochraceous chimney bee). **Sites: Cameron Co.** Laguna Atascosa NWR (*2021*) Nov. 7 (♀). Resaca de la Palma SP (*2021*) Nov. 23 (♀). **Hidalgo Co.** El Mesteño (*2021*) May 24 (♀). NBC (*2021*) May 2 (♀), May 7 (♀); (*2019*) Nov. 2 (♀), Nov. 10 (♀); (*2018*) Sept. 6 (♀), Nov. 16 (♀). Oleander Acres (*2021*) May 31 (♀). **Flora visited:** *Abutilon hypoleucum, Cirsium texanum, Herissantia crispa, Malvastrum americanum, Ruellia nudiflora* var. *runyonii.*

Diadasia (Dasiapis) tropicalis (Cockerell, 1918) (tropical chimney bee). **Sites: Cameron Co.** Resaca de la Palma SP (*2021*) Nov. 12 (♂). **Hidalgo Co.** NBC (*2021*) April 24 (♂), May 2 (♂), May 7 (♂); (*2019*) April 21 (♂), July 2 (♂). **Starr Co.** Campos Viejos (*2021*) June 21 (♂). La Puerta (*2021*) May 16 (♂). **Flora visited:** *Abutilon abutiloides, Abutilon hypoleucum, Antigonon leptopus, Chromolaena odorata, Malvastrum americanum, Melochia tomentosa.*

Diadasia (Diadasia) enavata (Cresson, 1872) (sunflower chimney bee). **Sites: Hidalgo Co.** MNC (*2021*) June 10 (♂). NBC (*2021*) June 27 (♂); (*2019*) June 30 (♀); (*2018*) June 30 (♀), Sept. 7 (♂), Sept. 14 (♂). **Starr Co.** Campos Viejos (*2021*) June 21 (♂). **Flora visited:** *Helianthus annuus, Verbesina encelioides.*

Epeolus pusillus Cresson, 1864 (dwarf epeolus). **Sites: Hidalgo Co.** MNC (*2018*) Nov. 17 (♀). NBC (*2018*) Nov. 27 (♀). **Flora visited:** *Symphyotrichum subulatum, Viguiera stenoloba.*

Eucera (Xenoglossodes) fasciatella (LaBerge, 1970) (fasciatella longhorn bee). **Sites: Starr Co.** Falcon SP (*2021*) May 29 (♂), June 3 (♂). **Flora visited:** *Salvia ballotiflora.*

Eucera (Xenoglossodes) paenalbata (LaBerge, 2001) (woolly-white longhorn bee). **Sites: Starr Co.** Campos Viejos (*2021*) June 21 (♀). **Flora visited:** *Ratibida columnifera.*

Eucera (Xenoglossodes) wilmattae (Cockerell, 1917) (Wilmatte's longhorn bee). **Sites: Cameron Co.** Hugh Ramsey (*2021*) June 6 (♀♂), June 11 (♀♂). Sabal Palm Sanctuary (*2021*) June 11 (♂). **Hidalgo Co.** El Mesteño (*2021*) June 2 (♀♂). NBC (*2021*) May 2 (♀), May 7 (♀), Nov. 5 (♀♂); (*2019*) April 21 (♀♂), Oct. 19 (♀); (*2018*) Nov. 8 (♂). Quinta Mazatlán (*2021*) May 4 (♂), Nov. 6 (♂). **Starr Co.** Campos Viejos (*2021*) June 21 (♀). Falcon SP (*2021*) June 12 (♀), June 25 (♀). **Flora visited:** *Bidens alba, Gaillardia pulchella, Melampodium cinereum, Palafoxia texana, Ratibida columnifera, Thymophylla tenuiloba, Verbesina encelioides, Viguiera stenoloba, Wedelia acapulcensis* var. *hispida.*

Exomalopsis (Exomalopsis) analis Spinola, 1853 (ringed exomalopsis). **Sites: Hidalgo Co.** Hub of Prosperity (*2021*) May 10 (♀). MNC May 22 (♀). NBC (*2021*) Nov. 5 (♀), Nov. 24 (♀); (*2018*) Nov. 17 (♀). Quinta Mazatlán (*2021*) Nov. 6 (♀). **Flora visited:** *Aloysia gratissima, Aloysia macrostachya, Chromolaena odorata, Heliotropium angiospermum, Solanum elaeagnifolium.*

Exomalopsis (Exomalopsis) mellipes Cresson, 1878 (honey-footed exomalopsis). **Sites: Cameron Co.** Resaca de la Palma SP (*2021*) Nov. 23 (♀). **Hidalgo Co.** El Mesteño (*2021*) May 20 (♀♂), May 21 (♀♂). NBC (*2021*) May 7 (♀), June 27 (♀), Nov. 5 (♀♂), Nov. 24 (♀); (*2019*) Oct. 27 (♂); (*2018*) Sept. 12 (♀), Sept. 20 (♀). **Starr Co.** Campos Viejos (*2021*) June 21 (♀). Falcon SP (*2021*) May 26 (♀), May 29 (♂), June 3 (♀♂). La Puerta (*2021*) May 16 (♂). **Flora visited:** *Aloysia gratissima, Aloysia macrostachya, Bidens alba, Chromolaena odorata, Clematis drummondii, Ebenopsis ebano, Leucophyllum frutescens, Malvastrum coromandelianum, Salvia ballotiflora, Solanum rostratum, Solidago sempervirens, Verbesina microptera.*

Exomalopsis (Exomalopsis) similis Cresson, 1865 (similar exomalopsis). **Sites: Hidalgo Co.** NBC (*2021*) Nov. 5 (♀♂); (*2019*) Oct. 22 (♂), Nov. 15 (♀). **Flora visited:** *Bidens alba, Solidago sempervirens, Viguiera stenoloba.*

Exomalopsis (Phanomalopsis) snowi Cockerell, 1906 (Snow's exomalopsis). **Sites: Hidalgo Co.** NBC (*2021*) Nov. 5 (♀), Nov. 24; (*2018*) Nov. 9 (♀♂). **Flora visited:** *Aloysia gratissima, Bidens alba, Chromolaena odorata, Ebenopsis ebano, Solidago sempervirens.*

Exomalopsis (Stilbomalopsis) birkmanni Cockerell, 1922 (Birkmann's exomalopsis). **Sites: Hidalgo Co.** Hub of Prosperity (*2021*) May 10 (♀). MNC (2021) May 22 (♀). NBC (*2021*) Nov. 5 (♀♂), Nov. 24 (♀); (*2019*) Oct. 18 (♀), Oct. 31(♂); (*2018*) Sept. 12 (♀), Sept. 20 (♀♂). Quinta Mazatlán (*2021*) May 13 (♀). **Starr Co.** Campos Viejos (*2021*) June 21 (♀). Falcon SP (*2021*) May 29 (♀), June 3 (♀♂), June 12 (♀). **Flora visited:** *Condalia hookeri, Leucophyllum frutescens, Melampodium cinereum, Oenothera* sp., *Salvia ballotiflora, Sideroxylon celastrinum, Solanum elaeagnifolium, Solanum rostratum, Symphyotrichum subulatum, Tridax procumbens, Verbesina microptera.*

Exomalopsis (Stilbomalopsis) solani Cockerell, 1896 (nightshade exomalopsis). **Sites: Hidalgo Co.** NBC (*2021*) Nov. 5 (♀); (*2019*) Oct. 17 (♀), Oct. 31 (♂). **Flora visited:** *Chromolaena odorata, Salvia ballotiflora, Solanum elaeagnifolium.*

Florilegus condignus (Cresson, 1898) (worthy longhorn bee). **Sites: Cameron Co.** Hugh Ramsey (*2021*) June 6 (♀♂), June 11. (♀♂). Sabal Palm Sanctuary (*2021*) June 11 (♂). **Hidalgo Co.** El Mesteño (*2021*) June 9 (♀). Estero Llano Grande SP (*2021*) June 15 (♂). Hub of Prosperity (*2021*) May 10 (♀♂). MNC (*2021*) May 22 (♀♂), June 10 (♀♂). NBC (*2019*) April 2 (♂), April 4 (♂), April 8 (♂), July 2 (♂). Oleander Acres (*2021*) May 31 (♀). Quinta Mazatlán (*2021*) May 5 (♀♂), May 13 (♀♂). **Starr Co.** Campos Viejos (*2021*) June 17 (♂), June 21 (♂). Falcon SP (*2021*) May 26 (♀♂), May 29 (♀♂). Salineño (*2021*) May 26 (♀♂). **Flora visited:** *Aloysia gratissima, Aloysia macrostachya, Bidens alba, Cordia boissieri, Duranta erecta, Eysenhardtia texana, Helianthus annuus, Lantana achyranthifolia, Melochia tomentosa, Ratibida columnifera, Ruellia nudiflora* var. *runyonii, Salvia ballotiflora, Salvia greggii, Solanum elaeagnifolium, Verbesina encelioides, Verbesina microptera, Wedelia acapulcensis* var. *hispida.*

Melissodes (Eumelissodes) tristis Cockerell, 1894 (mournful longhorn bee). **Sites: Hidalgo Co.** NBC (*2019*) April 24 (♀). Quinta Mazatlán (*2021*) May 4 (♀). **Flora visited:** *Prosopis glandulosa, Salvia ballotiflora.*

Melissodes (Melissodes) tepaneca Cresson, 1898 (Tepanec longhorn bee). **Sites: Cameron Co.** Hugh Ramsey (*2021*) June 6 (♀♂), June 9 (♀♂). Resaca de la Palma SP (*2021*) June 6 (♀). **Hidalgo Co.** El Mesteño (*2021*) May 20 (♀), May 24 (♀). Estero Llano Grande SP (*2021*) June 15 (♀). MNC (*2021*) June 10 (♀♂). NBC (*2021*) May 31 (♀♂), Nov. 5 (♀♂); (*2019*) April 9 (♀); (*2018*) Sept. 11 (♂). Oleander Acres (*2021*) May 31 (♂). Quinta Mazatlán (*2021*) May 4 (♂), May 28 (♂). **Starr Co.** Campos Viejos (*2021*) June 17 (♀). Falcon SP (*2021*) May 26 (♀), June 12 (♀). La Puerta (*2021*) May 6 (♀), May 7 (♀). Salineño (*2021*) May 26 (♀). **Flora visited:** *Chromolaena odorata, Citharexylum berlandieri, Duranta erecta, Helianthus annuus, Herissantia crispa, Melampodium cinereum, Melochia tomentosa, Opuntia engelmannii* var. *lindheimeri, Parkinsonia aculeata, Parkinsonia texana, Ratibida columnifera, Salvia ballotiflora, Tecoma stans, Verbesina encelioides, Viguiera stenoloba, Wedelia acapulcensis* var. *hispida, Zornia bracteata.*

Melissodes (Tachymelissodes) opuntiellus Cockerell, 1911 (little prickly pear longhorn bee). **Sites: Hidalgo Co.** El Mesteño (*2021*) May 18 (♀♂). Mission Hike and Bike (*2019*) April 19; MNC (*2019*) April 17; NBC (*2019*) March 25 (♂), April 18 (♀). **Starr Co.** Campos Viejos (*2021*) June 17 (♂), June 21 (♂). La Puerta (*2021*) May 9 (♀). **Flora visited:** *Abutilon abutiloides, Argemone sanguinea, Florestina tripteris, Opuntia engelmannii* var. *lindheimeri, Parkinsonia texana.*

Melissoptila pinguis (Cresson, 1878) (portly longhorn bee). **Sites: Hidalgo Co.** NBC (*2021*) Nov. 10 (♀♂), Nov. 24 (♀♂); (*2019*) Nov. 14 (♀♂); (*2018*) Nov. 8 (♂), Nov. 26 (♀). **Flora visited:** *Abutilon* sp., *Malvastrum amerianum, Verbesina encelioides.*

Melitoma marginella (Cresson, 1872) (marginate chimney bee). **Sites: Hidalgo Co.** NBC (*2021*) Nov. 10 (♀), Nov. 24 (♀); (*2019*) April 24 (♀); (*2018*) Nov. 8 (♀♂). **Flora visited:** *Merremia dissecta, Tecoma stans, Turnera ulmifolia.*

Melitoma taurea Say, 1837 (bull chimney bee). **Sites: Starr Co.** Falcon SP (*2021*) June 12 (♀), June 25 (♂). **Flora visited:** *Tecoma stans.*

Nomada texana Cresson, 1872 (Texas nomad bee). **Sites: Cameron Co.** Laguna Atascosa NWR (*2021*) Nov. 7 (♀). **Hidalgo Co.** NBC (*2021*) Nov. 10 (♀♂); (*2019*) Oct. 26 (♀), Nov. 1 (♀). **Flora visited:** *Bidens alba, Gaillardia pulchella, Ratibida columnifera, Solidago sempervirens, Symphyotrichum subulatum, Trixis inula.*

Nomada vierecki Cockerell, 1903 (Viereck's nomad bee). **Sites: Starr Co.** Falcon SP (*2021*) May 29 (♀♂). **Flora visited:** *Aphanostephus skirrhobasis, Leucophyllum frutescens, Melampodium cinereum.*

Svastra (Epimelissodes) atripes (Cresson, 1872) (black-legged longhorn bee). **Sites: Hidalgo Co.** El Mesteño (*2021*) June 16 (dark morph ♀). **Starr Co.** Campos Viejos (*2021*) June 21 (♂). **Flora visited:** *Proboscidea louisianica, Salvia ballotiflora.*

Svastra (Epimelissodes) petulca (Cresson, 1878) (frisky longhorn bee). **Sites: Cameron Co.** Resaca de la Palma SP (*2021*) Nov. 12 (♂). **Hidalgo Co.** El Mesteño (*2021*) June 2 (♀). MNC (*2021*) June 1 (♀). NBC (*2021*) Nov. 5 (♂); (*2019*) April 19 (♀♂). **Starr Co.** Campos Viejos (*2021*) June 17 (♀♂). Falcon SP (*2021*) June 3 (♀), June 12 (♀♂), June 25 (♀♂). La Puerta (*2021*) May 16 (♂). Salineño (*2021*) May 26 (♂). **Flora visited:** *Clematis drummondii, Eysenhardtia texana, Helianthus annuus, Malvastrum americanum, Ratibida columnifera, Senegalia berlandieri, Trixis inula, Verbesina encelioides, Viguiera stenoloba, Wedelia acapulcensis* var. *hispida.*

Svastra (Epimelissodes) sabinensis (Cockerell, 1924) (Sabine longhorn bee). **Sites: Hidalgo Co.** NBC (*2019*) April 19 (♀♂). **Starr Co.** Campos Viejos (*2021*) June 17 (♂). **Flora visited:** *Verbesina encelioides, Viguiera stenoloba.*

Svastra (Idiomelissodes) duplocincta (Cockerell, 1905) (barrel cactus longhorn bee). **Sites: Starr Co.** Campos Viejos (*2021*) June 17 (♀), June 21 (♀). **Flora visited:** *Ferocactus wislizeni.*

Triepeolus rufoclypeus (Fox, 1891) (red-faced triepeolus). **Sites: Cameron Co.** Hugh Ramsey (*2021*) June 6 (♀♂), Nov. 11 (♀). **Hidalgo Co.** El Mesteño (*2021*) May 21 (♀). NBC (*2021*) May 7 (♀); (*2019*) April 13 (♂), Oct. 28 (♀). Quinta Mazatlán (*2021*) May 13 (♂). **Starr Co.** Campos Viejos (*2021*) June 17 (♀). Falcon SP (*2021*) May 29 (♀). **Flora visited:** *Gaillardia pulchella, Helianthus annuus, Solidago sempervirens, Verbesina encelioides, Viguiera stenoloba, Wedelia acapulcensis* var. *hispida.*

Triepeolus vernus Rightmyer, 2008 (spring triepeolus). **Sites: Hidalgo Co.** NBC (*2019*) April 13 (♀). **Flora visited:** *Opuntia engelmannii* var. *lindheimeri, Thelocactus setispinus.*

Triepeolus zacatecus (Cresson, 1878) (Zacatec triepeolus). **Sites: Starr Co.** La Puerta (*2021*) May 16 (♂). **Flora visited:** *Senegalia berlandieri.*

Xylocopa (Neoxylocopa) griswoldi Mérida, Hinojosa-Díaz & Ayala, 2022 (Griswold's carpenter bee). **Sites: Cameron Co.** Hugh Ramsey (*2021*) Nov. 11 (♀). Resaca de la Palma SP (*2021*) Nov. 11 (♀), Nov. 23 (♀). South Padre Island BNC (*2021*) June 5 (♀). **Hidalgo Co.** Bentsen–Rio Grande Valley SP (*2021*) April 9 (♀). Estero Llano Grande SP (*2021*) June 11 (♀). NBC (*2021*) June 27 (♀♂), Nov. 5 (♀); (*2019*) March 27 (♀); (*2018*) Sept. 6 (♀), Nov. 15 (♀). Quinta Mazatlán (*2021*) May 5 (♀). **Starr Co.** La Puerta (*2021*) May 16 (♀). **Flora visited:** *Anisacanthus quadrifidus, Antigonon leptopus, Ehretia anacua, Malvaviscus arboreus* var. *drummondii, Parkinsonia aculeata, Parkinsonia texana, Passiflora incarnata, Prosopis glandulosa, Solanum elaeagnifolium, Tecoma stans, Verbesina encelioides.*

Xylocopa (Notoxylocopa) tabaniformis parkinsoniae Cockerell, 1917 (parkinsonia carpenter bee). **Sites: Cameron Co.** Resaca de la Palma SP (*2021*) June 11 (♀). South Padre Island BNC (*2021*) June 5 (♀). **Hidalgo Co.** Bentsen–Rio Grande Valley SP (*2020*) Feb. 27 (♀). NBC (*2021*) Nov. 5 (♀), Nov. 10; (*2019*) June 30 (♀♂), Oct. 27 (♀). **Flora visited:** *Anisacanthus quadrifidus, Hamelia patens, Monarda citriodora, Parkinsonia aculeata, Passiflora incarnata, Tecoma stans.*

Xylocopa (Schonnherria) micans Lepeletier, 1841 (southern carpenter bee). **Sites: Cameron Co.** Hugh Ramsey (*2021*) June 6 (♀). Resaca de la Palma SP (*2021*) June 6 (♀). Sabal Palm Sanctuary (*2021*) June 11 (♀). South Padre Island BNC (*2021*) June 5 (♀). **Hidalgo Co.** El Mesteño (*2021*) May 24 (♀). Estero Llano Grande SP (*2021*) May 22 (♀), June 15 (♀). MNC (*2021*) June 17 (♀). NBC (*2021*) May 31 (♀), Nov. 10 (♀). Oleander Acres (*2021*) May 31 (♀). Quinta Mazatlán (*2021*) May 13 (♀). **Starr Co.** Campos Viejos (*2021*) June 17 (♀). Salineño (*2021*) May 26 (♂). **Flora visited:** *Anisacanthus quadrifidus, Cordia boissieri, Justicia brandegeeana, Karwinskia humboldtiana, Malvaviscus arboreus* var. *drummondii, Melochia tomentosa, Passiflora incarnata, Salvia greggii.*

Xylocopa (Stenoxylocopa) strandi Dusmet y Alonso, 1924 (Strand's carpenter bee). **Sites: Hidalgo Co.** NBC (*2018*) Sept. 12 (♀). **Flora visited:** *Hamelia patens, Malvaviscus arboreus* var. *drummondii.*

FAMILY COLLETIDAE

Colletes birkmanni Swenk, 1906 (Birkmann's cellophane bee). **Sites: Hidalgo Co.** NBC (*2019*) Oct. 29 (♂), Nov. 4 (♀); (*2018*) Nov. 16 (♀). **Flora visited:** *Croton capitatus, Solidago sempervirens.*

Colletes intermixtus Swenk, 1905 (intermixed cellophane bee). **Sites: Hidalgo Co.** NBC (*2021*) May 4 (♀). **Flora visited:** *Solanum elaeagnifolium.*

Colletes mandibularis Smith, 1853 (pale gold cellophane bee). **Sites: Cameron Co.** Bahia Grande (*2021*) Nov. 9 (♀). **Hidalgo Co.** El Mesteño (*2021*) June 2 (♀), Nov. 20 (♀), Nov. 21 (♀). **Starr Co.** Campos Viejos (*2021*) June 21 (♀). **Flora visited:** *Amphiachyris dracunculoides, Aphanostephus skirrhobasis, Verbesina encelioides.*

Colletes swenki Stephen, 1954 (Swenk's cellophane bee). **Sites: Hidalgo Co.** NBC (*2019*) April 11 (♀). **Flora visited:** *Quincula lobata.*

Colletes texanus Cresson, 1872 (Texas cellophane bee). **Sites: Hidalgo Co.** NBC (*2020*) March 5 (♂). **Flora visited:** *Physalis pubescens.*

***Colletes* sp.** (undescribed species). **Sites: Hidalgo Co.** Quinta Mazatlán (*2021*) May 4 (♀). **Flora visited:** *Salvia ballotiflora.*

Hylaeus (Hylaeana) panamensis Michener, 1954 (Panamanian masked bee). **Sites: Hidalgo Co.** MNC (*2019*) Nov. 4 (♀). NBC (*2019*) Nov. 5 (♀). **Flora visited:** *Palafoxia texana.*

Hylaeus (Prosopis) affinis (Smith, 1853) (eastern masked bee). **Sites: Cameron Co.** Hugh Ramsey (*2021*) Nov. 11 (♀). **Hidalgo Co.** MNC (*2019*) Nov. 4 (♀). NBC (*2019*) April 30 (♀). **Flora visited:** *Palafoxia texana.*

Hylaeus (Prosopis) modestus Say, 1837 (modest masked bee). **Sites: Cameron Co.** Hugh Ramsey (*2021*) Nov. 11 (♀♂). Laguna Atascosa NWR (*2021*) Nov. 7 (♂). **Hidalgo Co.** NBC (*2020*) March 3 (♂); (*2019*) April 30 (♀), Dec. 22 (♀). **Flora visited:** *Croton humilis, Florestina tripteris, Teucrium cubense.*

Ptiloglossa mexicana (Cresson, 1878) (Mexican feather-tongued bee). **Sites: Hidalgo Co.** Garden of William Rich (*2021*) May 18 (♀), July 26 (♂). NBC (*2021*) May 31 (♀). **Flora visited:** *Asclepias texana, Passiflora foetida* var. *gossypiifolia, Passiflora incarnata, Solanum elaeagnifolium.*

FAMILY HALICTIDAE

Agapostemon (Agapostemon) melliventris Cresson, 1874 (honey-tailed agapostemon). **Sites: Cameron Co.** Bahia Grande (*2021*) Nov. 16 (♀♂). Las Palomas (*2021*) Nov. 12 (♀♂). **Hidalgo Co.** El Mesteño (*2021*) June 2 (♀♂). MNC (*2021*) Nov. 5 (♂). NBC (*2021*) Nov. 5 (♂); (*2018*) Sept. 12 (♀♂). Quinta Mazatlán (*2021*) Nov. 6. **Starr Co.** Campos Viejos (*2021*) June 17 (♀♂). Falcon SP (*2021*) May 26 (♂), June 12 (♀). **Flora visited:** *Aloysia gratissima, Bidens alba, Chromolaena odorata, Citharexylum berlandieri, Cucurbita foetidissima, Duranta erecta, Guaiacum angustifolium, Helianthus annuus, Leucophyllum frutescens, Ruellia nudiflora* var. *runyonii, Solidago sempervirens, Verbesina encelioides.*

Agapostemon (Agapostemon) texanus Cresson, 1872 (Texas agapostemon). **Sites: Hidalgo Co.** NBC (*2021*) Nov. 18; (*2018*) Nov. 14 (♀♂). Quinta Mazatlán (*2021*) May 3 (♀♂), June 2 (♂), June 27 (♀♂); (*2018*) Sept. 9 (♂). **Flora visited:** *Chromolaena odorata, Helianthus annuus, Ratibida columnifera, Salvia ballotiflora, Verbesina encelioides.*

Agapostemon (Agapostemon) tyleri Cockerell, 1917 (Tyler's agapostemon). **Sites: Starr Co.** Campos Viejos (*2021*) June 17 (♀♂), June 21 (♀♂). **Flora visited:** *Cordia boissieri*, *Ferocactus wislizeni*, *Leucophyllum frutescens*.

Augochlora (Augochlora) azteca (Vachal, 1911) (Aztec sweat bee). **Sites: Cameron Co.** Bahia Grande (*2021*) Nov. 16 (♀). Laguna Atascosa NWR (*2021*) Nov. 7 (♀). Resaca de la Palma SP (*2021*) Nov. 16 (♀♂), Nov. 23 (♀♂). South Padre Island BNC (*2021*) Nov. 16 (♀). **Hidalgo Co.** El Mesteño (*2021*) Nov. 20 (♀). NBC (*2021*) Nov. 10 (♀); (*2019*) April 17 (♀); (*2018*) Sept. 19 (♀). **Starr Co.** Campos Viejos (*2021*) June 17 (♂), June 21 (♀♂). **Flora visited:** *Abutilon hypoleucum*, *Anisacanthus quadrifidus*, *Chamaecrista fasciculata*, *Leucophyllum frutescens*, *Melochia tomentosa*, *Merremia dissecta*, *Opuntia engelmannii* var. *lindheimeri*, *Tecoma stans*.

Augochlora (Oxystoglossella) aurifera Cockerell, 1897 (golden-green sweat bee). **Sites: Cameron Co.** Laguna Atascosa NWR (*2021*) Nov. 7 (♀♂), Nov. 16 (♀♂). Resaca de la Palma SP (*2021*) June 6 (♀), Nov. 12 (♀♂). Sabal Palm Sanctuary (*2021*) June 11 (♀). **Hidalgo Co.** El Mesteño (*2021*) June 9 (♀). Hub of Prosperity (*2021*) May 10 (♀). MNC (*2021*) May 22 (♂). NBC (*2021*) Nov. 5 (♀), Nov. 24 (♀♂); (*2018*) Nov. 24 (♀♂). Oleander Acres (*2021*) May 31 (♀). Quinta Mazatlán (*2021*) May 4 (♀♂), May 5 (♀♂), May 13 (♀♂), Nov. 6 (♀). Santa Ana NWR (*2021*) April 9 (♀♂). **Starr Co.** Campos Viejos (*2021*) June 21 (♀). **Flora visited:** *Abutilon trisulcatum*, *Bidens alba*, *Chromolaena odorata*, *Cirsium texanum*, *Leucophyllum frutescens*, *Malvastrum americanum*, *Nama undulatum*, *Oenothera* sp., *Solanum elaeagnifolium*, *Solidago sempervirens*, *Sphaeralcea angustifolia*, *Teucrium cubense*, *Tridax procumbens*, *Verbesina encelioides*, *Viguiera stenoloba*, *Wedelia acapulcensis* var. *hispida*.

Augochlorella bracteata Ordway, 1966 (gilded augochlorella). **Sites: Cameron Co.** Bahia Grande (*2021*) Nov. 16 (♀♂). Resaca de la Palma SP (*2021*) Nov. 12 (♂), Nov. 23 (♀♂). Sabal Palm Sanctuary (*2021*) June 11 (♀). **Hidalgo Co.** El Mesteño (*2021*) June 16 (♀). Estero Llano Grande SP (*2021*) June 15 (♀). NBC (*2019*) Oct. 27 (♀). **Flora visited:** *Agastache breviflora*, *Amphiachyris dracunculoides*, *Antigonon leptopus*, *Heliotropium angiospermum*, *Sida ciliaris*, *Viguiera stenoloba*, *Wedelia acapulcensis* var. *hispida*.

Augochloropsis (Paraugochloropsis) metallica (Fabricius, 1793) (metallic-green sweat bee). **Sites: Cameron Co.** Laguna Atascosa NWR (*2021*) Nov. 7 (♀). **Hidalgo Co.** El Mesteño (*2021*) June 16 (♀), Nov. 20 (♀). NBC (*2018*) Sept. 18 (♀), Nov. 23 (♀♂), Dec. 12 (♀). Oleander Acres (*2021*) May 31 (♀). **Starr Co.** Campos Viejos (*2021*) June 21 (♀). **Flora visited:** *Boerhavia erecta*, *Salvia ballotiflora*, *Sida ciliaris*, *Solanum elaeagnifolium*, *Solanum rostratum*, *Solidago sempervirens*, *Symphyotrichum subulatum*, *Vitex agnus-castus* var. *alba* (nonnative).

Dieunomia (Epinomia) nevadensis bakeri (Cockerell, 1898) (Baker's dieunomia). **Sites: Hidalgo Co.** NBC (*2019*) Oct. 20 (♂). **Starr Co.** Falcon SP (*2021*) June 12 (♀). **Flora visited:** *Melampodium cinereum*, *Solidago sempervirens*.

Halictus (Odontolictus) ligatus Say, 1837 (ligated furrow bee). **Sites: Cameron Co.** Hugh Ramsey (*2021*) June 6 (♀), June 11 (♀). Resaca de la Palma SP (*2021*) Nov. 12 (♂). South Padre Island BNC (*2021*) June 5 (♀). **Hidalgo Co.** El Mesteño (*2021*) June 2 (♀). Hub of Prosperity (*2021*) May 10 (♀♂). MNC (*2021*) April 28 (♀♂), May 22 (♀♂). NBC (*2021*) May 7 (♀♂), May 31 (♀♂), Nov. 6 (♀♂), Nov. 24 (♀♂); (*2019*) Dec. 20 (♂); (*2018*) Nov. 25 (♀♂). Oleander Acres (*2021*) May 31 (♀♂). Quinta Mazatlán (*2021*) May 4 (♀♂), Nov. 6 (♀♂). Santa Ana NWR (*2021*) April 9 (♀). **Starr Co.** Campos Viejos (*2021*) June 17 (♀), June 21 (♀). Falcon SP (*2021*) May 26 (♀♂), May 29 (♀), June 3 (♀♂), June 12 (♀♂). Salineño (*2021*) May 26 (♀). **Flora visited:** *Aphanostephus ramosissimus, Bidens alba, Cirsium texanum, Gaillardia pulchella, Helianthus annuus, Larrea tridentata, Melampodium cinereum, Prosopis glandulosa, Ratibida columnifera, Symphyotrichum subulatum, Tridax procumbens, Verbesina encelioides, Viguiera stenoloba, Wedelia acapulcensis* var. *hispida.*

Lasioglossum (Dialictus) coactum (Cresson, 1872) (constricted metallic furrow bee). **Sites: Cameron Co.** Bahia Grande (*2021*) Nov. 16 (♀). Laguna Atascosa NWR (*2021*) Nov. 9 (♀). Las Palomas (*2021*) Nov. 12 (♀). **Hidalgo Co.** Quinta Mazatlán (*2021*) Nov. 6 (♀). **Flora visited:** *Amphiachyris dracunculoides, Argemone* sp., *Carlowrightia parviflora, Chromolaena odorata, Euphorbia cyathophora, Rayjacksonia phyllocephala, Salvia ballotiflora, Symphyotrichum subulatum, Verbesina encelioides, Viguiera stenoloba.*

***Lasioglossum (Dialictus)* sp.,** possibly *Lasioglossum hunteri* (Cockerell, 1932). **Sites: Starr Co.** Campos Viejos (*2021*) June 18 (♀). **Flora visited:** *Ferocactus wislizeni.*

Nomia (Acunomia) tetrazonata Cockerell, 1910 (four-banded nomia). **Sites: Hidalgo Co.** El Mesteño (*2021*) May 20 (♀♂), May 24 (♀). NBC (*2019*) March 30 (♀). **Starr Co.** Campos Viejos (*2021*) June 17 (♀♂). La Puerta (*2021*) May 6 (♀), May 9 (♀), May 16 (♀♂). **Flora visited:** *Citharexylum berlandieri, Colubrina texensis, Condalia hookeri, Guaiacum angustifolium, Karwinskia humboldtiana, Leucophyllum frutescens, Sideroxylon celastrinum, Zizyphus obtusifolia.*

Sphecodes heraclei heraclei Robertson, 1897 (Hercules sphecodes). **Sites: Hidalgo Co.** MNC (*2020*) Feb. 28 (♀). NBC (*2018*) Nov. 18 (♀). **Flora visited:** *Palafoxia texana, Ratibida columnifera, Symphyotrichum subulatum, Verbesina encelioides, Viguiera stenoloba.*

FAMILY MEGACHILIDAE

Ashmeadiella (Ashmeadiella) cactorum (Cockerell, 1897) (cactus ashmeadiella). **Sites: Hidalgo Co.** El Mesteño (*2021*) June 24 (♂). **Starr Co.** La Puerta (*2021*) May 9 (♀). **Flora visited:** *Aphanostephus skirrhobasis, Herissantia crispa, Parkinsonia texana.*

Ashmeadiella (Ashmeadiella) maxima Michener, 1936 (greater ashmeadiella). **Sites: Hidalgo Co.** Pixie Preserve (*2021*) May 11 (♂). **Flora visited:** *Ziziphus obtusifolia.*

Coelioxys (Glyptocoelioxys) totonacus Cresson, 1878 (Totonac cuckoo leafcutter). **Sites: Hidalgo Co.** Quinta Mazatlán (*2021*) Nov. 6 (♀). **Flora visited:** *Chromolaena odorata.*

Coelioxys (Leuraspidia) aztecus Cresson, 1878 (Aztec cuckoo leafcutter). **Sites: Cameron Co.** Hugh Ramsey (*2021*) Nov. 11 (♂). **Hidalgo Co.** NBC (*2021*) Nov. 5 (♂); (*2019*) Nov. 3 (♂), Dec. 12 (♂); (*2018*) Nov. 24. (♀). **Flora visited:** *Abutilon hypoleucum*, *Bidens alba*, *Salvia ballotiflora*, *Viguiera stenoloba.*

Coelioxys (Neocoelioxys) slossoni arenicola Crawford, 1914 (sand-dwelling cuckoo leafcutter). **Sites: Cameron Co.** Hugh Ramsey (*2021*) June 6 (♀), June 9 (♀), June 11 (♀♂). **Hidalgo Co.** El Mesteño (*2021*) May 21 (♀). NBC (*2021*) May 31 (♀), June 27 (♀); (*2019*) April 17 (♀♂); (*2018*) Sept. 6 (♀♂). Oleander Acres (*2021*) May 31 (♀). Quinta Mazatlán (*2021*) May 28 (♀). **Starr Co.** Campos Viejos (*2021*) June 21 (♂). Falcon SP (*2021*) May 29 (♀), June 3 (♀), June 12 (♀). La Puerta (*2021*) May 16 (♀). Salineño (*2021*) May 26 (♂). **Flora visited:** *Aloysia gratissima*, *Citharexylum berlandieri*, *Croton humilis*, *Helianthus annuus*, *Lantana achyranthifolia*, *Phyla nodiflora*, *Senegalia berlandieri*, *Viguiera stenoloba*, *Wedelia acapulcensis* var. *hispida.*

Coelioxys (Synocoelioxys) texanus Cresson, 1872 (Texas cuckoo leafcutter). **Sites: Cameron Co.** Resaca de la Palma SP (*2021*) June 6 (♂). **Hidalgo Co.** NBC (*2019*) Oct. 17 (♂), Oct. 21 (♀). **Starr Co.** Campos Viejos (*2021*) June 21 (♂). Falcon SP (*2021*) June 3 (♂). **Flora visited:** *Florestina tripteris*, *Salvia ballotiflora*, *Solidago sempervirens*, *Viguiera stenoloba*, *Wedelia acapulcensis* var. *hispida.*

Coelioxys (Xerocoelioxys) editus Cresson, 1872 (scaly cuckoo leafcutter). **Sites: Hidalgo Co.** NBC (*2018*) Sept. 9 (♀), Sept. 15 (♂). **Flora visited:** *Abutilon hypoleucum*, *Bidens alba*, *Helianthus annuus*, *Ratibida columnifera.*

Dianthidium (Dianthidium) discors Timberlake, 1948 (discordant pebble bee). **Sites: Starr Co.** Campos Viejos (*2021*) June 17, (♀), June 21 (♀♂). La Puerta (*2021*) May 6 (♂). **Flora visited:** *Ferocactus wislizeni*, *Guaiacum angustifolium.*

Heriades (Neotrypetes) variolosa (Cresson, 1872) (variegated resin bee). **Sites: Cameron Co.** Hugh Ramsey (*2021*) June 9 (♀♂), June 10 (♀♂), June 11 (♀♂), Nov. 11 (♀♂). Resaca de la Palma SP (*2021*) Nov. 12 (♀♂), Nov. 23 (♀). **Hidalgo Co.** NBC (*2019*) Oct. 25 (♀). **Starr Co.** Falcon SP (*2021*) May 26 (♀), June 12 (♀♂). **Flora visited:** *Bidens alba*, *Croton humilis*, *Palafoxia texana*, *Symphyotrichum subulatum*, *Teucrium cubense*, *Verbesina microptera*, *Viguiera stenoloba*, *Wedelia acapulcensis* var. *hispida.*

Lithurgopsis littoralis (Cockerell, 1917) (littoral cactus woodborer). **Sites: Cameron Co.** Bahia Grande (*2019*) April 27 (♀). **Hidalgo Co.** El Mesteño (*2021*) May 2 (♀), May 24 (♀). La Sal del Rey (*2021*) April 26 (♀♂). Mission Hike and Bike (*2021*) April 26 (♀); (*2019*) April 14 (♀♂). MNC (*2021*) April 24 (♀♂); (*2019*) March 17 (♀), April 20 (♀). Pixie Preserve (*2019*) May 18 (♀♂). **Starr Co.** Falcon SP (*2021*) May 26 (♀♂). La Puerta (*2021*) May 7 (♂). **Flora visited:** *Argemone aenea*, *Argemone sanguinea*, *Opuntia engelmannii* var. *lindheimeri.*

Megachile (Argyropyle) parallela Smith, 1853 (parallel leafcutter). **Sites: Hidalgo Co.** NBC (*2021*) June 27 (♀); (*2019*) March 24 (♂). **Flora visited:** *Bidens alba, Gaillardia pulchella, Verbesina encelioides.*

Megachile (Chelostomoides) exilis Cresson, 1872 (slender resin bee). **Sites: Cameron Co.** Hugh Ramsey (*2021*) June 6 (♀♂), June 9 (♀♂). **Hidalgo Co.** NBC (*2021*) June 27 (♀♂); (*2019*) April 25 (♀♂); (*2018*) Sept. 15 (♀), Sept. 17 (♂). **Flora visited:** *Dalea scandens, Prosopis glandulosa, Rhynchosia senna* var. *texana, Salvia ballotiflora.*

Megachile (Chelostomoides) prosopidis Cockerell, 1900 (mesquite resin bee). **Sites: Starr Co.** Falcon SP (*2021*) May 29 (♀), June 3 (♂). **Flora visited:** *Parkinsonia aculeata, Prosopis glandulosa, Salvia ballotiflora.*

Megachile (Chelostomoides) texensis Mitchell, 1956 (Texan resin bee). **Sites: Hidalgo Co.** Teniente (*2021*) April 16 (♀). **Flora visited:** Found flying over ground: *Prosopis glandulosa, Senegalia berlandieri,* and *Vachellia tortuosa* in bloom in area.

Megachile (Litomegachile) brevis Say, 1837 (common little leafcutter). **Sites: Cameron Co.** Laguna Atascosa NWR (*2021*) Nov. 7 (♀). South Padre Island BNC (*2021*) Nov. 16 (♀). **Hidalgo Co.** NBC (*2021*) Nov. 5 (♀); (*2019*) April 2 (♀), July 2 (♂). **Flora visited:** *Havardia pallens, Heliotropium angiospermum, Medicago sativa, Senegalia berlandieri, Symphyotrichum subulatum, Vachellia rigidula, Viguiera stenoloba.*

***Megachile (Litomegachile)* sp.,** possibly *Megachile coquilletti* Cockerell, 1915. **Sites: Hidalgo Co.** NBC (*2019*) April 2 (♀). **Flora visited:** *Cirsium texanum.*

Megachile (Megachile) montivaga Cresson, 1878 (silver-tailed petal cutter). **Sites: Hidalgo Co.** NBC (*2021*) April 24 (♀); (*2019*) July 1 (♀). **Flora visited:** *Argemone aenea, Oenothera speciosa.*

Megachile (Neochelynia) chichimeca Cresson, 1878 (Chichimec leafcutter). **Sites: Cameron Co.** Laguna Atascosa NWR (*2021*) Nov. 7 (♀). Resaca de la Palma SP (*2021*) June 6 (♀), June 11 (♀), Nov. 12 (♀♂), Nov. 23 (♀). **Hidalgo Co.** El Mesteño (*2021*) June 2 (♀♂). Hub of Prosperity (*2021*) May 10. NBC (*2021*) May 31 (♀), June 27 (♀), Nov. 5 (♀); (*2019*) July 1 (♀♂); (*2018*) Nov. 8 (♀♂). Oleander Acres (*2021*) May 31 (♀). Quinta Mazatlán (*2021*) Nov. 6 (♀). **Starr Co:** La Puerta (*2021*) May 16 (♀). **Flora visited:** *Aloysia gratissima, Bidens alba, Croton humilis, Ebenopsis ebano, Salvia ballotiflora, Tridax procumbens, Trixis inula, Viguiera stenoloba.*

Megachile (Pseudocentron) sidalceae Cockerell, 1897 (sidalcea leafcutter). **Sites: Hidalgo Co.** Estero Llano Grande SP (*2019*) June 15 (♀). NBC (*2019*) July 2 (♂). **Starr Co.** Campos Viejos (*2021*) June 17 (♂). Falcon SP (*2021*) May 29 (♀♂), June 12 (♀), June 25 (♂) (♀ in nest). **Flora visited:** *Aloysia gratissima, Eysenhardtia texana, Palafoxia texana, Parkinsonia aculeata, Viguiera stenoloba.*

Megachile (Sayapis) frugalis Cresson, 1872 (frugal leafcutter). **Sites: Hidalgo Co.** El Mesteño (*2021*) June 2 (♂), June 9 (♀). **Flora visited:** *Aphanostephus skirrhobasis, Salvia ballotiflora.*

Megachile (Sayapis) inimica inimica Cresson, 1872 (hostile leafcutter). **Sites: Hidalgo Co.** NBC (*2019*) April 4 (♀♂). **Starr Co.** La Puerta (*2021*) May 16 (♀). **Flora visited:** *Senegalia berlandieri, Viguiera stenoloba.*

Megachile (Sayapis) policaris Say, 1831 (policaris leafcutter). Sites: **Cameron Co.** Laguna Atascosa NWR (*2021*) Nov. 7 (♀). **Hidalgo Co.** Hub of Prosperity (*2021*) May 10 (♂). MNC (*2019*) March 28 (♂). NBC (*2021*) June 27 (♀♂); (*2019*) March 28 (♀), April 15 (♀). Pixie Preserve (*2019*) May 18 (♀). **Flora visited:** *Cirsium texanum, Helianthus annuus, Larrea tridentata, Senegalia greggii, Tridax procumbens.*

Megachile (Sayapis) zaptlana Cresson, 1878 (Zaptlan leafcutter). **Sites: Cameron Co.** Hugh Ramsey (*2021*) June 6 (♂), Nov. 11 (♂). **Hidalgo Co.** El Mesteño (*2021*) May 24 (♀). Estero Llano Grande SP (*2021*) June 15 (♂). NBC (*2021*) May 31 (♀♂); (*2018*) Sept. 6 (♂). **Starr Co.** Campos Viejos (*2021*) June 17 (♂). **Flora visited:** *Aloysia gratissima, Croton humilis, Herissantia crispa, Prosopis glandulosa, Ratibida columnifera, Verbesina encelioides.*

Megachile (Tylomegachile) cf. toluca Cresson, 1878 (Toluca leafcutter). **Sites: Hidalgo Co.** NBC (*2018*) Sept. 15 (♂). **Flora visited:** *Rhynchosia senna* var. *texana.*

Osmia (Diceratosmia) subfasciata Cresson, 1872 (faintly banded mason bee). **Sites: Hidalgo Co.** MNC (*2019*) March 29 (♀), April 3 (♀). NBC (*2020*) March 1 (♂). Pixie Preserve (*2021*) May 18 (♀). **Flora visited:** *Coursetia axillaris, Larrea tridentata, Parkinsonia aculeata, Prosopis glandulosa, Vicia ludoviciana.* These bees use plant materials from *Opuntia engelmannii* var. *lindheimeri* to construct nests.

Osmia (Helicosmia) chalybea Smith, 1853 (steel-blue mason bee). **Sites: Hidalgo Co.** NBC (*2021*) April 24 (♀). **Flora visited:** *Cirsium texanum.*

Stelis (Dolichostelis) costalis Cresson, 1872 (ribbed stelis cuckoo bee). **Sites: Cameron Co.** Hugh Ramsey (*2021*) June 6 (♀♂). **Flora visited:** *Wedelia acapulcensis* var. *hispida.*

References

Ackerman AJ. 1916. The carpenter bees of the United States of the genus *Xylocopa*. *Journal of the New York Entomological Society* 24(3): 196–232.

Alcock J. 1975. A day in the life of *Protoxaea gloriosa*. *Insect World Digest* 2(5): 10–13.

———. 1990. Body size and territorial behavior in the bee *Protoxaea gloriosa* (Fox) (Hymenoptera: Oxaeidae). *Pan-Pacific Entomologist* 66(2): 157–161.

———. 1998. Sleeping aggregations of the bee *Idiomelissodes duplocincta* (Cockerell) (Hymenoptera: Anthophorini) and their possible function. *Journal of the Kansas Entomological Society* 7(1): 74–84.

Almeida EAB. 2008. Revision of the Brazilian species of *Pseudaugochlora* Michener 1954 (Hymenoptera: Halictidae: Augochlorini). *Zootaxa* 1679:1–38.

Alves-dos-Santos I. 1999. Aspectos morfológicos e comportamentais dos machos de *Ancyloscelis* Latreille (Anthophoridae, Apoidea). *Revista brasileira de zoologia* 16(2): 37–43.

Armbrust E. 2004. Resource use and nesting behavior of *Megachile prosopidis* and *M. chilopsidis* with notes on *M. discorhina* (Hymenoptera: Megachilidae). *Journal of the Kansas Entomological Society* 77(2): 89–98.

Ascher JS, Pickering J. Discover Life bee species guide and world checklist (Hymenoptera: Apoidea: Anthophila). https://www.discoverlife.org/mp/20q?guide = Apoidea_species

Ashmead WH. 1898. Some new genera of bees. *Psyche* 8:282–285.

———. 1899a. Classification of the bees, or the superfamily Apoidea. *Transactions of the American Entomological Society* 26(1): 49–100.

———. 1899b. A generic table of the family Panurgidae: A reply to Mr. Cockerell's critique on the segregation of *Perdita* Cockerell (1899). *Psyche* 9:372–376.

Ayala R, Engel MS. 2014. A new species of *Ptiloglossa* from Mexico, with new records of *Ptiloglossa cyaniventris* from Panama and Costa Rica (Hymenoptera: Colletidae). *Journal of Melittology* 35:1–13.

Ayala R, Griswold TL, Bullock S. 1993. The native bees of Mexico. In: Ramamoorthy TP, Bye R, Lot A, Fa J (eds.), *Biological Diversity of Mexico: Origins and Distribution*. Oxford: Oxford University Press: 179–227.

Ayala R, Griswold TL, Yanega D. 1996. Apoidea (Hymenoptera). In: Llorente-Bousquets J, García-Aldrete AN, González-Soriano E (eds.), *Biodiversidad, taxonomía y biogeografía de artrópodos de México: Hacia una síntesis de su conocimiento*. Mexico City: UNAM/ CONABIO: 423–464.

Baker JR. 1975. Taxonomy of five Nearctic subgenera of *Coelioxys* (Hymenoptera; Megachilidae). *University of Kansas Science Bulletin* 50(12): 649–730.

Baker JR, Kuhn ED, Bambara SB. 1985. Nests and immature stages of leafcutter bees (Hymenoptera: Megachilidae). *Journal of the Kansas Entomological Society* 58(2): 290–313.

Barrows EM. 1973. Soil nesting by wood-inhabiting halictine bees, *Augochlora pura* and *Lasioglossum coeruleum* (Hymenoptera: Halictidae). *Journal of the Kansas Entomological Society* 46(4): 496–499.

Bartholomew CS, Prowell D, Griswold TL. 2006. An annotated checklist of bees (Hymenoptera: Apoidea) in longleaf pine savannas of southern Louisiana and Mississippi. *Journal of the Kansas Entomological Society* 79(2): 184–198.

Berger LA, Moffet JO, Rummel DR. 1985. Seasonal cycles of *Agapostemon angelicus* Cockerell relative to hybrid cottonseed production in Texas (Hymenoptera: Halictidae). *Journal of the Kansas Entomological Society* 58(1): 1–8.

Blair AW, Williamson PS. 2008. Effectiveness and importance of pollinators to star cactus (*Astrophytum asterias*). *Southwestern Naturalist* 54(4): 423–430.

Bossert S, Murray EA, Almeida EAB, Brady SG, Blaimer BB, Danforth BN. 2019. Combining transcriptomes and ultraconserved elements to illuminate the phylogeny of Apidae. *Molecular Phylogenetics and Evolution* 130:121–131.

Brush J. 2021, August 23. Personal communication regarding presence of *Protoxaea gloriosa* nests at La Puerta Tract of the Lower Rio Grande Valley National Wildlife Refuge.

Buchmann SL. 1987. The ecology of oil flowers and their bees. *Annual Review of Ecology and Systematics* 18:343–369.

———. 2021. *Bombus sonorus*, the Sonoran bumblebee. USDA Forest Service. https://www.fs.usda.gov/wildflowers/pollinators/pollinator-of-the-month/SonoranBumblebee.shtml

Buchmann SL, Cane JH. 1989. Bees assess pollen returns while sonicating *Solanum* flowers. *Oecologia* 81(3): 289–294.

Buchmann SL, Nabhan GP. 1996. *The Forgotten Pollinators*. Washington, DC: Island Press.

BugGuide. Apoidea (clade Anthophila) – Bees. Iowa State University Department of Plant Pathology, Entomology, and Microbiology. https://bugguide.net/node/view/8267/tree

Butler GD, Jr. 1965. The distribution and host plants of leafcutter bees in Arizona. *University of Arizona Agricultural Experiment Station Technical Bulletin* 167:1–19.

Cameron SA, Lozier JD, Strange JP, Koch JB, Cordes N, Solter LF, Griswold TL. 2011. Patterns of widespread decline in North American bumble bees. *Proceedings of the National Academy of Sciences of the United States of America* 108(2): 662–667.

Camillo E, Garofalo CA, Serrano JC. 1988. Observações preliminarias sobre a biologia da *Melitoma segmentaria* (Hymenoptera, Anthophoridae). *Ciência e cultura* 40(7): 914.

Cane JH. 1995. Notes on the nesting biology of *Svastra atripes atrimitra* (LaBerge) (Hymenoptera: Apidae). *Journal of the Kansas Entomological Society* 68(2): 238–240.

Cane JH, Griswold TL, Parker FD. 2007. Substrates and materials used for nesting by North American *Osmia* bees (Hymenoptera: Apiformes: Megachilidae). *Annals of the Entomological Society of America* 100(3): 350–358.

Carril OM, Wilson JS. 2021. *Common Bees of Eastern North America*. Princeton, NJ: Princeton University Press.

Carril OM, Wilson JS. 2023. *Common Bees of Western North America*. Princeton, NJ: Princeton University Press.

Cazier MA, Linsley EG. 1963. Territorial behavior among males of *Protoxaea gloriosa* (Fox) (Hymenoptera: Andrenidae). *Canadian Entomologist* 95(5): 547–556.

———. 1974. Foraging behavior of some bees and wasps at *Kallstroemia grandiflora* flowers in southern Arizona and New Mexico. *American Museum Novitates* 2546:1–20.

Chandler LD. 1961. Light trap collections of the nocturnal bee *Sphecodogastra texana* (Cresson) (Hymenoptera, Halictidae). *Proceedings of the Indiana Academy of Science* 71:124–128.

Cockerell TDA. 1895. New bees of the genus *Halictus* from New Mexico. *Annals and Magazine of Natural History*, series 6, 16:63–69.

———. 1896. The bees of the genus *Perdita* F. Smith. *Proceedings of the Academy of Natural Sciences of Philadelphia* 48:25–107.

———. 1897a. New and little-known bees. *Transactions of the American Entomological Society* 24:144–162.

———. 1897b. On the Mexican bees of the genus *Augochlora*. *Canadian Entomologist* 29:4–70.

———. 1898. On some panurgine and other bees. *Transactions of the American Entomological Society* 25:185–198.

———. 1899. Catálogo de las abejas de México. Mexico City: Secretária de Fomento.

———. 1900a. Descriptions of new bees collected by Mr. H. H. Smith in Brazil: I. *Proceedings of the Academy of Natural Sciences of Philadelphia* 52:356–377.

———. 1900b. The New Mexico bees of the genus *Megachile* and a new *Andrena*. *Annals and Magazine of Natural History*, series 7, 6:7–20.

———. 1901. A peculiar type of halictine bees. *Entomological News* 12:208–209.

———. 1903a. New American Hymenoptera, mostly of the genus *Nomada*. *Annals and Magazine of Natural History* 64(3): 211–212.

———. 1903b. The North American bees of the genus *Nomada*. *Proceedings of the Academy of Natural Sciences of Philadelphia* 55:580–614.

———. 1904a. New records of bees. *The Entomologist* 37:231–236.

———. 1904b. Notes on some bees in the British Museum. *Canadian Entomologist* 36(10): 301–304.

———. 1905a. *Diadasia* Patton: A genus of bees. *American Naturalist* 39(466): 741–745.

———. 1905b. Notes on some bees in the British Museum. *Transactions of the American Entomological Society (1890–)* 31(4): 309–364.

———. 1905c. Tables for the separation of some bees of the genera *Coelioxys* and *Colletes*. *Psyche* 12:85–90.

———. 1906. The North American bees of the family Anthophoridæ. *Transactions of the American Entomological Society (1890–)* 32(1): 63–116.

———. 1907. Descriptions and records of bees—XIV. *Annals and Magazine of Natural History*, series 7, 19(114): 531–540.

———. 1910a. Descriptions and records of bees—XXXII. *Annals and Magazine of Natural History*, series 8, 6:272–284.

———. 1910b. The North American bees of the genus *Nomia*. *Proceedings of the United States National Museum* 38:289–298.

———. 1911. Some new bees from flowers of Cactaceæ. *Canadian Entomologist* 43:131–132.

———. 1912a. Descriptions and records of bees—XLI. *Annals and Magazine of Natural History*, series 8, 9(49): 554–568.

———. 1912b. New records of bees—XLI. *Entomological News* 23:444–447.

———. 1918. Neotropical bees, principally collected by Professor Bruner in Argentina. *Transactions of the American Entomological Society (1890–)* 44(1): 25–38.

———. 1921a. Some parasitic megachilid bees of the western United States. *American Museum Novitates* 21:1–11.

———. 1921b. Western bees obtained by the American Museum expeditions. *American Museum Novitates* 24:1–15.

———. 1923. Some bees from Victoria, Mexico. *Proceedings of the United States National Museum* 63(2476): 1–5.

———. 1931. Descriptions and records of bees—CXXX. *Annals and Magazine of Natural History*, series 10, 8(47): 537–553.

———. 1933. The excessive abundance of certain bees. *American Naturalist* 67(710): 286–288.

———. 1934. Records of western bees. *American Museum Novitates* 697:1–15.

Cockerell TDA, Cockerell WP. 1899. Contributions from the New Mexico Biological Station—VII. Observations on bees, with descriptions of new genera and species. *Annals and Magazine of Natural History* 4(8): 403–421.

Cockerell WP. 1912. Collecting bees at Gualan, Guatemala. *Canadian Entomologist* 44:277–282.

———. 1917. Collecting bees in southern Texas. *Journal of the New York Entomological Society* 25(3): 187–193.

Coelho BWT. 2004. A review of the bee genus *Augochlorella* (Hymenoptera: Halictidae: Augochlorini). *Systematic Entomology* 29(3): 282–323.

Cortopassi-Laurino M, Boff S, Taniguchi M. 2010. Sobre ninhos de *Melitoma segmentaria* (Apidae: Apinae: Emphorini) em forno artisanal de pão. In *Anais do IX encontro sobre abelhas.* Ribeirão Preto, São Paulo.

Crawford JC. 1915. Descriptions of new Hymenoptera, no. 9. *Proceedings of the United States National Museum* 48:577–586.

Cresson ET. 1865. On the Hymenoptera of Cuba. *Proceedings of the Entomological Society of Philadelphia* 4:1–200.

———. 1872. Hymenoptera texana. *Transactions of the American Entomological Society* 4:153–292.

———. 1878a. Descriptions of new North American Hymenoptera in the collection of the American Entomological Society. *Transactions of the American Entomological Society* 7:61–136.

———. 1878b. Descriptions of new species of North American bees. *Proceedings of the Academy of Natural Sciences of Philadelphia* 30:181–222.

Cross EA, Bohart GE. 1960. The biology of *Nomia (Epinomia) triangulifera* with comparative notes on other species of *Nomia. University of Kansas Science Bulletin* 41:761–792.

Daly HV. 1973. Bees of the genus *Ceratina* in America north of Mexico (Hymenoptera: Apoidea). *University of California Publications in Entomology* 74:1–114.

Danforth BN. 1996. Phylogenetic analysis and taxonomic revision of the *Perdita* subgenera *Macrotera, Macroteropsis, Macroterella* and *Cockerellula* (Hymenoptera: Andrenidae). *University of Kansas Science Bulletin* 55(16): 635–692.

Danforth BN, Brady SG, Sipes SD, Pearson A. 2004. Single-copy nuclear genes recover Cretaceous-age divergences in bees. *Systematic Biology* 53(2): 309–326.

Danforth BN, Connal E, Packer L, Walker K, Pauly A, Fano R. 2008. Phylogeny of Halictidae with an emphasis on endemic African Halictinae. *Apidologie* 39:86–101.

Danforth BN, Eickwort GC. 2010. The evolution of social behavior in the augochlorine sweat bees (Hymenoptera: Halictidae) based on a phylogenetic analysis of the genera. In: Choe JC, Crespi BJ (eds.), *The Evolution of Social Behavior in Insects and Arachnids*. Cambridge: Cambridge University Press: 270–292.

Danforth BN, Minckley RL, Neff JL. 2019. *The Solitary Bees: Biology, Evolution, Conservation*. Princeton, NJ: Princeton University Press.

Devall MS, Thien LB. 1989. Factors influencing the reproductive success of *Ipomoea pes-caprae* (Convolvulaceae) around the Gulf of Mexico. *American Journal of Botany* 76(12): 1821–1831.

Dorchin A, Danforth BN, Griswold TL. 2018. A new genus of eucerine bees endemic to southwestern North America revealed in phylogenetic analyses of the *Eucera* complex (Hymenoptera: Apidae: Eucerini). *Arthropod Systematics and Phylogeny* 76(2): 215–234.

Dorchin A, López-Uribe MM, Praz CJ, Griswold TL, Danforth BN. 2018. Phylogeny, new generic-level classification, and historical biogeography of the *Eucera* complex (Hymenoptera: Apidae). *Molecular Phylogenetics and Evolution* 119:81–92.

Downing J, Borrero H, Liu H. 2016. Differential impacts from an extreme cold spell on subtropical vs. tropical specialist bees in southern Florida. *Ecosphere* 7(5): 1–9.

Downing J, Liu H. 2012. Friend or foe? Impacts of the introduced tropical oil bee *Centris nitida* on a threatened and specialized native mutualism in southern Florida. *Biological Invasions* 14(10): 2175–2185.

Droege S, Rightmyer MG, Sheffield CS, Brady SG. 2010. New synonymies in the bee genus *Nomada* from North America (Hymenoptera: Apidae). *Zootaxa* 2661:1–32.

Dunning DN. 1897. Another new species of *Protandrena*, Ckll. *Canadian Entomologist* 29(11): 1–264.

Dusmet y Alonso JM. 1924. Las *Xylocopa* (Hymenoptera: Apidae) en las colecciones de Madrid. *Trabajos del Museo Nacional de Ciencias Naturales (España), serie zoológica* 49:1–58.

Dyer JG, Shinn AF. 1978. Pollen collected by *Calliopsis andreniformis* Smith in North America (Hymenoptera: Andrenidae). *Journal of the Kansas Entomological Society* 51(4): 787–795.

Eickwort GC. 1969a. A comparative morphological study and generic revision of the augochlorine bees (Hymenoptera: Halictidae). *University of Kansas Science Bulletin* 48:325–524.

———. 1969b. Tribal positions of Western Hemisphere green sweat bees, with comments on their nest architecture (Hymenoptera: Halictidae). *Annals of the Entomological Society of America* 62:652–660.

———. 1981. Aspects of the nesting biology of five Nearctic species of *Agapostemon* (Hymenoptera: Halictidae). *Journal of the Kansas Entomological Society* 54(2): 337–351.

Eickwort GC, Eickwort KR. 1972. Aspects of the biology of Costa Rican halictine bees, IV. *Augochlora (Oxystoglossella)* (Hymenoptera: Halictidae). *Journal of the Kansas Entomological Society* 45(1): 18–45.

———. 1973. Notes on the nests of three wood-dwelling species of *Augochlora* from Costa Rica (Hymenoptera: Halictidae). *Journal of the Kansas Entomological Society* 46:17–22.

Eickwort GC, Eickwort KR, Linsley EG. 1977. Observations on nest aggregations of the bees *Diadasia olivacea* and *D. diminuta* (Hymenoptera: Anthophoridae). *Journal of the Kansas Entomological Society* 50(1): 1–17.

Eickwort GC, Matthews RW, Carpenter J. 1981. Observations on the nesting behavior of *Megachile rubi* and *M. texana* with a discussion of the significance of soil nesting in the evolution of megachilid bees (Hymenoptera: Megachilidae). *Journal of the Kansas Entomological Society* 54(3): 557–570.

Eickwort GC, Sakagami SF. 1979. A classification of nest architecture of bees in the tribe Augochlorini (Hymenoptera: Halictidae; Halictinae), with description of a Brazilian nest of *Rhinocorynura inflaticeps*. *Biotropica* 11:28–37.

Engel MS. 2000. Classification of the bee tribe Augochlorini (Hymenoptera: Halictidae). *Bulletin of the American Museum of Natural History* 250:1–89.

———. 2015. A review of the genera and subgenera of Oxaeinae (Hymenoptera: Andrenidae). *Journal of Melittology* 52:1–18.

Esquivel IL, Coulson RN, Brewer MJ. 2020. A native bee, *Melissodes tepaneca* (Hymenoptera: Apidae), benefits cotton production. *Insects* 11(8): 487.

Evans AV. 2007. *National Wildlife Federation Field Guide to Insects and Spiders of North America.* New York: Sterling.

Falk S. 2015. *Field Guide to the Bees of Great Britain and Ireland.* London: Bloomsbury.

Ferreira da Costa CC, Gonçalves RB. 2019. What do we know about neotropical trap-nesting bees? Synopsis about their nest biology and taxonomy. *Papéis avulsos de zoologia* 59:1–16.

Fischer RL. 1951. Observations on the nesting habits of megachilid bees. *Journal of the Kansas Entomological Society* 24(2): 46–50.

———. 1955. A nest of *Heriades variolosus* Cresson (Hymenoptera: Megachilidae). *Canadian Entomologist* 87(1): 33–36.

Fox WJ. 1893. Synopsis of the North American species of *Megacilissa*. *Psyche* 6:421–422.

Frankie GW, Coville RE, Pawele JC, Jadallah CC, Vinson SB, Martinez LES. 2018. Bee-flower-people relationships, field biologists, and conservation in northwest urban Costa Rica and beyond. *Zoosymposia* 12:18–28.

Frankie GW, Thorp RW, Coville RE, Ertter B. 2014. *California Bees and Blooms.* Berkeley, CA: Heyday Books.

Frankie GW, Vinson SB, Newstrom LE, Barthell JF. 1988. Nest site and habitat preferences of *Centris* bees in the Costa Rican dry forest. *Biotropica* 20(4): 301–310.

Freitas FV, Branstetter MG, Franceschini-Santos VH, Dorchin A, Wright KW, López-Uribe MM, Griswold TL, Silveira FA, Almeida EAB. 2023. UCE phylogenomics, biogeography, and classification of long-horned bees (Hymenoptera: Apidae: Eucerinae), with insights on using specimens with extremely degraded DNA. *Insect Systematics and Diversity* 7(4), 3: 1–21.

Freitas FV, Branstetter MG, Griswold TL, Almeida EAB. 2021. Partitioned gene-tree analyses and gene-based topology testing help resolve incongruence in a phylogenomic study of host-specialist bees (Apidae: Eucerinae). *Molecular Biology and Evolution* 38(3): 1090–1100.

Friese H. 1898. Monographie der Bienengattungen *Megacilissa, Caupolicana, Diphaglossa,* und *Oxaea. Annalen des Kaiserlich-*Königlichen *Naturhistorischen Hofmuseums* 13(1): 59–86.

———. 1899. Monographie der Bienengattungen *Megacilissa, Caupolicana, Diphaglossa,* und *Oxaea. Annalen des Kaiserlich-*Königlichen *Naturhistorischen Hofmuseums* 14(3): 239–246.

Furchenbiene. 2021, October 6. *Deutsches Bienen-Journal.* https://www.bienenjournal.de/glossar/furchenbiene/

Gardner J, Gibbs J. 2020. The "red-tailed" *Lasioglossum (Dialictus)* (Hymenoptera: Halictidae) of the western Nearctic. *European Journal of Taxonomy* 725(1): 1–242.

Garrett F. 2014, July 3. Mulher: Feliz, Danuncia Urban dedicou a vida à ciência: Zoóloga fala sobre escolhas e a satisfação que a ciência trouxe à sua vida. Rede Globo. http://redeglobo.globo.com/rpctv/noticia/2014/03/mulher-feliz-danuncia-urban-dedicou-vida-ciencia.html

Geiser SW. 1946. Notes on some workers in Texas entomology, 1839–1880. *Southwestern Historical Quarterly* 49(4): 593–598.

Genaro JA. 2001. Especie nueva de *Melissoptila* para Cuba y la Española (Hymenoptera: Apidae). *Avicennia* 14:61–64.

Gibbs J. 2010. Revision of the metallic species of *Lasioglossum (Dialictus)* in Canada (Hymenoptera, Halictidae, Halictini). *Zootaxa* 2591:1–382.

———. 2011. Revision of the metallic *Lasioglossum (Dialictus)* of eastern North America (Hymenoptera: Halictidae: Halictini). *Zootaxa* 3073:1–216.

———. 2017. Notes on the nests of *Augochloropsis metallica fulgida* and *Megachile mucida* in Central Michigan (Hymenoptera: Halictidae, Megachilidae). *Great Lakes Entomologist* 50:17–24.

Gibbs J, Packer L, Dumesh S, Danforth BN. 2013. Revision and reclassification of *Lasioglossum (Evylaeus), L. (Hemihalictus)* and *L. (Sphecodogastra)* in eastern North America (Hymenoptera: Apoidea: Halictidae). *Zootaxa* 3672:1–117.

Gonçalves RB. 2016. A molecular and morphological phylogeny of the extant Augochlorini (Hymenoptera, Apoidea) with comments on implications for biogeography. *Systematic Entomology* 41(2): 430–440.

———. 2019. Phylogeny of the *Augochlora* clade with the description of four new species (Hymenoptera, Apoidea). *Revista brasileira de zoologia* 63(1): 91–100.

Gonzalez J. 2021, November 16. Personal communication regarding presence of *Bombus pensylvanicus* at South Padre Island Birding and Nature Center.

Gonzalez VH, Moreno E, Richards MH. 2004. Nesting biology of a neotropical bee, *Ceratina mexicana currani* Schwarz (Hymenoptera: Apidae: Xylocopinae). *Journal of the Kansas Entomological Society* 77(1): 58–90.

Goulson D. 2010. *Bumblebees: Behavior, Ecology, and Conservation*. 2nd ed. Oxford: Oxford University Press.

Graenicher S. 1911. Bees of northwestern Wisconsin. *Bulletin of the Public Museum of the City of Milwaukee* 1:221–249.

Griswold TL, Michener CD. 1997. The classification of the Osmiini of the Eastern Hemisphere (Hymenoptera, Megachilidae). *Journal of the Kansas Entomological Society* 70(3): 207–253.

Griswold TL, Rightmyer MG. 2017. A revision of the subgenus *Osmia (Diceratosmia)*, with descriptions of four new species (Hymenoptera, Megachilidae). *Zootaxa* 4337(1): 1–37.

Hicks CH. 1935. Nesting habits of certain western bees. *Canadian Entomologist* 68(3): 47–52.

Hobbs GA, Lilly CE. 1954. Ecology of the species of *Megachile* Latreille in the mixed prairie region of southern Alberta with special reference to pollination of alfalfa. *Ecology* 35(4): 453–462.

Holm H. 2014. *Pollinators of Native Plants: Attract, Observe and Identify Pollinators and Beneficial Insects with Native Plants*. Minnetonka, MN: Pollination Press.

Holmberg EL. 1903. Delectus hymenopterologicus argentinus. *Anales del Museo Nacional de Historia Natural de Buenos Aires* 2(3): 377–468.

———. 1909. Sobre abejas de Mendoza. *Anales del Museo Nacional de Historia Natural de Buenos Aires* 1:40–94.

———. 1921. Apidae argentinae, generis *Psaenythia* Gerstaecker. *Anales del Museo Nacional de Historia Natural de Buenos Aires* 31:249–354.

Hurd PD, Jr. 1978. Bamboo-nesting carpenter bees (genus *Xylocopa* Latreille) of the subgenus *Stenoxylocopa* Hurd and Moure (Hymenoptera: Anthophoridae). *Journal of the Kansas Entomological Society* 51(4): 746–764.

Hurd PD, Jr., LaBerge WE, Linsley EG. 1980. Principal sunflower bees of North America with emphasis on the southwestern United States (Hymenoptera: Apoidea). *Smithsonian Contributions to Zoology* 310:1–158.

Hurd PD, Jr., Linsley EG. 1976. The bee family Oxaeidae with a revision of the North American species (Hymenoptera: Apoidea). *Smithsonian Contributions to Zoology* 220:1–75.

Jahrsdoerfer SE, Leslie DM. 1988. Tamaulipan brushland of the Lower Rio Grande Valley of South Texas: Description, human impacts and management options. *US Fish and Wildlife Service Biological Report* 88(36): 1–63.

Janjic J, Packer L. 2003. Phylogeny of the bee genus *Agapostemon* (Hymenoptera: Halictidae). *Systematic Entomology* 28:101–123.

Kearney TH. 1923. Self-fertilization and cross fertilization in pima cotton. *Bulletin of the US Department of Agriculture* 1134:1–67.

Kerfoot WB. 1964. Observations on the nests of *Nomia nevadensis bakeri* with comparative notes on *Nomia nevadensis arizonensis* (Hymenoptera: Halictidae). *Journal of the Kansas Entomological Society* 37(2): 152–157.

———. 1967. Nest architecture and associated behavior of the nocturnal bee, *Sphecodogastra texana* (Hymenoptera: Halictidae). *Journal of the Kansas Entomological Society* 40(1): 84–93.

———. 1993. Communal nesting in North American pearly-banded bee, *Nomia tetrazonata*, with notes on nesting behavior of *Dieunomia heteropoda* (Hymenoptera: Halictidae: Nomiinae). *Annals of the Entomological Society of America* 86(6): 813–821.

Koch J, Strange J, Williams P. 2012. *Bumblebees of the Western United States*. Washington, DC: USDA Forest Service / Pollinator Partnership.

Krombein KV. 1967. *Trap-Nesting Wasps and Bees: Life Histories, Nests and Associates*. Washington, DC: Smithsonian Institution Press.

———. 1970. Another note on the nesting habits of *Megachile texana* Cresson. *Proceedings of the Entomological Society of Washington* 72:415.

LaBerge WE. 1956a. A revision of the bees in the genus *Melissodes* in North and Central America. Part I (Hymenoptera, Apidae). *University of Kansas Science Bulletin* 37(2): 911–1194.

———. 1956b. A revision of the bees of the genus *Melissodes* in North and Central America. Part II (Hymenoptera, Apidae). *University of Kansas Science Bulletin* 38(1): 533–578.

———. 1957. The genera of bees of the tribe Eucerini in North and Central America (Hymenoptera, Apoidea). *American Museum Novitates* 1837:1–44.

———. 1958. Notes on the North and Central American bees of the genus *Svastra* Holmberg (Hymenoptera: Apidae). *Journal of the Kansas Entomological Society* 31(4): 266–273.

———. 1961. A revision of the bees of the genus *Melissodes* in North and Central America. Part III (Hymenoptera, Apidae). *University of Kansas Science Bulletin* 42:283–663.

———. 1963. Records of the genus *Svastra* in North America with the description of a new species (Hymenoptera: Apidae). *Journal of the Kansas Entomological Society* 36(1): 52–56.
———. 1964. Prodromus of American bees of the genus *Andrena* (Hymenoptera, Apoidea). *Bulletin of the University of Nebraska State Museum* 4:279–316.
———. 1970. A new genus with three new species of eucerine bees from Mexico (Hymenoptera: Anthophoridae). *Journal of the Kansas Entomological Society* 43:321–328.
———. 1989. A review of the bees of the genus *Pectinapis* (Hymenoptera: Anthophoridae). *Journal of the Kansas Entomological Society* 62:524–527.
———. 1994. Status of some species names of eucerine bees (Hymenoptera: Apoidea) proposed by Lepeletier in 1841. *Entomological News* 105:280–282.
———. 2001. Revision of the bees of the genus *Tetraloniella* in the New World (Hymenoptera: Apidae). *Illinois Natural History Survey Bulletin* 36:67–162.
LaBerge WE, Ribble DW. 1966a. Biology of *Florilegus condignus* (Hymenoptera: Anthophoridae), with a description of its larva, and remarks on its importance in alfalfa pollination. *Annals of the Entomological Society of America* 59:944–950.
———. 1966b. The nests and larvae of two species of *Agapostemon* (Hymenoptera: Halictidae). *Journal of the Kansas Entomological Society* 39:467–472.
Landry CL, Elliott NB, Vitale MR. 2014. Nesting ecology of *Megachile (Pseudocentron) alleni* Mitchell (Hymenoptera: Megachilidae) on San Salvador Island, the Bahamas. *Journal of the Kansas Entomological Society* 87(1): 37–46.
Lebuhn G. 2013. *A Field Guide to the Common Bees of California, Including Bees of the Western United States*. Berkeley: University of California Press.
Lee-Mader E., Xerxes Society. 2011. *Attracting Native Pollinators: Protecting North America's Bees and Butterflies*. North Adams, MA: Storey Publishing.
Linsley EG. 1960. Observations on some matinal bees at flowers of *Cucurbita*, *Ipomoea* and *Datura* in desert areas of New Mexico and southeastern Arizona. *Journal of the New York Entomological Society* 68(1): 13–20.
———. 1962. The colletid *Ptiloglossa arizonensis* Timberlake, a matinal pollinator of *Solanum* (Hymenoptera). *Pan-Pacific Entomologist* 38:75–82.
———. 1978. Temporal patterns of flower visitation by solitary bees, with particular reference to the southwestern United States. *Journal of the Kansas Entomological Society* 51(4): 531–546.
Linsley, EG, Cazier MA. 1963. Further observations on bees which take pollen from plants of the genus *Solanum*. *Pan-Pacific Entomologist* 39:1–18.
———. 1970. Some competitive relationships among matinal and late afternoon foraging activities of cupolicanine bees in southeastern Arizona (Hymenoptera, Colletidae). *Journal of the Kansas Entomological Society* 43(3): 251–261.
Linsley EG, MacSwain JW. 1958. The significance of floral constancy among bees of the genus *Diadasia* (Hymenoptera, Anthophoridae). *Evolution* 12(2): 219–223.
Linsley EG, MacSwain JW, Michener CD. 1980. *Nesting Biology and Associates of* Melitoma *(Hymenoptera: Anthophoridae)*. Berkeley: University of California Press.
Linsley EG, Michener CD. 1962. Brief notes on the habits of *Protoxaea* (Hymenoptera: Andrenidae). *Journal of the Kansas Entomological Society* 35(4): 385–389.

Litman JR, Praz CJ, Danforth BN, Griswold TL, Cardinal S. 2013. Origins, evolution and diversification of cleptoparasitic lineages in long-tongued bees. *Evolution* 67(10): 2982–2998.

MacPhail VJ, Richardson LL, Colla SR. 2019. Incorporating citizen science, museum specimens, and field work into the assessment of extinction risk of the American bumble bee (*Bombus pensylvanicus* De Geer 1773) in Canada. *Journal of Insect Conservation* 23:597–611.

Mahlman T, Hipólito J, de Oliveira FF. 2014. Male sleeping aggregation of multiple Eucerini bee genera (Hymenoptera: Apidae) in Chapada Diamantina, Bahia, Brazil. *Biodiversity Data Journal* 2014(2): 1–6.

Mamade Filho GF, Ramos MA, Oliveira JG. 1990. Contribuição à biologia da *Melitoma segmentaria* (Anthophoridae). *Revista brasileira de zoologia* 7(3): 217–221.

Marshall SA. 2006. *Insects: Their Natural History and Diversity; with a Photographic Guide to the Insects of North America*. Richmond Hill, Ontario: Firefly Books.

Martins AC, Santos LM, Dal Molin A. 2015. Danuncia Urban: A life devoted to entomology. In Aguiar AJC, Gonçalves RB, Ramos KS (eds.), *Ensaios sobre as abelhas da região neotropical: Homenagem aos 80 anos de Danuncia Urban*. Curitiba, Brazil: Editora UFPR: 11–47.

Matos MC, Silva SS, Teodoro AV. 2016. Seasonal population abundance of the assembly of solitary wasps and bees (Hymenoptera) according to land-use in Maranhão state, Brazil. *Revista brasileira de entomologia* 60: 171–177.

McGinley RJ. 2003. Studies of Halictinae (Halictidae), II. Revision of *Sphecodogastra* Ashmead, floral specialists of Onagraceae. *Smithsonian Contributions to Zoology* 610:1–54.

McIntosh ME. 2005. Pollination of two species of *Ferocactus*: Interactions between cactus-specialist bees and their host plants. *Functional Ecology* 19(4): 727–734.

Megachile frugalis. 2021, July 25. Minnesota Bee Atlas. University of Minnesota Extension. https://apps.extension.umn.edu/environment/citizen-science/bee-atlas/bee-blocks/SpeciesDetail?speciesId=156

Mérida-Rivas JA, Hinojosa-Díaz IA, Ayala-Barajas R, Barrientos-Villalobos S, Pozo C, Vandame R. 2022. Revision of carpenter bees of the subgenus *Neoxylocopa* Michener (Hymenoptera: Apidae) from Mexico and Mesoamerica. *Zootaxa* 5158(1): 1–67.

Michener CD. 1935. A revision of the genus *Megachile* in the Nearctic region. Part III. Taxonomy of subgenera *Anthemois* and *Delomegachile* (Hymenoptera: Megachilidae). *Transactions of the American Entomological Society (1890–)* 61(3): 155–205.

———. 1939. A revision of the genus *Ashmeadiella* (Hymen., Megachilidae). *American Midland Naturalist* 22(1): 1–84.

———. 1953. The biology of a leaf-cutter bee (*Megachile brevis*) and its associates. *University of Kansas Science Bulletin* 35(3): 1659–1748.

———. 1954. Bees of Panama. *Bulletin of the American Museum of Natural History* 104(1): 1–176.

———. 1964. Evolution of the nests of bees. *American Zoologist* 4(2): 227–239.

———. 1966. Evidence of cooperative provisioning of cells in *Exomalopsis* (Hymenoptera: Anthophoridae). *Journal of the Kansas Entomological Society* 39(2): 315–317.

———. 1973. The Brazilian honeybee. *BioScience* 23(9): 523–527.

———. 1974a. Further notes on nests of *Ancyloscelis* (Hymenoptera: Anthophoridae). *Journal of the Kansas Entomological Society* 47(1): 19–22.

———. 1974b. *The Social Behavior of the Bees*. Cambridge, MA: Harvard University Press.

———. 1975. Nests of *Paranthidium jugatorium* in association with *Melitoma taurea* (Hymenoptera: Megachilidae and Anthophoridae). *Journal of the Kansas Entomological Society* 48(2): 194–200.

———. 2007. *The Bees of the World*. 2nd ed. Baltimore: Johns Hopkins University Press.

Michener CD, Kerfoot WB. 1967. Nests and social behavior of three species of *Pseudaugochloropsis*. *Journal of the Kansas Entomological Society* 40:214–232.

Michener CD, Lange RB. 1958. Observations on the behavior of Brazilian halictid bees, III. *University of Kansas Science Bulletin* 39:473–505.

Michener CD., McGinley RJ, Danforth BN. 1994. *The Bee Genera of North and Central America (Hymenoptera: Apoidea)*. Washington, DC: Smithsonian Institution Press.

Milne LJ, Milne M. 1980. *National Audubon Society Field Guide to Insects and Spiders*. New York: Knopf.

Minckley RL. 1998. A cladistic analysis and classification of the subgenera and genera of the large carpenter bees, tribe Xylocopini (Hymenoptera: Apidae). *Scientific Papers, University of Kansas Natural History Museum* 9:1–47.

Mitchell TB. 1927. New megachild bees. *Psyche* 34(2): 104–121.

———. 1935. A revision of the genus *Megachile* in the Nearctic region. Part II. Morphology of the male sternites and genital armature and the taxonomy of the subgenera *Litomegachile*, *Neomegachile* and *Cressoniella* (Hymenoptera: Megachilidae). *Transactions of the American Entomological Society* 61(1): 1–44.

Mitchell TB. 1937a. A revision of the genus *Megachile* in the Nearctic region. Part VI. Taxonomy of subgenera *Argyropile*, *Leptorachis*, *Pseudocentron*, *Acentron* and *Melanosarus* (Hymenoptera: Megachilidae). *Transactions of the American Entomological Society (1890–)* 63(1): 45–83.

———. 1937b. A revision of the genus *Megachile* in the Nearctic region. Part VII. Taxonomy of the subgenus *Sayapis* (Hymenoptera: Megachilidae). *Transactions of the American Entomological Society (1890–)* 63(2): 175–206.

———. 1938. The genus *Lithurgus* (Hymenoptera: Megachilidae). *Psyche* 45(4): 146–155.

———. 1960. *Bees of the Eastern United States*. Vols. 1 and 2. Raleigh: North Carolina Agricultural Experiment Station.

Moffat L, Roig-Alsina A. 1992. Nidificación comunal en la abeja *Melissoptila pubescens* (Smith) (Hymenoptera: Anthophoridae). *Revista de la Sociedad Entomológica Argentina* 51(1–4): 103–106.

Morato EF. 2001. Efeitos da fragmentação florestal sobre vespas e abelhas solitárias na Amazônia Central. II. Estratificação vertical. *Revista brasileira de zoologia* 18(3): 737–747.

Morato EF, Martins RP. 2006. An overview of proximate factors affecting the nesting behavior of solitary wasps and bees (Hymenoptera: Aculeata) in preexisting cavities in wood. *Neotropical Entomology* 35(3): 285–298.

Mueller UG. 1996. Life history and social evolution of the primitively eusocial bee *Augochlorella striata* (Hymenoptera: Halictidae). *Journal of the Kansas Entomological Society* 69:116–138.

Mueller UG, Eickwort CG, Aquadro CF. 1994. DNA fingerprinting analysis of parent-offspring conflict in a bee *Augochlorella striata*. *Proceedings of the National Academy of Sciences of the United States of America* 91(11): 5143–5147.

Muesebeck CF, Krombein KV, Townes HK. 1951. *Hymenoptera of America North of Mexico: Synoptic Catalog*. Washington, DC: US Department of Agriculture.

Murray EA, Evanhoe L, Bossert S, Geber MA, Griswold TL, McCoshum SM. 2021. Phylogeny, phenology, and foraging breadth of *Ashmeadiella* (Hymenoptera: Megachilidae). *Insect Systematics and Diversity* 5(3): 1–29.

Navarro L, Medel R. 2009. Relationship between floral tube length and nectar robbing in *Duranta erecta* L. (Verbenaceae). *Biological Journal of the Linnean Society* 96(2): 392–398.

Neff JL. 2003. Nest and provisioning biology of the bee *Panurginus polytrichus* Cockerell (Hymenoptera: Andrenidae), with a description of a new *Holcopasites* species (Hymenoptera: Apidae), its probable nest parasite. *Journal of the Kansas Entomological Society* 76(2): 203–216.

———. 2004. Hooked hairs and not so narrow tubes: Two new species of *Colletes* Latreille from Texas (Hymenoptera: Apoidea: Colletidae). *Journal of Hymenoptera Research* 13(3): 250–261.

———. 2008. Components of nest provisioning behavior in solitary bees (Hymenoptera: Apoidea). *Apidologie* 39(1): 30–45.

Neff JL, Danforth BN. 1991. The nesting and foraging behavior of *Perdita texana* (Cresson) (Hymenoptera: Andrenidae). *Journal of the Kansas Entomological Society* 64(4): 394–405.

Neff JL, Simpson BB. 1981. Oil-collecting structures in the Anthophoridae (Hymenoptera): Morphology, function, and use in systematics. *Journal of the Kansas Entomological Society* 54(1): 95–123.

———. 1988. Vibratile pollen-harvesting by *Megachile mendica* Cresson. *Journal of the Kansas Entomological Society* 61(2): 242–244.

———. 1992a. Nest biology of *Osmia (Diceratosmia) subfasciata* Cresson in central Texas (Hymenoptera: Megachilidae). *Pan-Pacific Entomologist* 68:15–26.

———. 1992b. Partial bivoltinism in a ground-nesting bee: The biology of *Diadasia rinconis* in Texas (Hymenoptera, Anthophoridae). *Journal of the Kansas Entomological Society* 65(4): 377–392.

Norden BB, Scarbrough A. 1979. Nesting biology of *Andrena (Larandrena) miserabilis* Cresson and description of the prepupa (Hymenoptera: Andrenidae). *Brimleyana* 2:141–146.

North F, Lillywhite H. 1980. The function of burrow turrets in a gregariously nesting bee. *Southwestern Naturalist* 25(3): 373–378.

Onuferko TM. 2018. A revision of the cleptoparasitic bee genus *Epeolus* Latreille for Nearctic species, north of Mexico (Hymenoptera, Apidae). *ZooKeys* 755:1–185.

Ordway E. 1966a. The bionomics of *Augochlorella striata* and *A. persimilis* in eastern Kansas. *Journal of the Kansas Entomological Society* 39:270–313.

———. 1966b. Systemics of the genus *Augochlorella* (Hymenoptera, Halictidae) north of Mexico. *University of Kansas Science Bulletin* 46:509–624.

———. 1984. Aspects of the nesting behavior and nest structure of *Diadasia opuntiae* Cockerell (Hymenoptera: Anthophoridae). *Journal of the Kansas Entomological Society* 57(2): 216–230.

———. 1987. The life history of *Diadasia rinconis* Cockerell (Hymenoptera: Anthophoridae). *Journal of the Kansas Entomological Society* 60(1): 15–24.

Orr MC, Portman ZM, Griswold TL. 2015. *Megachile (Megachile) montivaga* (Hymenoptera: Megachilidae) nesting in live thistle (Asteraceae: *Cirsium*). *Journal of Melittology* 48:1–6.

Packer L. 1990. Solitary and eusocial nests in a population of *Augochlorella striata* (Provancher)

(Hymenoptera; Halictidae) at the northern edge of its range. *Behavioral Ecology and Sociobiology* 27:339–344.

Packer L, Genaro JA, Sheffield CS. 2007. The bee genera of eastern Canada. *Canadian Journal of Arthropod Identification* 3:1–32.

Packer L, Gravel AD, Gretchen L. 2007. Phenology and social organization of *Halictus (Seladonia) tripartitus* (Hymenoptera: Halictidae). *Journal of Hymenoptera Research* 16:281–292.

Packer L, Knerer G. 1986. An analysis of variation in the nest architecture of *Halictus ligatus* in Ontario. *Insectes sociaux* 33:190–205.

Papavero N, Ibanez-Bernal S. 2001. Contributions to a history of Mexican dipterology—Part I. Entomologists and their works before the Biologia Centrali-Americana. *Acta zoológica mexicana* 84:65–173.

Parker FD. 1986. Nesting, associates, and mortality of *Osmia sanrafaelae* Parker. *Journal of the Kansas Entomological Society* 59(2): 367–377.

Parker FD, Bohart GE. 1979. *Dolichostelis*, a new genus of parasitic bees (Hymenoptera: Megachilidae). *Journal of the Kansas Entomological Society* 52(1): 138–153.

Parker FD, Cane JH, Frankie GW, Vinson SB. 1987. Host records and nest entry by *Dolichostelis*, a kleptoparasitic anthidiine bee (Hymenoptera: Megachilidae). *Pan-Pacific Entomologist* 63(2): 172–177.

Parks HB. 1930. Notes on Texas bees. *Bulletin of the Brooklyn Entomological Society* 25:263–264.

Pellett FC. 1920. *American Honey Plants*. Hamilton, IL: American Bee Journal.

Pemberton, RW, Liu H. 2008a. Naturalization of the oil collecting bee *Centris nitida* (Hymenoptera, Apidae, Centrini), a potential pollinator of selected native, ornamental and invasive plants in Florida. *Florida Entomologist* 9:101–109.

———. 2008b. Potential of invasive and native solitary specialist bee pollinators to help restore the rare cowhorn orchid (*Cyrtopodium punctatum*) in Florida. *Biological Conservation* 141(7): 1758–1764.

Pisanty G, Richter R, Martin T, Dettman J, Cardinal S. 2020. Molecular phylogeny and historical biogeography of andrenine bees (Hymenoptera: Andrenidae). *Molecular Phylogenetics and Evolution* 170:107151.

Porter CC. 1981. Ecological notes on Lower Rio Grande Valley *Xylocopa* (Hymenoptera: Anthophoridae). *Florida Entomologist* 64(1): 175–183.

———. 1983. Ecological notes on Lower Rio Grande Valley *Augochloropsis* and *Agapostemon* (Hymenoptera: Halictidae). *Florida Entomologist* 66(3): 344–353.

Portman ZM, Neff JL, Griswold TL. 2016. Taxonomic revision of *Perdita* subgenus *Heteroperdita* Timberlake (Hymenoptera: Andrenidae), with descriptions of two ant-like males. *Zootaxa* 4214(1): 1–97.

Ramírez-Freire L, Alanis-Flores GJ, Ayala-Barajas R, Quiroz-Martínez H, Velazco-Macías CG. 2012. The large carpenter bees (Hymenoptera: Apidae: *Xylocopa* spp.) of Nuevo León, México. *Journal of Pollination Ecology* 7(1): 1–4.

Ramirez-Freire L, Glafiro GJA, Ayala-Barajas R, Quiroz-Martínez H, Velazco-Macías CG. 2012. Las abejas del género *Agapostemon* (Hymenoptera: Halictidae) del estado de Nuevo León, México. *Revista mexicana de biodiversidad* 83:63–72.

Rasmussen C. 2012. Joseph Vachal (1838–1911): French entomologist and politician. *Zootaxa* 3442:1–52.

Rasmussen C, Ascher JS. 2008. Heinrich Friese (1860–1948): Names proposed and notes on a pioneer melittologist (Hymenoptera, Anthophila). *Zootaxa* 1833:1–118.

Rasmussen C, Garcete-Barrett BR, Gonçalves RB. 2009. Curt Schrottky (1874–1937): South American entomology at the beginning of the 20th century (Hymenoptera, Lepidoptera, Diptera). *Zootaxa* 2282:1–50.

Rau P. 1924. Notes on captive colonies and homing of *Bombus pensylvanicus* De Geer. *Annals of the Entomological Society of America* 17(4): 368–381.

———. 1928. The nesting habits of the little carpenter bee *Ceratina calcarata*. *Annals of the Entomological Society of America* 21(3): 380–396.

———. 1929. The biology and behavior of mining bees, *Anthophora abrupta* and *Entechnia taurea*. *Psyche* 36:155–181.

Raw A. 1977. The biology of two *Exomalopsis* species (Hymenoptera: Anthophoridae) with remarks on sociality in bees. *Revista de biología tropical* 25(1): 1–11.

———. 1984. The nesting biology of nine species of Jamaican bees (Hymenoptera). *Revista brasileira de entomologia* 28(4): 497–506.

———. 2002. New combinations and synonymies of leafcutter and mason bees of the Americas (Megachile, Hymenoptera, Megachilidae). *Zootaxa* 71(1): 1–43.

———. 2007. An annotated catalog of the leafcutter and mason bees (*Megachile*) of the Neotropics. *Zootaxa* 1601:1–127.

Rehan SM. 2020. Small carpenter bees (*Ceratina*). In: Stark C, ed., *Encyclopedia of Social Insects*. New York: Springer: 1–4.

Rehan SM, Richards MH. 2010a. The influence of maternal quality on brood sex allocation in the small carpenter bee, *Ceratina calcarata*. *Ethology* 116(9): 876–887.

———. 2010b. Nesting biology and subsociality in *Ceratina calcarata*. *Canadian Entomologist* 142:65–73.

Ribble DW. 1967. The monotypic North American subgenus *Larandrena* of *Andrena* (Hymenoptera: Apoidea). *Bulletin of the University of Nebraska State Museum* 6(3): 27–42.

Richardson A. 1990. *Plants of the Rio Grande Delta*. Austin: University of Texas Press.

Richardson A, King K. 2011. *Plants of Deep South Texas*. College Station: Texas A&M University Press.

Rightmyer MG. 2008. A review of the cleptoparasitic bee genus *Triepeolus* (Hymenoptera: Apidae). *Zootaxa* 1710:1–170.

Roberts RB. 1969a. Biology of the bee genus *Agapostemon* (Hymenoptera: Halictidae). *University of Kansas Science Bulletin* 48(16): 689–719.

———. 1969b. Revision of the bee genus *Agapostemon* (Hymenoptera: Halictidae). *University of Kansas Science Bulletin* 49(9): 437–590.

———. 1972. Revision of the genus *Agapostemon* (Hymenoptera: Halictidae). *University of Kansas Science Bulletin* 59(9): 437–590.

———. 1973a. Bees of northwestern America: *Agapostemon* (Hymenoptera: Halictidae). *Oregon State University Agricultural Experiment Station Technical Bulletin* 125:1–23.

———. 1973b. Nest architecture and immature stages of the bee *Oxaea flavescens* and the status of Oxaeidae (Hymenoptera). *Journal of the Kansas Entomological Society* 46(4): 437–446.

———. 1978. The nesting biology, behavior and immature stages of *Lithurge chrysurus*, an adventitious wood-boring bee in New Jersey (Hymenoptera: Megachilidae). *Journal of the Kansas Entomological Society* 51(4): 735–745.

Roberts RB, Brooks RW. 1987. Agaposetmonine bees of Mesoamerica (Hymenoptera: Apoidea). *University of Kansas Science Bulletin* 53(7): 357–391.

Robertson C. 1897. North American bees—description and synonyms. *Transactions of the Academy of Science of St. Louis* 7(14): 315–356.

———. 1925. Heterotropic bees. *Ecology* 6(4): 412–436.

———. 1926. Phenology of inquiline and nest-making bees. *Psyche* 33(4–5): 116–120.

Rocha Filho LCD, Packer L. 2017. Phylogeny of the cleptoparasitic Megachilini genera *Coelioxys* and *Radoszkowskiana*, with the description of six new subgenera in *Coelioxys* (Hymenoptera: Megachilidae). *Zoological Journal of the Linnean Society* 180(2): 354–413.

Rodriguez MF, Roig-Alsina A. 2004. Especies de *Ancyloscelis* Latreille del sur de Sudamérica (Hymenoptera, Apidae) con aparato bucal no modificado: Nuevas especies y notas sinonímicas. *Revista del Museo Argentino de Ciencias Naturales* 6(2): 351–361.

Rojas-Nossa SV, Sánchez JM, Navarro L. 2016. Effects of nectar robbing on male and female reproductive success of a pollinator-dependent plant. *Annals of Botany* 117(2): 291–297.

Rozen JG, Jr. 1964. Phylogenetic-taxonomic significance of last instar of *Protoxaea gloriosa* Fox, with descriptions of first and last instars (Hymenoptera: Apoidea). *Journal of the New York Entomological Society* 72(4): 223–230.

———. 1983. Nesting biology of the bee *Svastra sabinensis* (Hymenoptera, Anthophoridae). *Journal of the New York Entomological Society* 91(3): 264–268.

———. 1984a. Comparative nesting biology of the bee tribe Exomalopsini (Apoidea, Anthophoridae). *American Museum Novitates* 2798:1–37.

———. 1984b. Nesting biology of diphaglossine bees (Hymenoptera, Colletidae). *American Museum Novitates* 2786:1–33.

Rozen JG, Jr., Favreau MS. 1968. Biological notes on *Colletes compactus* and its cuckoo bee, *Epeolus pusillus* (Hymenoptera: Colletidae and Anthophoridae). *Journal of the New York Entomological Society* 76(2): 106–111.

Rozen JG, Jr., Hall G. 2014. Nest site selection and nesting behavior of the bee *Lithurgopsis apicalis* (Megachilidae: Lithurginae). *American Museum Novitates* 3796:1–24.

Rozen JG, Jr., MacNeill CD. 1957. Biological observations on *Exomalopsis (Anthophorula) chionura* Cockerell, including a comparison of the biology of *Exomalopsis* with that of other anthophorid groups. *Annals of the Entomological Society of America* 50:522–529.

Rozen JG, Jr., Snelling R. 1986. Ethology of the bee *Exomalopsis nitens* and its cleptoparasite (Hymenoptera: Anthophoridae). *Journal of the New York Entomological Society* 94(4): 480–488.

Rozen MA, Rozen JG, Jr. 2010. The pupa of the bee *Protoxaea gloriosa* (Hymenoptera: Andrenidae: Oxaeinae). *Journal of the Kansas Entomological Society* 83(1): 76–79.

Sampson BJ, Danka RG, Stringer SJ. 2004. Nectar robbery by bees *Xylocopa virginica* and *Apis mellifera* contributes to the pollination of rabbiteye blueberry. *Journal of Economic Entomology* 97(3): 735–740.

Sandhouse GA. 1937. The bees of the genera *Augochlora*, *Augochloropsis* and *Augochlorella* (Hymenoptera; Apoidea) occurring in the United States. *Journal of the Washington Academy of Sciences* 27:65–79.

Sandhouse GA, Cockerell TDA. 1921. Expedition of the California Academy of Sciences to the Gulf of California in 1921. The bees (III). *Proceedings of the California Academy of Sciences* 13(4): 333–339.

Schaller A, Roig-Alsina A. 2021. A revision of the bee genus *Ancyloscelis* (Hymenoptera: Apidae) in Argentina. *Zootaxa* 4980(3): 521–540.

Schrottky C. 1913. As espécies brazileiras do gênero *Megachile*, supplemento ao "Ensaio sobre as abelhas solitárias do Brasil" (*Revista do Museu Paulista* 5 (1902): 330–613). *Revista do Museu Paulista* 9:134–223.

Schwartz HF. 1926. North American *Dianthidium*, *Anthidiellum* and *Paranthidium*. *American Museum Novitates* 226:1–25.

Sheffield CS. 2013. A new species of *Megachile* Latreille subgenus *Megachiloides* (Hymenoptera, Megachilidae). *ZooKeys* 283:43–58.

Sheffield CS, Ratti C, Packer L, Griswold TL. 2011. Leafcutter and mason bees of the genus *Megachile* Latreille (Hymenoptera: Megachilidae) in Canada and Alaska. *Canadian Journal of Arthropod Identification* 18:1–107.

Shelly TE, Villalobos EM, Buchmann SL, Cane JH. 1993. Temporal patterns of floral visitation for two bee species foraging on *Solanum*. *Journal of the Kansas Entomological Society* 66(3): 319–327.

Shinn AF. 1967. A revision of the genus *Calliopsis* and the biology and ecology of *C. andreniformis* (Hymenoptera: Andrenidae). *University of Kansas Science Bulletin* 46:753–936.

Silva-Pereira AI, Alves-dos-Santos I, Malagodi-Braga KS, Contrera FAL. 2003. Forrageamento de *Melissoptila thoracica* Smith (Hymenoptera, Eucerini, Apoidea) em flores de *Sida* (Malvaceae). *Revista brasileira de zoologia* 20(3): 427–432.

Simpson BB, Neff JL. 1981. Floral rewards: Alternatives to pollen and nectar. *Annals of the Missouri Botanical Garden* 68(2): 301–322.

Snelling RR. 1966. Studies on North American bees of the genus *Hylaeus*. 1. Distribution of the western species of the subgenus *Prosopis* with descriptions of new forms (Hymenoptera: Colletidae). *Contributions in Science* (Los Angeles, Calif.) 98:1–18.

———. 1968. Studies on the North American bees of the genus *Hylaeus*. 4. The subgenera *Cephalylaeus*, *Metziella* and *Hylaeana* (Hymenoptera: Colletidae). *Contributions in Science* (Los Angeles, Calif.) 144:1–6.

———. 1983. The North American species of the bee genus *Lithurge* (Hymenoptera: Megachilidae). *Contributions in Science* (Los Angeles, Calif.) 343:1–11.

———. 1984. Studies on the taxonomy and distributions of American centridine bees (Hymenoptera: Anthophoridae). *Contributions in Science* (Los Angeles, Calif.) 347:1– 69.

———. 1986. Contributions toward a revision of the New World nomadine bees: A partitioning of the genus *Nomada* (Hymenoptera: Anthophoridae). *Contributions in Science* (Los Angeles, Calif.) 376:1–32.

———. 1990. A review of the native North American bees of the genus *Chalicodoma* (Hymenoptera: Megachilidae). *Contributions in Science* (Los Angeles, Calif.) 421:1–39.

———. 1994. *Diadasia*, subgenus *Dasiapis*, in North America (Hymenoptera: Anthophoridae). *Contributions in Science* (Los Angeles, Calif.) 448:1–8.

Stabentheiner A, Press H, Papst T, Hrassnigg N, Crailsheim K. 2003. Endothermic heat production in honeybee winter clusters. *Journal of Experimental Biology* 206(2): 353–358.

Stephen WP. 1954. A revision of the bee genus *Colletes* in America north of Mexico (Hymenoptera, Colletidae). *University of Kansas Science Bulletin* 36(1): 149–527.

Stephen WP, Bohart G, Torchio PF. 1969. *The Biology and External Morphology of Bees—with a Synopsis of the Genera of Northwestern America.* Corvallis: Agricultural Experiment Station, Oregon State University.

Stockhammer KA. 1966. Nesting habits and life cycle of a sweat bee, *Augochlora pura* (Hymenoptera: Halictidae). *Journal of the Kansas Entomological Society* 39(2): 157–192.

Stucky JM, Beckmann RL. 1982. Pollination biology, self-incompatibility, and sterility in *Ipomoea pandurata* (L.) G. F. W. Meyer (Convolvulaceae). *American Journal of Botany* 69(6): 1022–1031.

Swenk MH. 1906. New bees of the genus *Colletes. Canadian Entomologist* 38(2): 39–44.

———. 1908. Specific characters in the bee genus *Colletes. University Studies of the University of Nebraska* 8(1): 43–102.

Thorp RW. 1969. Ecology and behavior of *Anthophora edwardsii* (Hymenoptera: Anthophoridae). *American Midland Naturalist* 82(2): 321–337.

Timberlake PH. 1928. Bees of the genus *Perdita* in the J. C. Bridwell Collection. *Proceedings of the Hawaiian Entomological Society* 7(1): 151–161.

———. 1943. Racial differentiation in Nearctic species of *Dianthidium* (Hymenoptera, Apoidea). *Journal of the New York Entomological Society* 51(2): 71–109.

———. 1947. A revision of the species of *Exomalopsis* inhabiting the United States (Hymenoptera, Apoidea). *Journal of the New York Entomological Society* 55(2): 85–106.

———. 1948. Additions and corrections to the list of Nearctic species of *Dianthidium* (Hymenoptera, Apoidea). *Journal of the New York Entomological Society* 56(3): 149–153.

———. 1953. Bees of the genus *Perdita* in the collection of the University of Kansas. *University of Kansas Science Bulletin* 35:961–985.

———. 1954. A revisional study of the bees of the genus *Perdita* F. Smith, with special reference to the fauna of the Pacific coast (Hymenoptera, Apoidea). Part I. *University of California Publications in Entomology* 9(6): 345–432.

———. 1955. A new genus for two new species of dufoureine bees from California. *Pan-Pacific Entomologist* 31:105–108.

———. 1961. A review of the genus *Conanthalictus. Pan-Pacific Entomologist* 37(3): 145–160.

———. 1965. Notes on caupolicanine bees of Arizona (Hymenoptera: Apoidea). *Journal of the New York Entomological Society* 73(1): 46–48.

———. 1968. A revisional study of the bees in the genus *Perdita* F. Smith with special reference to the fauna of the Pacific coast (Hymenoptera: Apoidea). Part VII (including index to parts I to VII). *University of California Publications in Entomology* 49:1–196.

———. 1980. *Review of North American* Exomalopsis *(Hymenoptera, Anthophoridae).* Berkeley: University of California Press.

Torretta JP, Durante SP, Guadalupe M, Basilio AM. 2012. Nesting biology of the leafcutting bee *Megachile (Pseudocentron) gomphrenoides* (Hymenoptera: Megachilidae) in an agro-ecosystem. *Apidologie* 43:624–633.

Torretta JP, Medan D, Roig-Alsina A, Montaldo MH. 2010. Visitantes flores diurnos del girasol (*Helianthus annuus*, Asterales: Asteraceae) en la Argentina. *Revista de la Sociedad Entomológica Argentina* 69(1–2): 17–32.

Tumlison R, Benjamin K. 2011. Occurrence of two rare prairie insects, *Tetraloniella albata* (Cresson) and *Microstylum morosum* (Loew), at Terre Noire Natural Area, Clark County, Arkansas. *Journal of the Arkansas Academy of Science* 65(29): 183–186.

Urban D. 1968. As espécies do género *Melissoptila* Holmberg, 1884 (Hymenoptera, Apoidea). *Revista brasileira de entomologia* 13:1–94.

———. 1970. As espécies do gênero *Florilegus* Robertson, 1900 (Hymenoptera, Apoidea). *Boletim da Universidade Federal do Paraná* 3:245–280.

———. 1998. Espécies novas de *Melissoptila* Holmberg da América do Sul e notas taxonômicas (Hymenoptera, Anthophoridae). *Revista brasileira de zoologia* 15(1): 1–46.

US Department of Agriculture, Natural Resources Conservation Service. 2022. PLANTS Database. Greensboro, NC: National Plant Data Team. http://plants.usda.gov

Vachal, J. 1911. Étude sur les *Halictus* d'Amérique (Hym.). Suite et fin. *Miscellanea entomologica* 19:9–24, 41–56, 107–116 [IV 53–94].

Velez-Ruiz RI. 2015. A revision of the bee genus *Ptiloglossa* Smith (Hymenoptera: Colletidae: Diphaglossinae). PhD diss., South Dakota State University. https://openprairie.sdstate.edu/cgi/viewcontent.cgi?article=2893&context=etd

Velez-Ruiz RI, Smith-Pardo AH. 2013. New species of *Exomalopsis* and its associated cleptoparasite *Nomada* from Colombia with description of the nest (Hymenoptera: Apoidea: Anthophila: Apidae). *ISRN Entomology* 2013:1–10.

Wallberg A, Han F, Wellhagen G, Dahle B, Kawata M, Haddad N, Luz Z, Simões P, Allsopp MH, Kandemir I, de la Rúa P, Pirk CW, Webster MT. 2014. A worldwide survey of genome sequence variation provides insight into the evolutionary history of the honeybee *Apis mellifera*. *Nature Genetics* 46(10): 1081–1088.

Warriner MD. 2010. A range extension for the large carpenter bee *Xylocopa micans* (Hymenoptera: Apidae) with notes on floral and habitat associations. *Journal of the Kansas Entomological Society* 83:267–269.

———. 2012. Bumble bees (Hymenoptera: Apidae) of Texas: Historical distributions. *Southwestern Naturalist* 57(4): 442–445.

Wcislo WT. 1993. Communal nesting in a North American pearly-banded bee, *Nomia tetrazonata*, with notes on nesting behavior of *Dieunomia heteropoda* (Hymenoptera: Halictidae: Nomiinae). *Annals of the Entomological Society of America* 86(6): 813–821.

———. 1999. Male territoriality and nesting behavior of *Calliopsis hondurasica* Cockerell. *Journal of the Kansas Entomological Society* 72:91–98.

Wcislo WT, Buchmann SL. 1995. Mating behavior in the bees, *Dieunomia heteropoda* and *Nomia tetrazonata*, with a review of courtship in Nomiinae (Hymenoptera: Halictidae). *Journal of Natural History* 29(4): 1015–1027.

Wcislo WT, Engel MS. 1996. Social behavior and nest architecture of nomiine bees (Hymenoptera: Halictidae; Nomiinae). *Journal of the Kansas Entomological Society* 69(4): 158–167.

Williams HJ, Vinson SB, Frankie GW, Coville RE, Ive GW. 1984. Description, chemical contents and function of the tibial gland in *Centris nitida* and *Centris trigonoides subtarsata* (Hymenoptera: Anthophoridae) in the Costa Rican dry forest. *Journal of the Kansas Entomological Society* 57:50–54.

Williams P, Thorp R, Richardson L, Colla S. 2014. *Bumble Bees of North America*. Princeton, NJ: Princeton University Press.

Wilson JS, Carril OM. 2016. *The Bees in Your Backyard: A Guide to North America's Bees*. Princeton, NJ: Princeton University Press.

Wittmann D, Blochtein B. 1995. Why males of leafcutter bees hold their females' antennae with their front legs during mating. *Apidologie* 26(3): 181–195.

Wojcik VA. 2011. Bees (Hymenoptera: Apoidea) utilizing *Tecoma stans* (L.) Juss. ex Kunth (Bignoniaceae) in urban landscapes: A comparison of occurrence patterns and community composition in three cities in northwestern Costa Rica. *Journal of the Kansas Entomological Society* 84(3): 197–208.

Wright KW, Miller KB, Song H. 2020. A molecular phylogeny of the long-horned bees in the genus *Melissodes* Latreille (Hymenoptera: Apidae: Eucerinae). *Insect Systematics and Evolution* 52(4): 428–443.

Yanega D. 1997. Demography and sociality in halictine bees (Hymenoptera: Halictidae). In: Choe JC, Crespi BJ (eds.), *The Evolution of Social Behavior in Insects and Arachnids*. Cambridge: Cambridge University Press: 293–315.

Yañez-Ordoñez O. 1997. Biología de *Lithurge (Lithurgopsis) littoralis* Cockerell 1917 (Apoidea Megachilidae) en la Reserva Ecológica del Pedregal de San Ángel. D. F. Thesis, Universidad Nacional Autónoma de México, Facultad de Ciencias.

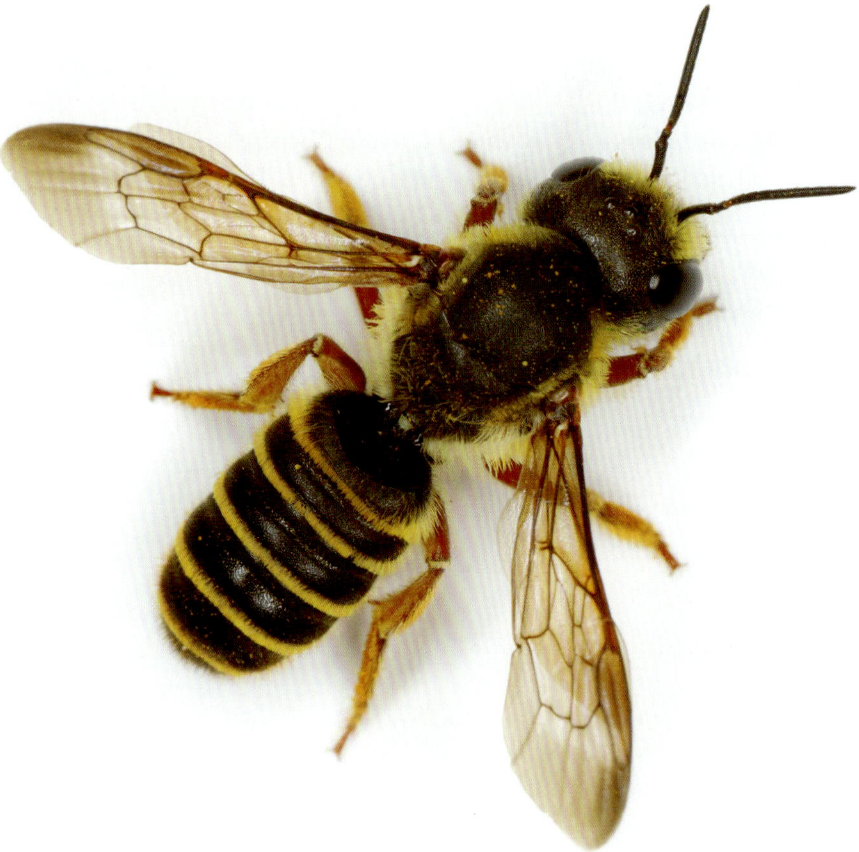

Subject Index

Note: Page numbers followed by *t* indicate tables and charts. Page numbers in *italics* indicate illustrations. Page numbers in **boldface** indicate definitions.

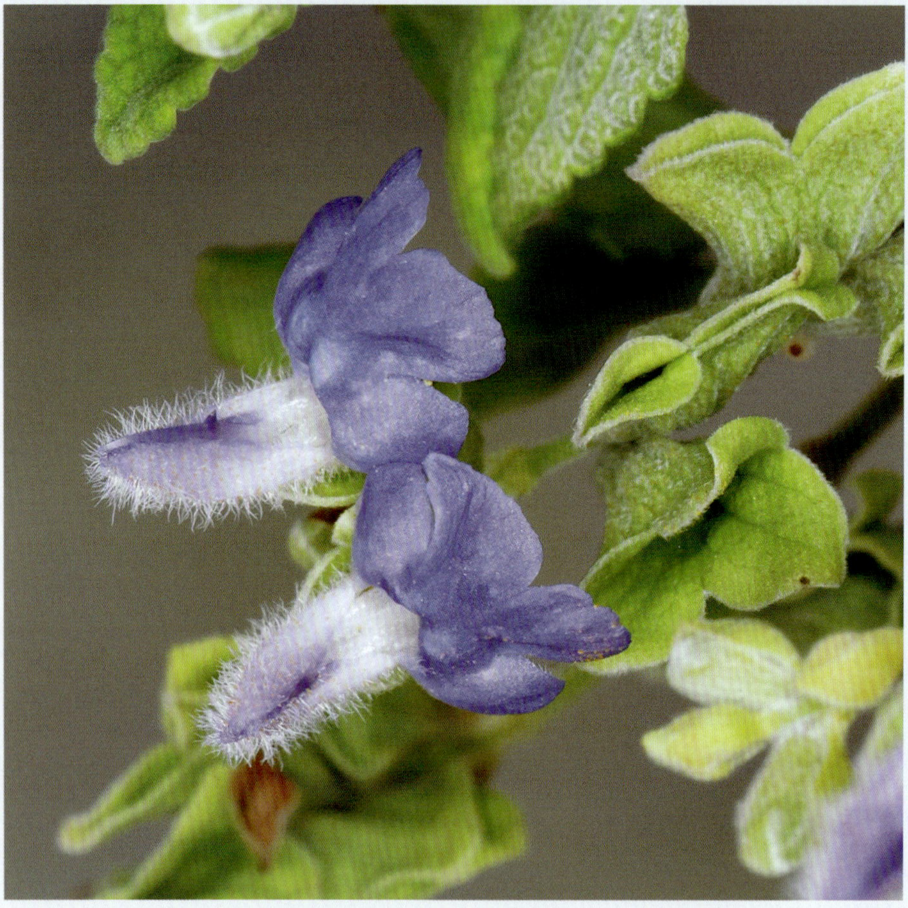

Plant Name Index

The USDA Plants Database is the primary source of scientific and common names for North American flora appearing here. Where USDA common names are not widely used in the Texas Rio Grande Valley, regional common names are provided as well, using the format (regional common name/ USDA common name). Regional common names are taken from Alfred Richardson and Ken King's comprehensive *Plants of Deep South Texas*.